AMERICAN MATHEMATICAL SOCIETY
COLLOQUIUM PUBLICATIONS
VOLUME XXIII

ORTHOGONAL POLYNOMIALS

BY

GABOR SZEGÖ

PROFESSOR OF MATHEMATICS
STANFORD UNIVERSITY

Published by the

American Mathematical Society
Providence, Rhode Island

1939

International Standard Serial Number 0065-9258
International Standard Book Number 0-8218-1023-5
Library of Congress Catalog Card Number 39-33497

First edition, 1939
Fourth edition, 1975
Reprinted 1985

Printed in the United States of America

TO MY WIFE

PREFACE

Recent years have seen a great deal of progress in the field of orthogonal polynomials, a subject closely related to many important branches of analysis. Orthogonal polynomials are connected with trigonometric, hypergeometric, Bessel, and elliptic functions, are related to the theory of continued fractions and to important problems of interpolation and mechanical quadrature, and are of occasional occurrence in the theories of differential and integral equations. In addition, they furnish comparatively general and instructive illustrations of certain situations in the theory of orthogonal systems. Recently, some of these polynomials have been shown to be of significance in quantum mechanics and in mathematical statistics.

The origins of the subject are to be found in the investigation of a certain type of continued fractions, bearing the name of Stieltjes. Special cases of these fractions were studied by Gauss, Jacobi, Christoffel, and Mehler, among others, while more general aspects of their theory were given by Tchebichef, Heine, Stieltjes, and A. Markoff.

Despite the close relationship between continued fractions and the problem of moments, and notwithstanding recent important advances in this latter subject, continued fractions have been gradually abandoned as a starting point for the theory of orthogonal polynomials. In their place, the orthogonal property itself has been taken as basic, and it is this point of view which has been adopted in the following exposition of the subject. Choosing this same basic property, we discuss certain special orthogonal polynomials, which have been treated in great detail independently of the general theory, and indeed, even before this theory existed at all. In this connection we add the names of Laplace, Legendre, Fourier, Abel, Laguerre, and Hermite to those previously mentioned.

As regards treatises on the subject, we note that the only systematic treatment thus far given is found in J. Shohat's monograph, *Théorie Générale des Polynomes Orthogonaux de Tchebichef*, Mémorial des Sciences Mathématiques, Paris, 1934. Limitations of space have compelled that work to be brief, and consequently, it does not enter into a detailed treatment of many problems which have been especially advanced in recent years. It has therefore seemed desirable to attempt a new and detailed development of the main ideas of this field, devoting, in particular, some space to recent investigations of the distribution of the zeros, of asymptotic representations, of expansion problems, and of certain questions of interpolation and mechanical quadrature.

In what follows, we are concerned partly with the general theory of orthogonal polynomials, and partly with the study of special classes of these polynomials. As might be expected, we have more exhaustive results for these special classes, and we cite as an instance the classical polynomials satisfying linear differential equations of the second order. Also, when the primary importance of these special classes in applications is taken into account, it should not be at all surprising that the present book is mainly devoted to their study. The general theory, however, as developed in Chapters XII and XIII, doubtless represents the most important progress made in recent years.

In the present work, no claim is made for completeness of treatment. On the contrary, the aim has purposely been to make the material suggestive rather than exhaustive. An attempt has been made to indicate the main and characteristic methods and to point out the relation of these to some general ideas in modern analysis. As a rule, preference has been given to those topics to which we were able to make some new, though modest, contributions, or which we could present in a new setting. Thus the book contains a number of results not previously published, some of which originated several years ago. For instance, we have included a discussion of the Cesàro summability of the Jacobi series at the end-points of the orthogonality interval (the method used here is of interest even in the classical case of Legendre series). Further, a new and simpler approach has been given to S. Bernstein's asymptotic formula for orthogonal polynomials. We also refer to certain details of minor importance, such as: simplifications and additions in the asymptotic investigation of Jacobi and Laguerre polynomials and in the discussion of the expansions in terms of these polynomials; the discussion of the cases in which the Jacobi differential equation has only polynomial solutions; the evaluation of the number of zeros of general Jacobi polynomials in the intervals $[-\infty, -1]$, $[-1, +1]$, $[+1, +\infty]$; a new proof of the Heine-Stieltjes theorem on linear differential equations of the second order with polynomial coefficients and polynomial solutions, and so on.

In general, we have preferred to discuss problems which may be stated and treated simply, and which could be presented in a more or less complete form. This was the main reason for devoting no space to the extremely interesting arithmetic and algebraic properties of orthogonal polynomials, such as, for instance, the recent important investigations of I. Schur concerning the irreducibility and related properties of Laguerre and Hermite polynomials. Furthermore, we have attached great importance to the idea of replacing incomplete and overlapping theorems, scattered in the literature, by complete results involving only intrinsic or necessary restrictions. We have also tried to exploit, as far as seemed to be at all possible, definite methods, such as, for instance, Sturm's methods in differential equations (see §§6.3, 6.31, 6.32, 6.83).

A complete treatment of Legendre polynomials was not feasible, and probably not desirable, in the framework of the general theory. Besides, there are already complete treatises on spherical and other harmonics.[1] We have selected and considered only those properties of Legendre polynomials which are the starting points of generalizations to ultraspherical, Jacobi, or to more general polynomials. Another subject which could not be included was Stieltjes'

[1] For instance, E. W. Hobson **1** (see bibliography).

problem of moments, which has been omitted in spite of its great interest; for this subject would have necessitated the development of a complicated apparatus of results and methods. Orthogonal polynomials of more than one variable also have not been treated.[2]

The book is based on a course given at Washington University during the academic year 1935–1936. Acquaintance with the general ideas and methods of the theory of functions of real and complex variables is naturally required. Occasionally, Stieltjes-Lebesgue and Lebesgue integrals are considered. In the greater part of the book, however, these integrals have been avoided, and, except in a very few places, no detailed properties of them were used.

The problems at the end of this book are, with few exceptions, not new, and they are not interconnected as are, for instance, those in Pólya-Szegö's *Aufgaben und Lehrsätze*. They are more or less supplementary in character and serve as illustrations and exercises; they sometimes differ widely from one another both as to subject and method.

The list of references is not complete; it contains only original memoirs, a few text books of primary importance, and monographs to which references are made in the text.

For the suggestion of preparing a book on orthogonal polynomials for the Colloquium Publications, I am indebted to Professor J. D. Tamarkin, who has also participated in the present work by offering a great number of valuable suggestions. It is with the greatest gratitude that I mention his friendly interest.

I have also received valuable advice from my friends and teachers L. Fejér (Budapest), and G. Pólya (Zürich). My colleagues P. Erdös (Manchester), G. Grünwald (Budapest), W. H. Roever (St. Louis), A. Ross (St. Louis), J. Shohat (Philadelphia), and P. Turán (Budapest) gave generously and unstintingly of their time. F. A. Butter, Jr. (at present in Los Angeles) collaborated with me in the preparation of the manuscript. This last aid was made possible through a grant from the Rockefeller Research Fund of Washington University (1936–1937). My student L. H. Kanter also rendered valuable assistance in the preparation of the manuscript.

My gratitude for the encouragement and help of these friends, colleagues, and institutions can hardly be measured by any formal acknowledgment. Lastly, I wish to express to the American Mathematical Society my great appreciation for the inclusion of the present book in its Colloquium Series.

G. SZEGÖ

WASHINGTON UNIVERSITY, 1938.

[2] Cf. the bibliography in Jackson **8**, p. 423.

PREFACE TO THE REVISED EDITION

The first printing of this book published in 1939 was about exhausted in 1948. Reprinting was arranged then but for various reasons no change in the text was made. During the past twenty years since the preparation of the original edition was completed, considerable progess was made in this field. A glance at the pertinent section of the Mathematical Reviews suggests that the interest in this topic is still very much alive. Systematic treatment of orthogonal polynomials has been incorporated in various modern texts published in the meantime. We refer only to the *Higher Transcendental Functions* published by the Bateman Manuscript Project Staff (cf. in particular, vol. 2, Chapter X, edited by Professor A. Erdélyi), and to the book of F. Tricomi, *Vorlesungen über Orthogonalreihen* (Chapters IV–VI).

Recently the council of the American Mathematical Society has authorized the author to prepare a revised edition of the book, adding a moderate amount of material in order to bring it up to date. Naturally, limitations of space and time did not allow including all new results (or, for that matter, the old ones which were missing from the original edition). Only a few particularly interesting new items have been added as well as some details which deserve attention because of elegance of the method or originality of ideas. We mention here in particular the important Pollaczek polynomials; they are treated in an Appendix. Further new material was incorporated in the form of Problems and Exercises. New bibliographic items have been included, again in a rather selective way. Finally, misprints have been corrected and numerous minor improvements and additions made.

The author recollects again, as was stated in the Preface of 1938, that the preparation of this book was suggested to him by the late Professor J. D. Tamarkin. Since his untimely death in 1945 his name is not too frequently mentioned. It is justified and probably necessary to remind the younger mathematical generation, in the rush of modern developments, how much American mathematics owes to his great energy and far-sighted intelligence.

STANFORD UNIVERSITY, 1958 G. SZEGÖ

PREFACE TO THE THIRD EDITION

The interest of the mathematical community for orthogonal polynomials, classical and non-classical, is still not entirely exhausted. During the past years I lectured about this subject several times at Stanford. The attendants of the course were upper division and graduate students, specializing in mathematics, mathematical statistics, calculus of probability, etc.

Only minor changes have been made in the text. I owe numerous improvements and corrections to various friends and colleagues. I mention particularly Professor Paul Turán (Budapest, Hungary) and Professor Lee Lorch (Edmonton, Canada). New references, published in the time interval 1958-1966, have been included.

STANFORD UNIVERSITY, 1966 G. SZEGÖ

PREFACE TO THE FOURTH EDITION

Again the American Mathematical Society has taken the initiative to reprint the present book, allowing some minor changes and new material. Among the persons interested in the field of orthogonal polynomials who have contributed to these changes and additions, I mention with particular indebtedness my friend and colleague Professor Richard Askey (Madison, Wisconsin) and the very active and original group of mathematicians around him. A very important set of lectures by Askey entitled, "Orthogonal Polynomials and Special Functions," reached me too late to be incorporated in the present edition.

Further material has been furnished by Professor Paul Turán (Budapest, Hungary) and Professor Lee Lorch (Toronto, Canada). New problems and exercises have also been included. Peter Szego (Redwood City, California) gave me valuable assistance in preparing the present manuscript.

My gratitude goes to all these friends and colleagues.

STANFORD UNIVERSITY, 1975 G. SZEGÖ

TABLE OF CONTENTS

CHAPTER I

PRELIMINARIES

1.1. Notation

Numbers in bold face type, like **1**, refer to the bibliography at the end of the book. The system of section numbering used is Peano's decimal system, and the numeration of formulas starts anew in each section. Thus, reference to §9.5 and (9.5.2) means section 9.5 in Chapter IX and formula (9.5.2) in the same section, respectively. A similar numeration has been used for the theorems.

We use the symbol: $\delta_{nm} = 0$ or 1, according as $n \neq m$, or $n = m$.

The closed real interval $a \leqq x \leqq b$ (a and b finite) will be denoted by $[a, b]$. The same symbol is used if either a or b is infinite or if both are; in this case the equality sign is excluded.

We often write for a real x

$$\text{sgn } x = -1, 0, +1, \tag{1.1.1}$$

according as x is negative, zero, or positive; more generally, for arbitrary complex x, $x \neq 0$, we write

$$\text{sgn } x = |x|^{-1} x. \tag{1.1.2}$$

The symbol $\bar{x}$ denotes the conjugate complex value, $\Re(x)$ the real part, and $\Im(x)$ the imaginary part of the complex number x.

If two sequences z_n and w_n of complex numbers have the property that $w_n \neq 0$ and $z_n/w_n \to 1$ as $n \to \infty$, we write $z_n \cong w_n$. If z_n and w_n are complex, $w_n \neq 0$, and the sequence $|z_n|/|w_n|$ has finite positive limits of indetermination, we write $z_n \sim w_n$.

Occasionally we make use of the notation

$$z_n = O(a_n), \qquad z_n = o(a_n) \tag{1.1.3}$$

if $a_n > 0$, to state that z_n/a_n is bounded, or tends to 0, respectively, as $n \to \infty$. A similar notation is used for a passage of limit other than $n \to \infty$.

A function $f(x)$ is called increasing (strictly increasing) if $x_1 < x_2$ implies $f(x_1) < f(x_2)$; it is called non-decreasing if $x_1 < x_2$ implies $f(x_1) \leqq f(x_2)$. An analogous terminology will be used for decreasing functions.

Let $p \geqq 1$, and let $\alpha(x)$ be a non-decreasing function in $[a, b]$ which is not constant. The class of functions $f(x)$ which are measurable with respect to $\alpha(x)$ and for which the Stieltjes-Lebesgue integral $\int_a^b |f(x)|^p \, d\alpha(x)$ exists (see §1.4) is called $L_\alpha^p(a, b)$. In case $\alpha(x) = x$ we use the notation $L^p(a, b)$; in case $p = 1$, $\alpha(x)$ arbitrary, the notation $L_\alpha(a, b)$ is used. If $f(x)$ and $g(x)$ belong to the class $L_\alpha^p(a, b)$, the same is true for $f(x) + g(x)$. (Cf. Kaczmarz-Steinhaus **1**, pp. 10–11.)

1.11. Inequalities

(1) *Cauchy's inequality.* Let $\{a_\nu\}$, $\{b_\nu\}$, $\nu = 1, 2, \cdots, n$, be two systems of complex numbers. Then

$$\left|\sum_{\nu=1}^{n} a_\nu b_\nu\right|^2 \leqq \sum_{\nu=1}^{n} |a_\nu|^2 \sum_{\nu=1}^{n} |b_\nu|^2. \tag{1.11.1}$$

The equality sign holds if and only if two numbers λ, μ, not both zero, exist such that $\lambda a_\nu + \mu \bar{b}_\nu = 0$, $\nu = 1, 2, \cdots, n$.

(2) *Schwarz's inequality.* Let $f(x)$ and $g(x)$ be two functions of class $L_\alpha^2(a, b)$. Then $f(x)g(x)$ is of class $L_\alpha(a, b)$, and

$$\left|\int_a^b f(x)g(x)\,d\alpha(x)\right|^2 \leqq \int_a^b |f(x)|^2\,d\alpha(x) \int_a^b |g(x)|^2\,d\alpha(x). \tag{1.11.2}$$

(3) *Inequality for the arithmetic and geometric mean.* If $f(x) > 0$, we have

$$\frac{\int_a^b f(x)\,d\alpha(x)}{\int_a^b d\alpha(x)} \geqq \exp\left\{\frac{\int_a^b \log f(x)\,d\alpha(x)}{\int_a^b d\alpha(x)}\right\}, \tag{1.11.3}$$

provided all integrals exist, and $\int_a^b d\alpha(x) > 0$. (Cf. Hardy-Littlewood-Pólya **1**, pp. 137–138.)

(4) *Abel's transformation and Abel's inequality.* From

$$\begin{aligned} &f_0g_0 + f_1g_1 + \cdots + f_ng_n \\ &\quad = (f_0 - f_1)G_0 + (f_1 - f_2)G_1 + \cdots + (f_{n-1} - f_n)G_{n-1} + f_nG_n, \end{aligned} \tag{1.11.4}$$

where

$$G_\nu = g_0 + g_1 + \cdots + g_\nu, \qquad \nu = 0, 1, 2, \cdots, n, \tag{1.11.5}$$

we obtain, assuming $f_0 \geqq f_1 \geqq \cdots \geqq f_n \geqq 0$, and $|G_\nu| \leqq G$, $\nu = 0, 1, \cdots, n$, the inequality

$$|f_0g_0 + f_1g_1 + \cdots + f_ng_n| \leqq f_0\, G. \tag{1.11.6}$$

(5) *Second mean-value theorem of the integral calculus.* Let $f(x) \geqq 0$ be a non-increasing function, and let $g(x)$ be continuous, $a \leqq x \leqq b$, a and b finite. Then

$$\int_a^b f(x)g(x)\,dx = f(a + 0) \int_a^\xi g(x)\,dx, \qquad a \leqq \xi \leqq b. \tag{1.11.7}$$

1.12. Polynomials and trigonometric polynomials

We shall consider polynomials in x of the form

$$\rho(x) = c_0 + c_1x + c_2x^2 + \cdots + c_mx^m, \tag{1.12.1}$$

with arbitrary complex coefficients $c_0, c_1, c_2, \cdots, c_m$. Here m is called the degree; and if $c_m \neq 0$, the precise degree of $\rho(x)$. In what follows an arbitrary polynomial of degree m will be denoted by π_m. If $\rho_0(x), \rho_1(x), \cdots, \rho_n(x)$ are arbitrary polynomials such that $\rho_m(x)$ has the precise degree m, every π_n can be represented as a linear combination of these polynomials with coefficients which are uniquely determined.

A trigonometric polynomial in θ of degree m has the form

$$g(\theta) = a_0 + a_1 \cos \theta + b_1 \sin \theta + \cdots + a_m \cos m\theta + b_m \sin m\theta, \tag{1.12.2}$$

with arbitrary complex coefficients. Here m is again called the degree of $g(\theta)$; m is the precise degree if $|a_m| + |b_m| > 0$. According as all the b_μ or all the a_μ vanish, $g(\theta)$ is referred to as a cosine or a sine polynomial.

The functions $\cos m\theta$ and $\sin (m + 1)\theta/\sin \theta$ are polynomials in $\cos \theta = x$ of the precise degree m and are called *Tchebichef polynomials of the first and second kind*, respectively. These polynomials play a fundamental rôle in subsequent considerations. Setting

$$\cos m\theta = T_m(\cos \theta) = T_m(x), \qquad \frac{\sin (m+1)\theta}{\sin \theta} = U_m(\cos \theta) = U_m(x), \tag{1.12.3}$$

we see that any cosine polynomial of degree m is a polynomial of the same degree in $\cos \theta = x$, and conversely. Any sine polynomial of degree m, divided by $\sin \theta$, furnishes a cosine polynomial of degree $m - 1$. Thus, a sine polynomial can be represented as the product of $\sin \theta = (1 - x^2)^{1/2}$ by a polynomial in $\cos \theta = x$.

The polynomials (1.12.3) are special cases of the so-called Jacobi polynomials (cf. Chapter IV). They contain only even or only odd powers of x according as m is even or odd. Thus $\cos (m + \frac{1}{2})\theta/\cos (\theta/2)$ and $\sin (m + \frac{1}{2})\theta/\sin(\theta/2)$ are cosine polynomials in θ of degree m; they are also connected with the Jacobi polynomials (see (4.1.8)).

We define the "reciprocal" polynomial of (1.12.1) by

$$\rho^*(x) = x^m \bar{\rho}(x^{-1}) = \bar{c}_m + \bar{c}_{m-1}x + \bar{c}_{m-2}x^2 + \cdots + \bar{c}_0 x^m. \tag{1.12.4}$$

If the zeros of $\rho(x)$ are $x_1, x_2, \cdots, x_m$, those of $\rho^*(x)$ are $x_1^*, x_2^*, \cdots, x_m^*$, where $x_\mu^* = \bar{x}_\mu^{-1}$ is the point which is obtained from x_μ by inversion with respect to the unit circle $|x| = 1$ in the complex x-plane. The zeros must be counted according to their multiplicity, and $0^* = \infty$, $\infty^* = 0$; ∞ as a zero of order k means that the coefficients of the k highest powers vanish.

1.2. Representation of non-negative trigonometric polynomials

THEOREM 1.2.1. *Let $g(\theta)$ be a trigonometric polynomial with real coefficients which is non-negative for all real values of θ. Then there exists a polynomial $\rho(z)$ of the same degree as $g(\theta)$ such that $g(\theta) = |\rho(z)|^2$, where $z = e^{i\theta}$. Conversely, if $z = e^{i\theta}$, the expression $|\rho(z)|^2$ always represents a non-negative trigonometric polynomial in θ of the same degree as the polynomial $\rho(z)$.*

See Fejér 5. The second part of the statement is obvious. The first part is easily derived from (1.12.2) by introducing $z^k + z^{-k}$ for $2 \cos k\theta$ and $z^k - z^{-k}$ for $2i \sin k\theta$. We then find $g(\theta) = z^{-m}G(z)$, where $G(z)$ is a π_{2m} for which $G^*(z) = G(z)$. Now those zeros of $G(z)$ which are different from 0 and ∞, and which do not have the absolute value 1, can be combined in pairs of the form z_μ, z_μ^*, $0 < |z_\mu| < 1$, where z_μ^* has a meaning similar to that in §1.12. Furthermore, every real zero θ_0 of $g(\theta)$ is of even multiplicity, and $e^{i\theta_0}$ is a zero of $G(z)$ of the same multiplicity. Thus

$$(1.2.1)\qquad G(z) = cz^\kappa \prod_{\mu=1}^{\sigma} (z - z_\mu)(z - z_\mu^*) \prod_{\nu=1}^{\tau} (z - \zeta_\nu)^2,$$

$$0 < |z_\mu| < 1, \qquad |\zeta_\nu| = 1; \qquad \kappa + \sigma + \tau = m.$$

Since $g(\theta) = |g(\theta)| = |G(z)|$, $z = e^{i\theta}$, and $|z - z_\mu| = |z_\mu| \, |z - z_\mu^*|$, $z = e^{i\theta}$, the theorem is established.

The representation in question is, however, not unique. Indeed, if α denotes an arbitrary zero of $\rho(z)$, the polynomial $\rho(z)\,(1 - \bar{\alpha}z)/(z - \alpha)$ furnishes another representation. Hence assuming $g(\theta) \not\equiv 0$, we can gradually remove all the zeros from $|z| < 1$ and obtain the following theorem:

Theorem 1.2.2. *Let $g(\theta)$ satisfy the condition of Theorem 1.2.1 and $g(\theta) \not\equiv 0$. Then a representation $g(\theta) = |h(e^{i\theta})|^2$ exists such that $h(z)$ is a polynomial of the same degree as $g(\theta)$, with $h(z) \neq 0$ in $|z| < 1$, and $h(0) > 0$. This polynomial is uniquely determined. If $g(\theta)$ is a cosine polynomial, $h(z)$ is a polynomial with real coefficients.*

A generalization of this normalized representation (its extension to a certain class of non-negative functions $g(\theta)$) is of great importance in the discussion of the asymptotic behavior of orthogonal polynomials. (See Chapters X-XIII.)

1.21. Theorem of Lukács concerning non-negative polynomials

(1) **Theorem** 1.21.1 (Theorem of Lukács). *Let $\rho(x)$ be a π_m non-negative in $[-1, +1]$. Then $\rho(x)$ can be represented in the form*

$$(1.21.1)\qquad \rho(x) = \begin{cases} \{A(x)\}^2 + (1 - x^2)\{B(x)\}^2 & \text{if } m \text{ is even,} \\ (1 + x)\{C(x)\}^2 + (1 - x)\{D(x)\}^2 & \text{if } m \text{ is odd.} \end{cases}$$

Here $A(x)$, $B(x)$, $C(x)$, and $D(x)$ are real polynomials such that the degrees of the single terms on the right-hand side do not exceed m.

The proof can be based on Theorem 1.2.2. We have

$$\rho(\cos\theta) = |h(e^{i\theta})|^2 = |e^{-im\theta/2} h(e^{i\theta})|^2,$$

where $h(z)$ is a π_m with real coefficients. Now the expressions

$$(1.21.2)\qquad \cos m\theta, \quad \frac{\sin (m + 1)\theta}{\sin\theta}, \quad \frac{\cos (m + \frac{1}{2})\theta}{\cos (\theta/2)}, \quad \frac{\sin (m + \frac{1}{2})\theta}{\sin (\theta/2)}$$

are all π_m in $\cos\theta$ (see §1.12), so that

$$e^{-im\theta/2}h(e^{i\theta}) = \begin{cases} A(\cos\theta) + i\sin\theta\, B(\cos\theta) & \text{if } m \text{ is even,} \\ 2^{\frac{1}{2}}\cos(\theta/2)\, C(\cos\theta) + i2^{\frac{1}{2}}\sin(\theta/2) D(\cos\theta) & \text{if } m \text{ is odd,} \end{cases}$$

where the degrees of $A(x)$, $B(x)$, $C(x)$, $D(x)$ are, respectively, $m/2$, $m/2 - 1$, $(m - 1)/2$, $(m - 1)/2$.

(2) The following theorem has a simpler character:

THEOREM 1.21.2. *Every polynomial in x, which is non-negative for all real values of x, can be represented in the form $\{A(x)\}^2 + \{B(x)\}^2$. Every polynomial which is non-negative for $x \geqq 0$, can be represented in the form $\{A(x)\}^2 + \{B(x)\}^2 + x[\{C(x)\}^2 + \{D(x)\}^2]$. Here $A(x)$, $B(x)$, $C(x)$, $D(x)$ are all real polynomials, and the degree of each term does not exceed the degree of the given polynomial.*

These representations can also be written in the form $|P(x)|^2$ and $|P(x)|^2 + x|Q(x)|^2$, respectively, where $P(x)$ and $Q(x)$ are polynomials with complex coefficients; for the degrees the same remark holds as before. In the case when $x \geqq 0$, $B(x)$ and $D(x)$ can be chosen to vanish identically. See Achieser **4**, [2.54], and Karlin-Studden **1**, Chapter V, Corollary 8.1.

In connection with this section see Pólya-Szegö **1**, vol. 2, pp. 82, 275, 276, problems 44, 45, 47.

1.22. Theorems of S. Bernstein

THEOREM 1.22.1. *If $g(\theta)$ is a trigonometric polynomial of degree m satisfying the condition $|g(\theta)| \leqq 1$, θ arbitrary and real, then $|g'(\theta)| \leqq m$.*

This theorem is due to S. Bernstein. (Cf. M. Riesz **1**.) The upper bound m cannot be replaced by a smaller one as is readily seen by taking $g(\theta) = \cos m\theta$. The following special case is worthy of notice:

THEOREM 1.22.2. *Let $\rho(z)$ be an arbitrary π_m satisfying the condition $|\rho(z)| \leqq 1$, where z is complex, and $|z| \leq 1$; then $|\rho'(z)| \leqq m$, $|z| \leqq 1$.*

With regard to this theorem see also Szász **1**, pp. 516–517. Finally we mention the following consequence of Theorem 1.22.1:

THEOREM 1.22.3. *Let $\rho(x)$ be a π_m satisfying the condition $|\rho(x)| \leqq 1$ in $-1 \leqq x \leqq +1$. Then*

$$|\rho'(x)| \leqq (1 - x^2)^{-\frac{1}{2}} m.$$

This follows by applying Theorem 1.22.1 to $g(\theta) = \rho(\cos\theta)$.

1.3. Approximation by polynomials

(1) THEOREM 1.3.1 (Theorem of Weierstrass). *A function, continuous in a finite closed interval, can be approximated with a preassigned accuracy by polynomials. A function of a real variable which is continuous and has the period 2π, can be approximated by trigonometric polynomials.*

For information concerning this theorem we refer to Jackson **4**. In the second part of the theorem let the function in question be even (odd); then the approximating trigonometric polynomials can be chosen as cosine (sine) polynomials.

THEOREM 1.3.2. *Let $\omega(\delta)$ be the modulus of continuity of a given function $f(x)$, continuous in the finite interval $[a, b]$,*

$$\omega(\delta) = \max |f(x') - f(x'')| \qquad \text{if } |x' - x''| \leqq \delta. \tag{1.3.1}$$

Then for each m we can find a polynomial $\rho(x)$ of degree m, such that in the given interval of length l we have

$$|f(x) - \rho(x)| < A\omega(l/m). \tag{1.3.2}$$

In the case of a periodic function $f(\theta)$ with period 2π, a trigonometric polynomial $g(\theta)$ of degree m can be found such that

$$|f(\theta) - g(\theta)| < B\omega(2\pi/m). \tag{1.3.3}$$

Here A and B are absolute constants.

In this connection see Jackson **4**, pp. 7, 15.

THEOREM 1.3.3. *Let $f(x)$ have a continuous derivative of order μ in the finite interval $[a, b]$, $\mu \geqq 1$, and let $\omega_\mu(\delta)$ be the modulus of continuity of $f^{(\mu)}(x)$. Then a polynomial $\rho(x)$ of degree $m + \mu$ exists such that*

$$\begin{aligned} |f(x) - \rho(x)| &< C(l/m)^{\mu}\omega_\mu(l/m), \\ |f'(x) - \rho'(x)| &< C(l/m)^{\mu-1}\omega_\mu(l/m), \qquad l = b - a. \end{aligned} \tag{1.3.4}$$

Here C is a constant depending only on μ.

Analogous inequalities can be obtained for all the derivatives $f(x)$, $f'(x)$, $\cdots, f^{(\mu)}(x)$.

For the first inequality see Jackson **4** (p. 18, Theorem VIII). To prove the second inequality we first establish the following lemma:

LEMMA. *Let $f(\theta)$ be a function of period 2π satisfying the Lipschitz condition*

$$|f(\theta_1) - f(\theta_2)| < \lambda |\theta_1 - \theta_2|, \tag{1.3.5}$$

where λ is a positive constant. Then there exist for each m trigonometric polynomials $g(\theta)$ of degree m such that

$$|f(\theta) - g(\theta)| < \frac{D'\lambda}{m}, \qquad |g'(\theta)| < D''\lambda, \tag{1.3.6}$$

where D' and D'' are absolute constants.

For the first inequality (1.3.6) see Jackson **4**, pp. 2–6. When we use his notation and argument, it suffices to show that $|\lambda^{-1}I'_m(\theta)|$ is less than an absolute constant. But

(1.3.7) $$I'_m(\theta) = -\frac{h_m}{2}\int_{-\pi/2}^{+\pi/2}\{f(\theta + 2u) - f(\theta)\}F'_m(u)\,du$$

and

$$\int_0^{\pi/2} u\,|\,F'_m(u)\,|\,du = 4\int_0^{\pi/2} u\left|\frac{\sin mu}{m\sin u}\right|^3\left|\frac{d}{du}\frac{\sin mu}{m\sin u}\right|du$$

(1.3.8) $$= O(1)\int_0^{\pi/2} u\left|\frac{\sin mu}{mu}\right|^3\left|\frac{d}{du}\frac{\sin mu}{mu}\right|du$$

$$+ O(1)\int_0^{\pi/2} u\left|\frac{\sin mu}{mu}\right|^3\left|\frac{\sin mu}{mu}\right|du,$$

since $u/\sin u$ is analytic in the closed interval $[0, \pi/2]$. On writing $mu = x$,

$$O(m^{-1})\int_0^{\infty} x\left|\frac{\sin x}{x}\right|^3\left|\frac{d}{dx}\frac{\sin x}{x}\right|dx + O(m^{-2})\int_0^{\infty} x\left|\frac{\sin x}{x}\right|^4 dx = O(m^{-1}).$$

Now we use (cf. loc. cit.) $h_m = O(m)$.

The analogue of the lemma for polynomials can be derived in the usual way. Then in the upper bound of the first inequality of (1.3.6) the factor $b - a = l$ appears. It is convenient to transform the interval $a \leqq x \leqq b$ into $-\frac{1}{2} \leqq y \leqq \frac{1}{2}$ (instead of $-1 \leqq y \leqq 1$, cf. Jackson, loc. cit., p. 14), defining the function in $[-1, -\frac{1}{2}]$ and $[\frac{1}{2}, 1]$ by a constant.

In order to prove Theorem 1.3.3, we apply Theorem VIII of Jackson (loc. cit., p. 18) to $f'(x)$. (For this argument cf. loc. cit., p. 16.) Thus

$$|\,f'(x) - q(x)\,| < K(l/m)^{\mu-1}\omega_\mu(l/m),$$

where $q(x)$ is a proper $\pi_{m+\mu-1}$. Applying the lemma to $f(x) - \int_a^x q(t)dt$, which satisfies a Lipschitz condition with

$$\lambda = K(l/m)^{\mu-1}\omega_\mu(l/m),$$

we obtain a π_m, say $\sigma(x)$, such that

$$\left|f(x) - \int_a^x q(t)\,dt - \sigma(x)\right| < K'(l/m)^{\mu}\omega_\mu(l/m), \qquad |\,\sigma'(x)\,| < K''(l/m)^{\mu-1}\omega_\mu(l/m).$$

If we write $\int_a^x q(t).dt + \sigma(x) = \rho(x)$, the statement is established.

The constants K, K', K'' in the last three inequalities depend only on μ.

(2) THEOREM 1.3.4 (Theorem of Runge-Walsh). *Let $f(x)$ be an analytic function regular in the interior of a Jordan curve C and continuous in the closed domain bounded by C. Then $f(x)$ can be approximated with an arbitrary accuracy by polynomials.*

See Walsh **1**, p. 36. This theorem has been proved by Runge in case $f(x)$ is analytic on C; the general case is due to Walsh.

We need also a supplement to the former theorem, due to Walsh (**1**, pp. 75–76). Let C be again a Jordan curve in the complex x-plane. Let $x = \phi(z)$ be the map function carrying over the exterior of C into $|z| > 1$ and preserving $x = z = \infty$. Then the circles $|z| = R$, $R > 1$, correspond to certain curves C_R, called *level curves.* We have

THEOREM 1.3.5. *Let $f(x)$ be analytic within and on C, and let C_R be the largest level curve in the interior of which $f(x)$ is regular. Then to an arbitrary r, $0 < r < R$, there corresponds a constant $M > 0$ such that, for each m, a polynomial $\rho_m(x)$ of degree m exists satisfying the inequality*

$$|f(x) - \rho_m(x)| < Mr^{-m}, \qquad x \text{ on } C. \tag{1.3.9}$$

This holds also if C is a Jordan arc, for example, the interval $-1 \leqq x \leqq +1$. In the latter case C_R is an ellipse with foci at ± 1, and R is the sum of the semi-axes (§1.9).

1.4. Orthogonality; weight function; vectors in function spaces

(1) *Let $\alpha(x)$ be a non-decreasing function in $[a, b]$ which is not constant.* If $a = -\infty$ (or $b = +\infty$), we require that $\alpha(-\infty) = \lim_{x \to -\infty} \alpha(x)$ ($\alpha(+\infty) = \lim_{x \to +\infty} \alpha(x)$) should be finite. The *scalar product* of two real functions $f(x)$ and $g(x)$, where x ranges over the real interval $[a, b]$, is defined by the Stieltjes-Lebesgue integral

$$(f, g) = \int_a^b f(x)g(x)\, d\alpha(x), \tag{1.4.1}$$

where we assume that $f(x)g(x)$ is of the class $L_\alpha(a, b)$. This is certainly the case if $f(x)$ and $g(x)$ are both continuous, or both of bounded variation, and $[a, b]$ is a finite interval. For a fixed function $\alpha(x)$ the *orthogonality* with respect to the "distribution" $d\alpha(x)$ may be defined by the relation

$$(f, g) = 0. \tag{1.4.2}$$

We shall also use the expression "$f(x)$ is orthogonal to $g(x)$."

If we permit $f(x)$ and $g(x)$ to be complex functions in general, definition (1.4.1) must be modified to read

$$(f, g) = \int_a^b f(x)\overline{g(x)}\, d\alpha(x). \tag{1.4.3}$$

With this change in the definition of (f, g), we retain (1.4.2) as the definition of orthogonality.

[For the definition of *Stieltjes-Lebesgue integrals* see, for instance, Hildebrandt **1**, pp. 185–194. This definition, given originally for a monotonic $\alpha(x)$, can easily be extended to the case where $\alpha(x)$ is of bounded variation. Hildebrandt **1**, pp. 177–178, may also be consulted for the definition of Riemann-Stieltjes integrals.

In what follows we sometimes need the formula for integration by parts:

$$\int_a^b f(x)\,d\alpha(x) + \int_a^b \alpha(x)\,df(x) = f(b)\alpha(b) - f(a)\alpha(a), \tag{1.4.4}$$

where a and b are finite, $\alpha(x)$ is of bounded variation, and $f(x)$ is continuous. The integrals are taken as Riemann-Stieltjes integrals.

The expression "distribution" used above arises from the classical interpretation of $d\alpha(x)$ as a continuous or discontinuous mass distribution in the interval $[a, b]$, the mass contributed by the interval $[x_1, x_2]$ of $[a, b]$ being $\alpha(x_2) - \alpha(x_1)$.]

(2) If $\alpha(x)$ is absolutely continuous, the scalar product (1.4.1) reduces to

$$(f, g) = \int_a^b f(x)g(x)w(x)\,dx, \tag{1.4.5}$$

where the integral is assumed to exist in Lebesgue's sense. *Here $w(x)$ is a non-negative function measurable in Lebesgue's sense for which $\int_a^b w(x)\,dx > 0$.* We shall call $w(x)$ the *weight function,* referring to a weight function of, or on, the given interval. Instead of "weight function" the term "norm function" is sometimes used in the literature.[3] In the case of a distribution $w(x)\,dx$ the total mass corresponding to the interval $[x_1, x_2]$ is obviously $\int_{x_1}^{x_2} w(x)\,dx$. In what follows we refer to distributions of the form $d\alpha(x)$ as *distributions of Stieltjes type.*

We use the same concept of distribution and weight function on a curve or on an arc in the complex plane, for example, on the unit circle. Then we replace the variable x by the real parameter which is used for the definition of the curve or arc in question. (See Chapters XI and XVI.)

(3) Let $d\alpha(x)$, or $w(x)\,dx$, $a \leqq x \leqq b$, be a fixed distribution, and consider a space of "vectors" defined by the set of real functions $f(x)$ which belong to the class $L_\alpha^2(a, b)$. The scalar product of two vectors (functions) $f(x)$ and $g(x)$ is defined by (1.4.1) and the length (magnitude, norm) of a vector $f(x)$ by $\|f\| = (f, f)^{\frac{1}{2}}$. Vectors (functions) with $\|f\| = 0$ are called zero-vectors (zero-functions); vectors (functions) with $\|f\| = 1$ are said to be normalized. When $f(x)$ is not a zero-function, $\lambda f(x)$ will be normalized provided $\lambda \neq 0$ is a proper constant, uniquely determined save possibly for sign. If the functions $\alpha(x)$ and $w(x)$ satisfy the conditions mentioned in (1) and (2), there exist functions of positive length for both cases. In the second case $f(x)$ is a zero-function if and only if $\{f(x)\}^2 w(x)$, or what amounts to the same thing, $f(x)w(x)$, vanishes everywhere in $[a, b]$ except on a set of measure zero. If $w(x)$ and $f(x)$ are integrable in Riemann's sense, $f(x)$ is a zero-function provided $f(x)w(x)$ vanishes at every point of continuity.

We note the inequality of Schwarz (cf. (1.11.2))

$$\|fg\| \leqq \|f\|\,\|g\|, \tag{1.4.6}$$

[3] Some corresponding German and French terms are: Belegungsfunktion, Gewichtsfunktion, fonction caractéristique (Stekloff), poids (S. Bernstein).

the equality sign holding if and only if $\lambda f(x) + \mu g(x)$ is a zero-function with λ and μ proper constants not both zero.

A finite set of functions $f_0(x), f_1(x), \cdots, f_l(x)$ is said to be linearly independent if the equation

$$|| \lambda_0 f_0(x) + \lambda_1 f_1(x) + \cdots + \lambda_l f_l(x) || = 0$$

can be true only for

$$\lambda_0 = \lambda_1 = \cdots = \lambda_l = 0.$$

Evidently no zero-function can be contained in such a system. An enumerable set of functions ($l = \infty$) is called linearly independent if the preceding condition is satisfied for every finite subset of the given set.

The extension of these considerations to complex vector spaces is not difficult. The scalar product is then defined as in (1.4.3).

Concerning the axiomatic foundation of these concepts see Stone **1**, Chapter I.

1.5. Closure; integral approximations

(1) DEFINITION. *Let $p \geqq 1$, and let $\alpha(x)$ be a non-decreasing function in $[a, b]$ which is not constant.*[4] *Let the functions*

$$f_0(x), f_1(x), f_2(x), \cdots, f_n(x), \cdots \tag{1.5.1}$$

be of the class $L_\alpha^p(a, b)$. The system (1.5.1) is called closed in $L_\alpha^p(a, b)$ if for every $f(x)$ of $L_\alpha^p(a, b)$ and for every $\epsilon > 0$ a function of the form

$$k(x) = c_0 f_0(x) + c_1 f_1(x) + \cdots + c_n f_n(x) \tag{1.5.2}$$

exists such that

$$\int_a^b | f(x) - k(x) |^p \, d\alpha(x) < \epsilon. \tag{1.5.3}$$

With regard to this definition see Kaczmarz-Steinhaus **1**, p. 49. These authors use the term "Abgeschlossenheit" for "closure."

(2) THEOREM 1.5.1. *Let p and $\alpha(x)$ have the same meaning as in the previous definition, and let the function $f(x)$ be of the class $L_\alpha^p(a, b)$, a and b finite. Then for every $\epsilon > 0$ a continuous function $F(x)$ can be determined such that*

$$\int_a^b | f(x) - F(x) |^p \, d\alpha(x) < \epsilon. \tag{1.5.4}$$

For a Riemann-integrable function with $\alpha(x) = x$, this follows by a well-known argument from the definition of the integral. In the general case, it is convenient to use the method of W. H. Young of approximating Stieltjes-Lebesgue integrals. (See Hildebrandt **1**, p. 190.)

Applying Weierstrass' theorem, we obtain the following:

[4] See the remark at the beginning of §1.4 (1).

THEOREM 1.5.2. *Let p, a, b, $\alpha(x)$, $f(x)$ satisfy the conditions of Theorem* 1.5.1. *For every $\epsilon > 0$ there exists a polynomial $\rho(x)$ such that*

$$\int_a^b |f(x) - \rho(x)|^p \, d\alpha(x) < \epsilon. \tag{1.5.5}$$

This means the closure of the system

$$\{x^n\}, \qquad n = 0, 1, 2, \cdots, \tag{1.5.6}$$

in the class $L_\alpha^p(a, b)$. In what follows, we shall use in particular the cases $p = 1$ and $p = 2$.

An analogous statement holds for the "mean approximation" of $f(x)$ by trigonometric polynomials, which is equivalent to the property of closure of the system

$$1, \cos x, \sin x, \cos 2x, \sin 2x, \cdots, \cos nx, \sin nx, \cdots \tag{1.5.7}$$

in $L_\alpha^p(-\pi, +\pi)$.

(3) A more precise form of Theorem 1.5.2 is often useful.

THEOREM 1.5.3. *Let p, a, b, $\alpha(x)$, $f(x)$ satisfy the conditions of Theorem* 1.5.1 *and let $f(x)$ be real-valued. Then we can find a polynomial $\rho(x)$ which satisfies* (1.5.5) *and is such that $\rho(x)$ remains between the upper and lower bounds of $f(x)$.*

We refer also to the following property of Riemann-Stieltjes integrals which plays a rôle in Chapter X.

THEOREM 1.5.4. *Let the real-valued function $f(x)$ be bounded in $[a, b]$, a and b finite, $\alpha(x)$ non-decreasing, and let the Riemann-Stieltjes integral $\int_a^b f(x)\, d\alpha(x)$ exist. For every $\epsilon > 0$ there exist polynomials $\rho(x)$ and $\mathrm{P}(x)$ such that*

$$\inf f(x) - \epsilon \leqq \rho(x) \leqq f(x) \leqq \mathrm{P}(x) \leqq \sup f(x) + \epsilon, \tag{1.5.8}$$

and

$$\int_a^b \{\mathrm{P}(x) - \rho(x)\} \, d\alpha(x) < \epsilon. \tag{1.5.9}$$

See (for $\alpha(x) = x$) Pólya-Szegö **1**, vol. 1, pp. 65, 228, problem 137.

Similar statements hold for approximations by trigonometric polynomials. If $f(x)$ is an even function, $-\pi \leqq x \leqq +\pi$, the approximating trigonometric polynomials can be chosen as cosine polynomials.

1.6. Linear functional operations

(1) Let $\mathfrak{U}(f)$ be an operation which makes a number $\mathfrak{U}(f)$ correspond to every function $f(x)$, continuous in the finite interval $[a, b]$. This operation is called *additive* if

$$\mathfrak{U}(c_1 f_1 + c_2 f_2) = c_1 \mathfrak{U}(f_1) + c_2 \mathfrak{U}(f_2) \tag{1.6.1}$$

whenever c_1 and c_2 are constants and $f_1(x)$ and $f_2(x)$ are arbitrary continuous functions in $[a, b]$. It is called *continuous* if $\mathfrak{U}(f_n) \to \mathfrak{U}(f)$ whenever $f_n(x) \to f(x)$ uniformly in $[a, b]$. Additive and continuous operations are called *linear*.

An operation $\mathfrak{U}(f)$ is called *limited* if there exists a constant M such that $| \mathfrak{U}(f) | \leqq M \max | f |$. The greatest lower bound of these constants M is the *norm* of $\mathfrak{U}(f)$. The class of additive and limited operations $\mathfrak{U}(f)$ coincides with that of the linear operations.

According to a theorem of F. Riesz (**1**), any linear operation can be written in the form

$$\mathfrak{U}(f) = \int_a^b f(x)\, d\alpha(x), \tag{1.6.2}$$

where $\alpha(x)$ is of bounded variation, defined in $[a, b]$ and independent of $f(x)$. It is obvious that (1.6.2) always represents a linear operation. In (1.6.2) the function $\alpha(x)$ can always be so normalized that $\alpha(x - 0) \leqq \alpha(x) \leqq \alpha(x + 0)$ or $\alpha(x + 0) \leqq \alpha(x) \leqq \alpha(x - 0)$ for $a < x < b$. Then the norm of $\mathfrak{U}(f)$ is given by $\int_a^b | d\alpha(x) |$, which is the total variation of $\alpha(x)$.

(2) Let $K(x)$ be a given function continuous in $[a, b]$. Then

$$\int_a^b f(x)K(x)\, dx \tag{1.6.3}$$

defines a linear operation. Dirichlet's integral

$$\frac{1}{2\pi} \int_{-\pi}^{+\pi} f(x) \frac{\sin \{(2n+1)(x - x_0)/2\}}{\sin \{(x - x_0)/2\}}\, dx, \tag{1.6.4}$$

where n is a non-negative integer and x_0 arbitrary, is a special case of (1.6.3). It represents the nth partial sum of the Fourier expansion of $f(x)$ at $x = x_0$. Another important example is Fejér's integral

$$\frac{1}{2\pi(n + 1)} \int_{-\pi}^{+\pi} f(x) \left(\frac{\sin \{(n + 1)(x - x_0)/2\}}{\sin \{(x - x_0)/2\}} \right)^2 dx, \tag{1.6.5}$$

which represents the nth Cesàro mean of the Fourier expansion of $f(x)$. A further illustration of linear operations is furnished by Lagrange's interpolation polynomial

$$L(x) = L(f; x) = f(x_0)l_0(x) + f(x_1)l_1(x) + \cdots + f(x_n)l_n(x), \tag{1.6.6}$$

where $l_0(x), l_1(x), \cdots, l_n(x)$ are the fundamental polynomials associated with the interpolation points $x_0, x_1, \cdots, x_n$ (see Chapter XIV). For a fixed value $x = \xi$, the expression $L(f; \xi)$ represents a linear operation on $f(x)$. Finally, the general mechanical quadrature formula,

$$Q(f) = \lambda_0 f(x_0) + \lambda_1 f(x_1) + \cdots + \lambda_n f(x_n) \tag{1.6.7}$$

is also an example; here $\lambda_0, \lambda_1, \cdots, \lambda_n$ are the so-called Cotes numbers (see Chapter XV).

(3) We consider the sequence of linear operations

$$\mathfrak{U}_n(f) = \int_a^b f(x)\, d\alpha_n(x), \qquad n = 0, 1, 2, \cdots, \tag{1.6.8}$$

and the operation

$$\mathfrak{U}(f) = \int_a^b f(x)\, d\alpha(x), \tag{1.6.9}$$

where $\alpha_n(x)$ are normalized as in (1.6.2). Then we have the following theorem:

THEOREM 1.6. *A necessary and sufficient condition that* $\lim_{n\to\infty}\mathfrak{U}_n(f) = \mathfrak{U}(f)$, *where* $f(x)$ *is an arbitrary continuous function, is that the following two relations be satisfied simultaneously:*

$$\begin{aligned} &\lim_{n\to\infty} \mathfrak{U}_n(x^k) = \mathfrak{U}(x^k), && k = 0, 1, 2, \cdots, \\ &\int_a^b |\, d\alpha_n(x)\,| < A, && n = 0, 1, 2, \cdots. \end{aligned} \tag{1.6.10}$$

Moreover, if the second condition (1.6.10) is not satisfied, a continuous function $f(x)$ exists such that the sequence $\{\mathfrak{U}_n(f)\}$ is unbounded.

This important theorem is due to E. Helly (**1**, pp. 268–271). See also Banach **1**, p. 123. The first condition (1.6.10) expresses the validity of the limiting relation for an arbitrary polynomial. The second condition (1.6.10) states that the total variations of the functions $\alpha_n(x)$ are bounded.

(4) Let $b - a = 2\pi$, and suppose that $f(x)$, $\alpha_n(x)$, and $\alpha(x)$ are functions with period 2π. Then the first condition (1.6.10) must be replaced by the following:

$$\begin{aligned} &\lim_{n\to\infty} \mathfrak{U}_n(\cos kx) = \mathfrak{U}(\cos kx), \\ &\lim_{n\to\infty} \mathfrak{U}_n(\sin kx) = \mathfrak{U}(\sin kx), \end{aligned} \qquad k = 0, 1, 2, \cdots \tag{1.6.11}$$

One of the most important applications of the preceding considerations is to the theory of "singular integrals" of Lebesgue:

$$\mathfrak{U}_n(f) = \int_a^b f(x) K_n(x)\, dx, \tag{1.6.12}$$

where $\{K_n(x)\}$ is a given sequence of continuous functions. In this case we are mainly interested in finding a necessary and sufficient condition that $\mathfrak{U}_n(f) \to f(x_0)$, where x_0 is a fixed point in $[a, b]$ and $f(x)$ an arbitrary continuous function. According to Helly's theorem, this must hold if $f(x)$ is an arbitrary polynomial (or trigonometric polynomial in the periodic case) and the so-called Lebesgue constants (which are the norms of $\mathfrak{U}_n(f)$) are bounded:

$$\int_a^b |\, K_n(x)\,|\, dx < A. \tag{1.6.13}$$

See Lebesgue **1**, **2**; in particular articles 45, 46, pp. 86–88. See also Haar **1**.

For Dirichlet's integral (1.6.4) this condition (1.6.13) is not satisfied; hence there exist continuous functions whose Fourier expansions are divergent at a preassigned point. (Du Bois-Reymond **1**; Lebesgue **2**, chapter IV, pp. 84–89.) This condition is, however, satisfied in the case of Fejér's integral (1.6.5) which implies the Cesàro summability of the Fourier expansion of a continuous function (Fejér **2**, in particular p. 60). The same holds for the Cesàro means of second order of the Legendre series (Fejér **4**).

Regarding applications of Helly's theorem to the theory of interpolation and mechanical quadrature, see Chapters XIV and XV.

1.7. The Gamma function

The Euler integral of the second kind

$$\Gamma(z) = \int_0^\infty e^{-t} t^{z-1}\, dt \tag{1.7.1}$$

defines the Gamma function $\Gamma(z)$ for $\Re(z) > 0$. By analytic continuation we obtain a meromorphic function without zeros and with simple poles at $z = 0, -1, -2, \cdots$. The functional equations

$$\Gamma(z+1) = z\Gamma(z), \qquad \Gamma(z)\Gamma(1-z) = \pi/\sin \pi z \tag{1.7.2}$$

hold. Another important formula is

$$\begin{aligned}\Gamma(z)\Gamma(z+1/n)\cdots\Gamma(z+(n-1)/n)\\ = n^{\frac{1}{2}-nz}(2\pi)^{(n-1)/2}\Gamma(nz),\end{aligned} \qquad n \text{ a positive integer.} \tag{1.7.3}$$

In what follows we use mainly the cases $n = 2$ and $n = 3$.

The Euler integral of the first kind

$$B(p, q) = \int_0^1 x^{p-1}(1-x)^{q-1}\, dx, \qquad p > 0, q > 0, \tag{1.7.4}$$

can be expressed in terms of the Gamma function thus:

$$B(p,\ ) = \frac{\Gamma(p)\Gamma(q)}{\Gamma(p+q)}. \tag{1.7.5}$$

The integral (1.7.4) exists also for complex p and q with positive real parts for which (1.7.5) remains valid. By means of (1.7.5) the definition of $B(p, q)$ can be extended to arbitrary complex p and q. (See Whittaker-Watson **1**, Chapter 12, pp. 237, 239, 240, 254.)

The special case $n = 2$ of (1.7.3) is as follows:

$$\Gamma(z)\,\Gamma(z+1/2) = 2^{1-2z}\pi^{1/2}\Gamma(2z). \tag{1.7.6}$$

We mention also the formula

$$\frac{1}{\Gamma(z)} = \frac{1}{2\pi i}\int_{-\infty}^{(0+)} e^t t^{-z}\, dt. \tag{1.7.7}$$

1.71. Bessel functions

(1) The Bessel function of the first kind of order α can be defined by

(1.71.1)
$$J_\alpha(z) = \sum_{\nu=0}^{\infty} \frac{(-1)^\nu (z/2)^{\alpha+2\nu}}{\nu!\, \Gamma(\nu + \alpha + 1)}.$$

Obviously, $z^{-\alpha} J_\alpha(z)$ is an even integral function. Here α is arbitrary real. If α is a negative integer, $\{\Gamma(\nu + \alpha + 1)\}^{-1}$ must be replaced by 0 whenever $\nu + \alpha + 1 \leqq 0$. We then obtain the relation $J_\alpha(z) = (-1)^\alpha J_{-\alpha}(z)$. If α is not an integer, $J_\alpha(z)$ and $J_{-\alpha}(z)$ are linearly independent. We notice the special cases

(1.71.2)
$$J_{-\frac{1}{2}}(z) = \left(\frac{2}{\pi z}\right)^{\frac{1}{2}} \cos z, \qquad J_{\frac{1}{2}}(z) = \left(\frac{2}{\pi z}\right)^{\frac{1}{2}} \sin z.$$

The function (1.71.1) satisfies Bessel's differential equation

(1.71.3)
$$y'' + z^{-1}y' + (1 - \alpha^2 z^{-2})y = 0, \qquad y = J_\alpha(z).$$

For non-negative integral values of α we introduce the Bessel functions of the second kind

(1.71.4)
$$\begin{aligned} Y_\alpha(z) = \frac{2}{\pi}\left(\gamma + \log \frac{z}{2}\right) J_\alpha(z) &- \frac{1}{\pi} \sum_{\nu=0}^{\alpha-1} \frac{(\alpha - \nu - 1)!\,(z/2)^{2\nu-\alpha}}{\nu!} \\ &- \frac{1}{\pi} \sum_{\nu=0}^{\infty} \frac{(-1)^\nu (z/2)^{2\nu+\alpha}}{\nu!\,(\nu + \alpha)!} \{1/1 + 1/2 + \cdots \\ &+ 1/\nu + 1/1 + 1/2 + \cdots + 1/(\nu + \alpha)\}, \qquad \alpha = 0, 1, 2, \cdots. \end{aligned}$$

Here γ is Euler's constant. The first sum is to be suppressed for $\alpha = 0$, and the curly brackets in the second sum are to be replaced by 1 for $\nu = 0$, $\alpha = 0$, and by $1/1 + 1/2 + \cdots + 1/\alpha$ for $\nu = 0$, $\alpha > 0$. This function furnishes a second solution of (1.71.3) independent of (1.71.1). (See Whittaker-Watson **1**, Chapter 17, pp. 370, 372.)

The formulas

(1.71.5)
$$J_{\alpha-1}(z) + J_{\alpha+1}(z) = 2\alpha z^{-1} J_\alpha(z), \qquad \frac{d}{dz}\{z^{-\alpha} J_\alpha(z)\} = -z^{-\alpha} J_{\alpha+1}(z),$$

follow directly from (1.71.1) on comparing the corresponding coefficients on both sides. The integral representation

(1.71.6)
$$J_\alpha(z) = \frac{(z/2)^\alpha}{\Gamma(\alpha + \frac{1}{2})\Gamma(\frac{1}{2})} \int_{-1}^{+1} (1 - t^2)^{\alpha - \frac{1}{2}} e^{izt}\, dt$$

holds for $\alpha > -\frac{1}{2}$. This can be verified by introducing the development of e^{izt} and integrating by means of (1.7.5).

(2) The following important asymptotic formula is used in various applications:

(1.71.7)
$$J_\alpha(z) = \left(\frac{2}{\pi z}\right)^{\frac{1}{2}} \cos(z - \alpha\pi/2 - \pi/4) + O(z^{-\frac{3}{2}}), \qquad z \to +\infty.$$

This is only a special case ($p = 1$) of the asymptotic expansion (see Whittaker-Watson **1**, p. 368):

$$J_\alpha(z) = \left(\frac{2}{\pi z}\right)^{\frac{1}{2}} \cos(z - \alpha\pi/2 - \pi/4)\left\{\sum_{\nu=0}^{p-1} a_\nu z^{-2\nu} + O(z^{-2p})\right\} \tag{1.71.8}$$
$$+ \left(\frac{2}{\pi z}\right)^{\frac{1}{2}} \sin(z - \alpha\pi/2 - \pi/4)\left\{\sum_{\nu=0}^{p-1} b_\nu z^{-2\nu-1} + O(z^{-2p-1})\right\};$$

here p is an arbitrary positive number, a_ν and b_ν certain constants depending only on ν, and $z \to +\infty$. Also, $a_0 = 1$.

This expansion holds also if z is complex, $|\arg z| \leqq \pi - \delta$, $\delta > 0$, if we agree that $z^{\frac{1}{2}} = \exp(\frac{1}{2}\log z)$ with $|\Im(\log z)| \leqq \pi - \delta$. We notice the following important consequence of this formula, valid for $-\pi/2 + \delta \leqq \arg z \leqq 3\pi/2 - \delta$, $\delta > 0$:

$$e^{\alpha\pi i/2} J_\alpha(e^{-i\pi/2}z) = (2\pi z)^{-\frac{1}{2}} e^z\{1 + O(|z|^{-1})\} \tag{1.71.9}$$
$$+ (2\pi z)^{-\frac{1}{2}} \exp[-z + (\alpha + \tfrac{1}{2})\pi i]\{1 + O(|z|^{-1})\}.$$

An asymptotic formula similar to (1.71.7) holds for Bessel functions $Y_\alpha(z)$ of the second kind, with the only change that cosine is to be replaced by sine. (See Whittaker-Watson **1**, p. 371.)

(3) It may be useful to notice the order of magnitude of $J_\alpha(z)$ and $Y_\alpha(z)$ for $z \to +0$ and $z \to +\infty$. From the preceding formulas we see that when $z \to +0$,

$$\begin{aligned} &J_\alpha(z) \sim z^\alpha, && \alpha \text{ real}, \alpha \neq -1, -2, -3, \cdots, \\ &Y_\alpha(z) \sim z^{-\alpha}, && \alpha = 1, 2, 3, \cdots, \\ &Y_0(z) \sim \log(1/z), \end{aligned} \tag{1.71.10}$$

while when $z \to +\infty$,

$$J_\alpha(z) = O(z^{-\frac{1}{2}}), \qquad Y_\alpha(z) = O(z^{-\frac{1}{2}}). \tag{1.71.11}$$

1.8. Differential equations

We shall make frequent use of certain elementary transformations of homogeneous linear differential equations of the second order.

(1) Let $K(x)$, $M(x)$, $N(x)$ be functions defined in the interval $a < x < b$ in which $K(x)$ and $M(x)$ have continuous derivatives and $K(x) \neq 0$; let $N(x)$ be continuous. If in

$$K(x)y'' + M(x)y' + N(x)y = 0 \tag{1.8.1}$$

we introduce $y = s(x)u(x)$, $u(x)$ being the new unknown function, $s(x)$ can be determined so that $u(x)$ satisfies an equation of the form

$$u'' + \lambda(x)u = 0. \tag{1.8.2}$$

Direct calculation gives

$$2Ks' + Ms = 0; \qquad s(x) = \exp\left\{-\int \frac{M\,dx}{2K}\right\}, \tag{1.8.3}$$

where the integration is extended from an arbitrary point x_0 to x. Then

$$\lambda(x) = -\frac{d}{dx}\left(\frac{M}{2K}\right) - \left(\frac{M}{2K}\right)^2 + \frac{N}{K}. \tag{1.8.4}$$

(2) If we introduce into (1.8.1) the new independent variable θ defined by $x = \sigma(\theta)$, we obtain

$$K(x)\sigma'(\theta)\frac{d^2y}{d\theta^2} + \{M(x)[\sigma'(\theta)]^2 - K(x)\sigma''(\theta)\}\frac{dy}{d\theta} + N(x)[\sigma'(\theta)]^3 y = 0. \tag{1.8.5}$$

If we apply the process in (1) to (1.8.5), the first derivative can be removed. We write $y = s^*u$; here, in view of (1.8.3),

$$s^* = \exp\left\{-\int \frac{M\sigma'^2 - K\sigma''}{2K\sigma'}\,d\theta\right\} = (\sigma')^{\frac{1}{2}}s, \tag{1.8.6}$$

where s has the same meaning as in (1.8.3). Hence, $y = (\sigma')^{\frac{1}{2}}su$, and u satisfies

$$\frac{d^2u}{d\theta^2} + \lambda^* u = 0, \tag{1.8.7}$$

with

$$\lambda^* = -\frac{d}{d\theta}\left(\frac{M\sigma'^2 - K\sigma''}{2K\sigma'}\right) - \left(\frac{M\sigma'^2 - K\sigma''}{2K\sigma'}\right)^2 + \frac{N}{K}\sigma'^2. \tag{1.8.8}$$

As an application of the above, we note the following transformations of Bessel's differential equation (1.71.3), $k \neq 0$,

$$\frac{d^2u}{dx^2} + \left(k^2 + \frac{\frac{1}{4} - \alpha^2}{x^2}\right)u = 0; \qquad u(x) = x^{\frac{1}{2}}J_\alpha(kx), \tag{1.8.9}$$

$$\frac{d^2u}{dx^2} + \left(\frac{k}{x} + \frac{1 - \alpha^2}{4x^2}\right)u = 0; \qquad u(x) = x^{\frac{1}{2}}J_\alpha\{2(kx)^{\frac{1}{2}}\}. \tag{1.8.10}$$

Another elementary formula, important for further exposition is the representation of a solution $y = y(x)$ of the *non-homogeneous* equation

$$K(x)y'' + M(x)y' + N(x)y = f(x) \tag{1.8.11}$$

in terms of a fundamental system $\{y_1(x), y_2(x)\}$ of the corresponding homogeneous equation (1.8.1). We have

$$y(x) = c_1y_1(x) + c_2y_2(x) + \int_{x_0}^{x} \frac{y_1(x)y_2(t) - y_2(x)y_1(t)}{y_1'(t)y_2(t) - y_2'(t)y_1(t)}\,\frac{f(t)}{K(t)}\,dt, \tag{1.8.12}$$

where x_0 is a fixed value and c_1, c_2 proper constants. Now

$$y_1'(x)y_2(x) - y_2'(x)y_1(x) = \text{const.}\exp\left\{-\int \frac{M}{K}\,dx\right\}. \tag{1.8.13}$$

When we take $M = 0$, this expression becomes a constant.

Applying this last remark to (1.8.9), we obtain the important formulas

$$(1.8.14)\qquad \begin{aligned} J'_\alpha(x)J_{-\alpha}(x) - J'_{-\alpha}(x)J_\alpha(x) &= \frac{2\sin\alpha\pi}{\pi x}, &\quad \alpha \text{ non-integral},\\ J'_\alpha(x)Y_\alpha(x) - Y'_\alpha(x)J_\alpha(x) &= -\frac{2}{\pi x}, &\quad \alpha = 0, 1, 2, \cdots. \end{aligned}$$

As regards the evaluation of the constants on the right side, see Watson **3**, p. 43, (2), p. 76, (1).

1.81. Airy's function

An interesting transformation of Bessel's differential equation (1.71.3) can be obtained in the special cases $\alpha = \pm 1/3$. If we use (1.8.7) and (1.8.8), there is no difficulty in showing that both integral functions

$$(1.81.1)\qquad \begin{aligned} k(x) &= \frac{\pi}{3}(x/3)^{\frac{1}{2}}J_{-\frac{1}{3}}\{2(x/3)^{\frac{3}{2}}\} = \frac{\pi}{3}\sum_{\nu=0}^{\infty}\frac{(-x/3)^{3\nu}}{\nu!\,\Gamma(\nu+2/3)},\\ l(x) &= \frac{\pi}{3}(x/3)^{\frac{1}{2}}J_{\frac{1}{3}}\{2(x/3)^{\frac{3}{2}}\} = \frac{\pi}{3}\frac{x}{3}\sum_{\nu=0}^{\infty}\frac{(-x/3)^{3\nu}}{\nu!\,\Gamma(\nu+4/3)}, \end{aligned}$$

satisfy the equation

$$(1.81.2)\qquad \frac{d^2y}{dx^2} + \tfrac{1}{3}xy = 0.$$

For negative x we have $k(x) > 0$, $l(x) < 0$. Using (1.71.9),[5] we obtain for $x < 0$, $x \to -\infty$

$$(1.81.3)\qquad k(x) \cong -l(x) \cong 2^{-1}3^{-\frac{1}{4}}\pi^{\frac{1}{2}}|x|^{-\frac{1}{4}}\exp\{2(|x|/3)^{\frac{3}{2}}\}.$$

Thus, but for a constant factor, the function

$$(1.81.4)\qquad A(x) = k(x) + l(x)$$

is the only possible particular solution of (1.81.2) which remains bounded if $x \to -\infty$. Indeed,

$$(1.81.5)\qquad A(x) \cong 2^{-1}3^{-\frac{1}{4}}\pi^{\frac{1}{2}}|x|^{-\frac{1}{4}}\exp\{-2(|x|/3)^{\frac{3}{2}}\}, \qquad x \to -\infty.$$

(See Watson **3**, pp. 188–190, 202.) This function $A(x)$ is called Airy's function; it can be considered as the standard solution of (1.81.2) and plays an important part in numerous questions in mathematical physics. The function $l(x)/k(x)$ is increasing (see Fig. 1) so that an arbitrary real solution of (1.81.2) has at most one negative zero and infinitely many positive zeros. In particular, $A(x)$

[5] If x is negative, we have

$$k(x) = \frac{\pi}{3}(|x|/3)^{1/2}e^{-i\pi/6}J_{-1/3}\{e^{-i\pi/2}2(|x|/3)^{3/2}\},$$

$$l(x) = -\frac{\pi}{3}(|x|/3)^{1/2}e^{i\pi/6}J_{1/3}\{e^{-i\pi/2}2(|x|/3)^{3/2}\}.$$

has no negative zero and infinitely many positive zeros. Since $A(x) > 0$ for $x < 0$, we see from (1.81.2) that $A''(x) > 0$ for $x < 0$; therefore $A'(x) \to 0$ as $x \to -\infty$.

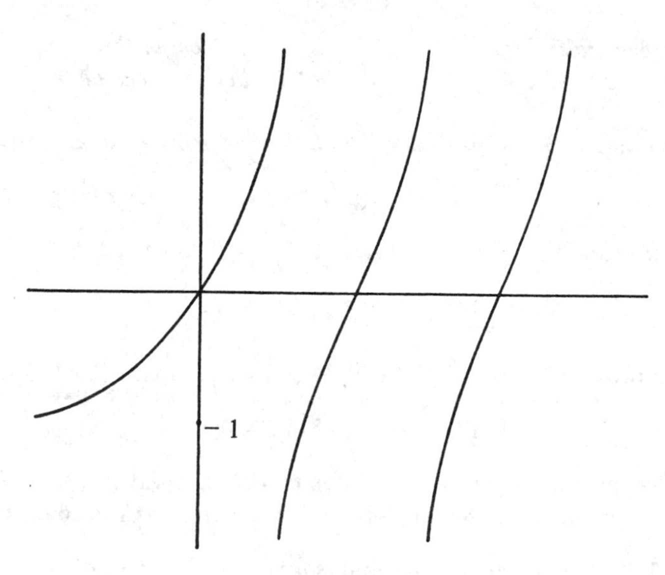

FIG. 1

1.82. Theorems of Sturm's type

The following "comparison theorems" of Sturm's type can be proved in the usual way (see Szegö **20**, pp. 3–4):

THEOREM 1.82.1. *Let $f(x)$ and $F(x)$ be functions continuous in $x_0 < x < X_0$ with $f(x) \leqq F(x)$. Let the functions $y(x)$ and $Y(x)$, both not identically zero, satisfy the differential equations*

$$y'' + f(x)y = 0, \qquad Y'' + F(x)Y = 0, \tag{1.82.1}$$

respectively. Let x' and x'', $x' < x''$, be two consecutive zeros of $y(x)$. Then the function $Y(x)$ has at least one variation of sign in the interval $x' < x < x''$ provided $f(x) \not\equiv F(x)$ in $[x', x'']$.

The statement holds also for $x' = x_0$ $[y(x_0 + 0) = 0]$ if the additional condition

$$\lim_{x \to x_0+0} \{y'(x)Y(x) - y(x)Y'(x)\} = 0 \tag{1.82.2}$$

is satisfied (similarly for $x'' = X_0$).

From Theorem 1.82.1 we readily derive (loc. cit., p. 4) the following theorem:

THEOREM 1.82.2. *Let $\phi(x)$ be continuous and decreasing in $x_0 < x < X_0$, and let y be a solution of*

$$y'' + \phi(x)y = 0 \tag{1.82.3}$$

which is not identically zero. Then $x' < x'' < x'''$ being three consecutive zeros of $y(x)$, we have $x'' - x' < x''' - x''$; that is, the sequence of the zeros of $y(x)$ is convex.

The last inequality holds also under the following more general condition:

$$\phi(x) > \phi(x'') > \phi(y), \qquad \text{for } x < x'' < y < x'''. \tag{1.82.4}$$

In addition, it holds also if $x' = x_0$ $[y(x_0 + 0) = 0]$ provided that

$$\lim_{x \to x_0+0} (x - x_0)y'(x) = 0. \tag{1.82.5}$$

In order to prove $x'' - x' = h < x''' - x''$, we compare (1.82.3) with

$$Y'' + \phi(x - h)Y = 0,$$

which has the solution $Y(x) = y(x - h)$ in the interval $x'' \leqq x \leqq x'''$.

Another very elementary remark of a related nature is the following:

THEOREM 1.82.3. *Let $f(x)$ be continuous and negative in $x_0 < x < X_0$. Then an arbitrary solution y of $y'' + f(x)y = 0$, for which $y \to 0$ if $x \to X_0$, cannot vanish in $x_0 \leqq x < X_0$.*

Suppose the contrary. Now between two consecutive zeros sgn y'' = sgn y is constant, say positive; then y is positive and convex, which is a contradiction.

A further remarkable result of Sturm's type is the following (Watson **3**, p. 518, Makai **2**):

THEOREM 1.82.4. *Let $f(x)$, $F(x)$, $y(x)$, $Y(x)$, x_0, X_0, x', x'' have the same meaning and satisfy the same conditions as in Theorem 1.82.1. We denote by ξ the first zero of $Y(x)$ to the right of x', $x' < \xi < x''$.*

Assuming that $y(x) > 0$; $Y(x) > 0$ in $x' < x < \xi$, and

$$\lim_{x \to x'+0} \frac{y(x)}{Y(x)} \geqq 1, \tag{1.82.6}$$

we have $y(x) > Y(x)$ in $x' < x < \xi$.

We conclude as usual that the function $y'(x)Y(x) - y(x)Y'(x)$ is increasing in $[x', \xi]$; it is zero at $x = x'(Y(x') = 0)$, thus positive in $x' < x < \xi$. Hence $y(x)/Y(x)$ is increasing. In view of (1.82.6) the assertion follows.

The statement holds also for $x' = x_0$ provided the condition (1.82.2) is satisfied.

We prove now the following important consequence of the last theorem (Hartman-Wintner **1**; Makai **2**):

THEOREM 1.82.5. *Let $\phi(x)$, x', x'', x''' have the same meaning as in Theorem* (1.82.2): $x'' - x' < x''' - x''$. *Let $y(x)$ describe a negative "wave" in $[x', x'']$ and a positive one in $[x'', x''']$. The first "wave" is then entirely contained in the second one.*

The meaning of the last assertion is the inequality:

$$0 < -y(2x'' - x) < y(x), \qquad x'' < x < 2x'' - x'.$$

The proof proceeds as usual, taking into account that, $Y(x) = -y(2x'' - x)$,

$$\lim_{x \to x''} \frac{y(x)}{-y(2x'' - x)} = \lim_{x \to x''} \frac{y'(x)}{y'(2x'' - x)} = 1.$$

1.9. An elementary conformal mapping

Let the complex variables x and z be connected by the relations

$$(1.9.1) \qquad x = \tfrac{1}{2}(z + z^{-1}), \qquad z = x + (x^2 - 1)^{\frac{1}{2}}.$$

The exterior of the unit circle, $|z| > 1$, as well as its interior, is mapped onto the whole x-plane except the closed interval $[-1, +1]$ (the so-called cut plane), with $z = \infty$ and $z = 0$, respectively, corresponding to $x = \infty$. If we take that branch of $x + (x^2 - 1)^{\frac{1}{2}}$ which becomes infinite at $x = \infty$, we obtain $|z| > 1$; if we take the other branch, which vanishes at $x = \infty$, we obtain $|z| < 1$.

The unit circle $z = e^{i\theta}$ is carried over into the closed segment $-1 \leqq x \leqq +1$ described twice since $x = \cos\theta$.

The circle $|z| = r$, or $|z| = r^{-1}$, $0 < r \neq 1$, corresponds to the ellipse with foci at -1, $+1$ and with semi-axes

$$\tfrac{1}{2}(r + r^{-1}), \qquad \tfrac{1}{2}|r - r^{-1}|.$$

Upon replacing z by e^{ζ} or by $e^{-\zeta}$, we obtain the representation

$$(1.9.2) \qquad x = \cosh \zeta.$$

It maps the half-strip

$$(1.9.3) \qquad \Re(\zeta) > 0, \qquad -\pi < \Im(\zeta) \leqq \pi$$

onto the same x-plane cut along $[-1, +1]$ as before; now, however, the point $x = \infty$ has to be removed.

1.91. The principle of argument; Rouché's theorem; sequences of analytic functions

THEOREM 1.91.1 (Principle of argument). *Let $f(x)$ be analytic both inside and on a Jordan curve C, and let $f(x) \neq 0$ on C. Then the variation of $\Im\{\log f(x)\} = \arg f(x)$, as x describes C in the positive sense is $2\pi i m$, where m is the number of the zeros of $f(x)$ in the interior of C, counted with the proper multiplicity.*

THEOREM 1.91.2 (Rouché's theorem). *Let $f(x)$ and $g(x)$ be analytic inside and on C, and let $|g(x)| < |f(x)|$ on C. Then $f(x) + g(x)$ and $f(x)$ have no zeros on C and the same number of zeros in the interior of C.*

THEOREM 1.91.3 (Theorem of Hurwitz). *Let $\{f_n(x)\}$ be a sequence of analytic functions regular in a region G, and let this sequence be uniformly convergent in every closed subset of G. Suppose the analytic function $\lim_{n\to\infty} f_n(x) = f(x)$ does not vanish identically. Then if $x = a$ is a zero of $f(x)$ of order k, a neighborhood $|x - a| < \delta$ of $x = a$ and a number N exist such that if $n > N$, $f_n(x)$ has exactly k zeros in $|x - a| < \delta$.*

The last theorem follows immediately from Theorem 1.91.2. Concerning the preceding theorems see Pólya-Szegö **1**, vol. 1, pp. 120-124.

In Theorem 1.91.3, let G be symmetric relative to the real axis, and let $f_n(x)$ be real if x is real. If $x = a$ is a simple *real* zero of $f(x)$, then for $n > N$ each $f_n(x)$ has exactly one *real* zero in $|x - a| < \delta$. For if $f_n(x_0) = 0$, then $f_n(\bar{x}_0)$ is also 0.

CHAPTER II

DEFINITION OF ORTHOGONAL POLYNOMIALS; PRINCIPAL EXAMPLES

2.1. Orthogonality

(1) In what follows $\alpha(x)$ is a fixed non-decreasing function which is not constant in the interval $a \leqq x \leqq b$. (See the remark at the beginning of §1.4 (1).)

DEFINITION. *An orthonormal set of functions* $\phi_0(x), \phi_1(x), \cdots, \phi_l(x)$, *$l$ finite or infinite, is defined by the relations*

$$(2.1.1) \qquad (\phi_n, \phi_m) = \int_a^b \phi_n(x)\phi_m(x)\, d\alpha(x) = \delta_{nm}, \qquad n, m = 0, 1, 2, \cdots, l.$$

Here $\phi_n(x)$ is real-valued and belongs to the class $L^2_\alpha(a, b)$.

Functions of this kind are necessarily linearly independent. If $\alpha(x)$ has only a finite number N of points of increase (that is, points in the neighborhood of which $\alpha(x)$ is not constant), l is necessarily finite and $l < N$.

THEOREM 2.1.1. *Let the real-valued functions*

$$(2.1.2) \qquad f_0(x), f_1(x), f_2(x), \cdots, f_l(x), \qquad l \text{ finite or infinite,}$$

be of the class $L^2_\alpha(a, b)$ and linearly independent. Then an orthonormal set

$$(2.1.3) \qquad \phi_0(x), \phi_1(x), \phi_2(x), \cdots, \phi_l(x)$$

exists such that, for $n = 0, 1, 2, \cdots, l$,

$$(2.1.4) \qquad \phi_n(x) = \lambda_{n0}f_0(x) + \lambda_{n1}f_1(x) + \cdots + \lambda_{nn}f_n(x), \qquad \lambda_{nn} > 0.$$

The set (2.1.3) is uniquely determined.

The procedure of deriving (2.1.3) from (2.1.2) is called *orthogonalization.* (Cf. Stone **1**, pp. 12–13.)

(2) For the orthonormal functions (2.1.4) the following explicit representation holds:

$$(2.1.5) \qquad \phi_n(x) = (D_{n-1}D_n)^{-\frac{1}{2}}D_n(x), \qquad n = 0, 1, 2, \cdots,$$

where, for $n \geqq 1$,

$$(2.1.6) \qquad D_n(x) = \begin{vmatrix} (f_0, f_0) & (f_0, f_1) & \cdots & (f_0, f_n) \\ (f_1, f_0) & (f_1, f_1) & \cdots & (f_1, f_n) \\ \cdots & \cdots & \cdots & \cdots \\ (f_{n-1}, f_0) & (f_{n-1}, f_1) & \cdots & (f_{n-1}, f_n) \\ f_0(x) & f_1(x) & \cdots & f_n(x) \end{vmatrix},$$

and, for $n \geqq 0$,

$$D_n = [(f_\nu, f_\mu)]_{\nu,\mu=0,1,2,\cdots,n} > 0. \tag{2.1.7}$$

We write $D_{-1} = 1$ and $D_0(x) = f_0(x)$. The determinant (2.1.7) corresponds to the positive definite quadratic form

$$\begin{aligned} &\| u_0 f_0 + u_1 f_1 + \cdots + u_n f_n \|^2 \\ &\quad = \int_a^b \{u_0 f_0(x) + u_1 f_1(x) + \cdots + u_n f_n(x)\}^2 \, d\alpha(x), \end{aligned} \tag{2.1.8}$$

so that $D_n > 0$ for each n.

Furthermore, the following integral representations can be established:

$$\begin{aligned} D_n(x) = \frac{1}{n!} \underbrace{\int_a^b \int_a^b \cdots \int_a^b}_{n} &\begin{vmatrix} f_0(x_0) & f_1(x_0) & \cdots & f_n(x_0) \\ f_0(x_1) & f_1(x_1) & \cdots & f_n(x_1) \\ \cdots & \cdots & \cdots & \cdots \\ f_0(x_{n-1}) & f_1(x_{n-1}) & \cdots & f_n(x_{n-1}) \\ f_0(x) & f_1(x) & \cdots & f_n(x) \end{vmatrix} \\ \cdot &\begin{vmatrix} f_0(x_0) & f_1(x_0) & \cdots & f_{n-1}(x_0) \\ f_0(x_1) & f_1(x_1) & \cdots & f_{n-1}(x_1) \\ \cdots & \cdots & \cdots & \cdots \\ f_0(x_{n-1}) & f_1(x_{n-1}) & \cdots & f_{n-1}(x_{n-1}) \end{vmatrix} d\alpha(x_0)\, d\alpha(x_1) \cdots d\alpha(x_{n-1}), \quad n \geqq 1, \end{aligned} \tag{2.1.9}$$

$$\begin{aligned} D_n = \frac{1}{(n+1)!} \\ \cdot \underbrace{\int_a^b \int_a^b \cdots \int_a^b}_{n+1} \begin{vmatrix} f_0(x_0) & f_1(x_0) & \cdots & f_n(x_0) \\ f_0(x_1) & f_1(x_1) & \cdots & f_n(x_1) \\ \cdots & \cdots & \cdots & \cdots \\ f_0(x_n) & f_1(x_n) & \cdots & f_n(x_n) \end{vmatrix}^2 d\alpha(x_0) d\alpha(x_1) \cdots d\alpha(x_n). \end{aligned} \tag{2.1.10}$$

(Cf. Kowalewski **1**, p. 326; Pólya-Szegö **1**, pp. 48–49, 208, problem 68.)

(3) Definition. *Let $\{\phi_n(x)\}$ be a given orthonormal set, finite or infinite. To an arbitrary real-valued function $f(x)$ let there correspond the formal Fourier expansion*

$$f(x) \sim f_0\phi_0(x) + f_1\phi_1(x) + \cdots + f_n\phi_n(x) + \cdots. \tag{2.1.11}$$

The coefficients f_n, called the Fourier coefficients of $f(x)$ with respect to the given system, are defined by

$$f_n = (f, \phi_n) = \int_a^b f(x)\phi_n(x) \, d\alpha(x), \qquad n = 0, 1, 2, \cdots. \tag{2.1.12}$$

Every finite section of the series (2.1.11) has the following important minimum property:

THEOREM 2.1.2. *Let $\phi_n(x)$, $f(x)$, f_n have the same meaning as in the previous definition. Let $l \geqq 0$ be a fixed integer, and $a_0, a_1, \cdots, a_l$ arbitrary real constants. If we write*

$$g(x) = a_0\phi_0(x) + a_1\phi_1(x) + \cdots + a_l\phi_l(x), \tag{2.1.13}$$

and the coefficients a_ν are variable, the integral

$$\int_a^b \{f(x) - g(x)\}^2 \, d\alpha(x) \tag{2.1.14}$$

becomes a minimum if and only if $a_\nu = f_\nu$, $\nu = 0, 1, 2, \cdots, l$.

The minimum itself is

$$\int_a^b \{f(x)\}^2 \, d\alpha(x) - \sum_{\nu=0}^{l} f_\nu^2 = \|f\|^2 - \sum_{\nu=0}^{l} f_\nu^2, \tag{2.1.15}$$

so that

$$f_0^2 + f_1^2 + \cdots + f_l^2 \leqq \|f\|^2, \tag{2.1.16}$$

and (Bessel's inequality)

$$f_0^2 + f_1^2 + f_2^2 + \cdots \leqq \|f\|^2 = \int_a^b \{f(x)\}^2 \, d\alpha(x). \tag{2.1.17}$$

If the left-hand side of (2.1.17) is an infinite series, it is convergent. The discussion of the equality sign in (2.1.17) leads to the concept of closure (§1.5).

A classical example of Fourier expansions of this kind is the ordinary Fourier series in terms of the trigonometric functions 1, $\cos nx$, $\sin nx$, $n = 1, 2, 3, \cdots$; $-\pi \leqq x \leqq +\pi$.

(4) Another important characterization of the orthonormal set (2.1.4) can be based on the preceding minimum property of the partial sums. Indeed, for variable real values of $\lambda_0, \lambda_1, \cdots, \lambda_{n-1}$ the expression

$$\| \lambda_0 f_0(x) + \lambda_1 f_1(x) + \cdots + \lambda_{n-1} f_{n-1}(x) + f_n(x) \| \tag{2.1.18}$$

becomes a minimum if and only if

$$\lambda_0 f_0(x) + \lambda_1 f_1(x) + \cdots + \lambda_{n-1} f_{n-1}(x) + f_n(x) = \lambda_{nn}^{-1} \phi_n(x). \tag{2.1.19}$$

The extension of these considerations to complex function spaces is not difficult. The scalar product of the functions $f(x)$ and $g(x)$ is then defined as in (1.4.3).

2.2. Orthogonal polynomials

(1) DEFINITION. *Let $\alpha(x)$ be a fixed non-decreasing function with infinitely many points of increase in the finite or infinite interval $[a, b]$, and let the "moments"*

$$c_n = \int_a^b x^n \, d\alpha(x), \qquad n = 0, 1, 2, \cdots, \tag{2.2.1}$$

exist. If we orthogonalize the set of non-negative powers of x:

$$1, x, x^2, \cdots, x^n, \cdots, \tag{2.2.2}$$

in the sense explained in §2.1 (the linear independence is shown below), *we obtain a set of polynomials*

$$p_0(x), p_1(x), p_2(x), \cdots, p_n(x), \cdots \tag{2.2.3}$$

uniquely determined by the following conditions:

(a) $p_n(x)$ *is a polynomial of precise degree* n *in which the coefficient of* x^n *is positive;*

(b) *the system* $\{p_n(x)\}$ *is orthonormal, that is,*

$$\int_a^b p_n(x)p_m(x)\, d\alpha(x) = \delta_{nm}, \qquad n, m, = 0, 1, 2, \cdots. \tag{2.2.4}$$

The existence of the moments (2.2.1) is equivalent to the fact that the functions x^n are of the class $L_\alpha(a, b)$.

A similar definition holds in the special case of a distribution of the type $w(x)\, dx$. Here we assume that $w(x)$ is non-negative and measurable in Lebesgue's sense and that $\int_a^b w(x)\, dx > 0$. Moreover, the moments must exist again.

We call $p_n(x)$ the *orthogonal polynomials*[6] associated with the distributions $d\alpha(x)$ and $w(x)\, dx$, respectively; in the latter case we also speak of the orthogonal polynomials associated with the weight function $w(x)$. The following chapters are devoted to the study of these polynomials. Evidently if the distribution is of the type $w(x)\, dx$, the system

$$\{[w(x)]^{\frac{1}{2}} p_n(x)\}, \qquad n = 0, 1, 2, \cdots, \tag{2.2.5}$$

is orthonormal in the usual sense.

The linear independence of the functions (2.2.2) can readily be shown. In fact if $\rho(x)$ is an arbitrary real polynomial, the relation

$$\|\rho\|^2 = \int_a^b \{\rho(x)\}^2\, d\alpha(x) = 0$$

is possible only if $\rho(x)$ vanishes at all points of increase of $\alpha(x)$. Since there are infinitely many such points, $\rho(x)$ must vanish identically.

If $\alpha(x)$ has only a finite number, say N, of points of increase, the functions $1, x, x^2, \cdots, x^{N-1}$ are still linearly independent. Through orthogonalization we obtain a finite system of polynomials $\{p_n(x)\}$, $n = 0, 1, 2, \cdots, N - 1$, satisfying similar conditions as required in the previous Definition. See §2.8 and §2.82.

(2) Using the general formulas (2.1.5) to (2.1.8), we obtain, for $n \geqq 1$,

[6] Sometimes these are called *Tchebichef* polynomials. We shall reserve this terminology for the important special cases (1.12.3).

$$(2.2.6)\qquad p_n(x) = (D_{n-1}D_n)^{-\frac{1}{2}} \begin{vmatrix} c_0 & c_1 & c_2 & \cdots & c_n \\ c_1 & c_2 & c_3 & \cdots & c_{n+1} \\ \cdots & \cdots & \cdots & \cdots & \cdots \\ c_{n-1} & c_n & c_{n+1} & \cdots & c_{2n-1} \\ 1 & x & x^2 & \cdots & x^n \end{vmatrix},$$

where for $n \geqq 0$

$$(2.2.7)\qquad D_n = [c_{\nu+\mu}]_{\nu,\mu=0,1,2,\cdots,n} > 0.$$

In addition to (2.2.6) we have $p_0(x) = D_0^{-\frac{1}{2}} = c_0^{-\frac{1}{2}}$. The determinant (2.2.7) is associated with the positive definite quadratic form

$$(2.2.8)\qquad \sum_{\nu=0}^{n}\sum_{\mu=0}^{n} c_{\nu+\mu}u_\nu u_\mu = \int_a^b (u_0 + u_1x + u_2x^2 + \cdots + u_nx^n)^2\,d\alpha(x),$$

which is called a form of *Hankel* or of *recurrent* type. (See Szegö **1**.)

The determinant in (2.2.6) can be transformed by multiplying the next to the last column by x, subtracting it from the last column, and repeating this operation for each of the preceding columns. In this way we obtain, $n \geqq 1$,

$$(2.2.9)\qquad p_n(x) = (D_{n-1}D_n)^{-\frac{1}{2}} \begin{vmatrix} c_0x - c_1 & c_1x - c_2 & \cdots & c_{n-1}x - c_n \\ c_1x - c_2 & c_2x - c_3 & \cdots & c_nx - c_{n+1} \\ \cdots & \cdots & \cdots & \cdots \\ c_{n-1}x - c_n & c_nx - c_{n+1} & \cdots & c_{2n-2}x - c_{2n-1} \end{vmatrix}.$$

Furthermore, according to (2.1.9) and (2.1.10), we have the following integral representations:

$$(2.2.10)\qquad \begin{aligned} p_n(x) = \frac{(D_{n-1}D_n)^{-\frac{1}{2}}}{n!} \underbrace{\int_a^b\int_a^b\cdots\int_a^b}_{n} &(x - x_0)(x - x_1)\cdots(x - x_{n-1}) \\ &\cdot \prod_{\substack{\nu,\mu=0,1,\cdots,n-1\\ \nu<\mu}} (x_\nu - x_\mu)^2\,d\alpha(x_0)\,d\alpha(x_1)\cdots d\alpha(x_{n-1}), \end{aligned}$$

and

$$(2.2.11)\qquad \begin{aligned} D_n = \frac{1}{(n+1)!}\underbrace{\int_a^b\int_a^b\cdots\int_a^b}_{n+1} &\prod_{\substack{\nu,\mu=0,1,\cdots,n\\ \nu<\mu}} (x_\nu - x_\mu)^2 \\ &\cdot d\alpha(x_0)\,d\alpha(x_1)\cdots d\alpha(x_n). \end{aligned}$$

For (2.2.10) and (2.2.11) see, for example, Heine **3**, vol. 1, p. 288. Formulas (2.2.6), (2.2.9), (2.2.10) are not suitable in general for derivation of properties of the polynomials in question. To this end we shall generally prefer the orthogonality property itself, or other representations derived by means of the orthogonality property.

(3) The Fourier expansion of an arbitrary function $f(x)$ in terms of the polynomials $\{p_n(x)\}$ has the form

$$f(x) \sim f_0p_0(x) + f_1p_1(x) + \cdots + f_np_n(x) + \cdots \tag{2.2.12}$$

with

$$f_n = \int_a^b f(x)p_n(x)\,d\alpha(x), \qquad n = 0, 1, 2, \cdots. \tag{2.2.13}$$

The partial sums have the minimum property formulated in Theorem 2.1.2. We notice as an important special case the following direct characterization of the orthogonal polynomials (see §2.1 (4)). Considering the set of all polynomials $\rho(x)$ of degree n with the coefficient of x^n unity, we find that the integral

$$\int_a^b \{\rho(x)\}^2\,d\alpha(x) \tag{2.2.14}$$

becomes a minimum if and only if $\rho(x) = \text{const.}\ p_n(x)$. Here the constant factor is determined by normalizing $\rho(x)$. If k_n denotes the highest coefficient of $p_n(x)$, the minimum is obviously k_n^{-2}. From (2.2.8) we find for this minimum the value D_n/D_{n-1}, so that

$$k_n = (D_{n-1}/D_n)^{\frac{1}{2}}, \tag{2.2.15}$$

which also follows directly from (2.2.6).

2.3. Further remarks

(1) The restriction (a) in the definition in §2.2 (1) concerning the highest coefficient, and the restriction (b) concerning the integral of the square, is only one of various possible ways of normalizing the polynomials in question. Sometimes other kinds of normalization are appropriate, such as fixing the value of $p_n(x)$ at $x = a$ or at $x = b$,[7] or fixing the highest coefficient of $p_n(x)$, and so on. Since $p_n(x)$ has the precise degree n, every π_n can be represented as a linear combination of $p_0(x), p_1(x), \cdots, p_n(x)$ (see §1.12). Therefore $p_n(x)$, $n \geqq 1$, is orthogonal to any π_{n-1}. In particular,

$$\int_a^b p_n(x)x^\nu\,d\alpha(x) = 0, \qquad \nu = 0, 1, 2, \cdots, n-1. \tag{2.3.1}$$

This condition determines $p_n(x)$ save for a constant factor. Frequently, this wider formulation of the orthogonality property is used. Observe also that if $\rho(x)$ is a π_n and

$$\int_a^b p_n(x)\rho(x)\,d\alpha(x) = c, \tag{2.3.2}$$

then the coefficient of x^n in $\rho(x)$ is ck_n.

[7] We have $p_n(a) \neq 0$, $p_n(b) \neq 0$ (see §3.3).

(2) Let $[a, b]$ be an interval symmetric with respect to the origin, that is, $a = -b$, and let us consider a distribution of the type $w(x)\,dx$ with an even weight function, that is, $w(-x) = w(x)$. Then $p_n(x)$ is an even or an odd polynomial according as n is even or odd:

$$p_n(-x) = (-1)^n p_n(x). \tag{2.3.3}$$

It can contain only those powers of x which are congruent to n (mod 2). Indeed, we have for $\nu = 0, 1, 2, \cdots, n-1$

$$\int_{-a}^{a} p_n(-x)x^\nu w(x)\,dx = (-1)^\nu \int_{-a}^{a} p_n(x)x^\nu w(x)\,dx = 0.$$

Consequently, $p_n(-x)$ possesses the same orthogonality property as $p_n(x)$ (in the wider sense). Therefore, comparing the coefficients of x^n, we obtain $p_n(-x) = \text{const.}\ p_n(x) = (-1)^n p_n(x)$.

The linear transformation $x = kx' + l$, $k \neq 0$, carries over the interval $[a, b]$ into an interval $[a', b']$ (or $[b', a']$), and the weight function $w(x)$ into $w(kx' + l)$. Then the polynomials

$$(\operatorname{sgn} k)^n \mid k \mid^{\frac{1}{2}} p_n(kx' + l) \tag{2.3.4}$$

are orthonormal on $[a', b']$ (or $[b', a']$) with the weight function $w(kx' + l)$.

2.4. The classical orthogonal polynomials

1. Let $a = -1$, $b = +1$, $w(x) = (1-x)^\alpha(1+x)^\beta$, $\alpha > -1$, $\beta > -1$. Then, except for a constant factor, the orthogonal polynomial $p_n(x)$ is the Jacobi polynomial $P_n^{(\alpha,\beta)}(x)$ (see §4.1).

2. Let $a = 0$, $b = +\infty$, $w(x) = e^{-x}x^\alpha$, $\alpha > -1$. In this case $p_n(x)$ is, except for a constant factor, the Laguerre polynomial $L_n^{(\alpha)}(x)$ (see §5.1).

3. Let $a = -\infty$, $b = +\infty$, $w(x) = e^{-x^2}$. In this case $p_n(x)$ is, save for a constant factor, the Hermite polynomial $H_n(x)$ (see §5.5).

Some special cases of 1, except for constant factors, are:

The ultraspherical polynomials, for $\alpha = \beta$.

The Tchebichef polynomials of the first kind, $T_n(x) = \cos n\theta$, $x = \cos\theta$, for $\alpha = \beta = -\frac{1}{2}$ (see (1.12.3)).

The Tchebichef polynomials of the second kind, $U_n(x) = \sin(n+1)\theta/(\sin\theta)$, $x = \cos\theta$, for $\alpha = \beta = +\frac{1}{2}$ (see (1.12.3)).

The polynomials $U_{2n}(\cos(\theta/2)) = \sin(n+\frac{1}{2})\theta/\sin(\theta/2)$ of $\cos\theta = x$, for $\alpha = -\beta = \frac{1}{2}$ (see §1.12).

The Legendre polynomials $P_n(x)$, for $\alpha = \beta = 0$.

A detailed investigation of these polynomials will be given in later chapters.

2.5. A formula of Christoffel

(1) THEOREM 2.5. *Let $\{p_n(x)\}$ be the orthonormal polynomials associated with the distribution $d\alpha(x)$ on the interval $[a, b]$. Also let*

$$\rho(x) = c(x - x_1)(x - x_2) \cdots (x - x_l), \qquad c \neq 0, \tag{2.5.1}$$

be a π_l which is non-negative in this interval. Then the orthogonal polynomials $\{q_n(x)\}$, associated with the distribution $\rho(x)\, d\alpha(x)$, can be represented in terms of the polynomials $p_n(x)$ as follows:

$$\rho(x)q_n(x) = \begin{vmatrix} p_n(x) & p_{n+1}(x) & \cdots & p_{n+l}(x) \\ p_n(x_1) & p_{n+1}(x_1) & \cdots & p_{n+l}(x_1) \\ \cdots & \cdots & \cdots & \cdots \\ p_n(x_l) & p_{n+1}(x_l) & \cdots & p_{n+l}(x_l) \end{vmatrix}. \tag{2.5.2}$$

In case of a zero x_k, of multiplicity m, $m > 1$, we replace the corresponding rows of (2.5.2) *by the derivatives of order* 0, 1, 2, $\cdots$, $m - 1$ *of the polynomials $p_n(x)$, $p_{n+1}(x)$, $\cdots$, $p_{n+l}(x)$ at $x = x_k$.*

This important result is due to Christoffel (see **1**, actually only in the special case $\alpha(x) = x$). The polynomials $q_n(x)$ are in general not normalized.

The proof is almost obvious. The right-hand member of (2.5.2) is a π_{n+l} which is evidently divisible by $\rho(x)$. Hence it has the form $\rho(x)q_n(x)$, where $q_n(x)$ is a π_n. Moreover, it is a linear combination of the polynomials $p_n(x)$, $p_{n+1}(x)$, $\cdots$, $p_{n+l}(x)$, so that if $q(x)$ is an arbitrary π_{n-1}, then

$$\int_a^b \rho(x)q_n(x)q(x)\, d\alpha(x) = \int_a^b q_n(x)q(x)\rho(x)\, d\alpha(x) = 0. \tag{2.5.3}$$

Finally, the right side of (2.5.2) is not identically zero. To show this, it suffices to prove that the coefficient of $p_{n+l}(x)$, that is, the determinant $[p_{n+\nu}(x_{\mu+1})]$, $\nu, \mu = 0, 1, 2, \cdots, l - 1$, does not vanish. Suppose it to vanish; then certain real constants $\lambda_0, \lambda_1, \lambda_2, \cdots, \lambda_{l-1}$ exist, not all zero, such that

$$\lambda_0 p_n(x) + \lambda_1 p_{n+1}(x) + \cdots + \lambda_{l-1} p_{n+l-1}(x) \tag{2.5.4}$$

vanishes for $x = x_1, x_2, \cdots, x_l$. Hence (2.5.4) is of the form $\rho(x)G(x)$, where $G(x)$ is a π_{n-1}. Since (2.5.4) is orthogonal to an arbitrary π_{n-1}, we have the relation

$$\int_a^b \rho(x)G(x) \cdot G(x)\, d\alpha(x) = 0;$$

whence $G(x) \equiv 0$, a contradiction.

(2) The representation (2.5.2) enables us, for instance, to reduce ultraspherical polynomials with $\alpha = \beta =$ an integer, or with $\alpha + \frac{1}{2} = \beta + \frac{1}{2} =$ an integer, to Legendre and Tchebichef polynomials, respectively [cf. §4.21 (3)]. Another illustration may be obtained in connection with the polynomials considered in §2.6.

By using some special properties of $d\alpha(x)$ or of $\rho(x)$, formula (2.5.2) can be simplified. For example, let $d\alpha(x) = w(x)\, dx$, $w(x)$ and $\rho(x)$ be even functions, and $a = -b$. Then, instead of (2.5.2), we have the representation (l even)

$$\rho(x)q_n(x) = \begin{vmatrix} p_n(x) & p_{n+2}(x) & p_{n+4}(x) & \cdots & p_{n+l}(x) \\ p_n(x_1) & p_{n+2}(x_1) & p_{n+4}(x_1) & \cdots & p_{n+l}(x_1) \\ \cdots & \cdots & \cdots & \cdots & \cdots \\ p_n(x_{l/2}) & p_{n+2}(x_{l/2}) & p_{n+4}(x_{l/2}) & \cdots & p_{n+l}(x_{l/2}) \end{vmatrix},$$

where $\{\pm x_1, \pm x_2, \cdots, \pm x_{l/2}\}$ is the total set of zeros of $\rho(x)$. For instance, the orthogonal polynomials $q_n(x)$ associated with the weight function $1 - x^2$ in $[-1, +1]$ can be determined from

$$(1 - x^2)q_n(x) = \begin{vmatrix} P_n(x) & P_{n+2}(x) \\ P_n(1) & P_{n+2}(1) \end{vmatrix} = P_n(x) - P_{n+2}(x).$$

(Cf. (4.7.27), $\lambda = \frac{1}{2}$.)

2.6. A class of polynomials considered by S. Bernstein and G. Szegö

Let $\rho(x)$ be a polynomial of precise degree l and positive in $[-1, +1]$. Then the orthonormal polynomials $p_n(x)$, which are associated with the weight functions

$$w(x) = \begin{cases} (1 - x^2)^{-\frac{1}{2}}\{\rho(x)\}^{-1}, \\ (1 - x^2)^{\frac{1}{2}}\{\rho(x)\}^{-1}, \\ \left(\dfrac{1 - x}{1 + x}\right)^{\frac{1}{2}}\{\rho(x)\}^{-1} \end{cases} \tag{2.6.1}$$

can be calculated explicitly provided $l < 2n$ in the first case, $l < 2(n + 1)$ in the second, and $l < 2n + 1$ in the third. The polynomials of the first case play an important rôle in the proof of Szegö's equiconvergence theorem (**9**; cf. Theorem 13.1.2). All three cases were later investigated by S. Bernstein (**3**) in connection with his asymptotic formula (**2**; cf. Theorem 12.1.4).

THEOREM 2.6. *Let $\rho(x)$ be a π_l of precise degree l and positive in $[-1, +1]$. Let $\rho(\cos\theta) = |h(e^{i\theta})|^2$ be the normalized representation of $\rho(\cos\theta)$ in the sense of Theorem* 1.2.2. *Writing $h(e^{i\theta}) = c(\theta) + is(\theta)$, $c(\theta)$ and $s(\theta)$ real, we have the following formulas:*

$$\begin{aligned} p_n(\cos\theta) &= (2/\pi)^{\frac{1}{2}}\Re\{e^{in\theta}\overline{h(e^{i\theta})}\} \\ &= (2/\pi)^{\frac{1}{2}}\{c(\theta)\cos n\theta + s(\theta)\sin n\theta\}, \\ w(x) &= (1 - x^2)^{-\frac{1}{2}}\{\rho(x)\}^{-1}, \qquad l < 2n; \end{aligned} \tag{2.6.2}$$

$$\begin{aligned} p_n(\cos\theta) &= (2/\pi)^{\frac{1}{2}}(\sin\theta)^{-1}\Im\{e^{i(n+1)\theta}\overline{h(e^{i\theta})}\} \\ &= (2/\pi)^{\frac{1}{2}}\left\{c(\theta)\frac{\sin(n+1)\theta}{\sin\theta} - s(\theta)\frac{\cos(n+1)\theta}{\sin\theta}\right\}, \\ w(x) &= (1 - x^2)^{\frac{1}{2}}\{\rho(x)\}^{-1}, \qquad l < 2(n + 1); \end{aligned} \tag{2.6.3}$$

$$p_n(\cos\theta) = \pi^{-\frac{1}{2}}(\sin(\theta/2))^{-1}\Im\{e^{i(n+\frac{1}{2})\theta}\overline{h(e^{i\theta})}\}$$

$$(2.6.4) \qquad = \pi^{-\frac{1}{2}}\left\{c(\theta)\frac{\sin(n+\frac{1}{2})\theta}{\sin(\theta/2)} - s(\theta)\frac{\cos(n+\frac{1}{2})\theta}{\sin(\theta/2)}\right\},$$

$$w(x) = \left(\frac{1-x}{1+x}\right)^{\frac{1}{2}}\{\rho(x)\}^{-1}, \qquad l < 2n+1.$$

These formulas must be modified for $l = 2n$, $l = 2(n+1)$, and $l = 2n+1$, respectively, by multiplying the right-hand member of (2.6.2) by $(1 + h_l/h_0)^{-\frac{1}{2}}$, and those of (2.6.3) and (2.6.4) by $(1 - h_l/h_0)^{-\frac{1}{2}}$, where $h_0 = h(0)$ and h_l is the coefficient of z^l in $h(z)$.

First we observe that the right-hand members of (2.6.2), (2.6.3), (2.6.4) are cosine polynomials with the highest terms

$$(2.6.5) \qquad (2/\pi)^{\frac{1}{2}} h_0 \cos n\theta, \qquad (2/\pi)^{\frac{1}{2}} h_0 \frac{\sin(n+1)\theta}{\sin\theta}, \qquad \pi^{-\frac{1}{2}} h_0 \frac{\sin(n+\frac{1}{2})\theta}{\sin(\theta/2)},$$

respectively. In the first of these expressions, if $l = 2n > 0$, h_0 must be replaced by $h_0 + h_l$; in the second and last, if $l = 2(n+1)$ and $l = 2n+1$, respectively, we have $h_0 - h_l$ in place of h_0.

We give the proof of (2.6.2). First we show that

$$\int_{-1}^{+1} p_n(x)x^\nu(1-x^2)^{-\frac{1}{2}}\{\rho(x)\}^{-1}\,dx = 0, \qquad \nu = 0, 1, \cdots, n-1,$$

or, what amounts to the same thing,

$$\int_0^\pi p_n(\cos\theta)\cos\nu\theta\{\rho(\cos\theta)\}^{-1}\,d\theta = 0, \qquad \nu = 0, 1, 2, \cdots, n-1.$$

Now,

$$\frac{(2/\pi)^{\frac{1}{2}}}{2}\Re\left\{\int_0^\pi e^{in\theta}\,\overline{h(e^{i\theta})}\,(e^{i\nu\theta} + e^{-i\nu\theta})\,|\,h(e^{i\theta})\,|^{-2}\,d\theta\right\}$$

$$= \frac{(2/\pi)^{\frac{1}{2}}}{4}\Re\left\{\int_{-\pi}^{+\pi}\frac{e^{i(n+\nu)\theta} + e^{i(n-\nu)\theta}}{h(e^{i\theta})}\,d\theta\right\} = \frac{(2/\pi)^{\frac{1}{2}}}{4}\Re\left\{\frac{1}{i}\int_{|z|=1}\frac{z^{n+\nu} + z^{n-\nu}}{zh(z)}\,dz\right\} = 0,$$

since the function $(z^{n+\nu} + z^{n-\nu})\{zh(z)\}^{-1}$ is regular for $|z| \leqq 1$. Furthermore,

$$\int_{-1}^{+1}\{p_n(x)\}^2(1-x^2)^{-\frac{1}{2}}\{\rho(x)\}^{-1}\,dx = \int_0^\pi\{p_n(\cos\theta)\}^2\{\rho(\cos\theta)\}^{-1}\,d\theta$$

$$= \int_0^\pi p_n(\cos\theta)(2/\pi)^{\frac{1}{2}}h_0\cos n\theta\{\rho(\cos\theta)\}^{-1}\,d\theta$$

$$= \frac{1}{4}(2/\pi)^{\frac{1}{2}}h_0(2/\pi)^{\frac{1}{2}}\Re\left\{\frac{1}{i}\int_{|z|=1}\frac{z^{2n}+1}{zh(z)}\,dz\right\} = \frac{1}{4}(2/\pi)h_0(2\pi/h_0) = 1.$$

The proofs of (2.6.3) and (2.6.4) are similar. In place of $\cos\nu\theta$ we use $\sin(\nu+1)\theta/\sin\theta$ and $\sin(\nu+\frac{1}{2})\theta/\sin(\theta/2)$, respectively. The modifications necessary for $l = 2n$, $l = 2(n + 1)$, and $l = 2n + 1$, in (2.6.2), (2.6.3), and (2.6.4), respectively, are also obvious. Finally, we notice that (2.6.2) arises from (2.6.4), (2.6.4) from (2.6.3), and (2.6.2) from (2.6.3) by replacing $\rho(x)$ by $(1 - x)\rho(x)$, $(1 + x)\rho(x)$, and $(1 - x^2)\rho(x)$, respectively.

2.7. Stieltjes-Wigert polynomials

Wigert (**2**, p. 7; also Stieltjes **11**, pp. 507–508) found a very elegant explicit representation for the orthonormal polynomials $p_n(x)$ associated with the weight function

$$(2.7.1)\qquad w(x) = \pi^{-\frac{1}{2}} k \exp(-k^2 \log^2 x) = \pi^{-\frac{1}{2}} k x^{-k^2 \log x}, \quad 0 < x < +\infty;\ k > 0.$$

Using the notation (cf. Gauss **1**, p. 16)

$$(2.7.2)\qquad \begin{bmatrix} n \\ \nu \end{bmatrix} = \frac{(1 - q^n)(1 - q^{n-1}) \cdots (1 - q^{n-\nu+1})}{(1 - q)(1 - q^2) \cdots (1 - q^\nu)}, \quad 0 < \nu < n, \begin{bmatrix} n \\ 0 \end{bmatrix} = \begin{bmatrix} n \\ n \end{bmatrix} = 1,$$

where

$$(2.7.3)\qquad q = \exp\{-(2k^2)^{-1}\},$$

we have

$$(2.7.4)\qquad p_n(x) = (-1)^n q^{n/2+\frac{1}{4}}\{(1 - q)(1 - q^2) \cdots (1 - q^n)\}^{-\frac{1}{2}} \sum_{\nu=0}^{n} \begin{bmatrix} n \\ \nu \end{bmatrix} q^{\nu^2}(-q^{\frac{1}{2}}x)^\nu.$$

For $n = 0$ the product in the braces must be replaced by 1.

The proof can be based on the identity of Gauss:

$$(2.7.5)\qquad \sum_{\nu=0}^{n} \begin{bmatrix} n \\ \nu \end{bmatrix} q^{\nu(\nu+1)/2} u^\nu = (1 + qu)(1 + q^2 u) \cdots (1 + q^n u).$$

See Szegö **12**, where other similar polynomials (related to the theory of theta functions) are also considered. Also see Hahn **5**.

2.8. Distributions of Stieltjes type; an analogue of Legendre polynomials

Tchebichef (**4**) investigated a remarkable finite set of orthogonal polynomials associated with the distribution $d\alpha(x)$ of Stieltjes type, where $\alpha(x)$ is a step function with jumps of one unit at the points $x = 0, 1, 2, \cdots, N - 1$ (N is a fixed positive integer). This is a distribution of the type mentioned at the end of §2.2 (1). The associated polynomials are, except for constant factors (see (2.8.3)),

$$(2.8.1)\qquad t_n(x) = n!\, \Delta^n \binom{x}{n}\binom{x - N}{n}, \qquad n = 0, 1, 2, \cdots, N - 1.$$

Indeed, Tchebichef shows (**4**, pp. 547, 552; see also A. Markoff **1**, pp. 21–22) that

$$(2.8.2)\qquad \int_{-\infty}^{+\infty} t_n(x) t_m(x)\, d\alpha(x) = \sum_{x=0,1,2,\cdots,N-1} t_n(x) t_m(x) = 0, \qquad \text{if } n \neq m,$$

and

$$\int_{-\infty}^{+\infty} \{t_n(x)\}^2 \, d\alpha(x) = \sum_{x=0,1,2,\cdots,N-1} \{t_n(x)\}^2$$

(2.8.3)

$$= \frac{N(N^2 - 1^2)(N^2 - 2^2) \cdots (N^2 - n^2)}{2n+1},$$

$$n, m = 0, 1, 2, \cdots, N-1.$$

These formulas hold for all non-negative values of n and m, but they are trivial for $n \geqq N$ or $m \geqq N$, since $t_n(x) = 0$ for $x = 0, 1, 2, \cdots, N-1$, if $n \geqq N$.

In (2.8.1) we used the symbols

$$\Delta f(x) = f(x+1) - f(x),$$

(2.8.4)

$$\Delta^n f(x) = \Delta\{\Delta^{n-1} f(x)\}$$

$$= f(x+n) - \binom{n}{1} f(x+n-1) + \cdots + (-1)^n f(x).$$

By the mean-value theorem

(2.8.5)

$$\Delta^n f(x) = f^{(n)}(x + \theta n), \qquad 0 < \theta < 1,$$

(see, for example, Pólya-Szegö **1**, vol. 2, pp. 55, 241, problem 98), we obtain for a fixed value of n the remarkable formula

(2.8.6)

$$\lim_{N\to\infty} N^{-n} t_n(Nx) = P_n(2x-1),$$

where $P_n(x)$ is the Legendre polynomial of degree n (see §4.1 (3)). The representation (2.8.1) is the "difference" analogue of (4.3.1), $\alpha = \beta = 0$. The proofs of (2.8.2) and (2.8.3) are analogous to those in §4.3.

Tchebichef also considers (**1**, **2**) the more general case in which the points $0, 1, 2, \cdots, N-1$ are replaced by an arbitrary set of N distinct points. In this connection he obtains an interpolation formula having a certain significance in mathematical statistics. (See Jordan **1**.)

2.81. Poisson-Charlier polynomials

These polynomials have become important in some recent investigations connected with the calculus of probability and statistics (see Doetsch **2** and the literature quoted in E. Schmidt **1**, also Meixner **1**, **2**). They belong to the distribution $d\alpha(x)$ where $\alpha(x)$ is a step function with the jump

(2.81.1)

$$j(x) = e^{-a} a^x (x!)^{-1} \text{ at the point } x, \qquad x = 0, 1, 2, \cdots; a > 0.$$

Obviously, the total variation of $\alpha(x)$ is

$$\alpha(+\infty) - \alpha(-\infty) = \sum_{x=0}^{\infty} j(x) = 1.$$

The corresponding orthonormal polynomials are:

$$p_n(x) = a^{n/2}(n!)^{-\frac{1}{2}} \sum_{\nu=0}^{n} (-1)^{n-\nu}\binom{n}{\nu}\nu!\, a^{-\nu}\binom{x}{\nu} \tag{2.81.2}$$
$$= a^{n/2}(n!)^{-\frac{1}{2}}(-1)^n\{j(x)\}^{-1}\Delta^n j(x-n).$$

A simple proof of (2.81.2) can be given by means of the method of generating functions (see §4.4; cf. Doetsch **2**, p. 260, and Meixner **1, 2**). Let, for a sufficiently small $|w|$,

$$G(x, w) = \sum_{n=0}^{\infty} a^{-n/2}(n!)^{-\frac{1}{2}} p_n(x) w^n = \sum_{n=0}^{\infty}\sum_{\nu=0}^{n} \frac{(-1)^{n-\nu}}{n!}\binom{n}{\nu}\nu! a^{-\nu}\binom{x}{\nu} w^n$$
$$= \sum_{\nu=0}^{\infty}\sum_{n=\nu}^{\infty} \frac{(-1)^{n-\nu}}{n!}\binom{n}{\nu}\nu! a^{-\nu}\binom{x}{\nu} w^n \tag{2.81.3}$$
$$= \sum_{\nu=0}^{\infty} a^{-\nu}\binom{x}{\nu} w^\nu e^{-w} = e^{-w}(1 + a^{-1}w)^x.$$

Then

$$\sum_{x=0,1,2,\cdots} j(x)G(x, u)G(x, v)$$
$$= \sum_{x=0,1,2,\cdots} e^{-a}a^x(x!)^{-1}e^{-u}(1 + a^{-1}u)^x e^{-v}(1 + a^{-1}v)^x \tag{2.81.4}$$
$$= e^{-a-u-v}e^{a(1+a^{-1}u)(1+a^{-1}v)} = e^{a^{-1}uv},$$

so that

$$\sum_{x=0,1,2,\cdots} j(x)a^{-n/2}(n!)^{-\frac{1}{2}}p_n(x)a^{-m/2}(m!)^{-\frac{1}{2}}p_m(x) = a^{-n}(n!)^{-1}\delta_{nm}, \tag{2.81.5}$$
$$n, m = 0, 1, 2, \cdots.$$

The polynomials (2.81.2) are connected with Laguerre polynomials (§5.1) by the relation

$$p_n(x) = a^{-n/2}(n!)^{\frac{1}{2}} L_n^{(x-n)}(a). \tag{2.81.6}$$

Concerning the expansion problem associated with Charlier-Poisson polynomials, we refer to E. Schmidt **1**.

2.82. Krawtchouk's polynomials

Considerations in the calculus of probability lead also to the following distribution $d\alpha(x)$.

Let $\alpha(x)$ be a step function with the jump, at the point x, of

$$j(x) = \binom{N}{x} p^x q^{N-x}, \qquad x = 0, 1, 2, \cdots, N. \tag{2.82.1}$$

Here N is a positive integer, $p > 0$, $q > 0$, and $p + q = 1$.

See Krawtchouk **1**. The total variation of $\alpha(x)$ is 1. The associated set of orthogonal polynomials is again finite as in §2.8.

(1) The method of generating functions yields the formula

$$p_n(x) = \left\{\binom{N}{n}\right\}^{-\frac{1}{2}} (pq)^{-n/2} \sum_{\nu=0}^{n} (-1)^{n-\nu} \binom{N-x}{n-\nu}\binom{x}{\nu} p^{n-\nu} q^{\nu}, \tag{2.82.2}$$

$$n = 0, 1, 2, \cdots, N.$$

Indeed, let

$$\begin{aligned} K(x, w) &= \sum_{n=0}^{N} \left\{\binom{N}{n}\right\}^{\frac{1}{2}} (pq)^{n/2} p_n(x) w^n \\ &= \sum_{n=0}^{N}\sum_{\nu=0}^{n} (-1)^{n-\nu} \binom{N-x}{n-\nu}\binom{x}{\nu} p^{n-\nu} q^{\nu} w^n. \end{aligned} \tag{2.82.3}$$

In case x is an integer, $0 \leqq x \leqq N$, the last summation can be extended over all values of ν, $n = 0, 1, 2, \cdots$; $\nu \leqq n$, since the general term vanishes if $N - x < n - \nu$ or if $x < \nu$. (From $N - x \geqq n - \nu$, $x \geqq \nu$, we have $N \geqq n$.) Thus

$$\begin{aligned} K(x, w) &= \sum_{\nu=0}^{\infty} \binom{x}{\nu} q^{\nu} w^{\nu} \sum_{n=\nu}^{\infty} (-1)^{n-\nu} \binom{N-x}{n-\nu} p^{n-\nu} w^{n-\nu} \\ &= \sum_{\nu=0}^{\infty} \binom{x}{\nu} q^{\nu} w^{\nu} (1 - pw)^{N-x} = (1 + qw)^x (1 - pw)^{N-x}, \end{aligned} \tag{2.82.4}$$

from which

$$\begin{aligned} &\sum_{x=0,1,2,\cdots,N} j(x) K(x, u) K(x, v) \\ &= \sum \binom{N}{x} p^x q^{N-x} (1 + qu)^x (1 - pu)^{N-x} (1 + qv)^x (1 - pv)^{N-x} \\ &= \{p(1 + qu)(1 + qv) + q(1 - pu)(1 - pv)\}^N = (1 + pquv)^N, \end{aligned} \tag{2.82.5}$$

so that in fact

$$\sum_{x=0,1,2,\cdots,N} j(x) \left\{\binom{N}{n}\right\}^{\frac{1}{2}} (pq)^{n/2} p_n(x) \left\{\binom{N}{m}\right\}^{\frac{1}{2}} (pq)^{m/2} p_m(x) = \binom{N}{n} (pq)^n \delta_{nm}, \tag{2.82.6}$$

$$n, m = 0, 1, 2, \cdots, N.$$

If $n > N$, obviously $p_n(x) = 0$ for $x = 0, 1, 2, \cdots, N$.

(2) Two other classes of polynomials can be derived from the polynomials (2.82.2) by two different limiting processes.

(a) Let z be real and let x denote the greatest integer less than or equal to $pN + z(2pqN)^{\frac{1}{2}}$ where p, q, z are fixed, and $N \to \infty$. Then for a fixed n

$$\lim_{N \to \infty} p_n(x) = (2^n n!)^{-\frac{1}{2}} H_n(z) \tag{2.82.7}$$

if $H_n(z)$ denotes the nth Hermite polynomial (§5.5). This follows readily from (2.82.3) and (2.82.4), since for x an integer, $0 < x < N$,

$$\lim_{N\to\infty}\sum_{n=0}^{N}\left\{\binom{N}{n}\right\}^{\frac{1}{2}}(pq)^{n/2}p_n(x)\{(2/N)^{\frac{1}{2}}w\}^n$$
$$= \lim_{N\to\infty}\{1+(2/N)^{\frac{1}{2}}qw\}^x\{1-(2/N)^{\frac{1}{2}}pw\}^{N-x}$$
$$= \lim_{N\to\infty}\exp\{(2/N)^{\frac{1}{2}}qwx-(2/N)^{\frac{1}{2}}pw(N-x)-N^{-1}q^2w^2x-N^{-1}p^2w^2(N-x)\}$$
$$= \exp\{2z(pq)^{\frac{1}{2}}w-pqw^2\} = \sum_{n=0}^{\infty}\frac{H_n(z)}{n!}\{(pq)^{\frac{1}{2}}w\}^n.$$

(Use Titchmarsh **1**, p. 95; see (5.5.7).) It is instructive to observe that the same limiting process applied to the given distribution $d\alpha(x)$ leads to the distribution $e^{-z^2}\,dz$ of the Hermite polynomials; more precisely,

$$j(x) \cong (2\pi pqN)^{-\frac{1}{2}}e^{-z^2}, \quad \text{or} \quad j(x)\,dx \cong \pi^{-\frac{1}{2}}e^{-z^2}\,dz. \tag{2.82.8}$$

(b) Let $pN = a$, where a is a fixed positive number, $N\to\infty$, $p\to 0$, $q\to 1$. Then for a fixed n and a fixed integer $x \geqq 0$, we find that $\lim_{N\to\infty}p_n(x)$ exists and is identical with the Poisson-Charlier polynomial (2.81.2). In fact ((2.81.3))

$$\begin{aligned}\lim_{N\to\infty}(1+qw)^x(1-pw)^{N-x} &= \lim_{p\to 0}(1+qw)^x(1-pw)^{-x}(1-pw)^{p^{-1}a}\\ &= (1+w)^xe^{-aw}\end{aligned} \tag{2.82.9}$$

2.9. Further special cases

Concerning other distributions $d\alpha(x)$ of Stieltjes type, see A. Markoff **1**, pp. 7–18; Stieltjes **11**, pp. 546–555; and Gottlieb **1**.

Markoff considers the case for which $\alpha(x)$ is a step function with the jump, at the point q^x, of $j(x) = q^x$, $x = 0, 1, 2, \cdots, N-1$, and $q > 0$, $q \neq 1$. This distribution is very much similar to that in §2.8. Analogues of (2.8.1) and (2.8.6) hold.

Stieltjes and Gottlieb investigate the case for which $\alpha(x)$ is a step function with the jump q^x at the point x, $x = 0, 1, 2, \cdots, 0 < q < 1$.

In addition to these "discrete" distributions as well as to those studied in §2.8 and §2.82, see Karlin-McGregor 2, Eagleson 1, and Gasper 6, 7.

A remarkable distribution can be defined by the weight function

$$w(x) = \{x(\alpha-x)(\beta-x)\}^{-\frac{1}{2}}, \qquad 0 < x \leqq \alpha,\ \alpha < \beta. \tag{2.9.1}$$

Heine (**3**, vol. 1, pp. 294–296) derives a second order linear differential equation for the associated orthogonal polynomials which are related to the Jacobian elliptic functions. Recently **Achieser (1)** investigated the orthogonal polynomials associated with the weight function

$$w(x) = \begin{cases}\{(1-x^2)(a-x)(b-x)\}^{-\frac{1}{2}}|c-x|, & -1 \leqq x \leqq a,\quad b \leqq x \leqq +1,\\ 0, & a < x < b,\end{cases} \tag{2.9.2}$$

where $-1 < a < b < +1$ and c depends in a proper way on a and b. These polynomials are also related to the elliptic functions.

In some cases the condition of the positiveness of the weight function can be removed to a certain extent. (Cf. Szegö **19**.)

Concerning the polynomials of Pollaczek (**1–4**), see Appendix.

CHAPTER III

GENERAL PROPERTIES OF ORTHOGONAL POLYNOMIALS

In this chapter we shall deal with properties of orthogonal polynomials which hold for distributions restricted only by certain conditions of integrability. Usually, we shall consider distributions of the Stieltjes type $d\alpha(x)$, but at times we shall be concerned with distributions of the special type $w(x)\,dx$. However, $\alpha(x)$ and $w(x)$ will always be taken subject to the conditions formulated in §2.2 (1).

3.1. Extremum properties; closure

(1) Let $f(x)$ be a given function of the class $L_\alpha^2(a, b)$, and let x^n belong to $L_\alpha(a, b)$ for $n = 0, 1, 2, \cdots$. Then it is evident that the integrals

$$\int_a^b |f(x)|^2\, d\alpha(x), \qquad \int_a^b f(x)x^n\, d\alpha(x), \qquad n = 0, 1, 2, \cdots, \tag{3.1.1}$$

exist in the Stieltjes-Lebesgue sense. Next, denoting by $\{p_n(x)\}$ the orthonormal set of polynomials associated with the distribution $d\alpha(x)$ in $[a, b]$, we state the following theorem:

THEOREM 3.1.1. *The weighted quadratic deviation*

$$\int_a^b |f(x) - \rho(x)|^2\, d\alpha(x), \tag{3.1.2}$$

where $\rho(x)$ ranges over the set of all π_n, becomes a minimum if and only if $\rho(x)$ is the nth partial sum of the Fourier expansion

$$\begin{aligned} f(x) &\sim f_0p_0(x) + f_1p_1(x) + f_2p_2(x) + \cdots + f_np_n(x) + \cdots, \\ f_n &= \int_a^b f(x)p_n(x)\, d\alpha(x), \qquad n = 0, 1, 2, \cdots \end{aligned} \tag{3.1.3}$$

See Theorem 2.1.2 and §2.2 (3). The minimum itself is

$$\int_a^b |f(x)|^2\, d\alpha(x) - \sum_{\nu=0}^{n} |f_\nu|^2. \tag{3.1.4}$$

This implies Bessel's inequality, that is,

$$|f_0|^2 + |f_1|^2 + |f_2|^2 + \cdots + |f_n|^2 + \cdots \leq \int_a^b |f(x)|^2\, d\alpha(x). \tag{3.1.5}$$

(2) On replacing n by $n - 1$ and taking $f(x) = x^n$, we obtain the following direct characterization of $p_n(x)$:

THEOREM 3.1.2. *The integral*

$$\int_a^b |\rho(x)|^2 \, d\alpha(x), \tag{3.1.6}$$

where $\rho(x)$ ranges over the set of all π_n with the highest term x^n, becomes a minimum if and only if $\rho(x)$ = const. $p_n(x)$.

See §2.2 (3). If k_n is the coefficient of x^n in $p_n(x)$, the minimum of (3.1.6) is attained for $\rho(x) = k_n^{-1} p_n(x)$.

(3) THEOREM 3.1.3. *Let x_0 be an arbitrary complex constant, $\rho(x)$ an arbitrary π_n with complex coefficients, normalized by the condition*

$$\int_a^b |\rho(x)|^2 \, d\alpha(x) = 1. \tag{3.1.7}$$

The maximum of $|\rho(x_0)|^2$ is given by the polynomials

$$\rho(x) = \epsilon\{K_n(x_0, x_0)\}^{-\frac{1}{2}} K_n(x_0, x), \qquad |\epsilon| = 1, \tag{3.1.8}$$

where

$$\begin{aligned} K_n(x_0, x) &= \overline{p_0(x_0)}p_0(x) + \overline{p_1(x_0)}p_1(x) + \cdots + \overline{p_n(x_0)}p_n(x). \\ &= p_0(\bar{x}_0)p_0(x) + p_1(\bar{x}_0)p_1(x) + \cdots + p_n(\bar{x}_0)p_n(x). \end{aligned} \tag{3.1.9}$$

The maximum itself is $K_n(x_0, x_0)$.

If we write $\rho(x) = \lambda_0 p_0(x) + \lambda_1 p_1(x) + \cdots + \lambda_n p_n(x)$, condition (3.1.7) becomes $|\lambda_0|^2 + |\lambda_1|^2 + \cdots + |\lambda_n|^2 = 1$, and by Cauchy's inequality it follows that

$$|\rho(x_0)|^2 \leq \sum_{\nu=0}^{n} |\lambda_\nu|^2 \sum_{\nu=0}^{n} |p_\nu(x_0)|^2 = K_n(x_0, x_0). \tag{3.1.10}$$

The latter bound is attained for $\lambda_\nu = \lambda \overline{p_\nu(x_0)}$, where λ is to be determined according to the condition

$$|\lambda|^2 \sum_{\nu=0}^{n} |p_\nu(x_0)|^2 = 1.$$

Thus the statement is established.

The "kernel polynomials" $K_n(x_0, x) = \overline{K_n(x, x_0)} = K_n(\bar{x}, \bar{x}_0)$ can be used for the representation of the nth partial sum $s_n(x)$ of the Fourier expansion (3.1.3) in the form of an integral. In fact we have

$$\begin{aligned} s_n(x) &= f_0 p_0(x) + f_1 p_1(x) + \cdots + f_n p_n(x) \\ &= \sum_{\nu=0}^{n} p_\nu(x) \int_a^b f(t) p_\nu(t) \, d\alpha(t) = \int_a^b f(t) K_n(t, x) \, d\alpha(t). \end{aligned} \tag{3.1.11}$$

As a consequence of (3.1.11) we obtain

$$\int_a^b K_n(t, x)\rho(t)\, d\alpha(t) = \rho(x), \tag{3.1.12}$$

where $\rho(x)$ is an arbitrary π_n. We may easily show that this is a characteristic property of $K_n(t, x)$ as a π_n in $\bar{t}$. As a further consequence we notice the following theorem:

Theorem 3.1.4. *Let a and x_0 be finite, $x_0 \leqq a$. Then the polynomials $\{K_n(x_0, x)\}$ are orthogonal with respect to the distribution $(x - x_0)\, d\alpha(x)$.*

This follows immediately from (3.1.12) by writing $x = x_0$, $\rho(t) = (t - x_0)r(t)$, where $r(t)$ is an arbitrary π_{n-1}. A similar result holds if b is finite.

(4) According to the previous results the expression (3.1.4) decreases as n increases, and consequently it tends to a non-negative limit as $n \to \infty$. We have Parseval's formula

$$|f_0|^2 + |f_1|^2 + |f_2|^2 + \cdots + |f_n|^2 + \cdots = \int_a^b |f(x)|^2\, d\alpha(x) \tag{3.1.13}$$

when and only when this limit is zero.

The validity of (3.1.13) is evidently equivalent to the fact that the integral (3.1.2) can be made arbitrarily small by a proper choice of the polynomial $\rho(x)$. This is, however, the same as the closure in $L^2_\alpha(a, b)$ of the system $\{p_n(x)\}$ or of the system $\{x^n\}$ (see the definition in §1.5 (1)). Thus, according to Theorem 1.5.2 we have the following:

Theorem 3.1.5. *The set of the orthogonal polynomials $\{p_n(x)\}$, $n = 0, 1, 2, \cdots$, associated with the distribution $d\alpha(x)$ on a finite interval $[a, b]$, is closed in $L^2_\alpha(a, b)$. More generally it is closed in $L^p_\alpha(a, b)$, $p \geqq 1$.*

For a function $f(x)$ of the class $L^2_\alpha(a, b)$ Parseval's formula (3.1.13) *holds.*

A function $f(x)$ of the class $L^2_\alpha(a, b)$, for which $f_n = 0$, $n = 0, 1, 2, \cdots$, is necessarily a zero-function.

The finiteness of the interval considered is an essential restriction. Some cases of infinite intervals will be studied later. (See §5.7.)

The assumption $f_n = 0$ in the last part of Theorem 3.1.5 is equivalent to the fact that

$$\int_a^b f(x)x^n\, d\alpha(x) = 0, \qquad n = 0, 1, 2, \cdots. \tag{3.1.14}$$

The discussion of this condition is closely connected with the uniqueness of Stieltjes' problem of moments. An example showing that Theorem 3.1.5 does not hold generally in case of an infinite interval is the following:

$$d\alpha(x) = \exp(-x^\mu \cos \mu\pi)\, dx, \quad f(x) = \sin(x^\mu \sin \mu\pi), \quad 0 < \mu < 1/2. \tag{3.1.15}$$

Here (3.1.14) is satisfied (cf. Pólya-Szegö **1**, vol. 1, pp. 114, 285, 286, problem 153), and yet $f(x)$ is not a zero-function. In the same case, if $\rho(x)$ is an arbitrary polynomial,

$$A = \int_0^\infty \{f(x)\}^2 d\alpha(x) = \int_0^\infty f(x)\{f(x) - \rho(x)\} d\alpha(x)$$

$$\leqq \int_0^\infty |f(x) - \rho(x)| d\alpha(x) \leqq \left\{\int_0^\infty |f(x) - \rho(x)|^2 d\alpha(x)\right\}^{\frac{1}{2}} \left\{\int_0^\infty d\alpha(x)\right\}^{\frac{1}{2}},$$

which shows that the integrals

$$\int_0^\infty |f(x) - \rho(x)| d\alpha(x) \quad \text{and} \quad \int_0^\infty |f(x) - \rho(x)|^2 d\alpha(x)$$

cannot be made arbitrarily small.

3.11. Generalizations

Numerous analogous problems arise if the weighted quadratic deviation (3.1.2) is replaced by other types of deviations. The most interesting cases are

$$\int_a^b |f(x) - \rho(x)|^p d\alpha(x), \tag{3.11.1}$$

p being a fixed positive number, and the limiting case $p \to \infty$[8] (called also the "Tchebichef deviation"):

$$\max_{a \leqq x \leqq b} \{|f(x) - \rho(x)| w(x)\}. \tag{3.11.2}$$

In the last case we assume that $f(x)$ and $w(x)$ are continuous. Similarly, the integral (3.1.6) might be replaced by the expression

$$\int_a^b |\rho(x)|^p d\alpha(x), \tag{3.11.3}$$

or by

$$\max_{a \leqq x \leqq b} \{|\rho(x)| w(x)\}. \tag{3.11.4}$$

The polynomials of fixed degree which minimize (3.11.1) and (3.11.2) represent a generalization of the nth partial sum of the expansion of $f(x)$ in terms of the orthogonal polynomials associated with $d\alpha(x)$ or $w(x)\,dx$. The polynomials of fixed degree, and with highest coefficient unity, which minimize (3.11.3) and (3.11.4) represent a generalization of the orthogonal polynomials themselves.

Since the number of investigations which can be classified under this general point of view is very considerable, only the most important aspects can be indicated here.

(1) For $p = 2$ the polynomials minimizing (3.11.1) are the partial sums of the

[8] Replacing $d\alpha(x)$ by $\{w(x)\}^p\,dx$, we have (a and b finite, $f(x)$ and $w(x)$ continuous)

$$\lim_{p\to\infty} \left[\int_a^b |f(x) - \rho(x)|^p \{w(x)\}^p dx\right]^{1/p} = \max_{a \leqq x \leqq b} \{|f(x) - \rho(x)| w(x)\}.$$

expansion of a given function $f(x)$ in a series of orthogonal polynomials (Theorem 3.1.1). In case $a = -1$, $b = +1$, $w(x) = (1 - x^2)^{-\frac{1}{2}}$, we obtain the expansion of $f(x) = f(\cos \theta)$ in a cosine series; in case $a = -1$, $b = +1$, $w(x) = 1$, we obtain the Legendre series. (Cf. Chapters IX and XIII.)

(2) In the general case (3.11.1) the existence and uniqueness of the minimizing polynomials have been investigated. See Jackson **1**, **2**, **3**; Shohat **1**, pp. 509–513, **4**, pp. 160–161. Both authors consider only distributions of the type $w(x)\,dx$. For general distributions see Tamarkin **1**, p. 118.

(3) Let a, b be finite, $d\alpha(x) = dx$, and $p = 1$. For problems (3.11.1) and (3.11.3) see S. Bernstein **2**, in particular pp. 135–137, where references are also given to the earlier literature. (Cf. also Geronimus **5**.) Recently Achyeser (**2**) discussed the problem of minimizing

$$\int_p^q |\,\rho(x)\,|\,dx + \int_r^s |\,\rho(x)\,|\,dx, \tag{3.11.5}$$

where $[p, q]$ and $[r, s]$ are given disjoint finite intervals and $\rho(x)$ ranges over all π_n with the highest term x^n. The minimizing polynomials can be represented in terms of elliptic functions.

(4) In the case where a and b are finite, $f(x)$ continuous, and $w(x) = 1$, the minimum problem corresponding to (3.11.2) leads to the closest approximation of continuous functions by polynomials. The connection between the closeness of this approximation (as $n \to \infty$) and the continuity properties of $f(x)$ has been investigated in great detail. (See Jackson **4**.)

(5) If a, b are finite, $f(x)$ continuous, and $d\alpha(x) = dx$, the minimizing polynomials of (3.11.1) (for a fixed n) tend to the minimizing polynomial of (3.11.2) as $p \to \infty$. (We have existence and uniqueness in both cases.) See Pólya **2**; also Shohat **1**, pp. 513–514, **4**, p. 171. Both authors consider only distributions of the form $w(x)\,dx$. For general distributions, see Tamarkin **1**, p. 125.

(6) If $a = -1$, $b = +1$, $w(x) = 1$, then problem (3.11.4) has the solution $\rho(x) = 2^{1-n}T_n(x)$ (see the notation in (1.12.3)). This is a classical result due to Tchebichef and is the starting point of various investigations of the highest interest. (Cf. S. Bernstein **1**.)

3.2. Recurrence formula; Christoffel-Darboux formula

(1) THEOREM 3.2.1. *The following relation holds for any three consecutive orthogonal polynomials:*

$$p_n(x) = (A_n x + B_n)p_{n-1}(x) - C_n p_{n-2}(x), \quad n = 2, 3, 4, \cdots. \tag{3.2.1}$$

Here A_n, B_n, and C_n are constants, $A_n > 0$ and $C_n > 0$. If the highest coefficient of $p_n(x)$ is denoted by k_n, we have

$$A_n = \frac{k_n}{k_{n-1}}, \qquad C_n = \frac{A_n}{A_{n-1}} = \frac{k_n k_{n-2}}{k_{r-1}^2}. \tag{3.2.2}$$

For the proof, we first determine A_n so that $p_n(x) - A_n x p_{n-1}(x)$ is a π_{n-1}.

This can be represented as a linear combination $\lambda_0 p_0(x) + \lambda_1 p_1(x) + \cdots + \lambda_{n-1}p_{n-1}(x)$, and because of the orthogonality it is readily seen that $\lambda_\nu = 0$ if $\nu < n - 2$. Therefore (3.2.1) follows. The first part of (3.2.2) is a consequence of (3.2.1); the second part follows from

$$\int_a^b p_n(x)p_{n-2}(x)\,d\alpha(x) = 0 = A_n \int_a^b x p_{n-1}(x)p_{n-2}(x)\,d\alpha(x) - C_n,$$

since the integral of the right-hand member is equal to

$$\int_a^b p_{n-1}(x)(k_{n-2}x^{n-1} + \cdots)\,d\alpha(x) = \frac{k_{n-2}}{k_{n-1}} \int_a^b \{p_{n-1}(x)\}^2\,d\alpha(x).$$

The recurrence formula (3.2.1) is valid also for $n = 1$ if we write $p_{-1}(x) = 0$, with the understanding that C_1 is arbitrary. The first formula in (3.2.2) then holds for $n = 1$.

Concerning a converse of Theorem 3.2.1, see Favard **1**.

(2) THEOREM 3.2.2. *We have*

$$p_0(x)p_0(y) + p_1(x)p_1(y) + \cdots + p_n(x)p_n(y) = \frac{k_n}{k_{n+1}} \frac{p_{n+1}(x)p_n(y) - p_n(x)p_{n+1}(y)}{x - y}. \tag{3.2.3}$$

For the special case $d\alpha(x) = dx$, see Christoffel **1**; see also Darboux **1**. This important identity can be easily derived from the recurrence formula. For we have

$$\begin{aligned} p_{n+1}(x)p_n(y) - p_n(x)p_{n+1}(y) &= \{(A_{n+1}x + B_{n+1})p_n(x) - C_{n+1}p_{n-1}(x)\}p_n(y) \\ &\quad - p_n(x)\{(A_{n+1}y + B_{n+1})p_n(y) - C_{n+1}p_{n-1}(y)\}, \\ &= A_{n+1}(x - y)p_n(x)p_n(y) + C_{n+1}\{p_n(x)p_{n-1}(y) - p_{n-1}(x)p_n(y)\}. \end{aligned}$$

By (3.2.2), this becomes

$$\frac{k_n}{k_{n+1}} \frac{p_{n+1}(x)p_n(y) - p_n(x)p_{n+1}(y)}{x - y} = p_n(x)p_n(y) + \frac{k_{n-1}}{k_n} \frac{p_n(x)p_{n-1}(y) - p_{n-1}(x)p_n(y)}{x - y},$$

which holds also for $n = 0$, with the understanding that k_{-1} is arbitrary. On replacing n by $0, 1, 2, \cdots, n$ and adding, we obtain (3.2.3).

We notice the special case $x = y$:

$$\{p_0(x)\}^2 + \{p_1(x)\}^2 + \cdots + \{p_n(x)\}^2 = \frac{k_n}{k_{n+1}} \{p'_{n+1}(x)p_n(x) - p'_n(x)p_{n+1}(x)\}. \tag{3.2.4}$$

(3) The left-hand member of (3.2.3) is identical with the "kernel" $K_n(\bar{x}, y) = K_n(\bar{y}, x)$ introduced in (3.1.9). Using (3.1.12), we may derive (3.2.3) in a different way by showing that the right-hand member of (3.2.3) (replacing y by t) satisfies (3.1.12). For we have

$$\frac{k_n}{k_{n+1}} \int_a^b \frac{p_{n+1}(x)p_n(t) - p_n(x)p_{n+1}(t)}{x - t} \rho(t)\, d\alpha(t)$$

$$= \frac{k_n}{k_{n+1}} \int_a^b \{p_{n+1}(x)p_n(t) - p_n(x)p_{n+1}(t)\} \frac{\rho(t) - \rho(x)}{x - t}\, d\alpha(t)$$

$$+ \frac{k_n}{k_{n+1}} \rho(x) \int_a^b p_n(t) \frac{p_{n+1}(x) - p_{n+1}(t)}{x - t}\, d\alpha(t)$$

$$+ \frac{k_n}{k_{n+1}} \rho(x) \int_a^b p_{n+1}(t) \frac{p_n(t) - p_n(x)}{x - t}\, d\alpha(t).$$

Here the first and third integrals of the right-hand member vanish (also for $n = 0$). The second term is $\rho(x)$ since

$$\frac{k_n}{k_{n+1}} \frac{p_{n+1}(x) - p_{n+1}(t)}{x - t} = k_n t^n + \cdots.$$

Another proof of (3.2.3) may be obtained by combining Theorem 3.1.4 with Theorem 2.5.

3.3. Elementary properties of the zeros

THEOREM 3.3.1. *The zeros of the orthogonal polynomials $p_n(x)$, associated with the distribution $d\alpha(x)$ on the interval $[a, b]$, are real and distinct and are located in the interior of the interval $[a, b]$.*

In special instances, particularly in the classical cases (see §2.4), we shall obtain later more exact information concerning the position of the zeros. (See Chapter VI.)

As a consequence of Theorem 3.3.1 we have $\alpha(a) < \alpha(x_1 - 0)$ and $\alpha(x_n + 0) < \alpha(b)$, where x_1 and x_n are the least and the greatest zeros of $p_n(x)$, respectively.

(1) The usual proof of the preceding theorem is based on the orthogonality property. From

$$\int_a^b p_n(x)\, d\alpha(x) = 0, \qquad n \geqq 1,$$

we are assured of the existence of at least one point in the interior of $[a, b]$ at which $p_n(x)$ changes sign. (The function $\alpha(x)$ has an infinite number of points of increase.) If $x_1, x_2, \cdots, x_l$ denote the abscissas of all such points, the product $p_n(x)(x - x_1)(x - x_2) \cdots (x - x_l)$ has a constant sign (that is, is non-negative or non-positive throughout $[a, b]$); we have $l \leqq n$. On the other hand, if $l < n$,

$$\int_a^b p_n(x)(x - x_1)(x - x_2) \cdots (x - x_l)\, d\alpha(x) = 0. \tag{3.3.1}$$

Since the integrand is not a zero-function, this is impossible. Therefore we have $l = n$.

(2) A slight variation of this argument may be made as follows. Let x_0 be an arbitrary zero of $p_n(x)$. The coefficients of $p_n(x)$ being real, we see that $p_n(x)/(x - \bar{x}_0)$ is a π_{n-1}. On the other hand

$$\int_a^b p_n(x) \frac{p_n(x)}{x - \bar{x}_0} d\alpha(x) = \int_a^b (x - x_0) \left| \frac{p_n(x)}{x - x_0} \right|^2 d\alpha(x) = 0, \tag{3.3.2}$$

so that

$$x_0 \int_a^b \left| \frac{p_n(x)}{x - x_0} \right|^2 d\alpha(x) = \int_a^b x \left| \frac{p_n(x)}{x - x_0} \right|^2 d\alpha(x). \tag{3.3.3}$$

In other words, x_0 is the centroid of a mass distribution on the interval $[a, b]$. The integral in the left-hand member of (3.3.3) being positive, x_0 is real. From (3.3.2) we see that $a < x_0 < b$.

If x_0 were a multiple zero, we should have

$$\int_a^b p_n(x) \frac{p_n(x)}{(x - x_0)^2} d\alpha(x) = \int_a^b \left\{ \frac{p_n(x)}{x - x_0} \right\}^2 d\alpha(x) = 0, \tag{3.3.4}$$

which is a contradiction.

(3) The statement concerning the location of the zeros (not their simplicity) follows also from the minimum property formulated in Theorem 3.1.2. Were a zero x_0 to lie outside $[a, b]$, the distance $| x - x_0 |$ could be diminished simultaneously for all x in $[a, b]$ by a proper displacement of x_0. Hence the corresponding integral (3.1.6) could not be a minimum.

For an extension of this argument to polynomials possessing an analogous or a more general minimum property in the real or complex region, see Fejér **7**, Szegö **5**.

From the orthogonality property of the kernel $K_n(x_0, x)$ (Theorem 3.1.4), we can similarly derive some theorems concerning the location of its zeros in x if x_0 is regarded as a parameter. (See Szegö **5**, pp. 241–244.)

(4) The reality and simplicity of the zeros (without the more exact statement concerning their location in $[a, b]$) follow from the recurrence formula by means of Sturm's theorem (Perron **4**, vol. 2, pp. 7–9). For, the polynomials

$$p_0(x), p_1(x), p_2(x), \cdots, p_n(x) \tag{3.3.5}$$

form a Sturmian sequence in $[a, b]$ since (a) if $p_\nu(x_0) = 0$, $\nu \geqq 1$, it follows from (3.2.1) that $p_{\nu-1}(x_0)p_{\nu+1}(x_0) < 0$; (b) $p_0(x)$ is a constant $\neq 0$, and $p_n(x)$ is of precise degree n; (c) at a point x_0 where $p_n(x_0) = 0$, we have $p_n'(x_0)p_{n-1}(x_0) > 0$. The latter fact follows from (3.2.4) if n is replaced by $n - 1$ and x by x_0 (see below). Now the number of variations of sign in (3.3.5) is n if $x < 0$ and $| x |$ is sufficiently large; it is 0 if $x > 0$ and sufficiently large. (Cf. §6.2 (1) and the footnote 32.)

(5) From (3.2.4) we obtain the important inequality

$$p_{n+1}'(x)p_n(x) - p_n'(x)p_{n+1}(x) > 0, \qquad x \text{ real}. \tag{3.3.6}$$

As a first consequence we point out that $p_n(x)$ and $p_{n+1}(x)$ cannot have common zeros. Furthermore, we obtain the following separation theorem:

THEOREM 3.3.2. *Let $x_1 < x_2 < \cdots < x_n$ be the zeros of $p_n(x)$, $x_0 = a$, $x_{n+1} = b$. Then each interval $[x_\nu, x_{\nu+1}]$, $\nu = 0, 1, 2, \cdots, n$, contains exactly one zero of $p_{n+1}(x)$.*

In fact if ξ and η, $\xi < \eta$, are two consecutive zeros of $p_n(x)$, we have $p_n'(\xi)p_n'(\eta) < 0$. On the other hand, (3.3.6) yields $-p_n'(\xi)p_{n+1}(\xi) > 0$, $-p_n'(\eta)p_{n+1}(\eta) > 0$, so that $p_{n+1}(\xi)p_{n+1}(\eta) < 0$. This indicates an odd number, that is, at least one zero of $p_{n+1}(x)$ in $\xi < x < \eta$. Now let $\xi = x_n$ be the greatest zero of $p_n(x)$; then $p_n'(\xi) > 0$, and (3.3.6) yields $p_{n+1}(\xi) < 0$. Since $p_{n+1}(b)$ is positive, we obtain at least one zero of $p_{n+1}(x)$ on the right of $\xi = x_n$, and similarly at least one on the left of the least zero x_1 of $p_n(x)$. Consequently, we can have only one zero of $p_{n+1}(x)$ between x_ν and $x_{\nu+1}$, $\nu = 0, 1, 2, \cdots, n$.

By interchanging the rôle of $p_n(x)$ and $p_{n+1}(x)$, we can prove as before the existence of at least one zero of $p_n(x)$ between two consecutive zeros of $p_{n+1}(x)$. This shows again that we cannot have more than one zero of $p_{n+1}(x)$ between two consecutive zeros of $p_n(x)$.

(6) THEOREM 3.3.3. *Between two zeros of $p_n(x)$ there is at least one zero of $p_m(x)$, $m > n$.*

See Stieltjes **11**, pp. 414-418. For the following proof see Popoviciu **1**.

Let $\xi_1, \xi_2, \cdots, \xi_m$ be the zeros of $p_m(x)$ in increasing order. According to Theorem 3.4.1 we have

$$\sum_{\mu=1}^{m} \lambda_\mu p_n(\xi_\mu)\rho(\xi_\mu) = \int_a^b p_n(x)\rho(x)\,d\alpha(x) = 0, \tag{3.3.7}$$

where $\{\lambda_\mu\}$ are the Christoffel numbers associated with $\{\xi_\mu\}$ (see §3.4) and $\rho(x)$ is an arbitrary π_{n-1}. Now an argument similar to that used in (1) shows that the sequence $\{p_n(\xi_1), p_n(\xi_2), \cdots, p_n(\xi_m)\}$ displays at least n, and therefore exactly n, variations of sign. Here $\operatorname{sgn} p_n(\xi_1) = (-1)^n$, $p_n(\xi_m) > 0$ Thus there are n distinct intervals

$$\xi_{\mu_\nu} < x < \xi_{\mu_\nu+1}, \qquad \nu = 1, 2, \cdots, n; \qquad 1 \leqq \mu_1 < \mu_2 < \cdots < \mu_{n+1} \leqq m,$$

containing exactly one zero of $p_n(x)$, respectively. This establishes the statement.

Other simple consequences of (3.3.6) are:

THEOREM 3.3.4. *Let c be an arbitrary real constant. Then the polynomial*

$$p_{n+1}(x) - cp_n(x) \tag{3.3.8}$$

has $n+1$ distinct real zeros. If $c > 0$ ($c < 0$), these zeros lie in the interior of $[a, b]$, with the exception of the greatest (least) zero which lies in $[a, b]$ only for $c \leqq p_{n+1}(b)/p_n(b)$, $[c \geqq p_{n+1}(a)/p_n(a)]$.

Indeed, the function $p_{n+1}(x)/p_n(x)$ increases from $-\infty$ to $+\infty$ in the intervals $x_\nu < x < x_{\nu+1}$, $\nu = 0, 1, 2, \cdots, n$, where $x_0 = -\infty$, $x_{n+1} = +\infty$.

THEOREM 3.3.5. *The following decomposition into partial fractions holds:*

$$\frac{p_n(x)}{p_{n+1}(x)} = \sum_{\nu=0}^{n} \frac{l_\nu}{x - \xi_\nu}, \qquad l_\nu > 0, \tag{3.3.9}$$

where $\{\xi_\nu\}$ *denote the zeros of* $p_{n+1}(x)$.

For we have

$$l_\nu = \frac{p_n(\xi_\nu)}{p'_{n+1}(\xi_\nu)} = \frac{p'_{n+1}(\xi_\nu)p_n(\xi_\nu) - p'_n(\xi_\nu)p_{n+1}(\xi_\nu)}{\{p'_{n+1}(\xi_\nu)\}^2} > 0. \tag{3.3.10}$$

3.4. The Gauss-Jacobi mechanical quadrature

(1) THEOREM 3.4.1. *If* $x_1 < x_2 < \cdots < x_n$ *denote the zeros of* $p_n(x)$, *there exist real numbers* $\lambda_1, \lambda_2, \cdots, \lambda_n$ *such that*

$$\int_a^b \rho(x)\, d\alpha(x) = \lambda_1 \rho(x_1) + \lambda_2 \rho(x_2) + \cdots + \lambda_n \rho(x_n), \tag{3.4.1}$$

whenever $\rho(x)$ *is an arbitrary* π_{2n-1}. *The distribution* $d\alpha(x)$ *and the integer* n *uniquely determine these numbers* λ_ν.

The set $\{x_\nu = x_{\nu n}\}$ of zeros, as well as the set of numbers $\{\lambda_\nu = \lambda_{\nu n}\}$, depends, of course, on n. Sometimes the numbers λ_ν are called Christoffel numbers. See Gauss **2**, Jacobi **1**, Christoffel **1**, Tchebichef **1**, Mehler **1**; Heine **3**, vol. 2, pp. 1–31.

It suffices to prove (3.4.1) for the special cases $\rho(x) = x^k$, $k = 0, 1, 2, \cdots, 2n-1$. These cases represent $2n$ conditions which uniquely determine, as we shall prove, the Christoffel numbers λ_ν and the points x_ν. (If the distinct points x_ν are given arbitrarily, the numbers λ_ν can be determined so that (3.4.1) holds for every π_{n-1}.)

To prove (3.4.1) we construct the Lagrange interpolation polynomial $L(x)$ of degree $n-1$ which coincides with $\rho(x)$ at the points x_ν, that is,

$$L(x) = \sum_{\nu=1}^{n} \rho(x_\nu) \frac{p_n(x)}{p'_n(x_\nu)(x - x_\nu)} = \sum_{\nu=1}^{n} \rho(x_\nu) l_\nu(x), \tag{3.4.2}$$

where the $l_\nu(x)$ are the fundamental polynomials associated with the abscissas $x_1, x_2, \cdots, x_n$ of the Lagrange interpolation (see §14.1). Now $\rho(x) - L(x)$ is divisible by $p_n(x)$, so that $\rho(x) - L(x) = p_n(x)r(x)$, where $r(x)$ is a π_{n-1}. Therefore

$$\begin{aligned} \int_a^b \rho(x)\, d\alpha(x) &= \int_a^b L(x)\, d\alpha(x) + \int_a^b p_n(x) r(x)\, d\alpha(x) \\ &= \int_a^b L(x)\, d\alpha(x) = \sum_{\nu=1}^{n} \rho(x_\nu) \int_a^b l_\nu(x)\, d\alpha(x). \end{aligned}$$

This establishes (3.4.1) with

$$\lambda_\nu = \int_a^b l_\nu(x)\,d\alpha(x) = \int_a^b \frac{p_n(x)}{p_n'(x_\nu)(x - x_\nu)}\,d\alpha(x), \qquad \nu = 1, 2, \cdots, n. \tag{3.4.3}$$

Conversely, let (3.4.1) hold for an arbitrary π_{2n-1}, called $\rho(x)$. Then we choose $\rho(x) = l(x)r(x)$, where $l(x) = (x - x_1)(x - x_2) \cdots (x - x_n)$ and $r(x)$ is an arbitrary π_{n-1}. We find from (3.4.1) that

$$\int_a^b l(x)r(x)\,d\alpha(x) = 0,$$

so that $l(x) = \text{const.}\ p_n(x)$.

The interpretation of the left-hand member of (3.4.1) as a mechanical quadrature is obvious. For an arbitrary function $f(x)$ defined in $[a, b]$ it may be written (cf. §15.1)

$$Q_n(f) = \lambda_1 f(x_1) + \lambda_2 f(x_2) + \cdots + \lambda_n f(x_n). \tag{3.4.4}$$

Then Theorem 3.4.1 can be formulated as follows: $Q_n(f) = \int_a^b f(x)\,d\alpha(x)$ provided $f(x)$ is an arbitrary π_{2n-1}. Further, from (3.4.3) the Christoffel numbers λ_ν are the values of $Q_n(f)$ for $f(x) = l_\nu(x)$. Also we can discuss for a fixed function $f(x)$ the convergence of the sequence $\{Q_n(f)\}$ as $n \to \infty$. (Compare Theorem 15.2.3 and also Problem 9 below.) Concerning mechanical quadrature formulas holding for an arbitrary π_{2n-k}, see Shohat **7**, p. 465.

(2) THEOREM 3.4.2. *The Christoffel numbers λ_ν are positive, and*

$$\lambda_1 + \lambda_2 + \cdots + \lambda_n = \int_a^b d\alpha(x) = \alpha(b) - \alpha(a). \tag{3.4.5}$$

The following representations hold:

$$\lambda_\nu = \int_a^b \left(\frac{p_n(x)}{p_n'(x_\nu)(x - x_\nu)}\right)^2 d\alpha(x), \tag{3.4.6}$$

$$\lambda_\nu = \frac{k_{n+1}}{k_n} \frac{-1}{p_{n+1}(x_\nu)p_n'(x_\nu)} = \frac{k_n}{k_{n-1}} \frac{1}{p_{n-1}(x_\nu)p_n'(x_\nu)}, \tag{3.4.7}$$

$$\begin{aligned}\lambda_\nu^{-1} &= \{p_0(x_\nu)\}^2 + \{p_1(x_\nu)\}^2 + \cdots + \{p_n(x_\nu)\}^2 \\ &= K_n(x_\nu, x_\nu).\end{aligned} \tag{3.4.8}$$

Here the previous notations are used.

Concerning (3.4.8) see Shohat **3**, p. 456. The special case $a = -1$, $b = +1$, $d\alpha(x) = dx$ is particularly important. Here the abscissas x_ν are the zeros of the nth Legendre polynomial, and the sum of the Christoffel numbers is 2, the length of the interval of integration. This is the case originally considered by

Gauss and Jacobi. Another important special case, namely $a = -1, b = +1$, $d\alpha(x) = (1 - x^2)^{-\frac{1}{2}}\, dx$, is due to Mehler (**1**).

The positiveness of λ_ν is clear from any of the representations (3.4.6), (3.4.7), (3.4.8). In case (3.4.7) we take into account (3.3.6). According to (3.4.5) the sum of the λ_ν is the total mass of the distribution $d\alpha(x)$ spread over the given interval.

The discussion of the representation (3.4.7) can be carried further in the case of the classical orthogonal polynomials. (Cf. §15.3 (1).)

For the proof of Theorem 3.4.2 we write $\rho(x) = \{l_\nu(x)\}^2$ in (3.4.1); this furnishes (3.4.6). Furthermore, writing $y = x_\nu$ in (3.2.3), multiplying by $d\alpha(x)$, and then integrating, we obtain, according to (3.4.3),

$$1 = \frac{k_n}{k_{n+1}} \int_a^b \frac{-p_n(x)p_{n+1}(x_\nu)}{x - x_\nu}\, d\alpha(x)$$

$$= -\frac{k_n}{k_{n+1}}\, p_{n+1}(x_\nu)p_n'(x_\nu)\lambda_\nu .$$

This establishes (3.4.7). Combining (3.4.7) with (3.2.4) for $x = x_\nu$, we get (3.4.8).

(3) As an application of (3.4.1) we obtain, for arbitrary real constants $u_0, u_1, \cdots, u_{n-1}$,

$$\begin{aligned} F(u) &= \int_a^b (u_0 + u_1 x + \cdots + u_{n-1}x^{n-1})^2\, d\alpha(x) \\ &= \sum_{\nu=1}^{n} \lambda_\nu (u_0 + u_1 x_\nu + \cdots + u_{n-1}x_\nu^{n-1})^2, \\ G(u) &= \int_a^b x(u_0 + u_1 x + \cdots + u_{n-1}x^{n-1})^2\, d\alpha(x) \\ &= \sum_{\nu=1}^{n} \lambda_\nu x_\nu (u_0 + u_1 x_\nu + \cdots + u_{n-1}x_\nu^{n-1})^2. \end{aligned} \tag{3.4.9}$$

Therefore, the characteristic values of the pencil

$$G(u) - \xi F(u) = \sum_{\nu=1}^{n} \lambda_\nu (x_\nu - \xi)(u_0 + u_1 x_\nu + \cdots + u_{n-1}x_\nu^{n-1})^2 \tag{3.4.10}$$

are precisely $\xi = x_1, x_2, \cdots, x_n$. With the notation (2.2.1), the quadratic form of the left-hand member becomes

$$\sum_{\nu=0}^{n-1}\sum_{\mu=0}^{n-1} (c_{\nu+\mu+1} - \xi c_{\nu+\mu})u_\nu u_\mu . \tag{3.4.11}$$

Its determinant is a π_n in ξ which vanishes for $\xi = x_1, x_2, \cdots, x_n$, and is therefore a constant multiple of $p_n(\xi)$. We thus arrive at a new proof of equation (2.2.9).

See also Problem 10.

3.41. Separation theorem of Tchebichef-A. Markoff-Stieltjes

In 1874 Tchebichef stated a very remarkable property of the Christoffel numbers (see **6, 8**), proofs of which were given independently by A. Markoff and Stieltjes. Let $n \geqq 2$. In view of the positiveness of λ_ν and of (3.4.5), there exist numbers $y_1 < y_2 < \cdots < y_{n-1}$, $a < y_1$, $y_{n-1} < b$,[9] such that

$$\lambda_\nu = \alpha(y_\nu) - \alpha(y_{\nu-1}), \qquad \nu = 1, 2, \cdots, n; \; y_0 = a, \; y_n = b. \tag{3.41.1}$$

THEOREM 3.41.1. *The zeros $x_1, x_2, \cdots, x_n$, arranged in increasing order, alternate with the numbers $y_1, y_2, \cdots, y_{n-1}$; that is,*

$$x_\nu < y_\nu < x_{\nu+1}; \tag{3.41.2}$$

more precisely

$$\begin{aligned} \alpha(x_\nu + 0) - \alpha(a) < \alpha(y_\nu) - \alpha(a) &= \lambda_1 + \lambda_2 + \cdots + \lambda_\nu \\ &< \alpha(x_{\nu+1} - 0) - \alpha(a), \quad \nu = 1, 2, \cdots, n-1. \end{aligned} \tag{3.41.3}$$

In view of (3.41.1) the quadrature formula (3.4.1) becomes

$$\int_a^b \rho(x)\, d\alpha(x) = \sum_{\nu=1}^{n} \rho(x_\nu)\{\alpha(y_\nu) - \alpha(y_{\nu-1})\}. \tag{3.41.4}$$

Since $y_{\nu-1} < x_\nu < y_\nu$, the right-hand member has the character of a "Riemann-Stieltjes sum."

As a further consequence of the inequalities (3.41.3) we notice that $\alpha(x_\nu + 0) < \alpha(x_{\nu+1} - 0)$. Thus we have proved the following:

THEOREM 3.41.2. *In the open interval $(x_\nu, x_{\nu+1})$, between two consecutive zeros of $p_n(x)$, the function $\alpha(x)$ cannot be constant.*

Or in other words: In an open interval in which $\alpha(x)$ is constant, $p_n(x)$ has at most one zero.

3.411. First proof of the separation theorem[10]

Let ν be an integer such that $1 \leqq \nu \leqq n - 1$. Choose for $\rho(x)$ in (3.4.1) a special π_{2n-2} subject to the following $2n - 1$ conditions:

$$\begin{aligned} \rho(x_k) &= \begin{cases} 1 & \text{if } k = 1, 2, \cdots, \nu, \\ 0 & \text{if } k = \nu + 1, \nu + 2, \cdots, n; \end{cases} \\ \rho'(x_k) &= 0 \quad \text{if } k = 1, 2, \cdots, \nu - 1, \nu + 1, \cdots, n. \end{aligned} \tag{3.411.1}$$

Then this polynomial is uniquely determined.

[9] To find such numbers y_ν it might be necessary to modify $\alpha(y)$ at some of its points of discontinuity, which has, of course, no influence on (3.4.1). It should be also observed that y_ν is in general not uniquely determined.

[10] Cf. A. Markoff **1, 2**, Stieltjes **1**, A. Markoff **3**.

By Rolle's theorem $\rho'(x)$ has at least one zero in each of the open intervals

(3.411.2) $$(x_1, x_2), (x_2, x_3), \cdots, (x_{\nu-1}, x_\nu); \qquad (x_{\nu+1}, x_{\nu+2}), \cdots, (x_{n-1}, x_n).$$

These zeros together with x_k, $1 \leqq k \leqq n$, $k \neq \nu$, furnish $(n-2)+(n-1) = 2n-3$ zeros for $\rho'(x)$. Since $\rho'(x)$ is a π_{2n-3}, it follows that these are the only zeros of $\rho'(x)$, and also that they are all simple zeros. Hence $\rho(x)$ is monotonic between any two consecutive zeros of $\rho'(x)$; in particular, it is monotonic between the zero in $(x_{\nu-1}, x_\nu)$ and $x_{\nu+1}$, and therefore also in $[x_\nu, x_{\nu+1}]$. Furthermore, $\rho(x)$ is decreasing in $[x_\nu, x_{\nu+1}]$ since $\rho(x_\nu) = 1$, $\rho(x_{\nu+1}) = 0$. From

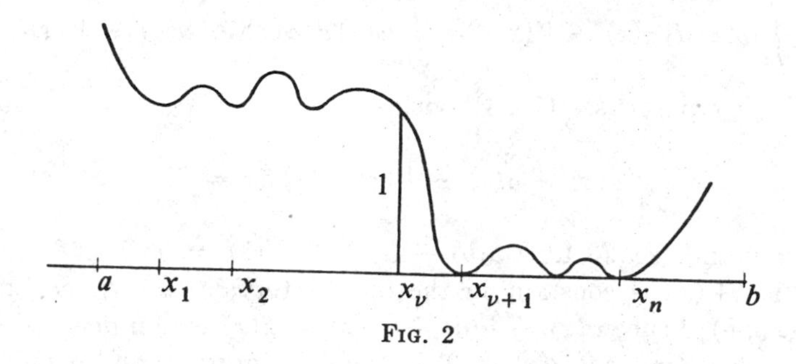

FIG. 2

these considerations the graph of $\rho(x)$ is easily seen to have the shape given in the figure. Therefore we have

(3.411.3) $$\begin{aligned} \rho(x) &\geqq 1 \quad \text{in} \quad a \leqq x \leqq x_\nu, \\ \rho(x) &\geqq 0 \quad \text{in} \quad x_\nu \leqq x \leqq b. \end{aligned}$$

For this special case the general formula (3.4.1) gives

$$\begin{aligned} \lambda_1 + \lambda_2 + \cdots + \lambda_\nu &= \int_a^b \rho(x)\, d\alpha(x) \\ &> \int_a^{x_\nu+0} \rho(x)\, d\alpha(x) > \int_a^{x_\nu+0} d\alpha(x), \end{aligned}$$

which establishes part of the inequalities (3.41.3).

To prove the remaining part we consider the distribution $d[-\alpha(-x)]$ in $[-b, -a]$. The associated orthonormal set is $\{(-1)^n p_n(-x)\}$ with the zeros $-x_n < -x_{n-1} < \cdots < -x_1$. In place of the numbers $y_1, y_2, \cdots, y_{n-1}$ we now have $-y_{n-1}, -y_{n-2}, \cdots, -y_1$. Then, according to the preceding result, $-\alpha(x_{n-\nu+1} - 0) < -\alpha(y_{n-\nu})$, or $\alpha(y_\nu) < \alpha(x_{\nu+1} - 0)$.

3.412. Second proof of the separation theorem[11]

Let the non-decreasing step-function $V(x)$ be defined by the following conditions:

[11] See Stieltjes **12**, especially pp. 588–592.

$$(3.412.1) \qquad V(x) = \begin{cases} 0, & a \leqq x < x_1, \\ \lambda_1, & x_1 \leqq x < x_2, \\ \lambda_1 + \lambda_2, & x_2 \leqq x < x_3, \\ \cdots\cdots\cdots\cdots & \\ \lambda_1 + \lambda_2 + \cdots + \lambda_n, & x_n \leqq x \leqq b. \end{cases}$$

Then (3.4.1) can be written in the form

$$(3.412.2) \qquad \int_a^b \rho(x)\, d\{\alpha(x) - V(x)\} = \int_a^b \rho(x)\, d\{\alpha(x) - \alpha(a) - V(x)\} = 0.$$

An integration by parts (see (1.4.4)) yields

$$(3.412.3) \qquad \int_a^b \{\alpha(x) - \alpha(a) - V(x)\}\rho'(x)\, dx = 0,$$

since $V(a) = 0$ and (see (3.4.5)) $\alpha(b) - \alpha(a) - V(b) = 0$.

The function $V(x)$ is constant in the open intervals (a, x_1), (x_1, x_2), $\cdots$, (x_{n-1}, x_n), (x_n, b); hence $\alpha(x) - \alpha(a) - V(x) = \beta(x)$ is non-decreasing there. We have $\beta(x) \geqq 0$ (but not $\beta(x) \equiv 0$) in the first interval, and $\beta(x) \leqq 0$ (but not $\beta(x) \equiv 0$) in the last interval. In the other intervals $(x_\nu, x_{\nu+1})$ the function $\beta(x)$ is either of constant sign (constantly non-negative or non-positive), or there exists a point y, $x_\nu < y < x_{\nu+1}$, such that $\beta(y - 0) < 0$ and $\beta(y + 0) > 0$. Thus the total interval $[a, b]$ can be subdivided in at most $2n$ intervals in which $\beta(x)$ is non-negative and non-positive alternately, without being identically zero. The end-points of these intervals are some of the zeros x_ν and some of the points y previously defined. Now from (3.412.3) we conclude, by means of an argument similar to that in §3.3 (1), that the number of these intervals is at least $2n$, and then exactly $2n$. Less precisely, $\beta(x)$ has exactly $2n$ variations of sign in $[a, b]$ which are located at the zeros x_ν as well as at the points y mentioned, whose number is $n - 1$. At the points y there is a transition from negative to positive values. Hence at the points x_ν there is a transition from positive to negative values.

Consequently,

$$(3.412.4) \qquad \beta(x_\nu - 0) > 0 > \beta(x_\nu + 0), \qquad \nu = 1, 2, \cdots, n,$$

which is equivalent to the statement of Theorem 3.41.1. (The first inequality is trivial for $\nu = 1$; the same holds for the second one for $\nu = n$.)

We can also prove the above statement by a slight modification of the argument used. Let y_ν denote the point in $(x_\nu, x_{\nu+1})$ with the variation of sign of the type y; then we have

$$(3.412.5) \qquad \beta(x_\nu + 0) < \beta(y_\nu) = 0 < \beta(x_{\nu+1} - 0).$$

(Cf. the footnote above.)

3.413. Third proof of the separation theorem[12]

Let the functions $\phi_k(x, t)$ be defined as follows

$$\phi_k(x, t) = \begin{cases} (x - t)^k & \text{if } x \leqq t, \\ 0 & \text{if } x > t, \end{cases} \tag{3.413.1}$$

where k is a non-negative integer, x and t arbitrary and real. Let[13]

$$F_k(t) = \int_a^b \phi_k(x, t)\, d\alpha(x) - \sum_{\nu=1}^{n} \lambda_\nu \phi_k(x_\nu, t). \tag{3.413.2}$$

Then according to (3.4.1)

$$F_k(a) = F_k(b) = 0, \qquad k = 0, 1, 2, \cdots, 2n - 1. \tag{3.413.3}$$

Also $F_k(t)$ is continuous if $k \geqq 1$; furthermore, we readily see that $F_0(t) = \beta(t)$, where $\beta(t)$ has the same meaning as in §3.412. Now

$$F_k'(t) = -kF_{k-1}(t), \qquad a < t < b,\ k \geqq 1. \tag{3.413.4}$$

[For $k = 1$, $t = x_\nu$, this means $F_1'(x_\nu \pm 0) = -F_0(x_\nu \pm 0)$.] If we take into account (3.413.3) and (3.413.4), Rolle's theorem furnishes for the number of zeros of $F_k(t)$ (including $t = a$ and $t = b$) the lower bound $2n + 1 - k$, $1 \leqq k \leqq 2n - 1$; this holds also for $k = 0$ in the sense that $F_0(t)$ has at least $2n - 1$ variations of sign. From this point on the statement follows by an argument similar to that in the second proof.

3.42. Another separation theorem

If $x_{1n} < x_{2n} < \cdots < x_{nn}$ denote the zeros of $p_n(x)$, we know (Theorem 3.3.2) that the system $\{x_{\nu n}\}$ alternates with the system $\{x_{\nu,n+1}\}$, that is,

$$x_{\nu-1,n} < x_{\nu,n+1} < x_{\nu n}, \qquad \nu = 1, 2, \cdots, n + 1;\ x_{0,n} = a,\ x_{n+1,n} = b. \tag{3.42.1}$$

Let now $\{\lambda_{\nu n}\}$ denote the system of Christoffel numbers associated with $p_n(x)$, and let $\{y_{\nu n}\}$ be the numbers y_ν defined in (3.41.1). Stieltjes showed (**12**) that in addition to Theorem 3.41.1 the following separation theorem holds:

THEOREM 3.42. *We may assert that*

$$y_{\nu-1,n} < y_{\nu,n+1} < y_{\nu n}; \tag{3.42.2}$$

or

$$\begin{aligned} \lambda_{1n} + \lambda_{2n} + \cdots + \lambda_{\nu-1,n} &< \lambda_{1,n+1} + \lambda_{2,n+1} + \cdots + \lambda_{\nu,n+1} \\ &< \lambda_{1n} + \lambda_{2n} + \cdots + \lambda_{\nu n}, \qquad \nu = 1, 2, \cdots, n. \end{aligned} \tag{3.42.3}$$

For $\nu = 1$ the inequalities involving y_{0n} and λ_{0n} must be disregarded.

[12] This proof is due to Pólya and Uspensky (written communication).

[13] In case $k = 0$, $t = a$ the integral in the right-hand member should be replaced by 0.

The analogy between (3.42.1) and (3.42.2) is obvious.

For the proof we use the same function $V(x)$ as in §3.412, denoting it now by $V_n(x)$, and introducing the corresponding function $V_{n+1}(x)$ associated with the system $\{\lambda_{\nu,n+1}\}$. Then by (3.412.3) we have

$$\int_a^b \{V_n(x) - V_{n+1}(x)\}\rho'(x)\, d\alpha(x) = 0, \tag{3.42.4}$$

where $\rho(x)$ is an arbitrary π_{2n-1}. Hence $V_n(x) - V_{n+1}(x)$ has at least $2n - 1$ variations of sign. Such variations can occur only at the points $x_{\nu n}$ and $x_{\nu,n+1}$. In the first case $V_{n+1}(x)$ is constant in the neighborhood of this point, and since $V_n(x)$ increases, the variation of sign is necessarily a transition from negative to positive values. The opposite is true for $x_{\nu,n+1}$. The total number of points $\{x_{\nu n}\}$ and $\{x_{\nu,n+1}\}$ is $2n + 1$. Now $V_n(x)$ and $V_{n+1}(x)$ are identical in the intervals $a \leqq x < x'$ and $x'' < x \leqq b$, where x' is the minimum and x'' the maximum of all the zeros $x_{\nu n}$ and $x_{\nu,n+1}$. Hence no variation of sign is possible at x' or x''. This means that a variation of sign actually occurs at each of the other zeros and is of the type described above.

As a first consequence of this, we again obtain (3.42.1), that is, Theorem 3.3.2, and as a second consequence, the inequalities

$$V_n(x_{\nu n} - 0) - V_{n+1}(x_{\nu n} - 0) < 0 < V_n(x_{\nu n} + 0) - V_{n+1}(x_{\nu n} + 0). \tag{3.42.5}$$

These are the same as the inequalities (3.42.3).

3.5. Continued fractions

Historically, the orthogonal polynomials $\{p_n(x)\}$ originated in the theory of continued fractions. This relationship is of great importance and is one of the possible starting points of the treatment of orthogonal polynomials. See Tchebichef **1-8**, Heine **3**, vol. 1, pp. 260-297, Stieltjes **11**.

(1) For an infinite continued fraction we use the notation

$$b_0 + \frac{a_1|}{|b_1} + \frac{a_2|}{|b_2} + \cdots + \frac{a_n|}{|b_n} + \cdots. \tag{3.5.1}$$

Here, as usual, the convergent R_n/S_n, $n = 0, 1, 2, \cdots$, is defined as the finite fraction obtained from (3.5.1) by stopping at the term b_n. (See, for example, Perron **3**.) We have

$$\begin{aligned} R_0 &= b_0, & R_1 &= b_0 b_1 + a_1, \cdots, \\ S_0 &= 1, & S_1 &= b_1, \cdots, \end{aligned} \tag{3.5.2}$$

and the recurrence formulas

$$R_n = b_n R_{n-1} + a_n R_{n-2}, \quad S_n = b_n S_{n-1} + a_n S_{n-2}, \quad n = 2, 3, 4, \cdots, \tag{3.5.3}$$

which hold also for $n = 1$ if we define $R_{-1} = 1$, $S_{-1} = 0$. Also, we easily obtain (see Perron, loc. cit., p. 16)

$$R_n S_{n-1} - R_{n-1} S_n = (-1)^{n-1} a_1 a_2 \cdots a_n,$$

or

$$\frac{R_n}{S_n} - \frac{R_{n-1}}{S_{n-1}} = \frac{(-1)^{n-1} a_1 a_2 \cdots a_n}{S_{n-1} S_n}, \qquad n = 1, 2, 3, \cdots. \tag{3.5.4}$$

(2) Let $\{p_n(x)\}$ be the orthonormal set of polynomials associated with the distribution $d\alpha(x)$ on $[a, b]$. The recurrence formula (3.2.1) then suggests the consideration of the continued fraction

$$\frac{1|}{|A_1 x + B_1} - \frac{C_2|}{|A_2 x + B_2} - \frac{C_3|}{|A_3 x + B_3} - \cdots - \frac{C_n|}{|A_n x + B_n} - \cdots. \tag{3.5.5}$$

Here A_n, B_n, C_n have the same meaning as in (3.2.1). Therefore,

$$b_0 = 0, \quad b_n = A_n x + B_n, \quad n \geqq 1; \quad a_1 = 1, \quad a_n = -C_n, \quad n \geqq 2. \tag{3.5.6}$$

Next we prove the following theorem:

THEOREM 3.5.1. *The convergents R_n/S_n of (3.5.5) are determined by the formulas*

$$\begin{aligned} R_n &= R_n(x) = c_0^{-\frac{3}{2}}(c_0 c_2 - c_1^2)^{\frac{1}{2}} \int_a^b \frac{p_n(x) - p_n(t)}{x - t} \, d\alpha(t), \\ S_n &= S_n(x) = c_0^{\frac{1}{2}} p_n(x), \qquad n = 0, 1, 2, \cdots. \end{aligned} \tag{3.5.7}$$

Here c_n has the same meaning as in (2.2.1).

Accordingly, the orthogonal polynomials are identical with the denominators of the convergents of the continued fraction (3.5.5).

The second part of the statement follows immediately by comparing (3.2.1) with (3.5.3) for $n \geqq 2$ and observing that the statement is true for $n = 0$ and $n = 1$. As regards the first part, we notice that it holds also for $n = 0$ and for $n = 1$. (Since $p_1(x) = k_1 x + \text{const.}$, the corresponding integral becomes $k_1 c_0$. Then we use (2.2.15) and (2.2.7).) Finally, if $n \geqq 2$, we have

$$\begin{aligned} &c_0^{\frac{3}{2}}(c_0 c_2 - c_1^2)^{-\frac{1}{2}}(R_n - b_n R_{n-1} - a_n R_{n-2}) \\ &\quad = \int_a^b \left\{ \frac{p_n(x) - p_n(t) - (A_n x + B_n)\{p_{n-1}(x) - p_{n-1}(t)\}}{x - t} \right. \\ &\qquad\qquad \left. + \frac{C_n\{p_{n-2}(x) - p_{n-2}(t)\}}{x - t} \right\} d\alpha(t) \\ &\quad = \int_a^b \frac{-(A_n t + B_n) p_{n-1}(t) + (A_n x + B_n) p_{n-1}(t)}{x - t} \, d\alpha(t) \\ &\quad = A_n \int_a^b p_{n-1}(t) \, d\alpha(t) = 0, \end{aligned}$$

which establishes the statement.

Therefore, the numerators of the convergents are expressible in terms of $p_n(x)$. Obviously, R_n is a polynomial of degree $n - 1$ in x.

(3) On expanding the rational function $R_n(x)/S_n(x)$ in descending powers of x, we obtain for $n \geqq 1$

$$\frac{R_n(x)}{S_n(x)} = d_{0n}x^{-1} + d_{1n}x^{-2} + d_{2n}x^{-3} + \cdots . \tag{3.5.8}$$

According to (3.5.4) this expansion agrees with that of $R_{n-1}(x)/S_{n-1}(x)$ up to, and including, the term $x^{-(2n-2)}$. Whence there exists a power series

$$d_0x^{-1} + d_1x^{-2} + d_2x^{-3} + \cdots \tag{3.5.9}$$

such that, for $n \geqq 1$,

$$\frac{R_n(x)}{S_n(x)} = d_0x^{-1} + d_1x^{-2} + \cdots + d_{2n-1}x^{-2n} + \sum_{\nu=2n}^{\infty} d_{\nu n}x^{-\nu-1}. \tag{3.5.10}$$

This is generally true for the convergents of any continued fraction of the type (3.5.5).

Theorem 3.5.2. *The equality*

$$d_\nu = c_0^{-2}(c_0c_2 - c_1^2)^{\frac{1}{2}}c_\nu, \qquad \nu = 0, 1, 2, \cdots, \tag{3.5.11}$$

is valid.

In fact, if d_ν is defined by these equations and we use (3.5.7), we find

$$R_n(x) - S_n(x)(d_0x^{-1} + d_1x^{-2} + \cdots + d_{2n-1}x^{-2n})$$
$$= c_0^{-\frac{3}{2}}(c_0c_2 - c_1^2)^{\frac{1}{2}}\left\{\int_a^b \frac{p_n(x) - p_n(t)}{x - t}\,d\alpha(t) - p_n(x)(c_0x^{-1} + \cdots + c_{2n-1}x^{-2n})\right\}.$$

Here the expression in the braces can be written in the form

$$\int_a^b \frac{p_n(x) - p_n(t)}{x - t}\,d\alpha(t) - p_n(x)\int_a^b \frac{1 - x^{-2n}t^{2n}}{x - t}\,d\alpha(t)$$
$$= \int_a^b \frac{p_n(x) - p_n(t)}{x - t}\,x^{-2n}t^{2n}\,d\alpha(t) - \int_a^b p_n(t)\,\frac{1 - x^{-2n}t^{2n}}{x - t}\,d\alpha(t)$$
$$= x^{-2n}\int_a^b \frac{p_n(x) - p_n(t)}{x - t}\,t^{2n}\,d\alpha(t)$$
$$- x^{-2n}\int_a^b p_n(t)(x^{2n-1} + x^{2n-2}t + \cdots + xt^{2n-2} + t^{2n-1})\,d\alpha(t).$$

Since the first integral of the right-hand member is a π_{n-1}, the expansion of the first term starts with x^{-n-1}. Since the contributions of the powers $1, t, t^2, \cdots, t^{n-1}$ in the second integral vanish, the expansion of the second term starts with $x^{-2n}x^{n-1} = x^{-n-1}$. On dividing by $S_n(x)$, we obtain an expansion of the form (3.5.10). This requirement uniquely determines the numbers $d_0, d_1, \cdots, d_{2n-1}$, and therefore, the whole sequence $\{d_\nu\}$.

(4) THEOREM 3.5.3. *The following decomposition into partial fractions holds for the convergents of* (3.5.5):

$$\frac{R_n(x)}{S_n(x)} = c_0^{-2}(c_0c_2 - c_1^2)^{\frac{1}{2}} \sum_{\nu=1}^{n} \frac{\lambda_{\nu n}}{x - x_{\nu n}}, \tag{3.5.12}$$

where $\lambda_\nu = \lambda_{\nu n}$ *and* $x_\nu = x_{\nu n}$ *have the same meaning as in* §3.4.

For, we find from Theorem 3.5.1

$$\frac{R_n(x_{\nu n})}{S'_n(x_{\nu n})} = \frac{c_0^{-\frac{3}{2}}(c_0c_2 - c_1^2)^{\frac{1}{2}}}{c_0^{\frac{1}{2}}p'_n(x_{\nu n})} \int_a^b \frac{p_n(t)}{t - x_{\nu n}}\, d\alpha(t).$$

Now we can apply (3.4.3).

From (3.5.12) we see that the zeros of $R_n(x)$ are real and that they alternate with those of $S_n(x)$.

(5) Finally, we consider the special case in which $[a, b]$ is a finite interval. Then the expansion (3.5.9) represents the function

$$F(x) = c_0^{-2}(c_0c_2 - c_1^2)^{\frac{1}{2}} \int_a^b \frac{d\alpha(t)}{x - t} \tag{3.5.13}$$

provided $|x|$ is sufficiently large. In various special cases, a function $F(x)$ representable in the form (3.5.13) may be developed directly into a continued fraction of the type (3.5.5), the denominators of this fraction being the orthogonal polynomials associated with the distribution $d\alpha(t)$. Such an approach to these polynomials is essentially different from that used in Chapter II.

THEOREM 3.5.4. *Let* $[a, b]$ *be a finite interval. Then*

$$\lim_{n\to\infty} \frac{R_n(x)}{S_n(x)} = F(x), \tag{3.5.14}$$

if x *is an arbitrary point in the complex plane cut along the segment* $[a, b]$. *The convergence is uniform on every closed set having no points in common with* $[a, b]$.

This theorem is due to A. Markoff (**5**, p. 89).

If x be real, $x > b$, we may combine Theorem 3.5.3 with Problem 9 to get

$$F(x) - \frac{R_n(x)}{S_n(x)} = c_0^{-2}(c_0c_2 - c_1^2)^{\frac{1}{2}}(x - \xi)^{-2n-1}k_n^{-2}, \qquad a \leqq \xi \leqq b. \tag{3.5.15}$$

This tends to 0 as $n \to \infty$ (Problem 52) provided x is sufficiently large. On the other hand, the left-hand member of (3.5.15) is uniformly bounded in the exterior of an arbitrary closed curve containing $[a, b]$ in its interior, since $\lambda_{\nu n} > 0$ and (3.4.5) holds. Now the statement follows by use of Vitali's theorem (Titchmarsh **1**, p. 168). Another proof can be based on Theorem 15.2.3.

Concerning further properties of the convergents we refer to Sherman **1** and the bibliography given there. Regarding the relation of the continued fraction (3.5.5) and of the orthogonal polynomials to the problem of moments, see Hamburger **1**, **2**, M. Riesz **2**, and the bibliography quoted in these papers.

CHAPTER IV

JACOBI POLYNOMIALS

In this chapter we shall be concerned with the main properties of Jacobi polynomials, which include as special cases the ultraspherical polynomials, particularly, the Legendre polynomials. Among the topics which are not considered here, but which are reserved for later study, are the properties of the zeros, asymptotic expressions, expansion problems, and properties connected with interpolation and mechanical quadrature.

Addition theorems for Legendre and ultraspherical polynomials have also been omitted, as have the relations of these polynomials to spherical and surface harmonics of various dimensions. Limitations of space and the existence of exhaustive treatises on these subjects are the chief reasons for such an omission. The interested reader may well consult Whittaker-Watson (**1**, pp. 326–328, 335) and Hobson (**1**).

4.1. Definition; notation; special cases

(1) The definition of the Jacobi polynomials $P_n^{(\alpha,\beta)}(x)$ has been given in §2.4, 1; they are orthogonal on $[-1, +1]$ with the weight function $w(x) = (1-x)^\alpha(1+x)^\beta$. Assurance of the integrability of $w(x)$ is achieved by requiring $\alpha > -1$ and $\beta > -1$; the normalization of $P_n^{(\alpha,\beta)}(x)$ is effected by[14]

$$P_n^{(\alpha,\beta)}(1) = \binom{n+\alpha}{n}. \tag{4.1.1}$$

The orthogonal polynomials with the weight function $(b-x)^\alpha(x-a)^\beta$ on the finite interval $[a, b]$ can be expressed in the form

$$\text{const. } P_n^{(\alpha,\beta)}\left\{2\frac{x-a}{b-a}-1\right\} \tag{4.1.2}$$

(see the last remark in §2.3). The case $a = 0$, $b = 1$ is often used (Jacobi **3**; Jordan **1**, vol. 3, pp. 231–234; Courant-Hilbert **1**, pp. 76–77).[15]

Stieltjes (**6**, p. 75) writes α and β for $(\beta+1)/2$ and $(\alpha+1)/2$, respectively, in terms of our notation. The same notation is used by Fejér (**13**, p. 42). Jordan's function $Z_n(u)$ in our notation is

$$(-1)^n\left\{\binom{n+\gamma-1}{n}\right\}^{-1} P_n^{(\alpha-\gamma,\gamma-1)}(2u-1).$$

[14] According to §3.3 the zeros of $P_n^{(\alpha,\beta)}(x)$ are in $-1 < x < +1$, so that $P_n^{(\alpha,\beta)}(1) \neq 0$.

[15] The statement on p. 76 in Courant-Hilbert must be corrected so as to read

$$p(x) = x^{q-1}(1-x)^{p-q}, \qquad q > 0,\ p - q > -1.$$

Courant-Hilbert's function $G_n(p, q, u)$ is the same as $Z_n(u)$ with $p = \alpha$, $q = \gamma$.

The important identity

$$P_n^{(\alpha,\beta)}(x) = (-1)^n P_n^{(\beta,\alpha)}(-x) \tag{4.1.3}$$

is readily derived by means of the last remark in §2.3. Combining (4.1.3) with (4.1.1), we have

$$P_n^{(\alpha,\beta)}(-1) = (-1)^n \binom{n+\beta}{n}. \tag{4.1.4}$$

(2) For $\alpha = \beta$ we have the ultraspherical polynomials. They are even or odd polynomials according as n is even or odd (§2.3 (2)).

THEOREM 4.1. *The following formulas hold:*

$$\begin{aligned} P_{2\nu}^{(\alpha,\alpha)}(x) &= \frac{\Gamma(2\nu+\alpha+1)\Gamma(\nu+1)}{\Gamma(\nu+\alpha+1)\Gamma(2\nu+1)} P_\nu^{(\alpha,-\frac{1}{2})}(2x^2-1) \\ &= (-1)^\nu \frac{\Gamma(2\nu+\alpha+1)\Gamma(\nu+1)}{\Gamma(\nu+\alpha+1)\Gamma(2\nu+1)} P_\nu^{(-\frac{1}{2},\alpha)}(1-2x^2), \\ P_{2\nu+1}^{(\alpha,\alpha)}(x) &= \frac{\Gamma(2\nu+\alpha+2)\Gamma(\nu+1)}{\Gamma(\nu+\alpha+1)\Gamma(2\nu+2)} xP_\nu^{(\alpha,\frac{1}{2})}(2x^2-1) \\ &= (-1)^\nu \frac{\Gamma(2\nu+\alpha+2)\Gamma(\nu+1)}{\Gamma(\nu+\alpha+1)\Gamma(2\nu+2)} xP_\nu^{(\frac{1}{2},\alpha)}(1-2x^2). \end{aligned} \tag{4.1.5}$$

As a consequence of these important relations, Jacobi polynomials with α or $\beta = \pm\frac{1}{2}$ may be expressed by ultraspherical polynomials. In order to establish the first relation, it suffices to prove that

$$\int_{-1}^{+1} P_\nu^{(\alpha,-\frac{1}{2})}(2x^2-1)\rho(x)(1-x^2)^\alpha\,dx = 0,$$

where $\rho(x)$ is an arbitrary $\pi_{2\nu-1}$. This is trivial if $\rho(x)$ is odd. Let $\rho(x)$ be even and equal to $r(x^2)$, where $r(x)$ is an arbitrary $\pi_{\nu-1}$. Then we have

$$\begin{aligned} \int_{-1}^{+1} P_\nu^{(\alpha,-\frac{1}{2})}(2x^2-1)r(x^2)(1-x^2)^\alpha\,dx &= 2\int_0^1 P_\nu^{(\alpha,-\frac{1}{2})}(2x^2-1)r(x^2)(1-x^2)^\alpha\,dx \\ &= \int_0^1 P_\nu^{(\alpha,-\frac{1}{2})}(2x-1)r(x)(1-x)^\alpha x^{-\frac{1}{2}}\,dx \\ &= 2^{-\alpha-\frac{1}{2}}\int_{-1}^{+1} P_\nu^{(\alpha,-\frac{1}{2})}(x)r\{\tfrac{1}{2}(1+x)\}(1-x)^\alpha(1+x)^{-\frac{1}{2}}\,dx = 0. \end{aligned}$$

A similar argument may be used to prove the second relation. The constant factors are determined according to (4.1.1), (4.1.3).

The case $a = -1$, $b = +1$, $w(x) = |x|^{2k}$, $k > -\frac{1}{2}$, can also be reduced to Jacobi polynomials. The corresponding orthogonal polynomials are (see Szegö **2**, p. 349):

$$(4.1.6)\qquad p_n(x) = \begin{cases} \text{const. } P_\nu^{(0,k-\frac{1}{2})}(2x^2 - 1) & \text{if } n = 2\nu, \\ \text{const. } xP_\nu^{(0,k+\frac{1}{2})}(2x^2 - 1) & \text{if } n = 2\nu + 1, \end{cases}$$

where the constant factors are different from zero; they depend on ν and k. The proof is similar to the previous one.

(3) The simplest cases of ultraspherical polynomials are[16]

$$P_n^{(-\frac{1}{2},-\frac{1}{2})}(x) = \frac{1\cdot 3 \cdots (2n-1)}{2\cdot 4 \cdots 2n} T_n(x) = \frac{1\cdot 3 \cdots (2n-1)}{2\cdot 4 \cdots 2n} \cos n\theta,$$

$$(4.1.7)\qquad P_n^{(\frac{1}{2},\frac{1}{2})}(x) = 2\frac{1\cdot 3 \cdots (2n+1)}{2\cdot 4 \cdots (2n+2)} U_n(x)$$

$$= 2\frac{1\cdot 3 \cdots (2n+1)}{2\cdot 4 \cdots (2n+2)} \frac{\sin (n+1)\theta}{\sin \theta},$$

where $x = \cos\theta$, and $T_n(x)$ and $U_n(x)$ denote the Tchebichef polynomials of the first and second kind [(1.12.3)]. This follows from

$$\int_{-1}^{+1} T_n(x)T_m(x)(1 - x^2)^{-\frac{1}{2}}\, dx = \int_0^\pi \cos n\theta \cos m\theta\, d\theta = 0,$$

$$\int_{-1}^{+1} U_n(x)U_m(x)(1 - x^2)^{\frac{1}{2}}\, dx = \int_0^\pi \sin (n+1)\theta \sin (m+1)\theta\, d\theta = 0, \qquad n \neq m,$$

because of (4.1.1).

In this connection, two "mixed" cases of importance may be mentioned:[17]

$$(4.1.8)\qquad \begin{aligned} P_n^{(\frac{1}{2},-\frac{1}{2})}(x) &= \frac{1\cdot 3 \cdots (2n-1)}{2\cdot 4 \cdots 2n} \frac{\sin \{(2n+1)\theta/2\}}{\sin (\theta/2)}, \\ P_n^{(-\frac{1}{2},\frac{1}{2})}(x) &= \frac{1\cdot 3 \cdots (2n-1)}{2\cdot 4 \cdots 2n} \frac{\cos \{(2n+1)\theta/2\}}{\cos (\theta/2)}, \end{aligned} \qquad x = \cos\theta.$$

The proof is similar to that of (4.1.7) (or is obtained by setting $\alpha = \frac{1}{2}$ and $\alpha = -\frac{1}{2}$ in (4.1.5)). Formulas (4.1.7) and (4.1.8) also follow from §2.6 by putting $\rho(x) = 1$ there.

Another important ultraspherical case is $\alpha = \beta = 0$, that is, the Legendre polynomials $P_n^{(0,0)}(x) = P_n(x)$. Less elementary cases are $\alpha = \beta = -\frac{3}{4}$, and $\alpha = \beta = -\frac{2}{3}$, for which Koschmieder (**1**) gave representations in terms of elliptic functions.

4.2. Differential equation

(1) THEOREM 4.2.1. *The Jacobi polynomials $y = P_n^{(\alpha,\beta)}(x)$ satisfy the following linear homogeneous differential equation of the second order:*

$$(4.2.1)\qquad (1 - x^2)y'' + [\beta - \alpha - (\alpha + \beta + 2)x]y' + n(n + \alpha + \beta + 1)y = 0,$$

[16] In the first equation the coefficient of $T_n(x)$ is 1 for $n = 0$.

[17] For $n = 0$, the numerical factor on the right side is 1.

or

$$\frac{d}{dx}\{(1-x)^{\alpha+1}(1+x)^{\beta+1}y'\} + n(n+\alpha+\beta+1)(1-x)^{\alpha}(1+x)^{\beta}y = 0. \tag{4.2.2}$$

To prove this, we note that since y is a π_n, the expression

$$d\{(1-x)^{\alpha+1}(1+x)^{\beta+1}y'\}/dx$$

has the form $(1-x)^{\alpha}(1+x)^{\beta}z$, where z is also a π_n. In order to show that $z = \text{const.}\, y$, we prove the orthogonality relation

$$\int_{-1}^{+1}\frac{d}{dx}\{(1-x)^{\alpha+1}(1+x)^{\beta+1}y'\}\rho(x)\,dx = 0,$$

where $\rho(x)$ is an arbitrary π_{n-1}. An integration by parts reduces the left-hand member to

$$-\int_{-1}^{+1}(1-x)^{\alpha+1}(1+x)^{\beta+1}y'\rho'(x)\,dx,$$

since $\alpha+1$ and $\beta+1$ are positive. A second integration by parts gives

$$\int_{-1}^{+1}y\frac{d}{dx}\{(1-x)^{\alpha+1}(1+x)^{\beta+1}\rho'(x)\}\,dx.$$

In the last integrand the coefficient of y is of the form $(1-x)^{\alpha}(1+x)^{\beta}r(x)$, where $r(x)$ is a π_{n-1}. Hence this integral vanishes and the statement is established. The constant factor $-n(n+\alpha+\beta+1)$ may be determined by comparing the highest terms.

An alternative form of (4.2.1) is

$$\begin{gathered}(1-x^2)Y'' + [\alpha-\beta+(\alpha+\beta-2)x]Y' + (n+1)(n+\alpha+\beta)Y = 0,\\ Y = (1-x)^{\alpha}(1+x)^{\beta}y = (1-x)^{\alpha}(1+x)^{\beta}P_n^{(\alpha,\beta)}(x).\end{gathered} \tag{4.2.3}$$

(2) Replacing $n(n+\alpha+\beta+1)$ in (4.2.1) by γ, we may ask: For what values of γ has this equation a polynomial solution which is not identically zero?

THEOREM 4.2.2. *Let $\alpha > -1$, $\beta > -1$. The differential equation*

$$(1-x^2)y'' + [\beta-\alpha-(\alpha+\beta+2)x]y' + \gamma y = 0, \tag{4.2.4}$$

where γ is a parameter, has a polynomial solution not identically zero if and only if γ has the form $n(n+\alpha+\beta+1)$, $n = 0, 1, 2, \cdots$. This solution is const. $\cdot P_n^{(\alpha,\beta)}(x)$, *and no solution which is linearly independent of $P_n^{(\alpha,\beta)}(x)$ can be a polynomial.*

To prove this, substitute $y = \sum_{\nu=0}^{\infty} a_\nu (x-1)^\nu$ in (4.2.4). We find

$$-(x+1)\sum_{\nu=2}^{\infty}\nu(\nu-1)a_\nu(x-1)^{\nu-1}$$

$$-[2(\alpha+1)+(\alpha+\beta+2)(x-1)]\sum_{\nu=1}^{\infty}\nu a_\nu(x-1)^{\nu-1}+\gamma\sum_{\nu=0}^{\infty}a_\nu(x-1)^\nu=0,$$

which yields the recurrence formula

$$[\gamma-\nu(\nu+\alpha+\beta+1)]a_\nu-2(\nu+1)(\nu+\alpha+1)a_{\nu+1}=0, \qquad \nu=0,1,2,\cdots. \tag{4.2.5}$$

Assuming that y is a polynomial, let us suppose that a_n is the last nonzero coefficient. Then we see from (4.2.5) that for $\nu = n$ the coefficient of a_n must vanish, that is, $\gamma = n(n+\alpha+\beta+1)$. Conversely, if this condition is satisfied, then $a_{n+1} = a_{n+2} = \cdots = 0$ since the coefficient of $a_{\nu+1}$ never vanishes.

Now let $\gamma = n(n+\alpha+\beta+1)$, and let z be a second solution of (4.2.1) or (4.2.2). If we let $x \to \pm 1$ in the relation

$$(1-x)^{\alpha+1}(1+x)^{\beta+1}(y'z-yz')=\text{const.}, \tag{4.2.6}$$

we see that y and z cannot both be polynomials unless the constant in the right member is zero, that is, unless y and z are linearly dependent. This argument shows that z cannot even be regular at $x = -1$ or at $x = +1$, unless y and z are linearly dependent.

4.21. Hypergeometric functions

(1) Substitution of $x = 1 - 2x'$ in (4.2.1) yields

$$x'(1-x')\frac{d^2y}{dx'^2}+[\alpha+1-(\alpha+\beta+2)x']\frac{dy}{dx'}+n(n+\alpha+\beta+1)y=0, \tag{4.21.1}$$

which is the hypergeometric equation of Gauss. On account of the second part of Theorem 4.2.2, for $n \geqq 1$, we obtain the important representation:

$$\begin{aligned}P_n^{(\alpha,\beta)}(x) &= \binom{n+\alpha}{n}F\left(-n, n+\alpha+\beta+1;\alpha+1;\frac{1-x}{2}\right)\\ &= \frac{1}{n!}\sum_{\nu=0}^{n}\binom{n}{\nu}(n+\alpha+\beta+1)\cdots(n+\alpha+\beta+\nu)\\ &\qquad\cdot(\alpha+\nu+1)\cdots(\alpha+n)\left(\frac{x-1}{2}\right)^\nu.\end{aligned} \tag{4.21.2}$$[18]

[18] The general coefficient

$$\binom{n}{\nu}(n+\alpha+\beta+1)\cdots(n+\alpha+\beta+\nu)(\alpha+\nu+1)\cdots(\alpha+n)$$

is to be replaced by $(\alpha+1)(\alpha+2)\cdots(\alpha+n)$ for $\nu = 0$, and by
$(n+\alpha+\beta+1)(n+\alpha+\beta+2)\cdots(2n+\alpha+\beta)$ for $\nu = n$.

Here, and in what follows, $F(a, b; c; x)$ is the usual notation for the hypergeometric series

$$
\begin{aligned}
&F(a, b; c; x) \\
(4.21.3)\qquad &= 1 + \sum_{\nu=1}^{\infty} \frac{a(a+1)\cdots(a+\nu-1)}{1\cdot 2\cdots \nu}\, \frac{b(b+1)\cdots(b+\nu-1)}{c(c+1)\cdots(c+\nu-1)}\, x^{\nu},
\end{aligned}
$$

convergent for $|x| < 1$ and satisfying

$$
(4.21.4)\qquad x(1-x)\frac{d^2y}{dx^2} + [c - (a+b+1)x]\frac{dy}{dx} - aby = 0.
$$

(See Whittaker-Watson **1**, p. 283.) For latter reference we observe that (4.21.3) is without meaning if c is a non-positive integer. However, it is readily seen that if m is a positive integer,

$$
\begin{aligned}
&\lim_{c \to -(m-1)} (c + m - 1)F(a, b; c; x) \\
(4.21.5)\qquad &= (-1)^{m-1} \frac{a(a+1)\cdots(a+m-1)b(b+1)\cdots(b+m-1)}{m!(m-1)!} \\
&\quad \cdot x^m F(a + m, b + m; m + 1; x),
\end{aligned}
$$

and the function $x^m F(a + m, b + m; m + 1; x)$ satisfies the equation (4.21.4) with $c = -(m - 1)$.

(2) In the formula (4.21.2) the hypergeometric series stops with the term in x^n. The constant factor in the first part of (4.21.2) is determined by (4.1.1). Using (4.21.2), note that the coefficient $l_n^{(\alpha,\beta)}$ of the highest term x^n in $P_n^{(\alpha,\beta)}(x)$ is

$$
(4.21.6)\qquad l_n^{(\alpha,\beta)} = \lim_{x\to\infty} x^{-n} P_n^{(\alpha,\beta)}(x) = 2^{-n}\binom{2n+\alpha+\beta}{n}.
$$

(3) Another application of (4.21.2) is the useful formula

$$
(4.21.7)\qquad \frac{d}{dx}\{P_n^{(\alpha,\beta)}(x)\} = \tfrac{1}{2}(n + \alpha + \beta + 1)P_{n-1}^{(\alpha+1,\beta+1)}(x),
$$

which follows immediately when we expand both sides of (4.21.7) according to (4.21.2).

As an application of (4.21.7) we observe that the successive derivatives $T_n'(x)$, $T_n''(x)$, $T_n'''(x)$, $\cdots$ of the Tchebichef polynomial $T_n(x)$ are, but for constant factors, $P_{n-1}^{(\frac{1}{2},\frac{1}{2})}(x)$, $P_{n-2}^{(\frac{3}{2},\frac{3}{2})}(x)$, $P_{n-3}^{(\frac{5}{2},\frac{5}{2})}(x)$, $\cdots$. The first is, except for a constant factor, $U_{n-1}(x)$ (see (4.1.7)). We note also that the derivatives $P_n'(x)$, $P_n''(x)$, $\cdots$ of the Legendre polynomial $P_n(x)$ are constant multiples of $P_{n-1}^{(1,1)}(x)$, $P_{n-2}^{(2,2)}(x)$, $\cdots$, respectively.

4.22. Generalization

(1) The second formula (4.21.2) furnishes the extension of the polynomial $P_n^{(\alpha,\beta)}(x)$ to arbitrary complex values of the parameters α and β. It is a poly-

nomial in x, α, and β. In the following, we again denote this π_n by $P_n^{(\alpha,\beta)}(x)$. Many of the properties of $P_n^{(\alpha,\beta)}(x)$ may be extended to this general case. The polynomial $P_n^{(\alpha,\beta)}(x)$ satisfies the differential equation (4.2.1), and the formulas (4.1.1), (4.1.3), (4.1.4) hold. Some other results, however, (for instance, the theorem on the location of the zeros, cf. §6.72) must be essentially modified. Using (4.1.3), the representation of $P_n^{(\alpha,\beta)}(x)$ as a π_n in $x + 1$ can easily be derived.

(2) By comparison of the corresponding powers of $x - 1$, we obtain the identity

$$
\begin{aligned}
&P_n^{(\alpha,\beta)}(x) \\
&= \binom{2n+\alpha+\beta}{n}\left(\frac{x-1}{2}\right)^n F\left(-n, -n-\alpha; -2n-\alpha-\beta; \frac{2}{1-x}\right) \qquad (4.22.1)\\
&= \left(\frac{1-x}{2}\right)^n P_n^{(\alpha',\beta)}\left(\frac{x+3}{x-1}\right), \qquad \alpha' = -2n-\alpha-\beta-1.
\end{aligned}
$$

Furthermore,

$$
\binom{n}{l} P_n^{(-l,\beta)}(x) = \binom{n+\beta}{l}\left(\frac{x-1}{2}\right)^l P_{n-l}^{(l,\beta)}(x), \quad l \text{ an integer}, \quad 1 \leqq l \leqq n, \tag{4.22.2}
$$

and

$$
\binom{n}{k-1} P_n^{(\alpha,\beta)}(x) = \binom{n+\alpha}{n-k+1} P_{k-1}^{(\alpha,\beta)}(x), \tag{4.22.3}
$$

$$
n+\alpha+\beta+k = 0,\ k \text{ an integer},\ 1 \leqq k \leqq n.
$$

In connection with (4.22.2) see (4.21.5).

(3) Let $n \geqq 1$. A reduction of the degree of $P_n^{(\alpha,\beta)}(x)$ occurs if and only if $n + \alpha + \beta + k = 0$ for a certain integer k, $1 \leqq k \leqq n$. In this case "$x = \infty$ is a zero of order $n - k + 1$," this being the precise order unless $\alpha = -l$, l an integer, $k \leqq l \leqq n$. If

$$
n+\alpha+\beta+k = 0, \qquad \alpha = -l, \qquad 1 \leqq k \leqq l \leqq n, \tag{4.22.4}
$$

the polynomial $P_n^{(\alpha,\beta)}(x)$ vanishes identically.

By setting $n + \alpha + \beta + k = \epsilon$, $\alpha + l = \eta$, it can be shown that $P_n^{(\alpha,\beta)}(x) = \epsilon r(x) + \eta s(x)$, except for terms of higher order, if $\epsilon \to 0$, $\eta \to 0$. Here $r(x)$ and $s(x)$ are certain π_n independent of ϵ and η. In view of (4.22.2) and (4.22.3) it follows that, apart from constant nonzero factors,

$$
r(x) = (1-x)^l P_{n-l}^{(l,-n+l-k)}(x), \qquad s(x) = P_{k-1}^{(-l,-n+l-k)}(x),
$$

or

$$
r(x) = (1-x)^{-\alpha} P_{n+\alpha}^{(-\alpha,\beta)}(x), \qquad s(x) = P_{-n-\alpha-\beta-1}^{(\alpha,\beta)}(x), \tag{4.22.5}
$$

α, β, n integers, $\alpha \geqq -n$, $\beta \geqq -n$, $\alpha + \beta \leqq -n - 1$, $n \geqq 1$. Both polynomials $r(x)$, $s(x)$ are solutions of (4.2.1); they are linearly independent since $r(1) = 0$,

$s(1) \neq 0$. Also, they have the precise degrees n and $k - 1 = -n - \alpha - \beta - 1$, respectively. In this instance the general solution of (4.2.1) is a polynomial.

(4) Once more let $n \geqq 1$. We then see that $P_n^{(\alpha,\beta)}(1) \neq 0$ unless $\alpha = -l$, $1 \leqq l \leqq n$. If $\alpha = -l$, $x = 1$ is a zero of order l, and this is the precise order unless $n + \alpha + \beta + k = 0$, $1 \leqq k \leqq l$, in which event we recognize the exceptional case (4.22.4).

According to (4.1.3) we have $P_n^{(\alpha,\beta)}(-1) \neq 0$, unless $\beta = -l$, $1 \leqq l \leqq n$. In this case $x = -1$ is a zero of order l, and here this is the precise order unless $n + \alpha + \beta + k = 0$, $1 \leqq k \leqq l$, which is again essentially the case (4.22.4).

(5) Let $n \geqq 0$. From (3) a second case may be derived in which the general solution of (4.2.1) is a polynomial. If we replace n by $-n - \alpha - \beta - 1$, the differential equation (4.2.1) remains unchanged, which leads to the linearly independent polynomial solutions:

$$r_1(x) = (1 - x)^{-\alpha} P_{-n-\beta-1}^{(-\alpha,\beta)}(x), \qquad s_1(x) = P_n^{(\alpha,\beta)}(x),$$
$$\alpha, \beta, n \text{ integers}, \alpha < -n, \beta < -n, n \geqq 0. \tag{4.22.6}$$

4.23. Second solution

(1) According to the theory of hypergeometric functions, a second solution of (4.2.1) is given by

$$(1 - x)^{-\alpha} F\left(-n - \alpha, n + \beta + 1; 1 - \alpha; \frac{1 - x}{2}\right), \tag{4.23.1}$$

unless α is an integer. (See Whittaker-Watson **1**, p. 286; cf. in particular y_1 and y_2.[19] The functions (4.21.2) and (4.23.1) are then linearly independent.

Now let α be an integer. If $\alpha = -l$, $0 \leqq l \leqq n$, the function (4.23.1) is, but for a constant factor, identical with $P_n^{(\alpha,\beta)}(x)$ (see (4.22.2)). The same is true if $\alpha \to \alpha_0 =$ a positive integer, provided we multiply (4.23.1) by $\alpha - \alpha_0$ before passing to the limit $\alpha \to \alpha_0$ (see (4.21.5)).

Finally, for integral values of α, $\alpha < -n$, $P_n^{(\alpha,\beta)}(x)$ and (4.23.1) are linearly independent, since

$$P_n^{(\alpha,\beta)}(1) = \binom{n + \alpha}{n} \neq 0$$

and (4.23.1) vanishes for $x = 1$. The latter function is a polynomial if and only if $n + \beta + 1$ is a non-positive integer, that is, β is an integer less than $-n$. This is the case referred to in §4.22 (5).

(2) Numerous other representations are obtained for the solutions of (4.2.1) by using the classical transformation formulas of hypergeometric functions. The only singularities of this differential equation are at $x = +1, -1$, and ∞. Interchanging α and β, and replacing x by $-x$, we obtain the expansions about $x = -1$.

[19] This can be readily shown by introducing in (4.21.1), $y = x'^{-\alpha}z$. Analogous methods can be used in the cases (4.23.2), (4.23.3).

The expansions about $x = \infty$ are especially important. From Whittaker-Watson 1, p. 286,[20] we obtain the solutions

$$(4.23.2) \qquad (1-x)^n F\left(-n, -n-\alpha; -2n-\alpha-\beta; \frac{2}{1-x}\right),$$

$$(4.23.3) \quad (1-x)^{-n-\alpha-\beta-1} F\left(n+\alpha+\beta+1, n+\beta+1; 2n+\alpha+\beta+2; \frac{2}{1-x}\right).$$

The first function is, except for a constant factor, $P_n^{(\alpha,\beta)}(x)$ [cf. (4.22.1)]. The second function is obtained from the first by replacing n by $-n-\alpha-\beta-1$.

Apart from constant factors, the expressions (4.23.2) and (4.23.3) arise from (4.21.2) and (4.23.1), respectively, by replacing α by $-2n-\alpha-\beta-1$, $(1-x)/2$ by $2/(1-x)$, and then multiplying by $(1-x)^n$. Consequently, (4.23.2) and (4.23.3) are linearly independent unless $-2n-\alpha-\beta-1$ is an integer not less than $-n$.

THEOREM 4.23.1. *Let α, β be arbitrary, $n \geqq 0$ an integer. The general solution of (4.2.1) can be represented in the forms*

$$(4.23.4) \quad \begin{aligned} &AP_n^{(\alpha,\beta)}(x) + B(1-x)^{-\alpha} F\left(-n-\alpha, n+\beta+1; 1-\alpha; \frac{1-x}{2}\right) \\ &\qquad\qquad \text{if } \alpha \neq -n, -n+1, -n+2, \cdots, \\ &AP_n^{(\alpha,\beta)}(x) + B(1+x)^{-\beta} F\left(-n-\beta, n+\alpha+1; 1-\beta; \frac{1+x}{2}\right) \\ &\qquad\qquad \text{if } \beta \neq -n, -n+1, -n+2, \cdots, \\ &AP_n^{(\alpha,\beta)}(x) + B(1-x)^{-n-\alpha-\beta-1} F\Big(n+\alpha+\beta+1, n+\beta+1; \\ &\qquad 2n+\alpha+\beta+2; \frac{2}{1-x}\Big) \quad \text{if } \alpha+\beta \neq -n-1, -n-2, \cdots, \end{aligned}$$

respectively. Here A and B are arbitrary constants.

(3) The preceding results enable us to prove the following:

THEOREM 4.23.2. *If α and β are arbitrary and $n \geqq 0$ is an integer, then (4.22.5) and (4.22.6) are the only cases in which the general solution of (4.2.1) is a polynomial. They can be characterized by one of the following sets of conditions:*

$$(4.23.5) \quad \begin{aligned} &\text{(a) } \alpha, \beta \text{ negative integers}, \alpha \geqq -n, \beta \geqq -n, \alpha+\beta \leqq -n-1, n \geqq 1, \\ &\text{(b) } \alpha, \beta \text{ negative integers}, \alpha < -n, \beta < -n, n \geqq 0. \end{aligned}$$

We see from (4.2.6) that in the case in question α and β must be negative integers. Now, let $\alpha < -n$, $\beta \geqq -n$; then (4.23.1) is a non-polynomial solu-

[20] In particular see the functions denoted by y_{21} and y_{22}. (There is a misprint in the corresponding formulas: the exponent of $-x$ should be $-A$ in y_{21} and $-B$ in y_{22}.)

tion. The case $\alpha \geqq -n$, $\beta < -n$ can be excluded by making use of (4.1.3). Finally, let $\alpha \geqq -n$, $\beta \geqq -n$, $\alpha + \beta \geqq -n$. Then (4.23.3) is a non-polynomial solution.

(4) Let α be an integer. In the exceptional cases, excluded in Theorem 4.23.1, we can show that the second solution contains logarithmic terms in its representation about $x = +1$ (similarly for $x = -1$, $x = \infty$). (See (4.61.6).)

An extension of the preceding discussion to arbitrary values of n is also possible. However, in what follows, we shall confine ourselves to non-negative integral values of n.

The consideration of the second solution, properly normalized, will be resumed in §4.61, where some other representations will also be given.

4.24. Transformation of the differential equation

Applying §1.8 to (4.2.1), we obtain the following important transformations of the differential equation of Jacobi polynomials:

$$\frac{d^2u}{dx^2} + \left\{\frac{1}{4}\frac{1-\alpha^2}{(1-x)^2} + \frac{1}{4}\frac{1-\beta^2}{(1+x)^2} + \frac{n(n+\alpha+\beta+1) + (\alpha+1)(\beta+1)/2}{1-x^2}\right\} u = 0, \tag{4.24.1}$$

$$u = u(x) = (1-x)^{(\alpha+1)/2}(1+x)^{(\beta+1)/2} P_n^{(\alpha,\beta)}(x);$$

$$\frac{d^2u}{d\theta^2} + \left\{\frac{\frac{1}{4}-\alpha^2}{4\sin^2\dfrac{\theta}{2}} + \frac{\frac{1}{4}-\beta^2}{4\cos^2\dfrac{\theta}{2}} + \left(n + \frac{\alpha+\beta+1}{2}\right)^2\right\} u = 0, \tag{4.24.2}$$

$$u = u(\theta) = \left(\sin\frac{\theta}{2}\right)^{\alpha+\frac{1}{2}} \left(\cos\frac{\theta}{2}\right)^{\beta+\frac{1}{2}} P_n^{(\alpha,\beta)}(\cos\theta).$$

The special cases $\alpha = \pm\frac{1}{2}$, $\beta = \pm\frac{1}{2}$ are to be particularly noted.

4.3. Rodrigues' formula; the orthonormal set

(1) Given α and β arbitrary, we have

$$(1-x)^\alpha(1+x)^\beta P_n^{(\alpha,\beta)}(x) = \frac{(-1)^n}{2^n n!}\left(\frac{d}{dx}\right)^n \{(1-x)^{n+\alpha}(1+x)^{n+\beta}\}. \tag{4.3.1}$$

First, take both α and β greater than -1. A simple application of Leibniz' rule then shows that the right-hand member is of the form $(1-x)^\alpha(1+x)^\beta\rho(x)$, $\rho(x)$ being a π_n. To show that $\rho(x) = \text{const.}\ P_n^{(\alpha,\beta)}(x)$, it suffices to prove that

$$\int_{-1}^{+1}\left(\frac{d}{dx}\right)^n \{(1-x)^{n+\alpha}(1+x)^{n+\beta}\} r(x)\, dx = 0,$$

where $r(x)$ is an arbitrary π_{n-1}. But integration by parts n times yields the result

$$(-1)^n \int_{-1}^{+1} (1-x)^{n+\alpha}(1+x)^{n+\beta} r^{(n)}(x)\, dx,$$

which vanishes since $r^{(n)}(x) \equiv 0$. The constant factor can then be determined by setting $x = 1$ and using (4.1.1) (see (4.3.2)).

Since $P_n^{(\alpha,\beta)}(x)$ is a polynomial in α and β by (4.21.2), and since the same is true for the right-hand member of (4.3.1) when divided by $(1-x)^\alpha(1+x)^\beta$, it follows that (4.3.1) is valid for arbitrary α and β.

On calculating the nth derivative in (4.3.1) by Leibniz' rule, we obtain the important representation

$$\begin{aligned} P_n^{(\alpha,\beta)}(x) &= \sum_{\nu=0}^{n} \binom{n+\alpha}{n-\nu}\binom{n+\beta}{\nu}\left(\frac{x-1}{2}\right)^\nu\left(\frac{x+1}{2}\right)^{n-\nu} \\ &= \binom{n+\alpha}{n}\left(\frac{x+1}{2}\right)^n \sum_{\nu=0}^{n} \frac{n(n-1)\cdots(n-\nu+1)}{(\alpha+1)(\alpha+2)\cdots(\alpha+\nu)}\binom{n+\beta}{\nu}\left(\frac{x-1}{x+1}\right)^\nu \\ &= \binom{n+\alpha}{n}\left(\frac{x+1}{2}\right)^n F\left(-n, -n-\beta; \alpha+1; \frac{x-1}{x+1}\right). \end{aligned} \tag{4.3.2}$$

(2) The argument used in (1) readily leads to the formula

$$\begin{aligned} &\int_{-1}^{+1} (1-x)^\alpha(1+x)^\beta\{P_n^{(\alpha,\beta)}(x)\}^2\, dx \\ &= \frac{2^{\alpha+\beta+1}}{2n+\alpha+\beta+1}\frac{\Gamma(n+\alpha+1)\Gamma(n+\beta+1)}{\Gamma(n+1)\Gamma(n+\alpha+\beta+1)} = h_n^{(\alpha,\beta)}. \end{aligned} \tag{4.3.3}$$

(Here we have the inequalities $\alpha > -1$, $\beta > -1$, and for $n = 0$ the product $(2n+\alpha+\beta+1)\Gamma(n+\alpha+\beta+1)$ must be replaced by $\Gamma(\alpha+\beta+2)$.) In fact, because of (4.3.1) and (4.21.6), we have

$$\begin{aligned} \int_{-1}^{+1} (1-x)^\alpha(1+x)^\beta\{P_n^{(\alpha,\beta)}(x)\}^2\, dx &= l_n^{(\alpha,\beta)} \int_{-1}^{+1} (1-x)^\alpha(1+x)^\beta P_n^{(\alpha,\beta)}(x) x^n\, dx \\ &= \frac{(-1)^n}{2^n n!} l_n^{(\alpha,\beta)} \int_{-1}^{+1} \left(\frac{d}{dx}\right)^n \{(1-x)^{n+\alpha}(1+x)^{n+\beta}\}\, x^n\, dx \\ &= 2^{-n} l_n^{(\alpha,\beta)} \int_{-1}^{+1} (1-x)^{n+\alpha}(1+x)^{n+\beta}\, dx. \end{aligned}$$

Now we employ (4.21.6) and (1.7.5).

Using the notation (4.3.3), we obtain as the orthonormal set associated with the weight function $(1-x)^\alpha(1+x)^\beta$ in $[-1, +1]$

$$\begin{aligned} p_n(x) &= \{h_n^{(\alpha,\beta)}\}^{-\frac{1}{2}} P_n^{(\alpha,\beta)}(x) \\ &= \left\{\frac{2n+\alpha+\beta+1}{2^{\alpha+\beta+1}}\frac{\Gamma(n+1)\Gamma(n+\alpha+\beta+1)}{\Gamma(n+\alpha+1)\Gamma(n+\beta+1)}\right\}^{\frac{1}{2}} P_n^{(\alpha,\beta)}(x), \\ &\qquad n = 0, 1, 2, \cdots. \end{aligned} \tag{4.3.4}$$

4.4. Generating function

(1) Formula (4.3.2) can be written as follows:

$$(4.4.1) \qquad P_n^{(\alpha,\beta)}(x) = \frac{1}{2\pi i}\int\left(1+\frac{x+1}{2}z\right)^{n+\alpha}\left(1+\frac{x-1}{2}z\right)^{n+\beta} z^{-n-1}\,dz,$$

where we assume that $x \neq \pm 1$. The integration is extended in the positive sense along a closed curve around the origin, such that the points $-2(x \pm 1)^{-1}$ lie neither on it nor in its interior. (We define the first and second factors of the integrand to be 1 for $z = 0$.) Hence for sufficiently small values of $|w|$,

$$(4.4.2) \qquad \sum_{n=0}^{\infty} P_n^{(\alpha,\beta)}(x) w^n = \frac{1}{2\pi i}\int \frac{\left(1+\frac{x+1}{2}z\right)^{\alpha}\left(1+\frac{x-1}{2}z\right)^{\beta}}{z - w\left(1+\frac{x+1}{2}z\right)\left(1+\frac{x-1}{2}z\right)}\,dz.$$

The denominator is

$$(4.4.3) \qquad -\tfrac{1}{4}(x^2-1)wz^2 - z(xw-1) - w = \tfrac{1}{4}(1-x^2)w(z-z_0)(z-Z_0),$$

where

$$(4.4.4) \qquad z_0 = z_0(w) = \frac{2}{1-x^2}\,\frac{xw-1+R}{w}, \qquad R = R(w) = (1-2xw+w^2)^{\frac{1}{2}}.$$

For $Z_0 = Z_0(w)$ there is an analogous expression with $-R$ instead of R. Here z_0 and R are regular analytic functions of w provided $|w|$ is sufficiently small; we take $R(0) = 1$. At $w = 0$ the function z_0 has a zero, and the function Z_0 has a pole. For sufficiently small $|w|$, z_0 lies in the interior, and Z_0 in the exterior, of the integration curve of (4.4.2), so that by Cauchy's theorem

$$\sum_{n=0}^{\infty} P_n^{(\alpha,\beta)} w^n = \left[\frac{1}{4}(1-x^2)w\right]^{-1}\left(1+\frac{x+1}{2}z_0\right)^{\alpha}\left(1+\frac{x-1}{2}z_0\right)^{\beta}(z_0-Z_0)^{-1}.$$

Now, we readily get

$$1+\frac{x+1}{2}z_0 = 2(1-w+R)^{-1}, \qquad 1+\frac{x-1}{2}z_0 = 2(1+w+R)^{-1},$$

$$z_0 - Z_0 = 4w^{-1}(1-x^2)^{-1}R,$$

so that

$$(4.4.5) \qquad \begin{aligned}\sum_{n=0}^{\infty} P_n^{(\alpha,\beta)}(x) w^n &= 2^{\alpha+\beta}R^{-1}(1-w+R)^{-\alpha}(1+w+R)^{-\beta} \\ &= 2^{\alpha+\beta}(1-2xw+w^2)^{-\frac{1}{2}}\{1-w+(1-2xw+w^2)^{\frac{1}{2}}\}^{-\alpha} \\ &\quad \cdot\{1+w+(1-2xw+w^2)^{\frac{1}{2}}\}^{-\beta}.\end{aligned}$$

This is the generating function (series) of the Jacobi polynomials (Jacobi **3**, pp. 193–194) which may be established directly for $x = \pm 1$. The expressions $\{\ \}^{-\alpha}$ and $\{\ \}^{-\beta}$ must be taken positive for $w = 0$.

(2) A slight variation of this argument may be made by writing (4.3.1) in the form

$$P_n^{(\alpha,\beta)}(x) = \frac{1}{2\pi i}\int\left(\frac{1}{2}\frac{t^2-1}{t-x}\right)^n\left(\frac{1-t}{1-x}\right)^\alpha\left(\frac{1+t}{1+x}\right)^\beta\frac{dt}{t-x}. \tag{4.4.6}$$

Here $x \neq \pm 1$, and the integration is extended in the positive sense around a closed contour enclosing the point $t = x$, but not the points $t = \pm 1$. Also the functions $((1-t)/(1-x))^\alpha$ and $((1+t)/(1+x))^\beta$ are assumed to reduce to 1 for $t = x$. We next write

$$\frac{1}{2}\frac{t^2-1}{t-x} = w^{-1}, \quad t = w^{-1}\{1-(1-2xw+w^2)^{\frac{1}{2}}\} = x + \frac{1}{2}(x^2-1)w + \cdots. \tag{4.4.7}$$

Here that branch of $(1-2xw+w^2)^{\frac{1}{2}}$ must be taken which is equal to $+1$ for $w = 0$. Then if w describes a small closed curve around the origin, t describes a curve of the type mentioned above. Furthermore,

$$\frac{1-t}{1-x} = 2\{1-w+(1-2xw+w^2)^{\frac{1}{2}}\}^{-1}, \tag{4.4.8}$$

$$\frac{1+t}{1+x} = 2\{1+w+(1-2xw+w^2)^{\frac{1}{2}}\}^{-1}, \quad \frac{dt}{t-x} = (1-2xw+w^2)^{-\frac{1}{2}}w^{-1}\,dw,$$

so that

$$\begin{aligned} P_n^{(\alpha,\beta)}(x) = \frac{1}{2\pi i}\int & w^{-n}2^\alpha\{1-w+(1-2xw+w^2)^{\frac{1}{2}}\}^{-\alpha} \\ & \cdot 2^\beta\{1+w+(1-2xw+w^2)^{\frac{1}{2}}\}^{-\beta}(1-2xw+w^2)^{-\frac{1}{2}}w^{-1}\,dw, \end{aligned} \tag{4.4.9}$$

which is the desired result.

(3) A third method of deriving the generating function is based on the following remark. If the function $F(x, w)$ of the right-hand member of (4.4.5) is developed in a power series in w, it is seen that the coefficient of w^n is a polynomial of degree n in x. To identify this polynomial with $P_n^{(\alpha,\beta)}(x)$ we show that

$$\int_{-1}^{+1}(1-x)^\alpha(1+x)^\beta F(x,u)F(x,v)\,dx, \tag{4.4.10}$$

considered as a function of u and v, is a function only of the product uv, which is equivalent to the orthogonality property. For $x = 1$, the identity (4.4.5) can be proved directly, and this procedure furnishes the normalization of the coefficients.

In Legendre's case: $\alpha = \beta = 0$, the integral can be calculated explicitly (Legendre **1**, p. 250). In the general case, Tchebichef (**5**) transformed this integral into the form

$$2^{\alpha+\beta+1}\int_0^1 t^\beta(1-t)^\alpha(1-uvt)^{-\alpha}(1-uvt^2)^{-1}\,dt, \tag{4.4.11}$$

from which the statement follows.

Concerning a fourth method based on Lagrange series, see Pólya-Szegö **1**, vol. 1, pp. 127, 303, problem 219.

4.5. Recurrence formula

(1) In the present case the general formula (3.2.1) becomes

$$
\begin{aligned}
&2n(n+\alpha+\beta)(2n+\alpha+\beta-2)P_n^{(\alpha,\beta)}(x)\\
&\quad=(2n+\alpha+\beta-1)\{(2n+\alpha+\beta)(2n+\alpha+\beta-2)x+\alpha^2-\beta^2\}P_{n-1}^{(\alpha,\beta)}(x)\\
&\qquad-2(n+\alpha-1)(n+\beta-1)(2n+\alpha+\beta)P_{n-2}^{(\alpha,\beta)}(x),\ n=2,3,4,\cdots;\\
&P_0^{(\alpha,\beta)}(x)=1,\qquad P_1^{(\alpha,\beta)}(x)=\tfrac{1}{2}(\alpha+\beta+2)x+\tfrac{1}{2}(\alpha-\beta).
\end{aligned}
\tag{4.5.1}
$$

Here, the coefficient of $xP_{n-1}^{(\alpha,\beta)}(x)$ may first be verified by means of (4.21.6); then, by alternately setting $x = +1$ and $x = -1$, the coefficients of $P_{n-1}^{(\alpha,\beta)}(x)$ and $P_{n-2}^{(\alpha,\beta)}(x)$ may be calculated. Actually, the formula is but a special case of the relations between contiguous Riemann P-functions (see Whittaker-Watson **1**, pp. 294–296).

(2) Using the notation (4.3.3), we obtain the following expression for the "kernel" (cf. (3.2.3)):

$$
\begin{aligned}
K_n^{(\alpha,\beta)}(x,y)&=\sum_{\nu=0}^{n}\{h_\nu^{(\alpha,\beta)}\}^{-1}P_\nu^{(\alpha,\beta)}(x)P_\nu^{(\alpha,\beta)}(y)\\
&=\frac{2^{-\alpha-\beta}}{2n+\alpha+\beta+2}\frac{\Gamma(n+2)\Gamma(n+\alpha+\beta+2)}{\Gamma(n+\alpha+1)\Gamma(n+\beta+1)}\\
&\qquad\cdot\frac{P_{n+1}^{(\alpha,\beta)}(x)P_n^{(\alpha,\beta)}(y)-P_n^{(\alpha,\beta)}(x)P_{n+1}^{(\alpha,\beta)}(y)}{x-y}.
\end{aligned}
\tag{4.5.2}
$$

In particular, for $y = 1$:

$$
\begin{aligned}
K_n^{(\alpha,\beta)}(x,1)&=K_n^{(\alpha,\beta)}(x)\\
&=\sum_{\nu=0}^{n}\frac{2\nu+\alpha+\beta+1}{2^{\alpha+\beta+1}}\frac{\Gamma(\nu+\alpha+\beta+1)}{\Gamma(\alpha+1)\Gamma(\nu+\beta+1)}P_\nu^{(\alpha,\beta)}(x)\\
&=2^{-\alpha-\beta}\frac{n+\alpha+1}{2n+\alpha+\beta+2}\frac{\Gamma(n+\alpha+\beta+2)}{\Gamma(\alpha+1)\Gamma(n+\beta+1)}\\
&\qquad\cdot\frac{P_n^{(\alpha,\beta)}(x)-\dfrac{n+1}{n+\alpha+1}P_{n+1}^{(\alpha,\beta)}(x)}{1-x}\\
&=2^{-\alpha-\beta-1}\frac{\Gamma(n+\alpha+\beta+2)}{\Gamma(\alpha+1)\Gamma(n+\beta+1)}P_n^{(\alpha+1,\beta)}(x).
\end{aligned}
\tag{4.5.3}
$$

The last representation in (4.5.3) is a consequence of Theorem 3.1.4, since $(1-x)^\alpha(1+x)^\beta(1-x) = (1-x)^{\alpha+1}(1+x)^\beta$. We also note that

$$
\begin{aligned}
P_n^{(\alpha+1,\beta)}(x)&=\frac{2}{2n+\alpha+\beta+2}\frac{(n+\alpha+1)P_n^{(\alpha,\beta)}(x)-(n+1)P_{n+1}^{(\alpha,\beta)}(x)}{1-x},\\
P_n^{(\alpha,\beta+1)}(x)&=\frac{2}{2n+\alpha+\beta+2}\frac{(n+\beta+1)P_n^{(\alpha,\beta)}(x)+(n+1)P_{n+1}^{(\alpha,\beta)}(x)}{1+x}.
\end{aligned}
\tag{4.5.4}
$$

The second formula follows from the first one if we interchange α and β and use (4.1.3). Finally, by using (4.21.7) and the last formulas, we obtain

$$(4.5.5)\qquad \begin{aligned}(1-x^2)\frac{d}{dx}\{P_n^{(\alpha,\beta)}(x)\} &= \tfrac{1}{2}(n+\alpha+\beta+1)(1-x^2)P_{n-1}^{(\alpha+1,\beta+1)}(x)\\ &= AP_{n-1}^{(\alpha,\beta)}(x) + BP_n^{(\alpha,\beta)}(x) + CP_{n+1}^{(\alpha,\beta)}(x),\end{aligned}$$

where

$$(4.5.6)\qquad \begin{aligned}A &= \frac{2(n+\alpha)(n+\beta)(n+\alpha+\beta+1)}{(2n+\alpha+\beta)(2n+\alpha+\beta+1)},\\ B &= (\alpha-\beta)\frac{2n(n+\alpha+\beta+1)}{(2n+\alpha+\beta)(2n+\alpha+\beta+2)},\\ C &= -\frac{2n(n+1)(n+\alpha+\beta+1)}{(2n+\alpha+\beta+1)(2n+\alpha+\beta+2)}.\end{aligned}$$

Here $P_{n+1}^{(\alpha,\beta)}(x)$ [or $P_{n-1}^{(\alpha,\beta)}(x)$] can be expressed by means of (4.5.1) in terms of $xP_n^{(\alpha,\beta)}(x)$, $P_n^{(\alpha,\beta)}(x)$, $P_{n-1}^{(\alpha,\beta)}(x)$ (or $P_{n+1}^{(\alpha,\beta)}(x)$). This yields

$$(4.5.7)\qquad \begin{aligned}&(2n+\alpha+\beta)(1-x^2)\frac{d}{dx}\{P_n^{(\alpha,\beta)}(x)\}\\ &\quad = -n\{(2n+\alpha+\beta)x+\beta-\alpha\}P_n^{(\alpha,\beta)}(x) + 2(n+\alpha)(n+\beta)P_{n-1}^{(\alpha,\beta)}(x),\\ &(2n+\alpha+\beta+2)(1-x^2)\frac{d}{dx}\{P_n^{(\alpha,\beta)}(x)\}\\ &\quad = (n+\alpha+\beta+1)\{(2n+\alpha+\beta+2)x+\alpha-\beta\}P_n^{(\alpha,\beta)}(x)\\ &\qquad\qquad - 2(n+1)(n+\alpha+\beta+1)P_{n+1}^{(\alpha,\beta)}(x).\end{aligned}$$

In addition, we notice the following consequence of the last formula (4.5.3):

$$(4.5.8)\qquad \begin{aligned}K_n^{(\alpha,\beta)}(1,1) &= K_n^{(\alpha,\beta)}(1)\\ &= 2^{-\alpha-\beta-1}\frac{\Gamma(n+\alpha+\beta+2)\Gamma(n+\alpha+2)}{\Gamma(\alpha+1)\Gamma(\alpha+2)\Gamma(n+1)\Gamma(n+\beta+1)}.\end{aligned}$$

4.6. Integral representations in general

The representation (4.3.1) and its integral form (4.4.6) are closely related to a classical method used for the integration of the hypergeometric equation and others of similar type. We again start from the formula (4.4.6):

$$(4.6.1)\qquad (1-x)^\alpha(1+x)^\beta P_n^{(\alpha,\beta)}(x) = \frac{(-\frac{1}{2})^n}{2\pi i}\int (1-t)^{n+\alpha}(1+t)^{n+\beta}(t-x)^{-n-1}\,dt,$$

where $x \neq \pm 1$. The integration is extended in the positive sense over a closed curve enclosing x, but not the points $t = \pm 1$. Using an idea of Euler (1), we try to integrate (4.2.1) and (4.2.3) by means of

$$(4.6.2)\qquad Y = (1-x)^\alpha(1+x)^\beta y = \int (1-t)^{n+\alpha}(1+t)^{n+\beta}(t-x)^{-n-1}\,dt,$$

with a proper choice of the contour of integration. Here $x \neq \pm 1$, and the path of integration must avoid the points -1, $+1$, and x. However, we allow -1 and $+1$ as end-points of a path provided the integrals (4.6.2) and (4.6.3) are convergent.

Substituting (4.6.2) in (4.2.3), we obtain

$$(1 - x^2)Y'' + [\alpha - \beta + (\alpha + \beta - 2)x]Y' + (n + 1)(n + \alpha + \beta)Y$$

$$(4.6.3)\quad \begin{aligned} &= \int (1 - t)^{n+\alpha}(1 + t)^{n+\beta}(t - x)^{-n-3}\{(n + 1)(n + 2)(1 - x^2) \\ &\qquad + (n + 1)[\alpha - \beta + (\alpha + \beta - 2)x](t - x) \\ &\qquad + (n + 1)(n + \alpha + \beta)(t - x)^2\}\, dt \\ &= -(n + 1) \int \frac{d}{dt} \{(1 - t)^{n+\alpha+1}(1 + t)^{n+\beta+1}(t - x)^{-n-2}\}\, dt. \end{aligned}$$

Therefore, we see that (4.6.2) satisfies (4.2.3), provided one of the following two conditions is fulfilled:

(a) The path of integration is a closed contour along which the expression $(1 - t)^{n+\alpha+1}(1 + t)^{n+\beta+1}(t - x)^{-n-2}$, or what amounts to the same thing, $(1 - t)^\alpha(1 + t)^\beta$ returns to its original value.

(b) The integration is extended along an arc, finite or infinite, such that the first expression mentioned vanishes at the end-points.

Specialization of the contour according to these restrictions yields numerous important integral representations for Jacobi polynomials as well as for other solutions of (4.2.1). (Cf. §4.61, §4.82.) For a special contour integral allowed in the sense of (a) and (b), we must first show that y is not identically zero; then y can be identified with a constant multiple of $P_n^{(\alpha,\beta)}(x)$, or with some other particular solution of (4.2.1); finally, the constant factor must be determined. The resulting integral representations hold, save for some exceptional values of α and β.

Further integral representations are obtained by replacing n by $-n - \alpha - \beta - 1$ in (4.6.2); this does not affect (4.2.3). Thus,

$$(4.6.4)\quad Y = (1 - x)^\alpha(1 + x)^\beta y = \int (1 - t)^{-n-\beta-1}(1 + t)^{-n-\alpha-1}(t - x)^{n+\alpha+\beta}\, dt,$$

where the contour is chosen as in (a) or (b) above. Instead of the first expression in (a) we now have

$$(1 - t)^{-n-\beta}(1 + t)^{-n-\alpha}(t - x)^{n+\alpha+\beta-1}.$$

Rodrigues' formula (4.3.1) is a special case of (4.6.2), the path of integration being a closed curve which encloses x but not ± 1. Condition (a) is then satisfied.

4.61. Application; functions of the second kind

(1) THEOREM 4.61.1. *Let x be arbitrary in the complex plane cut along the segment $[-1, +1]$. Let $\alpha > -1$, $\beta > -1$, $n \geqq 0$. Excluding the case $n = 0$, $\alpha + \beta + 1 = 0$, a solution $y = Q_n^{(\alpha,\beta)}(x)$ of the differential equation (4.2.1), which is linearly independent of $P_n^{(\alpha,\beta)}(x)$, can be obtained in the form*

$$(4.61.1)\quad \begin{aligned} &Q_n^{(\alpha,\beta)}(x) \\ &= 2^{-n-1}(x - 1)^{-\alpha}(x + 1)^{-\beta} \int_{-1}^{+1} (1 - t)^{n+\alpha}(1 + t)^{n+\beta}(x - t)^{-n-1}\, dt. \end{aligned}$$

In the exceptional case: $n = 0$, $\alpha + \beta + 1 = 0$, we have $Q_0^{(\alpha,\beta)}(x) =$ const.; then a non-constant solution is given by

$$Q^{(\alpha)}(x) = \log(x+1) + \pi^{-1}\sin\pi\alpha\,(x-1)^{-\alpha}(x+1)^{-\beta} \cdot \int_{-1}^{+1} \frac{(1-t)^\alpha(1+t)^\beta}{x-t}\log(1+t)\,dt. \tag{4.61.2}$$

The function $Q_n^{(\alpha,\beta)}(x)$ is called Jacobi's function of the second kind. We use this notation also if $n = 0$, $\alpha+\beta+1 = 0$, for the function $Q^{(\alpha)}(x)$. In the special case $\alpha = \beta = 0$ we write $Q_n^{(0,0)}(x) = Q_n(x)$ (Legendre's function of the second kind). (Cf. Jacobi **3**, pp. 195–197.)

Both (4.61.1) and (4.61.2) are multi-valued (except if α and β are integers). Both integrals are single-valued and regular in the complex plane cut along the segment $[-1, +1]$. Obviously, $Q_n^{(\alpha,\beta)}(x) \sim x^{-n-\alpha-\beta-1}$ as $x \to \infty$, which shows that $Q_n^{(\alpha,\beta)}(x)$ is linearly independent of $P_n^{(\alpha,\beta)}(x)$ (except if $n = 0$, $\alpha+\beta+1 = 0$, see below). The corresponding property of $Q^{(\alpha)}(x)$ is clear. We have, as is easily seen,

$$Q^{(\alpha)}(x) = 2\pi^{-1}\sin\pi\alpha\left\{\frac{\partial}{\partial\beta}Q_0^{(\alpha,\beta)}(x)\right\}_{\beta=-\alpha-1}. \tag{4.61.3}$$

The function $Q_n^{(\alpha,\beta)}(x)$ satisfies the differential equation (4.2.1); this follows from (4.6.2) since the segment $[-1, +1]$ which is the path of integration, satisfies the condition (b) of §4.6. Differentiating (4.2.1) with respect to β, and substituting $\beta = -\alpha-1$, we obtain the solution $Q^{(\alpha)}(x)$ since $Q_0^{(\alpha,\beta)}(x) = \text{const.}$ if $\beta = -\alpha-1$.

(2) THEOREM 4.61.2. *The following representations hold:*

$$Q_n^{(\alpha,\beta)}(x) = \tfrac{1}{2}(x-1)^{-\alpha}(x+1)^{-\beta}\int_{-1}^{+1}(1-t)^\alpha(1+t)^\beta\frac{P_n^{(\alpha,\beta)}(t)}{x-t}\,dt \tag{4.61.4}$$

$$= 2^{n+\alpha+\beta}\frac{\Gamma(n+\alpha+1)\Gamma(n+\beta+1)}{\Gamma(2n+\alpha+\beta+2)}(x-1)^{-n-\alpha-1}(x+1)^{-\beta} \cdot F\left(n+\alpha+1, n+1; 2n+\alpha+\beta+2; \frac{2}{1-x}\right). \tag{4.61.5}$$

Furthermore,

$$Q^{(\alpha)}(x) = \log(x+1) + \left(1-\frac{2}{1-x}\right)^{\alpha+1}\cdot\sum_{\nu=1}^{\infty}\binom{\alpha+\nu}{\nu}\left(\frac{1}{1}+\frac{1}{2}+\cdots+\frac{1}{\nu}\right)\left(\frac{2}{1-x}\right)^{\nu} + c, \tag{4.61.6}$$

$$c = \Gamma'(1) - \frac{\Gamma'(-\alpha)}{\Gamma(-\alpha)} - \log 2.$$

By use of Rodrigues' formula, (4.61.1) may be integrated by parts n times. This establishes (4.61.4). From (4.61.1) we readily obtain

$$(x-1)^\alpha(x+1)^\beta Q_n^{(\alpha,\beta)}(x)$$
$$= (-2)^{-n-1}\sum_{\nu=0}^{\infty}\binom{n+\nu}{n}(1-x)^{-n-\nu-1}\int_{-1}^{+1}(1-t)^{n+\nu+\alpha}(1+t)^{n+\beta}\,dt$$
$$= (-1)^{n+1}2^{\alpha+\beta-1}\sum_{\nu=0}^{\infty}\binom{n+\nu}{n}\frac{\Gamma(n+\nu+\alpha+1)\Gamma(n+\beta+1)}{\Gamma(2n+\nu+\alpha+\beta+2)}\left(\frac{2}{1-x}\right)^{n+\nu+1}.$$

This, in connection with the notation (4.21.3), yields (4.61.5).

Turning to the exceptional case, we first notice that for $n = 0, \alpha + \beta + 1 \neq 0$, (4.61.5) becomes

$$Q_0^{(\alpha,\beta)}(x) = 2^{\alpha+\beta} \frac{\Gamma(\alpha+1)\Gamma(\beta+1)}{\Gamma(\alpha+\beta+2)} (x-1)^{-\alpha-1}(x+1)^{-\beta} \cdot F\left(\alpha+1, 1; \alpha+\beta+2; \frac{2}{1-x}\right). \tag{4.61.7}$$

For $\alpha + \beta + 1 = 0$ this will, in fact, be a constant since ((1.7.2), second formula)

$$\frac{1}{2}\Gamma(\alpha+1)\Gamma(\beta+1)(x-1)^{-\alpha-1}(x+1)^{-\beta} F\left(\alpha+1, 1; 1; \frac{2}{1-x}\right)$$
$$= -\frac{1}{2}\frac{\pi}{\sin \pi\alpha}(x-1)^{-\alpha-1}(x+1)^{-\beta}\left(1 - \frac{2}{1-x}\right)^{-\alpha-1} = -\frac{1}{2}\frac{\pi}{\sin \pi\alpha}.$$

Now from (4.61.7), taking into account (4.61.3), we obtain (4.61.6).

(3) The case $n = 0$ may be treated in another way by means of the relation (4.2.6). For $z = 1$ this becomes

$$(1-x)^{\alpha+1}(1+x)^{\beta+1}y' = \text{const.} \tag{4.61.8}$$

This yields an integral representation for y.

THEOREM 4.61.3. *Let $\alpha > -1, \beta > -1$. We then have the following integral representations:*

$$Q_0^{(\alpha,\beta)}(x) = -2^{\alpha+\beta}\frac{\Gamma(\alpha+1)\Gamma(\beta+1)}{\Gamma(\alpha+\beta+1)}\int_\infty^x (t-1)^{-\alpha-1}(t+1)^{-\beta-1}\,dt \quad \text{if } \alpha+\beta+1 > 0, \tag{4.61.9}$$

$$Q_0^{(\alpha,\beta)}(x) = -2^{\alpha+\beta}\frac{\Gamma(\alpha+1)\Gamma(\beta+1)}{\Gamma(\alpha+\beta+1)}\left\{\int_\infty^x [(t-1)^{-\alpha-1}(t+1)^{-\beta-1} - t^{-\alpha-\beta-2}]\,dt - \frac{x^{-\alpha-\beta-1}}{\alpha+\beta+1}\right\} \quad \text{if } \alpha+\beta+1 < 0, \tag{4.61.10}$$

$$Q^{(\alpha)}(x) = \int_\infty^x \left[(t-1)^{-\alpha-1}(t+1)^{-\beta-1} - \frac{1}{t+1}\right] dt + \log(x+1) \quad \text{if } \alpha+\beta+1 = 0. \tag{4.61.11}$$

In the first case the integrand is $\sim t^{-\alpha-\beta-2}$, as $t \to \infty$, so that the integral is convergent. The constant factor in (4.61.9) can be obtained by comparing the principal terms in (4.61.7) and (4.61.9).

In the second case the principal term of the integrand is $(\alpha - \beta)t^{-\alpha-\beta-3}$, so that the integral is convergent. In the third case the principal term of the integrand is $2(\alpha+1)(t+1)^{-2}$, so that here too we have convergence.

(4) Another very general integral representation of a second solution of (4.2.1) is obtained by choosing the path of integration in (4.6.2) as in Jordan-Pochhammer's integral for the Gamma function (see the figure in Whittaker-Watson **1**, p. 257). This path can be defined by the scheme

(4.61.12) $\qquad (-1-), \quad (+1+), \quad (-1+), \quad (+1-).$

Condition (a) is then satisfied, $x \neq \pm 1$. The principal term is $x^{-n-\alpha-\beta-1}$ provided the integral $\int (1-t)^{n+\alpha}(1+t)^{n+\beta}\,dt$, extended over the contour in question, does not vanish. This is the case (see loc. cit., p. 257) unless one of the following conditions is satisfied:

$$n+\alpha = 0, 1, 2, \cdots; \qquad n+\beta = 0, 1, 2, \cdots;$$
$$2n+\alpha+\beta+2 = 0, -1, -2, \cdots. \tag{4.61.13}$$

In the special cases, where $\alpha+\beta$ or $\alpha-\beta$ is an integer, the contour can be simplified.

4.62. Further properties of the functions of the second kind

In the following considerations we again assume that $\alpha > -1$ and $\beta > -1$.

(1) The possible singular points of $Q_n^{(\alpha,\beta)}(x)$ are $+1, -1, \infty$. In order to discuss this function near $x = +1$ (and also for later purposes), we write (4.61.4) in the form

$$\begin{aligned} Q_n^{(\alpha,\beta)}(x) &= -\tfrac{1}{2}(x-1)^{-\alpha}(x+1)^{-\beta}\int_{-1}^{+1}(1-t)^\alpha(1+t)^\beta\,\frac{P_n^{(\alpha,\beta)}(x)-P_n^{(\alpha,\beta)}(t)}{x-t}\,dt \\ &\quad + \tfrac{1}{2}(x-1)^{-\alpha}(x+1)^{-\beta}P_n^{(\alpha,\beta)}(x)\int_{-1}^{+1}\frac{(1-t)^\alpha(1+t)^\beta}{x-t}\,dt \\ &= -\tfrac{1}{2}(x-1)^{-\alpha}(x+1)^{-\beta}\int_{-1}^{+1}(1-t)^\alpha(1+t)^\beta\,\frac{P_n^{(\alpha,\beta)}(x)-P_n^{(\alpha,\beta)}(t)}{x-t}\,dt \\ &\quad + P_n^{(\alpha,\beta)}(x)Q_0^{(\alpha,\beta)}(x). \end{aligned} \tag{4.62.1}$$

The last integral is a π_{n-1} in x (a constant multiple of the numerator $R_n(x)$ of the nth convergent of the continued fraction defined in §3.5; see the first part of (3.5.7)). Therefore, if x approaches $+1$, the behavior of $Q_n^{(\alpha,\beta)}(x)$ is to a certain extent determined by that of $Q_0^{(\alpha,\beta)}(x)$.

The discussion of $Q_0^{(\alpha,\beta)}(x)$ near $x = +1$ is not difficult. Expanding the factor $(t+1)^{-\beta-1}$ in the integrand of (4.61.9) into a power series in $t-1$, we obtain for $\alpha+\beta+1 > 0$, α not an integer,

$$Q_0^{(\alpha,\beta)}(x) = \text{const.} + (x-1)^{-\alpha}M\left(\frac{1-x}{2}\right);$$

here $M(u)$ is a power series of u, convergent for $|u| < 1$, and $M(0) \neq 0$. A similar representation holds if $\alpha+\beta+1 < 0$ [(4.61.10)]. (In the exceptional case $\alpha+\beta+1 = 0$, this is not true for $Q_0^{(\alpha,\beta)}(x)$; however, it is true for $Q^{(\alpha)}(x)$, cf. (4.61.11).) Now, let α be an integer; then we use (4.61.9) again. In view of

$$\begin{aligned}(t-1)^{-\alpha-1}(t+1)^{-\beta-1} &= 2^{-\beta-1}(t-1)^{-\alpha-1}\left(1-\frac{1-t}{2}\right)^{-\beta-1} \\ &= 2^{-\beta-1}(t-1)^{-\alpha-1}\left\{1+\cdots+\binom{\beta+\alpha}{\alpha}\left(\frac{1-t}{2}\right)^\alpha+\cdots\right\},\end{aligned}$$

the integration furnishes a logarithmic term. Thus,

$$(4.62.2)\quad Q_0^{(\alpha,\beta)}(x) = \begin{cases} \text{const.} + (x-1)^{-\alpha} M_1\left(\frac{1-x}{2}\right) \\ \qquad \text{if } \alpha > -1, \beta > -1; \alpha \neq 0, 1, 2, \cdots; \alpha + \beta + 1 \neq 0, \\ \frac{(-1)^\alpha}{2} \log \frac{1}{x-1} + (x-1)^{-\alpha} M_2\left(\frac{1-x}{2}\right) \\ \qquad \text{if } \alpha = 0, 1, 2, \cdots; \beta > -1. \end{cases}$$

Here $M_1(u)$ and $M_2(u)$ are power series convergent for $|u| < 1$ with $M_1(0) \neq 0$, $M_2(0) \neq 0$ (see below). A representation similar to the first one holds for $Q^{(\alpha)}(x)$.

We have, for instance,

$$(4.62.3)\qquad Q_0^{(0,0)}(x) = Q_0(x) = \int_\infty^x \frac{dt}{1-t^2} = \frac{1}{2} \log \frac{x+1}{x-1}.$$

[The statement $M_2(0) \neq 0$ requires further comment for $\alpha = 0$. Taking $x > 1$, and then integrating by parts, we have

$$\int_\infty^x (t-1)^{-1}(t+1)^{-\beta-1}\, dt$$

$$= (x+1)^{-\beta-1} \log (x-1) + (\beta+1) \int_\infty^x (t+1)^{-\beta-2} \log (t-1)\, dt,$$

so that

$$M_2(0) = \lim_{x \to 1+0} \{Q_0^{(\alpha,\beta)}(x) + \tfrac{1}{2} \log (x-1)\}$$

$$= -(\beta+1)2^\beta \int_\infty^1 (t+1)^{-\beta-2} \log (t-1)\, dt \neq 0.]$$

(2) Now we prove the following theorem:

THEOREM 4.62.1. *Let x be real, $x > 1$, and take $(x-1)^\alpha$, $(x+1)^\beta$ real and positive. We then have, for $x \to 1 + 0$,*

$$(4.62.4)\qquad Q_n^{(\alpha,\beta)}(x) \sim \begin{cases} (x-1)^{-\alpha}, & \alpha > 0, \\ \log(x-1), & \alpha = 0, \\ 1, & \alpha < 0. \end{cases}$$

More precisely,

$$(4.62.5)\qquad Q_n^{(\alpha,\beta)}(x) \cong \begin{cases} 2^{\alpha-1} \frac{\Gamma(\alpha)\Gamma(n+\beta+1)}{\Gamma(n+\alpha+\beta+1)} (x-1)^{-\alpha}, & \alpha > 0, \\ \frac{(-1)^\alpha}{2} \log \frac{1}{x-1}, & \alpha = 0. \end{cases}$$

The behavior near $x = -1$ of $Q_n^{(\alpha,\beta)}(x)$ is similar.

The case $\alpha > 0$ follows from (4.61.1):

$$Q_n^{(\alpha,\beta)}(x) \cong 2^{-n-\beta-1}(x-1)^{-\alpha} \int_{-1}^{+1} (1-t)^{\alpha-1}(1+t)^{n+\beta}\, dt.$$

In the case $\alpha = 0$ we use (4.62.1), (4.62.2), and $P_n^{(\alpha,\beta)}(1) = 1$. In the case $\alpha < 0$ the first term of the right-hand member of (4.62.1) vanishes as $x \to 1 + 0$.

Thus the statement is equivalent to $Q_0^{(\alpha,\beta)}(x) \sim 1$. This is immediately clear from (4.61.9) if $\alpha + \beta + 1 > 0$. If $\alpha + \beta + 1 < 0$, we use (4.61.10) and show that

$$(4.62.6) \qquad \int_1^\infty [(t-1)^{-\alpha-1}(t+1)^{-\beta-1} - t^{-\alpha-\beta-2}]\,dt + \frac{1}{\alpha+\beta+1} < 0.$$

On writing

$$(t-1)^{-\alpha-1}(t+1)^{-\beta-1} = (t-1)^{-1}(t+1)^{-\alpha-\beta-1}\left(\frac{t+1}{t-1}\right)^\alpha,$$

we see that (4.62.6), as a function of α and β, increases with α if $\alpha + \beta$ is constant. But when β approaches -1, we find for the left-hand member of (4.62.6) the result

$$\int_1^\infty [(t-1)^{-\alpha-1} - t^{-\alpha-1}]\,dt + \frac{1}{\alpha} = 0.$$

We have, as an instance, [(4.62.1), (4.62.3)]

$$(4.62.7) \qquad Q_n^{(0,0)}(x) = Q_n(x) = R(x) + \tfrac{1}{2}P_n(x)\log\frac{x+1}{x-1},$$

where $R(x)$ is a π_{n-1}. The logarithmic factor is chosen so that it tends to 0 as $x \to \infty$.

(3) THEOREM 4.62.2. *Let α be an integer, $\alpha \geqq 0$. We consider $Q_n^{(\alpha,\beta)}(x)$ (real and positive for $x > 1$) in the complex plane cut along the line $[-\infty, +1]$. Then*

$$(4.62.8) \qquad Q_n^{(\alpha,\beta)}(x+i0) - Q_n^{(\alpha,\beta)}(x-i0) = (-1)^{\alpha-1}\pi i P_n^{(\alpha,\beta)}(x),$$
$$-1 < x < +1.$$

This follows from (4.62.1) and (4.62.2).

On the other hand, the function

$$(4.62.9) \qquad \mathbf{Q}_n^{(\alpha,\beta)}(x) = \tfrac{1}{2}\{Q_n^{(\alpha,\beta)}(x+i0) + Q_n^{(\alpha,\beta)}(x-i0)\}$$

is analytic on $-1 < x < +1$ and satisfies the differential equation (4.2.1). As $x \to 1-0$, it displays a behavior similar to that of $Q_n^{(\alpha,\beta)}(x)$. In particular, we find for $\mathbf{Q}_n^{(0,0)}(x) = \mathbf{Q}_n(x)$

$$(4.62.10) \qquad \mathbf{Q}_n(x) = R(x) + \tfrac{1}{2}P_n(x)\log\frac{1+x}{1-x},$$

$$(4.62.11) \qquad \mathbf{Q}_n(-x) = (-1)^{n+1}\mathbf{Q}_n(x), \qquad -1 < x < +1,$$

$$(4.62.12) \qquad \lim_{x\to 1-0} \mathbf{Q}_n(x) = +\infty.$$

Here $R(x)$ has the same meaning as in (4.62.7).

In general, if α and β are both integers, the function $Q_n^{(\alpha,\beta)}(x)$ is regular and single-valued in the whole plane cut along $[-1, +1]$.

(4) The functions of the second kind satisfy the same recurrence formula as $P_n^{(\alpha,\beta)}(x)$ [(4.5.1)], that is,

$$
\begin{aligned}
&2n(n+\alpha+\beta)(2n+\alpha+\beta-2)Q_n^{(\alpha,\beta)}(x) \\
&\quad = (2n+\alpha+\beta-1)\{(2n+\alpha+\beta)(2n+\alpha+\beta-2)x+\alpha^2-\beta^2\} \\
&\quad \cdot Q_{n-1}^{(\alpha,\beta)}(x) - 2(n+\alpha-1)(n+\beta-1)(2n+\alpha+\beta)Q_{n-2}^{(\alpha,\beta)}(x), \\
&\qquad\qquad n = 2, 3, 4, \cdots.
\end{aligned}
\tag{4.62.13}
$$

This follows from (4.62.1) on account of (3.5.3) and Theorem 3.5.1. There is, however, an essential difference if $n = 1$. We then have, according to (4.62.1),

$$
\begin{aligned}
Q_1^{(\alpha,\beta)}(x) &= \tfrac{1}{2}\left[(\alpha+\beta+2)x+\alpha-\beta\right] Q_0^{(\alpha,\beta)}(x) \\
&\quad - 2^{\alpha+\beta-1}(\alpha+\beta+2)\frac{\Gamma(\alpha+1)\Gamma(\beta+1)}{\Gamma(\alpha+\beta+2)}(x-1)^{-\alpha}(x+1)^{-\beta}.
\end{aligned}
\tag{4.62.14}
$$

Therefore, both systems of functions [(4.3.3)]

$$
\begin{aligned}
p_n(x) &= \{h_n^{(\alpha,\beta)}\}^{-\frac{1}{2}} P_n^{(\alpha,\beta)}(x), \\
q_n(x) &= \{h_n^{(\alpha,\beta)}\}^{-\frac{1}{2}} Q_n^{(\alpha,\beta)}(x), \qquad n = 0, 1, 2, \cdots,
\end{aligned}
\tag{4.62.15}
$$

satisfy the same recurrence formula of the type (3.2.1) for $n \geqq 1$, provided we define

$$
p_{-1}(x) = 0, \qquad q_{-1}(x) = (x-1)^{-\alpha}(x+1)^{-\beta}.
\tag{4.62.16}
$$

Thus a procedure similar to that used in §3.2 (2) furnishes, for $n \geqq 1$,

$$
\begin{aligned}
&\frac{k_n}{k_{n+1}} \frac{p_{n+1}(x)q_n(y) - p_n(x)q_{n+1}(y)}{x-y} \\
&\quad = p_n(x)q_n(y) + \frac{k_{n-1}}{k_n}\cdot\frac{p_n(x)q_{n-1}(y) - p_{n-1}(x)q_n(y)}{x-y}.
\end{aligned}
\tag{4.62.17}
$$

Here k_n denotes the coefficient of x^n in the "normalized" polynomial $p_n(x)$. This formula also holds for $n = 0$ if we modify it as follows:

$$
\frac{k_0}{k_1}\frac{p_1(x)q_0(y) - p_0(x)q_1(y)}{x-y} = p_0(x)q_0(y) + \text{const.}\,\frac{q_{-1}(y)}{x-y}.
\tag{4.62.18}
$$

Adding, we obtain the important result:

$$
\begin{aligned}
&\sum_{\nu=0}^{n} \frac{2\nu+\alpha+\beta+1}{2^{\alpha+\beta+1}} \frac{\Gamma(\nu+1)\Gamma(\nu+\alpha+\beta+1)}{\Gamma(\nu+\alpha+1)\Gamma(\nu+\beta+1)} P_\nu^{(\alpha,\beta)}(x)Q_\nu^{(\alpha,\beta)}(y) \\
&\quad = \frac{1}{2}\frac{(y-1)^{-\alpha}(y+1)^{-\beta}}{y-x} + \frac{2^{-\alpha-\beta}}{2n+\alpha+\beta+2} \\
&\quad \cdot \frac{\Gamma(n+2)\Gamma(n+\alpha+\beta+2)}{\Gamma(n+\alpha+1)\Gamma(n+\beta+1)} \frac{P_{n+1}^{(\alpha,\beta)}(x)Q_n^{(\alpha,\beta)}(y) - P_n^{(\alpha,\beta)}(x)Q_{n+1}^{(\alpha,\beta)}(y)}{x-y}.
\end{aligned}
\tag{4.62.19}
$$

The constant $\frac{1}{2}$ in the right-hand member can be determined by substituting $n = 0$, multiplying by $y^{\alpha+\beta+1}$, then permitting $y \to \infty$, and finally using (4.61.5).

We shall return to this formula in §9.2, where it will be used in a classical manner for the expansion of an analytic function in terms of Jacobi polynomials or of Jacobi functions of the second kind.

4.7. Ultraspherical polynomials

(1) If $\alpha = \beta$, Jacobi's polynomial $P_n^{(\alpha,\beta)}(x)$ is called an ultraspherical polynomial.[21] The following is the customary notation and normalization:

$$(4.7.1)\qquad \begin{aligned} P_n^{(\lambda)}(x) &= \frac{\Gamma(\alpha+1)}{\Gamma(2\alpha+1)}\frac{\Gamma(n+2\alpha+1)}{\Gamma(n+\alpha+1)} P_n^{(\alpha,\alpha)}(x) \\ &= \frac{\Gamma(\lambda+\frac{1}{2})}{\Gamma(2\lambda)}\frac{\Gamma(n+2\lambda)}{\Gamma(n+\lambda+\frac{1}{2})} P_n^{(\lambda-\frac{1}{2},\lambda-\frac{1}{2})}(x), \qquad \alpha = \lambda - \tfrac{1}{2}. \end{aligned}$$

Here we assume first that $\alpha > -1$, or $\lambda > -\frac{1}{2}$. Some important special cases are (cf. (1.12.3))

$$(4.7.2)\qquad P_n^{(\frac{1}{2})}(x) = P_n(x), \qquad P_n^{(1)}(x) = U_n(x).$$

If $\alpha = -\frac{1}{2}$, or $\lambda = 0$, the polynomial $P_n^{(\lambda)}(x)$ vanishes identically for $n \geqq 1$. This case will be treated later. (Cf. (4.7.8).)

We next observe a number of formulas and theorems which can be obtained immediately from the theory of general Jacobi polynomials by setting $\alpha = \beta = \lambda - \frac{1}{2}$, $\lambda > -\frac{1}{2}$:

$$(4.7.3)\qquad P_n^{(\lambda)}(1) = \binom{n+2\lambda-1}{n};$$

$$(4.7.4)\qquad P_n^{(\lambda)}(-x) = (-1)^n P_n^{(\lambda)}(x);$$

$$(4.7.5)\qquad \begin{aligned} &(1-x^2)y'' - (2\lambda+1)xy' + n(n+2\lambda)y = 0, \qquad y = P_n^{(\lambda)}(x), \\ &(1-x^2)Y'' + (2\lambda-3)xY' + (n+1)(n+2\lambda-1)Y = 0, \\ &\qquad Y = (1-x^2)^{\lambda-\frac{1}{2}} P_n^{(\lambda)}(x); \end{aligned}$$

$$(4.7.6)\qquad \begin{aligned} P_n^{(\lambda)}(x) &= \binom{n+2\lambda-1}{n} F\left(-n, n+2\lambda; \lambda+\tfrac{1}{2}; \frac{1-x}{2}\right) \\ &= 2^n\binom{n+\lambda-1}{n}(x-1)^n \cdot F\left(-n, -n-\lambda+\tfrac{1}{2}; -2n-2\lambda+1; \frac{2}{1-x}\right). \end{aligned}$$

The last formulas define $P_n^{(\lambda)}(x)$ for all values of λ. If necessary, for some special values of λ, say $\lambda = \lambda_0$, the formulas may be interpreted as limits for $\lambda \to \lambda_0$.[22] For $\lambda = -m$, $m = 0, 1, 2, \cdots$, we obviously have (cf. the first formula (4.7.6)) $P_n^{(\lambda)}(x) \equiv 0$ if $n > 2m$. In this case

$$(4.7.7)\qquad \begin{aligned} \lim_{\lambda\to -m} \frac{P_n^{(\lambda)}(x)}{\lambda+m} &= \left\{\frac{d}{d\lambda} P_n^{(\lambda)}(x)\right\}_{\lambda=-m} \\ &= 2\frac{(2m)!(n-2m-1)!}{n!} F\left(-n, n-2m; -m+\tfrac{1}{2}; \frac{1-x}{2}\right) \end{aligned}$$

exists. These polynomials are, except for constant factors, again the Jacobi polynomials $P_n^{(\alpha,\alpha)}(x)$, $\alpha = -m - \frac{1}{2}$. For instance (cf. (1.12.3))

$$(4.7.8)\qquad \lim_{\lambda\to 0} \lambda^{-1} P_n^{(\lambda)}(x) = (2/n)T_n(x), \qquad n \geqq 1.$$

[21] They are sometimes called Gegenbauer's polynomials. See the papers of Gegenbauer (1–7). See also Heine 3, vol. 1, pp. 297-301, 449-464. Occasionally, the notation $C_n^{\lambda}(x)$ is used instead of $P_n^{(\lambda)}(x)$.

[22] This should also be done in all the subsequent formulas.

In case $\alpha = -l$, $(n+1)/2 \leq l \leq n$, the polynomial $P_n^{(\alpha,\alpha)}(x)$ vanishes identically (§4.22 (3)). However, the corresponding expression $P_n^{(\lambda)}(x)$, as a limit, has a meaning and does not vanish identically.

(2) Further formulas involving $P_n^{(\lambda)}(x)$ are

$$\lim_{x\to\infty} x^{-n}P_n^{(\lambda)}(x) = 2^n\binom{n+\lambda-1}{n}; \tag{4.7.9}$$

$$\frac{d^2u}{dx^2}+\left\{\frac{(n+\lambda)^2}{1-x^2}+\frac{\frac{1}{2}+\lambda-\lambda^2+x^2/4}{(1-x^2)^2}\right\}u=0, \quad u=(1-x^2)^{\lambda/2+\frac{1}{4}}P_n^{(\lambda)}(x); \tag{4.7.10}$$

$$\frac{d^2u}{d\theta^2}+\left\{(n+\lambda)^2+\frac{\lambda(1-\lambda)}{\sin^2\theta}\right\}u=0, \qquad u=(\sin\theta)^\lambda P_n^{(\lambda)}(\cos\theta); \tag{4.7.11}$$

$$(1-x^2)^{\lambda-\frac{1}{2}}P_n^{(\lambda)}(x) = \frac{(-2)^n}{n!}\frac{\Gamma(n+\lambda)\Gamma(n+2\lambda)}{\Gamma(\lambda)\Gamma(2n+2\lambda)}\left(\frac{d}{dx}\right)^n(1-x^2)^{n+\lambda-\frac{1}{2}}; \tag{4.7.12}$$

$$P_n^{(\lambda)}(x) = \binom{n+2\lambda-1}{n}\left(\frac{x+1}{2}\right)^n F\left(-n.-n-\lambda+\tfrac{1}{2};\lambda+\tfrac{1}{2};\frac{x-1}{x+1}\right); \tag{4.7.13}$$

$$\frac{d}{dx}P_n^{(\lambda)}(x) = 2\lambda P_{n-1}^{(\lambda+1)}(x); \tag{4.7.14}$$

$$\int_{-1}^{+1}(1-x^2)^{\lambda-\frac{1}{2}}\{P_n^{(\lambda)}(x)\}^2\,dx = 2^{1-2\lambda}\pi\{\Gamma(\lambda)\}^{-2}\frac{\Gamma(n+2\lambda)}{(n+\lambda)\Gamma(n+1)}, \tag{4.7.15}$$

$$\lambda > -\tfrac{1}{2},\ \lambda \neq 0;$$ [23]

$$\sum_{n=0}^{\infty}\frac{\Gamma(2\lambda)}{\Gamma(\lambda+\frac{1}{2})}\frac{\Gamma(n+\lambda+\frac{1}{2})}{\Gamma(n+2\lambda)}P_n^{(\lambda)}(x)w^n = 2^{\lambda-\frac{1}{2}}(1-2xw+w^2)^{-\frac{1}{2}}\{1-xw+(1-2xw+w^2)^{\frac{1}{2}}\}^{\frac{1}{2}-\lambda}; \tag{4.7.16}$$

$$nP_n^{(\lambda)}(x) = 2(n+\lambda-1)xP_{n-1}^{(\lambda)}(x)-(n+2\lambda-2)P_{n-2}^{(\lambda)}(x),\ n=2,3,4,\cdots; \quad P_0^{(\lambda)}(x)=1, \quad P_1^{(\lambda)}(x)=2\lambda x; \tag{4.7.17}$$

$$\sum_{\nu=0}^{n}(\nu+\lambda)P_\nu^{(\lambda)}(x) = \frac{1}{2}\,\frac{(n+2\lambda)P_n^{(\lambda)}(x)-(n+1)P_{n+1}^{(\lambda)}(x)}{1-x}. \tag{4.7.18}$$

(3) We obtain as a second solution of (4.7.5), which is linearly independent of $P_n^{(\lambda)}(x)$, (cf. (4.23.1), (4.23.3), (4.61.1), (4.61.4), (4.61.5))

$$y = (1-x)^{\frac{1}{2}-\lambda}F\left(-n-\lambda+\tfrac{1}{2}, n+\lambda+\tfrac{1}{2};\tfrac{3}{2}-\lambda;\frac{1-x}{2}\right), \tag{4.7.19}$$

$$\lambda \neq -n+1/2,\ -n+3/2,\ -n+5/2,\ \cdots;$$ [24]

[23] Use (1.7.3), $n = 2$. In the limiting case $\lambda \to 0$, $n \geq 1$, we multiply (4.7.15) by λ^{-2}, $\lambda \to 0$ [(4.7.8)].

[24] For $\lambda = -n-1/2, -n-3/2, \cdots$ this is a polynomial linearly independent of $P_n^{(\lambda)}(x)$ (cf. §4.23 (1)).

$$(4.7.20)\quad \begin{aligned} y &= (1-x)^{-n-2\lambda} F\left(n+2\lambda, n+\lambda+\tfrac{1}{2}; 2n+2\lambda+1; \frac{2}{1-x}\right), \\ &\qquad \lambda \neq -n/2, -n/2-1/2, -n/2-1, \cdots ;^{25} \end{aligned}$$

$$(4.7.21)\quad \begin{aligned} y &= (1-x^2)^{\frac{1}{2}-\lambda} \int_{-1}^{+1} (1-t^2)^{n+\lambda-\frac{1}{2}} (x-t)^{-n-1}\, dt \\ &= \text{const.}\ (1-x^2)^{\frac{1}{2}-\lambda} \int_{-1}^{+1} (1-t^2)^{\lambda-\frac{1}{2}} \frac{P_n^{(\lambda)}(t)}{x-t}\, dt, \quad \lambda > -1/2, \lambda \neq 0; \end{aligned}$$

$$(4.7.22)\quad \begin{aligned} y &= (1-x)^{-n-\lambda-\frac{1}{2}} (1+x)^{\frac{1}{2}-\lambda} F\left(n+\lambda+\tfrac{1}{2}, n+1; 2n+2\lambda+1; \frac{2}{1-x}\right), \\ &\qquad \lambda \neq -n/2, -n/2-1/2, -n/2-1, \cdots .^{26} \end{aligned}$$

According to Theorem 4.23.2 the general solution of (4.7.5) is a polynomial if and only if $\lambda - \frac{1}{2}$ is an integer and $\lambda \leqq -n/2$.

(4) Another generating function, essentially different from (4.7.16) (cf. Problem 16), and much simpler, is often used as a definition of the ultraspherical polynomials, namely:

$$(4.7.23)\quad \begin{aligned} P_0^{(\lambda)}(x) + P_1^{(\lambda)}(x)w + P_2^{(\lambda)}(x)w^2 + \cdots + P_n^{(\lambda)}(x)w^n + \cdots \\ = (1-2xw+w^2)^{-\lambda}. \end{aligned}$$

For the proof, we consider the recurrence formula (4.7.17), from which

$$\sum_{n=1}^{\infty} nP_n^{(\lambda)}(x)w^{n-1} = 2x\sum_{n=1}^{\infty}(n+\lambda-1)P_{n-1}^{(\lambda)}(x)w^{n-1} - \sum_{n=1}^{\infty}(n+2\lambda-2)P_{n-2}^{(\lambda)}(x)w^{n-1}$$

and in which we define $P_{-1}^{(\lambda)}(x) = 0$. If the left-hand member of (4.7.23) be denoted by $h(w)$, the last equation may be written in the form

$$\begin{aligned} h'(w) &= 2xw^{1-\lambda}(w^{\lambda}h(w))' - w^{2-2\lambda}(w^{2\lambda}h(w))' \\ &= 2x\{\lambda h(w) + wh'(w)\} - \{2\lambda w h(w) + w^2 h'(w)\}; \end{aligned}$$

or $h'(w)/h(w) = 2\lambda(x-w)(1-2xw+w^2)^{-1}$. Since $h(0) = P_0^{(\lambda)}(x) = 1$, (4.7.23) follows by integration.

On differentiating (4.7.23) with respect to λ, we obtain

$$(4.7.24)\qquad -(1-2xw+w^2)^{-\lambda}\log(1-2xw+w^2)$$

which, for $\lambda = -m$, $m = 0, 1, 2, \cdots$, gives the generating function of the polynomials (4.7.7), provided only those terms are considered for which $n > 2m$. These polynomials are, as mentioned in (1), constant multiples of the Jacobi polynomials $P_n^{(\alpha,\alpha)}(x)$ with $\alpha = -m - \frac{1}{2}$. The polynomials of degree $n \leqq 2m$, defined by this new generating function (4.7.24), are essentially different from

[25] Cf. §4.23 (2).

[26] Formula (4.61.5) has been considered only under the restriction $\alpha > -1$, $\beta > -1$. We see immediately, however, that except for the indicated values of λ, the function (4.7.22) is a solution which is $\sim x^{-n-2\lambda}$ as $x \to \infty$, so that it cannot be a polynomial.

the $P_n^{(\lambda)}(x)$ which are given in this case by the corresponding terms of (4.7.23). We have, for instance, [(4.1.7)],

$$(4.7.25)\quad \lambda = 0 \begin{cases} P_0^{(0)}(x) = 1, \quad P_n^{(0)}(x) = 0, \quad n \geqq 1, \\ -\log(1 - 2xw + w^2) = \sum_{n=1}^{\infty} (2/n)T_n(x)w^n \\ \qquad = \sum_{n=1}^{\infty} \frac{2}{n} \frac{2\cdot 4 \cdots 2n}{1\cdot 3 \cdots (2n-1)} P_n^{(-\frac{1}{2},-\frac{1}{2})}(x)w^n; \end{cases}$$

[27]

$$(4.7.26)\quad \lambda = -1 \begin{cases} P_0^{(-1)}(x) = 1, \quad P_1^{(-1)}(x) = -2x, \quad P_2^{(-1)}(x) = 1, \\ \qquad P_n^{(-1)}(x) = 0, \quad n \geqq 3, \\ -(1 - 2xw + w^2)\log(1 - 2xw + w^2) = 2xw - (2x^2 + 1)w^2 \\ \qquad - 32 \cdot \sum_{n=3}^{\infty} \frac{2\cdot 4 \cdots (2n-6)}{1\cdot 3 \cdots (2n-3)} P_n^{(-\frac{1}{2},-\frac{1}{2})}(x)w^n. \end{cases}$$

(5) From (4.5.5), (4.5.6), (4.5.7), (4.7.14), we obtain the relations

$$(4.7.27)\quad \begin{aligned} &(1 - x^2)\frac{d}{dx}\{P_n^{(\lambda)}(x)\} \\ &= [2(n+\lambda)]^{-1}\{(n + 2\lambda - 1)(n + 2\lambda)P_{n-1}^{(\lambda)}(x) - n(n+1)P_{n+1}^{(\lambda)}(x)\} \\ &= -nxP_n^{(\lambda)}(x) + (n + 2\lambda - 1)P_{n-1}^{(\lambda)}(x) \\ &= (n + 2\lambda)xP_n^{(\lambda)}(x) - (n+1)P_{n+1}^{(\lambda)}(x) \\ &= 2\lambda(1 - x^2)P_{n-1}^{(\lambda+1)}(x), \qquad n \geqq 0,\ P_{-1}^{(\lambda)}(x) = 0. \end{aligned}$$

We can then derive the following identities:

$$(4.7.28)\quad \begin{aligned} nP_n^{(\lambda)}(x) &= x\frac{d}{dx}\{P_n^{(\lambda)}(x)\} - \frac{d}{dx}\{P_{n-1}^{(\lambda)}(x)\}; \\ (n + 2\lambda)P_n^{(\lambda)}(x) &= \frac{d}{dx}\{P_{n+1}^{(\lambda)}(x)\} - x\frac{d}{dx}\{P_n^{(\lambda)}(x)\}. \end{aligned}$$

Adding these formulas we find, by use of (4.7.14),

$$(4.7.29)\quad \begin{aligned} \frac{d}{dx}\{P_{n+1}^{(\lambda)}(x) - P_{n-1}^{(\lambda)}(x)\} &= 2(n+\lambda)P_n^{(\lambda)}(x) \\ &= 2\lambda\{P_n^{(\lambda+1)}(x) - P_{n-2}^{(\lambda+1)}(x)\}, \\ &\qquad n \geqq 1,\ P_{-1}^{(\lambda)}(x) = 0. \end{aligned}$$

(6) Finally, we give some special formulas involving hypergeometric functions. Combining (4.1.5) and (4.21.2), we obtain

$$(4.7.30)\quad \begin{aligned} P_{2\nu}^{(\lambda)}(x) &= \binom{2\nu + 2\lambda - 1}{2\nu} F(-\nu, \nu + \lambda; \lambda + \tfrac{1}{2}; 1 - x^2) \\ &= (-1)^\nu \binom{\nu + \lambda - 1}{\nu} F(-\nu, \nu + \lambda; \tfrac{1}{2}; x^2), \end{aligned}$$

[27] The first of these identities can be derived directly by writing $1 - 2w\cos\theta + w^2 = (1 - we^{i\theta})(1 - we^{-i\theta})$.

$$(4.7.30)\qquad \begin{aligned} P_{2\nu+1}^{(\lambda)}(x) &= \binom{2\nu+2\lambda}{2\nu+1} xF(-\nu, \nu+\lambda+1; \lambda+\tfrac{1}{2}; 1-x^2) \\ &= (-1)^\nu 2\lambda \binom{\nu+\lambda}{\nu} xF(-\nu, \nu+\lambda+1; \tfrac{3}{2}; x^2). \end{aligned}$$

The constant factors can be calculated by substituting $x = 1$ and comparing the highest powers [or by computing $P_{2\nu}^{(\lambda)}(0)$ and $P_{2\nu+1}^{(\lambda)\prime}(0)$ from (4.7.23)].

Another way of writing the second part of (4.7.30) is the following:

$$(4.7.31)\quad P_n^{(\lambda)}(x) = \sum_{m=0}^{[n/2]} (-1)^m \frac{\Gamma(n-m+\lambda)}{\Gamma(\lambda)\Gamma(m+1)\Gamma(n-2m+1)} (2x)^{n-2m}.$$

This last expression is an explicit representation of the ultraspherical polynomials. (Cf. Problem 15.)

The formulas (4.1.5) may be proved by a more general consideration. For instance, in the first case, we may start from the corresponding differential equations

$$(4.7.32)\qquad \begin{aligned} &(1-x^2)y'' - 2(\alpha+1)xy' + 2\nu(2\nu+2\alpha+1)y = 0, \\ &(1-x^2)z'' - \{\alpha+\tfrac{1}{2}+(\alpha+\tfrac{3}{2})x\}z' + \nu(\nu+\alpha+\tfrac{1}{2})z = 0 \end{aligned}$$

and show the relation $y(x) = z(2x^2-1)$ between their general solutions. This argument furnishes at the same time relations between the non-polynomial solutions of (4.7.32) similar to (4.1.5). To be specific, let us replace the quantities n, α, β, x in the expression (4.23.3) first by $\nu, -\frac{1}{2}, \alpha, 1-2x^2$, respectively, and then by $\nu, +\frac{1}{2}, \alpha, 1-2x^2$, respectively; we then find (in the second case after multiplying by x)

$$(4.7.33)\qquad \begin{aligned} y &= x^{-n-2\alpha-1}F([(n+1)/2]+\alpha+1/2, [n/2]+\alpha+1; n+\alpha+\tfrac{3}{2}; x^{-2}) \\ &= x^{-n-2\lambda}F((n+1)/2+\lambda, n/2+\lambda; n+\lambda+1; x^{-2}), \end{aligned}$$

as a second solution of (4.7.5), which is linearly independent of $P_n^{(\lambda)}(x)$, provided $\lambda \neq -[(n+1)/2]-k; k = 0, 1, 2, \cdots$.

Starting from the first formula in (4.7.21), we find, save for a constant factor:

$$(4.7.34)\qquad \begin{aligned} y &= (1-x^2)^{\frac{1}{2}-\lambda}x^{-n-1}\sum_{\nu=0}^{\infty}\binom{n+2\nu}{2\nu}x^{-2\nu}\int_{-1}^{+1}(1-t^2)^{n+\lambda-\frac{1}{2}}t^{2\nu}\,dt \\ &= (1-x^2)^{\frac{1}{2}-\lambda}x^{-n-1}\sum_{\nu=0}^{\infty}\binom{n+2\nu}{2\nu}\frac{\Gamma(n+\lambda+\frac{1}{2})\Gamma(\nu+\frac{1}{2})}{\Gamma(n+\lambda+\nu+1)}x^{-2\nu} \\ &= \frac{\Gamma(n+\lambda+\frac{1}{2})\Gamma(\frac{1}{2})}{\Gamma(n+\lambda+1)}(1-x^2)^{\frac{1}{2}-\lambda}x^{-n-1} \\ &\qquad\qquad \cdot F((n+1)/2, 1+n/2; n+\lambda+1; x^{-2}). \end{aligned}$$

For further properties of ultraspherical polynomials the reader may consult Whittaker-Watson **1**, pp. 329, 330, 335, and the literature quoted there. See also Wangerin **1**, pp. 730–731. The function $C_n^\nu(x)$ of these authors is identical with $P_n^{(\nu)}(x)$ in our notation. The Legendre associated functions $P_n^m(x)$, m an integer (cf. Hobson **1**, p. 90), can be represented in the form (cf. (4.21.7))

$$(4.7.35)\qquad P_n^m(x) = \text{const.}\ (1-x^2)^{m/2}P_{n-m}^{(m+\frac{1}{2})}(x).$$

4.8. Integral representations for Legendre polynomials

In the important case $\alpha = \beta = 0$, that is, for Legendre polynomials, the method of §4.6 leads to various integral representations. According to (4.6.1),

$$P_n(x) = \frac{1}{2\pi i} \int \left(\frac{1}{2}\frac{t^2 - 1}{t - x}\right)^n \frac{dt}{t - x}. \tag{4.8.1}$$

Here the path of integration encloses the point x. At present, the position of this contour with reference to the points $t = \pm 1$ is immaterial.

(1) *Integrals of Dirichlet-Mehler.* (Cf. Dirichlet **1**, Mehler **5**.) Take x in the interior of the interval $[-1, +1]$, so that $x = \cos\theta$, $0 < \theta < \pi$. For the contour in question we choose the circle

$$|t - 1| = |e^{i\theta} - 1| = 2 \sin\frac{\theta}{2}, \qquad t = 1 + 2\sin\frac{\theta}{2}\cdot e^{i\psi}, \tag{4.8.2}$$

so that

$$\frac{1}{2}\frac{t^2 - 1}{t - \cos\theta} = \frac{1 + \sin\frac{\theta}{2}\cdot e^{i\psi}}{1 + \sin\frac{\theta}{2}\cdot e^{-i\psi}}. \tag{4.8.3}$$

Next let ψ vary from $-\pi$ to $+\pi$; then $1 + \sin\frac{1}{2}\theta\cdot e^{i\psi}$ describes the small circle in the figure. The expression in the right-hand member of (4.8.3) has the

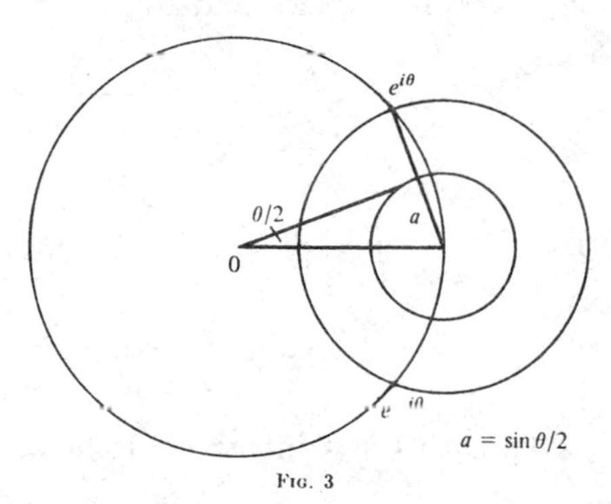

FIG. 3

absolute value 1 and an argument twice that of the numerator. Thus, if we write

$$\frac{1}{2}\frac{t^2 - 1}{t - \cos\theta} = e^{i\phi}, \tag{4.8.4}$$

the quantity ϕ varies from $-\theta$ to $+\theta$ and again from $+\theta$ to $-\theta$. Solving (4.8.4), we obtain

$$t = e^{i\phi} + e^{i\phi/2}(2\cos\phi - 2\cos\theta)^{\frac{1}{2}}. \tag{4.8.5}$$

Here the positive value of $(\quad)^{\frac{1}{2}}$ corresponds to $|\psi| < (\pi + \theta)/2$, that is, to the "exterior" arc, while the negative value corresponds to $|\psi| > (\pi + \theta)/2$, that is, to the "interior" arc. Furthermore, from (4.8.4) it follows that

$$t\,dt = e^{i\phi}\,dt + (t - \cos\theta)ie^{i\phi}\,d\phi;$$

whence

$$\frac{dt}{t - \cos\theta} = \frac{ie^{i\phi}\,d\phi}{t - e^{i\phi}} = \frac{ie^{i\phi/2}\,d\phi}{(2\cos\phi - 2\cos\theta)^{\frac{1}{2}}}.$$

Finally,

$$P_n(\cos\theta) = \frac{1}{2\pi}\int_{-\theta}^{+\theta} e^{in\phi}\frac{e^{i\phi/2}\,d\phi}{(2\cos\phi - 2\cos\theta)^{\frac{1}{2}}} + \frac{1}{2\pi}\int_{+\theta}^{-\theta} e^{in\phi}\frac{e^{i\phi/2}\,d\phi}{-(2\cos\phi - 2\cos\theta)^{\frac{1}{2}}};$$

or,

$$P_n\,(\cos\theta) = \frac{2}{\pi}\int_0^{\theta}\frac{\cos\,(n + \frac{1}{2})\phi}{(2\cos\phi - 2\cos\theta)^{\frac{1}{2}}}\,d\phi, \tag{4.8.6}$$

where the square root of $2\cos\phi - 2\cos\theta$ must be taken with positive sign. This is the first formula of Dirichlet-Mehler. Substituting $\pi - \theta$ for θ we obtain, because of (4.1.3), the second formula

$$P_n\,(\cos\theta) = \frac{2}{\pi}\int_{\theta}^{\pi}\frac{\sin\,(n + \frac{1}{2})\phi}{(2\cos\theta - 2\cos\phi)^{\frac{1}{2}}}\,d\phi. \tag{4.8.7}$$

(2) *First integral of Laplace.* (Cf. Whittaker-Watson **1**, pp. 312–313.) Let x be different from ± 1, and choose the circle

$$|\,t - x\,| = |\,x^2 - 1\,|^{\frac{1}{2}} \tag{4.8.8}$$

as the contour of integration. Writing $t = x + (x^2 - 1)^{\frac{1}{2}}e^{i\phi}$ (with an arbitrary but fixed determination of $(x^2 - 1)^{\frac{1}{2}}$), we find that

$$\frac{1}{2}\frac{t^2 - 1}{t - x} = x + (x^2 - 1)^{\frac{1}{2}}\cos\phi, \qquad \frac{dt}{t - x} = i\,d\phi. \tag{4.8.9}$$

Consequently,

$$\begin{aligned} P_n(x) &= \frac{1}{2\pi}\int_{-\pi}^{+\pi}\{x + (x^2 - 1)^{\frac{1}{2}}\cos\phi\}^n\,d\phi \\ &= \pi^{-1}\int_0^{\pi}\{x + (x^2 - 1)^{\frac{1}{2}}\cos\phi\}^n\,d\phi, \end{aligned} \tag{4.8.10}$$

an expression known as Laplace's first integral. It holds for arbitrary values of x.

(3) *Second integral of Laplace.* (Cf. Whittaker-Watson **1**, p. 314; Jacobi **2**, p. 153.) This integral is given by

$$\begin{aligned} P_n(x) &= \frac{1}{2\pi}\int_{-\pi}^{+\pi}\{x + (x^2 - 1)^{\frac{1}{2}}\cos\phi\}^{-n-1}\,d\phi \\ &= \pi^{-1}\int_0^{\pi}\{x + (x^2 - 1)^{\frac{1}{2}}\cos\phi\}^{-n-1}\,d\phi. \end{aligned} \tag{4.8.11}$$

It may be derived from the first integral of Laplace in the following manner. Let $0 < r < 1$ and

$$\begin{aligned} x &= \frac{1 + r^2}{1 - r^2}, \qquad (x^2 - 1)^{\frac{1}{2}} = \frac{2r}{1 - r^2}, \\ x + (x^2 - 1)^{\frac{1}{2}}\cos\phi &= (1 - r^2)^{-1}\,|\,1 + rz\,|^2, \qquad z = e^{i\phi}, \end{aligned} \tag{4.8.12}$$

whence,

$$P_n(x) = \frac{(1 - r^2)^{-n}}{2\pi} \int_{|z|=1} |1 + rz|^{2n} |dz|. \tag{4.8.13}$$

Now, substituting

$$z = \frac{w - r}{1 - rw}, \qquad 1 + rz = \frac{1 - r^2}{1 - rw}, \qquad \frac{dz}{dw} = \frac{1 - r^2}{(1 - rw)^2}, \tag{4.8.14}$$

we obtain[28]

$$P_n(x) = \frac{(1 - r^2)^{-n}}{2\pi} \int_{|w|=1} \frac{(1 - r^2)^{2n}}{|1 - rw|^{2n}} \frac{1 - r^2}{|1 - rw|^2} |dw|. \tag{4.8.15}$$

Substituting $w = -e^{i\phi'}$, we obtain (4.8.11). Using analytic continuation, this formula may be immediately extended to arbitrary complex values of x.

(4) As a special case of (4.4.9) we note the representation

$$P_n(\cos\theta) = \frac{1}{2\pi i} \int w^{-n-1}(1 - 2w\cos\theta + w^2)^{-\frac{1}{2}} dw. \tag{4.8.16}$$

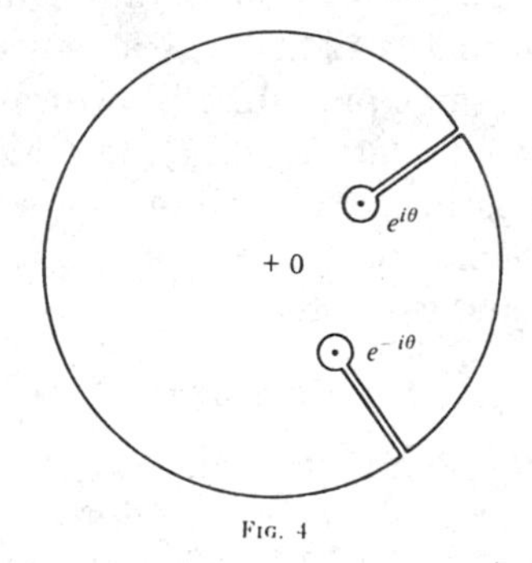

Fig. 4

The integration is extended in the positive sense along a contour enclosing the origin but neither of the points $e^{\pm i\theta}$. By a proper choice of the contour, formulas (4.8.6) and (4.8.10) can be derived again from (4.8.16) (see Pólya-Szegö **1**, vol. 1, pp. 115, 287–288, problem 157).

(5) *Integral representation of Stieltjes.* (Stieltjes **8**.) Suppose $0 < \theta < \pi$. This important representation can be obtained from (4.8.16) by using the contour in the figure. (The derivation from (4.8.1) seems too complicated.) Since the integrand is $O(w^{-n-2})$ as $w \to \infty$, the contribution of the large arcs tends to zero as the radius becomes infinite. The same is true of the contributions of the small arcs around $e^{\pm i\theta}$ (the integrand is $O\{|w - e^{\pm i\theta}|^{-\frac{1}{2}}\}$ there). Thus, we have

$$P_n(\cos\theta) = 2\Re \frac{1}{2\pi i} \int w^{-n-1}(1 - 2w\cos\theta + w^2)^{-\frac{1}{2}} dw,$$

extended twice over the straight line $w = t^{-1}e^{i\theta}$, where t increases from 0 to 1 and then decreases from 1 to 0. We have, then,

$$(1 - 2w\cos\theta + w^2)^{-\frac{1}{2}} = \pm\, te^{-i\theta}(1 - t)^{-\frac{1}{2}}(1 - te^{-2i\theta})^{-\frac{1}{2}},$$

[28] Concerning this argument see Szegö **21**.

where the signs + and − correspond to the first and second cases, respectively $[(1-t)^{-\frac{1}{2}} = (1-te^{-2i\theta})^{-\frac{1}{2}} = 1$ for $t = 0]$. Therefore,

$$
(4.8.17)\quad
\begin{aligned}
P_n(\cos\theta) &= 4\Re \frac{1}{2\pi i}\int_0^1 t^{n+1}e^{-i(n+1)\theta}te^{-i\theta}(1-t)^{-\frac{1}{2}}(1-te^{-2i\theta})^{-\frac{1}{2}}(-t^{-2}e^{i\theta})\,dt\\
&= \frac{2}{\pi}\Im\left\{e^{i(n+1)\theta}\int_0^1 t^n(1-t)^{-\frac{1}{2}}(1-te^{2i\theta})^{-\frac{1}{2}}\,dt\right\}.
\end{aligned}
$$

(6) Further integral representations are obtained by replacing n in (4.8.1) by $-n-1$ (cf. (4.6.4)) and observing the conditions (a), (b) formulated in §4.6 (here the expression $(1-t^2)^{-n}(t-x)^{n-1}$ must be considered). We then have

$$
(4.8.18)\qquad y = \frac{1}{2\pi i}\int\left(\frac{1}{2}\frac{t^2-1}{t-x}\right)^{-n-1}\frac{dt}{t-x}.
$$

Integrals of this type may represent Legendre polynomials as well as Legendre functions of the second kind, or a proper linear combination of these, according to the special choice of the path of integration.

By using the contour in (1), we again obtain Dirichlet-Mehler's formula (4.8.6), since the expression in the right-hand member does not change if we replace n by $-n-1$. The same procedure transforms Laplace's first integral (4.8.10) into Laplace's second integral (4.8.11), and conversely. Thus, choosing the contour in (4.8.18) as in (2), we obtain the second integral. The expression in the right-hand member of (4.8.11) cannot represent a solution other than $P_n(x)$, since it is finite and equal to $+1$ at $x = +1$.

4.81. Legendre functions of the second kind

(1) Let x be in the complex plane cut along the segment $[-1, +1]$, and $x = \frac{1}{2}(z+z^{-1})$, $|z| < 1$. We deform the path of integration in (4.61.1) into the circular arc through $-1, z, +1$ (cf. Whittaker-Watson **1**, p. 320, example 1). This deformation is permitted since $\operatorname{sgn}\Im x = -\operatorname{sgn}\Im z$. Then

$$
(4.81.1)\quad
\begin{aligned}
t &= \frac{(z+1)e^\tau + (z-1)}{(z+1)e^\tau - (z-1)},\\
\frac{1}{2}\frac{t^2-1}{t-x} &= \frac{dt}{d\tau}\frac{1}{x-t} = \{x + (x^2-1)^{\frac{1}{2}}\cosh\tau\}^{-1}.
\end{aligned}
$$

The new variable τ is real and ranges from $-\infty$ to $+\infty$; furthermore $(x^2-1)^{\frac{1}{2}} \cong x$ as $x \to \infty$. This furnishes the following integral representation for $Q_n^{(0,0)}(x) = Q_n(x)$, which is very similar to Laplace's second integral:

$$
(4.81.2)\qquad Q_n(x) = \frac{1}{2}\int_{-\infty}^{+\infty}\{x + (x^2-1)^{\frac{1}{2}}\cosh\tau\}^{-n-1}\,d\tau.
$$

(2) Let x be real, $x > 1$; we write $x = \cosh\zeta$, $\zeta > 0$. Introducing in (4.81.2)

$$
(4.81.3)\quad
\begin{aligned}
&\cosh\zeta + \sinh\zeta\cosh\tau = e^\theta,\\
&\frac{d\tau}{d\theta} = \frac{e^\theta}{\sinh\zeta\sinh\tau} = \frac{e^{\theta/2}}{(2\cosh\theta - 2\cosh\zeta)^{\frac{1}{2}}},
\end{aligned}
$$

we obtain

$$(4.81.4) \qquad Q_n(\cosh \zeta) = \int_\zeta^\infty \frac{e^{-(n+\frac{1}{2})\theta}}{(2\cosh\theta - 2\cosh\zeta)^{\frac{1}{2}}}\, d\theta.$$

In this formula (Watson **2**, p. 154) we first assume that $\zeta > 0$. By analytic continuation it can be extended to the half-strip (1.9.3). The path of integration is the horizontal line $\Re(\theta) \geqq \Re(\zeta)$, $\Im(\theta) = \Im(\zeta)$. This formula corresponds to Dirichlet-Mehler's formula.

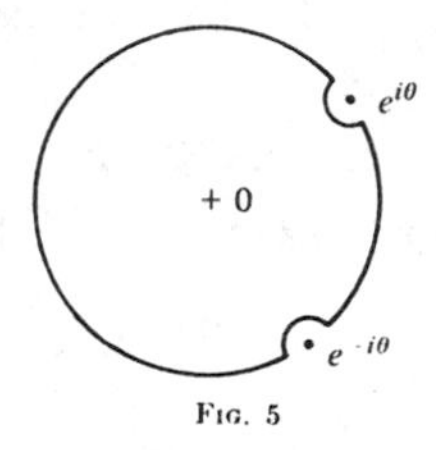

Fig. 5

4.82. Generalizations

(1) *Generalization of the Dirichlet-Mehler integral.* (See Fejér **12**.) Suppose $0 < \theta < \pi$ and $0 < \lambda < 1$. According to the generating function (4.7.23) we have

$$(4.82.1) \qquad P_n^{(\lambda)}(\cos\theta) = \frac{1}{2\pi i}\int w^{-n-1}(1 - 2w\cos\theta + w^2)^{-\lambda}\, dw.$$

Here we take the unit circle $|w| = 1$ as the path of integration, avoiding the singular points $w = e^{\pm i\theta}$, in the neighborhood of which the integrand is $O\{|w - e^{\pm i\theta}|^{-\lambda}\}$. (See Fig. 5.) For $w = e^{i\phi}$, $0 \leqq \phi < \theta$, we have

$$(1 - 2w\cos\theta + w^2)^{-\lambda} = e^{-i\lambda\phi}(w^{-1} - 2\cos\theta + w)^{-\lambda} = e^{-i\lambda\phi}(2\cos\phi - 2\cos\theta)^{-\lambda},$$

and for $w = e^{i\phi}$, $\theta < \phi \leqq \pi$,

$$(1 - 2w\cos\theta + w^2)^{-\lambda} = e^{-i\lambda(\phi-\pi)}(2\cos\theta - 2\cos\phi)^{-\lambda},$$

so that

$$P_n^{(\lambda)}(\cos\theta) = 2\Re\left\{\frac{1}{2\pi i}\int_0^\theta e^{-i(n+1)\phi}e^{-i\lambda\phi}(2\cos\phi - 2\cos\theta)^{-\lambda} ie^{i\phi}\, d\phi \right.$$
$$\left. + \frac{1}{2\pi i}\int_\theta^\pi e^{-i(n+1)\phi}e^{-i\lambda(\phi-\pi)}(2\cos\theta - 2\cos\phi)^{-\lambda} ie^{i\phi}\, d\phi\right\};$$

whence

$$P_n^{(\lambda)}(\cos\theta) = \pi^{-1}\int_0^\pi F(\phi)\,|2\cos\phi - 2\cos\theta|^{-\lambda}\, d\phi,$$

$$(4.82.2) \qquad F(\phi) = \begin{cases} \cos(n+\lambda)\phi & \text{if } 0 \leqq \phi < \theta, \\ \cos[(n+\lambda)\phi - \lambda\pi] & \text{if } \theta < \phi \leqq \pi. \end{cases}$$

The expression resulting from (4.82.2) for $\lambda = 1/2$, is the sum of the expressions (4.8.6) and (4.8.7).

(2) *Generalization of the integral of Stieltjes.* (Stieltjes; Hermite-Stieltjes **1**, vol. 2, p. 122, no. 284.) If we choose in (4.82.1) the same contour as in §4.8 (5), an argument similar to that used there leads to the representation

$$(4.82.3)\quad \begin{aligned} P_n^{(\lambda)}(\cos\theta) &= (2/\pi)\sin\lambda\pi\,\Im\left\{e^{il(\theta)}\int_0^1 t^{n+2\lambda-1}(1-t)^{-\lambda}(1-te^{2i\theta})^{-\lambda}\,dt\right\},\\ l(\theta) &= (n+2\lambda)\theta + (\tfrac{1}{2}-\lambda)\pi; \qquad 0<\theta<\pi,\ 0<\lambda<1. \end{aligned}$$

The formulas in §4.8 (1) and (2) can be extended to Jacobi polynomials $P_n^{(\alpha,\beta)}(x)$ for integral values of α. The resulting representations, however, are rather complicated. Finally, we notice the following generalization of (4.81.2) arising from (4.61.1) in a manner similar to that used in §4.81 (1) (notation the same as there):

$$(4.82.4)\quad \begin{aligned} Q_n^{(\alpha,\beta)}(x) = \tfrac{1}{2}\left(\frac{4z}{1-z}\right)^{\alpha}\left(\frac{4z}{1+z}\right)^{\beta}\int_{-\infty}^{+\infty} e^{\beta\tau}\{(1+z)e^{\tau}+1-z\}^{-\alpha-\beta}\\ \cdot\{x+(x^2-1)^{\frac{1}{2}}\cosh\tau\}^{-n-1}\,d\tau,\\ x = \tfrac{1}{2}(z+z^{-1}),\ |z|<1;\ \alpha>-1,\ \beta>-1, \end{aligned}$$

with a proper determination of the functions involved.

4.9. Trigonometric representations

(1) *Finite cosine expansion of Legendre polynomials.* From the generating function (4.7.23) we obtain, for $\lambda = \frac{1}{2}$,

$$(4.9.1)\quad \begin{aligned} (1-2w\cos\theta+w^2)^{-\frac{1}{2}} &= (1-we^{i\theta})^{-\frac{1}{2}}(1-we^{-i\theta})^{-\frac{1}{2}}\\ &= \sum_{m=0}^{\infty} g_m w^m e^{im\theta}\sum_{m=0}^{\infty} g_m w^m e^{-im\theta}, \end{aligned}$$

where

$$(4.9.2)\quad g_0 = 1,\qquad g_m = \frac{1\cdot 3\cdots(2m-1)}{2\cdot 4\cdots 2m},\qquad m = 1, 2, 3, \cdots,$$

so that

$$(4.9.3)\quad \begin{aligned} P_n(\cos\theta) &= \sum_{m=0}^{n} g_m e^{im\theta} g_{n-m} e^{-i(n-m)\theta}\\ &= \sum_{m=0}^{n} g_m g_{n-m} e^{i(2m-n)\theta} = \sum_{m=0}^{n} g_m g_{n-m}\cos(n-2m)\theta\\ &= 2g_0g_n\cos n\theta + 2g_1g_{n-1}\cos(n-2)\theta + \cdots\\ &\quad + \begin{cases} 2g_{(n-1)/2}g_{(n+1)/2}\cos\theta, & n \text{ odd},\\ g_{n/2}^2, & n \text{ even}. \end{cases} \end{aligned}$$

Consequently, $P_n(\cos\theta)$ is a trigonometric cosine polynomial with non-negative coefficients.

Relation (4.9.3), as an identity in θ, can be written as follows in terms of Tchebichef polynomials:

$$(4.9.4)\quad \begin{aligned} P_n(x) &= 2g_0g_nT_n(x) + 2g_1g_{n-1}T_{n-2}(x) + \cdots\\ &\quad + \begin{cases} 2g_{(n-1)/2}g_{(n+1)/2}T_1(x), & n \text{ odd},\\ g_{n/2}^2, & n \text{ even}. \end{cases} \end{aligned}$$

(2) *Infinite sine expansion of Legendre polynomials.* (Heine **3**, vol. 1, pp. 19, 89.) We have

$$P_n(\cos\theta) = \frac{4}{\pi}\frac{2\cdot 4\cdots 2n}{3\cdot 5\cdots(2n+1)}\{f_0\sin(n+1)\theta + f_1\sin(n+3)\theta + \cdots$$
$$(4.9.5)\qquad + f_\nu\sin(n+2\nu+1)\theta + \cdots\},$$
$$f_0 = 1;\quad f_\nu = f_{\nu n} = g_\nu\frac{(n+1)(n+2)\cdots(n+\nu)}{(n+\frac{3}{2})(n+\frac{5}{2})\cdots(n+\nu+\frac{1}{2})},\quad \nu = 1,2,3,\cdots.$$

For $n = 0$ the factor $(2\cdot 4\cdots 2n)/(3\cdot 5\cdots(2n+1))$ must be replaced by 1. Abel's transformation (§1.11 (4)) shows this expansion to converge for the values $0 < \theta < \pi$, and even uniformly for $\epsilon \leqq \theta \leqq \pi - \epsilon$, $0 < \epsilon < \pi/2$. The convergence may be deduced also from the elements of the theory of Fourier series. Indeed, (4.9.5) is the formal expansion of the function defined by $P_n(\cos\theta)$ for $0 < \theta < \pi$, and by $-P_n(\cos\theta)$ for $-\pi < \theta < 0$ (see below). It is a generalization of the classical expansion ($n = 0$)

$$(4.9.6)\qquad \pi/4 = (\sin\theta)/1 + (\sin 3\theta)/3 + (\sin 5\theta)/5 + \cdots.$$

First proof (Heine, loc. cit.; cf. also Fejér **20**, pp. 24–26). Using the notation (1.12.3), we have

$$(4.9.7)\qquad \int_0^\pi P_n(\cos\theta)\sin(m+1)\theta\,d\theta = \int_{-1}^{+1} P_n(t)U_m(t)\,dt.$$

This integral vanishes if $m < n$, or if $m \geqq n$, and $m - n$ odd. Now write $m = n + 2\nu$; using (4.9.3), we find

$$\int_0^\pi P_n(\cos\theta)\sin(m+1)\theta\,d\theta = \sum_{k=0}^n g_k g_{n-k}\int_0^\pi \cos(n-2k)\theta\sin(m+1)\theta\,d\theta$$
$$(4.9.8)\qquad = \sum_{k=0}^n g_k g_{n-k}\left(\frac{1}{m+1+n-2k} + \frac{1}{m+1-n+2k}\right)$$
$$= 2\sum_{k=0}^n \frac{g_k g_{n-k}}{m+1-n+2k} = \sum_{k=0}^n \frac{g_k g_{n-k}}{\nu+k+\frac{1}{2}}.$$

Considering ν as a continuous variable momentarily, we easily verify (by calculating the residues) that

$$(4.9.9)\qquad \sum_{k=0}^n \frac{g_k g_{n-k}}{\nu+k+\frac{1}{2}} = \frac{(\nu+1)(\nu+2)\cdots(\nu+n)}{(\nu+\frac{1}{2})(\nu+\frac{3}{2})\cdots(\nu+n+\frac{1}{2})} = 2\frac{2\cdot 4\cdots 2n}{3\cdot 5\cdots(2n+1)}f_\nu,$$

which establishes the statement.

Second proof (see Fejér **19**, pp. 202–203). Expansion of the last factor in the integrand of (4.8.17) gives

$$(4.9.10)\qquad P_n(\cos\theta) = \frac{2}{\pi}\,\Im\left\{e^{i(n+1)\theta}\sum_{m=0}^\infty g_m e^{2im\theta}\int_0^1 t^{n+m}(1-t)^{-\frac{1}{2}}\,dt\right\},$$

where g_m has the same meaning as in (4.9.2). The last integral can be calculated by means of (1.7.5), and the statement is thus established.

Third proof. We use mathematical induction with respect to n. According to the recurrence formula of $P_n(x)$ ((4.7.17), $\lambda = \frac{1}{2}$), it suffices to show that

$$(4.9.11)\qquad \frac{n^2}{n^2-\frac{1}{4}}f_{\nu n} = f_{\nu,n-1} + f_{\nu+1,n-1} - f_{\nu+1,n-2},$$
$$\nu = 0, 1, 2, \cdots, n = 1, 2, 3, \cdots;\ f_{\nu+1,-1} = 0.$$

For $n = 0$ we have (4.9.6). The identity (4.9.11) can be verified by direct calculation.

Remark. Formula (4.9.5) is closely related to certain expansions of the functions of second kind. We find from (4.61.4), $\alpha = \beta = 0$,

$$Q_n^{(0,0)}(x) = Q_n(x) = \frac{1}{2}\int_{-1}^{+1} \frac{P_n(t)}{x - t}\, dt. \tag{4.9.12}$$

If we substitute $x = \frac{1}{2}(w + w^{-1})$ (§1.9), the function $Q_n\{\frac{1}{2}(w + w^{-1})\}$ will be regular for $|w| \leqq 1$, $w \neq \pm 1$. Using (4.7.23), $\lambda = 1$, we obtain for $|w| < 1$,

$$Q_n\{\tfrac{1}{2}(w + w^{-1})\} = w \int_{-1}^{+1} \frac{P_n(t)}{1 - 2tw + w^2}\, dt = w \int_{-1}^{+1} P_n(t)\left\{\sum_{m=0}^{\infty} U_m(t) w^m\right\} dt,$$

so that, because of (4.9.7), (4.9.8), (4.9.9),

$$Q_n\{\tfrac{1}{2}(w + w^{-1})\} = 2\,\frac{2 \cdot 4 \cdots 2n}{3 \cdot 5 \cdots (2n+1)} \sum_{\nu=0}^{\infty} f_\nu w^{n+2\nu+1}, \qquad |w| < 1. \tag{4.9.13}$$

Now let $|w| < 1$, $w \to e^{i\theta}$, $0 < \theta < \pi$. Then $\frac{1}{2}(w + w^{-1}) \to \cos\theta - i0$, so that

$$Q_n(\cos\theta - i0) = 2\,\frac{2 \cdot 4 \cdots 2n}{3 \cdot 5 \cdots (2n+1)} \sum_{\nu=0}^{\infty} f_\nu e^{i(n+2\nu+1)\theta}, \qquad 0 < \theta < \pi. \tag{4.9.14}$$

(Here use was made of the convergence of the last series and of Abel's continuity theorem; see Titchmarsh **1**, pp. 9–10.) Replacing i by $-i$, we obtain

$$Q_n(\cos\theta + i0) = 2\,\frac{2 \cdot 4 \cdots 2n}{3 \cdot 5 \cdots (2n+1)} \sum_{\nu=0}^{\infty} f_\nu e^{-i(n+2\nu+1)\theta}, \qquad 0 < \theta < \pi. \tag{4.9.15}$$

From here (4.9.5) follows again because of (4.62.8), $\alpha = \beta = 0$. By use of (4.62.9), $\alpha = \beta = 0$, we find (Heine **3**, vol. 1, p. 130)

$$\mathbf{Q}_n(\cos\theta) = 2\,\frac{2 \cdot 4 \cdots 2n}{3 \cdot 5 \cdots (2n+1)} \sum_{\nu=0}^{\infty} f_\nu \cos(n + 2\nu + 1)\theta, \qquad 0 < \theta < \pi. \tag{4.9.16}$$

Another variation of these considerations (see Hobson **1**, pp. 57–58) is to introduce $x = \frac{1}{2}(w + w^{-1})$ and $y = w^{n+1}z$ into (4.2.1). Then z, as a function of w^2, satisfies a hypergeometric differential equation from which (4.9.13) (therefore also (4.9.5)) again follows.

By means of (4.9.9) we see that *the sequence $\{f_\nu\}$ is completely monotonic* (cf. §6.5 (4)).

(3) *Another trigonometric representation of Legendre polynomials.* (Stieltjes **7, 8**.) Starting again from (4.8.17), we write

$$(1 - te^{2i\theta})^{-\frac{1}{2}} = e^{i(\pi/4 - \theta/2)}(2\sin\theta)^{-\frac{1}{2}}\left\{1 - (1 - t)\,\frac{e^{i(\theta - \pi/2)}}{2\sin\theta}\right\}^{-\frac{1}{2}}.$$

Consequently, for $2\sin\theta > 1$

$$P_n(\cos\theta) = \frac{2}{\pi}\,\Im\left\{e^{i(n+1)\theta} e^{i(\pi/4 - \theta/2)}(2\sin\theta)^{-\frac{1}{2}} \sum_{\nu=0}^{\infty} g_\nu \frac{e^{i\nu(\theta - \pi/2)}}{(2\sin\theta)^\nu} \int_0^1 t^n(1 - t)^{\nu - \frac{1}{2}}\, dt\right\}, \text{or}$$

$$P_n(\cos\theta) = \frac{4}{\pi}\,\frac{2 \cdot 4 \cdots 2n}{3 \cdot 5 \cdots (2n+1)} \sum_{\nu=0}^{\infty} h_\nu \frac{\cos\{(n + \nu + \frac{1}{2})\theta - (\nu + \frac{1}{2})\pi/2\}}{(2\sin\theta)^{\nu+\frac{1}{2}}}, \tag{4.9.17}$$

where

$$h_0 = 1; \quad h_\nu = h_{\nu n} = g_\nu \frac{\frac{1}{2}\,\frac{3}{2}\cdots(\nu - \frac{1}{2})}{(n+\frac{3}{2})(n+\frac{5}{2})\cdots(n+\nu+\frac{1}{2})}, \qquad \nu = 1, 2, 3, \cdots. \tag{4.9.18}$$

Expansion (4.9.17) is convergent if $\pi/6 < \theta < 5\pi/6$. Its importance will be more fully realized in discussing the asymptotic behavior of $P_n(\cos\theta)$ for large values of n (§8.5). Concerning the connection of this representation with that of Heine [(4.9.5)], see Stieltjes **8**, p. 244.

(4) *Ultraspherical polynomials.* From (4.7.21) we obtain as in (1):

$$P_n^{(\lambda)}(\cos\theta) = 2\alpha_0\alpha_n \cos n\theta + 2\alpha_1\alpha_{n-1}\cos(n-2)\theta + \cdots + \begin{cases} 2\alpha_{(n-1)/2}\alpha_{(n+1)/2}\cos\theta, & n \text{ odd}, \\ \alpha_{n/2}^2, & n \text{ even}, \end{cases} \tag{4.9.19}$$

where

$$(1-u)^{-\lambda} = \alpha_0 + \alpha_1 u + \alpha_2 u^2 + \cdots + \alpha_n u^n + \cdots. \tag{4.9.20}$$

Therefore,

$$\alpha_n = \binom{n+\lambda-1}{n}, \qquad n = 0, 1, 2, \cdots. \tag{4.9.21}$$

Here the cases $\lambda = 0, -1, -2, \cdots$ are to be excluded. In particular, if $\lambda > 0$, the coefficients α_n are positive, so that $P_n^{(\lambda)}(\cos\theta)$ as a trigonometric cosine polynomial again has *non-negative coefficients.*

Expansion (4.9.5) can likewise be extended. We have (Szegö **19**, pp. 508–509) for $\lambda > 0$, $\lambda \neq 1, 2, 3, \cdots$, $0 < \theta < \pi$,

$$\begin{gathered}(\sin\theta)^{2\lambda-1}P_n^{(\lambda)}(\cos\theta) = \frac{2^{2-2\lambda}}{\Gamma(\lambda)}\frac{\Gamma(n+2\lambda)}{\Gamma(n+\lambda+1)}\sum_{\nu=0}^{\infty} f_\nu^{(\lambda)}\sin(n+2\nu+1)\theta, \\ f_0^{(\lambda)} = 1; \quad f_\nu^{(\lambda)} = f_{\nu n}^{(\lambda)} \\ = \frac{(1-\lambda)(2-\lambda)\cdots(\nu-\lambda)}{1\cdot 2\cdots\nu}\,\frac{(n+1)(n+2)\cdots(n+\nu)}{(n+\lambda+1)(n+\lambda+2)\cdots(n+\lambda+\nu)}, \\ \nu = 1, 2, 3, \cdots.\end{gathered} \tag{4.9.22}$$

The generalization of the third proof given in (2) is particularly simple. The special case $n = 0$ is

$$(\sin\theta)^{2\lambda-1} = 2\pi^{-\frac{1}{2}}\Gamma(\lambda+\tfrac{1}{2})\sum_{\nu=0}^{\infty}\frac{(1-\lambda)(2-\lambda)\cdots(\nu-\lambda)}{\Gamma(\lambda+\nu+1)}\sin(2\nu+1)\theta. \tag{4.9.23}$$

It can be verified in various ways. (Cf. Whittaker-Watson **1**, p. 263, problem 40; multiply this formula by $\cos x$, and substitute $x = \pi/2 - \theta$, $s = 2\lambda - 2$.)

Another extension of (4.9.5) is the following:

$$\begin{gathered}P_n^{(\lambda)}(\cos\theta) = \frac{2}{\Gamma(\lambda)}\sum_{\nu=0}^{\infty}\alpha_\nu\frac{\Gamma(n+\nu+2\lambda)}{\Gamma(n+\nu+\lambda+1)}\cos\{(n+2\nu+2\lambda)\theta - \lambda\pi\}, \\ 0 < \lambda < 1, 0 < \theta < \pi.\end{gathered} \tag{4.9.24}$$

It follows readily from (4.82.3) by means of the argument used in the second proof given in (2).

The extension of (4.9.17) (Szegö **17**, pp. 57–60) is

$$P_n^{(\lambda)}(\cos\theta) = (2/\pi)\sin\lambda\pi\,\frac{\Gamma(n+2\lambda)}{\Gamma(\lambda)}\sum_{\nu=0}^{\infty}\frac{\Gamma(\nu+\lambda)\Gamma(\nu-\lambda+1)}{\nu!\,\Gamma(n+\nu+\lambda+1)} \tag{4.9.25}$$
$$\cdot\frac{\cos\{(n+\nu+\lambda)\theta-(\nu+\lambda)\pi/2\}}{(2\sin\theta)^{\nu+\lambda}},\quad 0<\lambda<1;\ \pi/6<\theta<5\pi/6.$$

It follows from (4.82.3) by the same argument as used in (3).

4.10. Further properties of Jacobi polynomials

(1) *Rodrigues' formula.* A generalization of Rodrigues' formula (4.3.1) is the following:

$$(1-x)^{\alpha}(1+x)^{\beta}P_n^{(\alpha,\beta)}(x) = \frac{(-1)^m}{2^m n(n-1)\cdots(n-m+1)}\left(\frac{d}{dx}\right)^m\{(1-x)^{m+\alpha}(1+x)^{m+\beta}P_{n-m}^{(m+\alpha,m+\beta)}(x)\}. \tag{4.10.1}$$

Here $n \geqq m$. The proof is similar to that given in 4.3 (1).

(2) *Integral representations for ultraspherical polynomials.* (a) Combining (4.1.5) and (4.3.2) we find

$$\begin{aligned} P_{2\nu}^{(\alpha,\alpha)}(x) &= P_{2\nu}^{(\alpha,\alpha)}(1)\cdot x^{2\nu}F(-\nu,-\nu+\tfrac{1}{2};\alpha+1;1-x^{-2}),\\ P_{2\nu+1}^{(\alpha,\alpha)}(x) &= P_{2\nu+1}^{(\alpha,\alpha)}(1)\cdot x^{2\nu+1}F(-\nu,-\nu-\tfrac{1}{2};\alpha+1;1-x^{-2}) \end{aligned} \tag{4.10.2}$$

so that, in view of (4.7.1) and (4.7.3),

$$\begin{aligned} P_n^{(\lambda)}(x) &= P_n^{(\lambda)}(1)\cdot x^nF(-n/2,-n/2+\tfrac{1}{2};\lambda+\tfrac{1}{2};1-x^{-2})\\ &= \frac{2^{1-2\lambda}}{\Gamma(\lambda)}\frac{\Gamma(n+2\lambda)}{n!}\sum_{k=0}^{[n/2]}(-1)^k\binom{n}{2k}\frac{\Gamma(k+\frac{1}{2})}{\Gamma(\lambda+k+\frac{1}{2})}x^{n-2k}(1-x^2)^k\\ &= \frac{2^{1-2\lambda}}{[\Gamma(\lambda)]^2}\frac{\Gamma(n+2\lambda)}{n!}\int_0^{\pi}\{x+(x^2-1)^{\frac{1}{2}}\cos\varphi\}^n\sin^{2\lambda-1}\varphi\,d\varphi. \end{aligned} \tag{4.10.3}$$

We used (1.7.3), $n = 2$, and (1.7.5). This generalization of (4.8.10) holds provided that $\lambda > 0$. See Gegenbauer **1**, Seidel-Szász **1**.

(b) Another remarkable integral representation valid for non-negative integral values of $\lambda - \frac{1}{2}$ is the following:

$$(x^2-1)^{\lambda/2-1/4}P_n^{(\lambda)}(x) = 2^{\lambda-\frac{1}{2}}\frac{\Gamma(\lambda+\frac{1}{2})}{\Gamma(2\lambda)}\frac{\Gamma(n+2\lambda)}{\Gamma(n+\lambda+\frac{1}{2})}\cdot\frac{1}{\pi}\int_0^{\pi}\{x+(x^2-1)^{\frac{1}{2}}\cos\varphi\}^{n+\lambda-\frac{1}{2}}\cdot\cos(\lambda-\tfrac{1}{2})\varphi\,d\varphi. \tag{4.10.4}$$

For the sake of simplicity we assume here that $x > 1$, $(x^2-1)^{\frac{1}{2}} > 0$.

For the proof we use (4.4.6) with the circle (4.8.8) as contour of integration. Since

$$t^2-1 = 2(x^2-1)^{\frac{1}{2}}e^{i\varphi}\{x+(x^2-1)^{\frac{1}{2}}\cos\varphi\}$$

we have, $\alpha = \lambda - \frac{1}{2}$,

$$\begin{aligned} (x^2-1)^{\alpha}P_n^{(\alpha,\alpha)}(x) &= \frac{1}{2\pi i}\int\left(\frac{1}{2}\frac{t^2-1}{t-x}\right)^n(t^2-1)^{\alpha}\frac{dt}{t-x}\\ &= \frac{1}{2\pi}\int_{-\pi}^{+\pi}\{x+(x^2-1)^{\frac{1}{2}}\cos\varphi\}^{n+\alpha}\{2(x^2-1)^{\frac{1}{2}}e^{i\varphi}\}^{\alpha}\,d\varphi \end{aligned}$$

which yields easily (4.10.4).

(3) *Generating functions.* (a) A slight modification of the argument in 4.4 (3) (proving the formula (4.4.5) for the generating function $F(x, w)$) is the following. Let $-1 < w < 1$; we write (4.4.7) as follows:

$$x = t + \tfrac{1}{2}w(1 - t^2) \tag{4.10.5}$$

so that x and t describe the interval $[-1, +1]$ at the same time. Hence, using the formula (4.4.7) and the two first formulas of (4.4.8), we find

$$(1 - x)^\alpha(1 + x)^\beta F(x, w)\, dx = (1 - t)^\alpha(1 + t)^\beta\, dt$$

so that for any non-negative integer n

$$\int_{-1}^{+1} (1 - x)^\alpha(1 + x)^\beta F(x, w) x^n\, dx = \int_{-1}^{+1} (1 - t)^\alpha(1 + t)^\beta[t + \tfrac{1}{2}w(1 - t^2)]^n\, dt,$$

i.e., a polynomial of degree n in w. This shows that the coefficients of $F(x, w)$ must be, apart from constant factors, the Jacobi polynomials. These factors can be determined by writing $x = 1$.

(b) The following generating function holds for ultraspherical polynomials:

$$\begin{aligned} \sum_{n=0}^{\infty} \frac{P_n^{(\lambda)}(x)}{P_n^{(\lambda)}(1)} \frac{w^n}{n!} &= \sum_{n=0}^{\infty} \frac{\Gamma(2\lambda)}{\Gamma(n + 2\lambda)} P_n^{(\lambda)}(x) w^n \\ &= 2^{\lambda - \frac{1}{2}}\Gamma(\lambda + \tfrac{1}{2})\, e^{xw}[(1 - x^2)^{\frac{1}{2}}w]^{\frac{1}{2} - \lambda} J_{\lambda - \frac{1}{2}}[(1 - x^2)^{\frac{1}{2}}w]. \end{aligned} \tag{4.10.6}$$

It is different from (4.7.16) and (4.7.23). Setting $\lambda = \frac{1}{2}$ we obtain for the Legendre polynomials:

$$\sum_{n=0}^{\infty} \frac{P_n(x)}{n!} w^n = e^{xw} J_0[(1 - x^2)^{\frac{1}{2}}w]. \tag{4.10.7}$$

For the proof of (4.10.6) we use (4.10.3); we obtain for the left-hand side of (4.10.6):

$$\begin{aligned} &\frac{2^{1-2\lambda}\Gamma(2\lambda)}{[\Gamma(\lambda)]^2} \int_0^\pi \exp\,[w\{x + (x^2 - 1)^{\frac{1}{2}}\cos\varphi\}] \cdot \sin^{2\lambda - 1}\varphi\, d\varphi \\ &\qquad = \frac{2^{1-2\lambda}\Gamma(2\lambda)}{[\Gamma(\lambda)]^2} e^{xw} \sum_{m=0}^{\infty} \frac{[(x^2 - 1)^{\frac{1}{2}}w]^{2m}}{(2m)!} \int_0^\pi \cos^{2m}\varphi \sin^{2\lambda - 1}\varphi\, d\varphi \end{aligned}$$

which is easy to identify with the right-hand expression in (4.10.6).

Concerning further formal properties of the Legendre and ultraspherical polynomials, see Bateman Manuscript Project, vol. 1, Chapter 3 and vol. 2, Chapter 10. Cf. also Problems and Exercises, 61–66, 69–71, 84.

(4) The following remarkable identity, involving ultraspherical polynomials, is due to Feldheim 5, p. 278:

$$\begin{aligned} \frac{\Gamma(2\lambda)}{\Gamma(n + 2\lambda)} P_n^{(\lambda)}(\cos\theta) &= \frac{2\Gamma(\lambda + \frac{1}{2})}{\Gamma(\mu + \frac{1}{2})\Gamma(\lambda - \mu)} \\ &\cdot \frac{\Gamma(2\mu)}{\Gamma(n + 2\mu)} \int_0^{\pi/2} \sin^{2\mu}\varphi \cos^{2\lambda - 2\mu - 1}\varphi (1 - \sin^2\theta\cos^2\varphi)^{n/2} \\ &\cdot P_n^{(\mu)}\left(\frac{\cos\theta}{(1 - \sin^2\theta\cos^2\varphi)^{1/2}}\right) d\varphi;\ \lambda > \mu > -\tfrac{1}{2},\ \lambda \neq 0,\ \mu \neq 0,\ 0 \leqq \theta \leqq \pi. \end{aligned} \tag{4.10.8}$$

It is an easy consequence of the generating function (4.10.6). Indeed, multiplying both sides of (4.10.8) by w^n and extending the summation over $n = 0, 1, 2, \cdots$ we obtain

$$2^{\lambda-1/2}\Gamma(\lambda+\tfrac{1}{2})e^{w\cos\theta}(w\sin\theta)^{1/2-\lambda}J_{\lambda-1/2}(w\sin\theta)$$
$$= \frac{2\Gamma(\lambda+\frac{1}{2})}{\Gamma(\mu+\frac{1}{2})\Gamma(\lambda-\mu)}\cdot\int_0^{\pi/2}\sin^{2\mu}\varphi\cos^{2\lambda-2\mu-1}\varphi\cdot 2^{\mu-1/2}\Gamma(\mu+\tfrac{1}{2})e^{w\cos\theta}$$
$$\cdot(w\sin\theta\sin\varphi)^{1/2-\mu}J_{\mu-1/2}(w\sin\theta\sin\varphi)\,d\varphi,$$

or

$$J_{\lambda-1/2}(w\sin\theta) = \frac{(w\sin\theta)^{\lambda-\mu}}{2^{\lambda-\mu-1}\Gamma(\lambda-\mu)}\int_0^{\pi/2}J_{\mu-1/2}(w\sin\theta\sin\varphi)\sin^{\mu+1/2}\varphi\cos^{2\lambda-2\mu-1}\varphi\,d\varphi. \tag{4.10.9}$$

This identity is due to Sonine (Watson **3**, p. 373, (1)).

(5) *Positivity of certain sums.* The first result of this kind is due to L. Fejér **4**, p. 83:

$$P_0(x) + P_1(x) + P_2(x) + \cdots + P_n(x) \geqq 0 \text{ for } -1 \leqq x \leqq 1. \tag{4.10.10}$$

This fact is important in the study of summability of the Laplace series. It follows from (4.8.7) by observing the identity

$$\sum_{\nu=0}^{n}\frac{\sin(2\nu+1)\theta}{\sin\theta} = \left(\frac{\sin(n+1)\theta}{\sin\theta}\right)^2.$$

Generalizations of (4.10.10) are due to Fejér **12**, Feldheim **5**, Szegö **25**.

(6) The formulas (4.8.6) and (4.8.7) of Dirichlet-Mehler can be extended to

$$(1-x)^{\alpha+\mu}\frac{P_n^{(\alpha+\mu,\beta-\mu)}(x)}{P_n^{(\alpha+\mu,\beta-\mu)}(1)} = \frac{\Gamma(\alpha+\mu+1)}{\Gamma(\alpha+1)\Gamma(\mu)}\int_x^1(1-y)^\alpha\frac{P_n^{(\alpha,\beta)}(y)}{P_n^{(\alpha,\beta)}(1)}(y-x)^{\mu-1}dy,$$
$$\alpha > -1,\ \mu > 0,\ -1 < x < 1, \tag{4.10.11}$$

and

$$(1+x)^{\beta+\mu}\frac{P_n^{(\alpha-\mu,\beta+\mu)}(x)}{P_n^{(\beta+\mu,\alpha-\mu)}(1)} = \frac{\Gamma(\beta+\mu+1)}{\Gamma(\beta+1)\Gamma(\mu)}\int_{-1}^x(1+y)^\beta\frac{P_n^{(\alpha,\beta)}(y)}{P_n^{(\beta,\alpha)}(1)}(x-y)^{\mu-1}dy,$$
$$\beta > -1,\ \mu > 0,\ -1 < x < 1. \tag{4.10.12}$$

Feldheim's integral (4.10.8) follows by using the quadratic transformations (4.1.5) and the integral

$$\frac{(1-x)^{\alpha+\mu}}{(1+x)^{n+\alpha+1}}\frac{P_n^{(\alpha+\mu,\beta)}(x)}{P_n^{(\alpha+\mu,\beta)}(1)} = \frac{2^\mu\Gamma(\alpha+\mu+1)}{\Gamma(\alpha+1)\Gamma(\mu)}\int_x^1\frac{(1-y)^\alpha(y-x)^{\mu-1}}{(1+y)^{n+\alpha+\mu+1}}\frac{P_n^{(\alpha,\beta)}(y)}{P_n^{(\alpha,\beta)}(1)}dy, \tag{4.10.13}$$

$\alpha > -1$, $\mu > 0$, $-1 < x < 1$. These three integrals are all special cases of known integrals connecting hypergeometric functions which were found by Bateman **2**. See Askey-Fitch **2** for other useful integrals. Among the consequences of these integrals is the following theorem on positive sums.

THEOREM 4.10.1. *If* $0 \leqq \mu \leqq \nu$ *and* $\beta > -1$ *and if*

$$\sum_{k=0}^{n} a_k \frac{P_k^{(\alpha,\beta)}(x)}{P_k^{(\beta,\alpha)}(1)} \geqq 0, \qquad -1 \leqq x \leqq 1, \tag{4.10.14}$$

then

$$\sum_{k=0}^{n} a_k \frac{P_k^{(\alpha-\mu,\beta+\nu)}(y)}{P_k^{(\beta+\nu,\alpha-\mu)}(1)} \geqq 0, \qquad -1 \leqq y \leqq 1. \tag{4.10.15}$$

Feldheim's integral (4.10.8) can be used to prove a related theorem.

THEOREM 4.10.2. *If* $\nu > 0$, $\alpha > -1$ *and if*

$$\sum_{k=0}^{n} a_k \frac{P_k^{(\alpha,\alpha)}(x)}{P_k^{(\alpha,\alpha)}(1)} \geqq 0, \qquad -1 \leqq x \leqq 1, \tag{4.10.16}$$

then

$$\sum_{k=0}^{n} a_k \frac{P_k^{(\alpha+\nu,\alpha+\nu)}(y)}{P_k^{(\alpha+\nu,\alpha+\nu)}(1)} \geqq 0, \qquad -1 \leqq y \leqq 1. \tag{4.10.17}$$

As Feldheim 5 pointed out, (4.10.10) implies

$$\sum_{k=0}^{n} \frac{P_k^{(\alpha,\alpha)}(x)}{P_k^{(\alpha,\alpha)}(1)} > 0, \qquad -1 < x \leqq 1,\ \alpha > 0. \tag{4.10.18}$$

Askey-Gasper 4 and Askey-Steinig 2 have obtained generalizations of (4.10.18). They proved

$$\sum_{k=0}^{n} \frac{P_k^{(\alpha,\beta)}(x)}{P_k^{(\beta,\alpha)}(1)} > 0, \qquad -1 < x < 1, \tag{4.10.19}$$

for

(i) $\alpha+\beta \geq -2$, $\beta \geqq 0$ (when $\alpha = -2$, $\beta = 0$, assume $n \neq 1$),
(ii) $-\beta \leqq \alpha \leqq \beta+1$ ($\alpha = \frac{1}{2}$, $\beta = -\frac{1}{2}$ omitted),
(iii) $-\beta+1 \leqq \alpha \leqq 3/2$,
(iv) $-\beta+2 \leqq \alpha \leqq \beta+3$.

The case $\alpha = \frac{1}{2}$, $\beta = -\frac{1}{2}$ of the sum (4.10.19) is Fejér's sum (6.4.3), which is nonnegative, while the case $\alpha = 3/2$, $\beta = -\frac{1}{2}$ is equivalent to

$$\frac{d}{d\theta} \sum_{k=0}^{n} \frac{\sin(k+1)\theta}{(k+1)\sin(\theta/2)} < 0, \qquad 0 < \theta < \pi.$$

It is also equivalent to

$$\sum_{k=0}^{n} (k+1)^2 \frac{P_{2k}^{(2)}(x)}{P_{2k}^{(2)}(1)} > \frac{1}{4} \sum_{k=0}^{n} \frac{P_{2k}^{(2)}(x)}{P_{2k}^{(2)}(1)},$$

and, since the right-hand side of this inequality is the even part of (4.10.18) when $\alpha = 3/2$, it is nonnegative. The positivity of the left-hand side is equivalent to the positivity of $K_n^{(2)}(x)$ in (15.5.1).

(7) Jacobi polynomials satisfy the addition formula

$$P_n^{(\alpha,\beta)}(2|\cos\theta_1\cos\theta_2+re^{i\varphi}\sin\theta_1\sin\theta_2|^2-1)$$

(4.10.20)
$$=\sum_{k=0}^{n}\sum_{m=0}^{k} c_{n,k,m}^{(\alpha,\beta)}(\sin\theta_1\sin\theta_2)^{k+m}(\cos\theta_1\cos\theta_2)^{k-m}$$
$$\cdot P_{n-k}^{(\alpha+k+m,\beta+k-m)}(\cos 2\theta_1)\,P_{n-k}^{(\alpha+k+m,\beta+k-m)}(\cos 2\theta_2)$$
$$\cdot P_m^{(\alpha-\beta-1,\beta+k-m)}(2r^2-1)\,r^{k-m}\,\frac{\beta+k-m}{\beta}\,C_{k-m}^{\beta}(\cos\varphi),$$

where

$$c_{n,k,m}^{(\alpha,\beta)}=\frac{(k+m+\alpha)\,\Gamma(n+k+\alpha+\beta+1)\,\Gamma(k+\alpha)\,\Gamma(\beta+1)\,\Gamma(n+\beta+1)\,\Gamma(n-k+1)}{\Gamma(n+\alpha+\beta+1)\,\Gamma(n+m+\alpha+1)\,\Gamma(k+\beta+1)\,\Gamma(n-m+\beta+1)}$$

and the limit relation

$$\lim_{\beta\to 0}\frac{\beta+n}{\beta}\,C_n^{\beta}(\cos\varphi)=\begin{cases}2\cos n\varphi, & n=1,2,\cdots,\\ 1, & n=0,\end{cases}$$

is used when $\beta=0$.

When $r=1$ and $\alpha=\beta$ this formula is the addition theorem of Gegenbauer **1** for ultraspherical polynomials. It was discovered by Šapiro **1** in the case $\beta=0$ and independently by Koornwinder **1** in the general case. For other proofs see Koornwinder **2, 3, 4**.

Among the special results contained in (4.10.20) are

(4.10.21)
$$\frac{P_n^{(\alpha,\beta)}(\cos 2\theta_1)}{P_n^{(\alpha,\beta)}(1)}\,\frac{P_n^{(\alpha,\beta)}(\cos 2\theta_2)}{P_n^{(\alpha,\beta)}(1)}=\int_0^1\int_0^\pi \frac{P_n^{(\alpha,\beta)}(2|\cos\theta_1\cos\theta_2+re^{i\varphi}\sin\theta_1\sin\theta_2|^2-1)}{P_n^{(\alpha,\beta)}(1)}\,dm_{\alpha,\beta}(\varphi,r),$$

$\alpha>\beta>-\frac{1}{2}$, where

$$dm_{\alpha,\beta}(\varphi,r)=A_{\alpha,\beta}(1-r^2)^{\alpha-\beta-1}r^{2\beta+1}(\sin\varphi)^{2\beta}d\varphi\,dr,$$

and

$$A_{\alpha,\beta}^{-1}=\int_0^1\int_0^\pi(1-r^2)^{\alpha-\beta-1}r^{2\beta+1}(\sin\varphi)^{2\beta}d\varphi\,dr,$$

and also

(4.10.22)
$$\frac{P_n^{(\alpha,\beta)}(\cos 2\theta)}{P_n^{(\alpha,\beta)}(1)}=\int_0^1\int_0^\pi[\cos^2\theta-r^2\sin^2\theta+ir\cos\varphi\sin 2\theta]^n dm_{\alpha,\beta}(\varphi,r).$$

When $\alpha\to\beta$, (4.10.22) has (4.10.3) as a limit. In addition to the papers listed above, see Gasper **3, 4** and Askey **10**.

Another useful formula was found by Bateman **1**:

(4.10.23)
$$\frac{P_n^{(\alpha,\beta)}\left(\dfrac{1+xy}{x+y}\right)}{P_n^{(\alpha,\beta)}(1)}\left(\frac{x+y}{2}\right)^n=\sum_{k=0}^{n}c_{k,n}P_k^{(\alpha,\beta)}(x)\,\frac{P_k^{(\alpha,\beta)}(y)}{P_k^{(\alpha,\beta)}(1)}$$

where the $c_{k,n}$'s are defined by

(4.10.24)
$$\left(\frac{1+x}{2}\right)^n = \sum_{k=0}^{n} c_{k,n} P_k^{(\alpha,\beta)}(x).$$

The $c_{k,n}$ can be explicitly computed by use of Rodrigues' formula and orthogonality and they were given in Bateman **1**. However, in many applications only the positivity of the $c_{k,n}$ is needed (see e.g. Horton **1**). It can easily be proved by induction. Bateman **3** also discovered the inverse to (4.10.23):

(4.10.25)
$$\frac{P_n^{(\alpha,\beta)}(x)\,P_n^{(\alpha,\beta)}(y)}{P_n^{(\alpha,\beta)}(1)} = \sum_{k=0}^{n} b_{k,n} \frac{P_k^{(\alpha,\beta)}\left(\dfrac{1+xy}{x+y}\right)}{P_k^{(\alpha,\beta)}(1)} \left(\frac{x+y}{2}\right)^k$$

where the $b_{k,n}$'s are defined by

(4.10.26)
$$P_n^{(\alpha,\beta)}(x) = \sum_{k=0}^{n} b_{k,n} \left(\frac{1+x}{2}\right)^k.$$

This has been used by Koornwinder **3**. Again the specific form of the $b_{k,n}$'s was not needed.

(8) Gegenbauer **6** generalized (4.9.19) to obtain

(4.10.27)
$$P_n^{(\mu)}(x) = \sum_{k=0}^{[n/2]} a_{k,n} P_{n-2k}^{(\lambda)}(x)$$

where

(4.10.28)
$$a_{k,n} = \frac{\Gamma(\lambda)(n-2k+\lambda)\,\Gamma(k+\mu-\lambda)\,\Gamma(n-k+\mu)}{\Gamma(\mu)\,k!\,\Gamma(\mu-\lambda)\,\Gamma(n-k+\lambda+1)}.$$

Proofs are given in Hua **1** and Askey **2**. This can be inverted to give

(4.10.29)
$$(1-x^2)^{\mu-1/2} P_n^{(\mu)}(x) = \sum_{k=0}^{\infty} d_{k,n} P_{n+2k}^{(\lambda)}(x)(1-x^2)^{\lambda-1/2}, \quad \mu > (\lambda-1)/2,$$

where

(4.10.30)
$$d_{k,n} = \frac{\Gamma(\lambda)\,2^{2\lambda-2\mu}(n+2k+\lambda)\,\Gamma(n+2k+1)\,\Gamma(n+2\mu)\,\Gamma(n+k+\lambda)\,\Gamma(k+\lambda-\mu)}{\Gamma(\lambda-\mu)\,\Gamma(\mu)\,\Gamma(n+1)\,\Gamma(k+1)\,\Gamma(n+k+\mu+1)\,\Gamma(n+2k+2\lambda)}.$$

See Askey **1**. When $\lambda = 1$, (4.10.29) reduces to (4.9.22). Generalizations of these formulas to Jacobi polynomials as well as generalizations of Problem 84 are given in Askey **4**, Askey-Gasper **1**, **2**, Gasper **1**, **2**. In the general case, the coefficients are much more complicated and one needs to obtain asymptotic formulas and positivity when it holds. For an application of (4.10.27) and Problem 84 in which only the positivity is used see Askey-Wainger **3**. The linearization result in Problem 84 can be used to define Toeplitz operators and much of the classical theory of Toeplitz operators and the newer work on finite sections of such operators can be extended to these more general operators. See Hirschman **2** and Davis-Hirschman **1** and further references given in these papers.

CHAPTER V

LAGUERRE AND HERMITE POLYNOMIALS

Many of the properties of the polynomials with which we shall deal in this chapter are very similar, and more or less analogous, to the properties of Jacobi polynomials. For this reason we shall be brief and omit details, unless essential differences in statement or proof make the contrary necessary. Here, as in the case of Jacobi polynomials, the treatment of some special problems (zeros, extrema, and so on) is reserved for later chapters.

5.1. Elementary properties of Laguerre polynomials

(1) We define the Laguerre polynomials $\{L_n^{(\alpha)}(x)\}$, for $\alpha > -1$, by the following conditions of orthogonality and normalization:

$$\int_0^\infty e^{-x}x^\alpha L_n^{(\alpha)}(x)L_m^{(\alpha)}(x)\,dx = \Gamma(\alpha+1)\binom{n+\alpha}{n}\delta_{nm}, \tag{5.1.1}$$

$$n, m = 0, 1, 2, \cdots.$$

In addition, we require that the coefficient of x^n in the polynomial $L_n^{(\alpha)}(x)$ of degree n have the sign $(-1)^n$. (This differs from condition (a) in the definition of §2.2.) We also write $L_n^{(0)}(x) = L_n(x)$.

Reference is here made to Lagrange **1**, Abel **1**, p. 284, Tchebichef **3**, pp. 506–508, and Laguerre **1**, pp. 78–81 (pp. 434–437), who, however, consider only the case $\alpha = 0$. Laguerre uses the notation $f_n(x) = n!L_n(-x)$. Hilbert-Courant (**1**, pp. 79–80) also considers only the case $\alpha = 0$; the function there called $L_n(x)$ is the same as $n!L_n(x)$ in our notation. Concerning the general case $L_n^{(\alpha)}(x)$ see Sonin **1**, pp. 41–42.

We have the differential equations

$$\begin{aligned}
&xy'' + (\alpha + 1 - x)y' + ny = 0, && y = L_n^{(\alpha)}(x),\\
&xz'' + (x+1)z' + \left(n + \frac{\alpha}{2} + 1 - \frac{\alpha^2}{4x}\right)z = 0, && z = e^{-x}x^{\alpha/2}L_n^{(\alpha)}(x),\\
&u'' + \left(\frac{n + (\alpha+1)/2}{x} + \frac{1-\alpha^2}{4x^2} - \frac{1}{4}\right)u = 0, && u = e^{-x/2}x^{(\alpha+1)/2}L_n^{(\alpha)}(x),\\
&v'' + \left(4n + 2\alpha + 2 - x^2 + \frac{\frac{1}{4} - \alpha^2}{x^2}\right)v = 0, && v = e^{-x^2/2}x^{\alpha+\frac{1}{2}}L_n^{(\alpha)}(x^2).
\end{aligned} \tag{5.1.2}$$

Once again let $\alpha > -1$. Then an argument analogous to that in §4.2 (2) shows that a necessary and sufficient condition that

$$xy'' + (\alpha + 1 - x)y' + \lambda y = 0 \tag{5.1.3}$$

have a polynomial solution is that $\lambda = n$. Also, $L_n^{(\alpha)}(x)$ is the only polynomial solution. The latter statement follows from the relation

$$u_1'(x)u_2(x) - u_1(x)u_2'(x) = \text{const.}, \tag{5.1.4}$$

which holds for two arbitrary solutions $u_1(x)$, $u_2(x)$ of the third equation in (5.1.2). Incidentally, this argument furnishes slightly more: for $\alpha > -1$, the polynomials $L_n^{(\alpha)}(x)$ are the only solutions of (5.1.3) which are analytic near $x = 0$.

The analogue of Rodrigues' formula is

$$e^{-x}x^\alpha L_n^{(\alpha)}(x) = \frac{1}{n!}\left(\frac{d}{dx}\right)^n (e^{-x}x^{n+\alpha}). \tag{5.1.5}$$

To determine the constant factor we apply Leibniz' formula (which leads to (5.1.6)) and calculate the highest term of the right-hand member.

Further, we have the explicit representation

$$L_n^{(\alpha)}(x) = \sum_{\nu=0}^{n} \binom{n+\alpha}{n-\nu} \frac{(-x)^\nu}{\nu!}, \tag{5.1.6}$$

the formula

$$L_n^{(\alpha)}(0) = \binom{n+\alpha}{n}, \tag{5.1.7}$$

and the expression

$$l_n^{(\alpha)} = \frac{(-1)^n}{n!} \tag{5.1.8}$$

for the coefficient $l_n^{(\alpha)}$ of x^n in $L_n^{(\alpha)}(x)$. As a generating function we obtain

$$\begin{aligned} &L_0^{(\alpha)}(x) + L_1^{(\alpha)}(x)w + \cdots + L_n^{(\alpha)}(x)w^n + \cdots \\ &\qquad = (1-w)^{-\alpha-1}\exp\left(-\frac{xw}{1-w}\right). \end{aligned} \tag{5.1.9}$$

The following recurrence formula holds:

$$\begin{aligned} nL_n^{(\alpha)}(x) &= (-x + 2n + \alpha - 1)L_{n-1}^{(\alpha)}(x) - (n + \alpha - 1)L_{n-2}^{(\alpha)}(x), \\ &\qquad n = 2, 3, 4, \cdots, \\ L_0^{(\alpha)}(x) &= 1, \qquad L_1^{(\alpha)}(x) = -x + \alpha + 1. \end{aligned} \tag{5.1.10}$$

For the "kernel polynomial" $K_n^{(\alpha)}(x, y)$ we find

$$\begin{aligned} \Gamma(\alpha+1)K_n^{(\alpha)}(x,y) &= \sum_{\nu=0}^{n} \left\{\binom{\nu+\alpha}{\nu}\right\}^{-1} L_\nu^{(\alpha)}(x)L_\nu^{(\alpha)}(y) \\ &= (n+1)\left\{\binom{n+\alpha}{n}\right\}^{-1} \frac{L_n^{(\alpha)}(x)L_{n+1}^{(\alpha)}(y) - L_{n+1}^{(\alpha)}(x)L_n^{(\alpha)}(y)}{x-y}. \end{aligned} \tag{5.1.11}$$

The special case $y = 0$ is particularly important:

$$x \sum_{\nu=0}^{n} L_\nu^{(\alpha)}(x) = (n + \alpha + 1)L_n^{(\alpha)}(x) - (n + 1)L_{n+1}^{(\alpha)}(x). \tag{5.1.12}$$

(Cf. Theorem 2.5.) Finally, by means of (5.1.6), or (5.1.9), we readily obtain

$$\begin{gathered} \sum_{\nu=0}^{n} L_\nu^{(\alpha)}(x) = L_n^{(\alpha+1)}(x), \\ L_n^{(\alpha)}(x) = L_n^{(\alpha+1)}(x) - L_{n-1}^{(\alpha+1)}(x), \end{gathered} \tag{5.1.13}$$

$$\frac{d}{dx} L_n^{(\alpha)}(x) = -L_{n-1}^{(\alpha+1)}(x) = x^{-1}\{nL_n^{(\alpha)}(x) - (n + \alpha)L_{n-1}^{(\alpha)}(x)\}. \tag{5.1.14}$$

(2) THEOREM 5.1. *Let J_α have the same meaning as in* §1.71. *Then*

$$\begin{aligned} &\sum_{n=0}^{\infty} \left\{\binom{n + \alpha}{n}\right\}^{-1} L_n^{(\alpha)}(x)L_n^{(\alpha)}(y)w^n \\ &\quad = \Gamma(\alpha + 1)(1 - w)^{-1} \exp\left\{-(x + y)\frac{w}{1 - w}\right\}(-xyw)^{-\alpha/2} J_\alpha\left\{\frac{2(-xyw)^{\frac{1}{2}}}{1 - w}\right\}, \end{aligned} \tag{5.1.15}$$

and

$$\sum_{n=0}^{\infty} \frac{L_n^{(\alpha)}(x)}{\Gamma(n + \alpha + 1)} w^n = e^w(xw)^{-\alpha/2} J_\alpha\{2(xw)^{\frac{1}{2}}\}. \tag{5.1.16}$$

See Sonin **1**, p. 41, Wigert **1**, Hille **2**, Hardy **1**, Kogbetliantz **12**, Watson **4**. The first formula is a generalization of (5.1.9) ($y = 0$). The second formula is obtained from the first one by replacing w by $-y^{-1}w$, $y \to \infty$.

Direct calculation leads readily to (5.1.16) on account of (5.1.6). The formula (5.1.15) follows from (5.1.16) when we introduce for $L_n^{(\alpha)}(y)$ the integral expression which results from (5.4.1), and then integrate term-by-term. Finally an integral formula involving Bessel functions (Watson **3**, p. 395, (1)) must be used.

5.2. Generalization

By means of (5.1.6) the definition of Laguerre polynomials can be extended to arbitrary complex values of α. No reduction in the degree ever occurs (see (5.1.8)). For $n \geqq 1$ we have $L_n^{(\alpha)}(0) = 0$ if and only if $\alpha = -k$, k integral, $1 \leqq k \leqq n$. In this case $x = 0$ is a zero of precise order k, and from (5.1.6)

$$\begin{aligned} L_n^{(-k)}(x) &= (-x)^k \frac{(n - k)!}{n!} \sum_{\nu=0}^{n-k} \binom{n}{n - k - \nu} \frac{(-x)^\nu}{\nu!} \\ &= (-x)^k \frac{(n - k)!}{n!} L_{n-k}^{(k)}(x). \end{aligned} \tag{5.2.1}$$

Formula (5.1.16) remains true for arbitrary real α. From here (5.2.1) follows again.

5.3. Confluent hypergeometric series; relation between Jacobi and Laguerre polynomials; second solution

(1) In the notation of Pochhammer-Barnes, the confluent hypergeometric series is

$$ {}_1F_1(\alpha;\gamma;x) = 1 + \sum_{\nu=1}^{\infty} \frac{\alpha(\alpha+1)\cdots(\alpha+\nu-1)}{\gamma(\gamma+1)\cdots(\gamma+\nu-1)}\frac{x^\nu}{\nu!}. \tag{5.3.1}$$

This is obtained from the ordinary hypergeometric series [(4.21.3)] by the limiting process

$$ \lim_{\beta\to\infty} F(\alpha,\beta;\gamma;\beta^{-1}x). \tag{5.3.2}$$

We have

$$ L_n^{(\alpha)}(x) = \binom{n+\alpha}{n} {}_1F_1(-n;\alpha+1;x), \tag{5.3.3}$$

and using (4.21.2), we obtain the following important relation between Laguerre and Jacobi polynomials:

$$ L_n^{(\alpha)}(x) = \lim_{\beta\to\infty} P_n^{(\alpha,\beta)}(1-2\beta^{-1}x). \tag{5.3.4}$$

This holds uniformly in every closed part of the complex x-plane. Concerning further properties of the confluent hypergeometric functions see Whittaker-Watson **1**, Chapter XVI. Compare equation (B) in Whittaker-Watson **1**, p. 337, with our third equation in (5.1.2).

(2) From (4.23.1), by a limiting process similar to that used in (5.3.4), we obtain as a second solution of the first equation (5.1.2)

$$ x^{-\alpha}{}_1F_1(-n-\alpha;1-\alpha;x). \tag{5.3.5}$$

For non-integral values of α the functions (5.3.3) and (5.3.5) are evidently linearly independent. The same is true if α is negative integer less than $-n$, since in this case (5.3.5) is an infinite series. However, if α is an integer not less than $-n$, these solutions are identical. (If $-n \leqq \alpha \leqq 0$, use (5.2.1). In case $\alpha = g$, $g \geqq 1$, it is necessary to multiply (5.3.5) through by $\alpha - g$ before letting $\alpha \to g$.)

The representation (5.3.3) makes possible an extension of the definition of $L_n^{(\alpha)}(x)$ to arbitrary values of n.

5.4. Integral representations

THEOREM 5.4. *The following representation of Laguerre polynomials in terms of Bessel functions holds:*

$$ e^{-x}x^{\alpha/2}L_n^{(\alpha)}(x) = \frac{1}{n!}\int_0^\infty e^{-t}t^{n+\alpha/2}J_\alpha\{2(tx)^{\frac{1}{2}}\}\,dt, \quad n=0,1,2,\cdots;\alpha>-1. \tag{5.4.1}$$

For this representation, the reader may consult E. Le Roy **1**, pp. 379–384,

Erdélyi **1**. The same representation is valid if $\alpha \leqq -1$, provided $n + \alpha > -1$. From this, (5.2.1) follows again.

Several proofs of this formula may be given. The special case $n = 0$, that is,

$$e^{-x}x^{\alpha/2} = \int_0^\infty e^{-t}t^{\alpha/2}J_\alpha\{2(tx)^{\frac{1}{2}}\}\,dt, \tag{5.4.2}$$

can be obtained by expanding $J_\alpha(z)$ as in (1.71.1) and integrating term-by-term. (The formula is due to Sonin; cf. Watson **3**, p. 394, (4).) The general formula then follows from calculation of the generating function of both sides of (5.4.1); here (5.1.9) and (5.4.2) must be used.

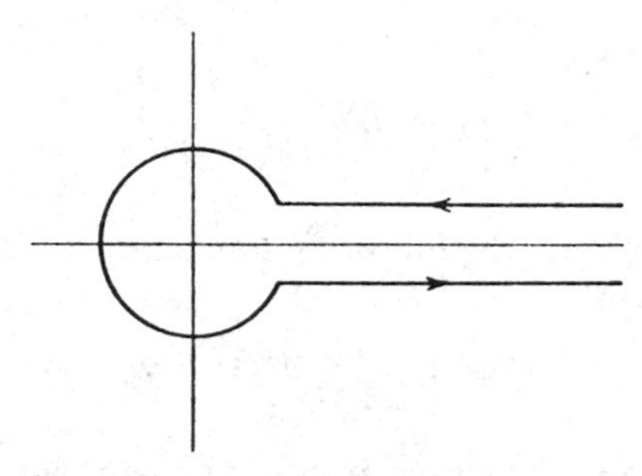

Fig. 6

Another proof, from a more general point of view, can be given as follows. We shall try to satisfy the second equation (5.1.2) by an integral of the form

$$z = z(x) = \int e^{-t}t^{n+\alpha/2}J_\alpha\{2(tx)^{\frac{1}{2}}\}\,dt \tag{5.4.3}$$

with a proper path of integration. Substituting this expression in the left-hand member of the equation mentioned, we obtain

$$\begin{aligned}\int e^{-t}t^{n+\alpha/2}\Big[&tJ''_\alpha\{2(tx)^{\frac{1}{2}}\} + (tx^{-1})^{\frac{1}{2}}(x+\tfrac{1}{2})J'_\alpha\{2(tx)^{\frac{1}{2}}\} \\ &+ \left(n + \frac{\alpha}{2} + 1 - \frac{\alpha^2}{4x}\right)J_\alpha\{2(tx)^{\frac{1}{2}}\}\Big]\,dt.\end{aligned} \tag{5.4.4}$$

In view of (1.71.3) the expression in the square brackets becomes

$$(tx)^{\frac{1}{2}}J'_\alpha\{2(tx)^{\frac{1}{2}}\} + (n + \alpha/2 + 1 - t)J_\alpha\{2(tx)^{\frac{1}{2}}\}, \tag{5.4.5}$$

so that the entire expression will be

$$\int \frac{d}{dt}[e^{-t}t^{n+\alpha/2+1}J_\alpha\{2(tx)^{\frac{1}{2}}\}]\,dt. \tag{5.4.6}$$

Hence (5.4.3) is a solution of (5.1.2) provided that

(a) the path of integration is a closed contour, and so chosen that $e^{-t}t^{n+\alpha/2+1}J_\alpha\{2(tx)^{\frac{1}{2}}\}$ resumes its initial value, or

(b) the path of integration is an arc, and the expression in (a) vanishes at its end-points.

For the interval $0 \leqq t < +\infty$, condition (b) is satisfied provided $n + \alpha + 1 > 0$. Assuming first that $\alpha > -1$, we notice that the function $x^{-\alpha/2}z$ is analytic near $x = 0$, so that (see the remark concerning (5.1.3)) $z = \text{const. } e^{-x}x^{\alpha/2}L_n^{(\alpha)}(x)$. The constant factor can be determined by comparing the "lowest terms," that is, the coefficients of $x^{\alpha/2}$. The restriction $\alpha > -1$ can then be removed by means of analytic continuation.

Another remarkable representation is obtained by choosing the contour as in the figure. Condition (b) is again satisfied, and the same argument as before yields $z = \text{const. } e^{-x}x^{\alpha/2}L_n^{(\alpha)}(x)$. The normalization of the integrand requires the determination of t^α at a certain point. We agree to take $\arg t = 0$ on the rectilinear part of the contour with $\Im t < 0$ (in its limiting position). Then the "lowest term" becomes, if $\alpha > -1$ and $\alpha \neq 0, 1, 2, \cdots$,

$$\frac{x^{\alpha/2}}{\Gamma(\alpha+1)}\int e^{-t}t^{n+\alpha}\,dt = \frac{x^{\alpha/2}}{\Gamma(\alpha+1)}(1 - e^{-2\pi i\alpha})\int_0^\infty e^{-t}t^{n+\alpha}\,dt$$

$$= 2\sin\pi\alpha\, e^{i\pi(\frac{1}{2}-\alpha)}\frac{\Gamma(n+\alpha+1)}{\Gamma(\alpha+1)}x^{\alpha/2},$$

so that on account of (5.1.7), we have

$$(5.4.7)\qquad e^{-x}x^{\alpha/2}L_n^{(\alpha)}(x) = \frac{(2\sin\pi\alpha)^{-1}e^{i\pi(\alpha-\frac{1}{2})}}{n!}\int_{+\infty}^{(0+)} e^{-t}t^{n+\alpha/2}J_\alpha\{2(tx)^{\frac{1}{2}}\}\,dt.$$

This formula can be extended to arbitrary non-integral values of α.

Further integral representations can be derived from (5.1.5) and (5.1.9) in a manner analogous to that used in the case of Jacobi polynomials (cf. (4.4.6), (4.4.9)). We have for instance, $x \neq 0$,

$$(5.4.8)\qquad e^{-x}x^\alpha L_n^{(\alpha)}(x) = \frac{1}{2\pi i}\int e^{-t}t^{n+\alpha}(t-x)^{-n-1}\,dt,$$

where the contour encloses $t = x$, but not $t = 0$.

5.5. Hermite polynomials

(1) These are defined by the conditions

$$(5.5.1)\qquad \int_{-\infty}^{+\infty} e^{-x^2}H_n(x)H_m(x)\,dx = \pi^{\frac{1}{2}}2^n n!\,\delta_{nm}, \quad n, m = 0, 1, 2, \cdots.$$

The coefficient of x^n in the nth polynomial is positive.

See the bibliography in Hille **1**. Our notation agrees with that of Hille. See also Hilbert-Courant **1**, pp. 77–79 where the same notation is used. In Pólya-Szegö **1** (vol. 2, pp. 94, 294, 295, problem 100), $H_n(x)$ is written for $(-1)^n(2^{n/2}n!)^{-1}H_n(2^{-\frac{1}{2}}x)$ in terms of our notation.

(2) The derivation of the following properties of Hermite polynomials presents no difficulty:

$$\begin{cases} y'' - 2xy' + 2ny = 0, \; y = H_n(x), \\ z'' + (2n + 1 - x^2)z = 0, \; z = e^{-x^2/2}H_n(x), \end{cases} \tag{5.5.2}$$

$$e^{-x^2}H_n(x) = (-1)^n \left(\frac{d}{dx}\right)^n e^{-x^2}, \tag{5.5.3}$$

$$\frac{H_n(x)}{n!} = \sum_{\nu=0}^{[n/2]} \frac{(-1)^\nu}{\nu!} \frac{(2x)^{n-2\nu}}{(n-2\nu)!}, \tag{5.5.4}$$

$$H_{2m}(0) = (-1)^m \frac{(2m)!}{m!}, \qquad H'_{2m+1}(0) = (-1)^m \frac{(2m+2)!}{(m+1)!}, \tag{5.5.5}$$

$$\lim_{x\to\infty} x^{-n}H_n(x) = 2^n, \tag{5.5.6}$$

$$\begin{aligned} H_0(x) + \frac{H_1(x)}{1!} w + \frac{H_2(x)}{2!} w^2 + \cdots + \frac{H_n(x)}{n!} w^n + \cdots \\ = \exp(2xw - w^2), \end{aligned} \tag{5.5.7}$$

$$\begin{aligned} H_n(x) = 2xH_{n-1}(x) - 2(n-1)H_{n-2}(x), \\ n = 2, 3, 4, \cdots; \; H_0(x) = 1, \; H_1(x) = 2x, \end{aligned} \tag{5.5.8}$$

$$\sum_{\nu=0}^{n} (2^\nu \nu!)^{-1} H_\nu(x)H_\nu(y) = (2^{n+1}n!)^{-1} \frac{H_{n+1}(x)H_n(y) - H_n(x)H_{n+1}(y)}{x - y}. \tag{5.5.9}$$

We notice the following "individual" properties:

$$H'_n(x) = 2nH_{n-1}(x), \qquad H_n(x) = 2xH_{n-1}(x) - H'_{n-1}(x). \tag{5.5.10}$$

From (5.5.7) we obtain

$$\sum_{\nu=0}^{n} \binom{n}{\nu} H_\nu(x)H_{n-\nu}(y) = 2^{n/2}H_n\{2^{-\frac{1}{2}}(x + y)\}, \tag{5.5.11}$$

and, by Cauchy's formula,

$$\frac{H_n(x)}{n!} = \frac{1}{2\pi i} \int w^{-n-1} \exp(2xw - w^2)\, dw, \tag{5.5.12}$$

where the contour encloses the origin.

5.6. Relation of Hermite polynomials to those of Laguerre

(1) Hermite polynomials can be entirely reduced to Laguerre polynomials with the parameters $\alpha = \pm\frac{1}{2}$, for we have

$$H_{2m}(x) = (-1)^m 2^{2m} m! L_m^{(-\frac{1}{2})}(x^2), \qquad H_{2m+1}(x) = (-1)^m 2^{2m+1} m! x L_m^{(\frac{1}{2})}(x^2). \tag{5.6.1}$$

These formulas are in some respects the analogues of (4.1.5). Their proofs are similar to those given there. Note the fourth equation in (5.1.2) ($\alpha = \pm\frac{1}{2}$), and the second equation in (5.5.2).

Combining (5.6.1) with (5.3.4), we obtain a representation of Hermite polynomials as limits of Jacobi, and in consequence of (4.1.5), of ultraspherical polynomials. This can be ascertained from (4.7.23) and (5.5.7). In fact, we have

$$\text{(5.6.2)} \qquad \exp(2xw - w^2) = \lim_{\lambda\to\infty}\left(1 - 2\frac{x}{\lambda}w + \frac{w^2}{\lambda}\right)^{-\lambda},$$

so that

$$\text{(5.6.3)} \qquad \frac{H_n(x)}{n!} = \lim_{\lambda\to\infty} \lambda^{-n/2} P_n^{(\lambda)}(\lambda^{-\frac{1}{2}}x).$$

From (5.6.1) the explicit representation (5.5.4) follows readily when we use (5.1.6). From Theorem 5.1 analogous expansions for $H_n(x)$ can be derived (see Watson **5**). The generating function (5.5.7) follows from (5.1.16), while the generating function arising from (5.1.9) for $\alpha = \pm\frac{1}{2}$ is of a different nature (see Problem 24).

By using (1.71.2) and (5.4.1), we obtain the integral representations

$$\text{(5.6.4)} \qquad \begin{aligned} e^{-x^2}H_n(x) &= (-1)^{[n/2]}2^{n+1}\pi^{-\frac{1}{2}}\int_0^\infty e^{-t^2}t^n\cos(2xt)\,dt, \qquad n \text{ even},\\ e^{-x^2}H_n(x) &= (-1)^{[n/2]}2^{n+1}\pi^{-\frac{1}{2}}\int_0^\infty e^{-t^2}t^n\sin(2xt)\,dt, \qquad n \text{ odd}.\end{aligned}$$

(2) Conversely, Laguerre polynomials can, to a certain extent, be reduced to Hermite polynomials. We have (Uspensky **1**, p. 604, (14))

$$\text{(5.6.5)} \quad L_n^{(\alpha)}(x) = \frac{(-1)^n\pi^{-\frac{1}{2}}}{\Gamma(\alpha+\frac{1}{2})}\frac{\Gamma(n+\alpha+1)}{(2n)!}\int_{-1}^{+1}(1-t^2)^{\alpha-\frac{1}{2}}H_{2n}(x^{\frac{1}{2}}t)\,dt, \quad \alpha > -\tfrac{1}{2}.$$

This can be readily shown by using the explicit formulas (5.1.6), (5.5.4), and (1.7.5).

(3) We conclude these formal considerations with the following remark concerning the identities (4.21.7), (5.1.14), and (5.5.10) on Jacobi, Laguerre, and Hermite polynomials, respectively. If the orthogonal polynomials $\{p_n(x)\}$ associated with a distribution $d\alpha(x)$ have the property that $\{p_n'(x)\}$ is, save for constant factors, a system of the same kind (that is, associated with a certain distribution $d\beta(x)$), then $\{p_n(x)\}$ is (save for trivial linear transformations) one of the three special systems mentioned before (classical polynomials). (W. Hahn **3**, Krall **1**.) A similar statement holds if we replace $\{p_n'(x)\}$ by $\{p_n^{(k)}(x)\}$ (Krall **2**, W. Hahn **4**).

Another problem of a similar nature has been considered by Bochner (**1**). He determines all sets of polynomials $\{p_n(x)\}$, where $p_n(x)$ is of precise degree n, satisfying a differential equation of the form

(5.6.6) $$f_0(x)y'' + f_1(x)y' + [f_2(x) + \lambda]y = 0, \qquad y = p_n(x),\ \lambda = \lambda_n.$$

Bochner obtains, in addition to the classical polynomials, certain polynomials related to $J_{n+\frac{1}{2}}(x)$, n integral, as possible solutions, as well as polynomials of the trivial type $ax^n + bx^m$, where a and b are constants.

5.7. Closure

Here we prove the analogue of Theorem 3.1.5 for Laguerre and Hermite polynomials. The main difficulty of these cases is due to the fact that the orthogonality interval is infinite. Using the customary notation (§1.1), we have the following theorem:

THEOREM 5.7.1. *The system*

(5.7.1) $$e^{-x/2}x^{\alpha/2}x^n, \qquad \alpha > -1,\ n = 0, 1, 2, \cdots,$$

is closed in $L^2(0, +\infty)$; *the system*

(5.7.2) $$e^{-x^2/2}x^n, \qquad n = 0, 1, 2, \cdots,$$

is closed in $L^2(-\infty, +\infty)$.

This statement is equivalent to the closure of the systems $\{e^{-x/2}x^{\alpha/2}L_n^{(\alpha)}(x)\}$ and $\{e^{-x^2/2}H_n(x)\}$, respectively. Theorem 5.7.1 remains true, of course, if we replace $e^{-x/2}$ by e^{-x} and $e^{-x^2/2}$ by e^{-x^2}, respectively. The idea of the following proof is due to J. von Neumann (see Hilbert-Courant **1**, pp. 81–82).

(1) We start with the remark that for $\alpha > -1$ the system

(5.7.3) $$\left(\log \frac{1}{y}\right)^{\alpha/2} y^n, \qquad n = 0, 1, 2, \cdots,$$

is closed in $L^2(0, 1)$. This is a consequence of Theorem 3.1.5, $p = 2$, since $(\log (1/y))^\alpha$ is integrable in $[0, 1]$.

Now let $e^{-x/2}x^{\alpha/2}f(x)$ belong to $L^2(0, +\infty)$. Then $(\log (1/y))^{\alpha/2}f(\log (1/y))$ belongs to $L^2(0, 1)$, and it can be approximated in mean by functions of the form $(\log (1/y))^{\alpha/2}\rho(y)$, where $\rho(y)$ is a polynomial. Thus, corresponding to every $\epsilon > 0$ a polynomial $\rho(y)$ can be determined such that

(5.7.4) $$\int_0^\infty e^{-x}x^\alpha\{f(x) - \rho(e^{-x})\}^2\, dx < \epsilon.$$

Hence all that remains to be shown is that if m is a non-negative integer, there exists for every $\delta > 0$ a polynomial $p(x)$ such that

(5.7.5) $$\int_0^\infty e^{-x}x^\alpha\{e^{-mx} - p(x)\}^2\, dx < \delta.$$

(2) For this purpose we use (5.1.9), writing

$$w = \frac{m}{m+1}, \qquad m = \frac{w}{1-w}, \tag{5.7.6}$$

and choosing

$$p(x) = (1-w)^{\alpha+1} \sum_{n=0}^{N} L_n^{(\alpha)}(x) w^n. \tag{5.7.7}$$

Then

$$\int_0^\infty e^{-x} x^\alpha \{e^{-mx} - p(x)\}^2 dx = (1-w)^{2\alpha+2} \int_0^\infty e^{-x} x^\alpha \left\{ \sum_{n=N+1}^{\infty} L_n^{(\alpha)}(x) w^n \right\}^2 dx, \tag{5.7.8}$$

which, in view of (5.1.1), is equal to

$$(1-w)^{2\alpha+2} \Gamma(\alpha+1) \sum_{n=N+1}^{\infty} \binom{n+\alpha}{n} w^{2n}. \tag{5.7.9}$$

Term-by-term integration is permitted here since the series

$$\sum_{n,n'=N+1}^{\infty} \int_0^\infty e^{-x} x^\alpha \, | L_n^{(\alpha)}(x) | \, | L_{n'}^{(\alpha)}(x) | \, dx \cdot w^{n+n'}$$
$$\leqq \sum_{n,n'=N+1}^{\infty} \left\{ \int_0^\infty e^{-x} x^\alpha [L_n^{(\alpha)}(x)]^2 dx \right\}^{\frac{1}{2}} \left\{ \int_0^\infty e^{-x} x^\alpha [L_{n'}^{(\alpha)}(x)]^2 dx \right\}^{\frac{1}{2}} w^{n+n'}$$

is convergent. The expression (5.7.9) becomes arbitrarily small when N is sufficiently large, and this establishes the statement.

(3) Let $e^{-x^2/2} f(x)$ belong to $L^2(-\infty, +\infty)$. Then both functions

$$e^{-y/2} y^{-\frac{1}{4}} \frac{f(y^{\frac{1}{2}}) \pm f(-y^{\frac{1}{2}})}{2} \tag{5.7.10}$$

belong to $L^2(0, +\infty)$. Therefore, by the preceding result (first taking $\alpha = -\frac{1}{2}$, then $\alpha = +\frac{1}{2}$), for every $\epsilon > 0$ there exist polynomials $\rho_1(y)$, $\rho_2(y)$ which satisfy the inequalities

$$\begin{aligned} &\int_0^\infty \left\{ e^{-y/2} y^{-\frac{1}{4}} \frac{f(y^{\frac{1}{2}}) + f(-y^{\frac{1}{2}})}{2} - e^{-y/2} y^{-\frac{1}{4}} \rho_1(y) \right\}^2 dy < \epsilon, \\ &\int_0^\infty \left\{ e^{-y/2} y^{-\frac{1}{4}} \frac{f(y^{\frac{1}{2}}) - f(-y^{\frac{1}{2}})}{2} - e^{-y/2} y^{\frac{1}{4}} \rho_2(y) \right\}^2 dy < \epsilon, \end{aligned} \tag{5.7.11}$$

or

$$\begin{aligned} &2 \int_0^\infty e^{-x^2} \left\{ \frac{f(x) + f(-x)}{2} - \rho_1(x^2) \right\}^2 dx < \epsilon, \\ &2 \int_0^\infty e^{-x^2} \left\{ \frac{f(x) - f(-x)}{2} - x\rho_2(x^2) \right\}^2 dx < \epsilon, \end{aligned} \tag{5.7.12}$$

so that

$$\int_{-\infty}^{+\infty} e^{-x^2} \{ f(x) - \rho_1(x^2) - x\rho_2(x^2) \}^2 dx < 2\epsilon. \tag{5.7.13}$$

For another proof based on the theory of integral equations, see Weyl **1**, in particular pp. 58–61, 64. See also Hamburger **1**, pp. 200–205.

(4) The argument used in (2) leads to the following result:

THEOREM 5.7.2. *The system*

$$e^{-x}x^{\alpha+n}, \qquad \alpha > -1,\ n = 0, 1, 2, \cdots, \tag{5.7.14}$$

is closed in $L(0, +\infty)$; *the system*

$$e^{-x^2}x^n, \qquad n = 0, 1, 2, \cdots, \tag{5.7.15}$$

is closed in $L(-\infty, +\infty)$.

We use again Theorem 3.1.5, $p = 1$. By means of (5.1.1) and Schwarz's inequality we find

$$\int_0^\infty e^{-x}x^\alpha \,|\, L_n^{(\alpha)}(x) \,|\, dx = O(n^{\alpha/2}). \tag{5.7.16}$$

For later purposes (§9.5 (1)) we note the following theorem:

THEOREM 5.7.3. *The functions*

$$f_n(x) = \phi(x)x^n, \qquad n = 0, 1, 2, \cdots, \tag{5.7.17}$$

where

$$\phi(x) = \begin{cases} x^\alpha, & 0 < x < 1, \\ e^{-x}x^\beta, & x \geqq 1, \end{cases} \tag{5.7.18}$$

$\alpha > -1$, β *arbitrary and real, form a closed system in* $L(0, +\infty)$.

We apply Theorem 3.1.5 with $d\alpha(y) = y^{-1}\phi\,(\log\,(1/y))\,dy$, $0 < y < 1$, and $p = 1$. Furthermore, we need a bound for

$$\int_0^\infty \phi(x) \,|\, L_n^{(\alpha)}(x) \,|\, dx = O(1) \int_0^1 e^{-x}x^\alpha \,|\, L_n^{(\alpha)}(x) \,|\, dx + \int_1^\infty e^{-x}x^\beta \,|\, L_n^{(\alpha)}(x) \,|\, dx.$$

The first integral in the right-hand member is $O(n^{\alpha/2})$; for the second integral, Schwarz's inequality yields the bound

$$\left\{\int_1^\infty e^{-x}x^{2\beta-\alpha}\,dx\right\}^{\frac{1}{2}}\left\{\int_1^\infty e^{-x}x^\alpha[L_n^{(\alpha)}(x)]^2\,dx\right\}^{\frac{1}{2}} = O(n^{\alpha/2}).$$

Concerning further formal properties of the Laguerre and Hermite polynomials, see Bateman Manuscript Project, vol. 2, Chapter 10, pp. 188–196. Cf. also Problems and Exercises, 67, 68, 72–80.

For problems related to Laguerre polynomials and applications of Laguerre polynomials see Szegö **26**, Askey-Gasper **3**, Askey **9**, Peetre **1**, and Roosenraad **1**. For Hermite polynomials see de Bruijn **1**.

CHAPTER VI

ZEROS OF ORTHOGONAL POLYNOMIALS

In §3.3 it was proved that the zeros of orthogonal polynomials are all real, distinct, and lie in the interior of the orthogonality interval. We shall now present a further and more detailed investigation of the location of these zeros. Starting with certain theorems valid under very general conditions imposed on the weight function, we proceed to the zeros of the classical polynomials and point out various methods used in their investigation. A particularly important tool in the latter connection is Sturm's theorem (§1.82), which in various cases leads to rather exact information concerning the zeros of polynomials satisfying certain linear differential equations of the second order.

No claim of completeness is made for the present survey on the zeros. Concerning the literature about zeros of Laguerre and Hermite polynomials, we refer to W. Hahn's "Bericht" (**2**).

The methods of this Chapter are quite elementary. In particular, no systematic use is made of the asymptotic properties of orthogonal polynomials of a special and general kind (Chapters VIII and XII), from which important information concerning zeros can also be derived.

6.1. Density of zeros

(1) THEOREM 6.1.1. *Let $d\alpha(x)$ be a distribution on the finite segment $[a, b]$, and let $\{p_n(x)\}$ denote the associated orthonormal set of polynomials. Let $[a', b']$ be a subinterval of $[a, b]$ such that $\int_{a'}^{b'} d\alpha(x) > 0$. Then if n is sufficiently large, every polynomial $p_n(x)$ has at least one zero in $[a', b']$.*

For the proof we shall use the Gauss-Jacobi mechanical quadrature (§3.4). Let $\rho(x)$ be an arbitrary π_m which is not greater than 0 in $[a, b]$, except possibly in $[a', b']$. Assuming that the polynomial $p_n(x)$ has no zeros x_ν in $[a', b']$, and taking $2n - 1 \geqq m$, we obtain

$$(6.1.1) \qquad \int_a^b \rho(x)\, d\alpha(x) = \sum_{\nu=1}^{n} \lambda_\nu \rho(x_\nu) \leqq 0.$$

Hence, when we apply the theorem of Weierstrass (Theorem 1.3.1), it follows that

$$(6.1.2) \qquad \int_a^b f(x)\, d\alpha(x) \leqq 0,$$

where $f(x)$ is continuous in $[a, b]$ and not greater than 0 in $[a, b]$, except possibly in $[a', b']$. If we define

$$
(6.1.3) \quad f(x) = \begin{cases} 0 & \text{in } a \leqq x \leqq a' \text{ and } b' \leqq x \leqq b, \\ (x - a')(b' - x) & \text{in } a' \leqq x \leqq b', \end{cases}
$$

we reach a contradiction.

(2) THEOREM 6.1.2. *Theorem* 6.1.1 *remains valid for infinite intervals* $[a, b]$ *provided the zero of the greatest modulus of* $p_n(x)$ *is* $o(n)$.

This remark is due to W. Hahn (**1**, pp. 215–217).[29] If we write

$$
\max |x_\nu - a'| |b' - x_\nu| = M_n, \qquad \nu = 1, 2, \cdots, n,
$$

the assumption means that $M_n = o(n^2)$. We now choose

$$
\rho(x) = \{T_k[2M_n^{-1}(x - a')(b' - x) + 1]\}^2, \qquad k = [n/2 - 1/4],
$$

where $T_k(x)$ denotes Tchebichef's polynomial (1.12.3). If we next assume that $[a', b']$ contains no zeros, we have

$$
-1 \leqq M_n^{-1}(x_\nu - a')(b' - x_\nu) \leqq 0,
$$

so that $\rho(x_\nu) \leqq 1$. It then follows that

$$
\int_{a'}^{b'} \rho(x)\, d\alpha(x) \leqq \int_a^b \rho(x)\, d\alpha(x) = \sum_{\nu=1}^{n} \lambda_\nu \rho(x_\nu) \leqq \sum_{\nu=1}^{n} \lambda_\nu = \int_a^b d\alpha(x).
$$

Now $T_k'(x)$ is increasing for $x \geqq 1$, so that $T_k(\xi + 1) > T_k'(1)\xi = k^2\xi$ for $\xi > 0$; thus we have in $[a', b']$

$$
\rho(x) \geqq k^4\{2M_n^{-1}(x - a')(b' - x)\}^2; \qquad \int_{a'}^{b'} \rho(x)\, d\alpha(x) > Ck^4M_n^{-2},
$$

where C is a positive constant independent of n. Hence $n^2M_n^{-1} = O(1)$, which is a contradiction.

Concerning the distribution of the zeros for large values of n, see Theorem 12.7.2.

6.11. Distance between consecutive zeros

Here and in the subsequent sections we consider distributions of the type $w(x)\, dx$.

(1) THEOREM 6.11.1. *Let* $w(x)$ *be a weight function on the finite interval* $[a, b]$, *bounded from zero:* $w(x) \geqq \mu > 0$. *Let* $x_1 > x_2 > \cdots > x_n$ *be the zeros of the associated orthonormal polynomial* $p_n(x)$ *in decreasing order.*[30] *On writing*

$$
(6.11.1) \quad x_\nu = \tfrac{1}{2}(a + b) + \tfrac{1}{2}(b - a) \cos \theta_\nu, \qquad 0 < \theta_\nu < \pi, \nu = 1, 2, \cdots, n,
$$

[29] He states (without a satisfactory proof) that if either a or b is finite, the subsequent argument succeeds even with $o(n^2)$.

[30] Of course, each x_ν depends on ν and n, $x_\nu = x_{\nu n}$.

we have

$$\theta_{\nu+1} - \theta_\nu < K \frac{\log n}{n}, \qquad \nu = 0, 1, 2, \cdots, n; \theta_0 = 0, \theta_{n+1} = \pi. \tag{6.11.2}$$

Here the constant K depends only on μ, a, and b.

See Krawtchouk **2**; Erdös-Turán (written communication). Erdös-Turán require the existence of $\int_a^b \{w(x)\}^{-1} dx$ instead of $w(x) \geqq \mu > 0$. Their proof (cf. **2**) is based on the study of the distribution of the interpolation points for which the associated fundamental polynomials (Chapter XIV) satisfy certain conditions. The following proof for the special case $w(x) \geqq \mu > 0$ has been prepared by B. Lengyel; he eliminated all references to the theory of interpolation from the argument of Erdös-Turán.

(2) Let ν be a fixed integer, $0 \leqq \nu \leqq n$, and $\gamma = (\theta_\nu + \theta_{\nu+1})/2$. We then define $\rho(x) = \rho\{\frac{1}{2}(a + b) + \frac{1}{2}(b - a) \cos \theta\}$ by

$$2\rho(x) = \left(\frac{\sin \{N(\gamma + \theta)/2\}}{N \sin \{(\gamma + \theta)/2\}}\right)^{2m} + \left(\frac{\sin \{N(\gamma - \theta)/2\}}{N \sin \{(\gamma - \theta)/2\}}\right)^{2m}, \tag{6.11.3}$$

where N and m denote certain positive integers. The single terms in the right-hand member represent the same trigonometric polynomial of degree $m(N - 1)$, taken alternately with arguments $\gamma + \theta$ and $\gamma - \theta$. Therefore, the sum is a cosine polynomial of the same degree $m(N - 1)$. If $m(N - 1) \leqq 2n - 1$, Theorem 3.4.1 can be applied.

We then have

$$(\theta_{\nu+1} - \theta_\nu)/4 \leqq \gamma/2 < (\gamma + \pi)/2 \leqq \pi - (\theta_{\nu+1} - \theta_\nu)/4; \tag{6.11.4}$$

whence for all values of k, $1 \leqq k \leqq n$,

$$0 < (\theta_{\nu+1} - \theta_\nu)/4 < (\gamma + \theta_k)/2 < \pi - (\theta_{\nu+1} - \theta_\nu)/4, \tag{6.11.5}$$

so that

$$\left(\sin \frac{\gamma + \theta_k}{2}\right)^{-1} < \left(\sin \frac{\theta_{\nu+1} - \theta_\nu}{4}\right)^{-1}, \qquad k = 1, 2, \cdots, n. \tag{6.11.6}$$

The same inequality, with $\leqq$ instead of $<$, holds for $|\sin (\gamma - \theta_k)/2|^{-1}$, since $|\gamma - \theta_k| \geqq (\theta_{\nu+1} - \theta_\nu)/2$. Thus,

$$\rho(x_k) \leqq \left(N \sin \frac{\theta_{\nu+1} - \theta_\nu}{4}\right)^{-2m}, \qquad k = 1, 2, \cdots, n. \tag{6.11.7}$$

By using (3.4.1) and (3.4.5), we obtain

$$\int_a^b \rho(x) w(x)\, dx \leqq \left(N \sin \frac{\theta_{\nu+1} - \theta_\nu}{4}\right)^{-2m} \int_a^b w(x)\, dx. \tag{6.11.8}$$

On the other hand, the value of $\rho(x)$ for $\theta = \gamma$ is not less than $\frac{1}{2}$, so that if we write $x_0 = \frac{1}{2}(a + b) + \frac{1}{2}(b - a) \cos \gamma$, we obtain from a later result (cf. Theorem 7.7)

$$(6.11.9) \qquad \int_a^b \rho(x)w(x)\, dx \geqq \mu \int_a^b \rho(x)\, dx \geqq \mu C n^{-2} \rho(x_0) \geqq \tfrac{1}{2}\mu C n^{-2},$$

where C is a positive constant depending on a and b only. Comparing (6.11.8) and (6.11.9), we now have

$$\tfrac{1}{2}\mu C n^{-2} \leqq \left(N \sin \frac{\theta_{\nu+1} - \theta_\nu}{4}\right)^{-2m} \int_a^b w(x)\, dx,$$

or

$$(6.11.10) \qquad \sin \frac{\theta_{\nu+1} - \theta_\nu}{4} \leqq N^{-1}\{(\mu C/2)^{-1} n^2\}^{1/(2m)} \left\{\int_a^b w(x)\, dx\right\}^{1/(2m)}$$

By substituting $N = [n/\log n]$, $m = [\log n]$, we see that the condition $m(N - 1) \leqq 2n - 1$ is satisfied for large values of n, and (6.11.2) follows immediately.

The same inequality (6.11.2) holds without essential change if $w(x) \geqq \mu(1 - x)^\alpha(1 + x)^\beta$, $\mu > 0$, where α and β are greater than -1. In this case, the last remark in §7.71 (4) must be used. The constant K in (6.11.2) now depends on μ, a, b, α, and β.

(3) We note the following simple result:

THEOREM 6.11.2. *Let $w(x)$ be a weight function on the interval $[-1, +1]$, and suppose*

$$(6.11.11) \qquad A \leqq (1 - x^2)^{\frac{1}{2}} w(x) \leqq B, \qquad -1 \leqq x \leqq +1,$$

where A and B are positive constants. If $x_\nu = \cos \theta_\nu$, $0 < \theta_\nu < \pi$, $\nu = 1, 2, \cdots, n$, stand for the zeros of the orthonormal polynomial $p_n(x)$ associated with $w(x)$, in decreasing order, we have

$$(6.11.12) \qquad \theta_{\nu+1} - \theta_\nu < \frac{4\pi B}{A} \frac{1}{n}, \qquad \nu = 0, 1, 2, \cdots, n;\ \theta_0 = 0,\ \theta_{n+1} = \pi.$$

This remark is due also to Erdös-Turán (written communication). The proof can be based on Theorem 3.41.1. We denote by λ_ν the Christoffel number corresponding to x_ν. Then

$$(6.11.13) \qquad A(\theta_{\nu+1} - \theta_\nu) \leqq \int_{\theta_\nu}^{\theta_{\nu+1}} w(\cos\theta) \sin\theta\, d\theta = \int_{x_{\nu+1}}^{x_\nu} w(x)\, dx \leqq \lambda_\nu + \lambda_{\nu+1},$$
$$\nu = 0, 1, 2, \cdots, n;\ \lambda_0 = \lambda_{n+1} = 0.$$

On the other hand,

$$\rho(x) = \rho(\cos\theta) = \left(\frac{\sin\{n(\theta - \theta_\nu)/2\}}{n \sin\{(\theta - \theta_\nu)/2\}}\right)^2 + \left(\frac{\sin\{n(\theta + \theta_\nu)/2\}}{n \sin\{(\theta + \theta_\nu)/2\}}\right)^2$$

is a π_{n-1} in $x = \cos\theta$. Furthermore, $\rho(x_\nu) = \rho(\cos\theta_\nu) \geqq 1$, so that (cf. (1.6.5))

$$\lambda_\nu \leqq \sum_{k=1}^{n} \lambda_k \rho(x_k) = \int_{-1}^{+1} \rho(x) w(x)\, dx \leqq B \int_0^\pi \rho(\cos\theta)\, d\theta$$

(6.11.14)

$$= B \int_{-\pi}^{+\pi} \left(\frac{\sin\{n\theta/2\}}{n \sin\{\theta/2\}} \right)^2 d\theta = \frac{2\pi B}{n}.$$

On combining this with (6.11.13), the statement is seen to be true.

Erdös-Turán proved also (cf. **2**) that if $0 < A \leqq w(x) \leqq B$, $-1 \leqq x \leqq +1$, *and* $0 < \epsilon < \pi/2$, *then, with the notations of Theorem* 6.11.2,

$$\frac{K_1}{n} < \theta_{\nu+1} - \theta_\nu < \frac{K_2}{n} \tag{6.11.15}$$

provided $\epsilon \leqq \theta_\nu \leqq \pi - \epsilon$. *Here* K_1 *and* K_2 *depend on* A, B, *and* ϵ.

6.12. Variation of the zeros with a parameter

A. Markoff proved (**4**) an important statement concerning the dependence of the zeros of $p_n(x)$ on a parameter τ which appears in the weight function $w(x) = w(x, \tau)$.

(1) THEOREM 6.12.1. *Let* $w(x, \tau)$ *be a weight function on the interval* $[a, b]$ *depending on a parameter* τ *such that* $w(x, \tau)$ *is positive and continuous for* $a < x < b$, $\tau_1 < \tau < \tau_2$. *Also, assume the existence and continuity of the partial derivative* $w_\tau(x, \tau)$ *for* $a < x < b$, $\tau_1 < \tau < \tau_2$, *and the convergence of the integrals*

$$\int_a^b x^\nu w_\tau(x, \tau)\, dx, \qquad \nu = 0, 1, 2, \cdots, 2n - 1, \tag{6.12.1}$$

uniformly in every closed interval $\tau' \leqq \tau \leqq \tau''$ *of the open segment* τ_1, τ_2. *If the zeros of* $p_n(x) = p_n(x, \tau)$ *be denoted by* $x_1(\tau) > x_2(\tau) > \cdots > x_n(\tau)$, *the* ν*th zero* $x_\nu(\tau)$ (*for a fixed value of* ν) *is an increasing function of* τ *provided that* w_τ/w *is an increasing function of* x, $a < x < b$.

The integrals (2.2.1) for the moments $c_\nu[d\alpha(x) = w(x, \tau)\, dx]$ converge uniformly in $\tau' \leqq \tau \leqq \tau''$, and the relations (2.2.1) may be differentiated with respect to τ; $\nu = 0, 1, 2, \cdots, 2n - 1$. Let $a < a' < b' < b$. For $a' \leqq x \leqq b'$, $\tau' \leqq \tau \leqq \tau''$, the function $w(x, \tau)$ has a positive minimum; whence the determinants D_{n-1} are uniformly bounded from zero if $\tau' \leqq \tau \leqq \tau''$ [(2.2.11)]. According to (2.2.6) the coefficients of $p_n(x)$, therefore also the zeros $x_\nu(\tau)$ (which are all distinct), possess continuous derivatives for $\tau_1 < \tau < \tau_2$.

Let $\rho(x)$ be a fixed π_{2n-1}. Apply Theorem 3.4.1 with $d\alpha(x) = w(x)\, dx$. The Christoffel numbers $\lambda_\nu = \lambda_\nu(\tau)$ are obviously functions of τ with a continuous derivative [(3.4.3)]. Differentiating (3.4.1) with respect to τ, we obtain

$$\int_a^b w_\tau(x, \tau)\rho(x)\, dx = \sum_{\nu=1}^{n} \lambda_\nu(\tau)\rho'(x_\nu)x'_\nu(\tau) + \sum_{\nu=1}^{n} \lambda'_\nu(\tau)\rho(x_\nu). \tag{6.12.2}$$

Now we substitute

$$\rho(x) = \frac{\{p_n(x)\}^2}{x - x_\nu}, \qquad \text{whence} \qquad \rho'(x_\nu) = \{p_n'(x_\nu)\}^2, \tag{6.12.3}$$

so that

$$\int_a^b w_\tau(x, \tau) \frac{\{p_n(x)\}^2}{x - x_\nu} dx = \lambda_\nu(\tau)\{p_n'(x_\nu)\}^2 x_\nu'(\tau), \tag{6.12.4}$$

since $\rho'(x_\mu) = 0$ if $\mu \neq \nu$. The left-hand member can be written in the form

$$\int_a^b \left\{w_\tau(x,\tau) - \frac{w_\tau(x_\nu, \tau)}{w(x_\nu, \tau)} w(x,\tau)\right\} \frac{\{p_n(x)\}^2}{x - x_\nu} dx, \tag{6.12.5}$$

the second term being zero because of the orthogonality. The difference

$$\frac{w_\tau(x, \tau)}{w(x, \tau)} - \frac{w_\tau(x_\nu, \tau)}{w(x_\nu, \tau)} \tag{6.12.6}$$

has the same sign as $x - x_\nu$ according to the assumption. This establishes the statement.[31]

(2) We point out the following consequence:

THEOREM 6.12.2. *Let $w(x)$ and $W(x)$ be two weight functions on $[a, b]$, both positive and continuous for $a < x < b$. Let $W(x)/w(x)$ be increasing. Then if $\{x_\nu\}$ and $\{X_\nu\}$ denote the zeros of the corresponding orthogonal polynomials of degree n in decreasing order, we have*

$$x_\nu < X_\nu, \qquad \nu = 1, 2, \cdots, n. \tag{6.12.7}$$

Defining $w(x, \tau) = (1 - \tau)w(x) + \tau W(x)$, $0 \leqq \tau \leqq 1$, we see that

$$\frac{w_\tau(x, \tau)}{w(x, \tau)} = \frac{W(x) - w(x)}{(1 - \tau)w(x) + \tau W(x)} = \tau^{-1} - \frac{\tau^{-1}}{1 - \tau + \tau W(x)/w(x)} \tag{6.12.8}$$

is an increasing function of x, $0 < \tau < 1$. We also have $w(x, 0) = w(x)$, $w(x, 1) = W(x)$.

Various applications of these results will be given in §6.21.

6.2. Location of the zeros of the classical polynomials

The discussion of this question, given in §3.3 for the general orthogonal polynomials, was based on the orthogonality property. In the particular cases called classical polynomials (§2.4) there exist various other approaches which are interesting from the point of view of method. It is assumed that $\alpha > -1$ and $\beta > -1$ in the Jacobi case, and $\alpha > -1$ in the Laguerre case; furthermore, $n \geqq 2$.

[31] This proof does not differ essentially from the original one due to A. Markoff, although the present arrangement is somewhat clearer.

(1) First of all, if we are dealing with the classical polynomials, a more precise form can be given to the argument indicated in §3.3 (4). In fact, for the polynomials in question the number of the sign variations in (3.3.5) for $x = a$ and $x = b$ can be readily calculated.[32] We use (4.1.1) and (4.1.4) for Jacobi polynomials, and (5.1.7) and (5.1.8) for Laguerre polynomials. In addition to the reality and distinctness of the zeros, the statement as to their location follows from the same argument.

In the Jacobi case the value of

$$\operatorname{sgn}\{P_{n-1}^{(\alpha,\beta)}(x)\}\cdot\frac{d}{dx}\{P_n^{(\alpha,\beta)}(x)\}$$

at the zeros of $P_n^{(\alpha,\beta)}(x)$ can also be determined by means of (4.5.7) (instead of the method in §3.3 (4) generally given). In the Laguerre case, (5.1.14) can be used. The situation is especially simple for Hermite polynomials; we have but to apply the first equation in (5.5.10).

(2) By Rolle's theorem the formulas (4.3.1), (5.1.5), (5.5.3) of Rodrigues' type furnish the statement again. We must bear in mind that the derivatives of $(1-x)^{n+\alpha}(1+x)^{n+\beta}$, $e^{-x}x^n$, and e^{-x^2} of the orders $0, 1, 2, \cdots, n-1$ vanish at $x = \pm 1$, $x = 0, +\infty$, and $x = \pm\infty$, respectively.

(3) The statement also follows from the differential equations (4.2.1), (5.1.2), and (5.5.2). To this end we first show that the zeros of Jacobi polynomials are different from -1, $+1$, and from one another. Differentiating (4.2.1) k times, we have

$$(1-x^2)y^{(k+2)} + [\beta - \alpha - (\alpha+\beta+2k+2)x]y^{(k+1)} + [n(n+\alpha+\beta+1) - k(k+\alpha+\beta+1)]y^{(k)} = 0.$$

Were y to vanish for $x = +1$ or $x = -1$, it would follow from (4.2.1) that $y' = 0$, whence from the equation just obtained ($k = 1$) $y'' = 0$, and so on; that is, $y \equiv 0$. (The coefficient of $y^{(k+1)}$ is different from zero for $x = \pm 1$.) Therefore, each of the zeros of $P_n^{(\alpha,\beta)}(x)$ is different from ± 1. Moreover, the zeros are all simple since (4.2.1) combined with $y = y' = 0$, $x \neq \pm 1$, yields $(1-x^2)y'' = 0$, or $y'' = 0$. Using the equation for y', we find $y''' = 0$ in the same way, and so on; that is, $y \equiv 0$ again.

Similarly, we show that the zeros of $L_n^{(\alpha)}(x)$ are simple and different from 0, and the zeros of $H_n(x)$ are simple.

We next apply the following theorem due to Laguerre (Pólya-Szegö **1**, vol. 2, pp. 59, 244–245, problem 118).

Let $f(x)$ be a π_n and x_0 one of its simple zeros. Then any circle through the points

$$x_0 \quad \textit{and} \quad x_0' = x_0 - 2(n-1)\frac{f'(x_0)}{f''(x_0)} \tag{6.2.1}$$

[32] Professor Pólya has kindly called to my attention the fact that the same can be done for the general orthogonal polynomials by using the determinant representation (2.2.6).

contains some zeros of $f(x)$ in both domains bounded by it, unless all the zeros lie on the circumference of this circle. The same is true if a straight line replaces the circle.[33]

The proof is as follows. Let $f(x) = (x - x_0)g(x)$, and let $x_1, x_2, \cdots, x_{n-1}$ denote the zeros of $g(x)$. Then $g(x_0) = f'(x_0)$ and $g'(x_0) = \frac{1}{2}f''(x_0)$, so that

$$\frac{1}{x_0 - x_1} + \frac{1}{x_0 - x_2} + \cdots + \frac{1}{x_0 - x_{n-1}} = \frac{g'(x_0)}{g(x_0)} = \frac{1}{2}\frac{f''(x_0)}{f'(x_0)}. \tag{6.2.2}$$

Hence (6.2.1) becomes

$$\frac{1}{n-1}\left(\frac{1}{x_0 - x_1} + \frac{1}{x_0 - x_2} + \cdots + \frac{1}{x_0 - x_{n-1}}\right) = \frac{1}{x_0 - x_0'}. \tag{6.2.3}$$

The linear transformation $X = (x_0 - x)^{-1}$ carries the points $x_1, x_2, \cdots, x_{n-1}$, and x_0' into certain points $X_1, X_2, \cdots, X_{n-1}$ and X_0'. Then we have

$$\frac{1}{n-1}(X_1 + X_2 + \cdots + X_{n-1}) = X_0', \tag{6.2.4}$$

so that any straight line through X_0' separates the points $X_1, X_2, \cdots, X_{n-1}$ from one another, unless they all lie on this straight line. Referred back to the x-plane, this yields the theorem.

For the Jacobi polynomials $y = P_n^{(\alpha,\beta)}(x)$ we obtain from (4.2.1)

$$\frac{y'}{y''} = \frac{1 - x^2}{\alpha - \beta + (\alpha + \beta + 2)x} \quad \text{if} \quad y = 0, \tag{6.2.5}$$

so that

$$x_0' = x_0 - \frac{2(n-1)}{\dfrac{\alpha+1}{1-x_0} - \dfrac{\beta+1}{1+x_0}}. \tag{6.2.6}$$

Let x_0 be a zero of y with the greatest imaginary part. If there were any non-real zeros, we should have $\mathfrak{I}(x_0) > 0$, and

$$\mathfrak{I}\left(\frac{\alpha+1}{1-x_0}\right) > 0, \qquad \mathfrak{I}\left(-\frac{\beta+1}{1+x_0}\right) > 0, \tag{6.2.7}$$

whence $\mathfrak{I}(x_0') > \mathfrak{I}(x_0)$. Therefore x_0' lies in the half-plane $\mathfrak{I}(x) > \mathfrak{I}(x_0)$, and a circle can be drawn through x_0 and x_0' which contains no zeros. (The zeros cannot all lie on this circle since they would then have to coincide with x_0, an impossibility.) Hence all the zeros are real. Now let x_0 be the greatest zero, $x_0 \neq \pm 1$. If we had $x_0 > 1$, (6.2.6) would give $x_0' > x_0$. Considering an arbitrary circle through x_0 and x_0', we are again led to a contradiction.

In the case of Laguerre polynomials,

[33] If $f''(x_0) = 0$, we have $x_0' = \infty$, and a straight line through x_0 must be considered.

$$x_0' = x_0 - \frac{2(n-1)}{1 - \dfrac{\alpha+1}{x_0}}. \tag{6.2.8}$$

From $\mathfrak{J}(x_0) > 0$ we obtain, as before, $\mathfrak{J}(x_0') > \mathfrak{J}(x_0)$. From $x_0 < 0$ we find $x_0' < x_0$.[34]

In the case of Hermite polynomials,

$$x_0' = x_0 - \frac{n-1}{x_0}, \tag{6.2.9}$$

so that $\mathfrak{J}(x_0') > \mathfrak{J}(x_0)$ if $\mathfrak{J}(x_0) > 0$.

(4) Laguerre's theorem also furnishes certain bounds for the zeros. Let x_1 be the largest and x_n the smallest of the zeros of $P_n^{(\alpha,\beta)}(x)$. Then $(\alpha+1)/(1-x_1) - (\beta+1)/(1+x_1) > 0$ and (cf. (6.2.6))

$$-1 < x_n \leqq x_1 - \frac{2(n-1)}{\dfrac{\alpha+1}{1-x_1} - \dfrac{\beta+1}{1+x_1}} < x_1, \tag{6.2.10}$$

so that

$$\frac{\alpha+1}{1-x_1} - \frac{\beta+1}{1+x_1} > \frac{2(n-1)}{1+x_1}, \qquad x_1 > \frac{\beta-\alpha+2n-2}{\beta+\alpha+2n}, \tag{6.2.11}$$

or for $\beta \geqq \alpha$

$$x_1 > \frac{n-1}{n+\alpha}. \tag{6.2.12}$$

In the ultraspherical case $x_n = -x_1$, so that

$$\frac{\alpha+1}{1-x_1} - \frac{\alpha+1}{1+x_1} > \frac{n-1}{x_1}, \qquad x_1 > \left(\frac{n-1}{n+2\alpha+1}\right)^{\frac{1}{2}}. \tag{6.2.13}$$

(If $n = 2$, the sign $>$ is to be replaced by $=$.) This bound is better than the preceding one. Both bounds have the form $1 - (\alpha+1)/n + O(n^{-2})$. Similarly, an upper bound can be obtained for x_n.

An analogous argument yields the bounds

$$x_0 > 2n + \alpha - 1, \qquad x_0 > \{\tfrac{1}{2}(n-1)\}^{\frac{1}{2}} \tag{6.2.14}$$

for the largest zeros x_0 of $L_n^{(\alpha)}(x)$ and of $H_n(x)$, respectively. (For $n = 2$ the second inequality becomes an equation.) These are very rough estimates (Theorem 6.32).

(5) Another proof of the reality and simplicity of the zeros (also furnishing $a < x_\nu < b$) can likewise be based on the differential equation by using the considerations of §6.7. It must be remembered that the polynomials in question are the only polynomial solutions of the corresponding differential equations (see §4.2 (2), §5.1 (1)).

[34] From (6.2.8) we also find that the least zero is $<\alpha + 1$.

(6) In this connection, we refer to a very elementary method due to Laguerre (3), which furnishes certain upper bounds for the zeros of the classical polynomials. Since $g''(x_0) = \frac{1}{3}f'''(x_0)$, we have from (6.2.2)

$$
\begin{aligned}
\frac{1}{(x_0 - x_1)^2} + \frac{1}{(x_0 - x_2)^2} + \cdots + \frac{1}{(x_0 - x_{n-1})^2} &= -\left(\frac{d}{dx_0}\right)\frac{g'(x_0)}{g(x_0)} \\
&= \frac{\{g'(x_0)\}^2 - g(x_0)g''(x_0)}{\{g(x_0)\}^2} = \frac{3\{f''(x_0)\}^2 - 4f'(x_0)f'''(x_0)}{12\{f'(x_0)\}^2},
\end{aligned}
\tag{6.2.15}
$$

and according to Cauchy's inequality

$$
\begin{aligned}
(n-1)\sum_{\nu=1}^{n-1}\frac{1}{(x_0 - x_\nu)^2} &- \left\{\sum_{\nu=1}^{n-1}\frac{1}{x_0 - x_\nu}\right\}^2 \\
&= (n-1)\frac{3\{f''(x_0)\}^2 - 4f'(x_0)f'''(x_0)}{12\{f'(x_0)\}^2} - \frac{\{f''(x_0)\}^2}{4\{f'(x_0)\}^2} \geqq 0,
\end{aligned}
$$

or

$$
3(n-2)\{f''(x_0)\}^2 - 4(n-1)f'(x_0)f'''(x_0) \geqq 0. \tag{6.2.16}
$$

This condition is necessary for each zero of a polynomial with real and distinct zeros.

In the case of Legendre polynomials,

$$
\begin{aligned}
(1 - x_0^2)f''(x_0) &= 2x_0f'(x_0), \\
(1 - x_0^2)f'''(x_0) &= 4x_0f''(x_0) - (n-1)(n+2)f'(x_0) \\
&= \frac{2 - n - n^2 + (6 + n + n^2)x_0^2}{1 - x_0^2}f'(x_0),
\end{aligned}
$$

so that

$$
3(n-2)4x_0^2 - 4(n-1)[2 - n - n^2 + (6 + n + n^2)x_0^2] \geqq 0;
$$

whence

$$
|x_0| \leqq (n-1)\left\{\frac{n+2}{n(n^2+2)}\right\}^{\frac{1}{2}} = 1 - \frac{5/2}{n^2} + \cdots. \tag{6.2.17}
$$

The "true" constant, as $n \to \infty$, in the second term is $j_1^2/2 = 2.891592\cdots$ (instead of $5/2$), where j_1 is the least positive zero of $J_0(x)$ (see (6.3.15)).

In the case of Hermite polynomials, from (5.5.2)

$$
\begin{gathered}
f''(x_0) = 2x_0f'(x_0), \qquad f'''(x_0) = 2(2x_0^2 - n + 1)f'(x_0), \\
3(n-2)x_0^2 - 2(n-1)(2x_0^2 - n + 1) \geqq 0;
\end{gathered}
$$

whence

$$
|x_0| \leqq \frac{2^{\frac{1}{2}}(n-1)}{(n+2)^{\frac{1}{2}}} = (2n+1)^{\frac{1}{2}} - \frac{9}{2}(2n+1)^{-\frac{1}{2}} + \cdots. \tag{6.2.18}
$$

This bound is better than that obtained by Sturm's method (cf. §6.31 (4)), although according to (6.32.5) the "true" order of the second term is $n^{-1/6}$

6.21. Inequalities for the zeros of the classical polynomials

A. Markoff's theorem (§6.12) furnishes several remarkable inequalities for the zeros of the classical polynomials.

(1) In discussing the zeros of Jacobi polynomials, we again enumerate the zeros $x_\nu = \cos\theta_\nu$ in decreasing order:

$$(6.21.1) \quad +1 > x_1 > x_2 > \cdots > x_n > -1; \quad 0 < \theta_1 < \theta_2 < \cdots < \theta_n < \pi.$$

THEOREM 6.21.1. *Let* $\{x_\nu = x_\nu(\alpha, \beta)\}$ *denote the zeros of the Jacobi polynomial* $P_n^{(\alpha,\beta)}(x)$ *in decreasing order. Then*

$$(6.21.2) \qquad \frac{\partial x_\nu}{\partial \alpha} < 0, \quad \frac{\partial x_\nu}{\partial \beta} > 0, \qquad \nu = 1, 2, \cdots, n.$$

In the ultraspherical case $\alpha = \beta$ *we have*

$$(6.21.3) \qquad \frac{\partial x_\nu}{\partial \alpha} < 0, \qquad \nu = 1, 2, \cdots, [n/2].$$

Applying Theorem 6.12.1 to $w(x, \tau) = (1 - x)^\alpha(1 + x)^\beta$ with $\alpha = \tau$, β fixed, or $\beta = \tau$, α fixed, we obtain the inequalities (6.21.2) (cf. A. Markoff **4**; Stieltjes **6**, p. 76). In the first case we have, in fact, $w_\tau/w = \log(1 - x)$, which is a decreasing function of x. The proof is similar in the second case.

The inequality (6.21.3) for the ultraspherical case is due to Stieltjes (**6**, p. 77). For the negative zeros the opposite inequality holds. This inequality does not follow directly from Theorem 6.12.1, since for $w(x, \tau) = (1 - x^2)^\tau$ the ratio $w_\tau/w = \log(1 - x^2)$ is not monotonic. However, it follows immediately from (6.21.2) by using (4.1.5).

[The proof of Stieltjes for (6.21.2) and (6.21.3) is entirely different from that of Markoff and is based on the differential equation (see below §6.22). Markoff also attempts a direct approach to (6.21.3) through a general theorem, but his proof is incorrect. In his notation (**4**, p. 181) the function

$$f(y) = \frac{(y - e)V(y, \xi)}{\dfrac{\partial V(y, \xi)}{\partial \xi}}$$

is equal to $y(\log 1/(1 - y^2))^{-1}$ in the ultraspherical case. This function approaches $+\infty$ as $y \to +0$ and $-\infty$ for $y \to -0$. Therefore, although $f'(y) < 0$, nothing can be said about the sign of the ratio $(f(y) - f(x_i))/(y - x_i)$. Incidentally, the condition $\partial V(y, \xi)/\partial\xi > 0$ of Markoff is not satisfied in the ultraspherical case at $y = e = 0$.

In the general case, $f(y) > 0$ for $y > e$, and $f(y) < 0$ for $y < e$. This fact is compatible with the decreasing property only if the denominator of $f(y)$ becomes

0 as $y \to e$. The function $f(y)$ is, however, in any case discontinuous at $y = e$, so that in the general case the same criticism applies as in the special case mentioned before.]

Apparently, Stieltjes was in possession of the general theorem of §6.12 (see **6**, p. 79, section 5, and the remark on p. 88).

(2) THEOREM 6.21.2. *Let the parameters α and β of the Jacobi polynomial $P_n^{(\alpha,\beta)}(x)$ be subject to the conditions*

$$-\tfrac{1}{2} \leqq \alpha \leqq +\tfrac{1}{2}, \qquad -\tfrac{1}{2} \leqq \beta \leqq +\tfrac{1}{2}. \tag{6.21.4}$$

Then we have for the zeros (notation as before)

$$\frac{2\nu - 1}{2n + 1}\pi \leqq \theta_\nu \leqq \frac{2\nu}{2n+1}\pi, \qquad \nu = 1, 2, \cdots, n, \tag{6.21.5}$$

with equality only in the special cases $\alpha = -\frac{1}{2}$, $\beta = +\frac{1}{2}$ and $\alpha = +\frac{1}{2}$, $\beta = -\frac{1}{2}$, respectively.

For Legendre polynomials, that is, for $\alpha = \beta = 0$, this result is due to Bruns (**1**). The general case is due to A. Markoff and Stieltjes. For the proof we observe that according to (6.21.2) the maximum and minimum of $x_\nu = \cos\theta_\nu$ are attained in the special cases mentioned above. Now we use (4.1.8).

(3) THEOREM 6.21.3. *In the ultraspherical case*

$$-\tfrac{1}{2} \leqq \alpha = \beta \leqq +\tfrac{1}{2} \tag{6.21.6}$$

we have the inequalities

$$(\nu - \tfrac{1}{2})\frac{\pi}{n} \leqq \theta_\nu \leqq \nu\frac{\pi}{n+1}, \qquad \nu = 1, 2, \cdots, [n/2], \tag{6.21.7}$$

with the equality sign valid only in the special cases $\alpha = \beta = -\frac{1}{2}$ and $\alpha = \beta = +\frac{1}{2}$, respectively.

The first proof of these inequalities, which are more precise than the corresponding inequalities (6.21.5) of "Bruns's type," is due to Stieltjes (**6**). Markoff's proof is not correct (see above). For the proof of Stieltjes see §6.22. Refer also to §6.3 (2) and (3). Corresponding inequalities for the negative zeros can readily be obtained from the symmetry relation $x_\nu + x_{n+1-\nu} = 0$.

We base our proof on Theorem 6.21.1. According to (6.21.3), the maximum and minimum of x_ν, $\nu \leqq [n/2]$, are attained if $\alpha = \beta = -\frac{1}{2}$ and $\alpha = \beta = +\frac{1}{2}$, respectively. Now the zeros of the polynomials (4.1.7) are

$$\cos(\nu - \tfrac{1}{2})\frac{\pi}{n} \quad \text{and} \quad \cos\nu\frac{\pi}{n+1}, \qquad \nu = 1, 2, \cdots, n. \tag{6.21.8}$$

(4) In the case of Laguerre polynomials we have $w(x, \tau) = e^{-x}x^\alpha$, with $\alpha = \tau$, and $w_\tau/w = \log x$ increasing. Hence the zeros of Laguerre polynomials are

increasing functions of the parameter α, $\alpha > -1$. Thus, on account of (5.6.1), we obtain the following theorem:

THEOREM 6.21.4. *If*

$$-\tfrac{1}{2} \leqq \alpha \leqq +\tfrac{1}{2}, \tag{6.21.9}$$

the zeros x_ν of $L_n^{(\alpha)}(x)$, arranged in increasing order, have the bounds

$$\xi_\nu^2 \leqq x_\nu \leqq \eta_\nu^2 . \tag{6.21.10}$$

Here ξ_ν and η_ν denote the νth positive zeros of the Hermite polynomials $H_{2n}(x)$ and $H_{2n+1}(x)$, respectively.

6.22. Proof of Stieltjes for the monotonic variation of the zeros of the classical polynomials

Stieltjes (**6**, pp. 73–77) gives a proof of Theorem 6.21.1 along the following lines. Substituting $x = x_\nu$ in (4.2.1), we have

$$\begin{aligned}\frac{1}{2}\frac{y''}{y'} + \frac{1}{2}\frac{\alpha+1}{x_\nu - 1} + \frac{1}{2}\frac{\beta+1}{x_\nu+1} = \frac{1}{x_\nu - x_1} + \cdots + \frac{1}{x_\nu - x_n} \\ + \frac{1}{2}\frac{\alpha+1}{x_\nu - 1} + \frac{1}{2}\frac{\beta+1}{x_\nu+1} = 0.\end{aligned} \tag{6.22.1}$$

Differentiation of this equation with respect to α yields

$$\begin{aligned}\frac{1}{(x_\nu - x_1)^2}\left(\frac{\partial x_\nu}{\partial\alpha} - \frac{\partial x_1}{\partial\alpha}\right) + \frac{1}{(x_\nu - x_2)^2}\left(\frac{\partial x_\nu}{\partial\alpha} - \frac{\partial x_2}{\partial\alpha}\right) + \cdots \\ + \frac{1}{(x_\nu - x_n)^2}\left(\frac{\partial x_\nu}{\partial\alpha} - \frac{\partial x_n}{\partial\alpha}\right) + \frac{1}{2}\frac{\alpha+1}{(x_\nu-1)^2}\frac{\partial x_\nu}{\partial\alpha} \\ + \frac{1}{2}\frac{\beta+1}{(x_\nu+1)^2}\frac{\partial x_\nu}{\partial\alpha} - \frac{1}{2}\frac{1}{x_\nu - 1} = 0,\end{aligned} \tag{6.22.2}$$

or

$$\sum_{\mu=1}^{n} a_{\nu\mu}\frac{\partial x_\mu}{\partial\alpha} = \frac{1}{2}\frac{1}{x_\nu - 1}, \qquad \nu = 1, 2, \cdots, n, \tag{6.22.3}$$

where

$$\begin{aligned}a_{\nu\nu} = \frac{1}{(x_\nu - x_1)^2} + \cdots + \frac{1}{(x_\nu - x_{\nu-1})^2} + \frac{1}{(x_\nu - x_{\nu+1})^2} + \cdots \\ + \frac{1}{(x_\nu - x_n)^2} + \frac{1}{2}\frac{\alpha+1}{(x_\nu - 1)^2} + \frac{1}{2}\frac{\beta+1}{(x_\nu+1)^2},\end{aligned} \tag{6.22.4}$$

and

$$a_{\nu\mu} = a_{\mu\nu} = -\frac{1}{(x_\nu - x_\mu)^2}, \qquad \nu \neq \mu. \tag{6.22.5}$$

The matrix $(a_{\nu\mu})$ is positive definite since

$$(6.22.6)\qquad \begin{aligned} K &= \sum_{\nu=1}^{n}\sum_{\mu=1}^{n} a_{\nu\mu}u_\nu u_\mu \\ &= \frac{1}{2}\sum_{\substack{\nu,\mu=1,2,\cdots,n\\ \nu\neq\mu}} \left(\frac{u_\nu - u_\mu}{x_\nu - x_\mu}\right)^2 + \frac{1}{2}\sum_{\nu=1}^{n}\left\{\frac{\alpha+1}{(x_\nu-1)^2} + \frac{\beta+1}{(x_\nu+1)^2}\right\}u_\nu^2. \end{aligned}$$

Stieltjes now uses the following theorem: If $A = (a_{\nu\mu})$ is a positive definite matrix with $a_{\nu\mu} < 0$, $\nu \neq \mu$, then the reciprocal matrix $(A)^{-1}$ has only positive elements.

[We can assume that $a_{\nu\nu} = 1$, $\nu = 1, 2, \cdots, n$, so that $K = E - L$, where E is the unit form, and the coefficients of L are non-negative. Then the absolute value of L is less than 1 if $E = 1$, and the reciprocal form of K can be written as follows:

$$(6.22.7)\qquad (K)^{-1} = E + L + L^2 + L^3 + \cdots .$$

All the forms of the right-hand member have non-negative coefficients, and the coefficients of $E + L$ are also positive.[35]]

By virtue of this theorem, the statement follows immediately from (6.22.3).

The proof for the second inequality of (6.21.2) is similar. The ultraspherical case (6.21.3) can either be treated by means of (4.1.5), or directly handled (cf. Stieltjes, loc. cit.). The same method applies to Laguerre polynomials (Theorem 6.21.4). In this case we have

$$(6.22.8)\qquad \begin{aligned} a_{\nu\nu} &= \frac{1}{(x_\nu - x_1)^2} + \cdots + \frac{1}{(x_\nu - x_{\nu-1})^2} + \frac{1}{(x_\nu - x_{\nu+1})^2} + \cdots \\ &\qquad + \frac{1}{(x_\nu - x_n)^2} + \frac{\alpha+1}{2x_\nu^2}, \\ a_{\nu\mu} &= -\frac{1}{(x_\nu - x_\mu)^2}, \qquad \nu \neq \mu, \end{aligned}$$

with $(2x_\nu)^{-1}$ on the right-hand side of the equations corresponding to (6.22.3).

6.3. Sturm's method; Jacobi polynomials

Sturm's method (see §1.82) leads very simply to certain inequalities for the zeros of Jacobi polynomials (see Szegö **20**, Buell **1**). In this way we not only confirm some of the results of §6.21, but we are also able to improve them to a considerable degree. We assume in this section that

$$(6.3.1)\qquad -\tfrac{1}{2} \leqq \alpha \leqq +\tfrac{1}{2}, \qquad -\tfrac{1}{2} \leqq \beta \leqq +\tfrac{1}{2},$$

[35] This argument is different from that of Stieltjes, which is likewise very simple and elementary. The present argument, however, furnishes similar theorems in more complicated cases.

excluding, in general, the case $\alpha^2 = \beta^2 = \frac{1}{4}$; we arrange the zeros $x_\nu = \cos\theta_\nu$ of $P_n^{(\alpha,\beta)}(x)$ in decreasing order:

$$+1 > x_1 > x_2 > \cdots > x_n > -1; \quad 0 < \theta_1 < \theta_2 < \cdots < \theta_n < \pi. \tag{6.3.2}$$

(1) THEOREM 6.3.1. *Under the conditions mentioned we have*

$$\theta_\nu - \theta_{\nu-1} < \frac{\pi}{n + (\alpha+\beta+1)/2}, \quad \nu = 1, 2, \cdots, n+1. \tag{6.3.3}$$

This holds for $\alpha^2 = \beta^2 = \frac{1}{4}$ with the $=$ sign instead of $<$. Here we define

$$\theta_0 = \begin{cases} 0 & \text{if } \alpha > -\frac{1}{2}, \\ -\theta_1 & \text{if } \alpha = -\frac{1}{2}, \end{cases} \quad \text{and} \quad \theta_{n+1} = \begin{cases} \pi & \text{if } \beta > -\frac{1}{2}, \\ 2\pi - \theta_n & \text{if } \beta = -\frac{1}{2}. \end{cases} \tag{6.3.4}$$

We notice that for $\alpha = \beta = -\frac{1}{2}$, $\alpha = \beta = +\frac{1}{2}$, $\alpha = -\beta = \frac{1}{2}$, $\alpha = -\beta = -\frac{1}{2}$,

$$\theta_\nu = (\nu - \tfrac{1}{2})\frac{\pi}{n}, \quad \nu\frac{\pi}{n+1}, \quad \nu\frac{\pi}{n+\frac{1}{2}}, \quad \frac{\nu-\frac{1}{2}}{n+\frac{1}{2}}\pi, \qquad \nu = 0, 1, \cdots, n+1, \tag{6.3.5}$$

respectively.

Inequality (6.3.3) follows immediately from Theorem 1.82.1 by comparing (4.24.2) with

$$\frac{d^2v}{d\theta^2} + \left(n + \frac{\alpha+\beta+1}{2}\right)^2 v = 0. \tag{6.3.6}$$

We consider the solution $v = \sin\{(n + (\alpha+\beta+1)/2)(\theta - \theta_{\nu-1})\}$. For $\alpha > -\frac{1}{2}$, the condition corresponding to (1.82.2) is satisfied at $x_0 = \theta_0 = 0$ (and similarly for $\beta > -\frac{1}{2}$ at $X_0 = \theta_{n+1} = \pi$).

(2) THEOREM 6.3.2. *Under the conditions mentioned we have*

$$\frac{\nu + (\alpha+\beta-1)/2}{n + (\alpha+\beta+1)/2}\pi < \theta_\nu < \frac{\nu}{n + (\alpha+\beta+1)/2}\pi, \quad \nu = 1, 2, \cdots, n, \tag{6.3.7}$$

whereas in the ultraspherical case $\alpha = \beta = \lambda - \frac{1}{2}$

$$\theta_\nu > \frac{\nu + \alpha/2 - \frac{1}{4}}{n + \alpha + \frac{1}{2}}\pi = \frac{\nu - (1-\lambda)/2}{n+\lambda}\pi, \quad \nu = 1, 2, \cdots, [n/2]. \tag{6.3.8}$$

The bounds (6.3.7) follow from (6.3.3) by addition and by using (4.1.3) (see Buell **1**, pp. 311–312). In case $\alpha = -\frac{1}{2}$, the factor ν of the upper bound can be replaced by $\nu - \frac{1}{2}$, while if $\beta = -\frac{1}{2}$, the factor $\nu + (\alpha+\beta-1)/2$ of the lower bound can be replaced by $\nu + (\alpha+\beta)/2$. In the cases (6.3.5) the same bounds hold, at times with $=$ replacing $<$. These inequalities are more precise than (6.21.5) provided $\alpha + \beta > 0$. In the case of Legendre polynomials ($\alpha = \beta = 0$) they are identical with (6.21.5), that is, with the inequalities of Bruns (**1**).

The inequality (6.3.8) holds only for the zeros $0 < \theta_\nu < \pi/2$; it becomes an equation for $\alpha = \beta = -\frac{1}{2}$ or $\alpha = \beta = +\frac{1}{2}$. It is more precise than the lower estimate of (6.21.7).

For the proof of (6.3.8) we observe that according to (6.3.3), the sequence

$$\theta'_\nu = \theta_\nu - \frac{\nu + \alpha/2 - \frac{1}{4}}{n + \alpha + \frac{1}{2}}\pi, \qquad \nu = 0, 1, 2, \cdots, [(n+1)/2], \tag{6.3.9}$$

is decreasing. Now for n odd, we have $\theta'_{(n+1)/2} = 0$. For n even it suffices to show that $\theta'_{n/2} > 0$. This follows from (6.3.3), since $\theta_{n/2} + \theta_{n/2+1} = \pi$.

A similar argument can be used to improve the left inequality in the general case (6.3.7) provided the zeros considered all lie in a certain preassigned part $0 \leqq \theta_\nu \leqq c$ of the interval $[0, \pi]$. For further proofs of (6.3.8) see (5) and §6.6 (2).

(3) In this connection we prove the following theorem:

THEOREM 6.3.3. *Let $n \geqq 2$. Under the condition $-\frac{1}{2} < \alpha = \beta < +\frac{1}{2}$ the sequence*

$$\theta_0, \theta_1, \theta_2, \cdots, \theta_{[n/2]+1} \tag{6.3.10}$$

of the zeros of $P_n^{(\alpha,\alpha)}(\cos\theta)$ is convex, that is, $\theta_\nu - \theta_{\nu-1}$ is increasing.

This follows by applying Theorem 1.82.2 to (4.7.11), $0 < \lambda < 1$. In fact, the coefficient of u in (4.7.11) decreases monotonically. In the cases $\alpha = \beta = \pm\frac{1}{2}$ the differences $\theta_\nu - \theta_{\nu-1}$ are constant. Here again $\theta_0 = 0$ if $\alpha > -\frac{1}{2}$, while $\theta_0 = -\theta_1$ if $\alpha = -\frac{1}{2}$. If n is even, the last term of (6.3.10) lies in $[\pi/2, \pi]$, and then (1.82.4) is used. For $\alpha > -\frac{1}{2}$ condition (1.82.5) is satisfied.

From Theorem 6.3.3 a similar convex property for the sequence $\{x_\nu\}$ can easily be derived (cf. Hille **4**).

By means of Theorem 6.3.3 the upper estimate of (6.21.7), more precisely

$$\theta_\nu < \frac{\nu}{n+1}\pi, \qquad -\tfrac{1}{2} \leqq \alpha < +\tfrac{1}{2}; \nu = 1, 2, \cdots, [n/2], \tag{6.3.11}$$

can be proved in another way (Szegö **20**, pp. 5–6, 8). For, let $-\frac{1}{2} < \alpha < +\frac{1}{2}$. The sequence

$$\theta''_\nu = \theta_\nu - \frac{\nu}{n+1}\pi, \qquad \nu = 0, 1, 2, \cdots, [n/2] + 1, \tag{6.3.12}$$

is convex; it therefore attains its maximum either for $\nu = 0$ or for $\nu = [n/2] + 1$. Now $\theta''_0 = \theta''_{[n/2]+1} = 0$ if n is odd. If n is even, we must bear in mind that $\theta''_{n/2} + \theta''_{n/2+1} = 0$.

(4) Finally, by means of Sturm's method, we derive certain inequalities which involve the zeros of Bessel functions. In some respects these are more precise than the preceding inequalities, although not so simple.

THEOREM 6.3.4. *Let $\alpha = \beta = \lambda - \frac{1}{2}, 0 < \lambda < 1$. Denoting by $j_1 < j_2 < j_3 < \cdots$ the positive zeros of Bessel's function $J_\alpha(x)$, we have*

(6.3.13)
$$\theta_\nu < \frac{j_\nu}{n+\lambda}, \qquad \nu = 1, 2, 3, \cdots, n.$$

For $\lambda = 0$ and $\lambda = 1$ the sign $<$ in (6.3.13) has to be replaced by the sign $=$. The statement follows by comparing (4.7.11) with (see (1.8.9))

(6.3.14)
$$\frac{d^2v}{d\theta^2} + \left\{(n+\lambda)^2 + \frac{\lambda(1-\lambda)}{\theta^2}\right\} v = 0, \quad v = \theta^{\frac{1}{2}} J_\alpha\{(n+\lambda)\theta\}, \alpha = \lambda - \tfrac{1}{2}.$$

The estimate (6.3.13) of θ_ν is the best possible in the sense that for a fixed ν and for n arbitrary the factor j_ν cannot be replaced by a smaller one since (Theorem 8.1.2)

(6.3.15)
$$\lim_{n\to\infty} n\theta_\nu = \lim_{n\to\infty} n\theta_{\nu n} = j_\nu.$$

Incidentally, for $0 < \theta_\nu \leqq \pi/2$, a similar lower bound for θ_ν can be obtained, namely,

(6.3.16)
$$\theta_\nu > j_\nu\{(n+\lambda)^2 + k\lambda(1-\lambda)\}^{-\frac{1}{2}},$$

where k is a positive numerical constant (Szegö **20**, p. 9). Then (6.3.15) follows from (6.3.13) and (6.3.16).

(5) Finally, another remarkable property of the zeros $\theta_\nu = \theta_{\nu n}$ of $P_n^{(\lambda)}(\cos\theta)$ can be proved by a proper application of Sturm's theorem. On substituting $\theta = \xi/(n+\lambda)$ in (4.7.11) we obtain

(6.3.17)
$$\frac{d^2u}{d\xi^2} + \left\{1 + \frac{\lambda(1-\lambda)}{(n+\lambda)^2\sin^2\{\xi/(n+\lambda)\}}\right\} u = 0.$$

If $0 < \lambda < 1$, then $(n+\lambda)^2 \sin^2\{\xi/(n+\lambda)\}$ increases with n provided ξ is fixed and $0 < \xi < (n+\lambda)\pi$. Thus $(n+\lambda)\theta_{\nu n}$ increases with n if ν is fixed.[36] From these considerations the estimate (6.3.13) follows again if (6.3.15) is known. Indeed, $(n+\lambda)\theta_{\nu n} < \lim_{n\to\infty}(n+\lambda)\theta_{\nu n} = j_\nu$.

As another application we can give also a new proof of (6.3.8) since

(6.3.18)
$$(n+\lambda)\theta_{\nu n} \geqq (2\nu - 1 + \lambda)\theta_{\nu,2\nu-1} = (2\nu - 1 + \lambda)\pi/2$$

for $n \geqq 2\nu - 1$. Cf. Problem 32.

6.31. Sturm's method; Laguerre and Hermite polynomials

Suppose $\alpha > -1$.

(1) THEOREM 6.31.1. *Let $x_\nu = x_{\nu n} = x_{\nu n}(\alpha)$, $\nu = 1, 2, \cdots, n$, be the zeros of $L_n^{(\alpha)}(x)$ in increasing order. Then*

(6.31.1)
$$x_\nu > \frac{(j_\nu/2)^2}{n + (\alpha+1)/2}, \qquad \nu = 1, 2, \cdots, n.$$

Here j_ν has the same meaning as in Theorem 6.3.4.

[36] By the separation theorem (Theorem 3.3.2) $\theta_{\nu n}$ decreases as n increases for fixed ν.

When we compare the third equation in (5.1.2) with

$$U'' + \left(\frac{n + (\alpha + 1)/2}{x} + \frac{1 - \alpha^2}{4x^2}\right) U = 0, \tag{6.31.2}$$

which has the solution

$$U = x^{\frac{1}{2}} J_\alpha \left\{2x^{\frac{1}{2}}\left(n + \frac{\alpha + 1}{2}\right)^{\frac{1}{2}}\right\} \tag{6.31.3}$$

(cf. (1.8.10)), the statement follows immediately. Condition (1.82.2), $x' = 0$, is satisfied in the present case.

An upper bound of x_ν of a similar kind can also be easily obtained. Let ω be a positive constant such that $\omega < 4n + 2(\alpha + 1)$. Compare the same equation (5.1.2) as before with

$$v'' + \left\{\frac{n + (\alpha + 1)/2 - \omega/4}{x} + \frac{1 - \alpha^2}{4x^2}\right\} v = 0, \tag{6.31.4}$$

where $0 < x \leqq \omega$. Then

$$x_\nu < \frac{(j_\nu/2)^2}{n + (\alpha + 1)/2 - \omega/4}, \tag{6.31.5}$$

if the expression on the right-hand side is not greater than ω. (For a fixed ν, this is the case, provided n is large enough.) The constant $(j_\nu/2)^2$ in the inequalities (6.31.1) and (6.31.5) is the best possible in the sense explained in §6.3 (4). For a fixed ν, we have, for the zero $x_\nu = x_{\nu n}$,

$$\lim_{n \to \infty} n x_{\nu n} = (j_\nu/2)^2. \tag{6.31.6}$$

The same results can be obtained by use of (1.8.9) and the fourth equation in (5.1.2). Condition (1.82.2), $x' = 0$, is again satisfied.

(2) Both equations (5.1.2) mentioned furnish upper bounds for the zeros if we use Theorem 1.82.3 and take into account the fact that the corresponding solutions vanish at $x = +\infty$. The bound which is obtained from the fourth equation is slightly better. It is given in the following theorem:

THEOREM 6.31.2. *The largest zero of $L_n^{(\alpha)}(x)$ satisfies the inequality*

$$x_n < 2n + \alpha + 1 + \{(2n + \alpha + 1)^2 + \tfrac{1}{4} - \alpha^2\}^{\frac{1}{2}} \cong 4n. \tag{6.31.7}$$

(3) Introducing $x = \{n + (\alpha + 1)/2\}^{-1}\xi$ in the third equation (5.1.2), we obtain

$$\frac{d^2u}{d\xi^2} + \left\{\frac{1}{\xi} + \frac{1 - \alpha^2}{4\xi^2} - \frac{1}{4}\left(n + \frac{\alpha + 1}{2}\right)^{-2}\right\} u = 0. \tag{6.31.8}$$

Since the coefficient of u increases with n, it follows that, for a fixed ν, the expression

$$\{n + (\alpha + 1)/2\}x_\nu = \{n + (\alpha + 1)/2\}x_{\nu n} \tag{6.31.9}$$

decreases with increasing n. The limit as $n \to \infty$ is $(j_\nu/2)^2$ (according to (6.31.6)). An interesting consequence of this decreasing property is

$$\{n + (\alpha + 1)/2\}x_{\nu n} \leqq \{\nu + (\alpha + 1)/2\}x_{\nu\nu}, \quad n = \nu, \nu + 1, \cdots . \tag{6.31.10}$$

Applying (6.31.7) and recalling (6.31.1), we obtain the following result:

THEOREM 6.31.3. *Let $\alpha > -1$; for the zeros $x_{\nu n}$ of $L_n^{(\alpha)}(x)$, arranged in increasing order, the following estimates hold:*

$$\frac{(j_\nu/2)^2}{n + (\alpha + 1)/2} < x_{\nu n} < \{\nu + (\alpha + 1)/2\} \cdot \frac{2\nu + \alpha + 1 + \{(2\nu + \alpha + 1)^2 + \frac{1}{4} - \alpha^2\}^{\frac{1}{2}}}{n + (\alpha + 1)/2}, \tag{6.31.11}$$
$$\nu = 1, 2, \cdots, n;\ n = 1, 2, \cdots .$$

In particular, for the least zero x_{1n} we have

$$\frac{(j_1/2)^2}{n + (\alpha + 1)/2} < x_{1n} \leqq \frac{(\alpha + 1)(\alpha + 3)}{2n + \alpha + 1}, \quad n = 1, 2, 3, \cdots . \tag{6.31.12}$$

Here j_ν has the same meaning as in Theorem 6.3.4.

If ν is large, $(j_\nu/2)^2 \cong \pi^2\nu^2/4$ (cf. (1.71.7)), while the coefficient in the right-hand member of (6.31.11) is $\cong 4\nu^2$. For $\nu = 1$ we do not need (6.31.7) since x_{11} can be calculated explicitly. In fact, $x_{11} = \alpha + 1$; whence (6.31.12) follows.

E. R. Neumann (**2**, p. 26) obtains for $\alpha = 0$ an inequality similar to (6.31.11) in a different manner. He finds in this case

$$x_{\nu n} = C_{\nu n} \frac{(\nu + 1)^2}{n + 1}, \quad \nu = 1, 2, \cdots, n;\ n = 1, 2, 3, \cdots, \tag{6.31.13}$$

where $\frac{1}{4} < C_{\nu n} < 4$. In view of the well-known estimate $j_\nu > (\nu - \frac{1}{4})\pi$ (cf. (8.1.4) and Problem 32), we can derive the following inequalities of the type mentioned from (6.31.11):

$$(3\pi/16)^2 < C_{\nu n} < 4. \tag{6.31.14}$$

(The upper bound 4 can not be diminished; cf. (8.9.15); also Problem 33.) W. Hahn (**1**, pp. 228–238) generalizes and extends Neumann's method to arbitrary real values of α.

A part of these results is more precise than those occurring in the literature (see W. Hahn **2**, pp. 228–230).

(4) The corresponding considerations for Hermite polynomials are very simple. To begin with, the second equation (5.5.2) furnishes the upper bound $(2n + 1)^{\frac{1}{2}}$ for the zeros. This is not as good as the bound (6.2.18). Furthermore, assuming $n \geqq 2$, we obtain the convexity of the sequence of the zeros

$$x_{0n} < x_{1n} < x_{2n} < \cdots \tag{6.31.15}$$

of $H_n(x)$, where $x_{0n} = 0$ if n is odd, and $x_{0n} = -x_{1n}$ if n is even. In all cases, $x_{1n}, x_{2n}, \cdots$ denote the positive zeros in increasing order. (See W. Hahn **1**, p. 244.)

On comparing the same equation with $Z'' + (2n+1)Z = 0$, we find[37]

$$x_{\nu n} > \begin{cases} \dfrac{\nu - \frac{1}{2}}{(2n+1)^{\frac{1}{2}}}\pi, \\[2ex] \dfrac{\nu}{(2n+1)^{\frac{1}{2}}}\pi, & \nu = 1, 2, \cdots, [n/2]. \end{cases} \tag{6.31.16}$$

(This follows also from (6.31.1) for $\alpha = \pm\frac{1}{2}$.) Now let ω be a fixed positive number, $\omega < (2n+1)^{\frac{1}{2}}$. Then

$$x_{\nu n} < \begin{cases} \dfrac{\nu - \frac{1}{2}}{(2n+1-\omega^2)^{\frac{1}{2}}}\pi, \\[2ex] \dfrac{\nu}{(2n+1-\omega^2)^{\frac{1}{2}}}\pi, & \nu = 1, 2, 3, \cdots, [n/2], \end{cases} \tag{6.31.17}$$

provided the right-hand members are not greater than ω. For a fixed ν we see that the constants $(\nu - \frac{1}{2})\pi$ and $\nu\pi$ are the best possible.

By introducing $x = (2n+1)^{-\frac{1}{2}}\xi$, the differential equation mentioned is transformed into

$$\begin{gathered} \frac{d^2 z}{d\xi^2} + \{1 - (2n+1)^{-2}\xi^2\}z = 0, \\ z = \exp\{-(2n+1)^{-1}\xi^2/2\} H_n\{(2n+1)^{-\frac{1}{2}}\xi\}. \end{gathered} \tag{6.31.18}$$

The coefficient of z increases with n; hence (for fixed ν) $(2n+1)^{\frac{1}{2}}x_{\nu n}$ decreases as n increases. Therefore, we have (cf. (3)) $(2n+1)^{\frac{1}{2}}x_{\nu n} \leqq (4\nu+1)^{\frac{1}{2}}x_{\nu,2\nu}$ or $(4\nu+3)^{\frac{1}{2}}x_{\nu,2\nu+1}$, respectively. Thus

$$\left.\begin{matrix} \dfrac{\nu - \frac{1}{2}}{(2n+1)^{\frac{1}{2}}}\pi \\[2ex] \dfrac{\nu}{(2n+1)^{\frac{1}{2}}}\pi \end{matrix}\right\} < x_{\nu n} < \begin{cases} \dfrac{4\nu+1}{(2n+1)^{\frac{1}{2}}}, \\[2ex] \dfrac{4\nu+3}{(2n+1)^{\frac{1}{2}}}, & \nu = 1, 2, \cdots, [n/2]. \end{cases} \tag{6.31.19}$$

For the least positive zero x_{1n}, we obtain ($x_{12} = 2^{-\frac{1}{2}}$, $x_{13} = (3/2)^{\frac{1}{2}}$)

$$\left.\begin{matrix} \dfrac{\pi/2}{(2n+1)^{\frac{1}{2}}} \\[2ex] \dfrac{\pi}{(2n+1)^{\frac{1}{2}}} \end{matrix}\right\} < x_{1n} \leqq \begin{cases} \left(\dfrac{5/2}{2n+1}\right)^{\frac{1}{2}}, \\[2ex] \left(\dfrac{21/2}{2n+1}\right)^{\frac{1}{2}}, & n \geqq 2. \end{cases} \tag{6.31.20}$$

[37] In this and subsequent formulas the upper line corresponds to the case n even, while the lower line to the case n odd.

The upper bounds are more precise than those resulting from (6.31.19) for $\nu = 1$.

From (6.31.20) we can derive bounds for the minimum distance d_n between consecutive zeros. The convex property mentioned above easily furnishes $d_n = x_{1n} - x_{0n}$, that is, $d_n = 2x_{1n}$ if n is even, and $d_n = x_{1n}$ if n is odd. It follows that

$$\frac{\pi}{(2n+1)^{\frac{1}{2}}} < d_n \leqq \begin{cases} \dfrac{10^{\frac{1}{2}}}{(2n+1)^{\frac{1}{2}}}, \\ \dfrac{(21/2)^{\frac{1}{2}}}{(2n+1)^{\frac{1}{2}}}, \end{cases} \qquad n \geqq 2. \tag{6.31.21}$$

In every case we have

$$\frac{\pi}{(2n+1)^{\frac{1}{2}}} < d_n \leqq \frac{(21/2)^{\frac{1}{2}}}{(2n+1)^{\frac{1}{2}}}. \tag{6.31.22}$$

Concerning the extensive literature on this subject, we refer to Laguerre (**2**, p. 105), Korous (**1**), Wiman (**1**), A. Brauer (**1**), Hille (**4**), and Winston (**1**). Hille obtains the same lower bounds as in (6.31.20) and (by a suitable choice of ω in (6.31.17)) the upper bounds

$$x_{1n} < \begin{cases} \dfrac{\pi/2}{(2n+1)^{\frac{1}{2}}}\left\{\frac{1}{2} + \frac{1}{2}\left[1 - \left(\dfrac{\pi}{2n+1}\right)^2\right]^{\frac{1}{2}}\right\}^{-\frac{1}{2}}, \\ \dfrac{\pi}{(2n+1)^{\frac{1}{2}}}\left\{\frac{1}{2} + \frac{1}{2}\left[1 - \left(\dfrac{2\pi}{2n+1}\right)^2\right]^{\frac{1}{2}}\right\}^{-\frac{1}{2}}. \end{cases} \tag{6.31.23}$$

These bounds are better than those resulting from (6.31.20), except for $n \leqq 6$. The results of the other authors are less precise than the preceding inequalities.[38]

6.32. Sturm's method; the largest zeros of Laguerre and Hermite polynomials

(1) Let $\alpha > -1$, and enumerate the zeros $x_\nu = x_{\nu n}$ of $L_n^{(\alpha)}(x)$ or $H_n(x)$ in *decreasing* order:

$$x_1 > x_2 > x_3 > \cdots. \tag{6.32.1}$$

Our purpose is to derive inequalities and asymptotic relations for x_{1n}, as well as for $x_{\nu n}$ for fixed values of ν as $n \to \infty$.

THEOREM 6.32. *Let $i_1 < i_2 < i_3 < \cdots$ be the real zeros of Airy's function $A(x)$ (§1.81, $i_1 > 0$). If $|\alpha| \geqq \frac{1}{4}$, $\alpha > -1$, the following inequalities hold for the zeros $\{x_\nu\}$ of $L_n^{(\alpha)}(x)$:*

$$x_\nu^{\frac{1}{2}} < (4n + 2\alpha + 2)^{\frac{1}{2}} - 6^{-\frac{1}{3}}(4n + 2\alpha + 2)^{-\frac{1}{6}} i_\nu, \tag{6.32.2}$$

[38] An exception is the lower bound in (6.31.20) for the special values $n = 2$ and $n = 3$ in which Wiman's expression furnishes the exact values. For $n = 3$ the upper bound of Wiman is the same as in (6.31.20).

whereas for the zeros $\{x_\nu\}$ *of* $H_n(x)$:

(6.32.3) $$x_\nu < (2n+1)^{\frac{1}{2}} - 6^{-\frac{1}{2}}(2n+1)^{-\frac{1}{6}} i_\nu.$$

Furthermore, we have for a fixed ν

(6.32.4) $$x_\nu^{\frac{1}{2}} = (4n+2\alpha+2)^{\frac{1}{2}} - 6^{-\frac{1}{2}}(4n+2\alpha+2)^{-\frac{1}{6}}\{i_\nu + \epsilon_n\},$$

(6.32.5) $$x_\nu = (2n+1)^{\frac{1}{2}} - 6^{-\frac{1}{2}}(2n+1)^{-\frac{1}{6}}\{i_\nu + \epsilon_n\},$$

in the Laguerre and Hermite cases, respectively, where $\lim_{n\to\infty} \epsilon_n = 0$.

These remarkable results possess an extended literature. (Zernike **1**, W. Hahn **1**, p. 227. See also Korous **1**, Bottema **1**, Van Veen **1**, and Spencer **1**.) In what follows, Sturm's method is used to prove (6.32.3). A similar argument can be applied for the proof of (6.32.2) (cf. the fourth equation in (5.1.2)). Formulas (6.32.5) and (6.32.4) follow from a certain asymptotic expansion of Hermite polynomials due to Plancherel and Rotach as well as from corresponding expansions for Laguerre polynomials; (6.32.4) holds for arbitrary real α. We discuss these formulas in Chapter VIII (cf. §8.9 (3)). They show that the constant i_ν in (6.32.2) and (6.32.3) is the best possible if ν is fixed and n is arbitrary.

We notice that the expressions

(6.32.6) $$\begin{gathered}\{(4n+2\alpha+2)^{\frac{1}{2}} - 6^{-\frac{1}{2}}(4n+2\alpha+2)^{-\frac{1}{6}} i_1\}^2,\\ (2n+1)^{\frac{1}{2}} - 6^{-\frac{1}{2}}(2n+1)^{-\frac{1}{6}} i_1\end{gathered}$$

are upper bounds for the zeros of $L_n^{(\alpha)}(x)$ and $H_n(x)$, respectively, $|\alpha| \geqq \frac{1}{4}$, $\alpha > -1$. Here the constant

(6.32.7) $$6^{-\frac{1}{2}} i_1 = 1.85575 \cdots$$

cannot be replaced by a smaller one. These bounds are more precise than those previously given.

Alternative forms of (6.32.4) and (6.32.5) are

(6.32.8) $$\begin{aligned} x_\nu^{\frac{1}{2}} &= (4n)^{\frac{1}{2}} - 6^{-\frac{1}{2}} i_\nu (4n)^{-\frac{1}{6}} + o(n^{-\frac{1}{6}}),\\ x_\nu &= (2n)^{\frac{1}{2}} - 6^{-\frac{1}{2}} i_\nu (2n)^{-\frac{1}{6}} + o(n^{-\frac{1}{6}}), \qquad n\to\infty,\end{aligned}$$

respectively. Also, the first formula can be written as follows:

(6.32.9) $$x_\nu = 4n - 2\cdot 6^{-\frac{1}{2}} i_\nu (4n)^{\frac{1}{3}} + o(n^{\frac{1}{3}}).$$

(2) Writing $h_n = (2n+1)^{\frac{1}{2}}$, we substitute $x = h_n - \xi$ in the second equation (5.5.2) and obtain

(6.32.10) $$\frac{d^2 z}{d\xi^2} + (2h_n\xi - \xi^2)z = 0.$$

Next we compare this equation with

$$\frac{d^2 Z}{d\xi^2} + 2h_n \xi Z = 0, \tag{6.32.11}$$

which has the solution $Z = A\{(6h_n)^{\frac{1}{3}}\xi\}$; here $A(x)$ is Airy's function defined in §1.81. We can then apply Theorem 1.82.1 in $[-\infty, +\infty]$; condition (1.82.2) is satisfied at $\xi = -\infty$ (cf. the last remark in §1.81). Thus we have $(6\, h_n)^{-\frac{1}{3}} i_\nu < h_n - x_\nu$, which establishes (6.32.3).

(3) We add a sketch of a direct proof of (6.32.5) based on Sturm's method. Let the real variable ξ be subject to the condition $|\xi| \leqq 2h_n\epsilon_n$, where $0 < \epsilon_n < 1$; we shall dispose of ϵ_n later. Then

$$2h_n\xi - \xi^2 \geqq \begin{cases} 2h_n\xi(1 - \epsilon_n) & \text{if} \quad \xi \geqq 0, \\ 2h_n\xi(1 + \epsilon_n) & \text{if} \quad \xi \leqq 0. \end{cases} \tag{6.32.12}$$

We now compare (6.32.10) with

$$\frac{d^2\zeta}{d\xi^2} + 2h_n\xi(1 \pm \epsilon_n)\zeta = 0, \tag{6.32.13}$$

where the signs $+$ and $-$ correspond to $\xi \leqq 0$ and $\xi \geqq 0$, respectively. Using the notation (1.81.1), we consider, for $-2h_n\epsilon_n \leqq \xi \leqq 0$, the solution

$$l\{(6h_n)^{\frac{1}{3}}(1 + \epsilon_n)^{\frac{1}{3}}\xi\} - k\{(6h_n)^{\frac{1}{3}}(1 + \epsilon_n)^{\frac{1}{3}}\xi\}\, \frac{l(-X_n)}{k(-X_n)} \tag{6.32.14}$$

with

$$X_n = (6h_n)^{\frac{1}{3}}(1 + \epsilon_n)^{\frac{1}{3}} 2h_n\epsilon_n . \tag{6.32.15}$$

This solution vanishes for $\xi = -2h_n\epsilon_n$. On the other hand, for $0 \leqq \xi \leqq 2h_n\epsilon_n$, we shall consider the solution

$$\left(\frac{1 + \epsilon_n}{1 - \epsilon_n}\right)^{\frac{1}{3}} l\{(6h_n)^{\frac{1}{3}}(1 - \epsilon_n)^{\frac{1}{3}}\xi\} - k\{(6h_n)^{\frac{1}{3}}(1 - \epsilon_n)^{\frac{1}{3}}\xi\}\, \frac{l(-X_n)}{k(-X_n)}. \tag{6.32.16}$$

At $\xi = 0$ it has the same value and the same derivative as (6.32.14) on account of (1.81.1). According to Sturm's theorem, $H_n(h_n - \xi)$ oscillates more rapidly in the interval $-2h_n\epsilon_n \leqq \xi \leqq + 2h_n\epsilon_n$ than the function $\zeta = \zeta(\xi)$ represented by (6.32.14) and (6.32.16).

The only negative zero of $\zeta(\xi)$ is $\xi = -2h_n\epsilon_n$. We now calculate the positive zeros of (6.32.16), that is, the values of ξ for which

$$\frac{l\{(6h_n)^{\frac{1}{3}}(1 - \epsilon_n)^{\frac{1}{3}}\xi\}}{k\{(6h_n)^{\frac{1}{3}}(1 - \epsilon_n)^{\frac{1}{3}}\xi\}} = \left(\frac{1 - \epsilon_n}{1 + \epsilon_n}\right)^{\frac{1}{3}} \frac{l(-X_n)}{k(-X_n)}. \tag{6.32.17}$$

If ϵ_n is small and X_n large, the right-hand member is nearly -1, and the νth zero in question is, for a given ν, near to $(6h_n)^{-\frac{1}{3}}(1 - \epsilon_n)^{-\frac{1}{3}} i_\nu$. If X_n is large and positive, we obtain from (1.81.3) and (1.81.5)

$$\frac{l(-X_n)}{k(-X_n)} + 1 = \frac{A(-X_n)}{k(-X_n)} \cong 3^{\frac{1}{2}} \exp\{-4(X_n/3)^{\frac{3}{2}}\}. \tag{6.32.18}$$

Now let ν_0 be a fixed positive integer and ϵ an arbitrary positive number. We choose

$$\epsilon_n = h_n^{-4/3}\omega, \qquad X_n = 2\cdot 6^{\frac{1}{3}}(1+\epsilon_n)^{\frac{1}{3}}\omega > 2.6^{\frac{1}{3}}\omega, \tag{6.32.19}$$

where ω is a fixed positive number, so large that $2\cdot 6^{\frac{1}{3}}\omega > i_{\nu_0} + 1$, and the left-hand member of (6.32.18) is less than ϵ. For sufficiently small values of ϵ, and as $n \to \infty$, the first ν_0 zeros of (6.32.17) then have the form

$$(6h_n)^{-\frac{1}{3}}(1-\epsilon_n)^{-\frac{1}{3}}(i_\nu + \delta), \qquad \nu = 1, 2, \cdots, \nu_0, \tag{6.32.20}$$

where $|\delta|$ is arbitrarily small with ϵ. (At any rate let $|\delta|$ be less than 1.) We see that for large n, the expressions (6.32.20) are less than $2h_n\epsilon_n$. Hence, when we apply Sturm's theorem in $-2h_n\epsilon_n \leqq \xi \leqq +2h_n\epsilon_n$, we have

$$h_n - x_\nu < (6h_n)^{-\frac{1}{3}}(1-\epsilon_n)^{-\frac{1}{3}}(i_\nu + \delta). \tag{6.32.21}$$

This latter relation combined with (6.32.3) establishes the statement (6.32.5).

6.4. Theorem of Pólya-Szegö on trigonometric polynomials with monotonic coefficients

THEOREM 6.4. *Let $a_0 > a_1 > \cdots > a_m > 0$. Then the functions*

$$f(t) = a_0 \cos mt + a_1 \cos(m-1)t + \cdots + a_{m-1}\cos t + a_m,$$

$$g(t) = a_0 \cos(m+\tfrac{1}{2})t + a_1 \cos(m-\tfrac{1}{2})t + \cdots + a_{m-1}\cos(3t/2) + a_m \cos(t/2), \tag{6.4.1}$$

have only real and simple zeros; there is, respectively, exactly one zero in each of the intervals

$$\frac{\mu-\frac{1}{2}}{m+\frac{1}{2}}\pi < t < \frac{\mu+\frac{1}{2}}{m+\frac{1}{2}}\pi \quad \text{and} \quad \frac{\mu-\frac{1}{2}}{m+1}\pi < t < \frac{\mu+\frac{1}{2}}{m+1}\pi, \tag{6.4.2}$$

where $\mu = 1, 2, \cdots, 2m$, and $\mu = 1, 2, \cdots, 2m+1$, respectively.

The first part of the statement is due to Pólya (**3**, p. 359); his proof uses the principle of argument (Theorem 1.91.1). The following proof furnishes Pólya's result again and, in addition, the inequalities (6.4.2) (Szegö **20**, pp. 9–11). It is based on Fejér's fundamental theorem (Fejér **1**; see Pólya-Szegö **1**, vol. 2, pp. 78, 269, problem 17) which asserts that the sine polynomials

$$\sigma_n(t) = \sin(t/2) + \sin(3t/2) + \cdots + \sin(n+\tfrac{1}{2})t, \qquad n = 0, 1, 2, \cdots; 0 < t < 2\pi, \tag{6.4.3}$$

are non-negative.

If $\tilde{f}(t)$ and $\tilde{g}(t)$ denote the conjugate functions of $f(t)$ and $g(t)$, respectively, we have

$$(6.4.4)\quad -\Im e^{-i(m+\frac{1}{2})t}\{f(t) + i\tilde{f}(t)\} = -\Im e^{-i(m+1)t}\{g(t) + i\tilde{g}(t)\} = \sum_{\mu=0}^{m} a_\mu \sin(\mu + \tfrac{1}{2})t,$$

which is positive for $0 < t < 2\pi$, according to Abel's transformation (1.11.4). Therefore,

$$(6.4.5)\quad \begin{aligned} &f(t)\sin(m + \tfrac{1}{2})t - \tilde{f}(t)\cos(m + \tfrac{1}{2})t > 0, \\ &g(t)\sin(m + 1)t - \tilde{g}(t)\cos(m + 1)t > 0, \qquad 0 < t < 2\pi; \end{aligned}$$

whence

$$(6.4.6)\quad \operatorname{sgn} f\left(\frac{\mu - \frac{1}{2}}{m + \frac{1}{2}}\pi\right) = \operatorname{sgn} g\left(\frac{\mu - \frac{1}{2}}{m + 1}\pi\right) = (-1)^{\mu+1}.$$

This shows the existence of at least one zero in each of the intervals (6.4.2). On the other hand, the functions (6.4.1) cannot have more than $2m$ and $2m + 1$ zeros, respectively, in $[0, 2\pi]$.

6.5. Fejér's generalization of Legendre polynomials

(1) Starting from the representation (4.9.3) of Legendre polynomials, Fejér (9) defines the "Legendre polynomials $F_n(x)$ associated with a given sequence $\alpha_0, \alpha_1, \alpha_2, \cdots$" in the following way:

$$(6.5.1)\quad \begin{aligned} F_n(\cos\theta) &= 2\alpha_0\alpha_n \cos n\theta + 2\alpha_1\alpha_{n-1}\cos(n-2)\theta + \cdots \\ &\quad + \begin{cases} 2\alpha_{(n-1)/2}\alpha_{(n+1)/2}\cos\theta, & \text{if } n \text{ odd}, \\ \alpha_{n/2}^2, & \text{if } n \text{ even}. \end{cases} \end{aligned}$$

The classical Legendre polynomials $P_n(x)$ are obtained if

$$(6.5.2)\quad \alpha_0 = g_0 = 1; \qquad \alpha_n = g_n = \frac{1\cdot 3 \cdots (2n-1)}{2\cdot 4 \cdots 2n}, \quad n = 1, 2, 3, \cdots,$$

the ultraspherical polynomials $P_n^{(\lambda)}(x)$ if (§4.9 (4))

$$(6.5.3)\quad \alpha_0 = 1; \qquad \alpha_n = \binom{n + \lambda - 1}{n}, \qquad n = 1, 2, 3, \cdots.$$

Various properties, well-known in these special cases, can be extended to the general polynomials $F_n(x)$ by imposing proper restrictions on the sequence $\{\alpha_n\}$. These restrictions concern certain properties of monotony and asymptotic behavior.

(2) THEOREM 6.5.1. *The zeros of $F_n(x)$ are real and simple and lie in the interval $-1 < x < +1$, provided $\alpha_n > 0$ and the sequence*

$$(6.5.4)\quad \alpha_1/\alpha_0, \; \alpha_2/\alpha_1, \; \cdots, \; \alpha_n/\alpha_{n-1}, \; \cdots$$

is increasing. More precisely, each interval

$$\frac{\nu - \frac{1}{2}}{n+1}\pi < \theta < \frac{\nu + \frac{1}{2}}{n+1}\pi, \qquad \nu = 1, 2, \cdots, n, \tag{6.5.5}$$

contains exactly one zero of $F_n(\cos\theta)$.

See Szegö **20**, pp. 15–17. Under the condition mentioned the coefficients of (6.5.1) are decreasing, and the statement follows immediately from Theorem 6.4 if we write $n = 2m$ or $n = 2m + 1$, according as n is even or odd, and $2\theta = t$. The condition in question is satisfied for Legendre polynomials and in the ultraspherical case for $0 < \lambda < 1$.

The inequalities (6.5.5) are not so precise as the Bruns inequalities (6.21.5). However, they hold for a comparatively general class of polynomials.

(3) THEOREM 6.5.2. *Let the sequence* $\{\alpha_n\}$, $\alpha_n > 0$, *be completely monotonic, that is, for all the differences,*[39]

$$\Delta^k \alpha_n = \alpha_n - \binom{k}{1}\alpha_{n+1} + \binom{k}{2}\alpha_{n+2} - \cdots + (-1)^k \alpha_{n+k} \geqq 0, \tag{6.5.6}$$
$$k, n = 0, 1, 2, \cdots.$$

Then the zeros $x_\nu = \cos\theta_\nu$, $0 < \theta_\nu < \pi$, *of* $F_n(x)$ *are not only real and lie in* $[-1, +1]$, *but they also satisfy the inequalities* (6.21.7) *of Stieltjes:*

$$(\nu - \tfrac{1}{2})\frac{\pi}{n} \leqq \theta_\nu \leqq \nu\frac{\pi}{n+1}, \qquad \nu = 1, 2, \cdots, [n/2]. \tag{6.5.7}$$

Here the signs of equality hold if and only if $F_n(x)$ *is Tchebichef's polynomial of the first* (see below) *or of the second kind, respectively.*

See Fejér **17**, pp. 311–312. According to an important theorem of Hausdorff (**1**) the class of completely monotonic sequences $\{\alpha_n\}$, $\alpha_n > 0$, is identical with the class of sequences which can be represented in the form

$$\alpha_n = \int_0^1 t^n \, d\alpha(t), \qquad n = 0, 1, 2, \cdots, \tag{6.5.8}$$

where $\alpha(t)$ is a non-decreasing function, not constant, with $2\alpha(t) = \alpha(t + 0) + \alpha(t - 0)$ for $0 < t < 1$. For sequences of this kind the condition of Theorem 6.5.1 is satisfied (Schwarz's inequality). The ultraspherical polynomials $P_n^{(\lambda)}(x)$, $0 < \lambda < 1$ are obtained [(1.7.2)] if

$$\alpha_n = \binom{n + \lambda - 1}{n} = \pi^{-1} \sin\lambda\pi \int_0^1 t^{n+\lambda-1}(1 - t)^{-\lambda}\, dt, \quad n = 0, 1, 2, \cdots. \tag{6.5.9}$$

Tchebichef's polynomials of the first kind are a *limiting case* of (6.5.9) since

[39] This definition of the differences of various orders is not the same as in (2.8.4).

$$\alpha_0 = 1, \qquad \lim_{\lambda \to 0} \lambda^{-1} \alpha_n = \int_0^1 t^{n-1}\,dt = \frac{1}{n}, \qquad n = 1, 2, 3, \cdots, \tag{6.5.10}$$

so that

$$\lim_{\lambda \to 0} \lambda^{-1} F_n(\cos\theta) = \frac{2}{n}\cos n\theta, \qquad n = 1, 2, 3, \cdots. \tag{6.5.11}$$

Tchebichef's polynomials of the second kind arise if $\alpha_n = r^n$, $0 < r \leqq 1$, that is, $\alpha(t)$ has only one point of increase in $0 < t \leqq 1$.

Fejér's argument is as follows:

$$F_n(\cos\theta) = \int_0^1 \int_0^1 \left\{ \sum_{k=0}^{n} t^k u^{n-k} \cos(n - 2k)\theta \right\} d\alpha(t)\, d\alpha(u). \tag{6.5.12}$$

An elementary transformation of the integrand gives

$$\frac{(t-u)(t^{n+1} - u^{n+1})}{t^2 - 2tu\cos 2\theta + u^2}\cos n\theta + \frac{2tu(t^n + u^n)\sin^2\theta}{t^2 - 2tu\cos 2\theta + u^2} \cdot \frac{\sin(n+1)\theta}{\sin\theta}, \tag{6.5.13}$$

so that

$$F_n(\cos\theta) = A_n(\theta)\cos n\theta + B_n(\theta)\frac{\sin(n+1)\theta}{\sin\theta}, \tag{6.5.14}$$

where $A_n(\theta)$ and $B_n(\theta)$ are positive functions in $0 < \theta < \pi$ provided $\alpha(t)$ has at least two points of increase. From this it follows that

$$\operatorname{sgn} F_n\left\{\cos\left(\nu - \tfrac{1}{2}\right)\frac{\pi}{n}\right\} = -\operatorname{sgn} F_n\left\{\cos \nu \frac{\pi}{n+1}\right\} = (-1)^{\nu+1}, \tag{6.5.15}$$
$$\nu = 1, 2, \cdots, [n/2],$$

which establishes the statement.

The last part of this argument is similar to that used in the proof of Theorem 6.4.

(4) Fejér considers (**20**, pp. 40–45) another remarkable generalization of Legendre polynomials. He starts from the representation (4.9.5). Let $\beta_m \downarrow 0$, and

$$\begin{aligned} G_n(\cos\theta) = \beta_0 \sin(n+1)\theta + \beta_1 \sin(n+3)\theta + \cdots \\ + \beta_m \sin(n + 2m + 1)\theta + \cdots. \end{aligned} \tag{6.5.16}$$

This series converges for $0 < \theta < \pi$ (see §4.9 (2)). Legendre polynomials are a special case, as well as the more general functions $(\sin\theta)^{2\lambda-1} P_n^{(\lambda)}(\cos\theta)$, $\lambda > 0$, $\lambda \neq 1, 2, 3, \cdots$ (see (4.9.22)). In these cases the sequence $\beta_m = f_m^{(\lambda)}$ (using the notation of (4.9.22)) is completely monotonic. In fact, we have

$$f_m^{(\lambda)} - f_{m+1}^{(\lambda)} = f_m^{(\lambda)}\left\{\frac{n\lambda}{n+\lambda}\frac{1}{m+1} + \frac{\lambda(n+2\lambda)}{n+\lambda}\frac{1}{n+\lambda+m+1}\right\} = f_m^{(\lambda)}\gamma_m. \tag{6.5.17}$$

The sequence $\{\gamma_m\}$ is completely monotonic and, on account of a well-known formula, we have

$$\Delta^{k+1} f_m^{(\lambda)} = \Delta^k \{f_m^{(\lambda)} - f_{m+1}^{(\lambda)}\} = \sum_{\nu=0}^{k} \binom{k}{\nu} \Delta^\nu f_m^{(\lambda)} \Delta^{k-\nu} \gamma_{m+\nu}. \tag{6.5.18}$$

Hence the statement follows by induction.[40]

Fejér shows (loc. cit.) that $G_n(\cos \theta)$ has at least one zero in each interval

$$(\nu - \tfrac{1}{2})\frac{\pi}{n} < \theta < \nu \frac{\pi}{n+1}, \qquad \nu = 1, 2, \cdots, [n/2], \tag{6.5.19}$$

provided

$$\beta_m \geqq 0, \Delta\beta_m \geqq 0, \Delta^2\beta_m \geqq 0, \Delta^3\beta_m \geqq 0, \quad m = 0, 1, 2, \cdots. \tag{6.5.20}$$

His proof is based on the positiveness of certain special trigonometric polynomials. We shall prove the following theorem:

THEOREM 6.5.3. *The function $G_n(\cos \theta)$ has at least one zero in each interval (6.5.19) provided $\{\beta_m\}$ is a completely monotonic sequence.*

This condition is more restrictive than that of Fejér. The proof is, however, very simple. Using (6.5.8), with β_n and $\beta(t)$ in place of α_n and $\alpha(t)$, respectively, we obtain

$$\begin{aligned} G_n(\cos \theta) &= \int_0^1 \left\{ \sum_{m=0}^{\infty} t^m \sin (n + 2m + 1)\theta \right\} d\beta(t) \\ &= \int_0^1 \frac{2t\, d\beta(t)}{1 - 2t \cos 2\theta + t^2} \sin \theta \cos n\theta + \int_0^1 \frac{(1 - t)\, d\beta(t)}{1 - 2t \cos 2\theta + t^2} \sin (n + 1)\theta. \end{aligned} \tag{6.5.21}$$

From this point the statement follows in the same way as in the proof of Theorem 6.5.2.

6.6. Recapitulation; additional remarks on ultraspherical polynomials

(1) We have obtained the following inequalities for the zeros $x_\nu = \cos \theta_\nu$, $0 < \theta_\nu < \pi$, (arranged in decreasing order) of the ultraspherical polynomial $P_n^{(\alpha,\alpha)}(x)$, $\alpha = \lambda - \frac{1}{2}$, provided $0 < \lambda < 1$:

(a) Inequalities (6.5.5), derived from the representation of $P_n^{(\alpha,\alpha)}(\cos \theta)$ as a cosine polynomial:

$$\frac{\nu - \frac{1}{2}}{n+1}\pi < \theta_\nu < \frac{\nu + \frac{1}{2}}{n+1}\pi, \qquad \nu = 1, 2, \cdots, n. \tag{6.6.1}$$

(b) Inequalities of the Bruns type:

$$\frac{\nu - \frac{1}{2}}{n + \frac{1}{2}}\pi < \theta_\nu < \frac{\nu}{n + \frac{1}{2}}\pi, \qquad \nu = 1, 2, \cdots, n, \tag{6.6.2}$$

[40] For $\lambda = 1/2$ this follows directly from (4.9.9).

(6.6.3)
$$\frac{\nu+\lambda-1}{n+\lambda}\pi < \theta_\nu < \frac{\nu}{n+\lambda}\pi, \qquad \nu = 1, 2, \cdots, n.$$

Inequalities (6.6.2) follow from (6.21.2), which was proved by A. Markoff and by Stieltjes in two different ways (cf. (6.21.5)); (6.6.3) is a special case of the more general inequalities (6.3.7) proved by Sturm's method. The inequalities (6.6.2) are more precise than those of (6.6.1) and those of (6.6.3) with $\lambda < \frac{1}{2}$; the opposite is true if $\lambda > \frac{1}{2}$.

(c) Inequalities of Stieltjes' type:

(6.6.4)
$$\frac{\nu-\frac{1}{2}}{n}\pi < \theta_\nu < \frac{\nu}{n+1}\pi, \qquad \nu = 1, 2, \cdots, [n/2].$$

These follow from (6.21.3) and were proved by Stieltjes. They can also be readily derived from (6.21.2) (which is due to A. Markoff and to Stieltjes). Fejér obtains them from (4.9.19) or (4.9.22) (cf. Theorems 6.5.2 and 6.5.3). An alternative proof for the upper bound is due to Szegö (Sturm's method, §6.3 (3)). The upper bound is better than that in each of the preceding inequalities; the lower bound is better than that in (6.6.2), and is better than that in (6.6.3) provided $\lambda < \frac{1}{2}$.

(d) Szegö's lower bound:

(6.6.5)
$$\theta_\nu > \frac{\nu-(1-\lambda)/2}{n+\lambda}\pi, \qquad \nu = 1, 2, \cdots, [n/2].$$

This follows by Sturm's method in two different ways (cf. 6.3 (2) and (5)). For a third method see (2) below. This lower bound is more precise than any of the preceding ones.

(2) By combining the integral representation (4.82.3) with the argument used in the proofs of Theorems 6.4, 6.5.2, 6.5.3,

(e) Fejér obtains (**19**, p. 208)

(6.6.6)
$$\frac{\nu-(1-\lambda)/2}{n+\lambda}\pi < \theta_\nu < \frac{\nu+\lambda-\frac{1}{2}}{n+2\lambda}\pi, \quad \nu = 1, 2, \cdots, [n/2].$$

The lower bound is the same as in (6.6.5). The upper bound is less or greater than that in (6.6.4) according as $\lambda < \frac{1}{2}$ or $\lambda > \frac{1}{2}$.

For the proof we substitute the bound in (6.6.5) for θ in (4.82.3), $0 < \theta < \pi/2$, and find that

(6.6.7)
$$\operatorname{sgn} P_n^{(\lambda)}(\cos\theta) = (-1)^\nu \operatorname{sgn} \mathfrak{J}\{e^{i\lambda(\theta-\pi/2)}(1-te^{2i\theta})^{-\lambda}\}.$$

It happens that the last sign is constant if t varies in $0 < t < 1$. Indeed, the argument of the expression in the braces lies between 0 and $-\lambda(\pi/2-\theta)$. Thus (6.6.7) becomes $(-1)^{\nu+1}$.

Upon substituting the upper bound of (6.6.6) in (4.82.3), we have

(6.6.8)
$$\operatorname{sgn} P_n^{(\lambda)}(\cos\theta) = (-1)^\nu \operatorname{sgn} \mathfrak{J}\{(1-te^{2i\theta})^{-\lambda}\} = (-1)^\nu,$$

which establishes the statement.

To recapitulate: the lower bound (6.6.5) is the best of all lower bounds given here; while in the case of the upper bounds, either (6.6.4) or (6.6.6) is best according as $\lambda > \frac{1}{2}$ or $\lambda < \frac{1}{2}$. Here we did not refer to inequalities which involve zeros of Bessel functions.

6.7. Electrostatic interpretation of the zeros of the classical polynomials

Stieltjes gave (**4**, **5**; **6**, pp. 75–76; cf., also, Schur **1**) a very interesting derivation of the differential equations of the classical polynomials, which is closely connected with the calculation of the discriminant of these polynomials (cf. §6.71) and can be interpreted as a problem of electrostatic equilibrium.

(1) PROBLEM. *Let p and q be two given positive numbers. If n unit "masses," $n \geqq 2$, at the variable points $x_1, x_2, x_3, \cdots, x_n$ in the interval $[-1, +1]$ and the fixed masses p and q at $+1$ and -1, respectively, are considered, for what position of the points $x_1, x_2, x_3, \cdots, x_n$ does the expression*

$$(6.7.1) \quad T(x_1, x_2, \cdots, x_n) = T(x) = \prod_{\kappa=1}^{n} (1 - x_\kappa)^p (1 + x_\kappa)^q \prod_{\substack{\nu,\mu=1,2,\cdots,n \\ \nu<\mu}} |x_\nu - x_\mu|$$

become a maximum?

Obviously, $\log (T^{-1})$ can be interpreted as the energy of the system of electrostatic masses just defined. They exert repulsive forces according to the law of logarithmic potential. The maximum position corresponds to the condition of electrostatic equilibrium. A maximum exists because T is a continuous function of $x_1, x_2, \cdots, x_n$ for $-1 \leqq x_\nu \leqq +1$, $\nu = 1, 2, \cdots, n$. It is clear that in the maximum position the x_ν are each different from ± 1 and from one another. In addition, this position is uniquely determined. To show this, let us suppose that (cf. Popoviciu **2**, p. 74)

$$(6.7.2) \quad \begin{aligned} &+1 > x_1 > x_2 > \cdots > x_n > -1, \\ &+1 > x_1' > x_2' > \cdots > x_n' > -1 \end{aligned}$$

are two positions of this kind; we write

$$(6.7.3) \quad y_\nu = (x_\nu + x_\nu')/2, \qquad \nu = 1, 2, \cdots, n.$$

Then

$$(6.7.4) \quad \begin{aligned} &|y_\nu - y_\mu| = \frac{|x_\nu - x_\mu| + |x_\nu' - x_\mu'|}{2} \geqq |x_\nu - x_\mu|^{\frac{1}{2}} |x_\nu' - x_\mu'|^{\frac{1}{2}}, \\ &|1 \pm y_\nu| \geqq |1 \pm x_\nu|^{\frac{1}{2}} |1 \pm x_\nu'|^{\frac{1}{2}}, \end{aligned}$$

so that $T(y) \geqq \{T(x)\}^{\frac{1}{2}}\{T(x')\}^{\frac{1}{2}}$, the equality sign being taken if and only if $x_\nu = x_\nu'$. This establishes the uniqueness.

THEOREM 6.7.1. *Let $p > 0$, $q > 0$, and let $\{x_\nu\}$, $-1 \leqq x_\nu \leqq +1$, be a system of values for which the expression* (6.7.1) *becomes a maximum. Then the $\{x_\nu\}$ are the zeros of the Jacobi polynomial $P_n^{(\alpha,\beta)}(x)$, where $\alpha = 2p - 1$, $\beta = 2q - 1$.*

From this fact the uniqueness of the maximum position follows again. For a maximum we have the conditions $\partial T/(\partial x_\nu) = 0$, or

$$(6.7.5)\quad \begin{aligned} &\frac{1}{x_\nu - x_1} + \cdots + \frac{1}{x_\nu - x_{\nu-1}} + \frac{1}{x_\nu - x_{\nu+1}} + \cdots \\ &\qquad + \frac{1}{x_\nu - x_n} + \frac{p}{x_\nu - 1} + \frac{q}{x_\nu + 1} = 0. \end{aligned}$$

If we introduce the polynomial $f(x) = (x - x_1)(x - x_2) \cdots (x - x_n)$, this becomes

$$(6.7.6)\quad \frac{1}{2}\frac{f''(x_\nu)}{f'(x_\nu)} + \frac{p}{x_\nu - 1} + \frac{q}{x_\nu + 1} = 0,$$

or

$$(1 - x_\nu^2)f''(x_\nu) + \{2q - 2p - (2q + 2p)x_\nu\}f'(x_\nu) = 0.$$

The last equation means that $(1 - x^2)f''(x) + \{\beta - \alpha - (\alpha + \beta + 2)x\}f'(x)$ is a π_n which vanishes for all the zeros of $f(x)$; whence this expression is equal to const. $f(x)$. By comparing the terms in x^n we obtain for the constant factor the value $-n(n + \alpha + \beta + 1)$. The resulting differential equation reduces to (4.2.1), so that according to Theorem 4.2.2, $f(x)$ must be a constant multiple of $P_n^{(\alpha,\beta)}(x)$.

See also Problem 37.

(2) The zeros of Laguerre and Hermite polynomials admit a similar interpretation.

THEOREM 6.7.2. *Let us consider the positive mass p at the fixed point $x = 0$ and unit masses at the variable points $x_1, x_2, \cdots, x_n$ in the interval $[0, +\infty]$ such that the "centroid" satisfies*

$$(6.7.7)\quad n^{-1}(x_1 + x_2 + \cdots + x_n) \leqq K,$$

where K is a preassigned positive number. Then the maximum of

$$(6.7.8)\quad U(x_1, x_2, \cdots, x_n) = \prod_{\kappa=1}^{n} x_\kappa^p \prod_{\substack{\nu,\mu=1,2,\cdots,n\\ \nu<\mu}} |x_\nu - x_\mu|$$

is attained if and only if the $\{x_\nu\}$ are the zeros of the Laguerre polynomial $L_n^{(\alpha)}(cx)$, where $\alpha = 2p - 1$, and $c = K^{-1}(n + \alpha)$.

THEOREM 6.7.3. *Let us consider a unit mass at each of the variable points $x_1, x_2, \cdots, x_n$ in the interval $[-\infty, +\infty]$ such that the "moment of inertia" satisfies*

$$(6.7.9)\quad n^{-1}(x_1^2 + x_2^2 + \cdots + x_n^2) \leqq L,$$

where L is a preassigned positive number. Then the maximum of

$$(6.7.10)\qquad V(x_1, x_2, \cdots, x_n) = \prod_{\substack{\nu,\mu=1,2,\cdots,n \\ \nu<\mu}} |x_\nu - x_\mu|$$

is attained if and only if the $\{x_\nu\}$ *are the zeros of the Hermite polynomial* $H_n(c'x)$, $c' = (2L)^{-\frac{1}{2}}(n - 1)^{\frac{1}{2}}$.

The existence and uniqueness of the position of maximum is clear in both cases. The corresponding x_ν are all different from one another; in the first case they are positive. It is clear, furthermore, that for the maximum position the sign of equality holds in (6.7.7) and (6.7.9). Hence, if ρ is a proper "multiplier," we have

$$(6.7.11)\qquad \begin{gathered} \frac{1}{x_\nu - x_1} + \frac{1}{x_\nu - x_2} + \cdots + \frac{1}{x_\nu - x_n} + \frac{p}{x_\nu} = \frac{\rho}{n}, \\ x_\nu f''(x_\nu) + \left(2p - \frac{2\rho}{n} x_\nu\right) f'(x_\nu) = 0 \end{gathered}$$

in the first case, and

$$(6.7.12)\qquad \begin{gathered} \frac{1}{x_\nu - x_1} + \frac{1}{x_\nu - x_2} + \cdots + \frac{1}{x_\nu - x_n} = \frac{2\rho}{n} x_\nu, \\ f''(x_\nu) - \frac{4\rho}{n} x_\nu f'(x_\nu) = 0 \end{gathered}$$

in the second. In both cases we have written

$$f(x) = (x - x_1)(x - x_2) \cdots (x - x_n).$$

If we replace x by cx, with c a proper constant factor, these conditions can easily be reduced to the first equation in (5.1.2) and to the first equation in (5.5.2), respectively. Therefore,

$$f(x) = \text{const. } L_n^{(\alpha)}(cx), \qquad \alpha = 2p - 1,\ c = 2\rho n^{-1},$$

in the first case, while $f(x) = \text{const. } H_n(c'x)$, $c' = (2\rho/n)^{\frac{1}{2}}$ in the second case. The constants c and c' can be determined from the conditions (6.7.7) and (6.7.9), in which the equality signs now hold. We observe that according to (5.1.6) the sum of the zeros of $L_n^{(\alpha)}(x)$ is equal to $n(n + \alpha)$; according to (5.5.4) the sum of the squares of the zeros of $H_n(x)$ is equal to $n(n - 1)/2$.

Cf. also Problem 38.

6.71. Discriminants of the classical polynomials

The maximum problems treated in the preceding section are closely related to the calculation of the discriminants of the classical polynomials (Hilbert **1**, Stieltjes **4**, **5**). The following method is due to I. Schur (**2**) (cf. Popoviciu **2**).

(1) Let $\{\rho_n(x)\}$ be a sequence of polynomials satisfying the recurrence formula

$$(6.71.1)\qquad \begin{aligned} \rho_n(x) &= (a_n x + b_n)\rho_{n-1}(x) - c_n\rho_{n-2}(x), \qquad n = 2, 3, 4, \cdots; \\ \rho_0(x) &= 1; \qquad \rho_1(x) = a_1 x + b_1. \end{aligned}$$

We suppose that $a_n c_n \neq 0$. Denoting by $\{x_{\nu n}\}$ the zeros of $\rho_n(x)$, we show that

$$(6.71.2)\qquad \Delta_n = \prod_{\nu=1}^{n} \rho_{n-1}(x_{\nu n}) = (-1)^{n(n-1)/2} \prod_{\nu=1}^{n} \{a_\nu^{n-2\nu+1} c_\nu^{\nu-1}\}, \qquad n = 1, 2, 3, \cdots.$$

Suppose $n \geqq 2$. The coefficient of x^{n-1} in $\rho_{n-1}(x)$ is $a_1 a_2 \cdots a_{n-1}$, so that

$$(6.71.3)\qquad \begin{cases} \Delta_n = (a_1 a_2 \cdots a_{n-1})^n \prod_{\nu=1}^{n} (x_{\nu n} - x_{1,n-1})(x_{\nu n} - x_{2,n-1}) \cdots (x_{\nu n} - x_{n-1,n-1}) \\ \quad = \dfrac{(a_1 a_2 \cdots a_{n-1})^n}{(a_1 a_2 \cdots a_n)^{n-1}} \rho_n(x_{1,n-1})\rho_n(x_{2,n-1}) \cdots \rho_n(x_{n-1,n-1}). \end{cases}$$

Using the recurrence formula, we obtain

$$(6.71.4)\qquad \begin{cases} \Delta_n = a_1 a_2 \cdots a_{n-1} \cdot a_n^{1-n}(-c_n)^{n-1} \prod_{\nu=1}^{n-1} \rho_{n-2}(x_{\nu,n-1}) \\ \quad = (-1)^{n-1} a_1 a_2 \cdots a_{n-1} \cdot a_n^{1-n} c_n^{n-1} \Delta_{n-1}, \end{cases}$$

which establishes the statement.

(2) THEOREM 6.71. *The discriminants of $P_n^{(\alpha,\beta)}(x)$, $L_n^{(\alpha)}(x)$, $H_n(x)$ are*

$$(6.71.5)\qquad D_n^{(\alpha,\beta)} = 2^{-n(n-1)} \prod_{\nu=1}^{n} \nu^{\nu-2n+2}(\nu + \alpha)^{\nu-1}(\nu + \beta)^{\nu-1}(n + \nu + \alpha + \beta)^{n-\nu},$$

$$(6.71.6)\qquad D_n^{(\alpha)} = \prod_{\nu=1}^{n} \nu^{\nu-2n+2}(\nu + \alpha)^{\nu-1},$$

$$(6.71.7)\qquad D_n = 2^{3n(n-1)/2} \prod_{\nu=1}^{n} \nu^\nu,$$

respectively.

We start from the familiar expression (cf., for instance, O. Perron **4**, vol. 1, p. 259, (12), (13); p. 260, (16))

$$(6.71.8)\qquad \begin{aligned} D_n^{(\alpha,\beta)} &= \{l_n^{(\alpha,\beta)}\}^{2n-2} \prod_{\substack{\nu,\mu=1,2,\cdots,n \\ \nu<\mu}} (x_{\nu n} - x_{\mu n})^2 \\ &= (-1)^{n(n-1)/2} \{l_n^{(\alpha,\beta)}\}^{n-2} \prod_{\nu=1}^{n} P_n^{(\alpha,\beta)\prime}(x_{\nu n}), \end{aligned}$$

where $l_n^{(\alpha,\beta)}$ has the same meaning as in (4.21.6), and $\{x_{\nu n}\}$ denotes the zeros of $P_n^{(\alpha,\beta)}(x)$. The discriminants $D_n^{(\alpha)}$ and D_n admit a similar representation. According to the first formula in (4.5.7), we have

$$(6.71.9)\quad (1-x^2)\frac{d}{dx}\{P_n^{(\alpha,\beta)}(x)\} = \frac{2(n+\alpha)(n+\beta)}{2n+\alpha+\beta}P_{n-1}^{(\alpha,\beta)}(x) \quad \text{if} \quad P_n^{(\alpha,\beta)}(x) = 0,$$

so that

$$(6.71.10)\quad \begin{aligned} D_n^{(\alpha,\beta)} &= (-1)^{n(n-1)/2}\{l_n^{(\alpha,\beta)}\}^{n-2}\left\{\frac{2(n+\alpha)(n+\beta)}{2n+\alpha+\beta}\right\}^n \\ &\quad \cdot\prod_{\nu=1}^{n}(1-x_{\nu n}^2)^{-1}P_{n-1}^{(\alpha,\beta)}(x_{\nu n}) \\ &= (-1)^{n(n+1)/2}\{l_n^{(\alpha,\beta)}\}^n\left\{\frac{2(n+\alpha)(n+\beta)}{2n+\alpha+\beta}\right\}^n \\ &\quad \cdot\{P_n^{(\alpha,\beta)}(1)P_n^{(\alpha,\beta)}(-1)\}^{-1}\prod_{\nu=1}^{n}P_{n-1}^{(\alpha,\beta)}(x_{\nu n}). \end{aligned}$$

The last factor can be calculated by means of (6.71.2), so that on account of (4.21.6), (4.1.1), (4.1.4), and (4.5.1), we obtain (6.71.5).

The expression (6.71.6) can be calculated in the same way by using (5.1.14), (5.1.8), (5.1.7), and (5.1.10); or even more simply, from (6.71.5) by using the limiting process of (5.3.4). Indeed, if $\{x_{\nu n} = x_{\nu n}(\beta)\}$ denotes the zeros of $P_n^{(\alpha,\beta)}(x)$ in decreasing order, we have, for fixed ν and n,

$$(6.71.11)\quad \lim_{\beta\to\infty}\beta(1-x_{\nu n}) = 2\xi_{\nu n},$$

where $\{\xi_{\nu n}\}$ are the zeros of $L_n^{(\alpha)}(x)$ in increasing order. Therefore,

$$(6.71.12)\quad \begin{aligned} D_n^{(\alpha)} &= \{l_n^{(\alpha)}\}^{2n-2}\prod_{\substack{\nu,\mu=1,2,\cdots,n\\ \nu<\mu}}(\xi_{\nu n}-\xi_{\mu n})^2 \\ &= \{l_n^{(\alpha)}\}^{2n-2}\lim_{\beta\to\infty}(\beta/2)^{n(n-1)}\prod_{\substack{\nu,\mu=1,2,\cdots,n\\ \nu<\mu}}(x_{\mu n}-x_{\nu n})^2 \\ &= \{l_n^{(\alpha)}\}^{2n-2}\lim_{\beta\to\infty}(\beta/2)^{n(n-1)}\{l_n^{(\alpha,\beta)}\}^{-2n+2}D_n^{(\alpha,\beta)}, \end{aligned}$$

which establishes the statement.

The discriminant D_n can also be obtained either directly, or from (6.71.5) by using (5.6.3), or from (6.71.6) by using (5.6.1). The first method is the simplest. By using (5.5.6), (5.5.10), (5.5.8), and (6.71.2), we find (6.71.7).

6.72. Distribution of the zeros of the general Jacobi polynomials

(1) Let α and β be arbitrary real numbers, $n \geqq 1$, and let $P_n^{(\alpha,\beta)}(x)$ denote the generalized Jacobi polynomials defined in §4.22. Then (6.71.5) still holds.

It follows from (4.1.1) or (4.21.2) that $x = +1$ is a zero of $P_n^{(\alpha,\beta)}(x)$ if and only if

$$(6.72.1)\quad \alpha = -1, -2, \cdots, -n.$$

(The multiplicity of this zero is $|\alpha|$; cf. (4.22.2).) Similarly, $x = -1$ is a zero if and only if (cf. (4.1.4))

$$\beta = -1, -2, \cdots, -n. \tag{6.72.2}$$

Finally, it follows from (4.21.6) that $x = \infty$ is a zero, if and only if

$$n + \alpha + \beta = -1, -2, \cdots, -n. \tag{6.72.3}$$

If such values of α and β are excluded, the zeros of $P_n^{(\alpha,\beta)}(x)$ are different from ± 1 and ∞; in addition, (6.71.5) shows that they are distinct. (This follows also from (4.2.1); cf. §6.2 (3).) Let N_1, N_2, N_3 be the number of zeros in $-1 < x < +1$, $-\infty < x < -1$, and $+1 < x < +\infty$, respectively. We shall now determine these numbers as functions of α and β.

Hilbert (**1**) calculated the number $N_1 + N_2 + N_3$ of the real zeros. A remark of Stieltjes (**5**, p. 444) indicates that he obtained the numbers N_1, N_2, N_3 three years before Hilbert's paper. The later results of Klein concerning the number of the zeros of the general hypergeometric function (**1**, pp. 562–567) readily lead to these numbers. (Cf. also Shibata **1**, Fujiwara **1**, Sen-Rangachariar **1**.) By use of Klein's symbol

$$E(u) = \begin{cases} 0 & \text{if } u \leqq 0, \\ [u] & \text{if } u > 0,\ u \text{ non-integral}, \\ u - 1 & \text{if } u = 1, 2, 3, \cdots, \end{cases} \tag{6.72.4}$$

we can formulate the following theorem:

THEOREM 6.72. *Let α, β be arbitrary real values, and set*

$$\begin{aligned} X &= X(\alpha, \beta) = E\{\tfrac{1}{2}(|2n + \alpha + \beta + 1| - |\alpha| - |\beta| + 1)\}, \\ Y &= Y(\alpha, \beta) = E\{\tfrac{1}{2}(-|2n + \alpha + \beta + 1| + |\alpha| - |\beta| + 1)\}, \\ Z &= Z(\alpha, \beta) = E\{\tfrac{1}{2}(-|2n + \alpha + \beta + 1| - |\alpha| + |\beta| + 1)\}. \end{aligned} \tag{6.72.5}$$

If we exclude the cases (6.72.1), (6.72.2), *and* (6.72.3), *the numbers of the zeros of $P_n^{(\alpha,\beta)}(x)$ in $-1 < x < +1$, $-\infty < x < -1$, $+1 < x < +\infty$, respectively, are*

$$N_1 = N_1(\alpha, \beta) = \begin{cases} 2[(X + 1)/2] & \textit{if } (-1)^n \binom{n+\alpha}{n}\binom{n+\beta}{n} > 0, \\ 2[X/2] + 1 & \textit{if } (-1)^n \binom{n+\alpha}{n}\binom{n+\beta}{n} < 0, \end{cases} \tag{6.72.6}$$

$$N_2 = N_2(\alpha, \beta) = \begin{cases} 2[(Y + 1)/2] & \textit{if } \binom{2n+\alpha+\beta}{n}\binom{n+\beta}{n} > 0, \\ 2[Y/2] + 1 & \textit{if } \binom{2n+\alpha+\beta}{n}\binom{n+\beta}{n} < 0, \end{cases} \tag{6.72.7}$$

$$(6.72.8) \qquad N_3 = N_3(\alpha, \beta) = \begin{cases} 2[(Z+1)/2] & \text{if } \binom{2n+\alpha+\beta}{n}\binom{n+\alpha}{n} > 0, \\ 2[Z/2]+1 & \text{if } \binom{2n+\alpha+\beta}{n}\binom{n+\alpha}{n} < 0. \end{cases}$$

We notice that the numbers $2[(X+1)/2]$, $2[X/2]+1$ can be characterized, respectively, as the even or odd of the numbers X and $X+1$, so that N_1 is either X or $X+1$. We also see that the conditions in (6.72.6) are equivalent to $P_n^{(\alpha,\beta)}(1)P_n^{(\alpha,\beta)}(-1) > 0$ or < 0, respectively. For instance, if $\operatorname{sgn} P_n^{(\alpha,\beta)}(1)P_n^{(\alpha,\beta)}(-1) = (-1)^X$ or $(-1)^{X+1}$, then $N_1(\alpha, \beta) = X$ or $X+1$, respectively. Similar remarks hold for N_2 and N_3.

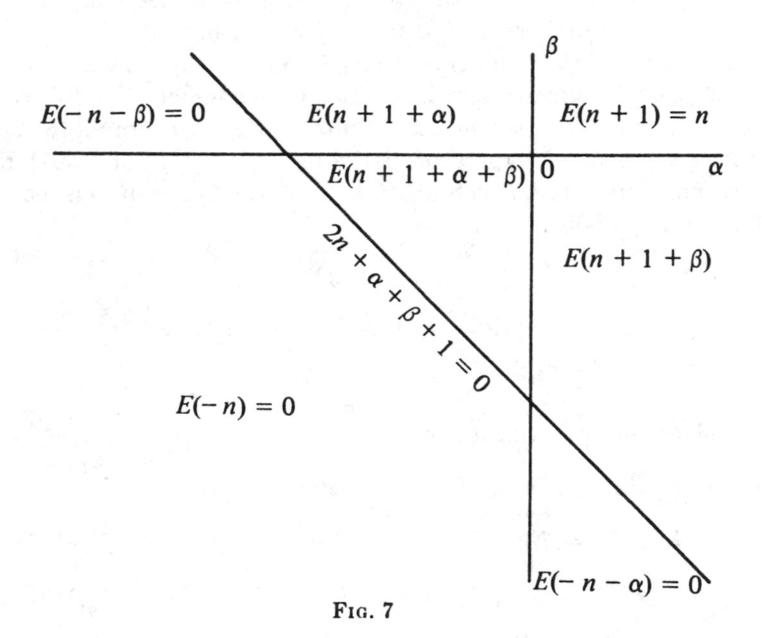

FIG. 7

It is sufficient to calculate N_1. On account of (4.22.1) we obtain N_2 by replacing α by $-2n-\alpha-\beta-1$ in (6.72.6), while N_3 is obtained from N_2 by interchanging α and β. The following proof is based on the continuity of the zeros as functions of α and β.

(2) For convenience, we introduce the notation $M(\alpha, \beta)$ for the function of the right-hand member of (6.72.6) so that we must prove $N_1(\alpha, \beta) = M(\alpha, \beta)$. If the point (α, β) varies, the function $N_1(\alpha, \beta)$ can change only if (α, β) crosses one of the straight lines (6.72.1) or (6.72.2). We first show that the function $M(\alpha, \beta)$ has the same property.

An easy calculation furnishes the value of $X(\alpha, \beta)$ in the seven regions bounded by the α- and β-axes and by the straight line $2n+\alpha+\beta+1 = 0$ (see figure 7). The only discontinuities of the function $E(u)$ are at the points

$u = 1, 2, 3, \cdots$; hence, apart from the straight lines (6.72.1) and (6.72.2), a jump of the function $M(\alpha, \beta)$ is possible only in the triangle where $X(\alpha, \beta) = E(n + 1 + \alpha + \beta)$ if $n + 1 + \alpha + \beta$ coincides with an integer k, $1 \leqq k \leqq n$. Let $\alpha = \alpha_0$ and $\beta = \beta_0$ be negative non-integers, $n + 1 + \alpha_0 + \beta_0 = k$, and let the integers p and q be chosen such that

$$(6.72.9) \qquad -p < \alpha_0 < -p + 1, \; -q < \beta_0 < -q + 1, \quad 1 \leqq p \leqq n, 1 \leqq q \leqq n.$$

Then we necessarily have $k = n + 2 - p - q$. We investigate $M(\alpha, \beta)$ for $n + 1 + \alpha + \beta = k \pm \epsilon$, $0 < \epsilon < 1$, where $|\alpha - \alpha_0|$ and $|\beta - \beta_0|$ are sufficiently small. Obviously,

$$\operatorname{sgn} (-1)^n \binom{n+\alpha}{n}\binom{n+\beta}{n} = (-1)^n(-1)^{p-1}(-1)^{q-1} = (-1)^k;$$

furthermore, $X(\alpha, \beta) = E(n + 1 + \alpha + \beta) = E(k \pm \epsilon) = k$ or $k - 1$, respectively. In other words $k = X$ or $k = X + 1$, respectively. Then (see the remark concerning Theorem 6.72) $M(\alpha, \beta) = k$ in both cases, so that no change occurs in $M(\alpha, \beta)$.

(3) We now show that when (α, β) crosses one of the lines (6.72.1) or (6.72.2), the jumps in $N_1(\alpha, \beta)$ and $M(\alpha, \beta)$ are the same. This will prove the statement (6.72.6), since, for $\alpha > 0$, $\beta > 0$, we have $N_1(\alpha, \beta) = n$ and $X(\alpha, \beta) = M(\alpha, \beta) = n$.

Because of the symmetry in α and β, it is sufficient to consider the case $\alpha = -k \pm \epsilon$, $\epsilon > 0$, $1 \leqq k \leqq n$, β not an integer, and to discuss the location of the zeros near $x = +1$.

From (4.21.2) we obtain for $\alpha = -k + \epsilon$

$$(6.72.10) \qquad \begin{aligned} &\frac{2^k k!\,(n-k)!}{(n+\alpha+\beta+1)\cdots(n+\alpha+\beta+k)(\alpha+k+1)\cdots(\alpha+n)} P_n^{(\alpha,\beta)}(x) \\ &\quad = c_0(\epsilon) + c_1(\epsilon)(x-1) + \cdots + c_{k-1}(\epsilon)(x-1)^{k-1} + (x-1)^k \\ &\qquad + c_{k+1}(\epsilon)(x-1)^{k+1} + \cdots + c_n(\epsilon)(x-1)^n. \end{aligned}$$

Here the coefficients are real rational functions of ϵ, regular for $\epsilon = 0$, and

$$(6.72.11) \qquad c_0(0) = c_1(0) = \cdots = c_{k-1}(0) = 0.$$

Furthermore,

$$(6.72.12) \qquad \begin{aligned} c_0(\epsilon) &= 2^k \left\{\binom{n}{k}\right\}^{-1} \frac{(\alpha+1)(\alpha+2)\cdots(\alpha+k)}{(n+\alpha+\beta+1)(n+\alpha+\beta+2)\cdots(n+\alpha+\beta+k)}; \\ c_0'(0) &= (-1)^{k-1} 2^k \left\{\binom{n}{k}\right\}^{-1} \frac{(k-1)!}{(n+\beta)(n+\beta-1)\cdots(n+\beta-k+1)} \neq 0. \end{aligned}$$

By means of a simple consideration from the theory of analytic functions (see below), we now obtain the following result. If δ is an arbitrarily small positive number ($\delta < \sin \pi/k$ for $k > 1$), then for sufficiently small values of $\epsilon > 0$, the function $P_n^{(\alpha,\beta)}(x)$, $\alpha = -k + \epsilon$, has exactly k distinct zeros in the neighborhood of $x = +1$. More precisely, if $\eta_1, \eta_2, \eta_3, \cdots, \eta_k$ denote the numbers satisfying the equation

$$\operatorname{sgn} c_0'(0) + \eta^k = 0, \tag{6.72.13}$$

these roots lie in the circles

$$x = 1 + (\epsilon \mid c_0'(0) \mid)^{1/k}(\eta_\nu + z), \; \mid z \mid < \delta, \qquad \nu = 1, 2, \cdots, k. \tag{6.72.14}$$

By replacing ϵ by $-\epsilon$ (that is, for $\alpha = -k - \epsilon$), the same result holds with the circles

$$x = 1 + (\epsilon \mid c_0'(0) \mid)^{1/k}(\zeta_\nu + z), \qquad \mid z \mid < \delta, \tag{6.72.15}$$

where $\zeta_1, \zeta_2, \cdots, \zeta_k$ are the roots of

$$-\operatorname{sgn} c_0'(0) + \zeta^k = 0. \tag{6.72.16}$$

The root corresponding to a real η_ν or ζ_ν is obviously real.

To prove the previous statement we introduce $x = 1 + (\epsilon \mid c_0'(0) \mid)^{1/k} y$ in (6.72.10) and obtain

$$\begin{aligned} &\epsilon c_0'(0) + \frac{\epsilon^2}{2!} c_0''(0) + \cdots + \frac{\epsilon^\nu}{\nu!} c_0^{(\nu)}(0) + \cdots + c_1(\epsilon)(\epsilon \mid c_0'(0) \mid)^{1/k} y \\ &+ \cdots + c_{k-1}(\epsilon)(\epsilon \mid c_0'(0) \mid)^{(k-1)/k} y^{k-1} + \epsilon \mid c_0'(0) \mid y^k \\ &+ c_{k+1}(\epsilon)(\epsilon \mid c_0'(0) \mid)^{(k+1)/k} y^{k+1} + \cdots + c_n(\epsilon)(\epsilon \mid c_0'(0) \mid)^{n/k} y^n = 0. \end{aligned} \tag{6.72.17}$$

If this be divided through by $\epsilon \mid c_0'(0) \mid$, the "principal terms" are $\operatorname{sgn} c_0'(0) + y^k$. Now we can apply Theorem 1.91.2 (Rouché's theorem); whence the statement follows immediately.

We are interested especially in the number of real zeros $x < +1$ near $x = +1$. From the preceding result we see that this number increases or decreases by one unit if we replace ϵ by $-\epsilon$, according as $(-1)^k c_0'(0)$ is positive or negative. With reference to (6.72.12) this is equivalent to the condition that

$$\binom{n+\beta}{k} < 0 \quad \text{or} \quad > 0,$$

respectively.

(4) On the other hand, we discuss the jump of $M(\alpha, \beta)$ as (α, β) crosses the line $\alpha = -k$ with β non-integral. First suppose $\beta > 0$, so that $X(\alpha, \beta) = E(n + 1 + \alpha)$. For $\alpha = -k \pm \epsilon$, $\epsilon > 0$, we have

$$X(\alpha, \beta) = E(n + 1 - k \pm \epsilon) = \begin{cases} n + 1 - k, \\ n - k, \end{cases} \tag{6.72.18}$$

respectively, and

$$(6.72.19)\quad \operatorname{sgn} (-1)^n \binom{n+\alpha}{n}\binom{n+\beta}{n} = \operatorname{sgn} (-1)^n \binom{n+\alpha}{n} = \begin{cases} (-1)^{n+k-1}, \\ (-1)^{n+k}, \end{cases}$$

respectively, which in both cases is $(-1)^{X(\alpha,\beta)}$. Hence (cf. the remark to Theorem 6.72) $M(\alpha, \beta)$ changes from $n + 1 - k$ to $n - k$, a loss of one unit. In this case

$$\binom{n+\beta}{k} > 0.$$

Now let β be negative and non-integral. Then $n + 1 + \alpha + \beta$ is non-integral near $\alpha = -k$, so that $X(\alpha, \beta)$ remains constant in this neighborhood. More precisely,

$$(6.72.20)\quad X(\alpha, \beta) = \begin{cases} n + 1 - k + [\beta] & \text{if } n + 1 - k + \beta > 0, \\ 0 & \text{if } n + 1 - k + \beta < 0. \end{cases}$$

In the first case we have for $\alpha = -k \pm \epsilon$, $\epsilon > 0$,

$$(6.72.21)\quad \operatorname{sgn} (-1)^n \binom{n+\alpha}{n}\binom{n+\beta}{n} = \begin{cases} (-1)^n(-1)^{k-1}(-1)^{[\beta]+1}, \\ (-1)^n(-1)^k(-1)^{[\beta]+1}, \end{cases}$$

respectively, which are $(-1)^{X(\alpha,\beta)+1}$ and $(-1)^{X(\alpha,\beta)}$, respectively. Hence (cf. the remark to Theorem 6.72), $M(\alpha, \beta)$ changes from $X(\alpha, \beta) + 1$ to $X(\alpha, \beta)$, a loss of one unit. In this case, again

$$\binom{n+\beta}{k} > 0.$$

Let us now consider the second case in (6.72.20). Then

$$\begin{aligned} &\operatorname{sgn} (-1)^n \binom{n+\alpha}{n}\binom{n+\beta}{n} \\ (6.72.22)\quad &= \operatorname{sgn} (-1)^n \binom{n+\alpha}{n}\binom{n+\beta}{k}\frac{k!}{n!}(\beta+1)(\beta+2)\cdots(\beta+n-k) \\ &= -\operatorname{sgn}\binom{n+\beta}{k} \quad \text{or} \quad \operatorname{sgn}\binom{n+\beta}{k}, \end{aligned}$$

respectively, for $\alpha = -k \pm \epsilon$, $\epsilon > 0$. For instance, if

$$\binom{n+\beta}{k} < 0,$$

$M(\alpha, \beta)$ changes from 0 to 1, a gain of one unit. The opposite is true if

$$\binom{n+\beta}{k} > 0.$$

This establishes Theorem 6.72.

6.73. Distribution of the zeros of the general Laguerre polynomials

The discriminant formula (6.71.6) is valid for general Laguerre polynomials $L_n^{(\alpha)}(x)$, α arbitrary and real, $n \geqq 1$. On account of (5.1.7), $x = 0$ is a zero of $L_n^{(\alpha)}(x)$ when and only when

$$\alpha = -1, -2, \cdots, -n. \tag{6.73.1}$$

(Its multiplicity is $|\alpha|$; cf. (5.2.1).) If such values of α are excluded, the zeros of $L_n^{(\alpha)}(x)$ are finite and different from 0; in addition, we conclude from (6.71.6) (or from the first differential equation (5.1.2)) that they are distinct. Let $n_1(\alpha)$ and $n_2(\alpha)$ denote the number of positive and negative zeros, respectively. By using Theorem 1.91.3 (Hurwitz's theorem) and (5.3.4), we see that if β is large, $P_n^{(\alpha,\beta)}(x)$ has at least $n_1(\alpha)$ zeros in $[-1, +1]$, at least $n_2(\alpha)$ zeros in $[+1, +\infty]$, and at least $n - n_1(\alpha) - n_2(\alpha)$ zeros which are not real. Therefore, using the notation of the previous section, we see that $n_1(\alpha) = N_1(\alpha, \beta)$ and $n_2(\alpha) = N_3(\alpha, \beta)$ if β is large; that is,

$$n_1(\alpha) = \lim_{\beta\to+\infty} N_1(\alpha, \beta); \qquad n_2(\alpha) = \lim_{\beta\to+\infty} N_3(\alpha, \beta). \tag{6.73.2}$$

Obviously, if $\alpha > -1$, then $n_1(\alpha) = n$, and $n_2(\alpha) = 0$.

Now suppose $\alpha < -1$, $\alpha \neq -2, -3, \cdots, -n$. From (6.72.5) we obtain

$$\lim_{\beta\to+\infty} X(\alpha, \beta) = E(n + \alpha + 1) = \begin{cases} n + [\alpha] + 1 & \text{if } \alpha > -n, \\ 0 & \text{if } \alpha < -n, \end{cases}$$

since the argument of E in the formula for $X(\alpha, \beta)$ is not a positive integer; furthermore,

$$\lim_{\beta\to+\infty} Z(\alpha, \beta) = E(-n) = 0.$$

Now

$$\operatorname{sgn} (-1)^n \binom{n+\alpha}{n} = \begin{cases} (-1)^{n+[\alpha]+1} & \text{if } \alpha > -n, \\ 1 & \text{if } \alpha < -n. \end{cases}$$

Thus, $N_1(\alpha, \beta) = n + [\alpha] + 1$ in the first case, and $N_1(\alpha, \beta) = 0$ in the second. Therefore,

$$n_1(\alpha) = \begin{cases} n + [\alpha] + 1 & \text{if } \alpha > -n, \\ 0 & \text{if } \alpha < -n. \end{cases} \tag{6.73.3}$$

Furthermore,

$$n_2(\alpha) = 0 \text{ or } 1, \tag{6.73.4}$$

according as

$$L_n^{(\alpha)}(0) = \binom{n+\alpha}{n} \gtrless 0.$$

THEOREM 6.73. *Let α be an arbitrary real number, $\alpha \neq -1, -2, \cdots, -n$. The number of the positive zeros of $L_n^{(\alpha)}(x)$ is n if $\alpha > -1$; it is $n + [\alpha] + 1$ if $-n < \alpha < -1$; it is 0 if $\alpha < -n$. The number of the negative zeros is 0 or 1.*

This result can also be obtained by a direct method similar to that used in §6.72. In fact, the numbers $n_1(\alpha)$, $n_2(\alpha)$ can change only if α passes one of the integers $-1, -2, \cdots, -n$. If α decreases through an odd value of this kind, a positive zero is lost and a negative zero gained. If the passed value is even, a positive and a negative zero are lost. For $\alpha < -n$, there are no real zeros if n is even and one real zero (which is negative) if n is odd. (Compare Lawton **1**, W. Hahn **1**.)

6.8. Polynomials which satisfy a second-order linear homogeneous differential equation with polynomial coefficients; theorem of Heine-Stieltjes

Heine (**3**, vol. 1, pp. 472–479) has studied the following problem:

PROBLEM. *Let $A(x)$ and $B(x)$ be given polynomials of degrees $p + 1$ and p, respectively. To determine a polynomial $C(x)$ of degree $p - 1$ such that the differential equation*

$$A(x)\frac{d^2y}{dx^2} + 2B(x)\frac{dy}{dx} + C(x)y = 0 \tag{6.8.1}$$

has a solution which is a polynomial of a preassigned degree n.

Heine asserts that, in general, there are exactly

$$\sigma = \sigma_{np} = \binom{n + p - 1}{n} \tag{6.8.2}$$

determinations of $C(x)$ of this kind.

The hypergeometric equations (4.2.1) and (4.21.1) are of this type with $p = 1$. Lamé functions satisfy an equation of the same type with $p \geqq 2$. These cases were the starting points of Heine's investigations on this subject.

Stieltjes (**3**) discusses only a special case of (6.8.1) which, however, is of primary importance. He obtains the following result:

THEOREM 6.8. *Let $A(x)$ and $B(x)$ be given polynomials of precise degree $p + 1$ and p, respectively, and let the highest coefficients of $A(x)$ and $B(x)$ have the same sign. If the zeros of $A(x)$ and $B(x)$ are real, distinct, and alternating with one another, there are exactly σ polynomials $C(x)$ of degree $p - 1$ such that the differential equation* (6.8.1) *has a solution which is a polynomial of degree n. Here σ has the meaning* (6.8.2).

The proof of Stieltjes uses a part of Heine's assertion, namely, that σ is an upper bound for the number of polynomials $C(x)$ in question. He obtains, however, not only the existence but also a characterization of the σ solutions mentioned, in the following way: The n zeros of these solutions are distributed in all possible ways in the p intervals defined by the $p + 1$ zeros of $A(x)$. (The number of such distributions is obviously σ.) The solutions are obtained by means of a maximum problem similar to those treated in §6.7.

The following proof of Stieltjes' theorem uses the latter's idea based on a maximum problem, but Heine's elimination process (which furnishes the upper bound σ) is replaced by certain elementary considerations related to Sturm's theorem (see §1.82). Our proof is, consequently, independent of Heine's work.

6.81. Preliminary remarks

We assume with Stieltjes that

$$A(x) = (x - a_0)(x - a_1) \cdots (x - a_p), \qquad a_0 < a_1 < \cdots < a_p, \tag{6.81.1}$$

and

$$\frac{B(x)}{A(x)} = \frac{\rho_0}{x - a_0} + \frac{\rho_1}{x - a_1} + \cdots + \frac{\rho_p}{x - a_p}, \quad \rho_\nu > 0, \nu = 0, 1, 2, \cdots, p. \tag{6.81.2}$$

This is equivalent to the assumption that the zeros of $A(x)$ alternate with those of $B(x)$ and that the highest coefficients of $A(x)$ and $B(x)$ have the same sign.

Let $C(x)$ be a given polynomial. Then (6.8.1) cannot have two polynomial solutions y and z linearly independent of each other. Otherwise, we should have for $x \neq a_\nu$

$$\begin{gathered} A(x)(y'z - yz')' + 2B(x)(y'z - yz') = 0, \\ y'z - yz' = \text{const.} \exp\left\{-\int \frac{2B(x)}{A(x)}\, dx\right\} = \text{const.} \prod_{\nu=0}^{p} |x - a_\nu|^{-2\rho_\nu}. \end{gathered} \tag{6.81.3}$$

Since the last product approaches ∞ as $x \to a_\nu$, this leads to a contradiction unless $y'z - yz' \equiv 0$.

Now let y be a polynomial solution, $y \not\equiv 0$. We show that $y \neq 0$ at $x = a_\nu$. If the contrary were true, substitution of $x = a_\nu$ in (6.8.1) would give $y' = 0$. Differentiation of (6.8.1) k times results in a differential equation of order $k + 2$ in y, which has the form

$$A(x)y^{(k+2)} + \{kA'(x) + 2B(x)\}y^{(k+1)} + \cdots = 0;$$

the further coefficients are again polynomials. Now from (6.81.2) we see that $B(a_\nu) = \rho_\nu A'(a_\nu)$, so that $kA'(a_\nu) + 2B(a_\nu) \neq 0$. Thus $y = y' = y'' = \cdots = y^{(k)} = 0$ would imply $y^{(k+1)} = 0$. In a similar way we can show that all the zeros of y are distinct.

Next we prove that the zeros of y lie in the interval $[a_0, a_p]$. Upon writing $y = f(x) = \text{const.}\,(x - x_1)(x - x_2) \cdots (x - x_n)$, we have, according to (6.8.1),

$$A(x_k)f''(x_k) + 2B(x_k)f'(x_k) = 0; \tag{6.81.4}$$

or, using (6.2.2) and (6.81.2),

$$\frac{1}{x_k - x_1} + \frac{1}{x_k - x_2} + \cdots + \frac{1}{x_k - x_{k-1}} + \frac{1}{x_k - x_{k+1}} + \cdots \tag{6.81.5}$$
$$+ \frac{1}{x_k - x_n} + \sum_{\nu=0}^{p} \frac{\rho_\nu}{x_k - a_\nu} = 0.$$

We assume that some of the zeros of $f(x)$ lie outside of $[a_0, a_p]$. Let x_k be one of these zeros such that the segment $[a_0, a_p]$ and the remaining zeros all lie in a closed half-plane with x_k as a boundary point, the segment itself lying in the open interior of the half-plane. Then the complex vectors

$$x_k - x_m, \qquad x_k - a_\nu$$

lie in an angle not greater than π for all values of ν and m, $m \neq k$, and the same is true of the reciprocals of these vectors. The vectors $x_k - a_\nu$ are directed into the interior of this angle. But then we see that (6.81.5) is a contradiction. (For this argument, see Pólya **1**.)

Let $n_1, n_2, \cdots, n_p$ be the number of zeros of y in $[a_0, a_1]$, $[a_1, a_2]$, $\cdots$, $[a_{p-1}, a_p]$, respectively. Then we say that y is of the type $\{n_1, n_2, \cdots, n_p\}$. Here $n_1 + n_2 + n_3 + \cdots + n_p = n$, and σ represents the number of all possible types. It is our intention to show the existence of exactly one polynomial solution of each type, corresponding to σ different determinations of the polynomial $C(x)$ of degree $p - 1$.

6.82. A maximum problem

Following Stieltjes (loc. cit.), we first show that a polynomial solution of each type exists.

Let $x_1, x_2, \cdots, x_n$ be variable points, each different from the a_ν and distributed in $[a_0, a_p]$ so that in each interval $[a_{\nu-1}, a_\nu]$ there lies a certain preassigned number, say n_ν, of these points, $\sum_{\nu=1}^{p} n_\nu = n$. Letting each x_k vary in a fixed interval, we consider the maximum of the product

$$W = \prod_{\substack{\kappa=1,2,\cdots,n \\ \nu=0,1,\cdots,p}} |x_\kappa - a_\nu|^{\rho_\nu} \prod_{\substack{\lambda,\mu=1,2,\cdots,n \\ \lambda<\mu}} |x_\lambda - x_\mu|. \tag{6.82.1}$$

The existence and positiveness of this maximum are clear. In the maximum position, the points x_k are different from the a_ν and from one another, and they are of a preassigned type. Furthermore, we have $\partial W/\partial x_k = 0$; whence (6.81.5) again follows. If $f(x)$ denotes the polynomial with the zeros x_k, the latter equation means that $A(x)f''(x) + 2B(x)f'(x)$ vanishes for $x = x_k$ and hence is divisible by $f(x)$. If this ratio is denoted by $-C(x)$, (6.8.1) follows. It is clear that $\log (W^{-1})$, apart from the constant terms $\rho_\mu \rho_\nu \log |a_\mu - a_\nu|^{-1}$, $\mu \neq \nu$, is the "energy" of the system of masses ρ_ν concentrated at a_ν and of unit masses concentrated at the x_k. (Cf. §6.7 (1).)

Incidentally, the argument used in §6.7 (1) shows that the system $\{x_k\}$ with the maximum property is uniquely determined. This is not the same as the uniqueness of the solution of a given type since (6.81.5) is not equivalent to the maximum property. It is, however, easy to show (cf. (6.22.6)) that (6.81.5) is equivalent to a relative maximum of W.

6.83. Uniqueness

Let $C(x)$, y and $D(x)$, z be two solutions of our problem of the same type, $C(x) \not\equiv D(x)$. Suppose both polynomials y and z have positive highest coefficients. By combining (6.8.1) with the corresponding equation for z, we find the relation

$$\text{(6.83.1)} \qquad \frac{d}{dx}(y'z - yz') + 2\sum_{\nu=0}^{p} \frac{\rho_\nu}{x - a_\nu}(y'z - yz') + \frac{C(x) - D(x)}{A(x)} yz = 0.$$

Introducing $H = \prod_{\nu=0}^{p} |x - a_\nu|^{2\rho_\nu}$, we obtain for $x \neq a_\nu$

$$\text{(6.83.2)} \qquad \frac{d}{dx}\{H(y'z - yz')\} = \frac{D(x) - C(x)}{A(x)} yzH.$$

Suppose the function $\{D(x) - C(x)\}/A(x)$ is non-negative in the fixed interval $a_{\nu-1} < x < a_\nu$. Then between two consecutive zeros α and β of y, $\alpha < \beta$, in this interval, z must change its sign at least once. Otherwise, yz would be permanently positive or negative, and therefore, $H(y'z - yz')$ increasing or decreasing in $\alpha < x < \beta$. For $x = \alpha + \epsilon$, $\epsilon > 0$, however, the last expression has the same sign as $y'z$ or yz, and for $x = \beta - \epsilon$, $\epsilon > 0$, the same sign as $y'z$ or $-yz$. Hence it passes from positive to negative values if $yz > 0$, and from negative to positive values if $yz < 0$. Either case contradicts the property of monotony of $H(y'z - yz')$ mentioned before.

Under the previous assumption concerning $\{D(x) - C(x)\}/A(x)$, the function z must change its sign also in $[a_{\nu-1}, \gamma]$ and in $[\delta, a_\nu]$ if γ and δ are, respectively, the first and last zeros of y in $[a_{\nu-1}, a_\nu]$. Otherwise, y and z would each have a constant sign in those intervals; moreover, y and z would have the same sign in a given interval because they are of the same type. Then $H(y'z - yz')$ would be an increasing function. It vanishes for $x \to a_{\nu-1} + 0$ and also for $x \to a_\nu - 0$, so that $y'z - yz'$ must be positive in $[a_{\nu-1}, \gamma]$ and negative in $[\delta, a_\nu]$. Now for $x = \gamma$, we have sgn $(y'z - yz')$ = sgn $y'z$, which is the same as sgn $(y'y)$ at $x = \gamma - \epsilon$, $\epsilon > 0$, that is, negative. This is a contradiction. The same argument can be used at $x = \delta$.

Finally, we remark, under the assumption made about $\{D(x) - C(x)\}/A(x)$, that the polynomial z must vanish in $[a_{\nu-1}, a_\nu]$, even if y has no zeros there. Otherwise, because of (6.83.2), the function $H(y'z - yz')$ would be monotonic. This is impossible since H vanishes for $x = a_{\nu-1}$ and for $x = a_\nu$.

To recapitulate: this argument would furnish at least one more zero for z in $[a_{\nu-1}, a_\nu]$ than for y, which is impossible. Therefore, $\{D(x) - C(x)\}/A(x)$ must be negative in some points of $a_{\nu-1} < x < a_\nu$. By interchanging $C(x)$, y with $D(x)$, z, it is seen that the same function must also be positive somewhere in $a_{\nu-1} < x < a_\nu$; whence $D(x) - C(x)$ must have at least one variation of sign in $[a_{\nu-1}, a_\nu]$. Since this is true for $\nu = 1, 2, \cdots, p$, the function $D(x) - C(x)$ has at least p variations of sign in $[a_0, a_p]$. However, this is impossible since $D(x) - C(x)$ is of degree $p - 1$.

6.9. Zeros of Legendre functions of the second kind; generalization

(1) In connection with Fejér's second generalization of Legendre polynomials (§6.5 (4)) we consider the function

$$H_n(\cos\theta) = \beta_0 \cos(n+1)\theta + \beta_1 \cos(n+3)\theta + \cdots + \beta_m \cos(n+2m+1)\theta + \cdots. \tag{6.9.1}$$

Here $\beta_m \downarrow 0$, so that the series converges for $0 < \theta < \pi$. This is the conjugate series of the function $G_n(\cos\theta)$ defined by (6.5.16). The function $\mathbf{Q}_n(\cos\theta)$ of (4.9.16) is a special case; then the sequence β_m is given by (4.9.5) and is completely monotonic. Now we prove the following theorem:

THEOREM 6.9.1. *Let $\beta_m > 0$ and let $\{\beta_m\}$ be a completely monotonic sequence. The function $H_n(\cos\theta)$ defined by* (6.9.1) *has at least one zero in each of the intervals*

$$\frac{\nu}{n+\frac{1}{2}}\pi < \theta < \frac{\nu+\frac{1}{2}}{n+\frac{1}{2}}\pi, \qquad \nu = 0, 1, 2, \cdots, n. \tag{6.9.2}$$

More precisely, $H_n(\cos\theta)$ has an odd number of zeros in each of these intervals. For the proof we again use Hausdorff's representation and obtain for (6.9.1)

$$\int_0^1 \left\{\sum_{m=0}^{\infty} t^m \cos(n+2m+1)\theta\right\} d\beta(t) = \int_0^1 \frac{\cos(n+1)\theta - t\cos(n-1)\theta}{1 - 2t\cos 2\theta + t^2}\, d\beta(t), \tag{6.9.3}$$

where $\beta(t)$ is a function of the same type as $\alpha(t)$ in (6.5.8). Upon substituting

$$\theta = \frac{\nu}{n+\frac{1}{2}}\pi \quad\text{and}\quad \theta = \frac{\nu+\frac{1}{2}}{n+\frac{1}{2}}\pi,$$

we find

$$\cos(n+1)\theta - t\cos(n-1)\theta = \begin{cases} (-1)^\nu\{\cos(\theta/2) - t\cos(3\theta/2)\}, \\ (-1)^{\nu+1}\{\sin(\theta/2) + t\sin(3\theta/2)\}, \end{cases} \tag{6.9.4}$$

respectively. Both expressions in the braces are positive; this establishes the statement. (For $\nu = 0$ and $\nu = n$ we must take into account that

$$\lim_{\theta\to+0} H_n(\cos\theta) = (-1)^{n+1} \lim_{\theta\to\pi-0} H_n(\cos\theta) > 0$$

($+\infty$ if $\sum_{m=0}^{\infty} \beta_m$ is divergent).)

(2) THEOREM 6.9.2. *The function* $\mathbf{Q}_n(\cos\theta)$ (§4.62 (3)) *has exactly* $n + 1$ *zeros in* $0 < \theta < \pi$, *which lie in the intervals* (6.9.2).

In this special case of Theorem 6.9.1 there cannot be more than one zero in each of the intervals (6.9.2); otherwise, $P_n(\cos\theta)$ would have more than n zeros by Sturm's theorem.

Inequalities (6.9.2) for the zeros of $\mathbf{Q}_n(\cos\theta)$ are due to Stieltjes (**8**, p. 252), whose proof is, however, different from that just given. Fejér obtains (**20**, pp. 51–52) less precise inequalities in a manner similar to that used above; his conditions concerning the sequence $\{\beta_m\}$, however, are more general.

(3) THEOREM 6.9.3. *Legendre's function of the second kind* $Q_n(x)$ (§4.61 (1)) *has no zeros in the complex plane cut along the segment* $[-1, +1]$, *except* $x = \infty$, *which is a zero of multiplicity* $n + 1$.

This theorem is due to Hermite (**3**) and Stieltjes (**9**) (cf. also Hermite-Stieltjes **1**, vol. 2, pp. 80–104, no. 267–274). The following argument is a slight modification of the second proof of Stieltjes.

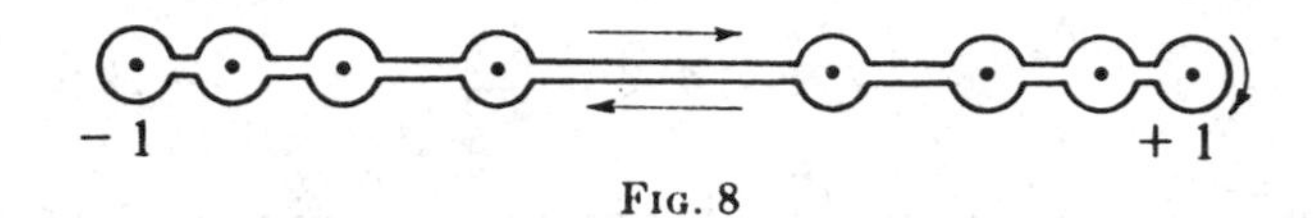

FIG. 8

We start from (4.62.10). This function $\mathbf{Q}_n(x)$ has $n + 1$ zeros in the interior of $[-1, +1]$, which alternate with the zeros of $P_n(x)$ (cf. Theorem 6.9.2). Let $+1 = x_0 > x_1 > \cdots > x_n > x_{n+1} = -1$ denote the zeros of $(1 - x^2)P_n(x)$ in decreasing order. Then

$$\text{(6.9.5)} \qquad \operatorname{sgn} \mathbf{Q}_n(x_\nu) = (-1)^\nu, \qquad \nu = 0, 1, 2, \cdots, n + 1.$$

Furthermore, $\mathbf{Q}_n(x)$ is a solution of (4.2.1), $\alpha = \beta = 0$, so that $\mathbf{Q}_n(x)/P_n(x)$ is increasing in the interval $-1 \leqq x \leqq +1$ (cf. (4.2.6)); it becomes infinite at each x_ν.

The curve in the figure encircles the points $x = \pm 1$ and avoids the zeros of $P_n(x)$ by means of semi-circles. Now we shall show that the variation of $\arg \{Q_n(x)/P_n(x)\}$ along this curve is equal to $2\pi(2n + 1)$. Then, according to Theorem 1.91.1 (principle of argument), the function $Q_n(x)/P_n(x)$ has exactly $2n + 1$ zeros exterior to this curve. But since it has a zero of multiplicity $2n + 1$ at $x = \infty$, the statement will follow.

Let ϵ be a sufficiently small positive number. As x encircles $+1$ in the negative (clockwise) direction from $1 + \epsilon$ to $1 - \epsilon$, the variation of the argument in question is approximately (cf. (4.62.7)) the variation of $\arg \{\log 1/(x - 1)\}$, a quantity which tends to 0 with ϵ. If x describes the segment from $x_\nu - \epsilon$ to $x_{\nu+1} + \epsilon$, $\nu = 0, 1, 2, \cdots, n$, along the "lower border" of $[-1, +1]$, we have

$$y = \frac{Q_n(x - i0)}{P_n(x)} = \frac{i\pi}{2} + \frac{\mathrm{Q}_n(x)}{P_n(x)}. \tag{6.9.6}$$

Then y describes a straight line $\mathfrak{J}y = \pi/2$ in the direction of decreasing abscissas. The variation of the argument of y is $+\pi$.

In the neighborhood of x_ν, $\nu = 1, 2, \cdots, n$, the function $\mathrm{Q}_n(x)/P_n(x)$ differs only by a bounded term from

$$\frac{\mathrm{Q}_n(x_\nu)}{P_n'(x_\nu)} \frac{1}{x - x_\nu}, \quad \text{where } \frac{\mathrm{Q}_n(x_\nu)}{P_n'(x_\nu)} < 0. \tag{6.9.7}$$

Therefore, the half-circle in the lower half-plane around x_ν will be carried over into a curve which approximates a large semi-circle in the lower half-plane; the argument of y then increases by $+\pi$.

Finally, if $x = -1$ is encircled in the negative sense, from $-1 + \epsilon$ to $-1 - \epsilon$, the variation in arg y is again a quantity which tends to zero with ϵ.

To recapitulate: While x moves along the lower border from $1 + \epsilon$ to $-1 - \epsilon$, the total increase in the argument of y is $(n + 1)\pi + n\pi = (2n + 1)\pi$. The same is true on the upper border as x varies from $-1 - \epsilon$ to $1 + \epsilon$. This establishes the statement.

6.10. Further results

(1) Let $\{\theta_{\nu n}\}$ denote the zeros of $P_n(\cos\theta)$ in the interval $[0, \pi]$, ordered in an increasing way. Turán (**1**) proved that the sequence $x_{\nu n} - x_{\nu,n-1}$, where $x_{\nu n} = \cos\theta_{\nu n}$, is increasing as ν runs from 1 to $[\frac{1}{2}(n - 1)]$. Szegö proved (in a correspondence with Turán, 1946) the same fact for the differences $\theta_{\nu,n-1} - \theta_{\nu n}$. Cf. Szegö-Turán **1**.

(2) Concerning the topics treated in §§6.8–6.83, see also Makai **3**.

(3) The argument of §6.9 (3) leads to a more general result concerning the number of the zeros of $Q_n(x) - aP_n(x)$ in the complex plane cut along the segment $[-1, +1]$ where a is a given complex constant (Hermite and Stieltjes, loc. cit.). This number is again $= 2n + 1$ if $-\pi/2 < \mathfrak{J}a < +\pi/2$, and $= n$ if $\mathfrak{J}a \geqq +\pi/2$ or $\mathfrak{J}a \leqq -\pi/2$.

Indeed, in the case $-\pi/2 < \mathfrak{J}a < +\pi/2$ no essential change in the reasoning is needed. Now let $\mathfrak{J}a > +\pi/2$. If x describes the segment from $x_\nu - \epsilon$ to $x_{\nu+1} + \epsilon$, $\nu = 0, 1, 2, \cdots, n$, along the lower border of $[-1, +1]$, the variation of the argument of $y - a$ is $-\pi$. There is no change in the contribution of the semi-circles around x_ν and of the whole upper border. Thus the total increase in the argument of $y - a$ is $-(n + 1)\pi + n\pi + (n + 1)\pi + n\pi = 2n\pi$.

Let $\mathfrak{J}a = +\pi/2$. On the segment from $x_\nu - \epsilon$ to $x_{\nu+1} + \epsilon$ there is a unique point ξ such that $\mathrm{Q}_n(\xi)/P_n(\xi) = \mathfrak{R}a$. We must use an indentation of the contour into the lower half-plane and take into account that for $\delta > 0$

$$\left(\frac{Q_n(x - i0)}{P_n(x)}\right)_{x=\xi-i\delta} = \left(\frac{Q_n(x - i0)}{P_n(x)}\right)_{x=\xi} - i\delta\left\{\frac{d}{dx}\frac{Q_n(x - i0)}{P_n(x)}\right\}_{x=\xi} + \cdots;$$

the imaginary part of the right-hand side is $\pi/2 - \delta' < \pi/2$ where $\delta' > 0$. The argument is similar if $\Im a = -\pi/2$.

(4) Makai-Turán (1) have proved the following theorem of the Picard-Landau type. Let $H_n(z)$ be Hermite's polynomial. There exists a positive (absolute) constant A such that every equation of the form

$$H_0(z) + H_1(z) + \gamma H_n(z) = 0$$

has a solution in the strip $|\Im z| \leqq A$; $n \geqq 2$, γ arbitrary real or complex. The exact value of A has been determined by Schmeisser **1**. The extremal polynomials are of third degree. The corresponding problems for quadrinomials and more general equations remain open.

(5) Szegö **20** improved the right-hand estimates in Theorem 6.4 when $2a_0 - a_1 > a_1 - a_2 \geqq a_2 - a_3 \geqq \cdots \geqq a_{n-1} - a_n \geqq a_n \geqq 0$. Both sides can be improved when $(2k-1)a_{k-1} \geqq 2ka_k > 0$, $k \geqq 1$. See Askey-Steinig **1**.

(6) The results mentioned in §6.10 (1) have been extended to ultraspherical polynomials in Szegö-Turán **1**.

(7) For the positive θ-zeros of the Legendre polynomials $P_n(\cos\theta)$ and also for the positive zeros of the Hermite and Laguerre polynomials, written in increasing order, the second differences of the respective sequences of consecutive zeros are all positive, as an immediate consequence of the Sturm theory (cf. Theorem 6.3.3). In L. Lorch-P. Szego **1**, **2**, it is conjectured that all *higher* differences are also positive, but this remains unresolved. Substantiating numerical evidence is cited there and in Davis-Rabinowitz **1**. The latter present also similar evidence connected with $P_n'(\cos\theta)$. See also Lorch-Muldoon-P. Szego **1**, **2**.

(8) If $\pi_n(x)$ is an arbitrary polynomial of degree less than or equal to n and if

$$\pi_n(x) = b_0 L_0(x) + \cdots + b_n L_n(x)$$

is its Laguerre expansion, then the number of sign changes of $\pi_n(x)$ for $x > 0$ is at least as great as the number of sign changes of the sequence of the differences

$$b_0,\ (b_0 - b_1),\ (b_0 - 2b_1 + b_2),\ \cdots.$$

See Turán **2**.

CHAPTER VII

INEQUALITIES

No inequalities, except trivial ones, are known for general orthogonal polynomials. However, inequalities involving an unspecified constant can easily be derived under certain conditions concerning the weight function $w(x)$. Still more precise estimates can be obtained if $w(x)$ is monotonic, and a great number of special inequalities follow from this added restriction.

Another very extensive class of inequalities can be derived for the classical orthogonal polynomials, and in the present chapter we intend to enumerate and compare the various methods used to obtain these inequalities. Aside from integral and series representations, the main tool is differential equations. As regards the latter, we remark that there is a special method for deriving inequalities for the solutions of certain differential equations (cf. Theorem 7.31.1). In recent years this method has been used in several special problems (not only for polynomials), with slight variations, primarily by G. N. Watson and S. Bernstein; it originated, however, in an idea of Sonin.[41]

At the end of this chapter we use the above mentioned inequalities in dealing with certain extremum problems which involve polynomials of a fixed degree.

The selection of the material treated in this chapter has been influenced by the needs of later chapters, especially by those of Chapters IX, XIV, and XV. Historically, the major part of the inequalities for classical polynomials arose from the discussion of the corresponding expansion problems.

We shall postpone till Chapter VIII the asymptotic calculation of certain maxima (which can also be expressed in terms of inequalities), since they require more intricate asymptotic considerations. However, we have found it necessary in the present chapter to anticipate certain asymptotic results of Chapter VIII.

7.1. Rough bounds for orthogonal polynomials

In this section we make essential use of the representation of positive functions discussed in §10.2. However, this does not play a rôle in the further course of Chapter VII.

(1) Let $w(x)$ be a weight function on the interval $[-1, +1]$ for which the integral

$$\int_{-1}^{+1} (1 - x^2)^{-\frac{1}{2}} \log w(x) \, dx \tag{7.1.1}$$

exists in Lebesgue's sense. (This implies that $w(x)$ cannot vanish on a whole

[41] Cf. Sonin **2**, pp. 23–24. I owe this reference to Professor J. A. Shohat.

segment.) Let $D(f; z) = D(z)$ be the analytic function associated with $f(\theta) = w(\cos\theta)\,|\sin\theta|$ in the sense of §10.2 (2).

The conformal representation $x = \frac{1}{2}(z + z^{-1})$ maps the unit circle $|z| < 1$ (or $|z| > 1$) onto the x-plane cut along the segment $-1 \leqq x \leqq +1$. For $z = e^{i\theta}$ we have $x = \cos\theta$. (Cf. §1.9.)

THEOREM 7.1.1. *Let $\{p_n(x)\}$ be the orthonormal set of polynomials associated with a weight function $w(x)$, $-1 \leqq x \leqq +1$, for which* (7.1.1) *exists; then*

$$|\pi^{\frac{1}{2}} D(z) p_n(x) z^n| < (1 - |z|^2)^{-\frac{1}{2}}, \qquad |z| < 1, \tag{7.1.2}$$

where $x = \frac{1}{2}(z + z^{-1})$ is an arbitrary point of the cut plane.

For, (cf. (10.2.9))

$$\begin{aligned} 1 &= \int_{-1}^{+1} \{p_n(x)\}^2 w(x)\, dx \\ &= \frac{1}{2}\int_{-\pi}^{+\pi} \{p_n(\cos\theta)\}^2 w(\cos\theta)|\sin\theta|\, d\theta \\ &= \lim_{r\to 1-0} \frac{1}{2}\int_{-\pi}^{+\pi} |p_n[\tfrac{1}{2}(z + z^{-1})]z^n|^2\, |D(z)|^2\, d\theta, \qquad z = re^{i\theta} \end{aligned} \tag{7.1.3}$$

Now if $f(z) = \sum_{m=0}^{\infty} c_m z^m$ is regular in $|z| < 1$, we have, according to Cauchy's inequality,

$$|f(z)|^2 \leqq \sum_{m=0}^{\infty} |c_m|^2 \sum_{m=0}^{\infty} |z^{2m}| = (1 - |z|^2)^{-1} \lim_{r\to 1-0} \frac{1}{2\pi}\int_{-\pi}^{+\pi} |f(re^{i\theta})|^2\, d\theta. \tag{7.1.4}$$

This establishes the statement.

The bound in (7.1.2) becomes infinite if x lies on the segment $[-1, +1]$; for all other values of x it furnishes a first appraisal of the magnitude of $p_n(x)$ under a rather general condition. This information is comparatively precise because we shall prove (cf. Theorem 12.1.2) that, for a fixed x, the left-hand member of (7.1.2) tends to $2^{-\frac{1}{2}}$ as $n \to \infty$. It is rather remarkable that no use of the orthogonal property has been made in deriving (7.1.2); only the normalization of $p_n(x)$ is employed.

(2) THEOREM 7.1.2. *Let $w(x)$ be bounded from zero, that is, $w(x) \geqq \mu > 0$. Then, if x is not on the segment $[-1, +1]$,*

$$|p_n(x)| < A\,|x + (x^2 - 1)^{\frac{1}{2}}|^n, \tag{7.1.5}$$

where $(x^2 - 1)^{\frac{1}{2}}$ is chosen so that $|x + (x^2 - 1)^{\frac{1}{2}}| > 1$. The constant A depends on x and μ but not on n; A is uniformly bounded in the exterior of any closed curve which contains $[-1, +1]$ in its interior.

We have in this case $|D(z)| > |\mu(1 - z^2)/2|^{\frac{1}{2}}$ (cf. (10.2.10)), so that from (7.1.2)

(7.1.6) $$|\pi\mu(1-z^2)/2|^{\frac{1}{2}}|p_n(x)z^n| < (1-|z|^2)^{-\frac{1}{2}}, \qquad |z|<1,$$

follows.

(3) In the same case $w(x) \geqq \mu > 0$, we can also easily obtain bounds for $p_n(x)$ on the orthogonality interval $-1 \leqq x \leqq +1$ itself. In fact, if $|z|<1$, we have, in view of (7.1.6), $(\pi\mu/2)^{\frac{1}{2}}(1-|z|^2)^{\frac{1}{2}}|p_n(x)z^n| < (1-|z|^2)^{-\frac{1}{2}}$. Let x be on the segment $[-1,+1]$, $x = \frac{1}{2}(z+z^{-1})$, where $|z|=1$. Then $p_n(x)z^n$ is a π_{2n} in z, and we have for $-1 \leqq x \leqq +1$, $|z|=1$, $r<1$,

$$|p_n(x)| = |p_n(x)z^n| < \max_{|\zeta|=r^{-1}} |p_n(\xi)\zeta^n|$$
$$= r^{-2n}\max_{|\zeta|=r}|p_n(\xi)\zeta^n| < r^{-2n}(\pi\mu/2)^{-\frac{1}{2}}(1-r^2)^{-1}.$$

Here $\xi = \frac{1}{2}(\zeta+\zeta^{-1})$. On putting $r^2 = 1-1/n$, for $n \geqq 2$, $-1 \leqq x \leqq +1$, we have

(7.1.7) $$|p_n(x)| < (\pi\mu/2)^{-\frac{1}{2}}(1-n^{-1})^{-n}n < 4(\pi\mu/2)^{-\frac{1}{2}}n.$$

For $-1 < x < +1$, we can reduce the exponent in (7.1.7) from 1 to $\frac{1}{2}$ (cf. (7.71.28)).

(4) The same elementary method gives an idea of the magnitude of Jacobi's polynomial if n is large. In this case $w(x) = (1-x)^\alpha(1+x)^\beta$, $\alpha > -1$, $\beta > -1$ and (cf. (10.2.13))

(7.1.8) $$D(z) = 2^{-(\alpha+\beta+1)/2}(1-z)^{\alpha+\frac{1}{2}}(1+z)^{\beta+\frac{1}{2}}.$$

Then by (7.1.2)

(7.1.9) $$\pi^{\frac{1}{2}}2^{-(\alpha+\beta+1)/2}|1-z|^{\alpha+\frac{1}{2}}|1+z|^{\beta+\frac{1}{2}}|p_n(x)z^n| < (1-|z|^2)^{-\frac{1}{2}},$$
$$x = \tfrac{1}{2}(z+z^{-1}),\ |z|<1.$$

Now assume $-1 \leqq x \leqq +1$; we obtain, as in (3),

(7.1.10) $$|p_n(x)| < Cr^{-2n}(1-r^2)^{-\frac{1}{2}}\max_{|\zeta|=r}|1-\zeta|^{-\alpha-\frac{1}{2}}|1+\zeta|^{-\beta-\frac{1}{2}},$$

where C depends only on α and β, and $0<r<1$. We choose again $r^2 = 1-1/n$. Discussing the right-hand member of (7.1.10) for $|\zeta|=r$, $\mathfrak{R}(\zeta) \geqq 0$, and for $|\zeta|=r$, $\mathfrak{R}(\zeta) \leqq 0$, we obtain

(7.1.11) $$|p_n(x)| < C'n^{\max(\alpha+1,\beta+1,\frac{1}{2})}, \qquad -1 \leqq x \leqq +1.$$

Here C' depends only on α and β. The "true" exponent is $\max(\alpha+\frac{1}{2}, \beta+\frac{1}{2}, 0)$ (cf. (7.32.2)).

For later purposes we give a formulation of (7.1.9) in terms of the Jacobi polynomials $P_n^{(\alpha,\beta)}(x)$ (cf. (4.3.4)). If x is exterior to the segment $[-1,+1]$, we have

(7.1.12) $$P_n^{(\alpha,\beta)}(x) = n^{-\frac{1}{2}}O(|z|^{-n}), \qquad x = \tfrac{1}{2}(z+z^{-1}),\ |z|<1;$$
$$\alpha > -1,\ \beta > -1,\ n \to \infty.$$

This holds uniformly in the exterior of any curve containing the segment $[-1, +1]$ in its interior. Inequality (7.1.12) follows, of course, immediately from the asymptotic formula (8.21.9).

(5) The following theorem is rather useful in obtaining bounds for orthonormal polynomials:

THEOREM 7.1.3. *Let $w(x)$ and $\tilde{w}(x)$ be two weight functions on the segment $[-1, +1]$, $w(x)/\tilde{w}(x) = k(x)$. Assume $k(x) \geqq k > 0$, and let $k(x)$ satisfy the Lipschitz condition*

$$(7.1.13) \qquad |k(x_1) - k(x_2)| < \lambda |x_1 - x_2|.$$

If $\{p_n(x)\}$ and $\{\tilde{p}_n(x)\}$ are the orthonormal polynomials associated with $w(x)$ and $\tilde{w}(x)$, respectively, we have

$$(7.1.14) \qquad |p_n(x)| \leqq k^{-\frac{1}{2}} |\tilde{p}_n(x)| + \lambda k^{-\frac{3}{2}}\{|\tilde{p}_n(x)| + |\tilde{p}_{n-1}(x)|\}.$$

This theorem is due to Korous (**3**). The proof follows from the identity ((3.2.3))

$$\begin{aligned} p_n(x) &= \int_{-1}^{+1} p_n(t)\left\{\sum_{\nu=0}^{n} \tilde{p}_\nu(x)\tilde{p}_\nu(t)\right\}\tilde{w}(t)\,dt \\ &= \frac{k_n}{\tilde{k}_n}\tilde{p}_n(x) + \int_{-1}^{+1} p_n(t)\left\{\sum_{\nu=0}^{n-1} \tilde{p}_\nu(x)\tilde{p}_\nu(t)\right\}\tilde{w}(t)\left(1 - \frac{k(t)}{k(x)}\right)dt \\ &= \frac{k_n}{\tilde{k}_n}\tilde{p}_n(x) + \frac{\tilde{k}_{n-1}}{\tilde{k}_n}\frac{1}{k(x)}\int_{-1}^{+1} p_n(t)\{\tilde{p}_n(x)\tilde{p}_{n-1}(t) - \tilde{p}_{n-1}(x)\tilde{p}_n(t)\} \\ &\qquad\qquad \cdot\tilde{w}(t)\,\frac{k(x) - k(t)}{x - t}\,dt, \end{aligned}$$

where k_n has the same meaning as in (2.2.15), and $\tilde{k}_n$ has the corresponding meaning for $\tilde{p}_n(x)$. Now, according to Schwarz's inequality

$$\begin{aligned} \frac{k_n}{\tilde{k}_n} = \int_{-1}^{+1} p_n(t)\tilde{p}_n(t)\tilde{w}(t)\,dt &\leqq \left\{\int_{-1}^{+1} [p_n(t)]^2\tilde{w}(t)\,dt\right\}^{\frac{1}{2}}\left\{\int_{-1}^{+1} [\tilde{p}_n(t)]^2\tilde{w}(t)\,dt\right\}^{\frac{1}{2}} \\ &= \left\{\int_{-1}^{+1} [k(t)]^{-1}w(t)[p_n(t)]^2\,dt\right\}^{\frac{1}{2}} \leqq k^{-\frac{1}{2}}, \end{aligned}$$

$$\frac{\tilde{k}_{n-1}}{\tilde{k}_n} = \int_{-1}^{+1} t\tilde{p}_{n-1}(t)\tilde{p}_n(t)\tilde{w}(t)\,dt \leqq \int_{-1}^{+1} |\tilde{p}_{n-1}(t)|\,|\tilde{p}_n(t)|\,\tilde{w}(t)\,dt \leqq 1,$$

$$\left|\int_{-1}^{+1} p_n(t)\tilde{p}_{n-1}(t)\tilde{w}(t)\,dt\right| \leqq \left\{\int_{-1}^{+1} [p_n(t)]^2\tilde{w}(t)\,dt\right\}^{\frac{1}{2}} \leqq k^{-\frac{1}{2}},$$

which establishes (7.1.14).

We mention two important special cases which follow immediately from (7.1.14) by use of the bounds of the Legendre and Tchebichef polynomials (concerning the first case, see (7.21.1) and (7.3.8)):

(a) If $w(x)$ is positive and satisfies a Lipschitz condition $|w(x_1) - w(x_2)| < \lambda|x_1 - x_2|$, we have

(7.1.15)
$$|p_n(x)| < \begin{cases} An^{\frac{1}{2}}, \\ A'(1 - x^2)^{-\frac{1}{4}}, \end{cases} \qquad -1 < x < +1.$$

Here the positive constants A and A' are independent of x and n.

(b) If $w(x) = (1 - x^2)^{-\frac{1}{2}} k(x)$ where $k(x)$ is positive and satisfies a Lipschitz condition $|k(x_1) - k(x_2)| < \lambda |x_1 - x_2|$, we have

(7.1.16)
$$|p_n(x)| < A, \qquad -1 < x < +1,$$

where A is independent of x and n.

For other elementary considerations of a similar nature, see Shohat **4**, pp. 165–166 and Jackson **6**, pp. 893–898. See, also, §7.71 (6).

7.2. Monotonic weight functions

THEOREM 7.2. *Let $w(x)$ be a weight function which is non-decreasing in the interval $[a, b]$, b finite. If $\{p_n(x)\}$ is the set of the corresponding orthogonal polynomials, the functions $\{w(x)\}^{\frac{1}{2}} |p_n(x)|$ attain their maximum in $[a, b]$ for $x = b$.*

See Szegö **3**. A corresponding statement holds for any subinterval $[x_0, b]$ of $[a, b]$ where $w(x)$ is non-decreasing.

The proof is based on the identity

$$w(b)\{p_n(b)\}^2 - w(x)\{p_n(x)\}^2 = 2\int_x^b w(t)p_n(t)p_n'(t)\,dt + \int_x^b \{p_n(t)\}^2\,dw(t),$$

which follows from (1.4.4). It suffices to show that this expression is non-negative in $a \leqq x \leqq b$. Denoting by $x_1 < x_2 < \cdots < x_n$ the zeros of $p_n(x)$ in increasing order, we have $p_n(t)p_n'(t) > 0$ for $t > x_n$ and $p_n(t)p_n'(t) < 0$ for $t < x_1$. Therefore, the statement is trivial for $x_n \leqq x \leqq b$ and follows from

$$\begin{aligned}\int_x^b w(t)p_n(t)p_n'(t)\,dt &= \int_a^b w(t)p_n(t)p_n'(t)\,dt - \int_a^x w(t)p_n(t)p_n'(t)\,dt \\ &= -\int_a^x w(t)p_n(t)p_n'(t)\,dt\end{aligned}$$

for $a \leqq x \leqq x_1$. Here we used the monotony of $w(x)$ only in $x \leqq t \leqq b$.

Now let $x_\nu \leqq x \leqq x_{\nu+1}$, $\nu = 1, 2, \cdots, n-1$, $n \geqq 2$. If we introduce the new weight function

$$W(x) = w(x)\{(x - x_1)(x - x_2) \cdots (x - x_\nu)\}^2,$$

the corresponding orthogonal polynomial of degree $n - \nu$ will be, save for a positive constant factor,

$$q_{n-\nu}(x) = \frac{p_n(x)}{(x - x_1)(x - x_2) \cdots (x - x_\nu)},$$

with the zeros $x_{\nu+1}, x_{\nu+2}, \cdots, x_n$. In fact, if $\rho(x)$ is an arbitrary $\pi_{n-\nu-1}$,

$$\int_a^b W(x)q_{n-\nu}(x)\rho(x)\,dx = \int_a^b w(x)p_n(x)(x - x_1)(x - x_2)\cdots(x - x_\nu)\rho(x)\,dx = 0.$$

But $W(x)$ increases monotonically for $x_\nu \leqq x \leqq b$, so that the preceding argument furnishes, for $x_\nu \leqq x \leqq x_{\nu+1}$,

$$W(b)\{q_{n-\nu}(b)\}^2 - W(x)\{q_{n-\nu}(x)\}^2 = w(b)\{p_n(b)\}^2 - w(x)\{p_n(x)\}^2 \geqq 0.$$

In addition we conclude that the equation $w(b)\{p_n(b)\}^2 = w(\xi)\{p_n(\xi)\}^2$ holds if and only if $\xi < x_1$ and $w(t)$ vanishes in $[a, \xi]$ (this condition has, of course, no significance if $a = \xi$), and is step-wise constant in $[\xi, b]$. Furthermore, $p_n(t)$ must vanish at the points of increase of $w(t)$. (The weight function $w(x)$ cannot vanish for $x \geqq x_1$.)

7.21. Applications

On putting $a = -1$, $b = +1$, and $w(x) = 1$, we obtain the important inequality

$$|P_n(x)| \leqq 1, \qquad -1 \leqq x \leqq +1, \tag{7.21.1}$$

for the Legendre polynomials $P_n(x)$, $P_n(1) = 1$. If $n > 0$, the equality sign holds only for $x = \pm 1$.

Another interesting case is $a = -1$, $b = +1$, and $w(x) = |x|^{2k}$, $k > 0$. On account of (4.1.6) we have $|x|^k |P_\nu^{(0,k-\frac{1}{2})}(2x^2 - 1)| \leqq P_\nu^{(0,k-\frac{1}{2})}(1) = 1$, so that

$$\{(1 - x)/2\}^{\alpha/2+\frac{1}{4}} |P_n^{(\alpha,0)}(x)| \leqq 1, \qquad -1 \leqq x \leqq +1,\ \alpha \geqq -\tfrac{1}{2}. \tag{7.21.2}$$

The equality sign holds only for $x = -1$. In the interval $0 \leqq x \leqq 1$ this is, for large n, less precise than the first inequality (7.32.6).

In case $a = 0$, $b = +\infty$, and $w(x) = e^{-x}$, we obtain for the Laguerre polynomials

$$e^{-x/2} |L_n(x)| \leqq 1, \qquad x \geqq 0, \tag{7.21.3}$$

the equality sign holding only for $x = 0$ if $n > 0$.

With regard to these special cases, see Szegö **2** and **3**.

7.3. Legendre polynomials

A second proof of (7.21.1) and various other important inequalities can be obtained by means of the differential equation for Legendre polynomials.

(1) THEOREM 7.3.1. *Let $n \geqq 2$. The successive relative maxima of $|P_n(x)|$, when x decreases from 1 to 0, form a decreasing sequence. More precisely, if $\mu_1, \mu_2, \cdots, \mu_{[n/2]}$ denote these maxima corresponding to decreasing values of x, we have*

$$1 > \mu_1 > \mu_2 > \cdots > \mu_{[n/2]}. \tag{7.3.1}$$

From this (7.21.1) follows again.

If n is even, we have

$$\mu_{n/2} = |P_n(0)| = \frac{1\cdot 3 \cdots (n-1)}{2\cdot 4 \cdots n}.$$

For the proof let

$$n(n+1)f(x) = n(n+1)\{P_n(x)\}^2 + (1-x^2)\{P_n'(x)\}^2. \tag{7.3.2}$$

Then we have $f(x) = \{P_n(x)\}^2$ if $P_n'(x) = 0$, or if $x = \pm 1$. Therefore,

$$\max_{-1 \leqq x \leqq +1} \{P_n(x)\}^2 \leqq \max_{-1 \leqq x \leqq +1} f(x). \tag{7.3.3}$$

Now, on account of (4.2.1),

$$\begin{aligned} n(n+1)f'(x) &= 2P_n'(x)\{n(n+1)P_n(x) - xP_n'(x) + (1-x^2)P_n''(x)\} \\ &= 2P_n'(x)\cdot xP_n'(x) = 2x\{P_n'(x)\}^2, \end{aligned} \tag{7.3.4}$$

so that $f(x)$ is decreasing for $x < 0$ and increasing for $x > 0$. This establishes the statement.

(2) THEOREM 7.3.2. *Let* $n \geqq 2$. *The successive relative maxima of* $(\sin\theta)^{\frac{1}{2}} | P_n(\cos\theta) |$ *when* θ *increases form* 0 *to* $\pi/2$, *form an increasing sequence.*

From (4.24.2) we obtain for $\alpha = \beta = 0$

$$\frac{d^2u}{d\theta^2} + \phi(\theta)u = 0; \qquad u = u(\theta) = (\sin\theta)^{\frac{1}{2}} P_n(\cos\theta),$$
$$\phi(\theta) = (2\sin\theta)^{-2} + (n + \tfrac{1}{2})^2. \tag{7.3.5}$$

Introducing

$$f(\theta) = \{u(\theta)\}^2 + \psi(\theta)\{u'(\theta)\}^2, \qquad \psi(\theta) = \{\phi(\theta)\}^{-1}, \tag{7.3.6}$$

we have

$$f'(\theta) = 2u'(\theta)\{u(\theta) + \psi(\theta)u''(\theta) + \tfrac{1}{2}\psi'(\theta)u'(\theta)\} = \psi'(\theta)\{u'(\theta)\}^2. \tag{7.3.7}$$

Now $\psi(\theta)$ is an increasing function in $[0, \pi/2]$, so that $f'(\theta) > 0$, and $f(\theta)$ is also increasing. But $f(\theta) = \{u(\theta)\}^2$ if $u'(\theta) = 0$; this proves the theorem.

(3) An important application of Theorem 7.3.2 is

THEOREM 7.3.3. *We have*

$$(\sin\theta)^{\frac{1}{2}} | P_n(\cos\theta) | < (2/\pi)^{\frac{1}{2}} n^{-\frac{1}{2}}, \qquad 0 \leqq \theta \leqq \pi. \tag{7.3.8}$$

Here the constant $(2/\pi)^{\frac{1}{2}}$ *cannot be replaced by a smaller one.*

The first proof of an inequality of the type (7.3.8), with a constant A instead of $(2/\pi)^{\frac{1}{2}}$, is due to Stieltjes (**8**, p. 241). Further proofs have been given by Gronwall (**1**, p. 221) and Fejér (**9**, pp. 289-291). The following proof is due to S. Bernstein (**2**, p. 236); it was the first leading to the precise constant $(2/\pi)^{\frac{1}{2}}$.

Let n be even. From Theorem 7.3.2

$$(7.3.9) \qquad (\sin\theta)^{\frac{1}{2}}\,|\,P_n(\cos\theta)\,| \leqq |\,P_n(0)\,|, \qquad 0 \leqq \theta \leqq \pi,$$

with the sign of equality holding if $\theta = \pi/2$. Now let n be odd. Then for $0 \leqq \theta \leqq \pi$

$$(7.3.10) \qquad \begin{aligned} (\sin\theta)^{\frac{1}{2}}\,|\,P_n(\cos\theta)\,| &< \max_{0\leqq\theta\leqq\pi} \{f(\theta)\}^{\frac{1}{2}} \\ &= \{f(\pi/2)\}^{\frac{1}{2}} = \{\tfrac{1}{4} + (n + \tfrac{1}{2})^2\}^{-\frac{1}{2}}\,|\,P_n'(0)\,|, \end{aligned}$$

where $f(\theta)$ is defined by (7.3.6). Using the notation (4.9.2), we have[42]

$$(7.3.11) \qquad \begin{aligned} |\,P_n(0)\,| &= g_{n/2} < (2/\pi)^{\frac{1}{2}} n^{-\frac{1}{2}}, \\ |\,P_n'(0)\,| &= (n + 1)g_{(n+1)/2} < (2/\pi)^{\frac{1}{2}}(n + 1)^{\frac{1}{2}}, \end{aligned}$$

according as n is even or odd. (In the second case we can use (4.7.31).) Now $\{\frac{1}{4} + (n + \frac{1}{2})^2\}^{-\frac{1}{2}}(n + 1)^{\frac{1}{2}} < n^{-\frac{1}{2}}$; therefore (7.3.8) follows.

That $(2/\pi)^{\frac{1}{2}}$ is the best possible constant is easily seen by considering $|\,P_n(0)\,|$, n even. Besides this we have (cf. (7.32.9), $\alpha = \beta = 0$)

$$(7.3.12) \qquad \max_{0\leqq\theta\leqq\pi} (\sin\theta)^{\frac{1}{2}}\,|\,P_n(\cos\theta)\,| \cong (2/\pi)^{\frac{1}{2}} n^{-\frac{1}{2}}, \qquad n \to \infty.$$

7.31. Theorem of Sonin; Bessel functions

The argument used in the preceding section can be generalized in various ways. For instance, the following important theorem holds.

THEOREM 7.31.1. *Let $y = y(x)$ satisfy the differential equation*

$$(7.31.1) \qquad y'' + \phi(x)y = 0,$$

where $\phi(x)$ is a positive function having a continuous derivative of a constant sign in $x_0 < x < X_0$. Then the successive relative maxima of $|\,y\,|$, as x increases from x_0 to X_0, form an increasing or decreasing sequence according as $\phi(x)$ decreases or increases.[43]

If we write

$$(7.31.2) \qquad f(x) = \{y(x)\}^2 + \{\phi(x)\}^{-1}\{y'(x)\}^2 = \{y(x)\}^2 + \psi(x)\{y'(x)\}^2,$$

we have, in fact, $f(x) = \{y(x)\}^2$ if $y'(x) = 0$, and

$$(7.31.3) \qquad f'(x) = 2y'(x)\{y(x) + \psi(x)y''(x) + \tfrac{1}{2}\psi'(x)y'(x)\} = \psi'(x)\{y'(x)\}^2.$$

That is, $\operatorname{sgn} f'(x) = -\operatorname{sgn} \phi'(x)$; whence the statement follows.

[42] The sequence $\{m^{1/2}g_m\}$ is increasing, so that $m^{1/2}g_m < \lim_{m\to\infty} m^{1/2}g_m = \pi^{-1/2}$.

[43] Professor Pólya has kindly pointed out to me the following generalization of this theorem: *Let $y(x)$ satisfy the differential equation*

$$\{k(x)y'\}' + \phi(x)y = 0,$$

where $k(x) > 0$, $\phi(x) > 0$, and both functions $k(x)$, $\phi(x)$ have a continuous derivative. Then the relative maxima of $|\,y\,|$ form an increasing or decreasing sequence according as $k(x)\phi(x)$ is decreasing or increasing. This was obtained independently by Butlewski 1.

As an illustration, we apply this result to (1.8.9) for $k = 1$, $x > 0$. In this case we have

$$\phi(x) = 1 + \frac{\frac{1}{4} - \alpha^2}{x^2}, \tag{7.31.4}$$

so that $\phi(x)$ is decreasing if $\alpha^2 < \frac{1}{4}$ and increasing if $\alpha^2 > \frac{1}{4}$. In the latter case x must be taken so large that $\phi(x) > 0$. We then conclude that the relative maxima of $x^{\frac{1}{2}} | J_\alpha(x) |$ form an increasing sequence if $\alpha^2 < \frac{1}{4}$, and a decreasing sequence if $\alpha^2 > \frac{1}{4}$. In the first case $x > 0$, and in the second $x > (\alpha^2 - \frac{1}{4})^{\frac{1}{2}}$.

According to (1.71.7), there are infinitely many such maxima, and they tend to $(2/\pi)^{\frac{1}{2}}$. Thus, we have the theorem:

Theorem 7.31.2. *If $J_\alpha(x)$ denotes Bessel's function of order α, we have*

$$\sup_{x \geq 0} \{x^{\frac{1}{2}} | J_\alpha(x) |\} = \begin{cases} (2/\pi)^{\frac{1}{2}} & \text{if } -\frac{1}{2} \leq \alpha \leq \frac{1}{2}, \\ \text{finite and} > (2/\pi)^{\frac{1}{2}} & \text{if } \qquad \alpha > \frac{1}{2}. \end{cases} \tag{7.31.5}$$

For $\alpha = \pm\frac{1}{2}$ we can use formulas (1.71.2). For $\alpha < -\frac{1}{2}$ we have $x^{\frac{1}{2}} J_\alpha(x) \to \infty$ as $x \to +0$ [(1.71.1)]. The second statement holds in this case provided the least upper bound in question is taken in an arbitrary interval $[x_0, +\infty]$ with $x_0 > 0$.

See Szegö **17**, pp. 40–41, and compare similar theorems in Watson **3**, pp. 488–489. See also §7.8.

7.32 Jacobi polynomials

(1) The consideration of §7.3 can easily be extended to ultraspherical polynomials. A treatment of this case will be given in §7.33. First, however, we discuss general Jacobi polynomials $P_n^{(\alpha,\beta)}(x)$. In applying the previous methods to $P_n^{(\alpha,\beta)}(x)$, the main difficulty lies in the fact that there is no special point $x = \xi$ interior to $[-1, +1]$ at which $P_n^{(\alpha,\beta)}(x)$ and its derivative have values which are easily calculated, as in the case of the ultraspherical, and especially of the Legendre, polynomials at $x = 0$. This is the reason that here we must anticipate certain comparatively simple results from Chapter VIII.[44] These are the following:

(a) The formula of the Mehler-Heine type [(8.1.1)],

$$\lim_{n \to \infty} n^{-\alpha} P_n^{(\alpha,\beta)}\left(\cos \frac{z}{n}\right) = (z/2)^{-\alpha} J_\alpha(z),$$

which holds uniformly for $| z | \leq R$, R fixed.

[44] After completing the manuscript I received a paper of Korous (**3**) in which some of the results of §7.32 are derived by use of the differential equation of Jacobi polynomials but *without* using the asymptotic formulas (8.1.1), (8.21.10).

(b) Darboux's formula [(8.21.10)],

$$P_n^{(\alpha,\beta)}(\cos\theta) = n^{-\frac{1}{2}}k(\theta)\cos(N\theta+\gamma) + O(n^{-\frac{3}{2}}),$$

$$k(\theta) = \pi^{-\frac{1}{2}}\left(\sin\frac{\theta}{2}\right)^{-\alpha-\frac{1}{2}}\left(\cos\frac{\theta}{2}\right)^{-\beta-\frac{1}{2}},$$

$$N = n + (\alpha+\beta+1)/2, \qquad \gamma = -(\alpha+\tfrac{1}{2})\pi/2, \qquad \epsilon \leqq \theta \leqq \pi - \epsilon, \qquad n \to \infty.$$

Here ϵ is a fixed positive number; the bound for the error term holds uniformly. In both cases, α and β are arbitrary real numbers.

Concerning the results of this section and subsequent ones, cf. Kogbetliantz **19**, p. 125, S. Bernstein **2**, and Szegö **17**.

(2) THEOREM 7.32.1. *Let $\alpha > -1$, $\beta > -1$,*

$$x_0 = \frac{\beta-\alpha}{\alpha+\beta+1}. \tag{7.32.1}$$

We have

$$\max_{-1\leqq x\leqq+1} |P_n^{(\alpha,\beta)}(x)| = \begin{cases} \binom{n+q}{n} \sim n^q & \text{if } q = \max(\alpha,\beta) \geqq -\frac{1}{2}, \\ |P_n^{(\alpha,\beta)}(x')| \sim n^{-\frac{1}{2}} & \text{if } q = \max(\alpha,\beta) < -\frac{1}{2}. \end{cases} \tag{7.32.2}$$

Here x' is one of the two maximum points nearest x_0.

The symbols $\sim$ refer to the limiting procedure $n \to \infty$. In the second case $-1 < x_0 < +1$; then we use (8.21.10). The subsequent argument furnishes the more precise result that the maximum of $|P_n^{(\alpha,\beta)}(x)|$ in the interval $0 \leqq x \leqq 1$ is of order $n^{\max(\alpha,-1/2)}$. A similar result holds for $-1 \leqq x \leqq 0$.

For the proof we generalize the argument of §7.3 (1) as follows. Let $n \geqq 1$, and let

$$n(n+\alpha+\beta+1)f(x) = n(n+\alpha+\beta+1)\{P_n^{(\alpha,\beta)}(x)\}^2 + (1-x^2)\left\{\frac{d}{dx}P_n^{(\alpha,\beta)}(x)\right\}^2. \tag{7.32.3}$$

Then by using (4.2.1), we obtain

$$n(n+\alpha+\beta+1)f'(x) = 2\{\alpha-\beta+(\alpha+\beta+1)x\}\left\{\frac{d}{dx}P_n^{(\alpha,\beta)}(x)\right\}^2. \tag{7.32.4}$$

Thus $f'(x)$ can change its sign only at $x = x_0$.

We see that the condition $-1 < x_0 < +1$ is equivalent to $(\alpha+\frac{1}{2})(\beta+\frac{1}{2}) > 0$. Now let $\alpha > -\frac{1}{2}$ and $\beta > -\frac{1}{2}$. Then the sequence formed by the relative maxima of $|P_n^{(\alpha,\beta)}(x)|$ in $-1 \leqq x \leqq x_0$ and by the value of this function at $x = -1$, is decreasing, while the sequence of the maxima in $x_0 \leqq x \leqq +1$ and of the value of the function at $x = +1$, is increasing. Therefore, $|P_n^{(\alpha,\beta)}(x)|$ attains its maximum in $[-1, +1]$ at one of the end-points.

In case $\alpha \geqq -\frac{1}{2}$ and $-1 < \beta \leqq -\frac{1}{2}$, the linear function $\alpha - \beta + (\alpha + \beta + 1)x$ is non-negative, so that the sequence of relative maxima in question is increasing in $[-1, +1]$, save for the case $\alpha = \beta = -\frac{1}{2}$, in which it is stationary; the situation is opposite in the case $\beta \geqq -\frac{1}{2}$, $-1 < \alpha \leqq -\frac{1}{2}$. Finally, let $-1 < \alpha < -\frac{1}{2}$, $-1 < \beta < -\frac{1}{2}$, so that again $-1 < x_0 < +1$; then the sequence of maxima is increasing in $[-1, x_0]$ and decreasing in $[x_0, +1]$, so that the absolute maximum of $| P_n^{(\alpha,\beta)}(x) |$ in $[-1, +1]$ is attained at the point of maximum nearest x_0 on the left or on the right.

See Problem 39.

(3) THEOREM 7.32.2. *Let α and β be arbitrary and real, and c a fixed positive constant, $n \to \infty$. Then*

$$P_n^{(\alpha,\beta)}(\cos\theta) = \begin{cases} \theta^{-\alpha-\frac{1}{2}} O(n^{-\frac{1}{2}}) & \text{if } cn^{-1} \leqq \theta \leqq \pi/2, \\ O(n^{\alpha}) & \text{if } 0 \leqq \theta \leqq cn^{-1}. \end{cases} \tag{7.32.5}$$

See S. Bernstein **2**, pp. 225–232 where Sonin's theorem is applied, but where the proof is perhaps slightly more complicated than ours. Szegö (**17**, p. 77) uses the asymptotic formula (8.21.17) which is, of course, a more complicated tool than (8.1.1) and (8.21.10) used below.

The bounds in (7.32.5) are precise as regards their orders in n. They follow also, as mentioned, from the more complicated asymptotic formula (8.21.17) of "Hilb's type." By use of (4.1.3) we can obtain similar bounds for the intervals $\pi/2 \leqq \theta \leqq \pi$.

We notice the useful inequalities

$$P_n^{(\alpha,\beta)}(\cos\theta) = \begin{cases} \theta^{-\alpha-\frac{1}{2}} O(n^{-\frac{1}{2}}), \\ O(n^{\alpha}), \end{cases} \qquad 0 < \theta \leqq \pi/2, \quad \alpha \geqq -\tfrac{1}{2}; \tag{7.32.6}$$

$$P_n^{(\alpha,\beta)}(\cos\theta) = O(n^{-\frac{1}{2}}), \qquad 0 < \theta \leqq \pi/2, \quad \alpha \leqq -\tfrac{1}{2}, \tag{7.32.7}$$

which follow from (7.32.5). Concerning the second bound in (7.32.6), and concerning (7.32.7), see §7.32 (2).

We observe that $\theta^{-\alpha-\frac{1}{2}} n^{-\frac{1}{2}} \sim n^{\alpha}$ if $\theta \sim n^{-1}$; thus it suffices to prove (7.32.5) for a special value of c. Apply Theorem 7.31.1 with [(4.24.2)]

$$\begin{aligned} &x = \theta, y = u_n(\theta) = \left(\sin\frac{\theta}{2}\right)^{\alpha+\frac{1}{2}} \left(\cos\frac{\theta}{2}\right)^{\beta+\frac{1}{2}} P_n^{(\alpha,\beta)}(\cos\theta), \\ &\phi(\theta) = \frac{\frac{1}{4} - \alpha^2}{4\sin^2\frac{\theta}{2}} + \frac{\frac{1}{4} - \beta^2}{4\cos^2\frac{\theta}{2}} + \left\{n + \frac{\alpha+\beta+1}{2}\right\}^2. \end{aligned} \tag{7.32.8}$$

First, let $\delta = \delta(\alpha, \beta)$ be a fixed positive number, sufficiently small. Then $\phi(\theta)$ is positive and decreasing in $0 < \theta \leqq \delta$ if $\alpha^2 < \frac{1}{4}$. It is positive and increasing in $kn^{-1} \leqq \theta \leqq \delta$ if $\alpha^2 > \frac{1}{4}$; here k is a fixed number, $k > (\alpha^2 - \frac{1}{4})^{\frac{1}{2}}$, and n is sufficiently large. Thus in both cases, the function $\phi(\theta)$ is positive and monotonic in $kn^{-1} \leqq \theta \leqq \delta$, where $k = k(\alpha, \beta)$, $\delta = \delta(\alpha, \beta)$, and n is sufficiently large. The same holds for $\alpha^2 = \frac{1}{4}, \beta^2 \neq \frac{1}{4}$. Therefore, the sequences of the relative maxima of $|u_n(\theta)|$ in the interval $kn^{-1} \leqq \theta \leqq \delta$ are increasing and decreasing, respectively, for large n, according as $\alpha^2 < \frac{1}{4}$ or $\alpha^2 > \frac{1}{4}$. According to (8.1.1) and (8.21.10), we find in both cases $u_n(\theta) = O(n^{-\frac{1}{2}})$. (We have $(\sin \theta/2)^{\alpha+\frac{1}{2}}(\cos \theta/2)^{\beta+\frac{1}{2}} \sim \theta^{\alpha+\frac{1}{2}}$.) This furnishes the first part of (7.32.5) with $c = k$. The second part follows immediately from (8.1.1).

In the case $\alpha^2 = \beta^2 = \frac{1}{4}$, excluded before, $\phi(\theta)$ is constant. Then we know $P_n^{(\alpha,\beta)}(\cos \theta)$ explicitly (cf. (4.1.7), (4.1.8)).

(4) THEOREM 7.32.3. *Let $u_n(\theta)$ have the same meaning as in* (7.32.8), *and $M_n = \max |u_n(\theta)|$ when $0 < \theta \leqq \pi/2$. We have*

$$\lim_{n\to\infty} n^{\frac{1}{2}} M_n = \begin{cases} \pi^{-\frac{1}{2}} & \text{if } -\frac{1}{2} \leqq \alpha \leqq +\frac{1}{2}, \\ \text{finite and } > \pi^{-\frac{1}{2}} & \text{if } \alpha > \frac{1}{2}. \end{cases} \tag{7.32.9}$$

Here β is greater than -1.

Cf. S. Bernstein **2**, pp. 225–232; Szegö **17**, pp. 79–80. Cf. Theorem 7.31.2.

The preceding argument needs only a slight modification. We have to discuss the maximum of $n^{\frac{1}{2}} |u_n(\theta)|$ for $\delta \leqq \theta \leqq \pi/2$ if $-\frac{1}{2} \leqq \alpha \leqq +\frac{1}{2}$, and for $\theta = n^{-1}z$, $0 \leqq z \leqq c$, if $\alpha > \frac{1}{2}$. The first is $\cong \pi^{-\frac{1}{2}}$ as $n \to \infty$ (according to (8.21.10)); the second is

$$\cong n^{\frac{1}{2}} \max_{0 \leqq z \leqq c} \left\{ \left(\frac{z}{2n}\right)^{\alpha+\frac{1}{2}} n^{\alpha}(z/2)^{-\alpha} |J_\alpha(z)| \right\} = \max_{0 \leqq z \leqq c} \{(z/2)^{\frac{1}{2}} |J_\alpha(z)|\}$$

(according to (8.1.1)). For sufficiently large c this is independent of c and greater than $\pi^{-\frac{1}{2}}$ (Theorem 7.31.2).

(5) Finally, as an application of (4.21.7) we point out the following generalization of Theorem 7.32.2:

THEOREM 7.32.4. *Let α and β be arbitrary and real, and c a fixed positive constant, $n \to \infty$. Then*

$$\left\{ \left(\frac{d}{dx}\right)^k P_n^{(\alpha,\beta)}(x) \right\}_{x=\cos\theta} = \begin{cases} \theta^{-\alpha-k-\frac{1}{2}} O(n^{k-\frac{1}{2}}) & \text{if } cn^{-1} \leqq \theta \leqq \pi/2, \\ O(n^{2k+\alpha}) & \text{if } 0 \leqq \theta \leqq cn^{-1}. \end{cases} \tag{7.32.10}$$

From this we find, uniformly in x, $-1 \leqq x \leqq +1$,

$$\left(\frac{d}{dx}\right)^k P_n^{(\alpha,\beta)}(x) = O(n^q), \qquad q = \max(2k + \alpha, 2k + \beta, k - \tfrac{1}{2}). \tag{7.32.11}$$

7.33. Ultraspherical polynomials

In the ultraspherical case the preceding considerations can be simplified.

(1) By the same argument as that in §7.32 (2) we find that

$$n(n+2\lambda)f(x) = n(n+2\lambda)\{P_n^{(\lambda)}(x)\}^2 + (1-x^2)\left\{\frac{d}{dx}P_n^{(\lambda)}(x)\right\}^2$$

is increasing or decreasing in $0 \leqq x \leqq 1$ according as $\lambda > 0$ or $\lambda < 0$, λ non-integral; we assume $n > 0$ in the first and $n > -2\lambda$ in the second case. Thus we obtain the following theorem:

THEOREM 7.33.1. *We have*

$$\max_{-1 \leqq x \leqq 1} |P_n^{(\lambda)}(x)| = \begin{cases} \binom{n+2\lambda-1}{n} & \text{if } \lambda > 0, \\ |P_n^{(\lambda)}(x')| & \text{if } \lambda < 0,\ \lambda \text{ non-integral.} \end{cases} \tag{7.33.1}$$

Here x' is one of the two maximum points nearest 0 if n is odd; $x' = 0$ if n is even.

In the first case (4.7.3) has been used (cf. also Theorem 7.4.1). In the second case we obtain, if n is even,

$$\max_{-1 \leqq x \leqq 1} |P_n^{(\lambda)}(x)| = |P_n^{(\lambda)}(0)| = \left|\binom{n/2+\lambda-1}{n/2}\right|; \tag{7.33.2}$$

whereas, for n odd,

$$\begin{aligned}\max_{-1 \leqq x \leqq 1} |P_n^{(\lambda)}(x)| &< \{f(0)\}^{\frac{1}{2}} = \{n(n+2\lambda)\}^{-\frac{1}{2}} |P_n^{(\lambda)\prime}(0)| \\ &= |2\lambda| \{n(n+2\lambda)\}^{-\frac{1}{2}} \left|\binom{\lambda + (n-1)/2}{(n-1)/2}\right|.\end{aligned} \tag{7.33.3}$$

Both bounds (7.33.2) and (7.33.3) are $\cong 2^{1-\lambda}|\Gamma(\lambda)|^{-1} n^{\lambda-1}$ as $n \to \infty$; the first bound is attained for $x = 0$, the second bound is precise in the asymptotic sense.

(2) By use of (4.7.11) we obtain in a manner similar to that in §7.3 (2), (3), the following:

THEOREM 7.33.2. *Let $0 < \lambda < 1$. Then we have for $0 \leqq \theta \leqq \pi$*

$$(\sin\theta)^\lambda |P_n^{(\lambda)}(\cos\theta)| \leqq \begin{cases} \alpha_{n/2} & \text{if } n \text{ is even}, \\ \{\lambda(1-\lambda) + (n+\lambda)^2\}^{-\frac{1}{2}}(n+1)\alpha_{(n+1)/2} & \text{if } n \text{ is odd}, \end{cases} \tag{7.33.4}$$

and

$$(\sin\theta)^\lambda |P_n^{(\lambda)}(\cos\theta)| < 2^{1-\lambda}\{\Gamma(\lambda)\}^{-1} n^{\lambda-1}. \tag{7.33.5}$$

Here the constant $2^{1-\lambda}\{\Gamma(\lambda)\}^{-1}$ cannot be replaced by a smaller one; α_n has the same meaning as in (4.9.21).

In (7.33.4) the sign of equality holds only for even n and $\theta = \pi/2$. Now $\alpha_n \cong \{\Gamma(\lambda)\}^{-1} n^{\lambda-1}$, and $\alpha_n < \{\Gamma(\lambda)\}^{-1} n^{\lambda-1}$ (since $\{n^{1-\lambda}\alpha_n\}$ is increasing[45]); moreover, $\{\lambda(1 - \lambda) + (n + \lambda)^2\}^{-\frac{1}{2}}(n + 1)^\lambda < n^{\lambda-1}$, [45] so that equation (7.33.5) follows.

Less precise (but more general) inequalities can be obtained from the general result (7.32.5) for $\alpha = \beta = \lambda - \frac{1}{2}$. We have

$$P_n^{(\lambda)}(\cos\theta) = \begin{cases} \theta^{-\lambda}O(n^{\lambda-1}), & cn^{-1} \leqq \theta \leqq \pi/2, \\ O(n^{2\lambda-1}), & 0 \leqq \theta \leqq cn^{-1}; \end{cases} \tag{7.33.6}$$

λ arbitrary and real, $\lambda \neq 0, -1, -2, \cdots; c > 0$.

(3) We point out an interesting special case of (7.33.6), namely, $\lambda = \frac{3}{2}$ (cf. Szegö **16**). We have $P_n'(x) = P_{n-1}^{(\frac{3}{2})}(x)$ (cf. (4.7.14)), so that

$$P_n'(\cos\theta) = \begin{cases} \theta^{-\frac{3}{2}}O(n^{\frac{1}{2}}), & cn^{-1} \leqq \theta \leqq \pi/2, \\ O(n^2), & 0 \leqq \theta \leqq cn^{-1}. \end{cases} \tag{7.33.7}$$

The first bound can be used for the whole interval $0 < \theta \leqq \pi/2$ [cf. (7.32.6)]. According to (7.33.1) the inequality

$$| P_n'(x) | \leqq n(n + 1)/2, \qquad -1 \leqq x \leqq +1, \tag{7.33.8}$$

holds, the equality sign being taken if $n = 0, 1$, or $n > 1$, $x = \pm 1$.

By using the first identity in (4.7.27), we find

$$(1 - x^2)P_n'(x) = \frac{n(n + 1)}{2n + 1}(P_{n-1}(x) - P_{n+1}(x)). \tag{7.33.9}$$

Thus we conclude from the first bound in (7.33.7) (which now holds for $0 < \theta \leqq \pi/2$, cf. the previous remark) the following:

THEOREM 7.33.3. *If $P_n(x)$ denotes Legendre's polynomial, we have for* $0 < \theta < \pi$

$$P_{n-1}(\cos\theta) - P_{n+1}(\cos\theta) = (\sin\theta)^{\frac{1}{2}}O(n^{-\frac{1}{2}}). \tag{7.33.10}$$

The bound of the factor $O(n^{-\frac{1}{2}})$ is independent of θ.

This result, without the factor $(\sin\theta)^{\frac{1}{2}}$, is due to Stieltjes (cf. Hermite-Stieltjes **1**, vol. 2, pp. 174–177; Fejér **9**, pp. 295–298). The present form of the theorem is implied in the previous more general results of S. Bernstein, Kogbetliantz, and Szegö; cf. Szegö **16**.

7.34. Bounds for integrals involving Jacobi polynomials

THEOREM 7.34. *Let α, β, μ be real numbers each greater than -1. Then as $n \to \infty$ (concerning the second part of the statement see below)*

[45] In view of the concavity of $\log x$, we have

$$(1 - \lambda)\log(n - 1) + \lambda \log n < \log(n + \lambda - 1),$$

$$(1 - \lambda)\log(n^2) + \lambda\log\{(n + 1)^2\} < \log\{(1 - \lambda)n^2 + \lambda(n + 1)^2\}.$$

$$(7.34.1)\qquad \int_0^1 (1-x)^\mu \,|\, P_n^{(\alpha,\beta)}(x) \,|\, dx \sim \begin{cases} n^{\alpha-2\mu-2}, & 2\mu < \alpha - \frac{3}{2}, \\ n^{-\frac{1}{2}} \log n, & 2\mu = \alpha - \frac{3}{2}, \\ n^{-\frac{1}{2}}, & 2\mu > \alpha - \frac{3}{2}. \end{cases}$$

See Szegö **17**, pp. 84–86, where the existence of the limits of the corresponding ratios is proved, and the limits are calculated. The proof of the second part of (7.34.1) requires a more complicated apparatus [(8.21.18)]; here we prove only that

$$(7.34.2)\qquad u_n = n^{\frac{1}{2}} \int_0^1 (1-x)^\mu \,|\, P_n^{(\alpha,\beta)}(x) \,|\, dx = O(\log n), \quad \text{and} \quad u_n \to \infty.$$

This is sufficient for later purposes (cf. §9.41 (5)).

We use (7.32.5); in fact, we have

$$(7.34.3)\qquad \begin{aligned} \int_0^1 (1-x)^\mu \,|\, P_n^{(\alpha,\beta)}(x) \,|\, dx &= O(1) \int_0^{\pi/2} \theta^{2\mu+1} \,|\, P_n^{(\alpha,\beta)}(\cos\theta) \,|\, d\theta \\ &= O(1) \int_0^{n^{-1}} \theta^{2\mu+1} n^\alpha \, d\theta + O(1) \int_{n^{-1}}^{\pi/2} \theta^{2\mu+1} \theta^{-\alpha-\frac{1}{2}} n^{-\frac{1}{2}} \, d\theta \\ &= O(n^{\alpha-2\mu-2}) + O(n^{-\frac{1}{2}})\{O(1) + O(n^{\alpha-2\mu-\frac{3}{2}})\}. \end{aligned}$$

If $2\mu - \alpha + \frac{3}{2} = 0$, the last term must be replaced by $O(\log n)$.

On the other hand,

$$(7.34.4)\qquad \int_0^1 (1-x)^\mu \,|\, P_n^{(\alpha,\beta)}(x) \,|\, dx > \begin{cases} A \int_0^{n^{-1}} \theta^{2\mu+1} \,|\, P_n^{(\alpha,\beta)}(\cos\theta) \,|\, d\theta, & 2\mu < \alpha - \frac{3}{2}, \\ A \int_{\pi/4}^{\pi/2} |\, P_n^{(\alpha,\beta)}(\cos\theta) \,|\, d\theta, & 2\mu > \alpha - \frac{3}{2}, \end{cases}$$

where A is a proper positive constant. According to (8.1.1), the first bound is

$$\cong A \int_0^1 (z/n)^{2\mu+1} n^\alpha (z/2)^{-\alpha} \,|\, J_\alpha(z) \,|\, n^{-1} \, dz \sim n^{\alpha-2\mu-2}.$$

According to (8.21.10), the second bound is

$$\sim n^{-\frac{1}{2}} \int_{\pi/4}^{\pi/2} |\cos(N\theta + \gamma)| \, d\theta \sim n^{-\frac{1}{2}}.$$

In case $2\mu = \alpha - \frac{3}{2}$, $\alpha > -\frac{1}{2}$, we have

$$\int_0^1 (1-x)^\mu \,|\, P_n^{(\alpha,\beta)}(x) \,|\, dx > A' \int_0^{\omega n^{-1}} \theta^{2\mu+1} \,|\, P_n^{(\alpha,\beta)}(\cos\theta) \,|\, d\theta,$$

where ω is a fixed positive number, and A' is independent of n and ω. From this inequality, according to (8.1.1),

$$\liminf_{n\to\infty} n^{\frac{1}{2}} \int_0^1 (1-x)^{\mu} \mid P_n^{(\alpha,\beta)}(x) \mid dx \geqq 2^{\alpha} A' \int_0^{\omega} z^{-\frac{1}{2}} \mid J_{\alpha}(z) \mid dz$$

follows. The last integral becomes arbitrarily large with ω; this furnishes the second part of (7.34.2).

7.4. Fejér's generalization of Legendre polynomials

(1) THEOREM 7.4.1. *Let $\{\alpha_n\}$ be a sequence with positive terms. Then the "Legendre polynomials" $F_n(x)$ (6.5.1) associated with the sequence $\{\alpha_n\}$ satisfy the inequalities*

$$\mid F_n(x) \mid \leqq F_n(1), \qquad -1 \leqq x \leqq +1. \tag{7.4.1}$$

The sign of equality holds only if $n = 0$, or $n > 0$ with $x = -1$ or $x = +1$.

This leads to a new proof for (7.21.1) and for the first part of (7.33.1).

(2) The inequalities (7.3.8) and (7.33.5), of the Stieltjes type, can likewise be extended to the polynomials $F_n(x)$, though with certain larger constants. We prove the following:

THEOREM 7.4.2. *Let*

$$\alpha_n > 0,\ \Delta\alpha_n = \alpha_n - \alpha_{n+1} > 0,\ \Delta^2\alpha_n = \alpha_n - 2\alpha_{n+1} + \alpha_{n+2} > 0, \qquad n = 0, 1, 2, \cdots, \tag{7.4.2}$$

and

$$f(z) = \alpha_0 + \alpha_1 z + \alpha_2 z^2 + \cdots + \alpha_n z^n + \cdots. \tag{7.4.3}$$

Then for the "Legendre polynomials" $F_n(x)$ associated with the sequence $\{\alpha_n\}$, we have

$$\mid F_n(\cos\theta) \mid \leqq 4\alpha_{[(n+1)/2]} \mid f(e^{2i\theta}) \mid, \qquad 0 < \theta < \pi;\ n = 0, 1, 2, \cdots. \tag{7.4.4}$$

See Fejér **9**, pp. 291–295; Szegö **11**, p. 179. Szegö obtains a larger bound under a more restrictive condition. The inequality (7.4.4) and the present proof are new.

Under the conditions (7.4.2) the function $f(z)$ is regular for $\mid z \mid < 1$ and continuous for $\mid z \mid \leqq 1$, $\mid z - 1 \mid \geqq \delta$, where δ is an arbitrarily small positive number. Indeed, we see that $\lim_{n\to\infty}\alpha_n = \alpha \geqq 0$ exists. If $\alpha = 0$, we use a well-known case of Abel's inequality (1.11.6); if $\alpha > 0$, we write $\alpha_n = (\alpha_n - \alpha) + \alpha$.

Now from (6.5.1),

$$F_n(\cos\theta) = z^{-n/2} \sum_{k=0}^{[n/2]}{}' \alpha_k\alpha_{n-k} z^k + z^{n/2} \sum_{k=0}^{[n/2]}{}' \alpha_k\alpha_{n-k} z^{-k}, \qquad z = e^{2i\theta}, \tag{7.4.5}$$

where the sign $\sum'$ indicates that for even n the last term $k = n/2$ has to be multiplied by $\frac{1}{2}$. Hence,

$$\mid F_n(\cos\theta) \mid \leqq 2 \left| \sum_{k=0}^{[n/2]}{}' \alpha_k\alpha_{n-k} z^k \right|, \qquad z = e^{2i\theta}. \tag{7.4.6}$$

By virtue of (1.11.6) we obtain

$$|F_n(\cos\theta)| \leqq 2\alpha_m \max_{0\leqq\nu\leqq[n/2]} \left|\sum_{k=\nu}^{[n/2]}{}' \alpha_k z^k\right|, \qquad m = n - [n/2],\ z = e^{2i\theta}. \tag{7.4.7}$$

Now, writing

$$\rho_n(z) = \alpha_n z^n + \alpha_{n+1} z^{n+1} + \alpha_{n+2} z^{n+2} + \cdots, \tag{7.4.8}$$

we have, according to a theorem of Fejér-Szegö (**1**),

$$|f(z)| \geqq |\rho_1(z)| \geqq |\rho_2(z)| \geqq \cdots, \qquad |z| \leqq 1,\ z \neq 1. \tag{7.4.9}$$

But

$$\sum_{k=\nu}^{[n/2]}{}' \alpha_k z^k = \begin{cases} \rho_\nu(z) - \frac{1}{2}\{\rho_{n/2}(z) + \rho_{n/2+1}(z)\}, & n \text{ even}, \\ \rho_\nu(z) - \rho_{(n+1)/2}(z), & n \text{ odd}, \end{cases} \tag{7.4.10}$$

so that the modulus of this sum is in both cases not greater than $2|f(z)|$; whence (7.4.4) follows.

We give here a sketch of the proof of (7.4.9). It suffices to show that $|f(z)| \geqq |\rho_1(z)|$, or $|f(z)| \geqq |f(z) - \alpha_0|$, or $\Re[f(z)] \geqq \alpha_0/2$ for $|z| < 1$. Now $\lim_{n\to\infty}\alpha_n = \alpha \geqq 0$ exists. If $f(z)$ is replaced by $f(z) - \alpha(1-z)^{-1}$, it is seen that we can assume $\alpha = 0$ from the start. Then for $|z| < 1$

$$\begin{aligned} f(z) &= \sum_{n=0}^{\infty} \Delta^2\alpha_n\{n + 1 + nz + (n-1)z^2 + \cdots + z^n\} \\ &= \frac{1}{2}\sum_{n=0}^{\infty} \Delta^2\alpha_n(n+1) + \sum_{n=0}^{\infty} \Delta^2\alpha_n\{(n+1)/2 + nz + (n-1)z^2 + \cdots + z^n\} \\ &= \alpha_0/2 + \sum_{n=0}^{\infty} \Delta^2\alpha_n\{(n+1)/2 + nz + (n-1)z^2 + \cdots + z^n\}. \end{aligned}$$

But for a real θ

$$\begin{aligned} &\Re\{(n+1)/2 + ne^{i\theta} + (n-1)e^{2i\theta} + \cdots + e^{ni\theta}\} \\ &= \sum_{\nu=0}^{n} (\tfrac{1}{2} + \cos\theta + \cos 2\theta + \cdots + \cos\nu\theta) \\ &= \sum_{\nu=0}^{n} \frac{\sin(\nu + \frac{1}{2})\theta}{2\sin\{\theta/2\}} = \frac{1}{2}\left(\frac{\sin\{(n+1)\theta/2\}}{\sin\{\theta/2\}}\right)^2 \end{aligned}$$

(cf. Fejér **1**; see also Pólya-Szegö **1**, vol. 2, pp. 78, 269, problem 17).

Fejér (**9**, pp. 295-298) also gives an extension of Stieltjes' theorem on $P_{n-1}(x) - P_{n+1}(x)$ (cf. Theorem 7.33.3) to the polynomials $F_n(x)$.

7.5. Recapitulation

In the last sections we gave various derivations of the important inequality (7.21.1):

(a) from the orthogonal property, by use of the general Theorem 7.2;
(b) from the differential equation (§7.3 (1));
(c) from the trigonometric representation (4.9.3).
Moreover it follows also
(d) from Laplace's integral representation (4.8.10); in fact, if $-1 \leqq x \leqq +1$,

$$|x + (x^2 - 1)^{\frac{1}{2}} \cos \phi| = |x + i(1 - x^2)^{\frac{1}{2}} \cos \phi|$$
$$= \{x^2 + (1 - x^2) \cos^2 \phi\}^{\frac{1}{2}} \leqq \{x^2 + (1 - x^2)\}^{\frac{1}{2}} = 1.$$

7.6. Laguerre and Hermite polynomials

(1) THEOREM 7.6.1. *Let α be arbitrary and real. The sequence formed by the relative maxima of $|L_n^{(\alpha)}(x)|$ and by the value of this function at $x = 0$, is decreasing for $x < \alpha + \frac{1}{2}$, and increasing for $x > \alpha + \frac{1}{2}$. The successive relative maxima of $|H_n(x)|$ form a decreasing sequence for $x \leqq 0$, and an increasing sequence for $x \geqq 0$.*

Indeed, the function

$$n\{L_n^{(\alpha)}(x)\}^2 + x\left\{\frac{d}{dx} L_n^{(\alpha)}(x)\right\}^2 \tag{7.6.1}$$

is decreasing for $x < \alpha + \frac{1}{2}$ and increasing for $x > \alpha + \frac{1}{2}$. The function

$$2n\{H_n(x)\}^2 + \{H_n'(x)\}^2 \tag{7.6.2}$$

is decreasing for $x < 0$ and increasing for $x > 0$. Both statements follow by differentiation as in §7.3 (1); we use the first differential equation in (5.1.2) and (5.5.2), respectively.

(2) THEOREM 7.6.2. *Let α be an arbitrary real number. The successive relative maxima of*

$$e^{-x/2} x^{(\alpha+1)/2} |L_n^{(\alpha)}(x)| \text{ and } e^{-x/2} x^{\alpha/2+\frac{1}{4}} |L_n^{(\alpha)}(x)| \tag{7.6.3}$$

form an increasing sequence provided $x > x_0$. In the first case

$$x_0 = \begin{cases} 0 & \text{if } \alpha^2 \leqq 1, \\ \dfrac{\alpha^2 - 1}{2n + \alpha + 1} & \text{if } \alpha^2 > 1. \end{cases} \tag{7.6.4}$$

In the second case

$$x_0 = \begin{cases} 0 & \text{if } \alpha^2 \leqq \frac{1}{4}, \\ (\alpha^2 - \frac{1}{4})^{\frac{1}{2}} & \text{if } \alpha^2 > \frac{1}{4}. \end{cases} \tag{7.6.5}$$

In the first case we take n so large that $2n + \alpha + 1 > 0$.

Sonin's theorem 7.31.1 applies to the functions u and v occurring, respectively, in the third and fourth equations of (5.1.2). The larger zero γ_n of the coefficient of u is an upper bound for the zeros of u (cf. Theorems 1.82.3 and 6.31.2).[46] The differential equation shows also that $x = \gamma_n$ is the last point of inflexion of u; thus γ_n is at the same time an upper bound for the points at which the relative extrema of u are attained. Similarly, if γ_n' denotes the larger zero of $4n + 2\alpha + 2 - x + (\frac{1}{4} - \alpha^2)x^{-1}$, we find in $(\gamma_n')^{\frac{1}{2}}$ an upper bound for the points at which v attains its relative extrema. (The bound of Theorem 6.31.2 is γ_n'.)

Now if x_0 is chosen according to (7.6.4) and (7.6.5), we have

$$\frac{n + (\alpha + 1)/2}{x} + \frac{1 - \alpha^2}{4x^2} - \tfrac{1}{4} > 0, \qquad -\frac{n + (\alpha + 1)/2}{x^2} - \frac{1 - \alpha^2}{2x^3} < 0 \qquad \text{for } x_0 < x < \gamma_n,$$

(7.6.6)

$$4n + 2\alpha + 2 - x + \frac{\frac{1}{4} - \alpha^2}{x} > 0, \qquad -1 - \frac{\frac{1}{4} - \alpha^2}{x^2} < 0 \qquad \text{for } x_0 < x < \gamma_n',$$

respectively. This establishes the statement.

Theorem 7.6.3. *The successive relative maxima of*

$$e^{-x^2/2}\,|H_n(x)| \tag{7.6.7}$$

form an increasing sequence for $x \geqq 0$.

Here the second equation in (5.5.2) can be used.

(3) The bounds analogous to (7.32.5) are readily obtained by means of the asymptotic formulas (8.1.8) and (8.22.1), which correspond to (8.1.1) and (8.21.10), respectively. It is convenient to use the fourth equation in (5.1.2). Then $4n + 2\alpha + 2 - x + (\frac{1}{4} - \alpha^2)x^{-1}$ is positive and decreasing in $0 < x \leqq \delta$ if $\alpha^2 \leqq \frac{1}{4}$; it is positive and increasing in $kn^{-1} \leqq x \leqq \delta$ provided $\alpha^2 > \frac{1}{4}$, $k > (\alpha^2 - \frac{1}{4})/4$ and n is sufficiently large. Here $\delta = \delta(\alpha)$ is a sufficiently small positive constant. Therefore, as in §7.32 (3), we obtain the following:

Theorem 7.6.4. *Let α be arbitrary and real, c and ω fixed positive constants, and let $n \to \infty$. Then*

$$L_n^{(\alpha)}(x) = \begin{cases} x^{-\alpha/2-\frac{1}{4}}O(n^{\alpha/2-\frac{1}{4}}) & \text{if } cn^{-1} \leqq x \leqq \omega, \\ O(n^{\alpha}) & \text{if } 0 \leqq x \leqq cn^{-1}. \end{cases} \tag{7.6.8}$$

These bounds are precise as regards their orders in n; they follow also from the more complicated formula (8.22.4) of Hilb's type.

For $\alpha \geqq -\frac{1}{2}$, both bounds hold in both intervals, that is,

$$L_n^{(\alpha)}(x) = \begin{cases} x^{-\alpha/2-\frac{1}{4}}O(n^{\alpha/2-\frac{1}{4}}), \\ O(n^{\alpha}), \end{cases} \qquad 0 < x \leqq \omega,\ \alpha \geqq -\tfrac{1}{2}. \tag{7.6.9}$$

[46] If this coefficient is constantly negative for $x > x_0$, $|u|$ has no zeros and no maxima. Similarly for v. These cases can be excluded.

On the other hand,

$$L_n^{(\alpha)}(x) = O(n^{\alpha/2-\frac{1}{4}}), \qquad 0 \leqq x \leqq \omega, \quad \alpha \leqq -\tfrac{1}{2}. \tag{7.6.10}$$

And generally, with α arbitrary and real,

$$L_n^{(\alpha)}(x) = O(n^{a}), \qquad a = \max(\tfrac{1}{2}\alpha - \tfrac{1}{4}, \alpha), \qquad 0 \leqq x \leqq \omega. \tag{7.6.11}$$

Finally, we obtain the following analogue of Theorem 7.32.3:

THEOREM 7.6.5. *Let* $M_n = \max_{0<x\leqq\omega} e^{-x/2}x^{\alpha/2+\frac{1}{4}} \mid L_n^{(\alpha)}(x) \mid$. *Then*

$$\lim_{n\to\infty} n^{-\alpha/2+\frac{1}{4}} M_n = \begin{cases} \pi^{-\frac{1}{2}} & \text{if } -\tfrac{1}{2} \leqq \alpha \leqq +\tfrac{1}{2}, \\ \text{finite and } > \pi^{-\frac{1}{2}} & \text{if } \alpha > \tfrac{1}{2}. \end{cases} \tag{7.6.12}$$

The proof is very similar to that in §7.32 (4); of course, (7.6.12) also follows directly from the deeper formula (8.22.4) combined with (7.31.5).

7.7. Problem of Lukács

(1) This problem (Lukács **1**) deals with a more precise form of the mean-value theorem

$$A \leqq \frac{1}{b-a}\int_a^b f(x)\,dx \leqq B, \tag{7.7.1}$$

where $A = \min_{a\leqq x\leqq b} f(x)$, $B = \max_{a\leqq x\leqq b} f(x)$, provided $f(x)$ is restricted to the set of all π_n with a fixed value of n.

THEOREM 7.7. *Let $f(x)$ be an arbitrary π_n with the minimum A and maximum B in $[a, b]$; then*

$$A + \frac{B-A}{\tau_n} \leqq \frac{1}{b-a}\int_a^b f(x)\,dx \leqq B - \frac{B-A}{\tau_n}, \tag{7.7.2}$$

where

$$\tau_n = \begin{cases} (m+1)^2 & \text{if } n = 2m, \\ (m+1)(m+2) & \text{if } n = 2m+1. \end{cases} \tag{7.7.3}$$

The number τ_n cannot be replaced by a smaller one.

This result is the analogue of an older theorem due to Fejér which deals with the analogous question for trigonometric polynomials of a fixed degree n, $b - a = 2\pi$. In this case $\tau_n = n + 1$. The proof of Fejér's theorem can be based on Theorem 1.2.1; however, various other methods have been used (cf. Pólya-Szegö **1**, vol. 2, pp. 83, 277–279, problem 50).

It suffices to prove the first inequality of (7.7.2); the second one follows when we replace $f(x)$ by $-f(x)$. In addition, there is no loss in generality in assuming $A = 0$. It is readily seen then that τ_n is the greatest possible value of $f(x)$ if x varies in $[a, b]$, and $f(x)$ ranges over the class of the π_n which are non-negative in

$[a, b]$ and satisfy

$$\frac{1}{b-a}\int_a^b f(x)\,dx = 1. \tag{7.7.4}$$

Let now $\max f(b) = M_n$ for the same set of π_n; we shall prove that $\tau_n = M_n$. It is clear that $\tau_n \geqq M_n$. On the other hand, if x_0 is an arbitrary point in $[a, b]$, we see, by means of a linear transformation, that

$$f(x_0) \leqq M_n \frac{1}{x_0 - a}\int_a^{x_0} f(x)\,dx, \qquad f(x_0) \leqq M_n \frac{1}{b - x_0}\int_{x_0}^b f(x)\,dx. \tag{7.7.5}$$

Multiplying the first inequality by $x_0 - a$, the second by $b - x_0$, and adding, we find

$$f(x_0) \leqq M_n \frac{1}{b-a}\int_a^b f(x)\,dx = M_n. \tag{7.7.6}$$

(2) *First method of calculating* M_n. See Pólya-Szegö **1**, vol. 2, pp. 96, 297, problem 108.

Assuming $a = -1$, $b = +1$, we use Theorem 1.21.1 and represent the polynomials in (1.21.1) as linear combinations of certain convenient polynomials. For arbitrary and real u_ν, v_ν (subject only to the normalization condition (7.7.8)), we write

$$f(x) = \begin{cases} \left\{\sum_{\nu=0}^{m} u_\nu P_\nu^{(0,0)}(x)\right\}^2 + (1-x^2)\left\{\sum_{\nu=0}^{m-1} v_\nu P_\nu^{(1,1)}(x)\right\}^2 & \text{if } n = 2m, \\ (1-x)\left\{\sum_{\nu=0}^{m} u_\nu P_\nu^{(1,0)}(x)\right\}^2 + (1+x)\left\{\sum_{\nu=0}^{m} v_\nu P_\nu^{(0,1)}(x)\right\}^2 & \\ & \text{if } n = 2m+1. \end{cases} \tag{7.7.7}$$

Because of the orthogonality of Jacobi polynomials, we have, in the notation of (4.3.3),

$$2 = \int_{-1}^{+1} f(x)\,dx = \begin{cases} \sum_{\nu=0}^{m} h_\nu^{(0,0)} u_\nu^2 + \sum_{\nu=0}^{m-1} h_\nu^{(1,1)} v_\nu^2 & \text{if } n = 2m, \\ \sum_{\nu=0}^{m} h_\nu^{(1,0)} u_\nu^2 + \sum_{\nu=0}^{m} h_\nu^{(0,1)} v_\nu^2 & \text{if } n = 2m+1. \end{cases} \tag{7.7.8}$$

But

$$f(1) = \begin{cases} \left\{\sum_{\nu=0}^{m} u_\nu P_\nu^{(0,0)}(1)\right\}^2 & \text{if } n = 2m \\ 2\left\{\sum_{\nu=0}^{m} v_\nu P_\nu^{(0,1)}(1)\right\}^2 & \text{if } n = 2m+1, \end{cases} \tag{7.7.9}$$

so that

$$M_n = \begin{cases} 2 \sum_{\nu=0}^{m} \{h_\nu^{(0,0)}\}^{-1} \{P_\nu^{(0,0)}(1)\}^2 & \text{if } n = 2m, \\ 4 \sum_{\nu=0}^{m} \{h_\nu^{(0,1)}\}^{-1} \{P_\nu^{(0,1)}(1)\}^2 & \text{if } n = 2m+1, \end{cases} \tag{7.7.10}$$

which is attained for u_ν = const. $\{h_\nu^{(0,0)}\}^{-1} P_\nu^{(0,0)}(1)$, $v_\nu = 0$, and $u_\nu = 0$, v_ν = const. $\{h_\nu^{(0,1)}\}^{-1} P_\nu^{(0,1)}(1)$, $\nu = 0, 1, 2, \cdots, m$, respectively. The corresponding polynomials are obviously the "kernel" polynomials $K_m^{(\alpha,\beta)}(x)$ (cf. (4.5.3)) for the cases $\alpha = \beta = 0$ and $\alpha = 0$, $\beta = 1$, that is, constant multiples of $P_m^{(1,0)}(x)$ and $P_m^{(1,1)}(x)$. In other words, we have $f(x)$ = const. $\{P_m^{(1,0)}(x)\}^2$, and $f(x)$ = const. $(1 + x)\{P_m^{(1,1)}(x)\}^2$, respectively. From (7.7.10) we now obtain (7.7.3) by direct calculation or, more easily, by mathematical induction, reasoning from m to $m + 1$.

(3) *Second method of calculating M_n.* See Lukács, loc. cit.[47] This method is based on certain mechanical quadrature formulas related to considerations similar to those in §3.4.

Let $n = 2m$, $x_0 = 1$, and let the zeros of $(1 - x)\, P_m^{(1,0)}(x)$ be denoted by $x_0, x_1, \cdots, x_m$. If $f(x)$ is a π_n and $L(x)$ the Lagrange interpolation polynomial of degree m which coincides with $f(x)$ at $x_0, x_1, \cdots, x_m$, then

$$f(x) - L(x) = (1 - x)\, P_m^{(1,0)}(x)\rho(x),$$

where $\rho(x)$ is a proper π_{m-1}. Therefore,

$$\int_{-1}^{+1} f(x)\, dx - \int_{-1}^{+1} L(x)\, dx = \int_{-1}^{+1} (1 - x) P_m^{(1,0)}(x)\rho(x)\, dx = 0, \tag{7.7.11}$$

so that as in (3.4.1)

$$\frac{1}{2}\int_{-1}^{+1} f(x)\, dx = \sum_{\nu=0}^{m} \lambda_\nu f(x_\nu), \tag{7.7.12}$$

where the coefficients λ_ν do not depend on $f(x)$. Upon writing

$$f(x) = \begin{cases} (1 - x)\left\{\dfrac{P_m^{(1,0)}(x)}{x - x_\nu}\right\}^2 & \text{if } \nu > 0, \\ \{P_m^{(1,0)}(x)\}^2 & \text{if } \nu = 0, \end{cases} \tag{7.7.13}$$

we show as in §3.4 (2) that the numbers λ_ν are positive. Now in view of $f(x) \geqq 0$ and of (7.7.4), we obtain from (7.7.12)

$$1 \geqq \lambda_0 f(1), \qquad f(1) \leqq \lambda_0^{-1}; \tag{7.7.14}$$

this is the precise bound of $f(1)$, attained when an only when $f(x_\nu) = 0$, $\nu = 1, 2, \cdots, m$; that is, when $f(x)$ = const. $\{P_m^{(1,0)}(x)\}^2$. In order to find λ_0, it is convenient to write $f(x) = \gamma\, P_m^{(1,0)}(x)$ in (7.7.12) (cf. (3.4.3)), where γ is a con-

[47] Lukács uses this second method in **1**; however, the first method was also in his possession (cf. **1**, p. 296).

stant; then, on account of (4.5.3), $\alpha = \beta = 0$,

$$\frac{1}{2}\int_{-1}^{+1} f(x)\,dx = \frac{\gamma}{2}\int_{-1}^{+1} P_m^{(1,0)}(x)\,dx = \frac{\gamma}{m+1}\int_{-1}^{+1} K_m^{(0,0)}(x)\,dx = \frac{\gamma}{m+1} = 1.$$

Consequently, $\lambda_0^{-1} = \gamma\, P_m^{(1,0)}(1) = (m+1)^2$.

Now let $n = 2m+1$, $x_0 = 1$, $x_{m+1} = -1$, and denote the zeros of the polynomial $(1-x^2)P_m^{(1,1)}(x)$ by $x_0, x_1, \cdots, x_{m+1}$. The same argument as before leads to

$$\frac{1}{2}\int_{-1}^{+1} f(x)\,dx = \sum_{\nu=0}^{m+1} \lambda_\nu f(x_\nu), \qquad \lambda_\nu > 0, \tag{7.7.15}$$

where $f(x)$ is an arbitrary π_n. The maximum in question is again λ_0^{-1}, which is attained when and only when $f(x_\nu) = 0$, $\nu = 1, 2, \cdots, m+1$; that is, $f(x) =$ const. $(1+x)\{P_m^{(1,1)}(x)\}^2$. In order to find λ_0, we write $f(x) = \gamma\,(1+x)\,P_m^{(1,1)}(x)$, so that on account of (4.5.3), $\alpha = 0$, $\beta = 1$,

$$\begin{aligned}\frac{1}{2}\int_{-1}^{+1} f(x)\,dx &= \frac{\gamma}{2}\int_{-1}^{+1} (1+x)P_m^{(1,1)}(x)\,dx \\ &= \frac{2\gamma}{m+2}\int_{-1}^{+1} (1+x)K_m^{(0,1)}(x)\,dx = \frac{2\gamma}{m+2} = 1,\end{aligned} \tag{7.7.16}$$

and we obtain

$$\lambda_0^{-1} = 2\gamma(m+1) = (m+1)(m+2).$$

7.71. Generalizations; applications

(1) Let $d\alpha(x)$ be an arbitrary distribution on the finite or infinite segment $[a, b]$, $\{p_n(x)\}$ the associated set of orthonormal polynomials, and x_0 an arbitrary but fixed point. Then if $\rho(x)$ is an arbitrary π_m with

$$\int_a^b |\,\rho(x)\,|^2\, d\alpha(x) = 1, \tag{7.71.1}$$

we have

$$|\,\rho(x_0)\,|^2 \leqq \sum_{\nu=0}^{m} |\,p_\nu(x_0)\,|^2, \tag{7.71.2}$$

with the sign of equality if and only if $\rho(x) =$ const. $\sum_{\nu=0}^{m} \overline{p_\nu(x_0)}\,p_\nu(x)$. This was proved in §3.1 (3).

In certain special cases we can calculate the maximum of (or some upper bounds for) the right-hand member of (7.71.2) if x_0 runs over a certain interval. The bounds obtained hold uniformly in that interval for the set of all $\rho(x)$ which are π_m and satisfy (7.71.1).

In what follows we consider various distributions of the form $d\alpha(x) = w(x)\,dx$; $\rho(x)$ denotes an arbitrary π_m which satisfies (7.71.1).

(2) Let $a = -1$, $b = +1$, $w(x) = 1$.

THEOREM 7.71.1. *Let $\rho(x)$ be an arbitrary π_m subject to the condition*

$$\int_{-1}^{+1} |\rho(x)|^2 dx = 1. \tag{7.71.3}$$

Then we have for $-1 \leqq x_0 \leqq +1$

$$|\rho(x_0)| \leqq \begin{cases} 2^{-\frac{1}{2}}(m+1), \\ A(1-x_0^2)^{-\frac{1}{4}} m^{\frac{1}{2}}. \end{cases} \tag{7.71.4}$$

Here A is an absolute constant.

The first bound is precise; it is attained for $x_0 = \pm 1$ if $\rho(x)$ is a proper π_m. These inequalities follow from (7.71.2) when we use (7.21.1) and (7.3.8).

Now let $a = -1$, $b = +1$, $w(x) = (1-x)^\alpha (1+x)^\beta$.

THEOREM 7.71.2. *Let $\alpha > -1$, $\beta > -1$, and $\rho(x)$ an arbitrary π_m subject to the condition*

$$\int_{-1}^{+1} (1-x)^\alpha (1+x)^\beta |\rho(x)|^2 dx = 1. \tag{7.71.5}$$

Then

$$\rho(\cos\theta) = \begin{cases} \theta^{-\alpha-\frac{1}{2}} O(m^{\frac{1}{2}}) & \text{if } cm^{-1} \leqq \theta \leqq \pi/2, \\ O(m^{\alpha+1}) & \text{if } 0 \leqq \theta \leqq cm^{-1}. \end{cases} \tag{7.71.6}$$

Here c is an arbitrary but fixed positive number, and the constants in the O-terms depend only on α, β, and c. Similar bounds hold in the interval $[\pi/2, \pi]$.

For the proof we notice that in this case $p_n(x) \sim n^{\frac{1}{2}} P_n^{(\alpha,\beta)}(x)$ [(4.3.3)], and according to (7.32.5),

$$\sum_{\nu=1}^{m} \nu \{P_\nu^{(\alpha,\beta)}(\cos\theta)\}^2 = \sum_{\nu\theta<c} \nu O(\nu^{2\alpha}) + \sum_{\nu\theta \geqq c} \nu\theta^{-2\alpha-1} O(\nu^{-1})$$
$$= O(\theta^{-2\alpha-2}) + \theta^{-2\alpha-1} O(m) = \theta^{-2\alpha-1} O(m),$$

if $cm^{-1} \leqq \theta \leqq \pi/2$. For the same sum we obtain the bound $\sum_{\nu=1}^{m} \nu O(\nu^{2\alpha}) = O(m^{2\alpha+2})$ if $0 \leqq \theta \leqq cm^{-1}$.

By use of (7.32.2) certain precise bounds can likewise be derived.

Let $a = 0$, $b = +\infty$, $w(x) = e^{-x}$. Then, according to (7.21.3),

$$e^{-x_0/2} |\rho(x_0)| \leqq (m+1)^{\frac{1}{2}}, \qquad x_0 \geqq 0, \qquad \int_0^\infty e^{-x} |\rho(x)|^2 dx = 1. \tag{7.71.7}$$

This bound is attained if $x_0 = 0$ and $\rho(x)$ is a proper π_n.

The case $a = 0$, $b = +\infty$, $w(x) = e^{-x} x^\alpha$, $\alpha > -1$, can be treated by a method similar to that used in the Jacobi case discussed above (cf. (7.6.8)).

(3) Now let $[a, b]$ be a finite interval and $f(x)$ an arbitrary π_n which is non-negative in $[a, b]$ and satisfies the condition

$$\int_a^b f(x)w(x)\,dx = 1. \tag{7.71.8}$$

We intend to determine the maximum of $|f(x_0)|$, where x_0 is a fixed real or complex value.

By Theorem 1.21.1 we can write

$$f(x) = \begin{cases} \left\{\sum_{\nu=0}^{m} u_\nu p_\nu(x)\right\}^2 + (x-a)(b-x)\left\{\sum_{\nu=0}^{m-1} v_\nu q_\nu(x)\right\}^2 & \text{if } n = 2m, \\ (x-a)\left\{\sum_{\nu=0}^{m} u_\nu r_\nu(x)\right\}^2 + (b-x)\left\{\sum_{\nu=0}^{m} v_\nu s_\nu(x)\right\}^2 & \text{if } n = 2m+1, \end{cases} \tag{7.71.9}$$

where $\{p_\nu(x)\}$, $\{q_\nu(x)\}$, $\{r_\nu(x)\}$, $\{s_\nu(x)\}$ are the orthonormal sets of polynomials associated with the weight functions

$$w(x),\ (x-a)(b-x)w(x),\ (x-a)w(x),\ (b-x)w(x),\ a \leqq x \leqq b, \tag{7.71.10}$$

respectively. The third and fourth sets are special cases ($x_0 = a$, $x_0 = b$) of the "kernel" polynomials (Theorem 3.1.4); the second set can be calculated by means of Theorem 2.5. In both cases, $n = 2m$ and $n = 2m + 1$, we have for the real numbers u_ν, v_ν

$$\int_a^b f(x)w(x)\,dx = \sum_{\nu=0}^{m} u_\nu^2 + \sum_{\nu=0}^{m} v_\nu^2 = 1, \qquad v_m = 0 \quad \text{if} \quad n = 2m, \tag{7.71.11}$$

so that according to Cauchy's inequality

$$|f(x_0)| \leqq \begin{cases} \max\left\{\sum_{\nu=0}^{m} |p_\nu(x_0)|^2,\ |x_0 - a|\,|b - x_0| \sum_{\nu=0}^{m-1} |q_\nu(x_0)|^2\right\} & \text{if } n = 2m, \\ \max\left\{|x_0 - a| \sum_{\nu=0}^{m} |r_\nu(x_0)|^2,\ |b - x_0| \sum_{\nu=0}^{m} |s_\nu(x_0)|^2\right\} & \text{if } n = 2m+1. \end{cases} \tag{7.71.12}$$

In case $a \leqq x_0 \leqq b$, the absolute value signs can be omitted. The right-hand side of (7.71.12) represents the maximum required.

(4) Let $a = -1$, $b = +1$, $w(x) = (1-x)^\alpha(1+x)^\beta$, α and $\beta > -1$. Then the four sets of polynomials mentioned in (3) are, respectively, constant multiples of

$$\{P_\nu^{(\alpha,\beta)}(x)\}, \qquad \{P_\nu^{(\alpha+1,\beta+1)}(x)\}, \qquad \{P_\nu^{(\alpha,\beta+1)}(x)\}, \qquad \{P_\nu^{(\alpha+1,\beta)}(x)\}. \tag{7.71.13}$$

In the special case $x_0 = 1$, for the maximum of $f(1)$ we obtain (cf. (4.5.3)):

$$\begin{aligned} &\sum_{\nu=0}^{m} \{p_\nu(1)\}^2 = K_m^{(\alpha,\beta)}(1) && \text{if } n = 2m, \\ &2\sum_{\nu=0}^{m} \{r_\nu(1)\}^2 = 2K_m^{(\alpha,\beta+1)}(1) && \text{if } n = 2m+1. \end{aligned} \tag{7.71.14}$$

Therefore, the following theorem holds:

THEOREM 7.71.3. *Let $f(x)$ be an arbitrary π_n which is non-negative in $[-1, +1]$ and satisfies the condition*

$$\int_{-1}^{+1} f(x)(1-x)^\alpha(1+x)^\beta\,dx = 1, \quad \alpha > -1, \beta > -1. \tag{7.71.15}$$

Then

$$f(1) \leqq \tag{7.71.16}$$
$$\begin{cases} 2^{-\alpha-\beta-1}\dfrac{\Gamma(m+\alpha+2)\Gamma(m+\alpha+\beta+2)}{\Gamma(\alpha+1)\Gamma(\alpha+2)\Gamma(m+1)\Gamma(m+\beta+1)} & \text{if } n = 2m, \\[2ex] 2^{-\alpha-\beta-1}\dfrac{\Gamma(m+\alpha+2)\Gamma(m+\alpha+\beta+3)}{\Gamma(\alpha+1)\Gamma(\alpha+2)\Gamma(m+1)\Gamma(m+\beta+2)} & \text{if } n = 2m+1. \end{cases}$$

These bounds are precise; both are $\sim m^{2\alpha+2}$ as $m \to \infty$.

Cf. Pólya-Szegö **1**, vol. 2, pp. 96-97, 298, problem 110. Upon permuting α and β, we obtain the corresponding bounds for $f(-1)$. In general, we find, under the same conditions as in Theorem 7.71.3 (cf. Theorem 7.71.2),

$$f(\cos\theta) = \begin{cases} \theta^{-2\alpha-1}O(m) & \text{if } cm^{-1} \leqq \theta \leqq \pi/2, \\ O(m^{2\alpha+2}) & \text{if } 0 \leqq \theta \leqq cm^{-1}. \end{cases} \tag{7.71.17}$$

The bounds for $f(\cos\theta)$ are similar in $\pi/2 \leqq \theta \leqq \pi$. Further, a bound of the form $O(m^c)$ holds uniformly in $0 \leqq \theta \leqq \pi$, where $c = \max(2\alpha+2, 2\beta+2, 1)$. The constants of all these O-terms depend only on α, β, and c.

(5) By means of Theorem 1.21.2 we can treat the following problem. Let $f(x)$ be an arbitrary π_n, non-negative for $x \geqq 0$, and

$$\int_0^\infty e^{-x}x^\alpha f(x)\,dx = 1, \qquad \alpha > -1. \tag{7.71.18}$$

What is $\max f(0)$?

We write (cf. (5.1.1))

$$f(x) = \begin{cases} \left|\sum_{\nu=0}^{m} u_\nu\left\{\Gamma(\alpha+1)\binom{\nu+\alpha}{\nu}\right\}^{-\frac{1}{2}} L_\nu^{(\alpha)}(x)\right|^2 \\ \qquad + x\left|\sum_{\nu=0}^{m-1} v_\nu\left\{\Gamma(\alpha+2)\binom{\nu+\alpha+1}{\nu}\right\}^{-\frac{1}{2}} L_\nu^{(\alpha+1)}(x)\right|^2 \\ \qquad\qquad \text{if } n = 2m, \\ \left|\sum_{\nu=0}^{m} u_\nu\left\{\Gamma(\alpha+1)\binom{\nu+\alpha}{\nu}\right\}^{-\frac{1}{2}} L_\nu^{(\alpha)}(x)\right|^2 \\ \qquad + x\left|\sum_{\nu=0}^{m} v_\nu\left\{\Gamma(\alpha+2)\binom{\nu+\alpha+1}{\nu}\right\}^{-\frac{1}{2}} L_\nu^{(\alpha+1)}(x)\right|^2 \\ \qquad\qquad \text{if } n = 2m+1, \end{cases} \tag{7.71.19}$$

where the complex numbers u_ν, v_ν satisfy the condition

$$\sum_{\nu=0}^{m} |u_\nu|^2 + \sum_{\nu=0}^{m} |v_\nu|^2 = 1. \tag{7.71.20}$$

(In the second sum $v_m = 0$ if $n = 2m$.) Now in both cases we have (cf. (5.1.7))

$$\begin{aligned} f(0) &= \left| \sum_{\nu=0}^{m} u_\nu \left\{ \Gamma(\alpha+1)\binom{\nu+\alpha}{\nu} \right\}^{-\frac{1}{2}} L_\nu^{(\alpha)}(0) \right|^2 \\ &\leqq \sum_{\nu=0}^{m} \left\{ \Gamma(\alpha+1)\binom{\nu+\alpha}{\nu} \right\}^{-1} \{L_\nu^{(\alpha)}(0)\}^2 \\ &= \{\Gamma(\alpha+1)\}^{-1} \sum_{\nu=0}^{m} \binom{\nu+\alpha}{\nu} = \{\Gamma(\alpha+1)\}^{-1} \binom{m+\alpha+1}{m}, \end{aligned} \tag{7.71.21}$$

and this is the required maximum.

If $\alpha = 0$, we obtain

$$f(0) \leqq [n/2] + 1, \qquad \int_0^\infty e^{-x} f(x)\, dx = 1, \tag{7.71.22}$$

provided $f(x)$ is a π_n, non-negative for $x \geqq 0$. We can readily prove the more general inequality

$$e^{-x} f(x) \leqq [n/2] + 1, \tag{7.71.23}$$

where $f(x)$ is subject to the same condition as in (7.71.22). To this end, let x_0 be an arbitrary positive number. Applying (7.71.22) to

$$f(x + x_0) \left\{ \int_0^\infty e^{-x} f(x + x_0)\, dx \right\}^{-1}$$

which satisfies the required condition, we find

$$f(x_0) \left\{ \int_0^\infty e^{-x} f(x + x_0)\, dx \right\}^{-1} \leqq [n/2] + 1.$$

Whence

$$\begin{aligned} e^{-x_0} f(x_0) &\leqq ([n/2] + 1) e^{-x_0} \int_0^\infty e^{-x} f(x + x_0)\, dx \\ &= ([n/2] + 1) \int_{x_0}^\infty e^{-x} f(x)\, dx \leqq [n/2] + 1. \end{aligned}$$

See also Problem 42.

(6) Certain bounds for the orthonormal polynomials $\{p_n(x)\}$ can be derived from the preceding results, provided the weight function $w(x)$ satisfies an inequality of the type

$$w(x) \geqq \mu > 0, \qquad a \leqq x \leqq b; \tag{7.71.24}$$

or

$$w(x) \geqq \mu(x - a)^\alpha(b - x)^\beta, \qquad a < x < b, \alpha > -1, \beta > -1; \tag{7.71.25}$$

or

$$w(x) \geqq \mu x^\alpha, \qquad x > 0, \alpha > -1. \tag{7.71.26}$$

In the first and second cases a and b are finite.

For instance, under the condition (7.71.24) we obtain

$$\int_a^b \{p_n(x)\}^2 dx \leqq \mu^{-1} \int_a^b \{p_n(x)\}^2 w(x)\, dx = \mu^{-1}. \tag{7.71.27}$$

Consequently, according to Theorem 7.71.1,

$$| p_n(x_0) | < \begin{cases} Am, \\ A'[(x_0 - a)(b - x_0)]^{-\frac{1}{4}} m^{\frac{1}{2}}, \end{cases} \qquad a < x_0 < b. \tag{7.71.28}$$

Here A and A' are positive constants depending only on a, b and μ.

A similar argument applies to the cases (7.71.25) and (7.71.26).

If we assume that $w(x)$ satisfies a Lipschitz condition, the bounds m and $m^{\frac{1}{2}}$ in (7.71.28) can be replaced by $m^{\frac{1}{2}}$ and 1, respectively. (Cf. (7.1.15).)

(7) Finally, by use of Theorem 7.32.4, we obtain the following generalization of (7.71.28). Let $w(x) \geqq \mu > 0$, $a = -1$, $b = +1$, and $k \geqq 0$, an integer. Then

$$p_n^{(k)}(\cos\theta) = \begin{cases} (\sin\theta)^{-k-\frac{1}{2}} O(n^{k+\frac{1}{2}}), \\ O(n^{2k+1}), \end{cases} \qquad 0 < \theta < \pi. \tag{7.71.29}$$

The bounds of the O-terms depend only on μ and k.

For the proof we use an argument similar to that used in proving Theorem 7.71.2.

7.72. A problem of Tchebichef

(1) Problem: *Let $w(x)$ be a weight function on the interval $[a, b]$, and let $W(x)$ be a given real-valued function, defined on the same interval, and for which the integrals*

$$\int_a^b W(x) x^k\, dx, \qquad k = 0, 1, 2, \cdots, n, \tag{7.72.1}$$

exist. Let $f(x)$ be an arbitrary polynomial of fixed degree n, not identically zero, and non-negative in $a \leqq x \leqq b$. To determine the maximum and the minimum of the ratio

$$\int_a^b f(x) W(x)\, dx : \int_a^b f(x) w(x)\, dx. \tag{7.72.2}$$

See Tchebichef **7**. First let a and b be finite. By using the representation (7.71.9) again, we easily find that the quantities in question are the maximum and minimum of the following quadratic forms in $\{u_\nu\}$ and $\{v_\nu\}$:

$$\int_a^b \left\{\sum_{\nu=0}^{m} u_\nu p_\nu(x)\right\}^2 W(x)\,dx + \int_a^b \left\{\sum_{\nu=0}^{m-1} v_\nu q_\nu(x)\right\}^2 (x-a)(b-x)W(x)\,dx$$

(7.72.3) if $n = 2m$,

$$\int_a^b \left\{\sum_{\nu=0}^{m} u_\nu r_\nu(x)\right\}^2 (x-a)W(x)\,dx + \int_a^b \left\{\sum_{\nu=0}^{m} v_\nu s_\nu(x)\right\}^2 (b-x)W(x)\,dx$$

if $n = 2m + 1$,

under the condition $\sum_{\nu=0}^m u_\nu^2 + \sum_{\nu=0}^m v_\nu^2 = 1$. (In the first case $v_m = 0$.) Here $\{p_\nu(x)\}$, $\{q_\nu(x)\}$, $\{r_\nu(x)\}$, $\{s_\nu(x)\}$ have the same meaning as in (7.71.9).

Let now a be finite and $b = +\infty$. Then we have to consider the maximum and minimum of the form

$$\int_a^\infty \left\{\sum_{\nu=0}^{[n/2]} u_\nu p_\nu(x)\right\}^2 W(x)\,dx + \int_a^\infty \left\{\sum_{\nu=0}^{[(n-1)/2]} v_\nu q_\nu(x)\right\}^2 (x-a)W(x)\,dx \tag{7.72.4}$$

under the condition $\sum u_\nu^2 + \sum v_\nu^2 = 1$. Here $\{p_\nu(x)\}$ and $\{q_\nu(x)\}$ are the orthonormal sets associated with $w(x)$ and $(x - a)w(x)$, respectively, $x \geqq a$.

In case $a = -\infty$, $b = +\infty$, we must consider the form

$$\int_{-\infty}^{+\infty} \left\{\sum_{\nu=0}^{[n/2]} u_\nu p_\nu(x)\right\}^2 W(x)\,dx, \qquad \sum_{\nu=0}^{[n/2]} u_\nu^2 = 1, \tag{7.72.5}$$

where $\{p_\nu(x)\}$ is associated with $w(x)$ in $[-\infty, +\infty]$.

Thus, in all these cases, the problem in question is reduced to the determination of the greatest and least characteristic values of a certain quadratic form. In dealing with the sum of two quadratic forms in $\{u_\nu\}$ and $\{v_\nu\}$, respectively, we determine the greatest characteristic value of the single forms, and the greater of these values is the maximum in question. A similar remark applies to the minimum. The actual application of this method is difficult, however, and certain mechanical quadrature formulas (see below) are often preferable.

Similar considerations apply if the integrals (7.72.2) are replaced by Stieltjes integrals.

(2) Let $a = -1$, $b = +1$, $W(x) = xw(x)$. It suffices to determine the maximum and minimum of

$$\int_{-1}^{+1} \{\rho(x)\}^2 x w(x)\,dx : \int_{-1}^{+1} \{\rho(x)\}^2 w(x)\,dx, \tag{7.72.6}$$

if $\rho(x)$ is an arbitrary π_m, not identically zero with real coefficients. Having done this, we must replace $w(x)$ by $(1 - x^2)w(x)$, $(1 \pm x)w(x)$, respectively; see below. Let $x_0, x_1, \cdots, x_m$ be the zeros of the orthogonal polynomial $p_{m+1}(x)$ associated with $w(x)$; according to (3.4.1), we find for the ratio (7.72.6), the representation

$$\sum_{\nu=0}^{m} \lambda_\nu \{\rho(x_\nu)\}^2 x_\nu : \sum_{\nu=0}^{m} \lambda_\nu \{\rho(x_\nu)\}^2, \tag{7.72.7}$$

where λ_ν denote the Christoffel numbers. Therefore, the maximum and minimum in question coincide with the greatest and least zero of $p_{m+1}(x)$, respectively. (Cf. §3.4 (3).) If $\hat{p}$ denotes the greatest zero of $p(x)$, it is seen from (7.72.3) that the maximum of (7.72.2) is, in this special case,

$$\tag{7.72.8} \begin{aligned} &\max (\hat{p}_{m+1}, \hat{q}_m) &&\text{if} \quad n = 2m, \\ &\max (\hat{r}_{m+1}, \hat{s}_{m+1}) &&\text{if} \quad n = 2m + 1. \end{aligned}$$

The result for the minimum is similar.

(3) Here the general discussion of Tchebichef ends (cf. **7**, p. 395). We can prove, however, that the expressions (7.72.8) are $\hat{p}_{m+1}$ and $\hat{r}_{m+1}$, respectively, so that the following theorem holds:

THEOREM 7.72.1. *Let $w(x)$ be a weight function on the interval $[-1, +1]$. Let $f(x)$ be an arbitrary π_n, not identically zero, and non-negative in $[-1, +1]$. Then the maximum of*

$$\tag{7.72.9} \int_{-1}^{+1} f(x)xw(x)\, dx : \int_{-1}^{+1} f(x)w(x)\, dx$$

is the greatest zero of $p_{m+1}(x)$ if $n = 2m$, and the greatest zero of $p_{m+2}(-1)p_{m+1}(x) - p_{m+1}(-1)p_{m+2}(x)$ if $n = 2m + 1$. Here $\{p_n(x)\}$ is the set of the orthonormal polynomials associated with $w(x)$ in the interval $[-1, +1]$.

According to Theorem 2.5,

$$\tag{7.72.10} \begin{aligned} (1 - x^2)q_m(x) &= \text{const.} \begin{vmatrix} p_m(x) & p_{m+1}(x) & p_{m+2}(x) \\ p_m(-1) & p_{m+1}(-1) & p_{m+2}(-1) \\ p_m(1) & p_{m+1}(1) & p_{m+2}(1) \end{vmatrix}, \\ (1 + x)r_m(x) &= \text{const.} \begin{vmatrix} p_m(x) & p_{m+1}(x) \\ p_m(-1) & p_{m+1}(-1) \end{vmatrix}, \\ (1 - x)s_m(x) &= \text{const.} \begin{vmatrix} p_m(x) & p_{m+1}(x) \\ p_m(1) & p_{m+1}(1) \end{vmatrix}. \end{aligned}$$

First, let $\xi_0 > \xi_1 > \cdots > \xi_m$ be the zeros of $p_{m+1}(x)$ in decreasing order. We show that the first determinant in the right-hand member of (7.72.10) is non-zero 0 if $\xi_0 < x < 1$. Indeed, according to (3.2.1)

$$\begin{vmatrix} p_m(x) & p_{m+1}(x) & p_{m+2}(x) \\ p_m(-1) & p_{m+1}(-1) & p_{m+2}(-1) \\ p_m(1) & p_{m+1}(1) & p_{m+2}(1) \end{vmatrix} = A_{m+2} \begin{vmatrix} p_m(x) & p_{m+1}(x) & xp_{m+1}(x) \\ p_m(-1) & p_{m+1}(-1) & -p_{m+1}(-1) \\ p_m(1) & p_{m+1}(1) & p_{m+1}(1) \end{vmatrix}$$

$$= A_{m+2}\, p_{m+1}(x)p_{m+1}(-1)p_{m+1}(1) \begin{vmatrix} h(x) & 1 & x \\ h(-1) & 1 & -1 \\ h(1) & 1 & 1 \end{vmatrix},$$

where $h(x) = p_m(x)/p_{m+1}(x)$. Now, by using (3.3.9), we see that the last determinant is positive in $\xi_0 < x < 1$, since

$$\begin{vmatrix} (x-\xi_\nu)^{-1} & 1 & x \\ (-1-\xi_\nu)^{-1} & 1 & -1 \\ (1-\xi_\nu)^{-1} & 1 & 1 \end{vmatrix} = \frac{2(1-x^2)}{(1-\xi_\nu^2)(x-\xi_\nu)} > 0.$$

On the other hand, $h(x)$ decreases from $+\infty$ to $-\infty$ between ξ_1 and ξ_0, and from $+\infty$ to $h(1)$ between ξ_0 and 1 (cf. the proof of Theorem 3.3.4). Furthermore, $h(-1) < 0$, $h(1) > 0$. Thus the greatest zero of $r_m(x)$, or $h(x) - h(-1)$, is greater than the greatest zero of $s_m(x)$, or $h(x) - h(1)$, $x < 1$.

(4) Tchebichef discusses in detail the case

$$(7.72.11)\qquad a = -1, \qquad b = +1, \qquad w(x) = 1, \qquad W(x) = x.$$

(Problem of the "centroid," see loc. cit., p. 399; cf. also Szegö **13**, pp. 627–629.) Save for constant factors, we now have

$$\begin{aligned} p_{m+1}(x) &= P_{m+1}(x); \qquad q_m(x) = P_m^{(1,1)}(x) = \text{const. } P'_{m+1}(x), \\ (7.72.12)\quad r_{m+1}(x) &= P_{m+1}^{(0,1)}(x) = \text{const. } \{P_{m+1}(x) + P_{m+2}(x)\}(1+x)^{-1}; \\ s_{m+1}(x) &= P_{m+1}^{(1,0)}(x) = \text{const. } \{P_{m+1}(x) - P_{m+2}(x)\}(1-x)^{-1}. \end{aligned}$$

This yields the following theorem:

THEOREM 7.72.2. *Let $f(x)$ be an arbitrary π_n, not identically zero, and non-negative for $-1 \leqq x \leqq 1$. Then the maximum of*

$$(7.72.13)\qquad \int_{-1}^{+1} xf(x)\,dx : \int_{-1}^{+1} f(x)\,dx$$

is the greatest zero of $P_{m+1}(x)$ if $n = 2m$; if $n = 2m+1$, it is the greatest zero of $P_{m+1}(x) + P_{m+2}(x)$.

The distance from the maximum value to 1 is $\sim n^{-2}$ as $n \to \infty$. The minimum is obviously the corresponding negative value.

For other refinements of the mean-value theorems, the reader is referred to Tchakaloff **1**. Cf. Problem 43. Concerning other extremum problems for polynomials and connected inequalities, see Geronimus **2**, **3**, **4**, and Shohat **2**.

7.8. Further results

(1) From Theorem 1.82.5 we conclude the following refinement of Sonin's Theorem 7.31.1. Let $y = y(x)$ satisfy the differential equation (7.31.1) and let $y(x)$ have an infinite set $\{x_m\}$ of zeros ordered in the increasing way: $x_1 < x_2 < x_3 < \cdots$. The function $\phi(x)$ should be positive, continuous, and decreasing. Let p be a fixed positive number. Then the integrals

$$\int_{x_\nu}^{x_{\nu+1}} |y(x)|^p\,dx$$

are increasing.

A similar statement holds if $\phi(x)$ is increasing.

For $p \to 0$ this reduces to the assertion of Theorem 1.82.2. Taking the pth root of the integral, for $p \to \infty$ we obtain the assertion of Sonin's theorem. Hence, the assertion above is a common generalization of both theorems (Makai **2**).

In Makai **2** one finds various remarkable special cases of this useful theorem.

(2) Let $\mu_{r,n}$ be the successive relative maxima of $|P_n(x)|$ when x decreases from $+1$ to -1. We have (Theorem 7.3.1):

$$1 > \mu_{1,n} > \mu_{2,n} > \cdots > \mu_{h,n} \quad \text{where} \quad h = [n/2].$$

Now for a fixed r, we have (Szegö **23**)

$$\mu_{r,n} > \mu_{r,n+1}, \qquad n \geq r+1. \tag{7.8.1}$$

These inequalities also hold for the relative maxima of $|P_n^{(\alpha,\beta)}(x)|/P_n^{(\alpha,\beta)}(1)$ when $\alpha = \beta > -\frac{1}{2}$, Szász **2**, and from this for $\alpha > \beta = -\frac{1}{2}$ by use of (4.1.5). The same result for $\alpha > \beta > -\frac{1}{2}$ is probable, but still open. Somewhat surprisingly these inequalities fail for

$$P_n^{(0,-1)}(x) = \frac{P_n(x) + P_{n-1}(x)}{2}$$

and graphical evidence suggests that the inequalities (7.8.1) are reversed for this function. See Askey-Gasper **4** for a problem which would be solved by these inequalities.

For the orthonormal Hermite functions defined by

$$\mathscr{H}_n(x) = \frac{H_n(x)e^{-x^2/2}}{\pi^{1/4}(2^n n!)^{1/2}},$$

Szász **3** proved inequalities similar to (7.8.1) and used them to prove

$$|\mathscr{H}_n(x)| \leqq \mathscr{H}_0(x) = \pi^{-1/4}.$$

Stronger inequalities in which the right-hand side goes to zero in n are known. See Askey-Wainger **1** and Muckenhoupt **3**. They also obtain refinements similar to that given in Problem 40 and similar inequalities for Laguerre polynomials.

(3) Many inequalities, involving in particular the classical orthogonal polynomials, have been investigated by Karlin-Szegö **1**. Problem 70 (Turán's inequality) is a special case. See Gasper **5** for a Turán type inequality for Jacobi polynomials.

(4) Theorems 7.31.2, 7.32.3, and 7.6.5 have been sharpened for $\alpha > -\frac{1}{2}$ by Lorch **3**.

(5) The inequalities (7.71.12) have been refined in Schoenberg-Szegö **1**.

(6) Turán **3** considers the problem of maximizing the Markoff-type functional

$$\frac{\int_0^\infty [\pi_n'(x)]^2 e^{-x}dx}{\int_0^\infty [\pi_n(x)]^2 e^{-x}dx}$$

over polynomials π_n of degree $\leqq n$. He shows that the exact maximum is

$$\frac{1}{2\sin \pi/(4n+2)}.$$

CHAPTER VIII

ASYMPTOTIC PROPERTIES OF THE CLASSICAL POLYNOMIALS

The consideration of the asymptotic properties of the orthogonal polynomials $\{p_n(x)\}$, $n \to \infty$, leads to two fundamental problems: the asymptotic behavior of the polynomials in question outside the orthogonality interval, especially in the non-real domain, and the asymptotic behavior on the orthogonality interval itself. In general, the second problem is deeper and more difficult than the first one. In our treatment we start with a discussion of Legendre polynomials, obtaining various important asymptotic formulas for them. We intend not only to give a survey of results, but also to point out the various methods used. The extension to ultraspherical and general Jacobi polynomials will also be indicated. The asymptotic investigation of Laguerre and Hermite polynomials, in general, requires new considerations, although essentially the same methods as before can also be applied to these cases.

The simplest special case,

$$T_n(x) = \tfrac{1}{2}(z^n + z^{-n}), \qquad x = \tfrac{1}{2}(z + z^{-1}),$$

the case of Tchebichef polynomials of the first kind, furnishes a good illustration of the characteristic features of our results. If x is located outside the interval $[-1, +1]$, we can take $|z| > 1$, and we then see that

$$T_n(x) \cong z^n/2, \qquad n \to \infty.$$

On the interval $[-1, +1]$, we write $z = e^{i\theta}$, $T_n(x) = \cos n\theta$. Here the polynomials have an oscillatory behavior.

These results need only a slight modification for Legendre, and even for Jacobi, polynomials as long as $x \neq \pm 1$. A new difficulty will, however, arise in the vicinity of the end-points ± 1, which are in some respects exceptional. This is ultimately due to the fact that the coefficient of $d\theta$ in

$$(1 - x)^\alpha(1 + x)^\beta dx = -(1 - \cos\theta)^\alpha(1 + \cos\theta)^\beta \sin\theta\, d\theta$$

vanishes, in general, or becomes infinite at $\theta = 0$ and $\theta = \pi$. When this occurs, functions of the type $\cos n\theta$ are not suitable for the approximation of the polynomials in question in the neighborhood of $x = \pm 1$. For this purpose we shall use certain Bessel functions.

Usually, the problems and results for Laguerre and Hermite polynomials are similar. But it is rather curious that, in the corresponding asymptotic expressions, the quantity $n^{\frac{1}{2}}$ appears instead of n. In the general Laguerre case, Bessel functions are needed near $x = 0$, whereas for the Hermite polynomials the origin $x = 0$ does not play an exceptional rôle. In both cases new difficulties arise due to the fact that the interval of integration is infinite. For the expansion problem it is of great importance to have asymptotic formulas that hold in intervals which become infinite as $n \to \infty$.

8.1. The formulas of Mehler-Heine type

These important formulas are elementary in character, and we shall discuss them briefly before entering into the general considerations of §§8.21–8.23.

(1) THEOREM 8.1.1. *Let α and β be arbitrary real numbers. Then*

$$\lim_{n\to\infty} n^{-\alpha} P_n^{(\alpha,\beta)}\left(\cos\frac{z}{n}\right) = \lim_{n\to\infty} n^{-\alpha} P_n^{(\alpha,\beta)}\left(1 - \frac{z^2}{2n^2}\right) = (z/2)^{-\alpha} J_\alpha(z), \tag{8.1.1}$$

where $J_\alpha(z)$ has the same meaning as in (1.71.1). *This formula holds uniformly in every bounded region of the complex z-plane.*

For Legendre polynomials, $\alpha = \beta = 0$, formula (8.1.1) is due to Mehler (**3**, p. 140) and Heine (**3**, vol. 1, p. 184). Concerning further literature we refer to Watson **3**, p. 155. In case $\alpha = \beta = 0$, a very simple proof follows from the first integral of Laplace [(4.8.10), (1.71.6)]. For $\alpha = \pm\frac{1}{2}$ the function in the right-hand member of (8.1.1) is a constant multiple of $z^{-1}\sin z$ and $\cos z$, respectively (cf. (1.71.2)). Formula (8.1.1) is trivial for the elementary cases (4.1.7) and (4.1.8).

The proof can be based on (4.21.2). In fact, we have for the $(\nu + 1)$st term of (4.21.2) if $x = \cos(z/n)$, z and ν fixed and $n \to \infty$, the following asymptotic expression:

$$\begin{aligned} &\frac{1}{\nu!(n-\nu)!}\frac{\Gamma(n+\alpha+\beta+\nu+1)}{\Gamma(n+\alpha+\beta+1)}\frac{\Gamma(n+\alpha+1)}{\Gamma(\nu+\alpha+1)}\left(-\sin^2\frac{z}{2n}\right)^\nu \\ &\qquad\cong \frac{n^\alpha}{\nu!\,\Gamma(\nu+\alpha+1)}\left(-\frac{z^2}{4}\right)^\nu. \end{aligned} \tag{8.1.2}$$

Here we exclude the case of a negative integer α. Passing to the limit under the summation sign is valid because of the existence of a dominant for the total sum which is readily derived. Indeed, we have, if n is large enough,

$$\begin{aligned} &\frac{n^{-\alpha}}{(n-\nu)!}\frac{\Gamma(n+\alpha+\beta+\nu+1)}{\Gamma(n+\alpha+\beta+1)}\frac{\Gamma(n+\alpha+1)}{2^\nu n^{2\nu}} \\ &\qquad\leqq \frac{n^{-\alpha}}{n!}n^\nu(2n+\alpha+\beta)^\nu\frac{\Gamma(n+\alpha+1)}{2^\nu n^{2\nu}} = O(1), \end{aligned}$$

uniformly in ν, $0 \leqq \nu \leqq n$. The argument needs only a slight modification if α is a negative integer.

Formula (8.1.1) gives a complete characterization of the function $P_n^{(\alpha,\beta)}(\cos\theta)$ for $\theta = O(n^{-1})$. As an important consequence we note the following:

THEOREM 8.1.2. *Let $x_{1n} > x_{2n} > \cdots$ be the zeros of $P_n^{(\alpha,\beta)}(x)$ in $[-1, +1]$ in decreasing order (α, β real but not necessarily greater than -1). If we write*

$x_{\nu n} = \cos\theta_{\nu n}$, $0 < \theta_{\nu n} < \pi$, then for a fixed ν,

(8.1.3)
$$\lim_{n\to\infty} n\theta_{\nu n} = j_\nu,$$

where j_ν is the νth positive zero of $J_\alpha(z)$.

(2) Relation (8.1.1) enables us to derive some properties of Bessel functions from the corresponding properties of Jacobi or Legendre polynomials. We use the symbol (a) $\to$ (b) to indicate that in writing $x = \cos(z/n)$, a certain formula (a) is transformed into another formula (b) by the limiting process $n \to \infty$. Then we have the following relations:

(4.1.7) and (4.1.8) $\to$ (1.71.2),
(4.2.1) $\to$ (1.71.3),
(4.22.2) $\to J_{-l}(z) = (-1)^l J_l(z)$, l an integer,
(4.24.2) $\to$ (1.8.9),
each of (4.8.6), (4.8.10), (4.9.3) $\to$ (1.71.6) in the special case $\alpha = 0$, and
(4.9.19) $\to$ (1.71.6) in the general case.

See also Problem 44.

From Theorem 1.91.3 (Hurwitz's theorem) the reality of the zeros of $z^{-\alpha}J_\alpha(z)$ follows for $\alpha > -1$. From (6.6.5) and (6.6.3) (or (6.6.2)) we obtain for the positive zeros j_ν of $J_\alpha(z)$, $\lambda = \alpha + \frac{1}{2}$,

(8.1.4)
$$(\nu + \tfrac{1}{2}\alpha - \tfrac{1}{4})\pi \leqq j_\nu \leqq \nu\pi, \qquad \nu = 1, 2, 3, \cdots, -\tfrac{1}{2} \leqq \alpha \leqq +\tfrac{1}{2}.$$

The upper bound can be replaced by $(\nu + \alpha)\pi$ if $-\frac{1}{2} \leqq \alpha \leqq 0$ (cf. (6.6.6)). Furthermore (cf. Watson **3**, p. 49, (1); cf. (7.31.5)),

(8.1.5)
$$(7.33.1) \to \Gamma(\alpha+1)(z/2)^{-\alpha}\,|\,J_\alpha(z)\,| \leqq 1, \qquad z > 0,\ \alpha \geqq -\tfrac{1}{2},$$

(8.1.6)
$$(7.33.5) \to z^{1/2}\,|\,J_\alpha(z)\,| \leqq (2/\pi)^{1/2}, \qquad z > 0,\ -\tfrac{1}{2} \leqq \alpha \leqq +\tfrac{1}{2}.$$

The expansion (4.9.17), (8.21.5), and the inequality (8.21.6) of the remainder term furnish the important formula (Stieltjes **8**, p. 242):

(8.1.7)
$$J_0(z) = \left(\frac{2}{\pi z}\right)^{\frac{1}{2}} \sum_{\nu=0}^{p-1} \frac{\{1\cdot 3 \cdots (2\nu-1)\}^2}{2\cdot 4 \cdots 2\nu} \frac{\cos\{z - (\nu + \frac{1}{2})\pi/2\}}{2^{2\nu} z^\nu} + \epsilon_p(z),$$
$$|\,\epsilon_p(z)\,| \leqq \left(\frac{2}{\pi z}\right)^{\frac{1}{2}} \frac{\{1\cdot 3 \cdots (2p-1)\}^2}{2\cdot 4 \cdots 2p} \frac{1}{2^{2p} z^p}, \qquad z > 0.$$

Thus the error $\epsilon_p(z)$ is numerically less than the first neglected term (replacing cos by 1). For $p = 1$ we obtain the special case $\alpha = 0$ of (1.71.7) (with a numerical constant in the bound of the remainder).

(3) THEOREM 8.1.3. *Let α be arbitrary and real. Then for an arbitrary complex z*

(8.1.8)
$$\lim_{n\to\infty} n^{-\alpha} L_n^{(\alpha)}(z/n) = z^{-\alpha/2} J_\alpha(2z^{\frac{1}{2}}),$$

uniformly if z is bounded.

This formula is of a type similar to (8.1.1) and yields similar results. The proof can be given along the same lines as there. In the special cases $\alpha = \pm \frac{1}{2}$, we again obtain trigonometric functions. From this case an analogous formula for Hermite polynomials can be derived. See Problem 45.

Both formulas (8.1.1) and (8.1.8) can be extended to an asymptotic expansion. Concerning the case of Laguerre polynomials, $\alpha = 0$, see Moecklin **1**, p. 28.

8.21. Asymptotic formulas for Legendre and Jacobi polynomials

From the point of view of the asymptotic problem, the Legendre polynomials $P_n(x)$ represent the simplest non-trivial case. We start with an enumeration of some classical results concerning the behavior of $P_n(x)$ as $n \to \infty$. The proofs, based on various methods, are given in subsequent sections. In what follows, ϵ denotes a fixed number with $0 < \epsilon < \pi/2$, so that the interval $[\epsilon, \pi - \epsilon]$ lies wholly in the interior of $[0, \pi]$; p is a fixed positive integer.

(1) THEOREM 8.21.1 (Formula of Laplace-Heine; Heine **3**, vol. 1, p. 174). *Let x be an arbitrary real or complex number which does not belong to the closed segment $[-1, +1]$. Then as $n \to \infty$,*

$$P_n(x) \cong (2\pi n)^{-1/2}(x^2 - 1)^{-1/4} \{x + (x^2 - 1)^{1/2}\}^{n+1/2} \tag{8.21.1}$$

Here $(x^2 - 1)^{-1/4}$, $(x^2 - 1)^{1/2}$, and $\{x + (x^2 - 1)^{1/2}\}^{n+1/2}$ are real and positive if x is real and greater than 1. This formula holds uniformly in the exterior of an arbitrary closed curve which encloses the segment $[-1, +1]$, in the sense that the ratio tends uniformly to 1.

THEOREM 8.21.2 (Formula of Laplace; Heine **3**, vol. 1, p. 175).

$$P_n(\cos\theta) = 2^{1/2}(\pi n \sin\theta)^{-1/2} \cos\{(n + \tfrac{1}{2})\theta - \pi/4\} + O(n^{-3/2}), \qquad 0 < \theta < \pi. \tag{8.21.2}$$

The bound for the error term holds uniformly in the interval $\epsilon \leqq \theta \leqq \pi - \epsilon$.

THEOREM 8.21.3 (Generalization of Laplace-Heine's formula). *Let x be in the complex plane cut along the segment $[-1, +1]$; let $x = \frac{1}{2}(z + z^{-1})$, $|z| > 1$. Then*

$$P_n(x) = g_n z^n \sum_{\nu=0}^{p-1} g_\nu \frac{1\cdot 3 \cdots (2\nu - 1)}{(2n-1)(2n-3)\cdots(2n-2\nu+1)} z^{-2\nu}(1 - z^{-2})^{-\nu-\frac{1}{2}} + O(n^{-p-\frac{1}{2}}|z|^n). \tag{8.21.3}$$

Here g_ν has the same meaning as in (4.9.2), that is,

$$g_0 = 1; \qquad g_\nu = \frac{1\cdot 3 \cdots (2\nu - 1)}{2\cdot 4 \cdots 2\nu}, \qquad \nu = 1, 2, 3, \cdots.$$

Formula (8.21.3) holds uniformly in the same sense as in Theorem 8.21.1.

THEOREM 8.21.4 (Darboux's generalization of Laplace's formula; Darboux **1**, p. 39).

(8.21.4)
$$P_n(\cos\theta) = 2g_n \sum_{\nu=0}^{p-1} g_\nu \frac{1\cdot 3 \cdots (2\nu-1)}{(2n-1)(2n-3)\cdots(2n-2\nu+1)} \cdot \frac{\cos\{(n-\nu+\frac{1}{2})\theta - (\nu+\frac{1}{2})\pi/2\}}{(2\sin\theta)^{\nu+\frac{1}{2}}} + O(n^{-p-\frac{1}{2}}), \qquad 0<\theta<\pi.$$

Here g_n has the same meaning as in Theorem 8.21.3. *The bound for the error term again holds uniformly in $\epsilon \leqq \theta \leqq \pi - \epsilon$.*

THEOREM 8.21.5 (Stieltjes' generalization of Laplace's formula; see Stieltjes **7**, **8**).

(8.21.5)
$$P_n(\cos\theta) = \frac{4}{\pi}\frac{2\cdot 4\cdots 2n}{3\cdot 5\cdots(2n+1)} \sum_{\nu=0}^{p-1} h_\nu \frac{\cos\{(n+\nu+\frac{1}{2})\theta - (\nu+\frac{1}{2})\pi/2\}}{(2\sin\theta)^{\nu+\frac{1}{2}}} + R_p(\theta), \qquad 0<\theta<\pi.$$

Here h_ν has the same meaning as in (4.9.18), *that is,*

$$h_0 = 1; \qquad h_\nu = \frac{1\cdot 3\cdots(2\nu-1)}{2\cdot 4\cdots 2\nu}\,\frac{\frac{1}{2}\frac{3}{2}\cdots(\nu-\frac{1}{2})}{(n+\frac{3}{2})(n+\frac{5}{2})\cdots(n+\nu+\frac{1}{2})}, \qquad \nu = 1, 2, 3, \cdots.$$

We have

(8.21.6)
$$|R_p(\theta)| < \frac{4}{\pi}\frac{2\cdot 4\cdots 2n}{3\cdot 5\cdots(2n+1)} h_p \frac{M}{(2\sin\theta)^{p+\frac{1}{2}}}, \qquad M = \max(|\cos\theta|^{-1}, 2\sin\theta).$$

The factor M is between 1 *and* 2. *Thus the error is numerically less than twice the first neglected term* (replacing cos by 1).

THEOREM 8.21.6 (Formula of Hilb; Hilb **1**).

(8.21.7)
$$P_n(\cos\theta) = (\theta/\sin\theta)^{\frac{1}{2}} J_0\{(n+\tfrac{1}{2})\theta\} + O(n^{-\frac{3}{2}}),$$

uniformly for $0 \leqq \theta \leqq \pi - \epsilon$. More precisely, for the error term we have the bounds

(8.21.8)
$$\theta^{\frac{1}{2}} O(n^{-\frac{3}{2}}) \quad \textit{if} \quad cn^{-1} \leqq \theta \leqq \pi - \epsilon,$$
$$\theta^{2} O(1) \quad \textit{if} \quad 0 < \theta \leqq cn^{-1},$$

where c is a fixed positive constant.

(2) Some of these results can be extended to Jacobi polynomials. The extension of (8.21.1) is due to Darboux (**1**):

THEOREM 8.21.7. *Let α and β be arbitrary real numbers. Then*

$$P_n^{(\alpha,\beta)}(x) \cong (x-1)^{-\alpha/2}(x+1)^{-\beta/2}\{(x+1)^{\frac{1}{2}}+(x-1)^{\frac{1}{2}}\}^{\alpha+\beta} \cdot (2\pi n)^{-\frac{1}{2}}(x^2-1)^{-\frac{1}{4}}\{x+(x^2-1)^{\frac{1}{2}}\}^{n+\frac{1}{2}}, \tag{8.21.9}$$

where x is outside of the closed segment $[-1, +1]$. This formula holds uniformly in the same sense as in Theorem 8.21.1. The determination of the multivalued functions occurring in this formula is obvious.

The extension of (8.21.2) is also due to Darboux (**1**); this is the important formula to which we referred in §7.32:

THEOREM 8.21.8. *Let α and β be arbitrary real numbers. Then*

$$P_n^{(\alpha,\beta)}(\cos\theta) = n^{-\frac{1}{2}}k(\theta)\cos(N\theta+\gamma) + O(n^{-\frac{3}{2}}),$$

$$k(\theta) = \pi^{-\frac{1}{2}}\left(\sin\frac{\theta}{2}\right)^{-\alpha-\frac{1}{2}}\left(\cos\frac{\theta}{2}\right)^{-\beta-\frac{1}{2}}, \qquad N = n + (\alpha+\beta+1)/2, \tag{8.21.10}$$

$$\gamma = -(\alpha+\tfrac{1}{2})\pi/2, \quad 0 < \theta < \pi.$$

The bound for the error term holds uniformly in the interval $[\epsilon, \pi-\epsilon]$.

The extension of Theorems 8.21.3 and 8.21.4 to Jacobi polynomials is readily achieved. However, the law of the coefficients is, in this case, rather complicated.

THEOREM 8.21.9. *Let α and β be arbitrary real numbers. There exists a sequence of analytic functions $\phi_\nu(z) = \phi_\nu(\alpha, \beta; z)$ which are real for real z and regular for $|z| > 1$ and $|z| = 1$, $z \neq \pm 1$, such that*

$$z^{-n}P_n^{(\alpha,\beta)}(x) = \sum_{\nu=0}^{p-1}\phi_\nu(z)n^{-\nu-\frac{1}{2}} + O(n^{-p-\frac{1}{2}}); \quad x = \tfrac{1}{2}(z+z^{-1}), \ |z| > 1, \tag{8.21.11}$$

uniformly for $|z| \geqq R$, $R > 1$.

Furthermore,

$$P_n^{(\alpha,\beta)}(\cos\theta) = 2\Re\left\{e^{in\theta}\sum_{\nu=0}^{p-1}\phi_\nu(e^{i\theta})n^{-\nu-\frac{1}{2}}\right\} + O(n^{-p-\frac{1}{2}}), \qquad 0<\theta<\pi, \tag{8.21.12}$$

uniformly for $\epsilon \leqq \theta \leqq \pi - \epsilon$.

These extensions attain, in the ultraspherical case, the following more precise form:

THEOREM 8.21.10. *Let $x = \frac{1}{2}(z+z^{-1})$, $|z| > 1$, and $\lambda > 0$ or $\lambda < 0$, $\lambda \neq -1, -2, -3, \cdots$. Then*

$$P_n^{(\lambda)}(x) = \alpha_n z^n \sum_{\nu=0}^{p-1}\alpha_\nu \frac{(1-\lambda)(2-\lambda)\cdots(\nu-\lambda)}{(n+\lambda-1)(n+\lambda-2)\cdots(n+\lambda-\nu)} \cdot z^{-2\nu}(1-z^{-2})^{-\nu-\lambda} + O(n^{\lambda-p-1}|z|^n). \tag{8.21.13}$$

Furthermore,

$$
(8.21.14)\qquad
\begin{aligned}
P_n^{(\lambda)}(\cos\theta) &= 2\alpha_n \sum_{\nu=0}^{p-1} \alpha_\nu \frac{(1-\lambda)(2-\lambda)\cdots(\nu-\lambda)}{(n+\lambda-1)(n+\lambda-2)\cdots(n+\lambda-\nu)} \\
&\quad \cdot \frac{\cos\{(n-\nu+\lambda)\theta-(\nu+\lambda)\pi/2\}}{(2\sin\theta)^{\nu+\lambda}} + O(n^{\lambda-p-1}), \qquad 0<\theta<\pi.
\end{aligned}
$$

Here α_ν has the same meaning as in (4.9.21). *Regarding the uniformity, the same remark holds as in the previous theorem.*

The special case of an integral value of λ is discussed in §8.4 (5).

An extension of Theorem 8.21.5 to ultraspherical polynomials is the following:

THEOREM 8.21.11. *Let* $0<\lambda<1$. *We have*

$$
(8.21.15)\qquad
\begin{aligned}
P_n^{(\lambda)}(\cos\theta) &= (2/\pi)\sin\lambda\pi \frac{\Gamma(n+2\lambda)}{\Gamma(\lambda)} \sum_{\nu=0}^{p-1} \frac{\Gamma(\nu+\lambda)\Gamma(\nu-\lambda+1)}{\nu!\,\Gamma(n+\nu+\lambda+1)} \\
&\quad \cdot \frac{\cos\{(n+\nu+\lambda)\theta-(\nu+\lambda)\pi/2\}}{(2\sin\theta)^{\nu+\lambda}} + R_p(\theta), \qquad 0<\theta<\pi,
\end{aligned}
$$

where

$$
(8.21.16)\qquad |R_p(\theta)| < (2/\pi)\sin\lambda\pi \frac{\Gamma(n+2\lambda)}{\Gamma(\lambda)} \frac{\Gamma(p+\lambda)\Gamma(p-\lambda+1)}{p!\,\Gamma(n+p+\lambda+1)} \frac{M}{(2\sin\theta)^{p+\lambda}}.
$$

Here M has the same meaning as in Theorem 8.21.5.

(3) Finally, we mention the following formula of "Hilb's type" (cf. Szegö **17**, p. 77; Rau **2**, pp. 691–692).

THEOREM 8.21.12. *Let $\alpha > -1$, and let β be arbitrary and real. Then we have*

$$
(8.21.17)\qquad
\begin{aligned}
\left(\sin\frac{\theta}{2}\right)^\alpha \left(\cos\frac{\theta}{2}\right)^\beta P_n^{(\alpha,\beta)}(\cos\theta) &= N^{-\alpha}\frac{\Gamma(n+\alpha+1)}{n!}(\theta/\sin\theta)^{\frac{1}{2}} J_\alpha(N\theta) \\
&\quad + \begin{cases} \theta^{\frac{1}{2}}O(n^{-\frac{3}{2}}) & \text{if } cn^{-1} \leqq \theta \leqq \pi-\epsilon, \\ \theta^{\alpha+2}O(n^\alpha) & \text{if } 0<\theta\leqq cn^{-1}, \end{cases}
\end{aligned}
$$

where N has the same meaning as in (8.21.10); *c and ϵ are fixed positive numbers.*

Obviously, the remainder term is always $\theta^{\frac{1}{2}}O(n^{-\frac{3}{2}})$. If we use (4.1.3), a similar formula can be obtained in the intervals $\epsilon \leqq \theta \leqq \pi - cn^{-1}$ and $\pi - cn^{-1} \leqq \theta < \pi$ provided $\beta > -1$. In view of (1.71.7) this leads to the following important result:

THEOREM 8.21.13. *Let $\alpha > -1$, $\beta > -1$. We have, with the same notation as in* (8.21.10),

$$P_n^{(\alpha,\beta)}(\cos\theta) = n^{-\frac{1}{2}}k(\theta)\{\cos(N\theta+\gamma) + (n\sin\theta)^{-1}O(1)\}, \qquad cn^{-1} \leqq \theta \leqq \pi - cn^{-1}. \tag{8.21.18}$$

Here c is a fixed positive number.

This formula (Szegö **17,** p. 77) is more precise than (8.21.10); for $\alpha = \beta = \lambda - \frac{1}{2}$, $0 < \lambda < 1$, it follows from Theorem 8.21.11, $p = 1$. The restriction $\alpha > -1$, $\beta > -1$ is not essential (cf. Szegö, loc. cit.). See also Obrechkoff **2.**

(4) Analogous formulas hold for Jacobi's functions of the second kind $Q_n^{(\alpha,\beta)}(x)$ and, particularly, for Legendre's functions of the second kind $Q_n(x)$, if x is in the cut plane, as well as for $\mathbf{Q}_n(\cos\theta)$ if $0 < \theta < \pi$. Here we point out only the analogue of Laplace's formula (8.21.2):

THEOREM 8.21.14. *For $0 < \theta < \pi$*

$$\mathbf{Q}_n(\cos\theta) = \pi^{\frac{1}{2}}(2n\sin\theta)^{-\frac{1}{2}}\cos\{(n+\tfrac{1}{2})\theta + \pi/4\} + O(n^{-\frac{3}{2}}). \tag{8.21.19}$$

This holds uniformly in the interval $[\epsilon, \pi - \epsilon]$.

8.22. Asymptotic formulas for Laguerre and Hermite polynomials

Similar, but slightly more complicated, formulas hold for Laguerre and Hermite polynomials. In what follows $n \to \infty$; we denote by ϵ and ω fixed positive numbers, $\epsilon < \omega$, by p a positive integer.

(1) THEOREM 8.22.1 (Fejér's formula; Fejér **3**). *Let α be an arbitrary real number; we have*

$$L_n^{(\alpha)}(x) = \pi^{-\frac{1}{2}}e^{x/2}x^{-\alpha/2-\frac{1}{4}}n^{\alpha/2-\frac{1}{4}}\cos\{2(nx)^{\frac{1}{2}} - \alpha\pi/2 - \pi/4\} + O(n^{\alpha/2-\frac{3}{4}}), \qquad x > 0. \tag{8.22.1}$$

The bound for the remainder holds uniformly in $[\epsilon, \omega]$.

THEOREM 8.22.2 (Perron's generalization of Fejér's formula; Perron **2**, p. 78, (49)). *Let α be an arbitrary real number; we have for $x > 0$*

$$\begin{aligned} L_n^{(\alpha)}(x) = {} & \pi^{-\frac{1}{2}}e^{x/2}x^{-\alpha/2-\frac{1}{4}}n^{\alpha/2-\frac{1}{4}}\cos\{2(nx)^{\frac{1}{2}} - \alpha\pi/2 - \pi/4\} \\ & \cdot\left\{\sum_{\nu=0}^{p-1} A_\nu(x)n^{-\nu/2} + O(n^{-p/2})\right\} \\ & + \pi^{-\frac{1}{2}}e^{x/2}x^{-\alpha/2-\frac{1}{4}}n^{\alpha/2-\frac{1}{4}}\sin\{2(nx)^{\frac{1}{2}} - \alpha\pi/2 - \pi/4\} \\ & \cdot\left\{\sum_{\nu=0}^{p-1} B_\nu(x)n^{-\nu/2} + O(n^{-p/2})\right\}, \end{aligned} \tag{8.22.2}$$

where $A_\nu(x)$ and $B_\nu(x)$ are certain functions of x independent of n and regular for $x > 0$. The bound for the remainder holds uniformly in $[\epsilon, \omega]$.

We notice that $A_0(x) = 1$ and $B_0(x) = 0$.

THEOREM 8.22.3 (Perron's formula in the complex domain; loc. cit.). *Let α be an arbitrary real number. Then*

$$(8.22.3)\quad \begin{aligned} L_n^{(\alpha)}(x) = \tfrac{1}{2}\pi^{-\frac{1}{2}}e^{x/2}(-x)^{-\alpha/2-\frac{1}{4}}n^{\alpha/2-\frac{1}{4}}\exp\{2(-nx)^{\frac{1}{2}}\} \\ \cdot\left\{\sum_{\nu=0}^{p-1} C_\nu(x)n^{-\nu/2} + O(n^{-p/2})\right\}, \end{aligned}$$

where $C_\nu(x)$ is again independent of n; it is regular in the complex plane cut along the positive part of the real axis. Formula (8.22.3) holds if x is in the cut plane mentioned; $(-x)^{-\alpha/2-\frac{1}{4}}$ and $(-x)^{\frac{1}{2}}$ must be taken real and positive if $x < 0$. The bound for the remainder holds uniformly in every closed domain with no points in common with $x \geqq 0$.

Here we have $C_0(x) = 1$.

THEOREM 8.22.4 (Asymptotic formula of Hilb's type). *For $\alpha > -1$ we have*

$$(8.22.4)\quad \begin{aligned} e^{-x/2}x^{\alpha/2}L_n^{(\alpha)}(x) = N^{-\alpha/2}\frac{\Gamma(n+\alpha+1)}{n!}J_\alpha\{2(Nx)^{\frac{1}{2}}\} + O(n^{\alpha/2-\frac{3}{4}}), \\ N = n + (\alpha+1)/2,\ x > 0, \end{aligned}$$

the bound holding uniformly in $0 < x \leqq \omega$. More precisely, the following bounds are valid:

$$(8.22.5)\quad \begin{aligned} &x^{5/4}O(n^{\alpha/2-\frac{3}{4}}) && \text{if } cn^{-1} \leqq x \leqq \omega, \\ &x^{\alpha/2+2}O(n^{\alpha}) && \text{if } 0 < x \leqq cn^{-1}. \end{aligned}$$

In case $\alpha = 0$ the last bound is to be replaced by $x^2 \log(x^{-1}n^{-1})$; in (8.22.5) c is a fixed positive number.

Evidently, the remainder term (8.22.5) is equal to $x^{5/4}O(n^{\alpha/2-\frac{3}{4}})$ throughout $0 < x \leqq \omega$.

As a consequence of (8.22.4) we obtain the following analogue of (8.21.18), which is more precise than (8.22.1).

THEOREM 8.22.5. *Let $\alpha > -1$ and $cn^{-1} \leqq x \leqq \omega$; then*

$$(8.22.6)\quad \begin{aligned} &L_n^{(\alpha)}(x) \\ &= \pi^{-\frac{1}{2}}e^{x/2}x^{-\alpha/2-\frac{1}{4}}n^{\alpha/2-\frac{1}{4}}\{\cos[2(nx)^{\frac{1}{2}} - \alpha\pi/2 - \pi/4] + (nx)^{-\frac{1}{2}}O(1)\}. \end{aligned}$$

Here c and ω are fixed positive constants.

We observe that $N^{\frac{1}{2}} - n^{\frac{1}{2}} = O(n^{-\frac{1}{2}})$ where N has the same meaning as in (8.22.4).

(2) Substituting $\alpha = \pm\frac{1}{2}$ in (8.22.4), we find, by using (5.6.1) and (1.71.2), a formula of Hilb's type for Hermite polynomials. This is contained in the more general theorem:

THEOREM 8.22.6 (Asymptotic expansion for Hermite polynomials). *For a real x*

$$\lambda_n^{-1} e^{-x^2/2} H_n(x) = \cos (N^{\frac{1}{2}} x - n\pi/2) \sum_{\nu=0}^{p-1} u_\nu(x) N^{-\nu}$$

(8.22.7)

$$+ N^{-\frac{1}{2}} \sin (N^{\frac{1}{2}} x - n\pi/2) \sum_{\nu=0}^{p-1} v_\nu(x) N^{-\nu} + O(n^{-p}),$$

$$N = 2n + 1,$$

where

$$\lambda_n = \frac{\Gamma(n+1)}{\Gamma(n/2+1)} \quad or \quad \frac{\Gamma(n+2)}{\Gamma(n/2+3/2)} N^{-\frac{1}{2}},$$

according as n is even or odd. The coefficients $u_\nu(x)$ and $v_\nu(x)$ are polynomials depending on ν; they contain only even and odd powers of x, respectively. The bound for the error term holds uniformly in every finite real interval, whether it contains the origin or not.

For arbitrary n, we have $\lambda_n = (\Gamma(n+1)/\Gamma(n/2+1))\{1 + O(n^{-1})\}$, and $u_0(x) = 1$, $v_0(x) = x^3/6$, so that

$$\frac{\Gamma(n/2+1)}{\Gamma(n+1)} e^{-x^2/2} H_n(x) = \cos (N^{\frac{1}{2}} x - n\pi/2)$$

(8.22.8)

$$+ \frac{x^3}{6} N^{-\frac{1}{2}} \sin (N^{\frac{1}{2}} x - n\pi/2) + O(n^{-1}).$$

THEOREM 8.22.7 (Asymptotic expansion for Hermite polynomials in the complex domain). *The expansion (8.22.7) holds in the complex x-plane if we replace the remainder term by* exp $\{N^{\frac{1}{2}} | \mathfrak{J}(x) |\} O(n^{-p})$. *This is true uniformly for $|x| \leqq R$ where R is an arbitrary fixed positive number.*

(3) Finally we deal with another type of asymptotic formulas requiring a more elaborate consideration.

THEOREM 8.22.8 (Formulas of Plancherel-Rotach type for Laguerre polynomials). *Let α be arbitrary and real, ϵ and ω fixed positive numbers. We have*

(a) *for* $x = (4n + 2\alpha + 2) \cos^2 \phi$, $\epsilon \leqq \phi \leqq \pi/2 - \epsilon n^{-\frac{1}{2}}$,

(8.22.9)

$$e^{-x/2} L_n^{(\alpha)}(x) = (-1)^n (\pi \sin \phi)^{-\frac{1}{2}} x^{-\alpha/2 - \frac{1}{4}} n^{\alpha/2 - \frac{1}{4}}$$
$$\{\sin [(n + (\alpha+1)/2)(\sin 2\phi - 2\phi) + 3\pi/4] + (nx)^{-\frac{1}{2}} O(1)\};$$

(b) *for* $x = (4n + 2\alpha + 2) \cosh^2 \phi$, $\epsilon \leqq \phi \leqq \omega$,

(8.22.10)

$$e^{-x/2} L_n^{(\alpha)}(x) = \tfrac{1}{2}(-1)^n (\pi \sinh \phi)^{-\frac{1}{2}} x^{-\alpha/2 - \frac{1}{4}} n^{\alpha/2 - \frac{1}{4}}$$
$$\cdot \exp \{(n + (\alpha+1)/2)(2\phi - \sinh 2\phi)\}\{1 + O(n^{-1})\};$$

(c) *for* $x = 4n + 2\alpha + 2 - 2(2n/3)^{\frac{1}{3}} t$, t *complex and bounded,*

(8.22.11) $$e^{-x/2}L_n^{(\alpha)}(x) = (-1)^n\pi^{-1}2^{-\alpha-\frac{1}{3}}3^{\frac{1}{3}}n^{-\frac{1}{3}}\{A(t) + O(n^{-\frac{2}{3}})\},$$

where $A(t)$ is Airy's function defined in §1.81.

In all of these formulas the O-terms hold uniformly.

The corresponding formulas for Hermite polynomials (Plancherel-Rotach **1**) are given by the following:

THEOREM 8.22.9. *Let ϵ and ω be fixed positive numbers. We have*

(a) *for* $x = (2n+1)^{\frac{1}{2}}\cos\phi,\ \epsilon \leqq \phi \leqq \pi - \epsilon,$

(8.22.12) $$e^{-x^2/2}H_n(x) = 2^{n/2+\frac{1}{4}}(n!)^{\frac{1}{2}}(\pi n)^{-\frac{1}{4}}(\sin\phi)^{-\frac{1}{2}} \cdot \{\sin[(n/2+\tfrac{1}{4})(\sin 2\phi - 2\phi) + 3\pi/4] + O(n^{-1})\};$$

(b) *for* $x = (2n+1)^{\frac{1}{2}}\cosh\phi,\ \epsilon \leqq \phi \leqq \omega,$

(8.22.13) $$e^{-x^2/2}H_n(x) = 2^{n/2-\frac{3}{4}}(n!)^{\frac{1}{2}}(\pi n)^{-\frac{1}{4}}(\sinh\phi)^{-\frac{1}{2}} \cdot \exp[(n/2+\tfrac{1}{4})(2\phi - \sinh 2\phi)]\{1 + O(n^{-1})\};$$

(c) *for* $x = (2n+1)^{\frac{1}{2}} - 2^{-\frac{1}{2}}3^{-\frac{1}{3}}n^{-\frac{1}{6}}t$, *t complex and bounded,*

(8.22.14) $$e^{-x^2/2}H_n(x) = 3^{\frac{1}{3}}\pi^{-\frac{3}{4}}2^{n/2+\frac{1}{4}}(n!)^{\frac{1}{2}}n^{-1/12}\{A(t) + O(n^{-\frac{2}{3}})\}.$$

In all these formulas the O-terms hold uniformly.

Note that (8.22.12) holds uniformly in the vicinity of $x = 0$.

8.23. Remarks on the preceding results

(1) Of all the formulas enumerated in §8.21, formula (8.21.1) has the simplest character. It can be proved by various methods. An extension to an asymptotic series is given by (8.21.3). Corresponding formulas hold for Jacobi polynomials (§8.21 (2)). The following simple consequence of (8.21.9) is important for various purposes:

(8.23.1) $$|P_n^{(\alpha,\beta)}(x)|^{1/n} \cong |x + (x^2-1)^{\frac{1}{2}}|, \qquad n \to \infty;$$

here x is in the cut plane. The right-hand member is >1 and represents the sum of the semi-axes of the ellipse with foci at ±1 and passing through x. We compare (8.23.1) with the following formula for Jacobi's functions of the second kind:

(8.23.2) $$|Q_n^{(\alpha,\beta)}(y)|^{1/n} \cong |y - (y^2-1)^{\frac{1}{2}}|.$$

Here y is again in the cut plane and the right-hand member is <1. (Cf. (8.71.19).)

We shall give also several proofs for the classical formula (8.21.2) of Laplace. It can be similarly generalized in various directions.

Darboux's formula (8.21.4) is the most important illustration of the method due to him (**1**). This method furnishes asymptotic formulas for the **coefficients** of power series whose singularities on the circle of convergence have a certain simple character. The same method yields (8.21.3), as well as similar expansions of Jacobi polynomials. (Cf. §8.4.)

The significance of the formula of Stieltjes is due to its unrestricted validity in the interval $0 < \theta < \pi$ (although the bound for the remainder given by (8.21.6) becomes infinite if $\theta \to 0$ or $\theta \to \pi$). In view of this fact, it can be used in various cases not only in a fixed interval in the interior of $[0, \pi]$, but even in the vicinity of the end-points where the formulas of Laplace and Darboux fail in general. For $p = 0$ it holds in the sense that $P_n(\cos\theta) = R_0(\theta)$. Then (8.21.6) furnishes inequality (7.3.8) but with a larger factor on the right side (with $2(2/\pi)^{\frac{1}{2}}$ instead of $(2/\pi)^{\frac{1}{2}}$). From the formula of Stieltjes an arbitrary number of terms in (8.21.4) can readily be derived. However, it seems difficult to obtain the general law of the coefficients of (8.21.4) in this manner.

The importance of Hilb's formula also lies in its unrestricted validity in the neighborhood of $\theta = 0$, with the additional advantage that the remainder term tends to 0 uniformly in this neighborhood. It furnishes the Mehler-Heine formula (8.1.1) immediately and yields Laplace's formula (8.21.2) by means of (1.71.7). The bounds (8.21.8) are a slight improvement over Hilb's result. Our proof (§8.62) is essentially the same as that of Hilb. Szegö (**15**) gives an asymptotic expansion in terms of Bessel functions of increasing order and generalizes Hilb's result. Analogous formulas hold for Legendre's function of the second kind. Szegö also obtains (**15**, p. 450) an analogue of (8.21.1) which holds in the cut plane, arbitrarily near its boundary. This formula involves Bessel functions with imaginary arguments.

Another formula of a type similar to (8.21.7) has been given by Watson (**2**) with a numerical estimate of the remainder. It involves $J_0(z)$ and $Y_0(z)$.

Theorem 8.21.12 is the extension of Hilb's formula to Jacobi polynomials. The proof given in §8.63 follows the same line of argument as that in §8.62.

The proofs of Theorems 8.21.1–8.21.14 are based on the following methods:

(a) Explicit series or integral representations;

(b) Darboux's method;

(c) method of Liouville-Stekloff (method of the integro-differential equation);

(d) method of steepest descent.

A short survey of these methods will be given at the proper places.

(2) Fejér's proof of (8.22.1) is based on the generating function (5.1.9), which in this case has the essential singularity $w = 1$ on the circle of convergence $|w| = 1$. This argument is of a character similar to Darboux's method. The more complicated type of singularity in this case naturally requires a more careful discussion; it is carried out by Fejér by an elementary method similar to the second mean-value theorem of the integral calculus.

In Perron's first proof of (8.22.3) (in the special case $p = 1$, see **1**), complex integration is used. He obtains the complete expansions (8.22.2) and (8.22.3) by using certain general asymptotic results concerning confluent hypergeometric functions.

Further proofs of Fejér's formula (partly giving more exact bounds for the remainder and holding on certain segments the end-points of which tend to $+0$ and $+\infty$) have been given by Rotach (**1**), Szegö (**10**), and Kogbetliantz (**14**). They use either the method of steepest descent or similar arguments. Fejér's formula is contained in (8.22.4); the latter result follows from certain general asymptotic theorems of Hilb's type due to Wright (**1**, p. 261, however only for fixed x). We give a proof for (8.22.4) by using the Liouville-Stekloff method (§8.64).

The special cases $\alpha = \pm\frac{1}{2}$ are equivalent to Hermite polynomials (see (8.22.7)); in these cases, Fejér's theorem was known previously by Adamoff (**1**). Adamoff obtains the remainder term with certain numerical bounds.

We derive (8.22.7) by using the method of Liouville-Stekloff. Uspensky's formula (5.6.5) then leads immediately to a corresponding asymptotic expansion for Laguerre polynomials involving Bessel functions. We indicate this in §8.66. Concerning this expansion see Wright, loc. cit.; its first term is (8.22.4). From this expansion Perron's formulas (8.22.2) and (8.22.3) follow readily. A second proof of these formulas can be based on the method of steepest descent (§8.72).

The first term of the asymptotic expansion mentioned for Hermite polynomials furnishes (8.22.8). Adamoff's formula is less precise. (On the other hand it contains numerical constants.) Comparison of (8.22.8) with Uspensky's formula (**1**, p. 597, (6)) indicates that it is convenient to work with $N = 2n + 1$ instead of $N = 2n$.

A very detailed asymptotic investigation of Hermite polynomials is due to Watson (**1**, second paper).

We mention the following simple consequence of Theorems 8.22.3 and 8.22.7: Let x be in the complex plane cut along the non-negative real axis. Then

$$(8.23.3)\qquad n^{-\frac{1}{2}} \log | L_n^{(\alpha)}(x) | \to 2\Re\{(-x)^{\frac{1}{2}}\}, \qquad n \to \infty.$$

Here $(-x)^{\frac{1}{2}}$ is taken real and positive if $x < 0$. However, if x is non-real,

$$(8.23.4)\qquad (2n)^{-\frac{1}{2}} \log \left\{ \frac{\Gamma(n/2 + 1)}{\Gamma(n + 1)} \, | H_n(x) | \right\} \to | \Im(x) |, \qquad n \to \infty.$$

Theorems 8.22.8 and 8.22.9 are closely related to the important results of Plancherel-Rotach (**1**). These authors deal exclusively with Hermite polynomials and use the method of steepest descent; they obtain a complete asymptotic expansion in all three cases of Theorem 8.22.9. Their argument has been applied to $L_n(x)$ by Moecklin (**1**). We shall derive (§§8.73–8.75) only the principal terms of these expansions, however, for general Laguerre polynomials $L_n^{(\alpha)}(x)$, by using the method of steepest descent. Our argument is based on the generating function (5.1.16) and on the asymptotic expansion of Bessel functions in the complex domain.

The formulas (8.22.9)–(8.22.11) describe Laguerre polynomials in the "oscillating region," in the region beyond this, and in a certain vicinity of the largest zero, respectively. The same is true for (8.22.12)–(8.22.14).

Van Veen (**1, 2**) derives the asymptotic series corresponding to (8.22.12) with numerical estimates. Schwid (**1**) applies the method of Liouville-Stekloff (in the more precise form due to Langer (**1**)) to the asymptotic investigation of Hermite polynomials.

8.3. "Elementary" proof of the formulas of Laplace-Heine and Laplace

(1) We start from the representation (4.9.4) and first prove (8.21.1). Let $x = \frac{1}{2}(z + z^{-1})$, $|z| > 1$; then

$$P_n(x) = \sum_{m=0}^{n} g_m g_{n-m} z^{n-2m} = g_n z^n \sum_{m=0}^{n} \frac{g_{n-m}}{g_n} g_m z^{-2m}. \tag{8.3.1}$$

We next show that

$$\lim_{n\to\infty} \sum_{m=0}^{n} \left(\frac{g_{n-m}}{g_n} - 1\right) g_m z^{-2m} = 0, \tag{8.3.2}$$

uniformly for $|z| \geqq R$, $R > 1$. Indeed, the expression in the brackets tends to zero if m is fixed and $n \to \infty$. On the other hand, it is easy to find a dominant; we have, for instance,

$$0 \leqq \left(\frac{g_{n-m}}{g_n} - 1\right) g_m \leqq \frac{g_{n-m} g_m}{g_n}.$$

Now $(n + 1)^{\frac{1}{2}} g_n$ is bounded from zero and from infinity, and

$$\frac{(n+1)^{\frac{1}{2}}}{(n-m+1)^{\frac{1}{2}}(m+1)^{\frac{1}{2}}} \leqq 1, \qquad 0 \leqq m \leqq n.$$

(2) We can prove (8.3.2) in another way, by use of certain very elementary properties of the sequence $\{g_n\}$. Let $\delta > 0$ be arbitrary, and let M be a positive integer, such that

$$\sum_{m=M+1}^{\infty} R^{-2m} < \delta.$$

The numbers $g_{n-m}/g_n - 1$ are positive and increase with m; we therefore have, if $n > M$,

$$\sum_{m=0}^{M} \left(\frac{g_{n-m}}{g_n} - 1\right) g_m R^{-2m} \leqq \left(\frac{g_{n-M}}{g_n} - 1\right) \sum_{m=0}^{M} g_m R^{-2m} < \left(\frac{g_{n-M}}{g_n} - 1\right) \sum_{m=0}^{\infty} g_m R^{-2m}.$$

The last expression tends to 0 as $n \to \infty$ since $g_n/g_{n-1} \to 1$. On the other hand,

$$\sum_{m=M+1}^{n} \left(\frac{g_{n-m}}{g_n} - 1\right) g_m R^{-2m} < \sum_{m=M+1}^{n} \frac{g_{n-m} g_m}{g_n} R^{-2m}.$$

Now g_m/g_{m-1} is increasing, so that

$$\frac{g_{n-m}g_m}{g_{n-m+1}g_{m-1}} \leqq 1 \text{ or } \geqq 1$$

according as $m \leqq (n+1)/2$ or $m \geqq (n+1)/2$. Consequently, $g_{n-m}g_m$ attains its maximum, as m varies from 0 to n, for $m = 0$ or $m = n$; hence $g_{n-m}g_m \leqq g_n$, so that

$$\sum_{m=M+1}^{n} \left(\frac{g_{n-m}}{g_n} - 1\right) g_m R^{-2m} \leqq \sum_{m=M+1}^{n} R^{-2m} < \sum_{m=M+1}^{\infty} R^{-2m} < \delta.$$

This establishes (8.3.2).

Therefore the expression

$$(8.3.3) \qquad (g_n z^n)^{-1} P_n(x) - \sum_{m=0}^{n} g_m z^{-2m}$$

tends to 0 as $n \to \infty$, uniformly for $|z| \geqq R$, $R > 1$. The last sum tends to $(1 - z^{-2})^{-\frac{1}{2}}$; whence (8.21.1) is readily derived.

(3) For the proof of (8.21.2) we use (4.9.3). We have for $0 < \theta < \pi$

$$(8.3.4) \qquad \begin{aligned} P_n(\cos\theta) &= 2g_n \Re\left\{e^{-in\theta} \sum_{m=0}^{[n/2]}{}' \frac{g_{n-m}}{g_n} g_m e^{2im\theta}\right\} \\ &= 2g_n \Re\left\{e^{-in\theta} \sum_{m=0}^{[n/2]}{}' \left(\frac{g_{n-m}}{g_n} - 1\right) g_m e^{2im\theta}\right\} + 2g_n \Re\left\{e^{-in\theta} \sum_{m=0}^{[n/2]}{}' g_m e^{2im\theta}\right\}, \end{aligned}$$

where $\sum'$ has the same meaning as in (7.4.5). The sequence $\{g_m\}$ tends monotonically to 0, so that for $0 < \theta < \pi$

$$\sum_{m=0}^{\infty} g_m e^{2im\theta} = (1 - e^{2i\theta})^{-\frac{1}{2}} = e^{i(\pi/4 - \theta/2)} (2 \sin\theta)^{-\frac{1}{2}}.$$

This series converges uniformly for $\epsilon \leqq \theta \leqq \pi - \epsilon$.

Now, if δ is an arbitrary positive number, we determine the positive integer M so that

$$(8.3.5) \qquad \left|\sum_{m=M'}^{M''} g_m e^{2im\theta}\right| < \delta, \qquad M'' > M' > M.$$

The numbers $g_{n-m}/g_n - 1$ being positive and increasing with m if $n > 2M$, we have, according to Abel's inequality,

$$(8.3.6) \qquad \begin{aligned} &\left|\sum_{m=0}^{M} \left(\frac{g_{n-m}}{g_n} - 1\right) g_m e^{2im\theta}\right| \\ &\leqq \left(\frac{g_{n-M}}{g_n} - 1\right) \max_{0 \leqq \mu \leqq M} \left|\sum_{m=\mu}^{M} g_m e^{2im\theta}\right| < K\left(\frac{g_{n-M}}{g_n} - 1\right), \end{aligned}$$

where K is a fixed constant. On the other hand,

(8.3.7)
$$\left|\sum_{m=M+1}^{[n/2]}{}' \left(\frac{g_{n-m}}{g_n} - 1\right) g_m e^{2im\theta}\right| \leqq \left(\frac{g_{n-[n/2]}}{g_n} - 1\right) \max_{M<\mu\leqq[n/2]} \left|\sum_{m=\mu}^{[n/2]}{}' g_m e^{2im\theta}\right| < \delta \frac{g_{n-[n/2]}}{g_n}.$$

Since g_n/g_{2n} is bounded, we see that the first $\sum'$ in the right-hand member of (8.3.4) tends to zero. Hence we have

(8.3.8)
$$(2g_n)^{-1}P_n(\cos\theta) = \Re\{e^{-in\theta}\cdot e^{i(\pi/4-\theta/2)}(2\sin\theta)^{-\frac{1}{2}}\} + \delta_n,$$

where $\delta_n \to 0$ uniformly in $\epsilon \leqq \theta \leqq \pi - \epsilon$. This is the formula of Laplace with a remainder term $o(n^{-\frac{1}{2}})$.

(4) Although they do not lead to the remainder term $O(n^{-\frac{3}{2}})$ stated in (8.21.2), these elementary arguments are important since they use only some very simple properties of the sequence $\{g_n\}$. At the same time they yield certain asymptotic formulas for Fejér's polynomials $F_n(x)$, introduced in §6.5, valid in the cut plane and in $-1 < x < +1$, respectively, provided certain conditions regarding the sequence $\{\alpha_n\}$ are satisfied. We have the following:

THEOREM 8.3. *Let $\{\alpha_m\}$ be a positive sequence, $\alpha_m \to 0$, and let $\alpha_m/\alpha_{m-1} \uparrow 1$. Then the following asymptotic formula holds:*

(8.3.9)
$$F_n(x) \cong \alpha_n z^n \sum_{m=0}^{\infty} \alpha_m z^{-2m}, \qquad n \to \infty,$$

where x is in the cut plane, $x = \frac{1}{2}(z + z^{-1})$, $|z| > 1$. If in addition α_m/α_{2m} remains bounded, we have

(8.3.10)
$$F_n(\cos\theta) = 2\alpha_n\Re\left\{e^{in\theta}\sum_{m=0}^{\infty}\alpha_m e^{-2im\theta}\right\} + o(\alpha_n), \qquad n \to \infty,$$

where $0 < \theta < \pi$. Both formulas are valid uniformly in the same sense as (8.21.1) *and* (8.21.2), *respectively.*

The generalization (8.3.9) of the Laplace-Heine formula is new; concerning the generalization (8.3.10) of Laplace's formula, see Szegö **11**, pp. 186-187.

We observe that the series in (8.3.9) is convergent, and the series in (8.3.10) is uniformly convergent in $\epsilon \leqq \theta \leqq \pi - \epsilon$. With regard to the latter fact we note that $\alpha_m/\alpha_{m-1} \leqq 1$, so that α_m is decreasing. As an application of Theorem 8.3, we obtain the analogue of (8.21.1) and (8.21.2) (with a less precise estimate of the remainder in the second case) for the ultraspherical polynomials $P_n^{(\lambda)}(x)$ provided $\lambda > 0$.

8.4. Darboux's formula proved by Darboux's method

(1) We shall prove formula (8.21.4), as well as others, by means of an important method due to Darboux (**1**), and we begin with an illustration of this method. Supposing $0 < \theta < \pi$, let us consider the generating function of

Legendre polynomials (cf. (4.7.23))

$$h(w) = (1 - we^{-i\theta})^{-\frac{1}{2}}(1 - we^{i\theta})^{-\frac{1}{2}} \tag{8.4.1}$$

in the neighborhood of $e^{i\theta}$:

$$\begin{aligned} h(w) &= (1 - we^{-i\theta})^{-\frac{1}{2}}(1 - e^{2i\theta})^{-\frac{1}{2}}\left\{1 - \frac{e^{2i\theta}}{e^{2i\theta} - 1}(1 - we^{-i\theta})\right\}^{-\frac{1}{2}} \\ &= (1 - e^{2i\theta})^{-\frac{1}{2}} \sum_{\nu=0}^{\infty} g_\nu \left(\frac{e^{2i\theta}}{e^{2i\theta} - 1}\right)^\nu (1 - we^{-i\theta})^{\nu - \frac{1}{2}}. \end{aligned} \tag{8.4.2}$$

A similar representation holds in the vicinity of $e^{-i\theta}$. Denoting the Lth partial sums of these expansions (stopping at the terms $\nu = L$) by $s_L^{(1)}(w)$ and $s_L^{(2)}(w)$, respectively, let us consider the difference

$$H(w) = h(w) - s_L^{(1)}(w) - s_L^{(2)}(w). \tag{8.4.3}$$

We see immediately that the Lth derivative $H^{(L)}(w)$ possesses continuous boundary values in $|w| \leqq 1$. Thus if we expand $H(w)$ in a power series about $w = 0$, the coefficients of $H^{(L)}(w)$ tend to 0. This simple remark shows that the coefficients d_n of $H(w)$ satisfy the condition

$$\lim_{n\to\infty} n^L d_n = 0. \tag{8.4.4}$$

Each of the terms of the finite sums $s_L^{(1)}(w)$ and $s_L^{(2)}(w)$ has only one singularity on the unit circle. The νth term of $s_L^{(1)}(w)$ contributes to the coefficient of w^n in $h(w)$ (and therefore to $P_n(\cos\theta)$) an expression of the form

$$(1 - e^{2i\theta})^{-\frac{1}{2}} g_\nu \left(\frac{e^{2i\theta}}{e^{2i\theta} - 1}\right)^\nu \binom{\nu - \frac{1}{2}}{n} (-e^{-i\theta})^n. \tag{8.4.5}$$

The sum $s_L^{(2)}(w)$ contributes the conjugate of (8.4.5). Both terms are $O(n^{-\nu-\frac{1}{2}})$. For a fixed value of p the coefficient of w^n in $H(w)$ is of higher order than $n^{-p-\frac{1}{2}}$, provided L is sufficiently large.

By use of the same argument, the following general theorem can be obtained:

THEOREM 8.4. *Let $h(w)$ be regular for $|w| < 1$, and let it have a finite number of singularities*

$$e^{i\phi_1}, e^{i\phi_2}, \cdots, e^{i\phi_l}, \qquad e^{i\phi_\alpha} \neq e^{i\phi_\beta}, \qquad \alpha \neq \beta, \tag{8.4.6}$$

on the unit circle $|w| = 1$. Let

$$h(w) = \sum_{\nu=0}^{\infty} c_\nu^{(k)} (1 - we^{-i\phi_k})^{a_k + \nu b_k}, \qquad k = 1, 2, \cdots, l, \tag{8.4.7}$$

in the vicinity of $e^{i\phi_k}$, where $b_k > 0$. Then the expression

$$\sum_{\nu=0}^{\infty} \sum_{k=1}^{l} c_\nu^{(k)} \binom{a_k + \nu b_k}{n} (-e^{i\phi_k})^n \tag{8.4.8}$$

furnishes an asymptotic expansion for the coefficient of w^n *in* $h(w)$ *in the following sense: if* Q *is an arbitrary positive number, and if a sufficiently large number of terms is taken in the sum* $\sum_{\nu=0}^{\infty}$ *in* (8.4.8), *we obtain an expression which approximates the coefficient in question with an error equal to* $O(n^{-Q})$.

A simple discussion shows that it suffices to stop at the term $\nu = p - 1$ of the sum, where p is a positive integer such that

$$p \geqq \max_{1 \leqq k \leqq l} b_k^{-1}\{Q - \Re(a_k) - 1\}. \tag{8.4.9}$$

Proper logarithmic singularities of $h(w)$ can also be admitted.

(2) In the case of $P_n(\cos\theta)$, this asymptotic expansion becomes

$$2\Re\left\{\sum_{\nu=0}^{\infty} g_\nu(1 - e^{2i\theta})^{-\frac{1}{2}}\left(\frac{e^{2i\theta}}{e^{2i\theta} - 1}\right)^{\nu}\binom{\nu - \frac{1}{2}}{n}(-e^{-i\theta})^n\right\}. \tag{8.4.10}$$

The general term is $O(n^{-\nu-\frac{1}{2}})$; thus, if we stop at $\nu = p - 1$, the error is $O(n^{-p-\frac{1}{2}})$. This agrees with formula (8.21.4). It is also clear that the bound for the remainder holds uniformly in $\epsilon \leqq \theta \leqq \pi - \epsilon$.

The same method can be used for the proof of the expansion (8.21.3), which corresponds to the Laplace-Heine formula (8.21.1). (Actually, this case is simpler than the preceding one.) Indeed, let $|z| > 1$; then

$$\begin{aligned} h(w) &= (1 - zw)^{-\frac{1}{2}}(1 - z^{-1}w)^{-\frac{1}{2}} \\ &= (1 - z^{-2})^{-\frac{1}{2}} \sum_{\nu=0}^{\infty} g_\nu \left(\frac{z^{-2}}{z^{-2} - 1}\right)^{\nu}(1 - zw)^{\nu-\frac{1}{2}}; \end{aligned} \tag{8.4.11}$$

whence (8.21.3) readily follows.

(3) The infinite series which corresponds to Darboux's formula (8.21.4) is convergent in the ordinary sense and represents $P_n(\cos\theta)$ provided $2\sin\theta > 1$, that is, $\pi/6 < \theta < 5\pi/6$. In fact, the representation (8.4.2) holds uniformly near $w = 0$ if

$$|(1 - we^{-i\theta})(e^{2i\theta} - 1)^{-1}| < 1. \tag{8.4.12}$$

See **8.92** (4).

(4) Darboux's method also applies to the general Jacobi, and in particular to the ultraspherical, polynomials and leads to the Theorems 8.21.9 and 8.21.10.

Another method of deriving the expansions of Theorem 8.21.9 will be indicated in §8.71 (4) and (5).

(5) Finally we observe that the expansion (8.21.14) stops at $\nu = \lambda - 1$ if λ is a positive integer. Then we obtain the exact representation

$$\begin{aligned} P_n^{(\lambda)}(\cos\theta) = 2\alpha_n \sum_{\nu=0}^{\lambda-1} \alpha_\nu & \frac{(1-\lambda)(2-\lambda)\cdots(\nu-\lambda)}{(n+\lambda-1)(n+\lambda-2)\cdots(n+\lambda-\nu)} \\ & \cdot \frac{\cos\{(n-\nu+\lambda)\theta - (\nu+\lambda)\pi/2\}}{(2\sin\theta)^{\nu+\lambda}}, \\ n = 0, 1, 2, \cdots; \quad \lambda = 1, 2, 3, \cdots; & \quad \alpha_n = \binom{n+\lambda-1}{n}. \end{aligned} \tag{8.4.13}$$

In fact, the difference (8.4.3) (with $L = \lambda - 1$) is in this case a rational function which has no singularities for any w (including $w = \infty$) and which vanishes for $w \to \infty$. Hence it must be identically zero.

The analogous representation of $P_n^{(\lambda)}(x)$ for $|z| > 1$, λ a positive integer, is slightly more complicated; it can be readily derived from (8.4.13).

8.5. Proof of the formula of Stieltjes

(1) Stieltjes' formula (4.9.17), (4.9.18) has been derived in §4.9 (3) from the integral representation (4.8.17). This argument furnishes for the remainder $R_p(\theta)$ of (8.21.5) the representation

$$R_p(\theta) = \frac{2}{\pi}\Im\left\{e^{i(n+1)\theta}e^{i(\pi/4-\theta/2)}(2\sin\theta)^{-\frac{1}{2}}\int_0^1 t^n(1-t)^{-\frac{1}{2}}\rho_p(t)\,dt\right\}. \tag{8.5.1}$$

Here (g_ν has the same meaning as in Theorem 8.21.3)

$$\rho_p(t) = (1-z)^{-\frac{1}{2}} - \sum_{\nu=0}^{p-1} g_\nu z^\nu; \qquad z = (1-t)\frac{e^{i(\theta-\pi/2)}}{2\sin\theta}. \tag{8.5.2}$$

According to Stieltjes we have

$$\begin{aligned} g_\nu &= \pi^{-1}\int_0^\pi \sin^{2\nu}\phi\,d\phi, \\ (1-z)^{-\frac{1}{2}} - \sum_{\nu=0}^{p-1} g_\nu z^\nu &= \pi^{-1}\int_0^\pi \frac{z^p\sin^{2p}\phi}{1-z\sin^2\phi}\,d\phi. \end{aligned} \tag{8.5.3}$$

The last formula is obvious first under the assumption that $|z| < 1$; it then can be extended to the whole strip $0 \leqq \Re(z) \leqq \frac{1}{2}$ without restriction. Now, writing $(1-t)\sin^2\phi = r$, we find

$$|1 - z\sin^2\phi|^2 = |1 - r/2 + (ir/2)\cot\theta|^2 = [\sin\theta - r/(2\sin\theta)]^2 + \cos^2\theta.$$

The minimum of this expression is $\cos^2\theta$ or $(2\sin\theta)^{-2}$, according as $2\sin^2\theta \leqq 1$ or $2\sin^2\theta \geqq 1$, so that

$$|\rho_p(t)| \leqq g_p(1-t)^p(2\sin\theta)^{-p}M, \tag{8.5.4}$$

where M has the same meaning as in (8.21.6). Thus

$$\begin{aligned} |R_p(\theta)| &\leqq \frac{2}{\pi}(2\sin\theta)^{-\frac{1}{2}}\int_0^1 t^n(1-t)^{-\frac{1}{2}}(1-t)^p g_p(2\sin\theta)^{-p}M\,dt \\ &= \frac{2}{\pi}g_p\frac{\Gamma(n+1)\Gamma(p+\frac{1}{2})}{\Gamma(n+p+\frac{3}{2})}\frac{M}{(2\sin\theta)^{p+\frac{1}{2}}}, \end{aligned} \tag{8.5.5}$$

which is equivalent to (8.21.6).

The analogous formula of Theorem 8.21.11 for the ultraspherical polynomials $P_n^{(\lambda)}(\cos\theta)$, $0 < \lambda < 1$, results from (4.82.3) (cf. Szegö **17**, pp. 57–60). It is the expansion (4.9.25) completed with an estimate of the error if we stop at the term $\nu = p - 1$. The error is again less than twice the first neglected term in which cos is replaced by 1. The proof is the same as before; we use for the quantities α_ν of (4.9.21) the representation (cf. (6.5.9))

$$\alpha_\nu = \pi^{-1} \sin \lambda\pi \int_0^\pi |\tan \phi|^{2\lambda-1} \sin^{2\nu} \phi \, d\phi. \tag{8.5.6}$$

The same remark as in §8.4 (3) applies to the infinite series corresponding to Stieltjes' formula and to its generalization. (Cf. (4.9.17), (4.9.25).)

8.61. Method of Liouville-Stekloff; formula of Laplace

We shall now derive the formula of Laplace from the differential equation (7.3.5). The essential idea is the transformation of this equation into an integral equation of Volterra's type which permits a successive improvement of the asymptotic formula in question. The idea is very old, and appears in the investigations of Liouville on differential equations of the Sturm-Liouville type. Stekloff (**1**) applied this method to the asymptotic discussion of certain classical polynomials.

Recently, Langer (**1, 2, 3**) employed this method systematically and improved its efficiency considerably. He generally considers "singular" cases like (4.24.2), or any of the equations (5.1.2), in the neighborhood of $\theta = 0$ and $x = 0$, respectively, and obtains general asymptotic formulas of "Hilb's type." He takes up also applications of this method in the complex domain.

(1) We write the differential equation (7.3.5) in the form

$$\left(\frac{d}{d\theta}\right)^2 \{(\sin \theta)^{\frac{1}{2}} P_n(\cos \theta)\} + (n + \tfrac{1}{2})^2 (\sin \theta)^{\frac{1}{2}} P_n(\cos \theta) = -\frac{(\sin \theta)^{\frac{1}{2}} P_n(\cos \theta)}{4 \sin^2 \theta}. \tag{8.61.1}$$

Interpreting this relation as a non-homogeneous equation for $(\sin \theta)^{\frac{1}{2}} P_n(\cos \theta)$, we can apply (1.8.12); the corresponding homogeneous equation has the fundamental system $\{\cos (n + \frac{1}{2})\theta, \sin (n + \frac{1}{2})\theta\}$, so that with certain constants θ_0, c_1, c_2,

$$\begin{aligned}(\sin \theta)^{\frac{1}{2}} P_n(\cos \theta) &= c_1 \cos (n + \tfrac{1}{2})\theta + c_2 \sin (n + \tfrac{1}{2})\theta \\ &\quad - \frac{1}{n + \frac{1}{2}} \int_{\theta_0}^{\theta} \frac{\sin \{(n + \frac{1}{2})(\theta - t)\}}{4 \sin^2 t} (\sin t)^{\frac{1}{2}} P_n(\cos t) \, dt.\end{aligned} \tag{8.61.2}$$

If we assume $\theta_0 = \pi/2$, $0 < \theta < \pi$, the last integral and its derivative vanish for $\theta = \pi/2$. This remark enables us to determine c_1 and c_2. We find

$$\begin{aligned}(\sin \theta)^{\frac{1}{2}} P_n(\cos \theta) &= \lambda_n \cos \{(n + \tfrac{1}{2})\theta - \pi/4\} \\ &\quad - \frac{1}{n + \frac{1}{2}} \int_{\pi/2}^{\theta} \frac{\sin \{(n + \frac{1}{2})(\theta - t)\}}{4 \sin^2 t} (\sin t)^{\frac{1}{2}} P_n(\cos t) \, dt,\end{aligned} \tag{8.61.3}$$

where

$$\lambda_n = \begin{cases} g_{n/2} & \text{if } n \text{ is even,} \\ \dfrac{n+1}{n+\frac{1}{2}} g_{(n+1)/2} & \text{if } n \text{ is odd.} \end{cases} \tag{8.61.4}$$

This is the Volterra equation mentioned above.

If θ is confined to the interval $[\epsilon, \pi - \epsilon]$ and M_n denotes the maximum of the absolute value of the left-hand member of (8.61.3), we have

$$M_n \leqq \lambda_n + \frac{\pi}{2n+1} \frac{M_n}{4 \sin^2 \epsilon}. \tag{8.61.5}$$

Therefore, if n is sufficiently large,

$$M_n < 2\lambda_n = O(n^{-\frac{1}{2}}),$$

or

$$(\sin \theta)^{\frac{1}{2}} P_n(\cos \theta) = \lambda_n \cos \{(n + \tfrac{1}{2})\theta - \pi/4\} + O(n^{-\frac{3}{2}}), \tag{8.61.6}$$

which readily furnishes Laplace's formula.

(2) Successive application of (8.61.3) leads to an expansion of Darboux type, namely, to the formula

$$\begin{aligned} (\sin \theta)^{\frac{1}{2}} P_n(\cos \theta) &= \cos (n + \tfrac{1}{2})\theta \left\{ \sum_{\nu=0}^{p-1} A_\nu(\theta) n^{-\nu-\frac{1}{2}} \right\} \\ &+ \sin (n + \tfrac{1}{2})\theta \left\{ \sum_{\nu=0}^{p-1} B_\nu(\theta) n^{-\nu-\frac{1}{2}} \right\} + O(n^{-p-\frac{1}{2}}), \qquad \epsilon \leqq \theta \leqq \pi - \epsilon, \end{aligned} \tag{8.61.7}$$

where $A_\nu(\theta)$ and $B_\nu(\theta)$ are certain functions analytic in $0 < \theta < \pi$ and independent of n and p. The explicit determination of these functions, that is, the identification of (8.61.7) with the formulas of Darboux or Stieltjes, seems, however, to be rather difficult.

For the proof we use mathematical induction. Assuming (8.61.7), we obtain from (8.61.3),

$$\begin{aligned} &(\sin \theta)^{\frac{1}{2}} P_n(\cos \theta) \\ &= \lambda_n \cos \{(n + \tfrac{1}{2})\theta - \pi/4\} - \frac{1}{n + \frac{1}{2}} \int_{\pi/2}^{\theta} \frac{\sin \{(n + \frac{1}{2})(\theta - t)\}}{4 \sin^2 t} \\ &\quad \cdot \left\{ \cos (n + \tfrac{1}{2})t \sum_{\nu=0}^{p-1} A_\nu(t) n^{-\nu-\frac{1}{2}} \right. \\ &\quad \left. + \sin (n + \tfrac{1}{2})t \sum_{\nu=0}^{p-1} B_\nu(t) n^{-\nu-\frac{1}{2}} \right\} dt + O(n^{-p-\frac{3}{2}}). \end{aligned}$$

Now $2 \sin \{(n + \frac{1}{2})(\theta - t)\} \cos (n + \frac{1}{2})t = \sin (n + \frac{1}{2})\theta + \sin \{(n + \frac{1}{2})(\theta - 2t)\}$, and integration by parts yields

$$\int_{\pi/2}^{\theta} \sin \{(n + \tfrac{1}{2})(\theta - 2t)\} \frac{A_\nu(t)}{4 \sin^2 t}\, dt$$
$$= \cos (n + \tfrac{1}{2})\, \theta \left\{ \sum_{\mu=0}^{p-1} a_\mu(\theta) n^{-\mu} \right\} + \sin (n + \tfrac{1}{2})\, \theta \left\{ \sum_{\mu=0}^{p-1} b_\mu(\theta) n^{-\mu} \right\} + O(n^{-p}),$$

where $a_\mu(\theta)$ and $b_\mu(\theta)$ are certain functions of the same type as $A_\nu(\theta)$ and $B_\nu(\theta)$. The integrals involving $B_\nu(t)$ can be dealt with in the same way. This leads by use of Stirling's series (cf. Pólya-Szegö **1**, vol. 1, pp. 29, 193, problem 155), to a representation of the form

$$(\sin \theta)^{\frac{1}{2}} P_n(\cos \theta) = \cos (n + \tfrac{1}{2})\, \theta \left\{ \sum_{\nu=0}^{p} A_\nu^{(1)}(\theta) n^{-\nu-\frac{1}{2}} \right\}$$
$$+ \sin (n + \tfrac{1}{2})\, \theta \left\{ \sum_{\nu=0}^{p} B_\nu^{(1)}(\theta) n^{-\nu-\frac{1}{2}} \right\} + O(n^{-p-\frac{3}{2}}).$$

Comparing this with (8.61.7), we see that

$$A_\nu^{(1)}(\theta) = A_\nu(\theta), \qquad B_\nu^{(1)}(\theta) = B_\nu(\theta), \qquad \nu = 0, 1, 2, \cdots, p - 1.$$

This establishes the statement.

We have

$$(8.61.8) \qquad A_0(\theta) = (2/\pi)^{\frac{1}{2}} \cos (\pi/4) = \pi^{-\frac{1}{2}}, \qquad B_0(\theta) = (2/\pi)^{\frac{1}{2}} \sin (\pi/4) = \pi^{-\frac{1}{2}}.$$

The same method can be used for the asymptotic evaluation of $\mathbf{Q}_n(\cos \theta)$, (see Problem 18), as well as of the ultraspherical polynomials. In the first case we obtain Theorem 8.21.14. The application of the method to general Jacobi polynomials is more difficult because no explicit values for these polynomials are known at the point $\theta = \pi/2$ (or at any other fixed point in $0 < \theta < \pi$).[48]

8.62. Method of Liouville-Stekloff; formula of Hilb

(1) We again deduce an integral equation for $P_n(\cos \theta)$, different from (8.61.3) and involving Bessel functions. Writing (7.3.5) in the form

$$(8.62.1) \qquad \left(\frac{d}{d\theta}\right)^2 \{(\sin \theta)^{\frac{1}{2}} P_n(\cos \theta)\} + \left\{ \frac{1}{4\theta^2} + (n + \tfrac{1}{2})^2 \right\} (\sin \theta)^{\frac{1}{2}} P_n(\cos \theta)$$
$$= \left\{ \frac{1}{4\theta^2} - \frac{1}{4 \sin^2 \theta} \right\} (\sin \theta)^{\frac{1}{2}} P_n(\cos \theta),$$

we apply (1.8.12). The corresponding homogeneous equation is (1.8.9) ($\alpha = 0$, $k = n + \frac{1}{2}$) with the solutions

$$(8.62.2) \qquad \theta^{\frac{1}{2}} J_0\{(n + \tfrac{1}{2})\theta\} \quad \text{and} \quad \theta^{\frac{1}{2}} Y_0\{(n + \tfrac{1}{2})\theta\}.$$

[48] Cf., however, Korous **3**.

Consequently, with certain constants θ_0, c_1, c_2, we have

$$(8.62.3)\quad \begin{aligned}(\sin\theta)^{\frac{1}{2}}P_n(\cos\theta) &= c_1\theta^{\frac{1}{2}}J_0\{(n+\tfrac{1}{2})\theta\} + c_2\theta^{\frac{1}{2}}Y_0\{(n+\tfrac{1}{2})\theta\}\\ &+ \frac{\theta^{\frac{1}{2}}}{n+\frac{1}{2}}\int_{\theta_0}^{\theta} t^{-\frac{1}{2}}\frac{J_0\{(n+\frac{1}{2})\theta\}Y_0\{(n+\frac{1}{2})t\} - Y_0\{(n+\frac{1}{2})\theta\}J_0\{(n+\frac{1}{2})t\}}{J_0'\{(n+\frac{1}{2})t\}Y_0\{(n+\frac{1}{2})t\} - Y_0'\{(n+\frac{1}{2})t\}J_0\{(n+\frac{1}{2})t\}}\\ &\qquad\cdot\left(\frac{1}{4t^2} - \frac{1}{4\sin^2 t}\right)(\sin t)^{\frac{1}{2}}P_n(\cos t)\,dt.\end{aligned}$$

According to (1.8.14)

$$(8.62.4)\quad J_0'\{(n+\tfrac{1}{2})t\}Y_0\{(n+\tfrac{1}{2})t\} - Y_0'\{(n+\tfrac{1}{2})t\}J_0\{(n+\tfrac{1}{2})t\} = -\frac{2}{\pi(n+\frac{1}{2})t}.$$

Furthermore, $t^{-2} - (\sin t)^{-2}$ is analytic at $t = 0$; that is, we can take $\theta_0 = 0$, and we then obtain

$$\begin{aligned}(\sin\theta)^{\frac{1}{2}}P_n(\cos\theta) &= c_1\theta^{\frac{1}{2}}J_0\{(n+\tfrac{1}{2})\theta\} + c_2\theta^{\frac{1}{2}}Y_0\{(n+\tfrac{1}{2})\theta\}\\ &- \frac{\pi}{2}\theta^{\frac{1}{2}}\int_0^{\theta} t^{\frac{1}{2}}[J_0\{(n+\tfrac{1}{2})\theta\}Y_0\{(n+\tfrac{1}{2})t\} - Y_0\{(n+\tfrac{1}{2})\theta\}J_0\{(n+\tfrac{1}{2})t\}]\\ &\qquad\cdot\left(\frac{1}{4t^2} - \frac{1}{4\sin^2 t}\right)(\sin t)^{\frac{1}{2}}P_n(\cos t)\,dt.\end{aligned}$$

If this equation is divided through by $\theta^{\frac{1}{2}}$, and θ approaches 0, the last term tends to 0, whereas the left-hand member tends to 1. Hence (cf. (1.71.10), (1.71.1)) $c_2 = 0$, $c_1 = 1$, and we find for $0 < \theta < \pi$

$$(8.62.5)\quad \begin{aligned}\left(\frac{\sin\theta}{\theta}\right)^{\frac{1}{2}}P_n(\cos\theta) &= J_0\{(n+\tfrac{1}{2})\theta\} + \frac{\pi}{8}\int_0^{\theta}[J_0\{(n+\tfrac{1}{2})\theta\}Y_0\{(n+\tfrac{1}{2})t\}\\ &- Y_0\{(n+\tfrac{1}{2})\theta\}J_0\{(n+\tfrac{1}{2})t\}]t^{-1}\left\{\left(\frac{t}{\sin t}\right)^2 - 1\right\}\left(\frac{\sin t}{t}\right)^{\frac{1}{2}}P_n(\cos t)\,dt.\end{aligned}$$

This is the integral equation required.

(2) First assume $0 < n\theta \leqq 1$. Then according to (1.71.1) and (1.71.4)

$$(8.62.6)\quad \begin{aligned}&J_0\{(n+\tfrac{1}{2})\theta\}Y_0\{(n+\tfrac{1}{2})t\} - Y_0\{(n+\tfrac{1}{2})\theta\}J_0\{(n+\tfrac{1}{2})t\}\\ &= J_0\{(n+\tfrac{1}{2})\theta\}\left[\frac{2}{\pi}\log\{(n+\tfrac{1}{2})t\}J_0\{(n+\tfrac{1}{2})t\} + O(1)\right]\\ &\quad - J_0\{(n+\tfrac{1}{2})t\}\left[\frac{2}{\pi}\log\{(n+\tfrac{1}{2})\theta\}J_0\{(n+\tfrac{1}{2})\theta\} + O(1)\right]\\ &= -\frac{2}{\pi}J_0\{(n+\tfrac{1}{2})\theta\}J_0\{(n+\tfrac{1}{2})t\}\log\frac{\theta}{t} + O(1) = O(1)\log\frac{\theta}{t} + O(1).\end{aligned}$$

Therefore, the integral of the right-hand member of (8.62.5) is (by (7.21.1))

$$O(1)\int_0^{\theta} t\log\frac{\theta}{t}\,dt + O(1)\int_0^{\theta} t\,dt = O(1)\theta^2\int_0^1 t\log\frac{1}{t}\,dt + O(\theta^2) = O(\theta^2).$$

(3) On the other hand, let $n\theta \geqq 1$, $\theta \leqq \pi - \epsilon$. Then, according to (1.71.10) and (1.71.11), the contribution of the interval $0 < t \leqq n^{-1}$ to the integral previously considered, is

$$O\{(n\theta)^{-\frac{1}{2}}\} \int_0^{n^{-1}} t\,|\,Y_0(nt)\,|\,dt + O\{(n\theta)^{-\frac{1}{2}}\} \int_0^{n^{-1}} t\,|\,J_0(nt)\,|\,dt$$

$$= O\{(n\theta)^{-\frac{1}{2}}\}\cdot n^{-2} = O(\theta^{-\frac{1}{2}}n^{-\frac{5}{2}}) = (n\theta)^{-1}O(\theta^{\frac{1}{2}}n^{-\frac{3}{2}}) = O(\theta^{\frac{1}{2}}n^{-\frac{3}{2}}).$$

Finally, we find for the contribution of the interval $n^{-1} \leqq t \leqq \theta$, according to (7.3.8),

$$O\{(n\theta)^{-\frac{1}{2}}\} \int_{n^{-1}}^{\theta} (nt)^{-\frac{1}{2}} t (nt)^{-\frac{1}{2}}\, dt = O(\theta^{\frac{1}{2}} n^{-\frac{3}{2}}).$$

8.63. Method of Liouville-Stekloff; extension of Hilb's formula to Jacobi polynomials

By use of complex integration, Szegö (**17**) has extended Hilb's formula and the corresponding asymptotic expansion mentioned in §8.23 (1), to ultraspherical, and even to general Jacobi, polynomials. Following Rau (**2**), we deduce the principal term of this general expansion by means of the Liouville-Stekloff method and obtain formula (8.21.17). The bounds for the remainder are better than those in Szegö **17**, p. 77, (47), and in Rau **2**, pp. 691–692, (29), (30).

(1) Let $\alpha > -1$. We write (4.24.2) in the form

$$\frac{d^2u}{d\theta^2} + \left\{\frac{\frac{1}{4} - \alpha^2}{\theta^2} + N^2\right\}u = \left\{\frac{\beta^2 - \frac{1}{4}}{4\cos^2\frac{\theta}{2}} + (\tfrac{1}{4} - \alpha^2)\left(\frac{1}{\theta^2} - \frac{1}{4\sin^2\frac{\theta}{2}}\right)\right\}u,$$

(8.63.1)

$$u = \left(\sin\frac{\theta}{2}\right)^{\alpha+\frac{1}{2}}\left(\cos\frac{\theta}{2}\right)^{\beta+\frac{1}{2}} P_n^{(\alpha,\beta)}(\cos\theta), \qquad N = n + (\alpha+\beta+1)/2.$$

Again applying (1.8.12), we obtain, because of (1.8.9),

$$\left(\sin\frac{\theta}{2}\right)^{\alpha+\frac{1}{2}}\left(\cos\frac{\theta}{2}\right)^{\beta+\frac{1}{2}} P_n^{(\alpha,\beta)}(\cos\theta) = c_1\theta^{\frac{1}{2}}J_\alpha(N\theta) + c_2\theta^{\frac{1}{2}}J_{-\alpha}(N\theta)$$

$$+ \frac{\theta^{\frac{1}{2}}}{N}\int_{\theta_0}^{\theta} t^{-\frac{1}{2}} \frac{J_\alpha(N\theta)J_{-\alpha}(Nt) - J_{-\alpha}(N\theta)J_\alpha(Nt)}{J'_\alpha(Nt)J_{-\alpha}(Nt) - J'_{-\alpha}(Nt)J_\alpha(Nt)}$$

(8.63.2)

$$\cdot\left\{\frac{\beta^2 - \frac{1}{4}}{4\cos^2\frac{t}{2}} + (\tfrac{1}{4} - \alpha^2)\left(\frac{1}{t^2} - \frac{1}{4\sin^2\frac{t}{2}}\right)\right\}$$

$$\cdot\left(\sin\frac{t}{2}\right)^{\alpha+\frac{1}{2}}\left(\cos\frac{t}{2}\right)^{\beta+\frac{1}{2}} P_n^{(\alpha,\beta)}(\cos t)\, dt.$$

Here and in what follows $J_{-\alpha}(z)$ must be replaced by $Y_\alpha(z)$ if α is an integer. We refer again to the identity (1.8.14), by which

$$(8.63.3) \qquad J'_\alpha(Nt)J_{-\alpha}(Nt) - J'_{-\alpha}(Nt)J_\alpha(Nt) = \frac{2\sin\alpha\pi}{\pi Nt}.$$

(If α is an integer, $\sin\alpha\pi$ is to be replaced by -1.) Therefore,

$$\theta^{-\frac{1}{2}}\left(\sin\frac{\theta}{2}\right)^{\alpha+\frac{1}{2}}\left(\cos\frac{\theta}{2}\right)^{\beta+\frac{1}{2}} P_n^{(\alpha,\beta)}(\cos\theta) = c_1 J_\alpha(N\theta) + c_2 J_{-\alpha}(N\theta)$$

$$(8.63.4) \qquad + \int_{\theta_0}^{\theta} \{J_\alpha(N\theta)J_{-\alpha}(Nt) - J_{-\alpha}(N\theta)J_\alpha(Nt)\}t^{\frac{1}{2}}f(t)\left(\sin\frac{t}{2}\right)^{\alpha+\frac{1}{2}}\left(\cos\frac{t}{2}\right)^{\beta+\frac{1}{2}} \cdot P_n^{(\alpha,\beta)}(\cos t)\,dt,$$

where $f(t)$ is regular in $0 \leqq t < \pi$, and independent of n.

The last integral is convergent for $\theta_0 = 0$; as $\theta \to +0$ it becomes (n fixed)

$$O(1)\int_0^\theta (\theta^\alpha t^{-\alpha} + \theta^{-\alpha}t^\alpha)t^{\frac{1}{2}}t^{\alpha+\frac{1}{2}}\,dt = O(\theta^\alpha)\int_0^\theta t\,dt + O(\theta^{-\alpha})\int_0^\theta t^{2\alpha+1}\,dt = O(\theta^{\alpha+2}).$$

This is true whether α is an integer or not [(1.71.10)], except for $\alpha = 0$. Then we obtain

$$O(1)\int_0^\theta \left(\log\frac{1}{t} + \log\frac{1}{\theta}\right)t^{\frac{1}{2}}t^{\frac{1}{2}}\,dt = O\left(\theta^2\log\frac{1}{\theta}\right).$$

Dividing (8.63.4) by θ^α, $\theta \to +0$, we find [(1.71.1)] a relation of the form

$$2^{-\alpha-\frac{1}{2}}P_n^{(\alpha,\beta)}(1) + O(\theta^2) = c_1\left\{\frac{(N/2)^\alpha}{\Gamma(\alpha+1)} + O(\theta^2)\right\} + c_2\theta^{-\alpha}J_{-\alpha}(N\theta) + O(\theta^2).$$

(The last term must be modified for $\alpha = 0$.) Hence if $\alpha \geqq 0$,

$$c_2 = 0 \qquad \text{and} \qquad c_1 = 2^{-\frac{1}{2}}N^{-\alpha}\,\Gamma(n+\alpha+1)(n!)^{-1}.$$

The same result holds when $-1 < \alpha < 0$ if we take into consideration the fact that the "principal term" of $\theta^{-\alpha}J_{-\alpha}(N\theta)$ is $\theta^{-2\alpha}$. Thus, for $0 < \theta < \pi$,

$$\theta^{-\frac{1}{2}}\left(\sin\frac{\theta}{2}\right)^{\alpha+\frac{1}{2}}\left(\cos\frac{\theta}{2}\right)^{\beta+\frac{1}{2}} P_n^{(\alpha,\beta)}(\cos\theta) = 2^{-\frac{1}{2}}N^{-\alpha}\frac{\Gamma(n+\alpha+1)}{n!}J_\alpha(N\theta)$$

$$(8.63.5) \qquad + \int_0^\theta \{J_\alpha(N\theta)J_{-\alpha}(Nt) - J_{-\alpha}(N\theta)J_\alpha(Nt)\}t^{\frac{1}{2}}f(t)\left(\sin\frac{t}{2}\right)^{\alpha+\frac{1}{2}}\left(\cos\frac{t}{2}\right)^{\beta+\frac{1}{2}} \cdot P_n^{(\alpha,\beta)}(\cos t)\,dt.$$

(2) Now let $n \to \infty$. We find bounds for the last integral in a manner similar to that in §8.62. First, let $0 < n\theta \leqq 1$. Then for $\alpha \neq 0$, the integral is (cf. the second bound in (7.32.5))

$$O(1)\int_0^\theta \{(n\theta)^\alpha(nt)^{-\alpha} + (n\theta)^{-\alpha}(nt)^\alpha\}t^{\frac{1}{2}}t^{\alpha+\frac{1}{2}}n^\alpha\,dt = O(n^\alpha\theta^{\alpha+2}).$$

In case $\alpha = 0$, we reason as in §8.62 (2) and obtain the same bound, that is,

$O(\theta^2)$. On the other hand, let $n^{-1} \leqq \theta \leqq \pi - \epsilon$. The contribution of the interval $0 < t \leqq n^{-1}$ is (cf. the second bound in (7.32.5))

$$O(1) \int_0^{n^{-1}} (n\theta)^{-\frac{1}{2}}\{| J_{-\alpha}(Nt) | + | J_\alpha(Nt) |\}t^{\frac{1}{2}}t^{\alpha+\frac{1}{2}}n^\alpha \, dt$$
$$= O(1)(n\theta)^{-\frac{1}{2}}n^{-\alpha-2}n^\alpha = O(\theta^{-\frac{1}{2}}n^{-\frac{5}{2}}) = (n\theta)^{-1}O(\theta^{\frac{1}{2}}n^{-\frac{3}{2}}) = O(\theta^{\frac{1}{2}}n^{-\frac{3}{2}}).$$

(Here $J_{-\alpha}$ must be replaced by Y_0 if $\alpha = 0$.) For $t \geqq n^{-1}$, we use the first bound in (7.32.5) and obtain

$$O(1) \int_{n^{-1}}^{\theta} (n\theta)^{-\frac{1}{2}}(nt)^{-\frac{1}{2}}t^{\frac{1}{2}}n^{-\frac{1}{2}} \, dt = O(\theta^{\frac{1}{2}}n^{-\frac{3}{2}}).$$

8.64. Method of Liouville-Stekloff; asymptotic formula of Hilb's type for Laguerre polynomials

By use of the fourth equation in (5.1.2) this method readily leads to (8.22.4). The third equation could likewise be used, but the calculation would then be slightly more complicated. Concerning an extension of (8.22.4) to an asymptotic expansion (at least for $\alpha > -\frac{1}{2}$) see §8.66.

(1) Let $\alpha > -1$. Writing the equation in question as

$$v'' + \left(4N + \frac{\frac{1}{4} - \alpha^2}{x^2}\right) v = x^2 v; \quad v = e^{-x^2/2}x^{\alpha+\frac{1}{2}}L_n^{(\alpha)}(x^2), \tag{8.64.1}$$
$$N = n + (\alpha + 1)/2,$$

we can apply (1.8.12) and (1.8.9). Hence with certain constants x_0, c_1, c_2,

$$e^{-x^2/2}x^{\alpha+\frac{1}{2}}L_n^{(\alpha)}(x^2) = c_1 x^{\frac{1}{2}}J_\alpha(2N^{\frac{1}{2}}x) + c_2 x^{\frac{1}{2}}J_{-\alpha}(2N^{\frac{1}{2}}x)$$
$$+ \frac{x^{\frac{1}{2}}}{2N^{\frac{1}{2}}} \int_{x_0}^{x} \frac{J_\alpha(2N^{\frac{1}{2}}x)J_{-\alpha}(2N^{\frac{1}{2}}t) - J_{-\alpha}(2N^{\frac{1}{2}}x)J_\alpha(2N^{\frac{1}{2}}t)}{J'_\alpha(2N^{\frac{1}{2}}t)J_{-\alpha}(2N^{\frac{1}{2}}t) - J'_{-\alpha}(2N^{\frac{1}{2}}t)J_\alpha(2N^{\frac{1}{2}}t)} e^{-t^2/2}t^{\alpha+2}L_n^{(\alpha)}(t^2) \, dt.$$

Once again we use (8.63.3) and obtain

$$e^{-x^2/2}x^\alpha L_n^{(\alpha)}(x^2) = c_1 J_\alpha(2N^{\frac{1}{2}}x) + c_2 J_{-\alpha}(2N^{\frac{1}{2}}x)$$
$$+ \frac{\pi}{2 \sin \alpha\pi} \int_{x_0}^{x} \{J_\alpha(2N^{\frac{1}{2}}x)J_{-\alpha}(2N^{\frac{1}{2}}t) - J_{-\alpha}(2N^{\frac{1}{2}}x)J_\alpha(2N^{\frac{1}{2}}t)\} e^{-t^2/2}t^{\alpha+3}L_n^{(\alpha)}(t^2) \, dt.$$

If α is an integer, $J_{-\alpha}(z)$ must be replaced by $Y_\alpha(z)$ and $\sin \alpha\pi$ by -1.

Let $x_0 = 0$. For a fixed n, as $x \to +0$, the last term is

$$O(1) \int_0^x (x^\alpha t^{-\alpha} + x^{-\alpha}t^\alpha)t^{\alpha+3} \, dt = O(x^{\alpha+4}).$$

(If $\alpha = 0$, this bound must be multiplied by $\log (1/x)$.) Therefore, as in §8.63, $c_2 = 0$ and $L_n^{(\alpha)}(0) = c_1 N^{\alpha/2}\{\Gamma(\alpha + 1)\}^{-1}$; whence

$$c_1 = N^{-\alpha/2}\,\Gamma(n + \alpha + 1)(n!)^{-1}. \tag{8.64.2}$$

Consequently,

$$e^{-x^2/2}x^{\alpha}L_n^{(\alpha)}(x^2) = N^{-\alpha/2}\frac{\Gamma(n+\alpha+1)}{n!}J_\alpha(2N^{\frac{1}{2}}x)$$

(8.64.3)

$$+\frac{\pi}{2\sin\alpha\pi}\int_0^x\{J_\alpha(2N^{\frac{1}{2}}x)J_{-\alpha}(2N^{\frac{1}{2}}t)$$

$$-J_{-\alpha}(2N^{\frac{1}{2}}x)J_\alpha(2N^{\frac{1}{2}}t)\}\,e^{-t^2/2}t^{\alpha+3}L_n^{(\alpha)}(t^2)\,dt.$$

(2) As $n \to \infty$, the remainder term can be estimated in the same way as in the previous cases. However, we avoid here the use of bounds of the type (7.32.5). (In §7.6 (3) we derived such bounds as a consequence of (8.1.8) and of Fejér's formula (8.22.1); this is almost (8.22.4), which is just what we must prove now.) In the following proof we apply only the elementary formula (8.1.8) of the Mehler-Heine type, in particular only the second bound in (7.6.8).

First, let $0 < x \leqq n^{-\frac{1}{2}}$. Then we have $L_n^{(\alpha)}(t^2) = O(n^\alpha)$ for $0 \leqq t \leqq x$. It then follows that the integral term in (8.64.3) is

(8.64.4) $$O(1)\int_0^x (n^{\alpha/2}x^\alpha n^{-\alpha/2}t^{-\alpha} + n^{-\alpha/2}x^{-\alpha}n^{\alpha/2}t^\alpha)t^{\alpha+3}n^\alpha\,dt = O(x^{\alpha+4}n^\alpha);$$

for $\alpha = 0$ this bound must be multiplied $\log(x^{-1}n^{-\frac{1}{2}})$.

Now let $n^{-\frac{1}{2}} \leqq x \leqq \omega^{\frac{1}{2}}$, where ω is a fixed positive number. Let M_n be the maximum of $e^{-x^2/2}x^\alpha\,|\,L_n^{(\alpha)}(x^2)\,|$ in this interval. Then the contribution of the part $0 \leqq t \leqq n^{-\frac{1}{2}}$ of the integral term is, $\alpha \neq 0$,

(8.64.5) $$O(1)\int_0^{n^{-\frac{1}{2}}} (n^{-\frac{1}{4}}x^{-\frac{1}{2}}n^{-\alpha/2}t^{-\alpha} + n^{-\frac{1}{4}}x^{-\frac{1}{2}}n^{\alpha/2}t^\alpha)t^{\alpha+3}n^\alpha\,dt = O(x^{-\frac{1}{2}}n^{\alpha/2-9/4}).$$

The same result holds if we have $\alpha = 0$. The contribution of the other part $n^{-\frac{1}{2}} \leqq t \leqq x$ is

(8.64.6) $$O(1)\int_0^x n^{-\frac{1}{4}}x^{-\frac{1}{2}}n^{-\frac{1}{4}}t^{-\frac{1}{2}}t^3M_n\,dt = O(1)n^{-\frac{1}{2}}x^3M_n = M_n\cdot o(1).$$

Taking account of (8.64.3), and using the same argument as in §8.61, we find that

(8.64.7) $$M_n = O(n^{-\alpha/2})O(n^\alpha)O(n^{-\frac{1}{4}}x^{-\frac{1}{2}}) = O(x^{-\frac{1}{2}}n^{\alpha/2-\frac{1}{4}}).$$

(This is, of course, identical with the first bound in (7.6.8).) Therefore, in view of (8.64.5), (8.64.6), and (8.64.7), we obtain for the remainder term, if $n^{-\frac{1}{2}} \leqq x \leqq \omega^{\frac{1}{2}}$,

$$O(x^{-\frac{1}{2}}n^{\alpha/2-9/4}) + O(1)n^{-\frac{1}{2}}x^3O(x^{-\frac{1}{2}}n^{\alpha/2-\frac{1}{4}})$$

$$= O(x^{-\frac{1}{2}}n^{\alpha/2-9/4}) + O(x^{5/2}n^{\alpha/2-\frac{3}{4}}) = O(x^{5/2}n^{\alpha/2-\frac{3}{4}}).$$

8.65. Method of Liouville-Stekloff; Hermite polynomials

(1) The integral equation (8.64.3) assumes a particularly simple form in case $\alpha = \pm\frac{1}{2}$, that is, for the Hermite polynomials. Using (5.6.1) and (1.71.2), and writing $n = 2m$, $\alpha = -\frac{1}{2}$, and $n = 2m + 1$, $\alpha = +\frac{1}{2}$, respectively, we obtain

$$(8.65.1)\qquad \begin{aligned} e^{-x^2/2}H_n(x) &= \lambda_n \cos(N^{\frac{1}{2}}x - n\pi/2) \\ &\quad + N^{-\frac{1}{2}} \int_0^x \sin\{N^{\frac{1}{2}}(x - t)\} t^2 e^{-t^2/2} H_n(t)\, dt, \end{aligned}$$

where

$$(8.65.2)\qquad \lambda_n = |H_n(0)|, \quad \text{or} \quad |H_n'(0)| N^{-\frac{1}{2}},$$

according as n is even or odd, and $N = 2n + 1$. However, it is more convenient to deduce this directly from the second equation in (5.5.2).

We prove (8.22.7) by mathematical induction. The statement is true for $p = 0$, replacing both sums $\sum_{\nu=0}^{p-1}$ by 0. In fact, if M_n denotes the maximum of $e^{-x^2/2} | H_n(x) |$ in a fixed real interval, we find from (8.65.1) that

$$(8.65.3)\qquad M_n \leqq \lambda_n + O(n^{-\frac{1}{2}}) \cdot M_n .$$

Then $M_n = \lambda_n O(1)$.

(2) Now assuming (8.22.7) for an arbitrary p, we obtain from (8.65.1)

$$(8.65.4)\qquad \begin{aligned} e^{-x^2/2}H_n(x) &= \lambda_n \cos(N^{\frac{1}{2}}x - n\pi/2) \\ &\quad + \lambda_n N^{-\frac{1}{2}} \int_0^x \sin\{N^{\frac{1}{2}}(x - t)\} \Big\{ \cos(N^{\frac{1}{2}}t - n\pi/2) \sum_{\nu=0}^{p-1} t^2 u_\nu(t) N^{-\nu} \\ &\quad + N^{-\frac{1}{2}} \sin(N^{\frac{1}{2}}t - n\pi/2) \sum_{\nu=0}^{p-1} t^2 v_\nu(t) N^{-\nu} \Big\} dt + \lambda_n O(n^{-p-\frac{1}{2}}). \end{aligned}$$

The second term of the right-hand member contains expressions of the following type:

$$(8.65.5)\qquad \begin{aligned} &\lambda_n N^{-\nu-\frac{1}{2}} \int_0^x \sin\{N^{\frac{1}{2}}(x - t)\} \cos(N^{\frac{1}{2}}t - n\pi/2) t^k\, dt \\ &\quad = (2k + 2)^{-1} \lambda_n N^{-\nu-\frac{1}{2}} x^{k+1} \sin(N^{\frac{1}{2}}x - n\pi/2) \\ &\qquad + \tfrac{1}{2}\lambda_n N^{-\nu-\frac{1}{2}} \int_0^x t^k \sin\{N^{\frac{1}{2}}(x - 2t) + n\pi/2\}\, dt, \\ &\lambda_n N^{-\nu-1} \int_0^x \sin\{N^{\frac{1}{2}}(x - t)\} \sin(N^{\frac{1}{2}}t - n\pi/2) t^l\, dt \\ &\quad = -(2l + 2)^{-1} \lambda_n N^{-\nu-1} x^{l+1} \cos(N^{\frac{1}{2}}x - n\pi/2) \\ &\qquad + \tfrac{1}{2}\lambda_n N^{-\nu-1} \int_0^x t^l \cos\{N^{\frac{1}{2}}(x - 2t) + n\pi/2\}\, dt, \end{aligned}$$

where k is even and l is odd ($k \geqq 2$, $l \geqq 3$). Integration by parts furnishes

(8.65.6)

$$\int_0^x t^k \sin\{N^{\frac{1}{2}}(x-2t)+n\pi/2\}\,dt = \tfrac{1}{2}N^{-\frac{1}{2}}x^k\cos(N^{\frac{1}{2}}x-n\pi/2) - \tfrac{1}{2}kN^{-\frac{1}{2}}\int_0^x t^{k-1}\cos\{N^{\frac{1}{2}}(x-2t)+n\pi/2\}\,dt,$$

$$\int_0^x t^l \cos\{N^{\frac{1}{2}}(x-2t)+n\pi/2\}\,dt = \tfrac{1}{2}N^{-\frac{1}{2}}x^l\sin(N^{\frac{1}{2}}x-n\pi/2) + \tfrac{1}{2}lN^{-\frac{1}{2}}\int_0^x t^{l-1}\sin\{N^{\frac{1}{2}}(x-2t)+n\pi/2\}\,dt;$$

the second formula holds also for $l = 1$, and

(8.65.7)
$$\int_0^x \sin\{N^{\frac{1}{2}}(x-2t)+n\pi/2\}\,dt = \begin{cases} 0, \\ N^{-\frac{1}{2}}\cos(N^{\frac{1}{2}}x-n\pi/2), \end{cases}$$

according as n is even or odd.

This consideration leads to a formula of the type

$$e^{-x^2/2}H_n(x) = \lambda_n\left\{\cos(N^{\frac{1}{2}}x-n\pi/2)\sum_{\nu=0}^{p}u_\nu^{(1)}(x)N^{-\nu} + N^{-\frac{1}{2}}\sin(N^{\frac{1}{2}}x-n\pi/2)\sum_{\nu=0}^{p-1}v_\nu^{(1)}(x)N^{-\nu}+O(n^{-p-\frac{1}{2}})\right\}.$$

By repeated application of the same argument

$$e^{-x^2/2}H_n(x) = \lambda_n\left\{\cos(N^{\frac{1}{2}}x-n\pi/2)\sum_{\nu=0}^{p}u_\nu^{(2)}(x)N^{-\nu} + N^{-\frac{1}{2}}\sin(N^{\frac{1}{2}}x-n\pi/2)\sum_{\nu=0}^{p}v_\nu^{(2)}(x)N^{-\nu}+O(n^{-p-1})\right\}.$$

We readily see that $u_\nu^{(1)}(x)$, $u_\nu^{(2)}(x)$; $v_\nu^{(1)}(x)$, $v_\nu^{(2)}(x)$ are polynomials of the same type as $u_\nu(x)$; $v_\nu(x)$, respectively, and $u_\nu(x) = u_\nu^{(1)}(x) = u_\nu^{(2)}(x)$, $v_\nu(x) = v_\nu^{(1)}(x) = v_\nu^{(2)}(x)$, $\nu \leqq p-1$; whence (8.22.7) follows.

The proof of Theorem 8.22.7 can be given along these same lines.

8.66. Application to Laguerre polynomials

(1) The asymptotic expansion (8.22.7), combined with the formula (5.6.5) of Uspensky, readily furnishes an asymptotic expansion of Hilb's type for the Laguerre polynomials $L_n^{(\alpha)}(x)$, at least for $\alpha > -\frac{1}{2}$. We shall give only an outline of the proof.

Substituting (8.22.7) in (5.6.5), we obtain an asymptotic expansion of which the general term is, apart from trivial constant factors depending on n,

(8.66.1)
$$\int_{-1}^{+1}(1-t^2)^{\alpha-\frac{1}{2}}e^{xt^2/2}\{\cos[(4n+1)^{\frac{1}{2}}x^{\frac{1}{2}}t]\,u_\nu(x^{\frac{1}{2}}t) + (4n+1)^{-\frac{1}{2}}\sin[(4n+1)^{\frac{1}{2}}x^{\frac{1}{2}}t]\,v_\nu(x^{\frac{1}{2}}t)\}\,dt.$$

Here the $u_\nu(x)$ and the $v_\nu(x)$ are even and odd polynomials, respectively, and both are independent of n. The expression (8.66.1) is a linear combination of terms of the form

$$\int_{-1}^{+1} (1 - t^2)^{\alpha-\frac{1}{2}} e^{xt^2/2}(x^{\frac{1}{2}}t)^k \cos\{(4n+1)^{\frac{1}{2}}x^{\frac{1}{2}}t\}\,dt,$$

(8.66.2)

$$(4n+1)^{-\frac{1}{2}}\int_{-1}^{+1} (1 - t^2)^{\alpha-\frac{1}{2}} e^{xt^2/2}(x^{\frac{1}{2}}t)^l \sin\{(4n+1)^{\frac{1}{2}}x^{\frac{1}{2}}t\}\,dt,$$

where k and l are non-negative integers, k even and l odd. If we expand $e^{\tau/2}\tau^{k/2}$ in a power series about $\tau = x$, we obtain for the first integral terms of the type

$$\int_{-1}^{+1} (1 - t^2)^{\alpha+q-\frac{1}{2}} \cos\{(4n+1)^{\frac{1}{2}}x^{\frac{1}{2}}t\}\,dt,$$

with q integral, and $q \geqq 0$. This can be expressed in terms of Bessel functions [(1.71.6)]. A similar method furnishes for the second integral (8.66.2) terms of the type

$$\int_{-1}^{+1} (1 - t^2)^{\alpha+q-\frac{1}{2}}\, t \sin\{(4n+1)^{\frac{1}{2}}x^{\frac{1}{2}}t\}\,dt,$$

q again being integral, and $q \geqq 0$. These can also be expressed in terms of Bessel functions (combine the second formula (1.71.5) with (1.71.6)).

If we stop the expansion of the first expression (8.66.2) at a certain term, the remainder appears in the form

$$(8.66.3)\qquad \int_{-1}^{+1} (1 - t^2)^{\alpha-\frac{1}{2}} f[x(1 - t^2)] \cos\{(4n+1)^{\frac{1}{2}}x^{\frac{1}{2}}t\}\,dt,$$

where $f(t) = c_m\tau^m + c_{m+1}\tau^{m+1} + \cdots$ is an integral function with a zero of order m at $\tau = 0$; here m is an arbitrary integer. The remainder is of similar form for the second integral in (8.66.2). Writing

$$g(t) = (1 - t^2)^{\alpha-\frac{1}{2}} f[x(1 - t^2)],$$

we see that the functions $g(t)$, $g'(t)$, $g''(t)$, $\cdots$, $g^{(m-1)}(t)$ vanish at $t = \pm 1$; they are all $x^m O(1)$, where $O(1)$ is uniformly bounded in $-1 \leqq t \leqq +1$, and in a fixed finite interval $a \leqq x \leqq b$ containing the origin or not. Integrating by parts, we find for (8.66.3) a bound of the form $O(n^{-K})$, where K is arbitrarily large with m, uniformly in x, $a \leqq x \leqq b$.

Similar remarks hold for the second remainder.

(2) The first term of this expansion furnishes (8.22.4). We also obtain readily an extension of (8.22.4) to the complex domain.

Now assume $0 < \epsilon \leqq x \leqq \omega$. Then applying (1.71.8), we find Perron's expansion (8.22.2). (Cf. Uspensky **1**, pp. 608–610.) There is no difficulty in deriving the complex formula (8.22.3) of Perron in the same way.

In all these considerations we assumed $\alpha > -\frac{1}{2}$. An extension of Perron's formulas to arbitrary real values of α is possible by means of the second formula in (5.1.13). Concerning a second proof of Perron's formulas (by use of the method of steepest descent), see §8.72.

8.71. The method of steepest descent; Legendre polynomials and related functions

(1) This method can be used for approximating integrals of the form

$$\int \{F(t)\}^n g(t)\,dt = \int e^{nf(t)} g(t)\,dt, \tag{8.71.1}$$

extended over a certain arc or closed curve, where $F(t) = e^{f(t)}$ and $g(t)$ are given analytic functions regular in a certain part of the complex t-plane, and $n \to \infty$. According to Cauchy's theorem, a deformation of the contour is possible. In many cases it is convenient to make it pass through some of the points t_0 at which $f'(t_0) = 0$ (saddle point); in addition, the direction of the contour at t_0 (critical direction) must be determined according to the condition that (assuming $f''(t_0) \neq 0$) the expression

$$nf''(t_0)(t - t_0)^2/2 \tag{8.71.2}$$

is real and negative if t is sufficiently near t_0. Then

$$\{F(t)\}^n = \{F(t_0)\}^n \exp \{nf''(t_0)(t - t_0)^2/2 + nf'''(t_0)(t - t_0)^3/6 + \cdots\}.$$

Hence under proper conditions concerning the behavior of $f(t)$ on the complementary part of the path of integration, the neighborhood

$$t - t_0 = O(n^{\delta - \frac{1}{2}}), \qquad 0 < \delta < \tfrac{1}{2}, \tag{8.71.3}$$

of the saddle point furnishes the "principal" part of the integral as $n \to \infty$. Its contribution is of the form

$$e^{nf(t_0)} g(t_0) n^{-\frac{1}{2}} \int_{-n^\delta}^{+n^\delta} \exp(-a\rho^2)\,d\rho \cong e^{nf(t_0)} g(t_0)\left(\frac{\pi}{an}\right)^{\frac{1}{2}}; \quad a > 0,\ n \to \infty. \tag{8.71.4}$$

(Additional difficulties arise if $f''(t_0) = 0$; concerning a case of this type, cf. §8.75.)

The term "method of steepest descent" arises from the following considerations. Let $t = u + iv$. Then representing u, v, $\Re[f(t)]$ as cartesian coördinates in the ordinary euclidean space, we obtain a surface with a *saddle point* at $t = t_0$, and the curve with the critical direction is the "steepest" curve on the surface through this point.

There is, of course, considerable freedom in the choice of the contour; only its direction through the saddle point is restricted. However, the exact calculation of the saddle points t_0 from $f'(t_0) = 0$, and particularly that of the corresponding critical directions, might well be a complicated task in certain cases.

For this reason, the following observation may greatly simplify matters. Instead of the critical direction itself, any other direction through the saddle point can be taken for the path of integration, provided it forms an angle of less than $\pi/4$ with the critical direction. Then the constant a in (8.71.4) becomes complex with a positive real part. Geometrically, this condition means that in passing through the saddle point on the surface, no higher level is reached in the neighborhood of this point than at the point itself.

For our purposes, we shall prefer a contour satisfying the last condition, and along which $\Re[f(t)]$ varies *monotonically*. Then the discussion of the integrand exterior to the neighborhood (8.71.3) becomes comparatively simple.

Concerning the history, further details, and important applications of this method, the reader is referred to Watson **3**, pp. 235–236.

(2) As a first illustration let us consider the Legendre function of the second kind, that is, the special case $\alpha = \beta = 0$ of (4.61.1). Let $x = \cos\theta - i0$, $0 < \theta < \pi$. Then

(8.71.5) $$Q_n^{(0,0)}(\cos\theta - i0) = Q_n(\cos\theta - i0) = \tfrac{1}{2}\int\left(\frac{1}{2}\frac{t^2-1}{t-\cos\theta}\right)^n \frac{dt}{\cos\theta - t}.$$

The original path of integration, $-1 \leqq t \leqq +1$, can be deformed into the upper half of the unit circle described in the negative sense. The saddle point condition is

(8.71.6) $$\frac{d}{dt}\left(\frac{1}{2}\frac{t^2-1}{t-\cos\theta}\right) = \frac{1}{2}\frac{t^2 - 2t\cos\theta + 1}{(t-\cos\theta)^2} = 0, \text{ whence } t = e^{\pm i\theta},$$

and we see that the path of integration passes through the saddle point $t = e^{i\theta}$. If $t = e^{i\theta}$ we have

(8.71.7) $$\left(\frac{d}{dt}\right)^2\left(\frac{1}{2}\frac{t^2-1}{t-\cos\theta}\right) = \frac{1}{i\sin\theta},$$

so that near this point

(8.71.8) $$\frac{1}{2}\frac{t^2-1}{t-\cos\theta} = e^{i\theta} + \frac{(t-e^{i\theta})^2}{2i\sin\theta} + \cdots.$$

Along the critical direction, $e^{-i(\theta+\pi/2)}(t - e^{i\theta})^2$ must be real and negative; that is,

(8.71.9) $$\arg(t - e^{i\theta}) = \theta/2 + 3\pi/4 \quad\text{or}\quad \theta/2 - \pi/4.$$

Now, the angle between this line and the tangent to the unit circle at the point $e^{i\theta}$ is

(8.71.10) $$\arg\{e^{i(\theta/2+3\pi/4)} : e^{i(\theta+\pi/2)}\} = \pi/4 - \theta/2,$$

and $|\pi/4 - \theta/2| \leqq \pi/4 - \epsilon/2$ if $\epsilon \leqq \theta \leqq \pi - \epsilon$. We can therefore use the circle $|t| = 1$ as the path of integration.[49]

[49] The critical direction is given by that of the bisector of the acute angle between the tangent at the point $e^{i\theta}$ and the horizontal direction.

Substituting $t = e^{i\phi}$, $0 \leqq \phi \leqq \pi$, we find that

$$\left|\frac{1}{2}\frac{t^2-1}{t-\cos\theta}\right| = \frac{\sin\phi}{\{(\cos\phi-\cos\theta)^2+\sin^2\phi\}^{\frac{1}{2}}} = \left\{\left(\frac{\cos\phi-\cos\theta}{\sin\phi}\right)^2+1\right\}^{-\frac{1}{2}} \tag{8.71.11}$$

is an increasing function of ϕ for $0 \leqq \phi \leqq \theta$, and a decreasing function for $\theta \leqq \phi \leqq \pi$. We next consider the contribution of the arc

$$\phi = \theta + n^{-\frac{1}{2}}\rho, \qquad -n^{\delta} \leqq \rho \leqq +n^{\delta}, \tag{8.71.12}$$

where δ is a properly chosen positive number. On this arc we have

$$e^{i\phi} - e^{i\theta} = e^{i\theta}(\exp[in^{-\frac{1}{2}}\rho] - 1) = e^{i\theta}\left\{in^{-\frac{1}{2}}\rho + \frac{(in^{-\frac{1}{2}}\rho)^2}{2!} + \cdots\right\},$$

and, in view of (8.71.8),

$$\frac{1}{2}\frac{t^2-1}{t-\cos\theta} = e^{i\theta} - \frac{e^{2i\theta}}{2i\sin\theta}n^{-1}\rho^2\{1 + c_1n^{-\frac{1}{2}}\rho + c_2(n^{-\frac{1}{2}}\rho)^2 + \cdots\},$$

provided $\delta < \frac{1}{2}$. Here $c_1, c_2, \cdots$ are certain functions of θ independent of n, for which $c_m = O(A^m)$ holds uniformly in $\epsilon \leqq \theta \leqq \pi - \epsilon$, and in m; $A = A(\epsilon)$. Now

$$\left(\frac{1}{2}\frac{t^2-1}{t-\cos\theta}\right)^n = e^{in\theta}\exp\left\{-\frac{e^{i\theta}}{2i\sin\theta}\rho^2\right\}e^W,$$

where

$$W = -\frac{e^{i\theta}}{2i\sin\theta}\rho^2\{c_1n^{-\frac{1}{2}}\rho + c_2(n^{-\frac{1}{2}}\rho)^2 + \cdots\}$$
$$- \sum_{\nu=2}^{\infty}\frac{1}{\nu}\left(\frac{e^{i\theta}}{2i\sin\theta}\right)^{\nu}n^{1-\nu}\rho^{2\nu}(1 + c_1n^{-\frac{1}{2}}\rho + \cdots)^{\nu}.$$

Let $\delta < \frac{1}{6}$; then if M is an arbitrarily large positive integer, W and e^W can be reduced to a finite sum plus a remainder which is $O(n^{-M})$. This yields the relation

$$\left(\frac{1}{2}\frac{t^2-1}{t-\cos\theta}\right)^n = e^{in\theta}\exp\left\{-\frac{e^{i\theta}}{2i\sin\theta}\rho^2\right\} \cdot\{1 + u_1(\rho,\theta)n^{-\frac{1}{2}} + u_2(\rho,\theta)n^{-1} + u_3(\rho,\theta)n^{-\frac{3}{2}} + \cdots\}. \tag{8.71.13}$$

The series in the braces is an asymptotic expansion. If only m terms of this are taken, the error is less than an arbitrarily large power of n^{-1} provided m is sufficiently large; here $u_\nu(\rho, \theta)$ is a polynomial in ρ and a function of θ, analytic in $\epsilon \leqq \theta \leqq \pi - \epsilon$. In addition $u_\nu(-\rho, \theta) = (-1)^\nu u_\nu(\rho, \theta)$.

(3) Now multiply (8.71.13) by

$$\frac{1}{2}\frac{dt}{\cos\theta - t} = \frac{1}{2}\frac{ie^{i\phi}\,d\phi}{\cos\theta - e^{i\phi}} = \frac{1}{2}\frac{ie^{i\theta}e^{in^{-\frac{1}{2}}\rho}n^{-\frac{1}{2}}\,d\rho}{\cos\theta - e^{i\theta}e^{in^{-\frac{1}{2}}\rho}}$$

$$= -\frac{e^{i\theta}}{2\sin\theta}n^{-\frac{1}{2}}\{1 + c_1'n^{-\frac{1}{2}}\rho + c_2'(n^{-\frac{1}{2}}\rho)^2 + \cdots\}\,d\rho,$$

where $\{c_m'\}$ is a sequence similar to $\{c_m\}$; this does not change the essential character of equation (8.71.13). We obtain, as the contribution of the arc (8.71.12),

$$(8.71.14)\qquad -\frac{e^{i(n+1)\theta}}{2\sin\theta}n^{-\frac{1}{2}}\int_{+n^{\delta}}^{-n^{\delta}}\exp\left\{-\frac{e^{i\theta}}{2i\sin\theta}\rho^2\right\}$$
$$\cdot\{1 + v_1(\rho,\theta)n^{-\frac{1}{2}} + v_2(\rho,\theta)n^{-1} + v_3(\rho,\theta)n^{-\frac{3}{2}} + \cdots\}\,d\rho,$$

where $\{v_\nu(\rho,\theta)\}$ is a sequence of polynomials similar to $\{u_\nu(\rho,\theta)\}$. The last series is again an asymptotic expansion of the same type as (8.71.13).

At the end-points of the arc in question the modulus of the integrand of (8.71.14) is $O(e^{-c\rho^2})$, that is, $O(e^{-cn^{2\delta}})$, $c > 0$; the same is true of the contribution of the complementary arc because of the monotonic character of the function (8.71.11). Thus,

$$Q_n(\cos\theta - i0)$$
$$= \frac{e^{i(n+1)\theta}}{2\sin\theta}n^{-\frac{1}{2}}\int_{-n^{\delta}}^{+n^{\delta}}\exp\left\{-\frac{e^{i\theta}}{2i\sin\theta}\rho^2\right\}\{1 + v_1(\rho,\theta)n^{-\frac{1}{2}} + v_2(\rho,\theta)n^{-1} + \cdots\}\,d\rho$$
$$+ O(e^{-cn^{2\delta}}).$$

The terms corresponding to the odd powers of $n^{-\frac{1}{2}}$ vanish after integration. Upon completing the interval of integration, we obtain

$$Q_n(\cos\theta - i0) = \frac{e^{i(n+1)\theta}}{2\sin\theta}n^{-\frac{1}{2}}\int_{-\infty}^{+\infty}\exp\left\{-\frac{e^{i\theta}}{2i\sin\theta}\rho^2\right\}d\rho + O(n^{-\frac{3}{2}})$$

$$(8.71.15)\qquad = \frac{e^{i(n+1)\theta}}{2\sin\theta}n^{-\frac{1}{2}}\left\{\frac{e^{i(\theta-\pi/2)}}{2\sin\theta}\right\}^{-\frac{1}{2}}\pi^{\frac{1}{2}} + O(n^{-\frac{3}{2}})$$

$$= \left(\frac{\pi}{2n\sin\theta}\right)^{\frac{1}{2}}\exp\{i[(n+\tfrac{1}{2})\theta + \pi/4]\} + O(n^{-\frac{3}{2}}),$$

the bound for the remainder holding uniformly for $\epsilon \leqq \theta \leqq \pi - \epsilon$.

This method leads to a complete asymptotic expansion of $Q_n(\cos\theta - i0)$ of the type (8.61.7) for $\epsilon \leqq \theta \leqq \pi - \epsilon$, but seems to be too difficult for use in finding the general law of the coefficients.

From (8.71.15) we obtain the corresponding asymptotic expansion for $Q_n(\cos\theta + i0)$ by merely replacing i by $-i$. From this, and (4.62.8), we easily derive Laplace's formula. Formula (8.61.7) can also be derived in this way. At the same time we obtain a similar asymptotic expansion for $\mathbf{Q}_n^{(0,0)}(\cos\theta) = \mathbf{Q}_n(\cos\theta)$ with the principal term (8.21.19).

(4) The analogous considerations for the Jacobi polynomials $P_n^{(\alpha,\beta)}(\cos\theta)$ are immediate. Here α and β are arbitrary and real, $n \to \infty$, and θ again satisfies the condition $\epsilon \leqq \theta \leqq \pi - \epsilon$. According to (4.4.6) we have

$$(8.71.16)\qquad \begin{aligned} P_n^{(\alpha,\beta)}(\cos\theta) = 2\Re\left\{\frac{1}{2\pi i}\int\left(\frac{1}{2}\frac{t^2-1}{t-\cos\theta}\right)^n\right. \\ \left.\cdot\left(\frac{1-t}{1-\cos\theta}\right)^\alpha\left(\frac{1+t}{1+\cos\theta}\right)^\beta\frac{dt}{\cos\theta - t}\right\}, \end{aligned}$$

where the contour is the same as that used in (2).[50] The additional factor

$$(8.71.17)\qquad \frac{2}{\pi i}\left(\frac{1-t}{1-\cos\theta}\right)^\alpha\left(\frac{1+t}{1+\cos\theta}\right)^\beta$$

does not cause any new difficulty and furnishes for $t = e^{i\theta}$

$$(8.71.18)\qquad \frac{2}{\pi i}\left(\sin\frac{\theta}{2}\right)^{-\alpha}\left(\cos\frac{\theta}{2}\right)^{-\beta}\exp\left[i\{\alpha(\theta-\pi)/2 + \beta\theta/2\}\right].$$

Therefore, we find (8.21.10) and an expansion of the type (8.21.12).

(5) The same method can readily be applied to the functions $Q_n(x)$, $P_n(x)$, or more generally to $Q_n^{(\alpha,\beta)}(x)$ and $P_n^{(\alpha,\beta)}(x)$, where x is arbitrary real or complex but not on the segment $[-1, +1]$. This leads to (8.21.9) and also to the expansion (8.21.11). In the case of $Q_n^{(\alpha,\beta)}(x)$, we start from (4.61.1). It is convenient to replace the half-circle used above by the circular arc through ± 1 and $z = x - (x^2-1)^{\frac{1}{2}}$, $|z| < 1$, (x is in the cut plane), which was introduced in §4.81 (1). The resulting integral is the same as (4.82.4). Applying the method of steepest descent, we obtain a formula of the type

$$(8.71.19)\qquad (x-1)^\alpha(x+1)^\beta Q_n^{(\alpha,\beta)}(x) \cong n^{-\frac{1}{2}}\{x - (x^2-1)^{\frac{1}{2}}\}^{n+1}\phi(x),$$

where $|x - (x^2-1)^{\frac{1}{2}}| < 1$, and $\phi(x)$ is independent of n, and regular and nonzero in the cut plane.

8.72. Method of steepest descent; Perron's formulas for Laguerre polynomials

As a further application of this method we again prove the expansions (8.22.2) and (8.22.3). The present proof is based on the integral representation (5.4.1), which is valid for an arbitrary real α, provided n is sufficiently large.

(1) Let x be arbitrary but nonzero. We start from the asymptotic expansion (1.71.8) of the Bessel function $J_\alpha(z)$, z complex. The contribution of the segment $0 \leqq t \leqq 1$ in (5.4.1) is $(n!)^{-1}O(1)$. We can therefore confine our attention to values of $t \geqq 1$, so that (1.71.8) may be applied. Substitution of this expansion into (5.4.1) leads to integrals of the type

[50] We must avoid the points $t = \pm 1$ by means of small semi-circles.

$$(8.72.1)\qquad \begin{aligned} &\frac{1}{n!}\int_0^\infty e^{-t}t^{n+\alpha/2}(tx)^{-m/2-1/4}\cos\{2(tx)^{\frac{1}{2}}-\alpha\pi/2-\pi/4\}\,dt, \quad m \text{ even},\\ &\frac{1}{n!}\int_0^\infty e^{-t}t^{n+\alpha/2}(tx)^{-m/2-1/4}\sin\{2(tx)^{\frac{1}{2}}-\alpha\pi/2-\pi/4\}\,dt, \quad m \text{ odd},\\ &\qquad\qquad m = 0, 1, 2, \cdots. \end{aligned}$$

(Here the range of integration has again been completed to $0 \leqq t < \infty$.) The remainder term will be of the form

$$(8.72.2)\qquad \frac{1}{n!}\int_0^\infty e^{-t}t^{n+\alpha/2-q}\exp\{2t^{\frac{1}{2}}\,|\,\Re(-x)^{\frac{1}{2}}\,|\}\,dt,$$

with q a fixed positive number which can be chosen arbitrarily large; the determination of $(-x)^{\frac{1}{2}}$ is the same as in Theorem 8.22.3. If x is a fixed positive number, this is obviously $O(n^{\alpha/2-q})$. If x is complex, the discussion of this remainder term requires greater care. Subsequent considerations furnish a bound in this case also.

(2) For the sake of convenience we shall first discuss the integral

$$(8.72.3)\qquad \frac{1}{n!}\int_0^\infty e^{-t}t^n\exp[t^{\frac{1}{2}}\xi]\,dt = \frac{n^{n+1}e^{-n}}{n!}\int_0^\infty (e^{1-t}t)^n\exp[n^{\frac{1}{2}}t^{\frac{1}{2}}\xi]\,dt,$$

($\xi \neq 0$, arbitrary complex) to which the integrals (8.72.1) and (8.72.2) can be reduced; here $n \to +\infty$, but n is not necessarily an integer. In the last integral the saddle point is "essentially" $t = 1$, and the positive real axis corresponds to the critical direction.[51] If we write, as in (8.71.12),

$$(8.72.4)\qquad t = 1 + n^{-\frac{1}{2}}\rho, \qquad -n^{\delta} \leqq \rho \leqq +n^{\delta},$$

we obtain, for $0 < \delta < 1/6$,

$$(8.72.5)\qquad \begin{aligned} (e^{1-t}t)^n &= \exp\{n(1-t) + n\log[1+(t-1)]\}\\ &= e^{-\rho^2/2}\exp\left\{n\sum_{\nu=3}^{\infty}(-1)^{\nu-1}\frac{(n^{-\frac{1}{2}}\rho)^\nu}{\nu}\right\}\\ &= e^{-\rho^2/2}\{1 + u_1(\rho)n^{-\frac{1}{2}} + u_2(\rho)n^{-1} + \cdots\}, \end{aligned}$$

where the $u_\nu(\rho)$ are polynomials in ρ independent of n. This is an asymptotic expansion similar in character to that in (8.71.13). Furthermore,

$$(8.72.6)\qquad \begin{aligned} \exp[n^{\frac{1}{2}}t^{\frac{1}{2}}\xi] &= \exp\{n^{\frac{1}{2}}\xi + n^{\frac{1}{2}}\xi([1+(t-1)]^{\frac{1}{2}}-1)\}\\ &= \exp\left\{n^{\frac{1}{2}}\xi + \rho\xi/2 + n^{\frac{1}{2}}\xi\sum_{\nu=2}^{\infty}c_\nu(n^{-\frac{1}{2}}\rho)^\nu\right\}, \end{aligned}$$

where the c_ν are certain numerical constants. This furnishes $\exp(n^{\frac{1}{2}}\xi + \rho\xi/2)$

[51] This case is not a direct application of the method indicated in §8.71 (1) since the integrand has the form $\{F(t)\}^n\{G(t)\}^{n^{1/2}}$.

multiplied by an expression similar to that in the braces in (8.72.5). The coefficients corresponding to the $u_\nu(\rho)$ are in this case polynomials in ρ and ξ. We also see that the contribution of the range of integration complementary to (8.72.4) is $O(\exp[-cn^{2\delta}])$, $c > 0$. Therefore, we have in the same sense as in (8.71.14)

$$\int_0^\infty (e^{1-t}t)^n \exp[n^{\frac{1}{2}}t^{\frac{1}{2}}\xi]\,dt$$

$$= n^{-\frac{1}{2}}\exp[n^{\frac{1}{2}}\xi]\int_{-n^\delta}^{+n^\delta} \exp[-\rho^2/2 + \rho\xi/2]\{1 + v_1(\rho, \xi)n^{-\frac{1}{2}} + v_2(\rho, \xi)n^{-1} + \cdots\}\,d\rho$$

$$= n^{-\frac{1}{2}}\exp[n^{\frac{1}{2}}\xi]\int_{-\infty}^{+\infty} \exp[-\rho^2/2 + \rho\xi/2]\{1 + v_1(\rho, \xi)n^{-\frac{1}{2}} + v_2(\rho, \xi)n^{-1} + \cdots\}\,d\rho,$$

where the $v_\nu(\rho, \xi)$ are polynomials in ρ and ξ. Now if q is a non-negative integer, we have, in general,

$$\int_{-\infty}^{+\infty} \exp[-\rho^2/2 + \rho\xi/2]\,\rho^q\,d\rho = e^{\xi^2/8}\int_{-\infty}^{+\infty} e^{-\rho^2/2}(\rho + \xi/2)^q\,d\rho,$$

and the last integral is a π_q in ξ. (For $q = 0$ we get $(2\pi)^{\frac{1}{2}}e^{\xi^2/8}$.) According to Stirling's formula an expansion of the type

$$\exp(n^{\frac{1}{2}}\xi + \xi^2/8)\{1 + v_1(\xi)n^{-\frac{1}{2}} + v_2(\xi)n^{-1} + \cdots\}$$

results for (8.72.3), in which the $v_\nu(\xi)$ are polynomials.

By applying this result to the expressions in (8.72.1), (replacing n by $n + \alpha/2 - m/2 - 1/4$), we obtain the required expansions. The special case $\xi > 0$ yields the required bound for the remainder term (8.72.2).

8.73. Method of steepest descent; Laguerre polynomials for $1 \leqq x \leqq (4 - \eta)n$

Here and in the next two sections we derive formulas (8.22.9), (8.22.10), and (8.22.11) by the method of steepest descent. We notice that in the first case the condition $x = (4n + 2\alpha + 2)\cos^2\phi$, $\epsilon \leqq \phi \leqq \pi/2 - \epsilon n^{-\frac{1}{2}}$, means that x satisfies the inequality $x_0 \leqq x \leqq (4 - \eta)n$, where ϵ, x_0, and η are fixed positive numbers, $\epsilon < \pi/2$, $\eta < 4$, n large. The parameter α is arbitrary and real.

(1) We start with formula (5.1.16) (cf. the remark at the end of §5.2), in which we replace x by ξ^2 and w by $-w^2/4$; thus

$$\text{(8.73.1)}\qquad \sum_{n=0}^{\infty} \frac{L_n^{(\alpha)}(\xi^2)}{\Gamma(n + \alpha + 1)}(-w^2/4)^n = 2^\alpha e^{-w^2/4}(\xi w)^{-\alpha}e^{\alpha\pi i/2}J_\alpha(e^{-i\pi/2}\xi w).$$

Consequently,

$$\text{(8.73.2)}\qquad (-\tfrac{1}{4})^n\frac{L_n^{(\alpha)}(\xi^2)}{\Gamma(n + \alpha + 1)} = \frac{2^\alpha\xi^{-\alpha}}{2\pi i}\int e^{-w^2/4}w^{-2n-\alpha-1}e^{\alpha\pi i/2}J_\alpha(e^{-i\pi/2}\xi w)\,dw.$$

The integration is extended over a contour enclosing the origin. We choose it as a circle with center at the origin and radius to be determined later. The function (8.73.1) is real if w is real.

Since $\xi = x^{\frac{1}{2}}$,

$$\xi = l_n \cos \phi, \qquad l_n = (4n + 2\alpha + 2)^{\frac{1}{2}}, \qquad \epsilon \leqq \phi \leqq \pi/2 - \epsilon n^{-\frac{1}{2}}. \tag{8.73.3}$$

Then ξ is bounded from zero. According to (1.71.9), as $|w| \to \infty$, we have uniformly in $0 \leqq \arg w \leqq \pi$,

$$\begin{aligned} e^{\alpha\pi i/2} J_\alpha(e^{-i\pi/2}\xi w) &= (2\pi\xi w)^{-\frac{1}{2}} e^{\xi w}\{1 + O(\xi^{-1}|w|^{-1})\} \\ &\quad + (2\pi\xi w)^{-\frac{1}{2}} \exp[-\xi w + (\alpha + \tfrac{1}{2})\pi i]\{1 + O(\xi^{-1}|w|^{-1})\}. \end{aligned} \tag{8.73.4}$$

Hence from (8.73.2), for $\xi = l_n \cos\phi$, $w = l_n z$,

$$(-\tfrac{1}{4})^n \frac{L_n^{(\alpha)}(\xi^2)}{\Gamma(n + \alpha + 1)} = 2^\alpha \xi^{-\alpha}(2\pi\xi)^{-\frac{1}{2}} l_n^{-2n-\alpha-\frac{1}{2}} 2\Re\left\{\frac{1}{2\pi i} G + \frac{1}{2\pi i} H + K\right\}, \tag{8.73.5}$$

where

$$\begin{aligned} G &= \int z^{-\frac{1}{2}} \exp\{-\tfrac{1}{4} l_n^2 z^2 + l_n^2 z \cos\phi - \tfrac{1}{2} l_n^2 \log z\}\, dz, \\ H &= e^{(\alpha+\frac{1}{2})\pi i} \int z^{-\frac{1}{2}} \exp\{-\tfrac{1}{4} l_n^2 z^2 - l_n^2 z \cos\phi - \tfrac{1}{2} l_n^2 \log z\}\, dz, \\ K &= O(\xi^{-1} l_n^{-1}) \int |\exp\{-\tfrac{1}{4} l_n^2 z^2 \pm l_n^2 z \cos\phi - \tfrac{1}{2} l_n^2 \log z\}|\, |dz|. \end{aligned} \tag{8.73.6}$$

Here the integration is extended over the upper half-circle $|z| = 1$, and $\log z$ is zero if z is 1. In the last integral we choose the plus or minus of the ambiguous sign according to which gives the integral the larger value. If now we set

$$f(z) = -\tfrac{1}{4}z^2 + z\cos\phi - \tfrac{1}{2}\log z, \tag{8.73.7}$$

we find the saddle points of the first integral from the equation $f'(z) = -z/2 + \cos\phi - (2z)^{-1} = 0$, that is, $z = e^{\pm i\phi}$. Since $f''(e^{i\phi}) = \sin\phi\, e^{-i(\phi+\pi/2)}$, we have in the neighborhood of $e^{i\phi}$,

$$f(z) = f(e^{i\phi}) + \tfrac{1}{2}\sin\phi\, e^{-i(\phi+\pi/2)}(z - e^{i\phi})^2 + \cdots. \tag{8.73.8}$$

Therefore, the critical direction can be found in a manner similar to that used in §8.71 (2).

(2) On the circle $z = e^{i\psi}$

$$\Re\{f(e^{i\psi})\} = -\tfrac{1}{4}\cos 2\psi + \cos\psi\cos\phi \tag{8.73.9}$$

is increasing if $0 \leqq \psi \leqq \phi$, and decreasing if $\phi \leqq \psi \leqq \pi$. Consequently, it suffices to consider the contribution of the arc

$$\psi = \phi + l_n^{-1}\rho, \qquad -n^\delta \leqq \rho \leqq +n^\delta, \tag{8.73.10}$$

where δ is a fixed positive number, $\delta < 1/6$. Now we have

$$(8.73.11)\qquad \begin{aligned} z = e^{i\psi} &= e^{i\phi}\left\{1 + il_n^{-1}\rho + \frac{(il_n^{-1}\rho)^2}{2!} + \cdots\right\},\\ dz &= e^{i\phi}il_n^{-1}\left\{1 + il_n^{-1}\rho + \frac{(il_n^{-1}\rho)^2}{2!} + \cdots\right\} d\rho,\\ z - e^{i\phi} &= e^{i\phi}il_n^{-1}\rho\left\{1 + \frac{il_n^{-1}\rho}{2!} + \cdots\right\}. \end{aligned}$$

Whence from (8.73.8)

$$(8.73.12)\qquad f(z) = f(e^{i\phi}) - \tfrac{1}{2}\sin\phi\, e^{i(\phi-\pi/2)}l_n^{-2}\rho^2\{1 + c_1l_n^{-1}\rho + c_2(l_n^{-1}\rho)^2 + \cdots\},$$

where the c_ν are functions of ϕ, independent of n and ρ, and $c_m = O(A^m)$ uniformly in $\epsilon \leqq \phi \leqq \pi - \epsilon$, and uniformly as to m; $A = A(\epsilon)$. (We note that this condition for ϕ is more general than that in (8.22.9).) Hence, we have in the same sense as in (8.71.14), the following asymptotic expansion:

$$(8.73.13)\qquad \begin{aligned} G &= e^{-i\phi/2}\exp\{l_n^2 f(e^{i\phi})\}e^{i\phi}il_n^{-1}\\ &\quad\cdot\int_{-n^\delta}^{+n^\delta}\exp\{-\tfrac{1}{2}\sin\phi\, e^{i(\phi-\pi/2)}\rho^2\}\{1 + l_n^{-1}(c_1'\rho + c_1''\rho^3)\\ &\quad + l_n^{-2}(c_2'\rho^2 + c_2''\rho^4 + c_2'''\rho^6) + \cdots\}\, d\rho, \end{aligned}$$

where $c_1', c_1'', c_2', c_2'', c_2''', \cdots$ are constants. The principal term furnishes

$$(8.73.14)\qquad \begin{aligned} G &= e^{-i\phi/2}\exp\{l_n^2 f(e^{i\phi})\}\, e^{i\phi}il_n^{-1}(2\pi)^{\frac{1}{2}}(\sin\phi)^{-\frac{1}{2}}e^{-i(\phi/2-\pi/4)}\{1 + O(l_n^{-2})\}\\ &= \left(\frac{2\pi}{\sin\phi}\right)^{\frac{1}{2}}\exp\{-i(2n+\alpha+1)\phi + 3\pi i/4\}l_n^{-1}\\ &\qquad\cdot\exp\{\tfrac{1}{2}l_n^2\cos^2\phi + \tfrac{1}{4}l_n^2 + \tfrac{1}{2}il_n^2\sin\phi\cos\phi\}\{1 + O(l_n^{-2})\}, \end{aligned}$$

since the integrals with odd powers of ρ vanish. The bound for the error holds uniformly for $\epsilon \leqq \phi \leqq \pi - \epsilon$.

The integral in H is obtained by replacing ϕ by $\pi - \phi$, so that

$$(8.73.15)\qquad \begin{aligned} H &= \left(\frac{2\pi}{\sin\phi}\right)^{\frac{1}{2}}\exp\{(\alpha+\tfrac{1}{2})\pi i - i(2n+\alpha+1)(\pi-\phi) + 3\pi i/4\}\\ &\qquad\cdot l_n^{-1}\exp\{\tfrac{1}{2}l_n^2\cos^2\phi + \tfrac{1}{4}l_n^2 - \tfrac{1}{2}il_n^2\sin\phi\cos\phi\}\{1 + O(l_n^{-2})\}. \end{aligned}$$

Taking the absolute values of the integrands in G and H, we obtain, $\epsilon \leqq \phi \leqq \pi/2 - \epsilon n^{-\frac{1}{2}}$,

$$(8.73.16)\qquad K = O(\xi^{-1}l_n^{-1})l_n^{-1}\exp\{\tfrac{1}{2}l_n^2\cos^2\phi + \tfrac{1}{4}l_n^2\}.$$

We easily see that, except for terms of higher order, $H = -\bar{G}$. Therefore,

$$(8.73.17)\quad \begin{aligned}\Re\left\{\frac{1}{2\pi i}G + \frac{1}{2\pi i}H + K\right\} &= \left(\frac{2}{\pi\sin\phi}\right)^{\frac{1}{2}} l_n^{-1}\exp\{\tfrac{1}{2}l_n^2\cos^2\phi + \tfrac{1}{4}l_n^2\}\\ &\cdot\{\sin[\tfrac{1}{2}l_n^2\sin\phi\cos\phi - (2n+\alpha+1)\phi + 3\pi/4] + O(\xi^{-1}l_n^{-1})\},\end{aligned}$$

since $l_n^{-2} = O(\xi^{-1}l_n^{-1})$. Introducing this into (8.73.5), we obtain

$$\begin{aligned}(-\tfrac{1}{4})^n\frac{L_n^{(\alpha)}(\xi^2)}{\Gamma(n+\alpha+1)} &= \pi^{-1}(\sin\phi)^{-\frac{1}{2}}\xi^{-\alpha-\frac{1}{2}}2^{\alpha+1}l_n^{-2n-\alpha-\frac{1}{2}}\exp\{\tfrac{1}{2}l_n^2\cos^2\phi + \tfrac{1}{4}l_n^2\}\\ &\cdot\{\sin[\tfrac{1}{2}l_n^2\sin\phi\cos\phi - (2n+\alpha+1)\phi + 3\pi/4] + O(\xi^{-1}l_n^{-1})\}.\end{aligned}$$

Now

$$l_n = 2n^{\frac{1}{2}}\exp\left(\frac{\alpha+1}{4n}\right)\{1 + O(n^{-2})\},$$

$$4^n\Gamma(n+\alpha+1) = \pi^{\frac{1}{2}}2^{2n+\frac{1}{2}}n^{n+\alpha+\frac{1}{2}}e^{-n}\{1 + O(n^{-1})\}.$$

And since $n^{-1} = O(\xi^{-1}l_n^{-1})$, we have

$$\begin{aligned}L_n^{(\alpha)}(\xi^2) &= (-1)^n(\pi\sin\phi)^{-\frac{1}{2}}\xi^{-\alpha-\frac{1}{2}}n^{\alpha/2-1/4}\\ &\cdot\exp\{-n - (\alpha+1)/2\}\exp\{\tfrac{1}{2}\xi^2 + n + (\alpha+1)/2\}\\ &\cdot\{\sin[\tfrac{1}{2}l_n^2\sin\phi\cos\phi - (2n+\alpha+1)\phi + 3\pi/4] + O(\xi^{-1}l_n^{-1})\},\end{aligned}$$

$$(8.73.18)\quad \begin{aligned}e^{-\xi^2/2}L_n^{(\alpha)}(\xi^2) &= (-1)^n(\pi\sin\phi)^{-\frac{1}{2}}\xi^{-\alpha-\frac{1}{2}}n^{\alpha/2-\frac{1}{4}}\\ &\cdot\{\sin[(n+(\alpha+1)/2)\sin 2\phi - (2n+\alpha+1)\phi + 3\pi/4] + O(\xi^{-1}l_n^{-1})\}.\end{aligned}$$

This is identical with (8.22.9).

We observe that in the application of this result to Hermite polynomials certain simplifications are possible. For $\alpha = \pm\frac{1}{2}$ the O-terms in (1.71.9) and (8.73.4) vanish identically, so that K in (8.73.5) can be cancelled, and ξ can be arbitrarily near zero. Therefore, (8.22.12) follows readily by use of (5.6.1).

8.74. Method of steepest descent; Laguerre polynomials for $(4+\eta)n \leqq x \leqq An$

We start again from (8.73.2) and integrate along a proper circle about the origin. Using a notation analogous to that in the previous section, we assume that

$$(8.74.1)\quad \xi = l_n\cosh\phi,\qquad \epsilon \leqq \phi \leqq \omega,\qquad w = l_n z.$$

Then we have

$$(8.74.2)\quad \begin{aligned}(-\tfrac{1}{4})^n\frac{L_n^{(\alpha)}(\xi^2)}{\Gamma(n+\alpha+1)} &= 2^\alpha\xi^{-\alpha}(2\pi\xi)^{-\frac{1}{2}}l_n^{-2n-\alpha-\frac{1}{2}}\\ &\cdot\left\{\frac{1}{2\pi i}G_1' + \frac{1}{2\pi i}H_1' + 2\Re\left(\frac{1}{2\pi i}G_2'\right) + 2\Re\left(\frac{1}{2\pi i}H_2'\right) + K'\right\},\end{aligned}$$

where

$$G_1' = \int z^{-\frac{1}{2}} \exp\{-\tfrac{1}{4}l_n^2 z^2 + l_n^2 z \cosh\phi - \tfrac{1}{2}l_n^2 \log z\}\, dz,$$

$$(8.74.3) \qquad H_1' = e^{(\alpha+\frac{1}{2})\pi i} \int z^{-\frac{1}{2}} \exp\{-\tfrac{1}{4}l_n^2 z^2 - l_n^2 z \cosh\phi - \tfrac{1}{2}l_n^2 \log z\}\, dz,$$

$$K' = O(\xi^{-1} l_n^{-1}) \int |\exp\{-\tfrac{1}{4}l_n^2 z^2 \pm l_n^2 z \cosh\phi - \tfrac{1}{2}l_n^2 \log z\}|\, |dz|.$$

In each of G_1' and H_1', we integrate along both arcs

$$-\pi/4 \leqq \arg z \leqq +\pi/4, \qquad 3\pi/4 \leqq \arg z \leqq 5\pi/4,$$

of the circle $|z| = e^{-\phi}$, while in G_2' and H_2' (which have the same integrands as G_1' and H_1', respectively) we take $\pi/4 \leqq \arg z \leqq 3\pi/4$; in K' we take the arc $0 \leqq \arg z \leqq \pi$ of the preceding section. Then (8.73.4) can again be used [cf. (1.71.9)].

In this case we discuss

$$(8.74.4) \qquad f(z) = -\tfrac{1}{4}z^2 + z\cosh\phi - \tfrac{1}{2}\log z.$$

The condition $f'(z) = 0$ furnishes $z = e^{\pm\phi}$, and we have $f''(e^{-\phi}) = (e^{2\phi} - 1)/2 = e^{\phi}\sinh\phi$. Therefore, the circle $|z| = e^{-\phi}$ passes through the saddle point with the smaller modulus and has the critical direction at that point. For the second integral we obtain the saddle points $-e^{\pm\phi}$, from which it follows that the circle $|z| = e^{-\phi}$ can be used here again. Obviously, for $z = e^{-\phi+i\psi}$,

$$(8.74.5) \qquad \Re\{f(z)\} = -\tfrac{1}{4}e^{-2\phi}\cos 2\psi + e^{-\phi}\cosh\phi\cos\psi + \phi/2$$

decreases as ψ increases from 0 to π.

On writing $\psi = l_n^{-1}\rho$, $-n^{\delta} \leqq \rho \leqq +n^{\delta}$, $0 < \delta < 1/6$, we obtain

$$z = e^{-\phi+i\psi} = e^{-\phi}\left\{1 + il_n^{-1}\rho + \frac{(il_n^{-1}\rho)^2}{2!} + \cdots\right\},$$

$$(8.74.6) \qquad dz = e^{-\phi} il_n^{-1}\left\{1 + il_n^{-1}\rho + \frac{(il_n^{-1}\rho)^2}{2!} + \cdots\right\} d\rho,$$

$$z - e^{-\phi} = e^{-\phi} il_n^{-1}\rho\left\{1 + \frac{il_n^{-1}\rho}{2!} + \cdots\right\};$$

whence

$$(8.74.7) \qquad f(z) = f(e^{-\phi}) - \tfrac{1}{2}\sinh\phi\, e^{-\phi} l_n^{-2}\rho^2\{1 + c_1 l_n^{-1}\rho + c_2(l_n^{-1}\rho)^2 + \cdots\}.$$

Here the coefficients $c_1, c_2, \cdots$ (and also $c_1', c_1'', c_2', c_2'', c_2''', \cdots$ below) are analogous to those in §8.73. Hence, in the same sense as before, we have the asymptotic expansion

$$
\begin{aligned}
G_1' &= e^{\phi/2}\exp\{l_n^2 f(e^{-\phi})\}e^{-\phi}il_n^{-1}\int_{-n^\delta}^{+n^\delta}\exp\{-\tfrac{1}{2}\sinh\phi\, e^{-\phi}\rho^2\} \\
&\qquad\cdot\{1 + l_n^{-1}(c_1'\rho + c_1''\rho^3) + l_n^{-2}(c_2'\rho^2 + c_2''\rho^4 + c_2'''\rho^6) + \cdots\}\,d\rho \\
&= e^{\phi/2}\exp\{l_n^2 f(e^{-\phi})\}e^{-\phi}il_n^{-1}(2\pi)^{\frac{1}{2}}(\sinh\phi)^{-\frac{1}{2}}e^{\phi/2}\{1 + O(l_n^{-2})\} \\
&= \left(\frac{2\pi}{\sinh\phi}\right)^{\frac{1}{2}} e^{(2n+\alpha+1)\phi}il_n^{-1}\exp\{l_n^2\cosh^2\phi - \tfrac{1}{4}l_n^2 e^{2\phi}\}\{1 + O(l_n^{-2})\}.
\end{aligned}
\tag{8.74.8}
$$

The "principal parts" of G_1' and H_1' are given by the arcs

$$-\pi/4 \leqq \arg z \leqq +\pi/4, \qquad 3\pi/4 \leqq \arg z \leqq 5\pi/4,$$

respectively. If we replace z by $e^{i\pi}z$ in H_1', we see immediately that these principal parts are identical. Consequently, we have $G_1' = H_1'$ except for terms which are of higher order than the remainder term in (8.74.8); furthermore, G_2' and H_2' are of higher order than the same remainder term. Therefore, because of $\xi^{-1}l_n^{-1} = O(l_n^{-2})$, we have

$$
\begin{aligned}
&\frac{1}{2\pi i}G_1' + \frac{1}{2\pi i}H_1' + 2\Re\left(\frac{1}{2\pi i}G_2'\right) + 2\Re\left(\frac{1}{2\pi i}H_2'\right) + K' \\
&\quad = \left(\frac{2}{\pi\sinh\phi}\right)^{\frac{1}{2}} e^{(2n+\alpha+1)\phi}l_n^{-1}\exp\{l_n^2\cosh^2\phi - \tfrac{1}{4}l_n^2 e^{2\phi}\}\{1 + O(l_n^{-2})\},
\end{aligned}
\tag{8.74.9}
$$

so that

$$
\begin{aligned}
(-\tfrac{1}{4})^n\frac{L_n^{(\alpha)}(\xi^2)}{\Gamma(n+\alpha+1)} &= \pi^{-1}(\sinh\phi)^{-\frac{1}{2}}\xi^{-\alpha-\frac{1}{2}}2^\alpha l_n^{-2n-\alpha-\frac{1}{2}} \\
&\qquad\cdot\exp\{\xi^2 - \tfrac{1}{4}l_n^2 e^{2\phi} + (2n+\alpha+1)\phi\}\{1 + O(l_n^{-2})\};
\end{aligned}
\tag{8.74.10}
$$

or

$$
\begin{aligned}
e^{-\xi^2/2}L_n^{(\alpha)}(\xi^2) &= \tfrac{1}{2}(-1)^n(\pi\sinh\phi)^{-\frac{1}{2}}\xi^{-\alpha-\frac{1}{2}}n^{\alpha/2-\frac{1}{4}} \\
&\qquad\cdot\exp\{[n + (\alpha+1)/2](2\phi - \sinh 2\phi)\}\{1 + O(n^{-1})\}.
\end{aligned}
\tag{8.74.11}
$$

Returning to the variable x, we obtain (8.22.10). From this result (8.22.13) follows immediately.

8.75. Method of steepest descent; Laguerre polynomials for $x = 4n + O(n^{\frac{1}{3}})$

(1) First, let t be real and bounded. We write as before

$$x = \xi^2, \qquad \xi = l_n - (6l_n)^{-\frac{1}{3}}t, \qquad w = l_n z,$$

$$(-\tfrac{1}{4})^n\frac{L_n^{(\alpha)}(\xi^2)}{\Gamma(n+\alpha+1)} = 2^\alpha\xi^{-\alpha}(2\pi\xi)^{-\frac{1}{2}}l_n^{-2n-\alpha-\frac{1}{2}}2\Re\left\{\frac{1}{2\pi i}G'' + \frac{1}{2\pi i}H'' + K''\right\}, \tag{8.75.1}$$

where

$$G'' = \int z^{-\frac{1}{2}} \exp \{l_n^2 f_1(z) - 6^{-\frac{1}{3}} l_n^{\frac{4}{3}} tz\}\, dz,$$

(8.75.2) $$H'' = e^{(\alpha+\frac{1}{2})\pi i} \int z^{-\frac{1}{2}} \exp \{l_n^2 f_2(z) + 6^{-\frac{1}{3}} l_n^{\frac{4}{3}} tz\}\, dz,$$

$$K'' = O(l_n^{-2}) \int | \exp \{l_n^2 f_\nu(z) \mp 6^{-\frac{1}{3}} l_n^{\frac{4}{3}} tz\} |\, | dz |, \qquad \nu = 1, 2,$$

with

(8.75.3) $$f_1(z) = -\tfrac{1}{4} z^2 + z - \tfrac{1}{2} \log z, \qquad f_2(z) = -\tfrac{1}{4} z^2 - z - \tfrac{1}{2} \log z.$$

The integrals in (8.75.2) are extended over the upper half of a proper curve symmetric to the real axis for which $| z |$ and $| z |^{-1}$ are bounded.

The "saddle point condition" $f_1'(z) = -z/2 + 1 - (2z)^{-1} = 0$ furnishes $z = 1$, and we notice that $f_1''(1) = 0$ and $f_1'''(1) = -1$. Therefore, this saddle point is of a different character from the preceding ones.

We first integrate along the segment

(8.75.4) $$z = 1 + 6^{\frac{1}{3}} l_n^{-\frac{2}{3}} \rho e^{2\pi i/3}, \qquad 0 \leqq \rho \leqq n^{\delta},$$

where δ is a fixed positive number, $\delta < 1/6$, then along the segment symmetric to this one with respect to the imaginary axis, and finally along the arc of a

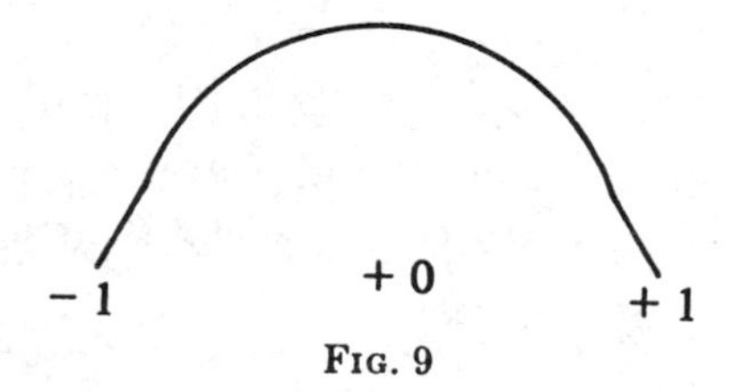

Fig. 9

circle with center at $z = 0$ which connects the ends of the segments mentioned (see Fig. 9). For certain constants $c_4, c_5, \cdots$, we have

(8.75.5) $$f_1(z) = f_1(1) + \frac{1}{3!}(z - 1)^3 f_1'''(1) + \cdots = \tfrac{3}{4} - l_n^{-2}\rho^3 + c_4(l_n^{-\frac{2}{3}}\rho)^4 + c_5(l_n^{-\frac{2}{3}}\rho)^5 + \cdots.$$

Furthermore, if r is the radius of the arc of the circle, then

(8.75.6) $$\Re\{f_1(re^{i\psi})\} = -\tfrac{1}{4}r^2 \cos 2\psi + r \cos \psi - \tfrac{1}{2} \log r$$

is decreasing for $0 \leqq \psi \leqq \pi$, since $r \cos \psi < 1$. We now obtain the following expansion

$$G'' = \exp\{\tfrac{3}{4}l_n^2 - 6^{-1/3}l^{4/3}t\}\cdot 6^{1/3}l_n^{-2/3}e^{2\pi i/3}$$

$$\cdot\int_0^{n^\delta} \exp\{-\rho^3 - \rho e^{2\pi i/3}t\}\{1 + (c_1'\rho + c_1''\rho^4)\, l_n^{-2/3} + (c_2'\rho^2 + c_2''\rho^5 + c_2'''\rho^8)l_n^{-4/3} + \cdots\}\, d\rho. \tag{8.75.7}$$

This is an asymptotic expansion in the usual sense. At the end-point $\rho = n^\delta$ we have the bound $O(\exp[-cn^{3\delta}])$, $c > 0$, for the integrand, which is at the same time a bound for the contribution of the remainder of the contour. (The slight modification of the circle around $z = -1$ is immaterial.) Thus

$$G'' = \exp\{\tfrac{3}{4}\, l_n^2 - 6^{-1/3}l_n^{4/3}t\}\cdot 6^{1/3}l_n^{-2/3}e^{2\pi i/3} \cdot\left\{\int_0^\infty \exp(-\rho^3 - \rho e^{2\pi i/3}t)\, d\rho + O(l_n^{-2/3})\right\}. \tag{8.75.8}$$

In the same way we see that the principal part of H'' is due to the small segment

$$z = -1 + 6^{1/3}l_n^{-2/3}\rho c^{\pi i/3}, \qquad 0 \leqq \rho \leqq n^\delta. \tag{8.75.9}$$

We find

$$\begin{aligned} f_2(z) &= f_2(-1) + \frac{1}{3!}(z+1)^3 f_2'''(-1) + \cdots \\ &= 3/4 - i\pi/2 - l_n^{-2}\rho^3 + \cdots . \end{aligned} \tag{8.75.10}$$

Therefore,

$$H'' = -\, e^{(\alpha+1/2)\pi i}e^{-i\pi/2}\exp\left\{\tfrac{3}{4}\, l_n^2 - \frac{i\pi}{2}\, l_n^2 - 6^{-1/3}l_n^{4/3}t\right\} \cdot 6^{1/3}l_n^{-2/3}e^{\pi i/3}\left\{\int_0^\infty \exp(-\rho^3 + \rho e^{\pi i/3}t)\, d\rho + O(l_n^{-2/3})\right\}, \tag{8.75.11}$$

and

$$K'' = O(l_n^{-2})l_n^{-2/3}\exp\{\tfrac{3}{4}\, l_n^2 - 6^{-1/3}l_n^{4/3}t\}. \tag{8.75.12}$$

Now we can readily show that $H'' = -\overline{G''}$ except for terms of higher order. Consequently,

$$(-\tfrac{1}{4})^n \frac{L_n^{(\alpha)}(\xi^2)}{\Gamma(n+\alpha+1)} = \pi^{-1/2}2^{\alpha+1/2}6^{1/3}\xi^{-\alpha-1/2}l_n^{-2n-\alpha-7/6}\exp\{\tfrac{3}{4}l_n^2 - 6^{-1/3}l_n^{4/3}t\} \cdot\left\{\Im\left[e^{2\pi i/3}\int_0^\infty \exp(-\rho^3 - \rho e^{2\pi i/3}t)\, d\rho\right] + O(l_n^{-2/3})\right\}. \tag{8.75.13}$$

The imaginary part in the braces is Airy's function $A(t)$ (Problem 2). Substituting the approximate values of $4^n\Gamma(n + \alpha + 1)$, $l_n^{-2n-\alpha-7/6}$, and $\xi^{-\alpha-1/2} = l_n^{-\alpha-1/2}\{1 + O(l_n^{-4/3})\}$, and observing that

$$\xi^2/2 = l_n^2/2 - 6^{-1/3}l_n^{4/3}t + O(l_n^{-2/3}), \tag{8.75.14}$$

we obtain

$$e^{-\xi^2/2} L_n^{(\alpha)}(\xi^2) = (-1)^n \pi^{-1} 2^{-\alpha-1/3} 3^{1/3} n^{-1/3}\{A(t) + O(l_n^{-2/3})\},$$

which is (8.22.9) with the less precise remainder term $O(n^{-1/3})$.

(2) Direct calculation of a further term in (8.75.5) leads to

$$c_4 = \frac{1}{4!} f_1^{(4)}(1) 6^{4/3} e^{8\pi i/3} = \tfrac{1}{8} 6^{4/3} e^{8\pi i/3}. \tag{8.75.15}$$

Furthermore, from (8.75.4)

$$z^{-1/3} = 1 - \tfrac{1}{2} 6^{1/3} e^{2\pi i/3} l_n^{-2/3} \rho + O(l_n^{-4/3} \rho^2). \tag{8.75.16}$$

Consequently, the expression in the braces in (8.75.8) can be written as follows

$$\int_0^\infty \exp(-\rho^3 - \rho e^{2\pi i/3} t)\, d\rho [1 + (-\tfrac{1}{2} 6^{1/3} e^{2\pi i/3} \rho + \tfrac{1}{8} 6^{4/3} e^{8\pi i/3} \rho^4) l_n^{-2/3} + O(l_n^{-4/3})].$$

The corresponding more precise form of (8.75.13) can be obtained by replacing the expression in the braces by

$$\begin{aligned} &\Im\left[e^{2\pi i/3} \int_0^\infty \exp(-\rho^3 - \rho e^{2\pi i/3} t)\, d\rho\right] \\ &\quad - \tfrac{1}{2} 6^{1/3} l_n^{-2/3} \Im\left[e^{4\pi i/3} \int_0^\infty \exp(-\rho^3 - \rho e^{2\pi i/3} t) \rho\, d\rho\right] \\ &\quad + \tfrac{1}{8} 6^{4/3} l_n^{-2/3} \Im\left[e^{10\pi i/3} \int_0^\infty \exp(-\rho^3 - \rho e^{2\pi i/3} t) \rho^4\, d\rho\right] + O(l_n^{-4/3}) \\ &\quad = A(t) + \{\tfrac{1}{2} 6^{1/3} A'(t) + \tfrac{1}{8} 6^{4/3} A^{(4)}(t)\} l_n^{-2/3} + O(l_n^{-4/3}) \\ &\quad = A(t) + \tfrac{1}{2} 6^{-2/3} l_n^{-2/3} t^2 A(t) + O(l_n^{-4/3}). \end{aligned}$$

Here the differential equation (1.81.2) has been used. Now (8.75.14) can be also written in the more precise form

$$\xi^2/2 - l_n^2/2 - 6^{-1/3} l_n^{2/3} t + \tfrac{1}{2} 6^{-2/3} l_n^{-2/3} t^2,$$

so that

$$e^{-\xi^2/2} = \exp[-l_n^2/2 + 6^{1/3} l_n^{2/3} t]\{1 - \tfrac{1}{2} 6^{-2/3} l_n^{-2/3} t^2 + O(l_n^{-4/3})\}.$$

From this, (8.22.9) readily follows with the remainder term $O(l_n^{-4/3}) = O(n^{-2/3})$.

(3) The case of a general complex t can be settled by a slight modification of the argument in (1).

The corresponding asymptotic formula for Hermite polynomials follows immediately by means of (5.6.1).

8.8. Differentiation of certain asymptotic formulas

Differentiation of an asymptotic formula with respect to a parameter occurring in that formula is in general not permitted. In some of our previous formulas, however, the permissibility of differentiation is readily established. We must, of course, modify the remainder in a proper way. We shall discuss the formulas (8.21.18) and (8.22.6), involving Jacobi and Laguerre polynomials, from this point of view, using the important identities (4.21.7) and (5.1.14).

(1) First let us consider (8.21.18) which contains Darboux's formula (8.21.10). We shall prove that

$$\frac{d}{d\theta}\{P_n^{(\alpha,\beta)}(\cos\theta)\} = n^{\frac{1}{2}}k(\theta)\{-\sin(N\theta+\gamma) + (n\sin\theta)^{-1}O(1)\}, \tag{8.8.1}$$

$$\alpha > -1,\ \beta > -1,\ cn^{-1} \leqq \theta \leqq \pi - cn^{-1},$$

with the same notation as in (8.21.10). We note that $k'(\theta) = k(\theta)(\sin\theta)^{-1}O(1)$.

In the proof, we write for the left-hand member of (8.8.1),

$$\begin{aligned}-\tfrac{1}{2}\sin\theta(n+\alpha+\beta+1)&P_{n-1}^{(\alpha+1,\beta+1)}(\cos\theta)\\ &= -\tfrac{1}{2}\sin\theta(n+\alpha+\beta+1)(n-1)^{-\frac{1}{2}}k(\theta)\\ &\qquad\cdot\left(\sin\frac{\theta}{2}\cos\frac{\theta}{2}\right)^{-1}\{\cos(N\theta+\gamma-\pi/2) + (n\sin\theta)^{-1}O(1)\},\end{aligned}$$

which establishes the statement. Evidently, the remainder term of (8.8.1) can be replaced by $\theta^{-\alpha-\frac{3}{2}}O(n^{-\frac{1}{2}})$ if $cn^{-1} \leqq \theta \leqq \pi - \epsilon$, and by $O(n^{-\frac{1}{2}})$ if $\epsilon \leqq \theta \leqq \pi - \epsilon$.

From (8.8.1) we derive the following important formulas:

$$\begin{aligned}&\frac{P_n^{(\alpha,\beta)}(\cos\theta_1) - P_n^{(\alpha,\beta)}(\cos\theta_2)}{\theta_1-\theta_2}\\ &\qquad = n^{-\frac{1}{2}}k(\theta_1)\frac{\cos(N\theta_1+\gamma)-\cos(N\theta_2+\gamma)}{\theta_1-\theta_2}\\ &\qquad\qquad + \theta_1^{-\alpha-\frac{3}{2}}O(n^{-\frac{1}{2}}) + (\pi-\theta_2)^{-\beta-\frac{3}{2}}O(n^{-\frac{1}{2}}),\end{aligned}$$

$$\begin{aligned}&\frac{P_n^{(\alpha,\beta)}(\cos\theta_1) - P_n^{(\alpha,\beta)}(\cos\theta_2)}{\cos\theta_1-\cos\theta_2}\\ &\qquad = n^{-\frac{1}{2}}k(\theta_1)\frac{\cos(N\theta_1+\gamma)-\cos(N\theta_2+\gamma)}{\cos\theta_1-\cos\theta_2}\\ &\qquad\qquad + \theta_1^{-\alpha-\frac{5}{2}}O(n^{-\frac{1}{2}}) + (\pi-\theta_2)^{-\beta-\frac{5}{2}}O(n^{-\frac{1}{2}}),\\ &\qquad\qquad \alpha > -1,\ \beta > -1,\ cn^{-1} \leqq \theta_1 < \theta_2 \leqq \pi - cn^{-1}.\end{aligned} \tag{8.8.2}$$

Similar formulas hold with $k(\theta_2)$ instead of $k(\theta_1)$ in the right-hand member.

For the proof we apply the mean-value theorem to

$$\phi(\theta) = P_n^{(\alpha,\beta)}(\cos\theta) - n^{-\frac{1}{2}}k(\theta)\cos(N\theta+\gamma).$$

Thus

$$\frac{\phi(\theta_1) - \phi(\theta_2)}{\theta_1 - \theta_2} = \phi'(\tau_1), \qquad \frac{\phi(\theta_1) - \phi(\theta_2)}{\cos\theta_1 - \cos\theta_2} = -\frac{\phi'(\tau_2)}{\sin\tau_2}, \qquad \begin{array}{l}\theta_1 < \tau_1 < \theta_2,\\ \theta_1 < \tau_2 < \theta_2.\end{array}$$

But, we have $\phi'(\tau_1) = n^{\frac{1}{2}}k(\tau_1)(n \sin \tau_1)^{-1}O(1)$, which is $\theta_1^{-\alpha-\frac{3}{2}}O(n^{-\frac{1}{2}})$ or $(\pi - \theta_2)^{-\beta-\frac{3}{2}}O(n^{-\frac{1}{2}})$, according as $\tau_1 \leqq \pi/2$ or $\tau_1 \geqq \pi/2$. Moreover,

$$\frac{k(\theta_1)\cos(N\theta_1 + \gamma) - k(\theta_2)\cos(N\theta_2 + \gamma)}{\theta_1 - \theta_2}$$
$$= k(\theta_1)\frac{\cos(N\theta_1 + \gamma) - \cos(N\theta_2 + \gamma)}{\theta_1 - \theta_2} + \cos(N\theta_2 + \gamma)\frac{k(\theta_1) - k(\theta_2)}{\theta_1 - \theta_2}.$$

The latter ratio is $k'(\tau) = \theta_1^{-\alpha-\frac{3}{2}}O(1) + (\pi - \theta_2)^{-\beta-\frac{3}{2}}O(1)$, $\theta_1 < \tau < \theta_2$. A similar argument leads to the second formula in (8.8.2).

If both θ_1 and θ_2 are confined to an interior interval $\epsilon \leqq \theta \leqq \pi - \epsilon$ of $[0, \pi]$, the remainders in the formulas (8.8.2) are simply $O(n^{-\frac{1}{2}})$.

(2) Next we consider (8.22.6), which contains Fejér's formula (8.22.1). We write (8.22.6) in the form

$$(8.8.3) \qquad \begin{gathered} L_n^{(\alpha)}(x) = k(x)n^{\alpha/2-\frac{1}{4}}\{\cos[2(nx)^{\frac{1}{2}} + \gamma] + (nx)^{-\frac{1}{2}}O(1)\},\\ k(x) = \pi^{-\frac{1}{2}}e^{x/2}x^{-\alpha/2-\frac{1}{4}}, \qquad \gamma = -(\alpha + \tfrac{1}{2})\pi/2,\\ \alpha > -1,\ cn^{-1} \leqq x \leqq \omega, \end{gathered}$$

where c and ω are fixed positive numbers. By means of (5.1.14) we now obtain

$$(8.8.4) \qquad \frac{d}{d(x^{\frac{1}{2}})}\{L_n^{(\alpha)}(x)\} = 2k(x)n^{\alpha/2+\frac{1}{4}}\{-\sin[2(nx)^{\frac{1}{2}} + \gamma] + (nx)^{-\frac{1}{2}}O(1)\},$$

noting that $d\{k(x)\}/d(x^{\frac{1}{2}}) = k(x)O(x^{-\frac{1}{2}})$, and that $(n-1)^{\frac{1}{2}} - n^{\frac{1}{2}} = O(n^{-\frac{1}{2}})$. The mean-value theorem furnishes, if both x_1 and x_2 belong to the interval $cn^{-1} \leqq x \leqq \omega$,

$$(8.8.5) \qquad \begin{aligned} &\frac{L_n^{(\alpha)}(x_1) - L_n^{(\alpha)}(x_2)}{x_1^{\frac{1}{2}} - x_2^{\frac{1}{2}}}\\ &\quad = k(x_1)n^{\alpha/2-\frac{1}{4}}\frac{\cos[2(nx_1)^{\frac{1}{2}} + \gamma] - \cos[2(nx_2)^{\frac{1}{2}} + \gamma]}{x_1^{\frac{1}{2}} - x_2^{\frac{1}{2}}}\\ &\quad\quad + x_1^{-\alpha/2-\frac{3}{4}}O(n^{\alpha/2-\frac{1}{4}}) + x_2^{-\alpha/2-\frac{3}{4}}O(n^{\alpha/2-\frac{1}{4}}), \qquad \alpha > -1. \end{aligned}$$

The proof is similar to that used in the case of Jacobi polynomials.

8.9. Applications; asymptotic properties of the zeros of Jacobi and Laguerre polynomials

(1) We shall first point out some consequences of the formulas (8.21.18) and (8.8.1), which will be important for the discussion of interpolation with Jacobi abscissas (Chapter XIV). Setting $c = 1$ in (8.21.18), and using the same notation as before, we have the result:

THEOREM 8.9.1. *Let $\alpha > -1$, $\beta > -1$, and let $0 < \theta_1 < \theta_2 < \cdots < \theta_n < \pi$ be the zeros of $P_n^{(\alpha,\beta)}(\cos\theta)$. Then*

$$\theta_\nu = n^{-1}\{\nu\pi + O(1)\}, \tag{8.9.1}$$

with $O(1)$ being uniformly bounded for all values of $\nu = 1, 2, \cdots n; n = 1, 2, 3, \cdots$.
Furthermore,

$$| P_n^{(\alpha,\beta)\prime}(\cos\theta_\nu) | \sim \nu^{-\alpha-\frac{3}{2}} n^{\alpha+2}, \qquad 0 < \theta_\nu \leqq \pi/2, \tag{8.9.2}$$

in the sense that the ratio of these expressions remains between certain positive bounds depending only on α and β.

Let ρ be a fixed number, $0 < \rho < \pi/2$. Substitute in (8.21.18)

$$N\theta = (\nu - \tfrac{1}{2})\pi - \gamma \pm \rho, \qquad \nu > 0, \nu \text{ an integer}, 0 < \theta \leqq \pi/2. \tag{8.9.3}$$

Then the first term on the right, that is, $\pm(-1)^\nu n^{-\frac{1}{2}}k(\theta)\sin\rho$, is the principal term provided the remainder $\theta^{-\alpha-\frac{3}{2}}O(n^{-\frac{3}{2}})$ is less than $n^{-\frac{1}{2}}k(\theta)\sin\rho$. This is so if ν and n are sufficiently large, $\nu \geqq M = M(\alpha, \beta, \rho)$. The same is true for the formula (8.8.1). We have also $\theta > 0$ if M is properly chosen. Furthermore, let $(\nu - \frac{1}{2})\pi - \gamma + \rho \leqq N\pi/2$, so that $\theta \leqq \pi/2$. Then for the values (8.9.3)

$$\operatorname{sgn} P_n^{(\alpha,\beta)}(\cos\theta) = (-1)^\nu \quad \text{and} \quad (-1)^{\nu+1}, \tag{8.9.4}$$

respectively. Hence $P_n^{(\alpha,\beta)}(\cos\theta)$ has at least one zero between the bounds given by the values (8.9.3) and, since (8.8.1) shows that it is monotonic in this interval, it has only one zero there. Also, we see that in the same interval

$$\frac{d}{d\theta}\{P_n^{(\alpha,\beta)}(\cos\theta)\} \sim n^{\frac{1}{2}}k(\theta) \sim n^{\frac{1}{2}}\theta^{-\alpha-\frac{1}{2}}, \tag{8.9.5}$$

and in the "complementary intervals" of $[0, \pi/2]$,

$$| P_n^{(\alpha,\beta)}(\cos\theta) | \sim n^{-\frac{1}{2}}k(\theta) \sim n^{-\frac{1}{2}}\theta^{-\alpha-\frac{1}{2}}. \tag{8.9.6}$$

The statements (8.9.5) and (8.9.6) hold in the sense that the ratio of the expressions in question remains bounded from zero and infinity, uniformly in θ if θ lies in the intervals mentioned; here n is sufficiently large, $\nu \geqq M(\alpha, \beta, \rho)$, and $(\nu - \frac{1}{2})\pi - \gamma + \rho \leqq N\pi/2$.

Thus $P_n^{(\alpha,\beta)}(\cos\theta)$ has no zeros in the "complementary intervals."

In the interval $0 \leqq \theta \leqq N^{-1}\{(M - \frac{1}{2})\pi - \gamma - \rho\}$ we have a bounded number of zeros which have the form $n^{-1}(j_k + \epsilon_n)$, where j_k denote the positive zeros of $J_\alpha(z)$ and $\epsilon_n \to 0$ (cf. (8.1.1) and (8.1.3)). This furnishes (8.9.1) for $0 < \theta_\nu \leqq \pi/2$. Taking into consideration the formula (4.1.3), the full statement (8.9.1) follows.

In addition, we find, by use of (8.9.5), that

$$| P_n^{(\alpha,\beta)\prime}(\cos\theta_\nu) | \sim \theta_\nu^{-1} n^{\frac{1}{2}} \theta_\nu^{-\alpha-\frac{1}{2}} = n^{\frac{1}{2}} \theta_\nu^{-\alpha-\frac{3}{2}}, \tag{8.9.7}$$

provided the conditions mentioned are satisfied. The extension to the total interval follows again by means of (8.1.1). Combining this result with (8.9.1), we obtain (8.9.2).

For the zeros θ_ν from a fixed interval $[a, b]$ in the interior of $[0, \pi]$, that is, $0 < a < b < \pi$, (8.9.1) can be written in the more precise form

$$\theta_\nu = N^{-1}\{(\nu - \tfrac{1}{2})\pi - \gamma + K\pi + \epsilon_n\}, \tag{8.9.8}$$

where K is a fixed integer (depending only on α, β, a, b), and $\epsilon_n \to 0$. In this case, (8.9.2) attains the simpler form

$$| P_n^{(\alpha,\beta)\prime}(\cos \theta_\nu) | \sim n^{\frac{3}{2}}. \tag{8.9.9}$$

(2) The analogous statements for Laguerre polynomials follow from (8.22.6), (8.1.8), and (8.8.4). We use here the same notation as before.

THEOREM 8.9.2. *Let $\alpha > -1$, and let $x_1 < x_2 < \cdots < x_n$ be the zeros of $L_n^{(\alpha)}(x)$; then we have for the zeros x_ν from a fixed interval $0 < x \leqq \omega$*

$$2x_\nu^{\frac{1}{2}} = n^{-\frac{1}{2}}\{\nu\pi + O(1)\}. \tag{8.9.10}$$

Moreover,

$$| L_n^{(\alpha)\prime}(x_\nu) | \sim x_\nu^{-\alpha/2-\frac{3}{4}} n^{\alpha/2+\frac{3}{4}} \sim \nu^{-\alpha-\frac{3}{2}} n^{\alpha+1}. \tag{8.9.11}$$

Let ρ be a fixed positive number. Introducing in (8.22.6)

$$2(nx)^{\frac{1}{2}} = (\nu - \tfrac{1}{2})\pi - \gamma \pm \rho, \qquad \nu > 0,\ \nu \text{ an integer},\ 0 < x \leqq \omega, \tag{8.9.12}$$

we obtain values of opposite signs provided $\nu \geqq M = M(\alpha, \rho)$, and n is sufficiently large. Then we see from (8.8.4) that $L_n^{(\alpha)}(x)$ is monotonic between the corresponding values of x, and hence $L_n^{(\alpha)}(x)$ has precisely one zero in the corresponding interval. Taking (8.1.8) into account, we find (8.9.10) as in (1). Again, from (8.8.4) and (8.1.8), $0 < x_\nu \leqq \omega$,

$$| L_n^{(\alpha)\prime}(x_\nu) | \sim x_\nu^{-\frac{1}{2}} k(x_\nu) n^{\alpha/2+\frac{3}{4}}.$$

For the zeros x_ν from a fixed positive interval $\epsilon \leqq x_\nu \leqq \omega$, $\epsilon > 0$, we obtain

$$| L_n^{(\alpha)\prime}(x_\nu) | \sim n^{\alpha/2+\frac{3}{4}}. \tag{8.9.13}$$

(3) Let α be arbitrary and real. The preceding results (especially formula (8.22.11)) enable us to discuss the largest zeros of $L_n^{(\alpha)}(x)$. Let $i_1 < i_2 < i_3 < \cdots$ be the zeros of Airy's function $A(t)$; then the only zeros of $L_n^{(\alpha)}(x)$ in the interval

$$x = 4n + 2\alpha + 2 - 2(2n/3)^{\frac{1}{3}} t, \qquad t \text{ real and bounded}, \tag{8.9.14}$$

are those corresponding to the values $t = i_\nu$. (Near $t = i_\nu$ we have precisely one zero if n is sufficiently large; this follows from the uniform validity of (8.22.11) for bounded complex t, by use of Theorem 1.91.3 (Theorem of Hurwitz).) On the other hand, the upper bound (6.31.7) of the zeros belongs to this interval.

Consequently, if we arrange the zeros $x_1 > x_2 > x_3 > \cdots$ of $L_n^{(\alpha)}(x)$ in decreasing order, we have

$$(8.9.15) \qquad x_\nu = x_{\nu n} = 4n + 2\alpha + 2 - 2(2n/3)^{\frac{1}{3}}(i_\nu + \epsilon_n), \quad \lim_{n\to\infty} \epsilon_n = 0, \ \nu \text{ fixed.}$$

This is identical with (6.32.9). An analogous result holds for Hermite polynomials, namely (6.32.5). It follows immediately from (8.22.14). This formula was proved in §6.32 by use of Sturm's method.

8.91. Applications; asymptotic properties of the maxima of Laguerre and Hermite polynomials

Another application is the discussion of the magnitude of the Laguerre and Hermite polynomials for non-negative and for real values of x, respectively.

(1) THEOREM 8.91.1. *Let α be arbitrary and real, $a > 0$, $0 < \eta < 4$. We have for $n \to \infty$*

$$(8.91.1) \quad \max e^{-x/2} x^{(\alpha+1)/2} \mid L_n^{(\alpha)}(x) \mid \sim \begin{cases} n^{\alpha/2} & \text{if } a \leqq x \leqq (4-\eta)n, \\ n^{\alpha/2+\frac{1}{4}} & \text{if } x \geqq a; \end{cases}$$

$$(8.91.2) \quad \max e^{-x/2} x^{\alpha/2+\frac{1}{4}} \mid L_n^{(\alpha)}(x) \mid \sim \begin{cases} n^{\alpha/2-\frac{1}{4}} & \text{if } a \leqq x \leqq (4-\eta)n, \\ n^{\alpha/2-1/12} & \text{if } x \geqq a. \end{cases}$$

These maxima are taken, respectively, in the intervals pointed out in the right-hand members.

These asymptotic formulas play an important rôle in the discussion of Laguerre series (Chapter IX).[52]

A more exact characterization of these maxima when $n \to \infty$ is also possible. We can replace $\sim$ by $\cong$ by introducing into the right-hand members the constant factors

$$(8.91.3) \qquad \begin{matrix} (2/\pi)^{\frac{1}{2}}(4\eta^{-1} - 1)^{\frac{1}{4}}, & \pi^{-1}(12)^{\frac{1}{4}} \max A(t); \\ \pi^{-\frac{1}{2}}(4\eta^{-1})^{\frac{1}{4}}, & \pi^{-1}(18)^{\frac{1}{6}} \max A(t), \end{matrix}$$

respectively. This follows readily from the subsequent proof. We note that these factors are independent of α and a. The first maximum of $A(t)$ is positive, and is actually the greatest value of $| A(t) |$ for all real values of t. (Cf. Theorem 7.31.1 and (1.81.2).)

(2) Let $t = j_1$ be the first maximum point of $A(t)$, $0 < j_1 < i_1$. Now choose two values t', t'' such that $0 < t' < t'' < j_1$ and denote the corresponding

[52] Kogbetliantz (**22**, p. 144; **23**, pp. 39, 51–53), states erroneously the appraisal $L_n^{(\alpha)}(x) = O(e^{x/2}x^{-\alpha/2-1/4}n^{\alpha/2-1/4})$, $x \geqq a$. The error is made in **22**, p. 154, where a certain bound valid for $H_n(x)$, if $| x | < cn^{1/2}$, is applied to an arbitrary $H_m(x)$, $m \leqq n$. Unfortunately, the main results of the paper **23** (and partly also those of **24**) are based on this erroneous statement.

values of x in (8. 9. 14) by x' and x'', respectively, so that $x' > x'' > x_{1n}$ if n is sufficiently large. Then $A(t') < A(t'')$, so that by (8.22.11)

$$(8.91.4) \qquad (-1)^n e^{-x'/2} L_n^{(\alpha)}(x') < (-1)^n e^{-x''/2} L_n^{(\alpha)}(x'').$$

Therefore, x' and x'' both cannot be on the left of the last extremum point of $e^{-x/2} L_n^{(\alpha)}(x)$, whence this extremum point lies in the interval (8.9.14).

The same consideration applies to $e^{-x/2} x^\lambda L_n^{(\alpha)}(x)$, where λ is an arbitrary but fixed real number. Indeed, $x^\lambda = (4n)^\lambda \{1 + O(n^{-\frac{2}{3}})\}$. Thus, a formula analogous to (8.22.11) holds for $e^{-x/2} x^\lambda L_n^{(\alpha)}(x)$ with the additional factor $(4n)^\lambda$ in the right-hand member.

(3) In the interval $a \leqq x \leqq (4 - \eta)n$ formula (8.22.9) can be applied. Since

$$x^{(\alpha+1)/2} (\sin \phi)^{-\frac{1}{2}} x^{-\alpha/2-\frac{1}{4}} n^{\alpha/2-\frac{1}{4}} = x^{\frac{1}{4}} (\sin \phi)^{-\frac{1}{2}} n^{\alpha/2-\frac{1}{4}} \sim n^{\alpha/2} (\cot \phi)^{\frac{1}{2}},$$

$$x^{\alpha/2+\frac{1}{4}} (\sin \phi)^{-\frac{1}{2}} x^{-\alpha/2-\frac{1}{4}} n^{\alpha/2-\frac{1}{4}} = (\sin \phi)^{-\frac{1}{2}} n^{\alpha/2-\frac{1}{4}},$$

the maxima in question are $\sim n^{\alpha/2}$ and $n^{\alpha/2-\frac{1}{4}}$, respectively.

In order to calculate the maximum in the interval $x \geqq a$, we use Theorem 7.6.2. The sequence of the relative maxima of

$$(8.91.5) \qquad e^{-x/2} x^{(\alpha+1)/2} \,|\, L_n^{(\alpha)}(x) \,| \quad \text{and} \quad e^{-x/2} x^{\alpha/2+\frac{1}{4}} \,|\, L_n^{(\alpha)}(x) \,| \; .$$

is increasing for $x > x_0$, where x_0 is a certain non-negative number depending on α, and n is sufficiently large. Therefore, on account of (1), the absolute maxima of the functions (8.91.5) for $x \geqq a$ are attained in the interval (8.9.14). According to (8.22.11) these maxima are $\sim n^{(\alpha+1)/2} n^{-\frac{1}{3}}$ and $n^{\alpha/2+\frac{1}{4}} n^{-\frac{1}{3}}$, respectively. Between a and x_0, Fejér's formula (8.22.1) must be used. This furnishes the complete proof.

(4) We also readily prove the following more general theorem:

THEOREM 8.91.2. *Let α and λ be arbitrary and real, $a > 0$, $0 < \eta < 4$. Then for $n \to \infty$*

$$(8.91.6) \qquad \max e^{-x/2} x^\lambda \,|\, L_n^{(\alpha)}(x) \,| \sim n^Q,$$

where

$$(8.91.7) \qquad Q = \begin{cases} \max (\lambda - \frac{1}{2}, \alpha/2 - \frac{1}{4}) & \text{if } \; a \leqq x \leqq (4 - \eta)n, \\ \max (\lambda - \frac{1}{3}, \alpha/2 - \frac{1}{4}) & \text{if } \; x \geqq a. \end{cases}$$

In the first case, that is to say, in the interval $a \leqq x \leqq (4 - \eta)n$, we again apply (8.22.9). Now

$$x^\lambda (\sin \phi)^{-\frac{1}{2}} x^{-\alpha/2-\frac{1}{4}} n^{\alpha/2-\frac{1}{4}} \sim n^{\lambda-\frac{1}{2}} (\cos \phi)^{2\lambda-\alpha-\frac{1}{2}} (\sin \phi)^{-\frac{1}{2}}.$$

This expression attains its maximum in the interval $\epsilon \leqq \phi \leqq \pi/2 - \epsilon n^{-\frac{1}{2}}$ for $\phi = \epsilon$ or $\phi = \pi/2 - \epsilon n^{-\frac{1}{2}}$ according as $\lambda \geqq \alpha/2 + \frac{1}{4}$ or $\lambda < \alpha/2 + \frac{1}{4}$. The maxima are $\sim n^{\lambda-\frac{1}{2}}$ and $n^{\lambda-\frac{1}{2}}(n^{-\frac{1}{2}})^{2\lambda-\alpha-\frac{1}{2}} = n^{\alpha/2-\frac{1}{4}}$, respectively. This establishes the first part of the statement. We also see that the maximum is attained near $x = (4 - \eta)n$ if $\lambda \geqq \alpha/2 + \frac{1}{4}$, and near $x = a$ if $\lambda < \alpha/2 + \frac{1}{4}$.

In the interval $(4 - \eta)n \leqq x \leqq 4n + O(n^{\frac{1}{3}})$, we have

$$\max e^{-x/2}x^{\lambda} \mid L_n^{(\alpha)}(x) \mid \sim n^{\lambda-(\alpha+1)/2} \max e^{-x/2}x^{(\alpha+1)/2} \mid L_n^{(\alpha)}(x) \mid.$$

Because of (8.91.1), this furnishes the second part of the statement. The maximum is attained near $x = 4n$ if $\lambda - \frac{1}{3} > \alpha/2 - \frac{1}{4}$, and near $x = a$ if $\lambda - \frac{1}{3} < \alpha/2 - \frac{1}{4}$.

(5) By use of (5.6.1), or directly from (8.22.12) and (8.22.14), we obtain the corresponding results for Hermite polynomials.

THEOREM 8.91.3. *Let λ be arbitrary and real, $a > 0$, $0 < \eta < 2$. Then*

$$\max e^{-x^2/2}x^{\lambda} \mid H_n(x) \mid \sim (2^n n!)^{\frac{1}{2}} n^S, \tag{8.91.8}$$

where

$$S = \begin{cases} \max(\lambda/2 - 1/4, -1/4) & \text{if} \quad a \leqq |x| \leqq [(2-\eta)n]^{\frac{1}{2}}, \\ \max(\lambda/2 - 1/12, -1/4) & \text{if} \quad |x| \geqq a;\ x \text{ real}. \end{cases} \tag{8.91.9}$$

We have, for instance (Theorem 7.6.3),

$$\max e^{-x^2/2} \mid H_n(x) \mid \cong (2^n n!)^{\frac{1}{2}} 2^{\frac{1}{4}} 3^{\frac{1}{2}} \pi^{-\frac{3}{4}} n^{-1/12} \max A(t). \tag{8.91.10}$$

Here x and t range over all real values. (Cf. Hille **1**, p. 436, (30).)

8.92. Further results

(1) Gatteschi (**1**, **2**) investigated various asymptotic formulas involving the classical polynomials and their zeros, and replaced the (usually unspecified) constants occurring in the O-terms by numerical values. As an illustration we mention the following remarkable refinements of Theorem 8.21.6.

We have

$$P_n(\cos\theta) = (\theta/\sin\theta)^{\frac{1}{2}} J_0\{(n+\tfrac{1}{2})\theta\} + \sigma$$

where

$$|\sigma| < 0.09\,\theta^2 \quad \text{if} \quad 0 < \theta \leqq \pi/(2n),$$

$$|\sigma| < 0.63\,\theta^{\frac{1}{2}} n^{-\frac{3}{2}} \quad \text{if} \quad \pi/(2n) < \theta \leqq \pi/2.$$

The following formula is even more informative:

$$\left(\frac{\sin\theta}{\theta}\right)^{\frac{1}{2}} P_n(\cos\theta) = J_0\{(n+\tfrac{1}{2})\theta\} - \frac{\theta}{24(n+\frac{1}{2})} J_1\{(n+\tfrac{1}{2})\theta\} + \sigma'$$

where

$$|\sigma'| < 0.03\,\theta^4 \quad \text{if} \quad 0 < \theta \leqq \pi/(2n),$$

$$|\sigma'| < 0.25\,\theta^{\frac{5}{2}} n^{-\frac{3}{2}} \quad \text{if} \quad \pi/(2n) < \theta \leqq \pi/2.$$

These formulas can be used for the asymptotic evaluation of the "first" zeros with specified constants in the error terms. See also Tricomi **4**.

(2) Further information concerning the asymptotic behavior of the Jacobi and Laguerre polynomials and their zeros can be found in Bateman Manuscript Project, vol. 2, Chapter 10, pp. 196–202; also Tricomi **5**, pp. 219–224, Thorne **1**, and Erdélyi-Swanson **1**. See also the literature quoted at these places.

Recently, Erdélyi **3** has obtained two new asymptotic formulas for the Laguerre polynomials $L_n^{(\alpha)}(x)$ valid for $x \leqq a(4n + 2\alpha + 2)$ and for $x \geqq b(4n + 2\alpha + 2)$ where a and b are fixed, $0 < b < a < 1$. These formulas involve Bessel functions and Airy's function, respectively, and hold on the above ranges which overlap and cover the entire real axis. This is an important result.

Erdélyi only proved these asymptotic formulas for $\alpha \geqq 0$. They were extended to $\alpha > -1$ by recurrence relations by Muckenhoupt **1**.

(3) The formula (8.21.1) of Laplace-Heine has been used in the estimation of the smallest eigenvalue of the truncated Hilbert matrix $(1/(i+j+1))_{i,j=0}^{N}$. See Szegö **27** and Widom-Wilf **1**. The size of this eigenvalue gives a measure of the degree of difficulty in finding the inverse of this matrix.

(4) The remark in §8.4 (3) is not quite correct. The infinite series corresponding to Darboux's formula (8.21.4) is convergent for $\pi/6 < \theta < 5\pi/6$, but it converges to $2P_n(\cos\theta)$ rather than $P_n(\cos\theta)$. See Olver **1**. While asymptotic series in the sense of Poincaré must converge to the "right" function if they converge this is not true of more general asymptotic expansions; so great care must be exercised when using them.

CHAPTER IX

EXPANSION PROBLEMS ASSOCIATED WITH THE CLASSICAL POLYNOMIALS

The formal expansion of a function in terms of general orthogonal polynomials having been defined in §3.1 (cf. (3.1.3)), we shall now turn our attention to expansions in terms of the classical polynomials. In this connection we shall deal with the following problems:

Expansion of an analytic function in series of Jacobi, of Laguerre, and of Hermite polynomials; discussion of the domain of the convergence.

Expansion of an "arbitrary" function in series of Jacobi, of Laguerre, and of Hermite polynomials; discussion of equiconvergence and summability theorems.

In the main, our principal concern will be with the second problem, and we agree that "arbitrary" function shall mean a function restricted only by certain conditions of integrability, or continuity, and by conditions involving the existence of certain integrals.

Two series $\sum_{n=0}^{\infty} u_n$ and $\sum_{n=0}^{\infty} v_n$ are called equiconvergent if the series $\sum_{n=0}^{\infty} (u_n - v_n)$ or, more generally, if

$$\sum_{n=0}^{\infty} (u_n - Av_n), \quad A \neq 0,$$

is convergent. We shall try to find simple trigonometric (Fourier) expansions equiconvergent with a given type of polynomial expansion. Such a procedure will enable us to reduce the discussion of the polynomial expansion to the discussion of trigonometric series under very general conditions.

Of the various methods of summability, we shall be primarily interested in that of Cesàro. A series $\sum_{n=0}^{\infty} u_n$ is called (C, k)-summable, $k > -1$, with the sum s if

$$\lim_{n \to \infty} \frac{s_n^{(k)}}{C_n^{(k)}} = s,$$

where

$$(1 - r)^{-k-1} \sum_{n=0}^{\infty} u_n r^n = \sum_{n=0}^{\infty} s_n^{(k)} r^n,$$

$$(1 - r)^{-k-1} = \sum_{n=0}^{\infty} C_n^{(k)} r^n = \sum_{n=0}^{\infty} \binom{n+k}{n} r^n.$$

Obviously, $k = 0$ corresponds to convergence in the ordinary sense. If $k' > k$, it is readily shown that (C, k)-summability involves (C, k')-summability with the same sum.

A necessary condition for (C, k)-summability, $k \geqq 0$, is $u_n = O(n^k)$.

Cesàro summability of any order k implies Abel summability; that is, the existence of the

$$\lim_{r \to 1-0} \sum_{n=0}^{\infty} u_n r^n$$

with the same sum.

9.1. Results

(1) The formal expansion of a function $f(x)$ in a Jacobi series is (compare (4.3.3))

$$(9.1.1)\qquad \begin{aligned} f(x) &\sim \sum_{n=0}^{\infty} a_n P_n^{(\alpha,\beta)}(x), \\ h_n^{(\alpha,\beta)} a_n &= \int_{-1}^{+1} (1-x)^{\alpha}(1+x)^{\beta} f(x) P_n^{(\alpha,\beta)}(x)\, dx. \end{aligned}$$

The expansion of $f(x)$ in a Laguerre series is (cf. (5.1.1))

$$(9.1.2)\qquad \begin{aligned} f(x) &\sim \sum_{n=0}^{\infty} a_n L_n^{(\alpha)}(x), \\ \Gamma(\alpha+1)\binom{n+\alpha}{n} a_n &= \int_0^{+\infty} e^{-x} x^{\alpha} f(x) L_n^{(\alpha)}(x)\, dx; \end{aligned}$$

and in an Hermite series it is (cf. (5.5.1))

$$(9.1.3)\qquad \begin{aligned} f(x) &\sim \sum_{n=0}^{\infty} a_n H_n(x), \\ \pi^{\frac{1}{2}} 2^n n! a_n &= \int_{-\infty}^{+\infty} e^{-x^2} f(x) H_n(x)\, dx. \end{aligned}$$

In all these cases $f(x)$ is a measurable function, and the existence of all integrals occurring is required.

In what follows we assume $\alpha > -1$, $\beta > -1$ in the Jacobi case and $\alpha > -1$ in the Laguerre case.

(2) In this connection the following results may be stated:

THEOREM 9.1.1 (Expansion of an analytic function in a Jacobi series). *Let $f(x)$ be analytic on the closed segment* $[-1, +1]$. *The expansion of $f(x)$ in a Jacobi series is convergent in the interior of the greatest ellipse with foci at ± 1, in which $f(x)$ is regular. The expansion is divergent in the exterior of this ellipse. Using the notation* (9.1.1), *we have the following representation for the sum R of the semi-axes of the ellipse of convergence*

$$(9.1.4)\qquad R = \liminf_{n \to \infty} |a_n|^{-1/n}.$$

THEOREM 9.1.2 (Equiconvergence theorem for Jacobi series in the interior of the interval $-1, +1$). *Let $f(x)$ be Lebesgue-measurable in $[-1, +1]$, and let the integrals*

$$\int_{-1}^{+1} (1-x)^{\alpha}(1+x)^{\beta}\,|f(x)|\,dx,$$
(9.1.5)
$$\int_{-1}^{+1} (1-x)^{\alpha/2-\frac{1}{4}}(1+x)^{\beta/2-\frac{1}{4}}\,|f(x)|\,dx$$

exist. If $s_n(x)$ denotes the nth partial sum of the expansion of $f(x)$ in a Jacobi series, and $\mathbf{s}_n(\cos\theta)$ the nth partial sum of the Fourier (cosine) series of

$$(1-\cos\theta)^{\alpha/2+\frac{1}{4}}(1+\cos\theta)^{\beta/2+\frac{1}{4}}f(\cos\theta), \tag{9.1.6}$$

then for $-1 < x < +1$,

$$\lim_{n\to\infty}\{s_n(x) - (1-x)^{-\alpha/2-\frac{1}{4}}(1+x)^{-\beta/2-\frac{1}{4}}\,\mathbf{s}_n(x)\} = 0, \tag{9.1.7}$$

uniformly in $-1+\epsilon \leqq x \leqq 1-\epsilon$, where ϵ is a fixed positive number, $\epsilon < 1$.

THEOREM 9.1.3 (Summability theorem for Jacobi series at the end-points $x = \pm 1$). *Let $f(x)$ be continuous on the closed segment $[-1, +1]$. The expansion of $f(x)$ in a Jacobi series is (C, k)-summable at $x = +1$, provided $k > \alpha + \frac{1}{2}$. This is in general not true if $k = \alpha + \frac{1}{2}$. An analogous statement holds for $x = -1$, α being replaced by β.*

THEOREM 9.1.4 (Generalized summability theorem for Jacobi series). *Let $f(x)$ be Lebesgue-measurable in $[-1, +1]$ and continuous at $x = +1$. Then if we assume the existence of the integral*

$$\int_{-1}^{+1} (1-x)^{\alpha}(1+x)^{\beta}\,|f(x)|\,dx, \tag{9.1.8}$$

the Jacobi series is (C, k)-summable, $k > \alpha + \frac{1}{2}$, at $x = +1$, provided that in the case

$$\beta > -\tfrac{1}{2}, \qquad \alpha + \tfrac{1}{2} < k < \alpha + \beta + 1, \tag{9.1.9}$$

the following additional "antipole condition" is satisfied: the integral

$$\int_{-1}^{0} (1+x)^{\beta/2-\frac{1}{4}}\,|f(x)|\,dx \tag{9.1.10}$$

exists. (For $k \geqq \alpha + \beta + 1$ no antipole condition is necessary.) For $k \leqq \alpha + \frac{1}{2}$ or for $k > \alpha + \frac{1}{2}$, but without the antipole condition, the statement is not true.

THEOREM 9.1.5 (Equiconvergence theorem for Laguerre series for $x > 0$). *Let $f(x)$ be Lebesgue-measurable in $[0, +\infty]$, and let the integrals*

$$\int_0^1 x^{\alpha}\,|f(x)|\,dx, \qquad \int_0^1 x^{\alpha/2-\frac{1}{4}}\,|f(x)|\,dx \tag{9.1.11}$$

exist. If the condition

$$\int_n^\infty e^{-x/2} x^{\alpha/2-13/12} \,|\, f(x) \,|\, dx = o(n^{-1/2}), \qquad n \to \infty, \tag{9.1.12}$$

is satisfied, and if $s_n(x)$ denotes the nth partial sum of the Laguerre series of $f(x)$, we have, for $x > 0$,

$$\lim_{n\to\infty} \left\{ s_n(x) - \pi^{-1} \int_{x^{1/2}-\delta}^{x^{1/2}+\delta} f(\tau^2) \frac{\sin \{2n^{1/2}(x^{1/2} - \tau)\}}{x^{1/2} - \tau} \, d\tau \right\} = 0, \tag{9.1.13}$$

where δ is a fixed positive number, $\delta < x^{1/2}$. This holds uniformly for every fixed positive interval $\epsilon \leqq x \leqq \omega$, $\delta < \epsilon^{1/2}$.

The same equiconvergence theorem (9.1.13) *is valid if the integrals* (9.1.11) *exist and condition* (9.1.12) *is replaced by the following:*

$$\begin{aligned} &\int_1^\infty e^{-x/2} x^{\alpha/2-3/4} \,|\, f(x) \,|\, dx \quad \textit{is convergent;} \\ &\int_n^\infty e^{-x} x^{\alpha-2} \,|\, f(x) \,|^2 \, dx = o(n^{-3/2}), \qquad n \to \infty. \end{aligned} \tag{9.1.14}$$

THEOREM 9.1.6 (Equiconvergence theorem for Hermite series, x arbitrary and real). *Let $f(x)$ be Lebesgue-measurable in $[-\infty, +\infty]$, and let the integral*

$$\int_{-a}^{+a} |\, f(x) \,|\, dx \tag{9.1.15}$$

exist for every $a > 0$. If the condition

$$\int_n^\infty e^{-x^2/2} x^{-5/3} \{|\, f(x) \,| + |\, f(-x) \,|\} \, dx = o(n^{-1}), \qquad n \to \infty, \tag{9.1.16}$$

is satisfied, and if $s_n(x)$ denotes the nth partial sum of the Hermite series of $f(x)$, we have, for an arbitrary and real x,

$$\lim_{n\to\infty} \left\{ s_n(x) - \pi^{-1} \int_{x-\delta}^{x+\delta} f(t) \frac{\sin \{(2n)^{1/2}(x - t)\}}{x - t} \, dt \right\} = 0, \tag{9.1.17}$$

where δ is a fixed positive number. Moreover, (9.1.17) *holds uniformly in every finite interval.*

The same equiconvergence theorem (9.1.17) *holds if the integral* (9.1.15) *exists and we replace condition* (9.1.16) *by the following:*

$$\begin{aligned} &\int_1^\infty e^{-x^2/2} x^{-1} \{|\, f(x) \,| + |\, f(-x) \,|\} \, dx \quad \textit{is convergent,} \\ &\int_n^\infty e^{-x^2} x^{-4} \{|\, f(x) \,|^2 + |\, f(-x) \,|^2\} \, dx = o(n^{-3}), \qquad n \to \infty. \end{aligned} \tag{9.1.18}$$

THEOREM 9.1.7 (Summability theorem for Laguerre series at $x = 0$). *Let*

$f(x)$ be Lebesgue-measurable in $[0, +\infty]$ and continuous at $x = 0$. If we assume the existence of the integral

$$\int_1^\infty e^{-x/2} x^{\alpha-k-\frac{1}{3}} |f(x)| \, dx, \tag{9.1.19}$$

the Laguerre series of $f(x)$ is (C, k)-summable at $x = 0$ with the sum $f(0)$ provided $k > \alpha + \frac{1}{2}$. This statement is not true for $k \leqq \alpha + \frac{1}{2}$.

9.11. Remarks

(1) Theorem 9.1.1 is well-known in the Tchebichef and Legendre cases.[53] The determination (9.1.4) of the ellipse of convergence is analogous to the well-known Cauchy-Hadamard formula.

(2) The importance of Theorem 9.1.2 is due to the possibility of applying it to convergence and summability problems for Jacobi series similar to those which are found in the classical theory of ordinary Fourier series. This theorem has been proved in the special case $\alpha = \beta = 0$ (Legendre series) by Haar (**2**) and W. H. Young (**1**) with conditions concerning $f(x)$ somewhat different from those arising from our general theorem in the special case $\alpha = \beta = 0$. W. H. Young considers also series proceeding in terms of Legendre polynomials, without being Fourier expansions in the ordinary sense. The proof of Theorem 9.1.2 given in §9.3 is due to Szegö (**17**, pp. 88–92). Recently, Obrechkoff (**2**) has treated the same problem, using as his main tool Darboux's formula in the more precise form (8.21.18).

Summability theorems for interior points were investigated earlier (in the ultraspherical case) by Adamoff (**2**) and (in the ultraspherical as well as in the general Jacobi case) by Kogbetliantz (**1, 2, 3, 7, 18, 19**). Concerning properties of these expansions analogous to those treated in Riemann's theory of trigonometric series (in particular theorems of uniqueness) see Kogbetliantz (**6, 20**) and Zygmund (**1**).

In setting up the expansion in Jacobi series, we must require the existence of the first integral (9.1.5). Using an obvious notation, we readily see that

$$\int_0^1 (1 - x)^\alpha |f(x)| \, dx \leftrightarrows \int_0^1 (1 - x)^{\alpha/2-\frac{1}{4}} |f(x)| \, dx, \tag{9.11.1}$$

according as $\alpha \geqq -\frac{1}{2}$ or $-1 < \alpha \leqq -\frac{1}{2}$. That the condition of the existence of the second integral cannot be improved upon for $\alpha > -\frac{1}{2}$ follows when we consider the special function $f(x) = (1 - x)^\mu$ with $\mu = -\alpha/2 - 3/4$ (cf. §9.3 (4)). The existence of both integrals (9.1.5) follows from that of

$$\int_{-1}^{+1} (1 - x)^\alpha (1 + x)^\beta \{f(x)\}^2 \, dx. \tag{9.11.2}$$

[53] The Legendre case is usually attributed to F. Neumann (cf. Whittaker-Watson **1**, pp. 322–323).

Instead of (9.1.6) the function

$$(1 - \cos\theta)^{\alpha+\frac{1}{2}}(1 + \cos\theta)^{\beta+\frac{1}{2}} f(\cos\theta) \tag{9.11.3}$$

may also be considered (cf. Szegö, loc. cit.). The change, necessary in (9.1.7) for this function, is at once apparent. Both functions (9.1.6) and (9.11.3) are integrable in $-\pi \leqq \theta \leqq +\pi$. The difference (9.1.7) may be written as

$$\begin{aligned}\int_{-1}^{+1} f(t)\{(1-t)^{\alpha}(1+t)^{\beta} K_n^{(\alpha,\beta)}(x,t) \\ - (1-x)^{-\alpha/2-\frac{1}{4}}(1+x)^{-\beta/2-\frac{1}{4}}(1-t)^{\alpha/2-\frac{1}{4}}(1+t)^{\beta/2-\frac{1}{4}} K_n^{(-\frac{1}{2},-\frac{1}{2})}(x,t)\}\, dt,\end{aligned} \tag{9.11.4}$$

where the notation (4.5.2) has been used. Comparison of $s_n(x) = s_n(\alpha, \beta; x)$ with $s_n(\gamma, \delta; x)$, where γ and δ are arbitrary, is also possible.

(3) Theorems 9.1.3 and 9.1.4 have an extensive literature. Gronwall (**1, 2**) has investigated the special case $\alpha = \beta = 0$ (Legendre series). His proofs have been simplified to a considerable extent by Lukács (**2**), Hilb (**1**), and Fejér (**8**). The ultraspherical case has been considered in great detail by Kogbetliantz (**2, 4, 19**; **21**, pp. 70–73) (cf. also Obrechkoff **1**). The method used in §§9.4–9.42 is new and comparatively simple. It is based on a peculiar relationship between the Jacobi polynomial $P_n^{(\alpha+k+1,\beta)}(x)$ and the "kernel" of the kth Cesàro mean of the Jacobi series which corresponds to the parameters α and β. (In the case of Laguerre series the corresponding relation is trivial (cf. §9.6).) The case of a Legendre series (if we consider the series at the end-point $x = +1$) is equivalent to a Laplace series; whence the term "antipole condition". At the end-point $x = -1$ a similar theorem holds; in this case the summability index k must exceed $\beta + \frac{1}{2}$, and an "antipole condition" must be satisfied near $x = +1$.

Theorem 9.1.3 furnishes the convergence of the Jacobi series at the end-point $x = +1$ provided $f(x)$ is continuous in $[-1, +1]$ and $-1 < \alpha < -\frac{1}{2}$ (cf. also Rau **1**; Szegö **17**, §20; Lorch **1, 2**).

In the case of Legendre series *the "kernels" of the Cesàro means of the second order are non-negative* (Fejér **4**). This fact has, of course, important consequences. Concerning similar theorems in case of ultraspherical expansions, cf. Kogbetliantz **19**.

In the special case of Legendre series Kogbetliantz (**16**) gave important refinements of Theorems 9.1.3 and 9.1.4. He studied, among others, the C-summability of proper order at the end-points $x = \pm 1$ if $f(x)$ becomes infinite of a certain order at the antipole.

As regards certain older (very complicated) results on the convergence of ultraspherical expansions at the end-points $x = \pm 1$, see Adamoff **2**.

(4) The equiconvergence problem for Laguerre and Hermite expansions has been treated (the first only in the special case $\alpha = 0$) by Rotach (**1**), and (in all cases) independently by Szegö (**10**). For the special Laguerre case $\alpha = 0$ the conditions of Szegö are more restrictive than those of Rotach, and the same

is true in the Hermite case. The conditions formulated in §9.1 are, however, slightly more general than those of Rotach.[54]

That the condition of the existence of the second integral (9.1.11) cannot be improved upon is shown by considering the expansion of the function $f(x) = x^{\mu}$, $\mu = -\alpha/2 - 3/4$, in a Laguerre series (§9.5 (6)).

Neither of the two sets of sufficient conditions formulated in Theorem 9.1.5 contains the other. Indeed, the function $f(x) = e^{x/2}x^{-\alpha/2-\frac{1}{4}}$ satisfies (9.1.14), but not (9.1.12). On the other hand, if

$$(9.11.5) \qquad e^{-x/2}x^{\alpha/2}f(x) = \begin{cases} 1, & m^2 \leqq x < m^2 + 1, \\ 0, & \text{otherwise}, \quad m = 1, 2, 3, \cdots, \end{cases}$$

then (9.1.12) is satisfied, but not the second part of condition (9.1.14).

Condition (9.1.12) implies, of course, that the integral on the left side exists. From the condition (9.1.12) the first part of (9.1.14) follows. In fact, if we write

$$(9.11.6) \qquad u(x) = \int_x^{\infty} e^{-t/2}t^{\alpha/2-13/12} \,|\, f(t) \,|\, dt,$$

we find from (1.4.4) that

$$(9.11.7) \qquad \int_1^{\omega} x^{\frac{1}{3}}\, du(x) = \omega^{\frac{1}{3}}u(\omega) - u(1) - \tfrac{1}{3}\int_1^{\omega} x^{-\frac{2}{3}}u(x)\, dx$$

is bounded as $\omega \to \infty$, since $u(x) = o(x^{-\frac{1}{3}})$.

A sufficient condition for the validity of (9.1.13) is that

$$(9.11.8) \qquad f(x) = O(e^{x/2}x^{-\alpha/2-\frac{1}{4}-\delta}), \qquad \delta > 0,\ x \to +\infty.$$

Then (9.1.14) is satisfied. On the other hand, for $f(x) = e^{x/2}x^{-\alpha/2+\frac{1}{4}}$ conditions (9.1.11) are satisfied, but not (9.1.12) and (9.1.14), and (as we are going to show in §9.5 (7)) the Laguerre series is divergent for $x > 0$.

A sufficient condition for the validity of (9.1.17) is that

$$(9.11.9) \qquad f(x) = O(e^{x^2/2} \,|\, x \,|^{-\delta}), \qquad \delta > 0,\ x \to \infty.$$

Then (9.1.18) holds. An analogous "Gegenbeispiel" here is $f(x) = xe^{x^2/2}$.

From Theorems 9.1.5 and 9.1.6 there follow the usual theorems on the convergence and the summability of Laguerre and Hermite expansions. Indeed, the integrals occurring in (9.1.13) and (9.1.17) are essentially the partial sums of order $[n^{\frac{1}{2}}]$ of a Fourier series (cf. (1.6.4)).

As early as 1907 Adamoff (**2**) obtained a convergence theorem for Hermite

[54] Rotach's second condition b_2' on p. 8 makes the first part of b_1' superfluous. His first set of conditions is equivalent to (9.1.11) plus (9.1.14) (for $\alpha = 0$), whereas his second set is more restrictive than (9.1.11) plus (9.1.12). In the theorem on p. 6, the second condition b_2 must be corrected to read "$\int_1^{\infty} e^{-z^2/4} \,|\, f(\pm z) \,|\, z^{3/2}\, dz$ exists." (I owe this to a written communication from Mr. Plancherel.) This, of course, implies b_1. Here the first set of conditions is more restrictive than (9.1.18), and the second set is more restrictive than (9.1.16). (Rotach's notation differs from ours; we must write $z = 2^{\frac{1}{2}}x$.)

expansions. There is an extensive further literature on this subject; E. R. Neumann (**1**), Galbrun (**1**), Wigert (**1**), Hille (**1**, **2**, **3**), Cramér (**1**), Uspensky (**1**), Korous (**1**, **2**), Stone (**1**), Müntz (**1**), and Kowallik (**1**) have all given direct treatments of Laguerre and Hermite expansions of an "arbitrary" function. (Concerning the expansion of an analytic function into Hermite series see Watson **1** (first paper).) These authors have obtained convergence and summability results but no equiconvergence theorems, except Korous (loc. cit.). The conditions which all these authors use at $x = \infty$ are more restrictive than those in our theorems. For instance, in case of Laguerre expansions, the convergence of

$$\int_0^\infty e^{-x/2} x^{\alpha/2-\frac{1}{4}} \, |f(x)| \, dx \tag{9.11.10}$$

is required by Korous; for Hermite expansions, convergence of the integrals

$$\int_1^\infty e^{-x^2} |f(\pm x)|^2 dx, \qquad \int_1^\infty e^{-x^2} x |f(\pm x)|^2 dx \tag{9.11.11}$$

are required by Uspensky and Kowallik, respectively. In addition, we mention the earlier treatment of Laguerre and Hermite expansions by means of the theory of integral equations (Myller-Lebedeff **1**, Weyl **1**).

Concerning Kogbetliantz **22**, **23**, **24**, see the footnote in §8.91. Concerning Theorem 9.1.7, see Szegö **10** and Kogbetliantz **10**. The condition regarding (9.1.19) is satisfied if

$$f(x) = O(e^{x/2} x^{k-\alpha-\frac{2}{3}-\delta}), \qquad \delta > 0, \; x \to +\infty. \tag{9.11.12}$$

On the other hand, the series is not (C, k)-summable for the function $f(x) = e^{x/2} x^{k-\alpha}$, $k > \alpha + 1/2$ (cf. §9.6 (3)).

Gibbs' phenomenon in the case of Hermite expansions has been studied by Jacob (**2**).

9.2. Expansion of an analytic function in Jacobi, Laguerre, and Hermite series

(1) Theorem 9.2.1. *Assume $\alpha > -1$, $\beta > -1$, and let $P_n^{(\alpha,\beta)}(x)$, $Q_n^{(\alpha,\beta)}(x)$, and $h_n^{(\alpha,\beta)}$ have the same meaning as in Chapter* IV *(see* §4.61 *and* (4.3.3)*); then*

$$\sum_{n=0}^{\infty} \{h_n^{(\alpha,\beta)}\}^{-1} P_n^{(\alpha,\beta)}(x) Q_n^{(\alpha,\beta)}(y) = \frac{1}{2} \frac{(y-1)^{-\alpha}(y+1)^{-\beta}}{y-x}, \tag{9.2.1}$$

where x lies in the interior of, and y in the exterior of, an arbitrary ellipse with foci at ± 1. This expansion holds uniformly if x and y belong to closed sets which are, respectively, in the interior and exterior of the ellipse mentioned.

The functions $(y-1)^{\alpha}(y+1)^{\beta} Q_n^{(\alpha,\beta)}(y)$ are single-valued and regular in the complex y-plane cut along $[-1, +1]$.

This important formula is well-known in the special case $\alpha = \beta = 0$ (Heine **3**, p. 78). The general formula is obtained from the identity (4.62.19) by letting n approach ∞. We write $x = \frac{1}{2}(z + z^{-1})$, $y = \frac{1}{2}(\zeta + \zeta^{-1})$; $1 < |z| < |\zeta|$. By use of (8.23.1) and (8.23.2) we find the "remainder term" of (4.62.19) to be

$$\frac{2^{-\alpha-\beta}}{2n+\alpha+\beta+2}\frac{\Gamma(n+2)\Gamma(n+\alpha+\beta+2)}{\Gamma(n+\alpha+1)\Gamma(n+\beta+1)} \cdot \frac{P_{n+1}^{(\alpha,\beta)}(x)Q_n^{(\alpha,\beta)}(y)-P_n^{(\alpha,\beta)}(x)Q_{n+1}^{(\alpha,\beta)}(y)}{x-y} = O(n)O[(|z|+\epsilon)^n]O[(|\zeta|^{-1}+\epsilon)^n] \tag{9.2.2}$$

as $n \to \infty$, with $\epsilon > 0$ arbitrarily small. This tends to 0 as $n \to \infty$ provided ϵ is sufficiently small.

(2) Now let $f(x)$ be regular if x is in the interior of the ellipse $|z| = R > 1$. On multiplying (9.2.1) by $(\pi i)^{-1}(y - 1)^\alpha(y + 1)^\beta f(y)$ and integrating over the ellipse $|\zeta| = R - \epsilon$, $0 < \epsilon < R/2$, we obtain an expansion of $f(x)$ in a Jacobi series which is uniformly convergent for $|z| \leqq R - 2\epsilon$. The usual term-by-term integration over the segment $[-1, +1]$ identifies this expansion with the expansion (9.1.1), and for the coefficients a_n we obtain the representation

$$a_n = \{\pi i h_n^{(\alpha,\beta)}\}^{-1} \int (y-1)^\alpha(y+1)^\beta Q_n^{(\alpha,\beta)}(y)f(y)\,dy, \qquad n = 0, 1, 2, \cdots, \tag{9.2.3}$$

where the integration is extended over the ellipse $|\zeta| = R - \epsilon$ in the positive sense.

(3) By means of (8.23.1) we can discuss formal series of the type

$$a_0P_0^{(\alpha,\beta)}(x) + a_1P_1^{(\alpha,\beta)}(x) + a_2P_2^{(\alpha,\beta)}(x) + \cdots + a_nP_n^{(\alpha,\beta)}(x) + \cdots, \tag{9.2.4}$$

which are not necessarily Fourier expansions in the ordinary sense. Let R have the same meaning as in (9.1.4), and assume $R > 1$. Then (9.2.4) has as its domain of convergence the ellipse of Theorem 9.1.1. In the interior of this ellipse it represents an analytic function.

(4) Expansions in terms of Jacobi's functions of the second kind, which are the analogues of Laurent expansions, can also be readily discussed.

THEOREM 9.2.2. *Assume $\alpha > -1$, $\beta > -1$, and let $F(y)$ be regular at $y = \infty$ with $F(\infty) = 0$. Then*

$$(y-1)^{-\alpha}(y+1)^{-\beta}F(y) = b_0Q_0^{(\alpha,\beta)}(y) + b_1Q_1^{(\alpha,\beta)}(y) + b_2Q_2^{(\alpha,\beta)}(y) + \cdots + b_nQ_n^{(\alpha,\beta)}(y) + \cdots. \tag{9.2.5}$$

This expansion is convergent in the exterior of the smallest ellipse with foci at ± 1, in the exterior of which $F(y)$ is regular. It is divergent in the interior of this ellipse. The sum of the semi-axes of this ellipse is given by

$$\rho = \limsup_{n\to\infty} |b_n|^{1/n}. \tag{9.2.6}$$

Obviously, $\rho \geqq 1$. In case $\rho = 1$ the statement needs a slight modification. The following representation holds for the coefficients:

$$(9.2.7)\qquad b_n = \{\pi i h_n^{(\alpha,\beta)}\}^{-1} \int F(x) P_n^{(\alpha,\beta)}(x)\, dx, \qquad n = 0, 1, 2, \cdots,$$

where the integration is extended over the ellipse $|z| = \rho + \epsilon$, $x = \frac{1}{2}(z + z^{-1})$, $\epsilon > 0$, in the positive sense. For the proof we again use (9.2.1).

Formal series proceeding in terms of Jacobi's functions of the second kind can also be discussed.

(5) The boundary of the convergence domain of a Laguerre series

$$(9.2.8)\qquad a_0 L_0^{(\alpha)}(x) + a_1 L_1^{(\alpha)}(x) + a_2 L_2^{(\alpha)}(x) + \cdots + a_n L_n^{(\alpha)}(x) + \cdots$$

can be characterized by the condition $\Re\{(-x)^{\frac{1}{2}}\} = \text{const.}$ Therefore, this boundary is a parabola with its focus at the origin. The series is convergent in the "interior" of this parabola and divergent in its "exterior". The analogue of Cauchy-Hadamard's formula holds. The proof is based on (8.23.3).

For an Hermite series

$$(9.2.9)\qquad a_0 H_0(x) + a_1 H_1(x) + a_2 H_2(x) + \cdots + a_n H_n(x) + \cdots$$

the corresponding condition is $|\Im(x)| = \text{const.}$, which defines a strip with the real axis as axis of symmetry (cf. (8.23.4)). The series is convergent and divergent in the "interior" and in the "exterior" of this strip, respectively. The analogue of Cauchy-Hadamard's formula holds again.

Concerning the expansion in an Hermite series of a given analytic function regular in the strip $|\Im(x)| \leqq a$, $a > 0$, see Watson **1** (first paper).

9.3. Proof of Theorem 9.1.2

(1) First let us replace $f(x)$ by a polynomial $\rho(x)$. Then $s_n(x) = \rho(x)$ if n exceeds the degree of $\rho(x)$. Furthermore, $(1 - x)^{-\alpha/2-\frac{1}{4}}(1 + x)^{-\beta/2-\frac{1}{4}} s_n(x) \to \rho(x)$ as $n \to \infty$, according to elementary tests for the convergence of Fourier series (see for instance Zygmund **2**, p. 25). Now the integral

$$(9.3.1)\qquad \int_{-1}^{+1} (1 - t)^a (1 + t)^b \,|f(t) - \rho(t)|\, dt, \qquad \begin{cases} a = \min(\alpha, \alpha/2 - 1/4), \\ b = \min(\beta, \beta/2 - 1/4), \end{cases}$$

can be made arbitrarily small by proper choice of $\rho(x)$ (cf. Theorem 1.5.2). It is therefore sufficient to show that the difference (9.11.4) admits an estimate

$$(9.3.2)\qquad O(1) \int_{-1}^{+1} (1 - t)^a (1 + t)^b \,|f(t)|\, dt + o(1), \qquad n \to \infty,$$

where both bounds hold uniformly in x, $-1 + \epsilon \leqq x \leqq 1 - \epsilon$, and $O(1)$ is independent of $f(x)$.

(2) In subsequent considerations we use Darboux's formula (8.21.10) for $P_n^{(\alpha,\beta)}(\cos\theta)$, as well as the second formula (8.8.2) for the difference ratio. In the latter case we assume that both arguments θ_1, θ_2 lie in an interval which is entirely in the interior of $[0, \pi]$.

According to (4.5.2), we have

$$K_n^{(\alpha,\beta)}(x,t)$$
$$(9.3.3)\qquad = \lambda_n\left\{P_n^{(\alpha,\beta)}(x)\,\frac{P_{n+1}^{(\alpha,\beta)}(t)-P_{n+1}^{(\alpha,\beta)}(x)}{t-x} - P_{n+1}^{(\alpha,\beta)}(x)\,\frac{P_n^{(\alpha,\beta)}(t)-P_n^{(\alpha,\beta)}(x)}{t-x}\right\},$$

where

$$(9.3.4)\qquad \lambda_n = 2^{-\alpha-\beta-1}\{n+O(1)\}.$$

Writing $x = \cos\theta$, $t = \cos\phi$, and using the notation of (8.21.10), we find for $-1 < x < +1$, $-1 < t < +1$,

$$K_n^{(\alpha,\beta)}(x,t) = 2^{-\alpha-\beta-1}k(\theta)k(\phi)\left\{\cos(N\theta+\gamma)\frac{\cos[(N+1)\phi+\gamma]-\cos[(N+1)\theta+\gamma]}{\cos\phi-\cos\theta}\right.$$
$$\left. - \cos[(N+1)\theta+\gamma]\frac{\cos(N\phi+\gamma)-\cos(N\theta+\gamma)}{\cos\phi-\cos\theta} + O(1)\right\}$$
$$= 2^{-\alpha-\beta-2}k(\theta)k(\phi)\left\{\frac{\cos[N(\phi+\theta)+\phi+2\gamma]+\cos[N(\phi-\theta)+\phi]}{\cos\phi-\cos\theta}\right.$$
$$(9.3.5)\qquad \left. - \frac{\cos[N(\phi+\theta)+\theta+2\gamma]+\cos[N(\phi-\theta)-\theta]}{\cos\phi-\cos\theta} + O(1)\right\}$$
$$= 2^{-\alpha-\beta-2}k(\theta)k(\phi)\left\{\frac{\sin[(N+\frac{1}{2})(\phi+\theta)+2\gamma]}{\sin\dfrac{\phi+\theta}{2}}\right.$$
$$\left. + \frac{\sin[(N+\frac{1}{2})(\phi-\theta)]}{\sin\dfrac{\phi-\theta}{2}} + O(1)\right\}.$$

Now assume $-1+\epsilon \leqq x \leqq 1-\epsilon$. Then

$$\int_{-1+\epsilon/2}^{1-\epsilon/2}(1-t)^\alpha(1+t)^\beta f(t)K_n^{(\alpha,\beta)}(x,t)\,dt$$
$$= \int_\eta^{\pi-\eta} 2^{\alpha+\beta+1}\left(\sin\frac{\phi}{2}\right)^{2\alpha+1}\left(\cos\frac{\phi}{2}\right)^{2\beta+1} f(\cos\phi)K_n^{(\alpha,\beta)}(x,t)\,d\phi,$$

where $\cos\eta = 1-\epsilon/2$. Next replace $K_n^{(\alpha,\beta)}(x,t)$ by (9.3.5). By Riemann's lemma (Zygmund 2, p. 18), the result will be

$$(9.3.6)\qquad \frac{1}{2\pi}\int_\eta^{\pi-\eta} f(\cos\phi)\frac{\sin(2n+1)\dfrac{\phi-\theta}{2}}{\sin\dfrac{\phi-\theta}{2}}\,d\phi + O(1)\int_{-1+\epsilon/2}^{1-\epsilon/2}|f(t)|\,dt + o(1),$$

as $n \to \infty$.

The bound of the term $O(1)$ is independent of $f(x)$. When we replace α and β by $-\frac{1}{2}$, and $f(t)$ by $(1-t)^{\alpha/2+\frac{1}{4}}(1+t)^{\beta/2+\frac{1}{4}}f(t)$, repeated application of Riemann's lemma yields

$$\int_{-1+\epsilon/2}^{1-\epsilon/2} (1-t)^{\alpha/2-\frac{1}{4}}(1+t)^{\beta/2-\frac{1}{4}} f(t) K_n^{(-\frac{1}{2},-\frac{1}{2})}(x,t)\,dt$$

$$= \frac{1}{2\pi}\int_{\eta}^{\pi-\eta} (1-\cos\phi)^{\alpha/2+\frac{1}{4}}(1+\cos\phi)^{\beta/2+\frac{1}{4}} f(\cos\phi)\,\frac{\sin (2n+1)\dfrac{\phi-\theta}{2}}{\sin\dfrac{\phi-\theta}{2}}\,d\phi$$

$$+ O(1)\int_{-1+\epsilon/2}^{1-\epsilon/2} |f(t)|\,dt + o(1) \tag{9.3.7}$$

$$= \frac{1}{2\pi}(1-\cos\theta)^{\alpha/2+\frac{1}{4}}(1+\cos\theta)^{\beta/2+\frac{1}{4}}\int_{\eta}^{\pi-\eta} f(\cos\phi)\,\frac{\sin (2n+1)\dfrac{\phi-\theta}{2}}{\sin\dfrac{\phi-\theta}{2}}\,d\phi$$

$$+ O(1)\int_{-1+\epsilon/2}^{1-\epsilon/2} |f(t)|\,dt + o(1).$$

Consequently, the part of (9.11.4) which corresponds to $-1+\epsilon/2 \leqq t \leqq 1-\epsilon/2$ has the form (9.3.2).

(3) We consider now the expressions

$$O(n)\,|P_n^{(\alpha,\beta)}(x)|\int_{1-\epsilon/2}^{1}\left|\frac{P_{n+1}^{(\alpha,\beta)}(t)-P_{n+1}^{(\alpha,\beta)}(x)}{t-x}\right|(1-t)^{\alpha}(1+t)^{\beta}|f(t)|\,dt, \tag{9.3.8}$$

$$O(n)\,|P_n^{(-\frac{1}{2},-\frac{1}{2})}(x)| \cdot\int_{1-\epsilon/2}^{1}\left|\frac{P_{n+1}^{(-\frac{1}{2},-\frac{1}{2})}(t)-P_{n+1}^{(-\frac{1}{2},-\frac{1}{2})}(x)}{t-x}\right|(1-t)^{\alpha/2-\frac{1}{4}}(1+t)^{\beta/2-\frac{1}{4}}|f(t)|\,dt, \tag{9.3.9}$$

and the corresponding integrals extended over $-1 \leqq t \leqq -1+\epsilon/2$. The first integral is

$$O(n)O(n^{-\frac{1}{2}})\int_{1-\epsilon/2}^{1}\{|P_{n+1}^{(\alpha,\beta)}(t)| + |P_{n+1}^{(\alpha,\beta)}(x)|\}(1-t)^{\alpha}(1+t)^{\beta}|f(t)|\,dt$$

$$= O(n)O(n^{-\frac{1}{2}})\int_0^{\eta}|P_{n+1}^{(\alpha,\beta)}(\cos\phi)|\,\phi^{2\alpha+1}|f(\cos\phi)|\,d\phi$$

$$+ O(n)O(n^{-\frac{1}{2}})O(n^{-\frac{1}{2}})\int_{-1}^{+1}(1-t)^{\alpha}(1+t)^{\beta}|f(t)|\,dt.$$

From (7.32.6) and (7.32.7) it follows that the first term of the right-hand member is

$$O(1)\int_0^{\eta}\phi^{-\alpha-\frac{1}{2}}\phi^{2\alpha+1}|f(\cos\phi)|\,d\phi = O(1)\int_{-1}^{+1}(1-t)^{\alpha/2-\frac{1}{4}}(1+t)^{\beta/2-\frac{1}{4}}|f(t)|\,dt,$$

or is

$$O(1) \int_0^\eta \phi^{2\alpha+1} |f(\cos \phi)| \, d\phi = O(1) \int_{-1}^{+1} (1-t)^\alpha (1+t)^\beta |f(t)| \, dt,$$

according as $\alpha \geqq -\frac{1}{2}$ or $\alpha \leqq -\frac{1}{2}$. In view of (4.1.7) the expression (9.3.9) is

$$O(n)O(n^{-\frac{1}{2}})O(n^{-\frac{1}{2}}) \int_{1-\epsilon/2}^{1} (1-t)^{\alpha/2-\frac{3}{4}}(1+t)^{\beta/2-\frac{3}{4}} |f(t)| \, dt$$

$$= O(1) \int_{-1}^{+1} (1-t)^{\alpha/2-\frac{3}{4}}(1+t)^{\beta/2-\frac{3}{4}} |f(t)| \, dt.$$

The integrals corresponding to $-1 \leqq t \leqq -1 + \epsilon/2$ are similarly treated.

(4) Finally, we consider the Jacobi series of the function (see, in the special case $\alpha = \beta$, Kogbetliantz **19**, p. 184, (62))

$$f(x) = (1-x)^\mu, \tag{9.3.10}$$

and we show that for proper values of μ the first integral in (9.1.5) exists, but that the second does not. Moreover, the Jacobi series is divergent in $-1 < x < +1$. Here $\alpha > -\frac{1}{2}$.

To this end we note that, according to (4.3.1) and (4.3.3),

$$\begin{aligned} \{h_n^{(\alpha,\beta)}\}^{-1} & \int_{-1}^{+1} (1-t)^{\mu+\alpha}(1+t)^\beta P_n^{(\alpha,\beta)}(t) \, dt \\ &= \{h_n^{(\alpha,\beta)}\}^{-1} \frac{(-1)^n \mu(\mu-1)\cdots(\mu-n+1)}{2^n n!} \\ &\qquad \int_{-1}^{+1} (1-t)^{n+\alpha}(1+t)^{n+\beta}(1-t)^{\mu-n} \, dt \\ &\cong 2^{-\alpha-\beta} n \frac{n^{-\mu-1}}{2^n \Gamma(-\mu)} 2^{n+\mu+\alpha+\beta+1} n^{-\mu-\alpha-1} \Gamma(\mu+\alpha+1) \\ &\sim n^{-2\mu-\alpha-1} \qquad \text{as } n \to \infty. \end{aligned} \tag{9.3.11}$$

If we assume that $\mu + \alpha > -1$, $\mu \neq 0, 1, 2, \cdots$, and take $0 < \theta < \pi$, the general term of the Jacobi series has the form $n^{-2\mu-\alpha-\frac{3}{2}} \cos(N\theta + \gamma)$. The required values of μ are those which satisfy the condition

$$-1 - \alpha < \mu \leqq -\alpha/2 - 3/4. \tag{9.3.12}$$

9.4. Proof of Theorem 9.1.3; preliminaries

(1) We first derive the following important identity:

$$\begin{aligned} P_n^{(\alpha+k+1,\beta)}(x) &= 2^{\alpha+\beta+1} \Gamma(\alpha+1) \frac{\Gamma(n+\beta+1)}{\Gamma(n+\alpha+\beta+k+2)} \\ &\cdot \sum_{\nu=0}^{n} \frac{\Gamma(n+\nu+\alpha+\beta+k+2)}{\Gamma(n+\nu+\alpha+\beta+2)} C_{n-\nu}^{(k)} \{h_\nu^{(\alpha,\beta)}\}^{-1} P_\nu^{(\alpha,\beta)}(1) P_\nu^{(\alpha,\beta)}(x), \end{aligned} \tag{9.4.1}$$

where $h_n^{(\alpha,\beta)}$ and $C_n^{(k)}$ have the usual meaning (see (4.3.3) and the introduction to

Chapter IX). This is a generalization of (4.5.3) (last identity), to which it reduces when $k = 0$.

For the proof we calculate

$$\int_{-1}^{+1} P_n^{(\alpha+k+1,\beta)}(x) P_\nu^{(\alpha,\beta)}(x)(1-x)^\alpha(1+x)^\beta\,dx. \tag{9.4.2}$$

According to (4.3.1), this last expression is equal to

$$\frac{(-1)^\nu}{2^\nu \nu!}\int_{-1}^{+1} P_n^{(\alpha+k+1,\beta)}(x)\left(\frac{d}{dx}\right)^\nu \{(1-x)^{\nu+\alpha}(1+x)^{\nu+\beta}\}\,dx$$

$$= \frac{1}{2^\nu \nu!}\int_{-1}^{+1} (1-x)^{\nu+\alpha}(1+x)^{\nu+\beta}\left(\frac{d}{dx}\right)^\nu \{P_n^{(\alpha+k+1,\beta)}(x)\}\,dx;$$

or, by (4.21.7) and (4.3.1), it is equal to

$$\frac{1}{2^\nu \nu!}\frac{1}{2^\nu}(n+\alpha+\beta+k+2)(n+\alpha+\beta+k+3)\cdots(n+\alpha+\beta+k+\nu+1)$$

$$\cdot\int_{-1}^{+1}(1-x)^{\nu+\alpha}(1+x)^{\nu+\beta}P_{n-\nu}^{(\alpha+k+\nu+1,\beta+\nu)}(x)\,dx$$

$$= \frac{(-1)^{n-\nu}}{2^{n+\nu}\nu!(n-\nu)!}\frac{\Gamma(n+\alpha+\beta+k+\nu+2)}{\Gamma(n+\alpha+\beta+k+2)}$$

$$\int_{-1}^{+1}(1-x)^{-k-1}\left(\frac{d}{dx}\right)^{n-\nu}\{(1-x)^{n+\alpha+k+1}(1+x)^{n+\beta}\}\,dx$$

$$= \frac{(k+1)(k+2)\cdots(k+n-\nu)}{2^{n+\nu}\nu!(n-\nu)!}\frac{\Gamma(n+\alpha+\beta+k+\nu+2)}{\Gamma(n+\alpha+\beta+k+2)}$$

$$\cdot\int_{-1}^{+1}(1-x)^{-k-1-n+\nu}(1-x)^{n+\alpha+k+1}(1+x)^{n+\beta}\,dx,$$

which, if (1.7.5) is taken into account, furnishes the statement.

(2) On substituting the explicit expressions for $P_\nu^{(\alpha,\beta)}(1)$ and $h_\nu^{(\alpha,\beta)}$ (see (4.1.1) and (4.3.3)), we obtain

$$\begin{aligned} P_n^{(\alpha+k+1,\beta)}(x) &= \frac{\Gamma(n+\beta+1)}{\Gamma(n+\alpha+\beta+k+2)}\sum_{\nu=0}^{n}\frac{\Gamma(n+\nu+\alpha+\beta+k+2)}{\Gamma(n+\nu+\alpha+\beta+2)} \\ &\quad\cdot C_{n-\nu}^{(k)}(2\nu+\alpha+\beta+1)\frac{\Gamma(\nu+\alpha+\beta+1)}{\Gamma(\nu+\beta+1)}P_\nu^{(\alpha,\beta)}(x). \end{aligned} \tag{9.4.3}$$

Since this is an identity in α, β, k, we may replace α by $\alpha + k + 1$ and k by $-k-2$. Whence the inversion formula

$$\begin{aligned} P_n^{(\alpha,\beta)}(x) &= \frac{\Gamma(n+\beta+1)}{\Gamma(n+\alpha+\beta+1)}\sum_{\nu=0}^{n}\frac{\Gamma(n+\nu+\alpha+\beta+1)}{\Gamma(n+\nu+\alpha+\beta+k+3)} \\ &\quad\cdot C_{n-\nu}^{(-k-2)}(2\nu+\alpha+\beta+k+2)\frac{\Gamma(\nu+\alpha+\beta+k+2)}{\Gamma(\nu+\beta+1)}P_\nu^{(\alpha+k+1,\beta)}(x) \end{aligned} \tag{9.4.4}$$

follows. Consequently,

$$S_n^{(k)}(x) \equiv \sum_{m=0}^{n} C_{n-m}^{(k)} \{h_m^{(\alpha,\beta)}\}^{-1} P_m^{(\alpha,\beta)}(1) P_m^{(\alpha,\beta)}(x)$$

$$(9.4.5) \qquad = \frac{2^{-\alpha-\beta-1}}{\Gamma(\alpha+1)} \sum_{\nu=0}^{n} G_\nu(n, k)(2\nu + \alpha + \beta + k + 2) \cdot \frac{\Gamma(\nu + \alpha + \beta + k + 2)}{\Gamma(\nu + \beta + 1)} P_\nu^{(\alpha+k+1,\beta)}(x),$$

with

$$(9.4.6) \quad G_\nu(n, k) = \sum_{m=\nu}^{n} C_{n-m}^{(k)} C_{m-\nu}^{(-k-2)} (2m + \alpha + \beta + 1) \frac{\Gamma(m + \nu + \alpha + \beta + 1)}{\Gamma(m + \nu + \alpha + \beta + k + 3)}.$$

The expression $S_n^{(k)}(x)$ represents the numerator of the nth Cesàro kernel of order k.

9.41. Continuation; the Lebesgue constants of order k

(1) For the proof of Theorem 9.1.3, according to Theorem 1.6 (Helly's theorem), it is sufficient to show that the sequence of the "Lebesgue constants"

$$(9.41.1) \qquad L_n^{(k)} = \{C_n^{(k)}\}^{-1} \int_{-1}^{+1} (1 - x)^\alpha (1 + x)^\beta \, | S_n^{(k)}(x) | \, dx,$$

is bounded if and only if $k > \alpha + 1/2$. Since for $k > \alpha + \frac{1}{2}$ (cf. (7.34.1), first and third case),

$$\int_{-1}^{+1} (1 - x)^\alpha (1 + x)^\beta \, | P_n^{(\alpha+k+1,\beta)}(x) | \, dx$$

$$(9.41.2) \qquad = O(1) \int_0^1 (1 - x)^\alpha \, | P_n^{(\alpha+k+1,\beta)}(x) | \, dx + O(1) \int_0^1 (1 - x)^\beta \, | P_n^{(\beta,\alpha+k+1)}(x) | \, dx$$

$$= O(n^{k-\alpha-1}) + O(n^{-\frac{1}{2}}) = O(n^{k-\alpha-1}),$$

we have from (9.4.5)

$$(9.41.3) \qquad L_n^{(k)} = O(n^{-k}) \sum_{\nu=0}^{n} | G_\nu(n, k) | \, \nu^{2k+1}.$$

The last factor ν^{2k+1} must be replaced by 1 for $\nu = 0$.

(2) First let k be a non-negative integer. Then in (9.4.6) we must consider only the terms $\nu \leqq m \leqq \nu + k + 1$, so that $G_\nu(n, k)$ can be written in the form $G(n - \nu - 1)$, where[55]

[55] If $\nu + k + 1 > n$, certain additional terms not permitted by (9.4.6) occur in $G(u)$. However, these terms vanish if we substitute $u = n - \nu - 1$, except if $\nu = n$ and $m = n + k + 1$. Now the contribution of this term to $G(u) = G(-1)$ is $O(m)(m + \nu)^{-k-2} = O(n^{-k-1})$. This multiplied by the factor $\nu^{2k+1} = n^{2k+1}$, which occurs in (9.41.3), furnishes a total contribution $O(n^{-k})O(n^k) = O(1)$.

(9.41.4)
$$G(u) = \sum_{m=\nu}^{\nu+k+1} (-1)^{m-\nu} \binom{u+k+1-m+\nu}{k} \cdot \binom{k+1}{m-\nu} (2m+\alpha+\beta+1) \frac{\Gamma(m+\nu+\alpha+\beta+1)}{\Gamma(m+\nu+\alpha+\beta+k+3)}.$$

This is a π_k in u, containing ν and k as parameters. Newton's formula shows us (cf. Markoff **5**, p. 15; notation as in (2.8.4)) that

(9.41.5)
$$G(u) = \sum_{\rho=0}^{k} \binom{u}{\rho} \Delta^\rho G(0);$$

and from (9.41.4) it follows that

$$\Delta^\rho G(0) = \sum_{m=\nu}^{\nu+\rho+1} (-1)^{m-\nu} \binom{k+1-m+\nu}{k-\rho} \binom{k+1}{m-\nu} \cdot (2m+\alpha+\beta+1) \frac{\Gamma(m+\nu+\alpha+\beta+1)}{\Gamma(m+\nu+\alpha+\beta+k+3)}$$

$$= \frac{1}{(k-\rho)!} \int_0^1 \sum_{m=\nu}^{\nu+\rho+1} \frac{(-1)^{m-\nu}(2m+\alpha+\beta+1)}{(m-\nu)!(\rho+1-m+\nu)!} t^{m+\nu+\alpha+\beta}(1-t)^{k+1}\, dt$$

(9.41.6)
$$= \frac{1}{(k-\rho)!} \int_0^1 \left\{ -2 \sum_{m=\nu+1}^{\nu+\rho+1} (-1)^{m-\nu-1} \frac{t^{m+\nu+\alpha+\beta}(1-t)^{k+1}}{(m-\nu-1)!(\rho+1-m+\nu)!} + (2\nu+\alpha+\beta+1) \sum_{m=\nu}^{\nu+\rho+1} (-1)^{m-\nu} \frac{t^{m+\nu+\alpha+\beta}(1-t)^{k+1}}{(m-\nu)!(\rho+1-m+\nu)!} \right\} dt$$

$$= \frac{1}{(k-\rho)!(\rho+1)!} \int_0^1 \left\{ -2(\rho+1)t^{2\nu+\alpha+\beta+1}(1-t)^{k+\rho+1} + (2\nu+\alpha+\beta+1)t^{2\nu+\alpha+\beta}(1-t)^{k+\rho+2} \right\} dt$$

$$= O(\nu^{-k-\rho-2}) + O(\nu)O(\nu^{-k-\rho-3}) = O(\nu^{-k-\rho-2}), \qquad \nu \geqq 1.$$

The last integral formula also shows that $\Delta^k G(0) = 0$, $\nu \geqq 1$; in particular, $G(u) \equiv 0$ for $k = 0$, $\nu \geqq 1$. Both facts hold also for $\nu = 0$, as analytic continuation with respect to $\alpha + \beta$ shows. This settles the case $k = 0$ in view of the remark in the last footnote. For $k > 0$,

(9.41.7)
$$G_\nu(n, k) = \sum_{\rho=0}^{k-1} (n-\nu)^\rho O(\nu^{-k-\rho-2}),$$

to which (according to the same remark) the term $O(n^{-k-1})$ must be added if $\nu = n$. In this case the factor $(n-\nu)^\rho$, in the case $\nu = 0$ the factor $\nu^{-k-\rho-2}$, must be replaced by 1. For $k > \alpha + \frac{1}{2}$, this provides, as desired,

(9.41.8)
$$L_n^{(k)} = O(n^{-k}) \sum_{\rho=0}^{k-1} n^\rho + O(n^{-k}) \sum_{\nu=1}^{n-1} \sum_{\rho=0}^{k-1} (n-\nu)^\rho \nu^{-k-\rho-2} \nu^{2k+1} + O(n^{-k}) \left\{ \sum_{\rho=0}^{k-1} n^{-k-\rho-2} n^{2k+1} + n^{-k-1} n^{2k+1} \right\} = O(1).$$

(3) We now consider the case $k > \alpha + \frac{1}{2}$, k not an integer. Then, according to the previous result,

$$L_n^{(k')} \leqq A, \qquad n = 0, 1, 2, \cdots, \tag{9.41.9}$$

where $k' = [k] + 1$, and A is a proper constant independent of n.

Let $\sigma > k'$. If we write $u_n(x) = \{h_n^{(\alpha,\beta)}\}^{-1} P_n^{(\alpha,\beta)}(1) P_n^{(\alpha,\beta)}(x)$, the definition (9.4.5) furnishes

$$\sum_{n=0}^{\infty} S_n^{(k')}(x) r^n = (1 - r)^{-k'-1} \sum_{n=0}^{\infty} u_n(x) r^n.$$

Thus,

$$\sum_{n=0}^{\infty} S_n^{(\sigma)}(x) r^n = (1 - r)^{k'-\sigma}(1 - r)^{-k'-1} \sum_{n=0}^{\infty} u_n(x) r^n = (1 - r)^{k'-\sigma} \sum_{n=0}^{\infty} S_n^{(k')}(x) r^n,$$

so that $S_n^{(\sigma)}(x) = \sum_{m=0}^{n} C_{n-m}^{(\sigma-k'-1)} S_m^{(k')}(x)$ and

$$L_n^{(\sigma)} \leqq \{C_n^{(\sigma)}\}^{-1} \sum_{m=0}^{n} C_{n-m}^{(\sigma-k'-1)} C_m^{(k')} L_m^{(k')} \leqq A\{C_n^{(\sigma)}\}^{-1} \sum_{m=0}^{n} C_{n-m}^{(\sigma-k'-1)} C_m^{(k')}.$$

Consequently,

$$L_n^{(\sigma)} \leqq A, \qquad \sigma \geqq k',\ n = 0, 1, 2, \cdots. \tag{9.41.10}$$

This holds in particular if $\sigma = k + 1, k + 2, \cdots$.

(4) Now we prove the identity

$$\begin{aligned} &\frac{\Gamma(n + \nu + \alpha + \beta + k + 2)}{\Gamma(n + \nu + \alpha + \beta + 2)} C_{n-\nu}^{(k)} = \frac{\Gamma(2n + \alpha + \beta + 2k + 3)}{\Gamma(2n + \alpha + \beta + k + 3)} \sum_{\rho=0}^{\infty} (-1)^\rho \\ &\cdot \binom{k+\rho}{\rho} C_{n-\nu}^{(k+\rho)} \frac{k(k-1)\cdots(k-\rho+1)}{(2n + \alpha + \beta + k + 3)\cdots(2n + \alpha + \beta + k + \rho + 2)}, \\ &\qquad 0 \leqq \nu \leqq n. \end{aligned} \tag{9.41.11}$$

The series on the right converges absolutely if $n \geqq 1$. (For $\rho = 0$, the last fraction is to be replaced by 1.) For the proof we use the following transformation of the left-hand member:

$$\begin{aligned} &\frac{\Gamma(2n + \alpha + \beta + 2k + 3)}{\Gamma(n + \nu + \alpha + \beta + 2)} \frac{1}{(n - \nu)!\,\Gamma(k + 1)} \int_0^1 t^{n-\nu+k}(1 - t)^{n+\nu+\alpha+\beta+k+1}\, dt \\ &= \frac{\Gamma(2n + \alpha + \beta + 2k + 3)}{\Gamma(n + \nu + \alpha + \beta + 2)} \frac{1}{(n - \nu)!\,\Gamma(k + 1)} \\ &\cdot \sum_{\rho=0}^{\infty} (-1)^\rho \binom{k}{\rho} \int_0^1 t^{n-\nu+k+\rho}(1 - t)^{n+\nu+\alpha+\beta+1}\, dt. \end{aligned}$$

Calculation of the last integral completes the proof. Term-by-term integration is permitted, since the terms are, save for a finite number of exceptions, all of the same sign.

Consequently, from (9.4.1) we have the representation

$$
\begin{aligned}
P_n^{(\alpha+k+1,\beta)}(x) &= 2^{\alpha+\beta+1}\Gamma(\alpha+1) \\
&\cdot \frac{\Gamma(n+\beta+1)\Gamma(2n+\alpha+\beta+2k+3)}{\Gamma(n+\alpha+\beta+k+2)\Gamma(2n+\alpha+\beta+k+3)} \sum_{\rho=0}^{\infty} (-1)^\rho \binom{k+\rho}{\rho} \\
&\cdot \frac{k(k-1)\cdots(k-\rho+1)}{(2n+\alpha+\beta+k+3)\cdots(2n+\alpha+\beta+k+\rho+2)} S_n^{(k+\rho)}(x),
\end{aligned}
\tag{9.41.12}
$$

or

$$
\begin{aligned}
S_n^{(k)}(x) &= \left\{ 2^{\alpha+\beta+1}\Gamma(\alpha+1) \right. \\
&\left. \cdot \frac{\Gamma(n+\beta+1)\Gamma(2n+\alpha+\beta+2k+3)}{\Gamma(n+\alpha+\beta+k+2)\Gamma(2n+\alpha+\beta+k+3)} \right\}^{-1} P_n^{(\alpha+k+1,\beta)}(x) \\
&- \sum_{\rho=1}^{\infty} (-1)^\rho \binom{k+\rho}{\rho} \\
&\cdot \frac{k(k-1)\cdots(k-\rho+1)}{(2n+\alpha+\beta+k+3)\cdots(2n+\alpha+\beta+k+\rho+2)} S_n^{(k+\rho)}(x).
\end{aligned}
\tag{9.41.13}
$$

On account of (9.41.2), this gives

$$
L_n^{(k)} = O(n^{-k})O(n^{\alpha+1})O(n^{k-\alpha-1}) + O(n^{-k})A_n = O(1) + O(n^{-k})A_n, \tag{9.41.14}
$$

where

$$
\begin{aligned}
A_n = \sum_{\rho=1}^{\infty} \binom{k+\rho}{\rho} & \frac{|k(k-1)\cdots(k-\rho+1)|}{(2n+\alpha+\beta+k+3)\cdots(2n+\alpha+\beta+k+\rho+2)} \\
& \cdot \int_{-1}^{+1} (1-x)^\alpha(1+x)^\beta \,|S_n^{(k+\rho)}(x)|\, dx.
\end{aligned}
\tag{9.41.15}
$$

Now, according to (9.41.10),

$$
\begin{aligned}
A_n &\leqq A \sum_{\rho=1}^{\infty} \binom{k+\rho}{\rho} \\
&\cdot \frac{|k(k-1)\cdots(k-\rho+1)|}{(2n+\alpha+\beta+k+3)\cdots(2n+\alpha+\beta+k+\rho+2)} C_n^{(k+\rho)} \\
&= O(1) \sum_{\rho=1}^{\infty} \left|\binom{k}{\rho}\right| \binom{n+k}{n} \\
&\cdot \frac{(n+k+1)\cdots(n+k+\rho)}{(2n+\alpha+\beta+k+3)\cdots(2n+\alpha+\beta+k+\rho+2)} \\
&= O(n^k) \sum_{\rho=1}^{\infty} \left|\binom{k}{\rho}\right| \frac{(k+2)\cdots(k+\rho+1)}{(\alpha+\beta+k+5)\cdots(\alpha+\beta+k+\rho+4)} = O(n^k),
\end{aligned}
\tag{9.41.16}
$$

since $(n+k+l)/(2n+\alpha+\beta+k+l+2)$ decreases as n increases provided $l > \alpha+\beta-k+2$. This completes the proof of the first part of the statement in Theorem 9.1.3.

(5) Finally, let $k = \alpha + \frac{1}{2}$. Using the previous notation, we have

$$L_n^{(k)} > \left\{2^{\alpha+\beta+1}\Gamma(\alpha+1)\frac{\Gamma(n+\beta+1)\Gamma(2n+\alpha+\beta+2k+3)}{\Gamma(n+\alpha+\beta+k+2)\Gamma(2n+\alpha+\beta+k+3)}\right\}^{-1}$$

$$\cdot\{C_n^{(k)}\}^{-1}\int_{-1}^{+1}(1-x)^\alpha(1+x)^\beta\,|\,P_n^{(\alpha+k+1,\beta)}(x)\,|\,dx - \{C_n^{(k)}\}^{-1}A_n$$

from (9.41.13). The first term on the right is $\sim n^{\alpha+1}n^{-k}n^{-\frac{1}{2}}\log n = \log n$ (according to the second part of (7.34.1)[56]); the second term is bounded, according to the result of (4). That is,

$$L_n^{(k)} > A\log n, \qquad A > 0, \quad k = \alpha + \tfrac{1}{2}, \tag{9.41.17}$$

so that the expansion of a continuous function in Jacobi series is, in general, not $(C, k = \alpha + \frac{1}{2})$-summable.

9.42. Proof of Theorem 9.1.4

(1) Let $f(x)$ be continuous at $x = 1$, let $f(1) = 0$, and assume $k > \alpha + \frac{1}{2}$. First we discuss the integral

$$\int_{-1}^{+1}|\,f(x)\,|\,(1-x)^\alpha(1+x)^\beta\,|\,P_n^{(\alpha+k+1,\beta)}(x)\,|\,dx \tag{9.42.1}$$

as $n \to \infty$. Denoting by ϵ an arbitrary positive number, $\epsilon < \pi/2$, we decompose the interval $0 \leqq \theta \leqq \pi$, $x = \cos\theta$, into

$$0 \leqq \theta \leqq \epsilon, \qquad \epsilon \leqq \theta \leqq \pi - \epsilon, \qquad \pi - \epsilon \leqq \theta \leqq \pi. \tag{9.42.2}$$

The corresponding integrals I, II, III, can be estimated as follows (cf. (9.41.2), (7.32.6), and (7.32.7)):

$$\begin{aligned}\mathrm{I} &= \max_{0\leqq\theta\leqq\epsilon}|\,f(\cos\theta)\,|\int_{\cos\epsilon}^{1}(1-x)^\alpha(1+x)^\beta\,|\,P_n^{(\alpha+k+1,\beta)}(x)\,|\,dx\\ &= \max_{0\leqq\theta\leqq\epsilon}|\,f(\cos\theta)\,|\,O(n^{k-\alpha-1}),\\ \mathrm{II} &= O(n^{-\frac{1}{2}}) = o(n^{k-\alpha-1}),\end{aligned}$$

$$\mathrm{III} = \begin{cases}\displaystyle\int_{\pi-\epsilon}^{\pi}|\,f(\cos\theta)\,|\,(\pi-\theta)^{2\beta+1}O(n^{-\frac{1}{2}})\,d\theta = O(n^{-\frac{1}{2}}) = o(n^{k-\alpha-1}),\\ \displaystyle\int_{\pi-\epsilon}^{\pi}|\,f(\cos\theta)\,|\,(\pi-\theta)^{2\beta+1}O(n^{\beta})\,d\theta\\ \qquad\displaystyle = O(n^{k-\alpha-1})\int_{\pi-\epsilon}^{\pi}|\,f(\cos\theta)\,|\,(\pi-\theta)^{2\beta+1}\,d\theta,\\ \displaystyle\int_{\pi-\epsilon}^{\pi}|\,f(\cos\theta)\,|\,(\pi-\theta)^{2\beta+1}(\pi-\theta)^{-\beta-\frac{1}{2}}O(n^{-\frac{1}{2}})\,d\theta = O(n^{-\frac{1}{2}}) = o(n^{k-\alpha-1}),\end{cases}$$

[56] If we use only (7.34.2), we obtain $L_n^{(k)} \to \infty$ as $n \to \infty$; this is sufficient for our purpose.

according as $-1 < \beta \leqq -\frac{1}{2}$, $-\frac{1}{2} < \beta \leqq k - \alpha - 1$, or $\beta > k - \alpha - 1$. In the last case we use the antipole condition of Theorem 9.1.4. Since ϵ is arbitrarily small, it appears that in all cases the integral (9.42.1) is $o(n^{k-\alpha-1})$.

(2) Next we introduce the constants

$$M_n^{(k)} = \{C_n^{(k)}\}^{-1} \int_{-1}^{+1} | f(x) | (1 - x)^\alpha (1 + x)^\beta | S_n^{(k)}(x) | \, dx, \tag{9.42.3}$$
$$n = 0, 1, 2, \cdots,$$

where $S_n^{(k)}(x)$ has the same meaning as in (9.4.5). We obtain the following analogue of (9.41.3):

$$M_n^{(k)} = O(n^{-k}) \sum_{\nu=0}^{n} | G_\nu(n, k) | \, o(\nu^{2k+1}).$$

The last factor $o(\nu^{2k+1})$ must be replaced by 1 for $\nu = 0$. Therefore, if k is any positive integer, we have as in (9.41.8),

$$\begin{aligned} M_n^{(k)} &= O(n^{-k}) \sum_{\rho=0}^{k-1} n^\rho + O(n^{-k}) \sum_{\nu=1}^{n-1} \sum_{\rho=0}^{k-1} (n - \nu)^\rho \nu^{-k-\rho-2} o(\nu^{2k+1}) \\ &\quad + O(n^{-k}) \left\{ \sum_{\rho=0}^{k-1} n^{-k-\rho-2} o(n^{2k+1}) + n^{-k-1} o(n^{2k+1}) \right\} = o(1). \end{aligned} \tag{9.42.4}$$

The same holds if $k = 0$, for then $G(u) \equiv 0$ (see the remark in the footnote of §9.41 (2)).

(3) For non-integral k we use (9.41.13) again. Let ϵ be an arbitrary positive number, and let n_0 be so chosen that, for $k' = [k] + 1$,

$$M_n^{(k')} \leqq \epsilon \quad \text{if} \quad n \geqq n_0 . \tag{9.42.5}$$

Then for $n \geqq n_0$, $\sigma > k'$ we find, as in §9.41 (3),

$$M_n^{(\sigma)} \leqq \{C_n^{(\sigma)}\}^{-1} \sum_{m=0}^{n} C_{n-m}^{(\sigma-k'-1)} C_m^{(k')} M_m^{(k')}.$$

We decompose the latter sum into the sums $\sum_{m=0}^{n_0-1}$ and $\sum_{m=n_0}^{n}$. In the second part (9.42.5) can be applied so that

$$M_n^{(\sigma)} \leqq \{C_n^{(\sigma)}\}^{-1} \sum_{m=0}^{n_0-1} C_{n-m}^{(\sigma-k'-1)} C_m^{(k')} M_m^{(k')} + \epsilon \{C_n^{(\sigma)}\}^{-1} \sum_{m=n_0}^{n} C_{n-m}^{(\sigma-k'-1)} C_m^{(k')}.$$

The second term of the right-hand member is less than ϵ. Therefore,

$$M_n^{(\sigma)} < \epsilon + A \{C_n^{(\sigma)}\}^{-1} C_n^{(\sigma-k'-1)}, \tag{9.42.6}$$

where A is a positive constant depending on ϵ and independent of σ. This holds in particular if $\sigma = k + 1, k + 2, \cdots$. We must replace $C_n^{(\sigma-k'-1)}$ by $C_{n-n_0+1}^{(\sigma-k'-1)}$ if $\sigma < k' + 1$, which occurs when $\sigma = k + 1$, but not when $\sigma \geqq k + 2$. Consequently, we obtain, as in (9.41.14),

$$M_n^{(k)} < O(n^{-k})O(n^{\alpha+1})o(n^{k-\alpha-1}) + O(n^{-k}) \sum_{\rho=1}^{\infty} \binom{k+\rho}{\rho}$$

(9.42.7)
$$\cdot \frac{|k(k-1)\cdots(k-\rho+1)|}{(2n+\alpha+\beta+k+3)\cdots(2n+\alpha+\beta+k+\rho+2)}$$
$$\cdot \{\epsilon C_n^{(k+\rho)} + A C_n^{(k+\rho-k'-1)}\}.$$

(For $\rho = 1$ we must replace $C_n^{(k+\rho-k'-1)}$ by $C_{n-n_0+1}^{(k+\rho-k'-1)}$.) The first term of the right-hand member is $o(1)$. The second term can be decomposed into two parts. The first part has the form $\epsilon O(1)$, according to §9.41 (4). To estimate the second part we use the fact that

$$\frac{C_n^{(k+\rho-k'-1)}}{C_n^{(k+\rho)}} = \frac{\Gamma(n+k+\rho-k')\Gamma(k+\rho+1)}{\Gamma(n+k+\rho+1)\Gamma(k+\rho-k')} < B\left(\frac{k+\rho}{n+k+\rho}\right)^{k'+1}$$

where $B > 0$ depends only on k. For the second part we therefore obtain the bound (cf. (9.41.16))

$$O(1) \sum_{\rho=1}^{\infty} \left|\binom{k}{\rho}\right| \frac{(n+k+1)\cdots(n+k+\rho)}{(2n+\alpha+\beta+k+3)\cdots(2n+\alpha+\beta+k+\rho+2)}$$
$$\cdot\left(\frac{k+\rho}{n+k+\rho}\right)^{k'+1}.$$

Decomposing this sum into the sums $\sum_{\rho=1}^{\mathrm{P}}$ and $\sum_{\rho=\mathrm{P}+1}^{\infty}$, where P is an arbitrary positive integer, we obtain the bound

$$O(1) \sum_{\rho=1}^{\mathrm{P}} \left(\frac{k+\rho}{n+k+\rho}\right)^{k'+1}$$
$$+ O(1) \sum_{\rho=\mathrm{P}+1}^{\infty} \left|\binom{k}{\rho}\right| \frac{(k+2)\cdots(k+\rho+1)}{(\alpha+\beta+k+5)\cdots(\alpha+\beta+k+\rho+4)}.$$

The first term tends to zero as $n \to \infty$. The second term is arbitrarily small if P is sufficiently large. Therefore $M_n^{(k)} = o(1)$ as $n \to \infty$.

Concerning the case $k = \alpha + \frac{1}{2}$, compare §9.41 (5).

(4) REMARK. *The continuity at $x = +1$ can be replaced by the more general condition*

$$\int_0^{\delta} |f(\cos\theta) - f(1)|\, d\theta = o(\delta), \qquad \delta \to +0. \tag{9.42.8}$$

Only the estimation of the integral I (cf. (9.42.2)) must be slightly modified. We have, by use of (7.32.5), if $f(1) = 0$,

$$\mathrm{I} = O(n^{\alpha+k+1}) \int_0^{n^{-1}} \theta^{2\alpha+1} |f(\cos\theta)|\, d\theta + O(n^{-\frac{1}{2}}) \int_{n^{-1}}^{\epsilon} \theta^{\alpha-k-\frac{1}{2}} |f(\cos\theta)|\, d\theta.$$

In both integrals we integrate by parts (cf. Fejér **8**, p. 280). Let

$$\int_0^\theta |f(\cos t)|\, dt = F(\theta).$$

Then we find

$$\begin{aligned}
\mathrm{I} &= O(n^{\alpha+k+1})n^{-2\alpha-1}F(n^{-1}) + O(n^{\alpha+k+1}) \int_0^{n^{-1}} \theta^{2\alpha}F(\theta)\,d\theta \\
&\quad + O(n^{-\frac{1}{2}})\{\epsilon^{\alpha-k-\frac{1}{2}}F(\epsilon) + n^{k-\alpha+\frac{1}{2}}F(n^{-1})\} + O(n^{-\frac{1}{2}}) \int_{n^{-1}}^{\epsilon} \theta^{\alpha-k-\frac{3}{2}}F(\theta)\,d\theta \\
&= o(n^{k-\alpha-1}) + o(n^{k-\alpha-1}) + O(n^{-\frac{1}{2}}) + o(n^{k-\alpha-1}) + O(n^{k-\alpha-1}) \max_{0<\theta\leqq\epsilon} \{\theta^{-1}F(\theta)\} \\
&= o(n^{k-\alpha-1}) + O(n^{k-\alpha-1}) \max_{0<\theta\leqq\epsilon} \{\theta^{-1}F(\theta)\}.
\end{aligned}$$

(5) Finally, we show that the assertion of Theorem 9.1.4 does not hold in general if the "antipole condition" is not satisfied. We consider the function (cf. §9.3 (4))

$$f(x) = (1 + x)^\mu. \tag{9.42.9}$$

Its expansion at the point $x = 1$ is

$$\begin{aligned}
&\sum_{n=0}^{\infty} \{h_n^{(\alpha,\beta)}\}^{-1}\binom{n+\alpha}{n}\int_{-1}^{+1} (1-x)^\alpha(1+x)^{\mu+\beta}P_n^{(\alpha,\beta)}(x)\,dx \\
&\quad = \sum_{n=0}^{\infty} (-1)^n\{h_n^{(\alpha,\beta)}\}^{-1}\binom{n+\alpha}{n}\int_{-1}^{+1} (1-x)^{\mu+\beta}(1+x)^\alpha P_n^{(\beta,\alpha)}(x)\,dx.
\end{aligned} \tag{9.42.10}$$

According to (9.3.11), up to a fixed constant nonzero factor, the principal part of the general term of (9.42.10) is

$$(-1)^n n^{\alpha-\beta-2\mu-1}, \quad \text{or} \quad (-1)^n C_n^{(\alpha-\beta-2\mu-1)}. \tag{9.42.11}$$

But (9.42.10) cannot be (C, k)-summable if $k \leqq \alpha - \beta - 2\mu - 1 = \lambda$. Indeed,

$$(1 - r)^{-k-1} \sum_{n=0}^{\infty} (-1)^n C_n^{(\lambda)} r^n = (1 - r)^{-k-1}(1 + r)^{-\lambda-1} \tag{9.42.12}$$

Darboux's method (§8.4) yields for the coefficient of r^n in the power series expansion of this function, the principal term

$$AC_n^{(k)} + B(-1)^n C_n^{(\lambda)}, \tag{9.42.13}$$

where A and B are fixed constants, different from zero.

Now let (9.1.9) be satisfied. If we take

$$-\beta - 1 < \mu \leqq \tfrac{1}{2}(\alpha - \beta - k - 1), \tag{9.42.14}$$

the Jacobi series exists, the integral (9.1.10) is divergent, (since we have $\frac{1}{2}(\alpha - \beta - k - 1) < -\beta/2 - 3/4$), and the series (9.42.10) is not (C, k)-summable.

9.5. Proof of Theorems 9.1.5 and 9.1.6

(1) We start from the representation

$$s_n(x) = \int_0^\infty e^{-t} t^\alpha f(t) K_n^{(\alpha)}(x, t) dt \tag{9.5.1}$$

of the nth partial sum of the Laguerre series of the function $f(x)$, where the "kernel" $K_n^{(\alpha)}(x, t)$ has the meaning in (5.1.11). We write the formula given there in the more convenient form

$$\begin{aligned}\Gamma(\alpha+1)K_n^{(\alpha)}(x,t) &= \frac{n+1}{\binom{n+\alpha}{n}} \frac{L_{n+1}^{(\alpha)}(x)L_{n+1}^{(\alpha-1)}(t) - L_{n+1}^{(\alpha-1)}(x)L_{n+1}^{(\alpha)}(t)}{x-t} \\ &= \frac{n+1}{\binom{n+\alpha}{n}} \left\{ L_{n+1}^{(\alpha)}(x) \frac{L_{n+1}^{(\alpha-1)}(t) - L_{n+1}^{(\alpha-1)}(x)}{x-t} \right. \\ &\qquad \left. - L_{n+1}^{(\alpha-1)}(x) \frac{L_{n+1}^{(\alpha)}(t) - L_{n+1}^{(\alpha)}(x)}{x-t} \right\}.\end{aligned} \tag{9.5.2}$$

This follows from (5.1.11) if for $L_n^{(\alpha)}(x)$ we substitute $L_{n+1}^{(\alpha)}(x) - L_{n+1}^{(\alpha-1)}(x)$, and similarly for $L_n^{(\alpha)}(t)$ (cf. (5.1.13)).

First we assume that the integrals (9.1.11) exist and that the condition (9.1.12) is satisfied. Then the first integral (9.1.14) exists (cf. the remark in §9.11 (4)). Let $f(x)$ be a polynomial $\rho(x)$; then the statement (9.1.13) is true. Therefore, according to the closure property pointed out in Theorem 5.7.3, it suffices to show that the difference in (9.1.13) admits an estimate of the form

$$O(1) \int_0^1 t^a \,|\, f(t) \,|\, dt + O(1) \int_1^\infty e^{-t/2} t^{\alpha/2 - \frac{3}{4}} \,|\, f(t) \,|\, dt + o(1), \quad n \to \infty, \tag{9.5.3}$$

where $a = \min(\alpha, \alpha/2 - 1/4)$, and the bounds $O(1)$ and $o(1)$ hold uniformly in x, $\epsilon \leqq x \leqq \omega$. Furthermore, both factors $O(1)$ are independent of $f(x)$.

(2) Let us consider the contribution to $s_n(x)$ of the interval $0 \leqq t \leqq \epsilon/2$. In accordance with the first formula (9.5.2) and (7.6.9), for $\alpha \geqq \frac{1}{2}$ this is

$$\begin{aligned} O(n^{1-\alpha}) &\int_0^{\epsilon/2} t^\alpha \,|\, f(t) \,|\, \{n^{\alpha/2-\frac{1}{4}} t^{-(\alpha-1)/2-\frac{1}{4}} n^{(\alpha-1)/2-\frac{1}{4}} + n^{(\alpha-1)/2-\frac{1}{4}} t^{-\alpha/2-\frac{1}{4}} n^{\alpha/2-\frac{1}{4}}\} dt \\ &= O(1) \int_0^\epsilon t^{\alpha/2+\frac{1}{4}} \,|\, f(t) \,|\, dt + O(1) \int_0^\epsilon t^{\alpha/2-\frac{1}{4}} \,|\, f(t) \,|\, dt \\ &= O(1) \int_0^\epsilon t^{\alpha/2-\frac{1}{4}} \,|\, f(t) \,|\, dt. \end{aligned} \tag{9.5.4}$$

If $-\frac{1}{2} \leqq \alpha < \frac{1}{2}$, we use (7.6.9) and (7.6.10); if $\alpha < -\frac{1}{2}$, we use (7.6.10). In the first case the result (9.5.4) remains valid, while in the second case we obtain $O(1) \int_0^\epsilon t^\alpha \,|\, f(t) \,|\, dt$.

(3) We next consider the contribution of the interval $\epsilon/2 \leqq t \leqq 2\omega$, and we apply (8.8.3) and (8.8.5) to the second formula (9.5.2). Here the variables are confined to a fixed positive interval; so the remainders in (8.8.3) and (8.8.5) depend only on n. We find (notation as in (8.8.3))

$$K_n^{(\alpha)}(x, t) = n^{1-\alpha} n^{\alpha/2-\frac{1}{2}} n^{(\alpha-1)/2-\frac{1}{4}} \frac{k(x)k(t)}{x^{\frac{1}{2}} + t^{\frac{1}{2}}}$$

$$(9.5.5) \qquad \cdot\Big\{t^{\frac{1}{2}} \cos [2(nx)^{\frac{1}{2}} + \gamma] \frac{\sin [2(nt)^{\frac{1}{2}} + \gamma] - \sin [2(nx)^{\frac{1}{2}} + \gamma]}{t^{\frac{1}{2}} - x^{\frac{1}{2}}}$$

$$- x^{\frac{1}{2}} \sin [2(nx)^{\frac{1}{2}} + \gamma] \frac{\cos [2(nt)^{\frac{1}{2}} + \gamma] - \cos [2(nx)^{\frac{1}{2}} + \gamma]}{t^{\frac{1}{2}} - x^{\frac{1}{2}}} + O(1)\Big\}.$$

The mean-value theorem allows us to replace $n' = n + 1$ by n. Suppose, for instance, that $\phi(m, t) = \cos (2m^{\frac{1}{2}}t + \gamma)$. Then

$$\frac{[\phi(n', t) - \phi(n, t)] - [\phi(n', x) - \phi(n, x)]}{(n'-n)(t-x)} = \frac{\partial^2 \phi}{\partial m \partial t}$$

taken at a proper place $\bar{m}$, $\bar{t}$, where $\bar{m}$ is between n and n', and $\bar{t}$ is between x and t. This readily furnishes

$$K_n^{(\alpha)}(x, t) = \frac{k(x)k(t)}{x^{\frac{1}{2}} + t^{\frac{1}{2}}} x^{\frac{1}{2}}\Big\{\cos [2(nx)^{\frac{1}{2}} + \gamma] \frac{\sin [2(nt)^{\frac{1}{2}} + \gamma] - \sin [2(nx)^{\frac{1}{2}} + \gamma]}{t^{\frac{1}{2}} - x^{\frac{1}{2}}}$$

$$(9.5.6) \qquad - \sin [2(nx)^{\frac{1}{2}} + \gamma] \frac{\cos [2(nt)^{\frac{1}{2}} + \gamma] - \cos [2(nx)^{\frac{1}{2}} + \gamma]}{t^{\frac{1}{2}} - x^{\frac{1}{2}}} + O(1)\Big\}$$

$$= \tfrac{1}{2}x^{\frac{1}{2}}\{k(x)\}^2 t^{-\frac{1}{4}} \frac{\sin \{2n^{\frac{1}{2}}(t^{\frac{1}{2}} - x^{\frac{1}{2}})\}}{t^{\frac{1}{2}} - x^{\frac{1}{2}}} + O(1).$$

Thus, according to Riemann's lemma, if $\epsilon < 1 < \omega$,

$$\int_{\epsilon/2}^{2\omega} e^{-t} t^{\alpha} f(t) K_n^{(\alpha)}(x, t)\, dt = \tfrac{1}{2}e^{-x} x^{\alpha+\frac{1}{2}}\{k(x)\}^2 \int_{\epsilon/2}^{2\omega} t^{-\frac{1}{4}} f(t) \frac{\sin \{2n^{\frac{1}{2}}(t^{\frac{1}{2}} - x^{\frac{1}{2}})\}}{t^{\frac{1}{2}} - x^{\frac{1}{2}}}\, dt$$

$$(9.5.7) \qquad + O(1) \int_{\epsilon/2}^{2\omega} |f(t)|\, dt$$

$$= \pi^{-1} \int_{x^{\frac{1}{2}}-\delta}^{x^{\frac{1}{2}}+\delta} f(\tau^2) \frac{\sin \{2n^{\frac{1}{2}}(\tau - x^{\frac{1}{2}})\}}{\tau - x^{\frac{1}{2}}}\, d\tau$$

$$+ O(1) \int_0^1 t^{\alpha} |f(t)|\, dt + O(1) \int_1^{\infty} e^{-t/2} t^{\alpha/2-\frac{3}{4}} |f(t)|\, dt + o(1).$$

(4) In the interval $2\omega \leqq t \leqq 3n$ (n large), we have, according to the first statement (8.91.2),

$$(9.5.8) \qquad e^{-t/2} t^{\alpha/2+\frac{1}{4}} | L_n^{(\alpha)}(t) | = O(n^{\alpha/2-\frac{1}{4}}).$$

Consequently, from the first formula (9.5.2) it follows that

(9.5.9)
$$\int_{2\omega}^{3n} e^{-t}t^{\alpha}f(t)K_n^{(\alpha)}(x,t)dt = O(n^{1-\alpha})n^{\alpha/2-\frac{1}{4}}\int_{2\omega}^{3n} e^{-t}t^{\alpha-1}|f(t)||L_{n+1}^{(\alpha-1)}(t)|\,dt$$
$$+ O(n^{1-\alpha})n^{(\alpha-1)/2-\frac{1}{4}}\int_{2\omega}^{3n} e^{-t}t^{\alpha-1}|f(t)||L_{n+1}^{(\alpha)}(t)|\,dt$$
$$= O(n^{1-\alpha})n^{\alpha/2-\frac{1}{4}}n^{(\alpha-1)/2-\frac{1}{4}}\int_{2\omega}^{3n} e^{-t/2}t^{\alpha/2-\frac{3}{4}}|f(t)|\,dt$$
$$+ O(n^{1-\alpha})n^{(\alpha-1)/2-\frac{1}{4}}n^{\alpha/2-\frac{1}{4}}\int_{2\omega}^{3n} e^{-t/2}t^{\alpha/2-\frac{3}{4}}|f(t)|\,dt,$$

which is equal to

(9.5.10)
$$O(1)\int_1^{\infty} e^{-t/2}t^{\alpha/2-\frac{3}{4}}|f(t)|\,dt.$$

In the interval $3n \leqq t < +\infty$, Theorem 8.91.2 ($x \geqq a$, $\lambda = \alpha/2 + 1/12$), tells us that

(9.5.11)
$$e^{-t/2}t^{\alpha/2+1/12}|L_n^{(\alpha)}(t)| = O(n^{\alpha/2-\frac{1}{4}}).$$

Therefore, from (5.1.11), on account of (9.1.12), we have

(9.5.12)
$$\int_{3n}^{\infty} e^{-t}t^{\alpha}f(t)K_n^{(\alpha)}(x,t)\,dt$$
$$= O(n^{1-\alpha})n^{\alpha/2-\frac{1}{4}}\int_{3n}^{\infty} e^{-t}t^{\alpha-1}|f(t)|\{|L_n^{(\alpha)}(t)| + |L_{n+1}^{(\alpha)}(t)|\}\,dt$$
$$= O(n^{1-\alpha})n^{\alpha/2-\frac{1}{4}}n^{\alpha/2-\frac{1}{4}}\int_{3n}^{\infty} e^{-t/2}t^{\alpha/2-13/12}|f(t)|\,dt = o(1).$$

(5) If the condition (9.1.12) is replaced by the requirements of (9.1.14), a treatment of the interval $2\omega \leqq t \leqq 3n$, the same as that given in (4), applies. In $3n \leqq t < +\infty$ we use Schwarz's inequality:

(9.5.13)
$$O(n^{\frac{3}{4}-\alpha/2})\int_{3n}^{\infty} e^{-t}t^{\alpha-1}|f(t)||L_n^{(\alpha)}(t)|\,dt$$
$$= O(n^{\frac{3}{4}-\alpha/2})\left\{\int_{3n}^{\infty} e^{-t}t^{\alpha-2}|f(t)|^2\,dt\right\}^{\frac{1}{2}}\left\{\int_{3n}^{\infty} e^{-t}t^{\alpha}[L_n^{(\alpha)}(t)]^2\,dt\right\}^{\frac{1}{2}}.$$

The last integral is $O(n^{\alpha})$ (see (5.1.1)), and this establishes the statement.

(6) Let

(9.5.14)
$$f(x) = x^{\mu}$$

(Blumenthal **1**, pp. 32-33). We shall show that for proper values of μ the first integral in (9.1.11) exists, but that the second does not; furthermore the Laguerre series is divergent for $x > 0$. Here $\alpha > -\frac{1}{2}$.

The coefficient a_n of the corresponding Laguerre series is given by

(9.5.15)
$$\Gamma(\alpha+1)\binom{n+\alpha}{n}a_n = \int_0^{\infty} e^{-x}x^{\alpha+\mu}L_n^{(\alpha)}(x)\,dx$$
$$= \frac{1}{n!}\int_0^{\infty} x^{\mu}\left(\frac{d}{dx}\right)^n\{e^{-x}x^{n+\alpha}\}\,dx$$

(see (5.1.5)). On integrating by parts, we have

(9.5.16)
$$a_n = \frac{\Gamma(\mu + \alpha + 1)}{\Gamma(-\mu)} \frac{\Gamma(n - \mu)}{\Gamma(n + \alpha + 1)}.$$

Here we assume $\mu + \alpha > -1$, $\mu \neq 0, 1, 2, \cdots$. Then $a_n \sim n^{-\mu-\alpha-1}$, and in view of Fejér's formula (8.22.1), the principal term of the Laguerre series behaves like

(9.5.17)
$$n^{-\mu-\alpha/2-5/4} \cos \{2(nx)^{\frac{1}{2}} - \alpha\pi/2 - \pi/4\}$$

for $x > 0$. This series is therefore divergent (cf. Problem 47) when and only when $\mu + \alpha/2 + \frac{5}{4} \leqq \frac{1}{2}$. If $\alpha > -\frac{1}{2}$ and

(9.5.18)
$$-\alpha - 1 < \mu \leqq -\alpha/2 - 3/4,$$

the first integral (9.1.11) exists, but the second does not, and the series is divergent for $x > 0$.

(7) We consider also the function

(9.5.19)
$$f(x) = e^{x/2}x^{\mu},$$

and we show that for a proper value of μ (particularly for $\mu = -\alpha/2 + 1/4$) the integrals (9.1.11) exist, but the conditions (9.1.12) and (9.1.14) are not satisfied; in addition, the Laguerre series is divergent for $x > 0$. Here $\alpha > -1$, $\alpha + \mu > -1$.

The integral

(9.5.20)
$$\Gamma(\alpha + 1)\binom{n + \alpha}{n} a_n = \int_0^\infty e^{-x/2} x^{\alpha+\mu} L_n^{(\alpha)}(x)\, dx$$

can be calculated by means of the generating function (see (5.1.9))

(9.5.21)
$$\begin{aligned}\sum_{n=0}^{\infty} \Gamma(\alpha + 1)\binom{n + \alpha}{n} a_n r^n &= (1 - r)^{-\alpha-1} \int_0^\infty e^{-x(\frac{1}{2}+r/(1-r))} x^{\alpha+\mu}\, dx \\ &= \Gamma(\alpha + \mu + 1)(1 - r)^{-\alpha-1}\left(\frac{1}{2} + \frac{r}{1 - r}\right)^{-\alpha-\mu-1} \\ &= \Gamma(\alpha + \mu + 1)2^{\alpha+\mu+1}(1 - r)^{\mu}(1 + r)^{-\alpha-\mu-1}.\end{aligned}$$

Thus, Darboux's method (§8.4) gives for a_n the principal term

(9.5.22)
$$An^{-\alpha-\mu-1} + B(-1)^n n^{\mu}, \qquad n \to \infty,$$

where A and B are fixed constants, $A \neq 0$, $B \neq 0$. We have $A = 0$ if $\mu = 0, 1, 2, \cdots$.

If $x > 0$, this shows, on account of Fejér's formula (8.22.1), that $a_n L_n^{(\alpha)}(x)$ does not tend to zero if $\mu = -\alpha/2 + 1/4$.

(8) Finally, we discuss the Hermite series

(9.5.23)
$$f(x) \sim a_0 H_0(x) + a_1 H_1(x) + \cdots + a_n H_n(x) + \cdots.$$

Writing $y = x^2$, we readily see that (see (5.6.1))

$$\sum_{m=0}^{\infty} a_{2m} H_{2m}(x) \quad \text{and} \quad \sum_{m=0}^{\infty} a_{2m+1} H_{2m+1}(x)$$

are the expansions of

$$\{f(x) + f(-x)\}/2 = \{f(y^{\frac{1}{2}}) + f(-y^{\frac{1}{2}})\}/2 \text{ and}$$
$$x^{-1}\{f(x) - f(-x)\}/2 = y^{-\frac{1}{2}}\{f(y^{\frac{1}{2}}) - f(-y^{\frac{1}{2}})\}/2$$

into a Laguerre series with the parameters $\alpha = -\frac{1}{2}$ and $\alpha = +\frac{1}{2}$, respectively. Applying Theorem 9.1.5 to these functions with $\alpha = -\frac{1}{2}$ and $\alpha = +\frac{1}{2}$, respectively, we obtain the conditions (9.1.11), (9.1.12), and (9.1.14) in the following form:

$$\int_0^1 y^{-\frac{1}{2}} \mid f(\pm y^{\frac{1}{2}}) \mid dy = 2 \int_0^1 \mid f(\pm x) \mid dx \text{ exists;}$$

$$\int_n^\infty e^{-y/2} y^{-\frac{3}{4}} \mid f(\pm y^{\frac{1}{2}}) \mid dy = o(n^{-\frac{1}{2}}), \text{ or } \int_{n}^\infty e^{-x^2/2} x^{-\frac{3}{2}} \mid f(\pm x) \mid dx = o(n^{-1});$$

$$\int_1^\infty e^{-y/2} y^{-1} \mid f(\pm y^{\frac{1}{2}}) \mid dy = 2 \int_1^\infty e^{-x^2/2} x^{-1} \mid f(\pm x) \mid dx \text{ exists;}$$

$$\int_n^\infty e^{-y} y^{-\frac{5}{2}} \mid f(\pm y^{\frac{1}{2}}) \mid^2 dy = o(n^{-\frac{3}{2}}), \text{ or } \int_n^\infty e^{-x^2} x^{-4} \mid f(\pm x) \mid^2 dx = o(n^{-3}).$$

This establishes (9.1.17), provided x belongs to an interval not containing the origin.

In order to accomplish the proof, if x lies in an interval of the form $[-\epsilon, +\epsilon]$, we have only to show that

$$(9.5.24) \quad e^{-t^2} \sum_{\nu=0}^{n} (2^\nu \nu! \pi^{\frac{1}{2}})^{-1} H_\nu(x) H_\nu(t) - \pi^{-1} \frac{\sin \{(2n)^{\frac{1}{2}}(x-t)\}}{x-t} = O(1)$$

uniformly if both x and t belong to $[-\epsilon, +\epsilon]$. The first member of this difference is, according to (5.5.9),

$$(9.5.25) \quad (2^{n+1} n! \pi^{\frac{1}{2}})^{-1} e^{-t^2} \left\{ H_{n+1}(x) \frac{H_n(t) - H_n(x)}{x-t} - H_n(x) \frac{H_{n+1}(t) - H_{n+1}(x)}{x-t} \right\}.$$

Now, by using the the notation of Theorem 8.22.6, we obtain

$$\lambda_n^{-1} e^{-x^2/2} H_n(x) = \cos(N^{\frac{1}{2}} x - n\pi/2) + O(n^{-\frac{1}{2}}),$$

$$\lambda_n^{-1} \frac{e^{-x^2/2} H_n(x) - e^{-t^2/2} H_n(t)}{x-t} = \frac{\cos(N^{\frac{1}{2}} x - n\pi/2) - \cos(N^{\frac{1}{2}} t - n\pi/2)}{x-t} + O(1).$$

The second formula follows by an argument similar to that used in §8.8 (cf. the first formula in (5.5.10)). Its left member can also be written in the form

$$\lambda_n^{-1} e^{-x^2/2} \frac{H_n(x) - H_n(t)}{x-t} + O(1).$$

Replacing $N^{\frac{1}{2}}$ by $(N+c)^{\frac{1}{2}}$, c a fixed constant, we find the error committed in the right-hand member to be $O(1)$.

These asymptotic formulas furnish for (9.5.25), by use of Stirling's formula,

$$(2^{n+1}n!\,\pi^{\frac{1}{2}})^{-1}\lambda_n\lambda_{n+1}e^{x^2-t^2}\Big\{\sin(N^{\frac{1}{2}}x - n\pi/2)\,\frac{\cos(N^{\frac{1}{2}}t - n\pi/2) - \cos(N^{\frac{1}{2}}x - n\pi/2)}{x-t}$$

$$- \cos(N^{\frac{1}{2}}x - n\pi/2)\,\frac{\sin(N^{\frac{1}{2}}t - n\pi/2) - \sin(N^{\frac{1}{2}}x - n\pi/2)}{x-t} + O(1)\Big\}$$

$$= \pi^{-1}\frac{\sin\{N^{\frac{1}{2}}(x-t)\}}{x-t} + O(1) = \pi^{-1}\frac{\sin\{(2n)^{\frac{1}{2}}(x-t)\}}{x-t} + O(1).$$

9.6. Proof of Theorem 9.1.7

(1) The Cesàro means of the Laguerre series at $x = 0$ can be represented in a particularly simple form. For, from (5.1.7),

$$\sum_{n=0}^{\infty} a_n L_n^{(\alpha)}(0) = \{\Gamma(\alpha+1)\}^{-1}\sum_{n=0}^{\infty}\int_0^{\infty} e^{-t}t^{\alpha}f(t)\,L_n^{(\alpha)}(t)\,dt. \tag{9.6.1}$$

Hence, by applying (5.1.9), the Cesàro means of order k are found to be

$$\{C_n^{(k)}\Gamma(\alpha+1)\}^{-1}\int_0^{\infty} e^{-t}t^{\alpha}f(t)L_n^{(\alpha+k+1)}(t)\,dt. \tag{9.6.2}$$

Assume $k > \alpha + \frac{1}{2}$, and subdivide the interval $[0, +\infty]$ into $[0, \epsilon]$, $[\epsilon, \omega]$, and $[\omega, +\infty]$. Then we find for (9.6.2)

$$\begin{aligned} &O(n^{-k})\max_{0\leq t\leq\epsilon}|f(t)|\int_0^{\epsilon} t^{\alpha}\,|L_n^{(\alpha+k+1)}(t)|\,dt \\ &\quad + O(n^{-k})\int_{\epsilon}^{\omega} e^{-t}t^{\alpha}\,|f(t)|\,|L_n^{(\alpha+k+1)}(t)|\,dt \\ &\quad\quad + O(n^{-k})\int_{\omega}^{\infty} e^{-t}t^{\alpha}\,|f(t)|\,|L_n^{(\alpha+k+1)}(t)|\,dt. \end{aligned} \tag{9.6.3}$$

By use of (7.6.8) the first integral becomes

$$O(1)\int_0^{n^{-1}} t^{\alpha}n^{\alpha+k+1}\,dt + O(1)\int_{n^{-1}}^{\epsilon} t^{\alpha}t^{-(\alpha+k+1)/2-\frac{1}{4}}n^{(\alpha+k+1)/2-\frac{1}{4}}\,dt = O(n^k). \tag{9.6.4}$$

The second integral is $O(n^{(\alpha+k+1)/2-\frac{1}{4}}) = o(n^k)$. Now, using Theorem 8.91.2, with α replaced by $\alpha + k + 1$, and $\lambda = k + \frac{1}{3}$, we see that $\lambda - \frac{1}{3} = k > (\alpha + k + 1)/2 - \frac{1}{4}$; whence the third integral is

$$O(1)\int_{\omega}^{\infty} e^{-t/2}t^{\alpha}\,|f(t)|\,t^{-k-\frac{1}{3}}n^k\,dt = O(n^k)\int_{\omega}^{\infty} e^{-t/2}t^{\alpha-k-\frac{1}{3}}\,|f(t)|\,dt. \tag{9.6.5}$$

Thus (9.6.2) can be represented in the form

$$O(1)\max_{0\leq t\leq\epsilon}|f(t)| + O(1)\int_{\omega}^{\infty} e^{-t/2}t^{\alpha-k-\frac{1}{3}}\,|f(t)|\,dt + o(1). \tag{9.6.6}$$

Here the bounds of the terms $O(1)$ are independent of ϵ and ω. If it is assumed that $f(0) = 0$, the statement is established.

(2) Let $k = \alpha + \frac{1}{2}$, and apply Theorem 1.6 (Helly's theorem) to the linear operations

$$\mathfrak{U}_n(f) = \{C_n^{(k)}\Gamma(\alpha + 1)\}^{-1} \int_0^1 e^{-t} t^\alpha f(t) L_n^{(\alpha+k+1)}(t)\, dt, \tag{9.6.7}$$
$$\mathfrak{U}(f) = f(0).$$

It suffices to show that the second condition in (1.6.10) is not satisfied. As a matter of fact, if this condition is not satisfied, then a continuous function $f(x)$, $0 \leqq x \leqq 1$, exists for which

$$\lim_{n\to\infty} \mathfrak{U}_n(f) = \mathfrak{U}(f)$$

does not hold. Extending the definition of this function to $x > 1$ by means of the condition $f(x) = 0$, we obtain the "Gegenbeispiel" required. (By use of the remark made in connection with Theorem 1.6, a continuous function $f(x)$ can be constructed for which (9.6.2) is unbounded when $n \to \infty$.)

Now if Ω is a positive constant independent of n,

$$\{C_n^{(k)}\Gamma(\alpha + 1)\}^{-1} \int_0^1 e^{-t} t^\alpha \,|\, L_n^{(\alpha+k+1)}(t) \,|\, dt \tag{9.6.8}$$
$$> An^{-k} \int_0^{\Omega/n} t^\alpha \,|\, L_n^{(\alpha+k+1)}(t) \,|\, dt,$$

where A is positive and independent of Ω and n. The last expression is, according to (8.1.8),

$$\sim \int_0^\Omega z^\alpha z^{-(\alpha+k+1)/2} \,|\, J_{\alpha+k+1}(2z^{\frac{1}{2}}) \,|\, dz = \int_0^\Omega z^{-\frac{3}{4}} \,|\, J_{\alpha+k+1}(2z^{\frac{1}{2}}) \,|\, dz.$$

This integral becomes arbitrarily large with Ω.

(3) Integral (9.1.19) exists if

$$f(x) = O(e^{x/2} x^{k-\alpha-\frac{1}{2}-\delta}), \qquad \delta > 0,\ x \to +\infty. \tag{9.6.9}$$

On the other hand, there is no difficulty in proving that the Laguerre series $\sum_{n=0}^\infty a_n L_n^{(\alpha)}(x)$ of $f(x) = e^{x/2}x^{k-\alpha}$ is not (C, k)-summable at $x = 0$. Here the condition $k > \alpha + \frac{1}{2}$ is satisfied. Indeed, from (9.5.21) we obtain, $\mu = k - \alpha$,

$$(1 - r)^{-k-1} \sum_{n=0}^\infty a_n L_n^{(\alpha)}(0) r^n = \frac{\Gamma(k + 1)}{\Gamma(\alpha + 1)} 2^{k+1}(1 - r)^{-\alpha-1}(1 + r)^{-k-1}. \tag{9.6.10}$$

Darboux's method furnishes for the coefficient of r^n in the expansion of this function the number

$$C(-1)^n n^k + o(n^k), \qquad C > 0,\ n \to \infty. \tag{9.6.11}$$

This establishes the statement.

We also notice that the Laguerre series is (C, k)-summable at $x = 0$ with an arbitrary $k > \alpha + 1/2$, provided $f(x)$ is continuous at $x = 0$, and

$$f(x) = O(e^{x/2} x^{-1/6}), \qquad x \to +\infty. \tag{9.6.12}$$

Again, the Laguerre series of the special function $f(x) = e^{x/2}x^{\frac{1}{3}}$ is not (C, k)-summable at $x = 0$ with *any* $k > \alpha + 1/2$.

9.7. Further results

(1) Using a generalized translation operator which can be defined by means of (4.10.21) it is possible to use Theorem 9.1.3 to show that the Jacobi series of a continuous function is uniformly (C, k) summable for $k > \max(\alpha + \frac{1}{2}, \beta + \frac{1}{2})$ when $\alpha, \beta \geqq -\frac{1}{2}$, or $\alpha + \beta \geqq -1$, and $\alpha, \beta > -1$. See Askey-Wainger **4** and Gasper **3**, **4**.

(2) For $\alpha = \beta \geqq -\frac{1}{2}$ Kogbetliantz **19** proved the positivity of the $(C, 2\alpha + 2)$ means of the formal reproducing kernel. A simple proof was given in Askey-Pollard **1**. For the general Jacobi case it is likely that the $(C, \alpha + \beta + 2)$ means are positive when $\alpha, \beta \geqq -\frac{1}{2}$. This has been proven for $\max(\beta - 1, -\beta) \leqq \alpha \leqq \beta + 1$ and $\max(\beta - 2, 3 - \beta) \leqq \alpha \leqq \beta + 2$. See Askey **8**.

(3) Theorem 9.1.2 can be used with the Carleson-Hunt theorem (Carleson **1**, Hunt **1**) to obtain almost everywhere convergence of Jacobi series when the function satisfies the conditions of both theorems.

(4) Slightly stronger equiconvergence theorems for Laguerre and Hermite series were obtained by Muckenhoupt **4**. Again almost everywhere convergence theorems follow from these results. These improvements and the mean convergence results to be given below use Erdélyi's relatively recent asymptotic formulas in an essential way. These estimates are described in 8.92 (2).

(5) If $\int_{-1}^{1} |f(x)|^p dx < \infty$, if $f(x)$ is expanded in a series of Legendre polynomials, and if $s_n(x)$ denotes the partial sums of this series, then Pollard **1** proved that

$$\lim_{n \to \infty} \int_{-1}^{1} |f(x) - s_n(x)|^p dx = 0$$

for $4/3 < p < 4$ and these bounds are best possible (Newman-Rudin **1**). This theorem has been extended to Jacobi series (Pollard **2**, **3**, Muckenhoupt **5**), and Laguerre and Hermite series (Askey-Wainger **1**, Muckenhoupt **2**, **3**). A slight simplification in the proof was given in Askey **6**, and this result and the positivity of some Cesàro mean was used in this paper to solve a problem of L^p convergence of Lagrange interpolation at the zeros of Jacobi polynomials.

(6) There are a number of other interesting positive summability methods, e.g. the analogues of Poisson-Abel summability (Bailey **1**); Gauss-Weierstrass summability (Bochner **2**, Karlin-McGregor **3**); and the de la Vallée-Poussin method which uses Bateman's formula (4.10.23). Horton **1** used this formula to prove that the de la Vallée-Poussin method is positive and variation diminishing.

(7) Muckenhoupt-Stein **1** have introduced a number of the deeper functionals from classical Fourier series into the study of orthogonal polynomials and were able to construct a theory of H^p spaces associated with singular Cauchy-Riemann equations and prove an analogue of the Marcinkiewicz multiplier theorem. Another proof of this theorem by mapping the theorem for Fourier series to Jacobi series was given by Askey-Wainger **2** and Askey **3**. Another extension of the Marcinkiewicz theorem was proved by Bonami-Clerc **1**. They were motivated by work of Coifman-G. Weiss **1**.

(8) Hirschman **1** has constructed a theory of variation diminishing transformations associated with orthogonal polynomial expansions which parallels Schoenberg's classical theory.

CHAPTER X

REPRESENTATION OF POSITIVE FUNCTIONS

In the present chapter we deal with an extension of Fejér's representation of non-negative trigonometric polynomials, given in §1.2, to certain general classes of non-negative functions. In particular, we are interested in the discussion of this representation if the given function is subjected to certain continuity conditions. Extensions of Fejér's theorem in this direction are important for the investigation of the asymptotic behavior of the general orthogonal polynomials associated with a distribution on a finite real interval or on the unit circle (Chapter XI). It seemed convenient to separate these considerations from the subject proper.

Concerning the results of this chapter see Szegö **6**, **7**, **8**, **9**. See also Grenander-Szegö **1**, 1.12–1.15.

10.1. Fatou's theorems

THEOREM 10.1.1. *Let $f(\theta)$ be integrable in Lebesgue's sense, and let*

$$f(r, \theta) = \frac{1}{2\pi}\int_{-\pi}^{+\pi} f(t)\,\frac{1 - r^2}{1 - 2r\cos(t - \theta) + r^2}\,dt \tag{10.1.1}$$

be the corresponding Poisson integral. Then we have almost everywhere in $-\pi \leqq \theta \leqq +\pi$,

$$\lim_{r \to 1-0} f(r, \theta) = f(\theta). \tag{10.1.2}$$

See Zygmund **2**, p. 54, §3.442.

THEOREM 10.1.2. *Let*

$$F(z) = c_0 + c_1 z + c_2 z^2 + \cdots + c_n z^n + \cdots \tag{10.1.3}$$

be regular for $|z| < 1$, and let the integral

$$\frac{1}{2\pi}\int_{-\pi}^{+\pi} |F(re^{i\theta})|^2\,d\theta \tag{10.1.4}$$

be bounded for $r < 1$. (This condition is equivalent to the convergence of

$$|c_0|^2 + |c_1|^2 + |c_2|^2 + \cdots + |c_n|^2 + \cdots.) \tag{10.1.5}$$

Then we have almost everywhere in $-\pi \leqq \theta \leqq +\pi$

$$\lim_{r \to 1-0} F(re^{i\theta}) = F(e^{i\theta}). \tag{10.1.6}$$

Furthermore, $F(e^{i\theta})$ is measurable, and $|F(e^{i\theta})|^2$ is integrable in Lebesgue's sense. The Fourier series of $F(e^{i\theta})$ is obtained by writing $z = e^{i\theta}$ in (10.1.3).

This theorem is due to Fatou (**1**). We say that $F(z)$ is of the class H_2. A function $F(z)$ which is regular and bounded in $|z| < 1$ belongs to this class. Concerning the more general classes H_δ, see F. Riesz **2** and Smirnoff **2**. In particular, if $F(z)$ is of the class H_1, the boundary values $F(e^{i\theta})$ exist (almost everywhere), $|F(e^{i\theta})|$ is integrable in Lebesgue's sense and Cauchy's theorem can be applied on $|z| = 1$. (Cf. F. Riesz, loc. cit., p. 94, c); Smirnoff, loc. cit., pp. 337–338.)

10.2. Generalization of Fejér's representation

Concerning this section, see Szegö **7**.

(1) Let $g(\theta)$ be a non-negative trigonometric polynomial not vanishing identically. According to Theorem 1.2.2, there exists a polynomial $h(z)$ of the same degree, uniquely determined by the following conditions:

(10.2.1)
(a) $g(\theta) = |h(z)|^2$, where $z = e^{i\theta}$;
(b) $h(z)$ is different from zero in $|z| < 1$;
(c) $h(0)$ is real and positive.

We obviously have

$$\log g(\theta) = 2\Re\{\log h(z)\}, \qquad z = e^{i\theta}. \tag{10.2.2}$$

The function $\log h(z)$ is regular for $|z| \leqq 1$ except at the points $z = e^{i\theta}$ which correspond to the zeros of $g(\theta)$, at which both functions $\log g(\theta)$ and $\log h(z)$ become logarithmically infinite; $\log h(0)$ is real. The function $2\Re\{\log h(z)\}$ is regular and harmonic for $|z| < 1$ and has absolutely integrable boundary values $\log g(\theta)$. Applying the mean-value theorem of Gauss, we obtain

$$\frac{1}{2\pi}\int_{-\pi}^{+\pi} \log g(\theta)\, d\theta = 2\Re\{\log h(0)\} = 2\log h(0), \tag{10.2.3}$$

so that

$$\{h(0)\}^2 = \exp\left\{\frac{1}{2\pi}\int_{-\pi}^{+\pi} \log g(\theta)\, d\theta\right\} = \mathfrak{G}(g). \tag{10.2.4}$$

The last expression is called the *geometric mean* of the function $g(\theta)$.

(2) The relation (10.2.2) enables us to extend this consideration to arbitrary non-negative functions $f(\theta)$ (instead of $g(\theta)$), defined in $[-\pi, +\pi]$ and integrable in Lebesgue's sense, provided $\mathfrak{G}(f) > 0$. This last condition is equivalent to the existence of the integral

$$\int_{-\pi}^{+\pi} \log f(\theta)\, d\theta = \int_{0 \leqq f(\theta) \leqq 1} \log f(\theta)\, d\theta + \int_{f(\theta) > 1} \log f(\theta)\, d\theta. \tag{10.2.5}$$

The existence of the second integral on the right follows from the integrability of $f(\theta)$. Thus, the condition $\mathfrak{G}(f) > 0$ is equivalent to the existence of the first integral on the right-hand side. This is a restriction on the "nearness" of $f(\theta)$ to 0. A consequence of this condition is that $f(\theta)$ can vanish only on a set of measure zero.

We now introduce the harmonic function $u(re^{i\theta})$ defined by Poisson's integral of $\frac{1}{2}\log f(\theta)$

$$(10.2.6)\qquad u(re^{i\theta}) = \frac{1}{4\pi}\int_{-\pi}^{+\pi} \log f(t)\,\frac{1-r^2}{1-2r\cos(\theta-t)+r^2}\,dt, \qquad 0 \leqq r < 1.$$

In the case of a continuous function $f(\theta) > 0$, we know that (see for instance Zygmund **2**, p. 51)

$$(10.2.7)\qquad \lim_{r\to 1-0} u(re^{i\theta}) = \tfrac{1}{2}\log f(\theta),$$

uniformly in θ. In the general case considered above, this is true only with the exception of a set of measure zero and without uniformity in general (see Theorem 10.1.2). If now u is completed to an analytic function $u + iv = k(z)$ with the condition that $k(0)$ is real, then $k(z)$ is uniquely determined. Writing $D(z) = e^{k(z)}$, we obtain the analogue (generalization) of the function $h(z)$ considered before. This function $D(z) = D(f; z)$ has the following properties (Szegö **7**, p. 237):

(a′) $D(z)$ is of the class H_2 (§10.1); almost everywhere in $-\pi \leqq \theta \leqq +\pi$,

$$(10.2.8)\qquad \lim_{r\to 1-0} D(re^{i\theta}) = D(e^{i\theta}) \text{ exists, and } f(\theta) = |D(e^{i\theta})|^2;$$

(b′) $D(z) \neq 0$ in $|z| < 1$;
(c′) $D(0)$ is real and positive.

We have again $\{D(0)\}^2 = \mathfrak{G}(f)$; this is obvious from (10.2.6). Furthermore, we show that for an arbitrary continuous function $F(\theta)$ of period 2π,

$$(10.2.9)\qquad \lim_{r\to 1-0}\int_{-\pi}^{+\pi} F(\theta)\,|D(re^{i\theta})|^2\,d\theta = \int_{-\pi}^{+\pi} F(\theta)f(\theta)\,d\theta.$$

If $D(z) = d_0 + d_1 z + d_2 z^2 + \cdots$, according to Schwarz's inequality, we have

$$\left\{\int_{-\pi}^{+\pi} \big|\,|D(e^{i\theta})|^2 - |D(re^{i\theta})|^2\big|\,d\theta\right\}^2$$
$$\leqq \int_{-\pi}^{+\pi} (|D(e^{i\theta})| - |D(re^{i\theta})|)^2\,d\theta \cdot \int_{-\pi}^{+\pi} (|D(e^{i\theta})| + |D(re^{i\theta})|)^2\,d\theta$$
$$\leqq 2\int_{-\pi}^{+\pi} |D(e^{i\theta}) - D(re^{i\theta})|^2\,d\theta \cdot \int_{-\pi}^{+\pi} (|D(e^{i\theta})|^2 + |D(re^{i\theta})|^2)\,d\theta$$
$$\leqq 8\pi\int_{-\pi}^{+\pi} |D(e^{i\theta})|^2\,d\theta \cdot \sum_{n=1}^{\infty} (1-r^n)^2\,|d_n|^2 \to 0$$

for $r \to 1 - 0$. Here (a′) is used. (See also Smirnoff **2**, p. 338.)

It should be observed that the function $D(z)$ is not uniquely determined by the conditions (a′), (b′), and (c′). For instance, we can multiply it by $\exp\{-(1+z)/(1-z)\}$. For this point see Szegö **7**, p. 241.

(3) By means of (10.2.6) we can obtain the following explicit representation of $D(z)$ in terms of $f(\theta)$:

$$D(z) = D(f; z) = \exp\left\{\frac{1}{4\pi}\int_{-\pi}^{+\pi} \log f(t)\, \frac{1 + ze^{-it}}{1 - ze^{-it}}\, dt\right\}, \qquad |z| < 1. \tag{10.2.10}$$

If $f(\theta)$ is an even function, the expansion of $D(z)$ around $z = 0$ has real coefficients. Let $f_1(\theta)$ and $f_2(\theta)$ be arbitrary functions satisfying the same conditions as $f(\theta)$, and let ρ be arbitrary and complex, $\rho \neq 0$. Then

$$D(f_1; z)D(f_2; z) = D(f_1f_2; z); \qquad \{D(f; z)\}^\rho = D(f^\rho; z). \tag{10.2.11}$$

As an example we mention the case $f(\theta) = g(\theta)$ considered in (1). We have

$$D(g; z) = h(z), \qquad D(g^{-1}; z) = \{h(z)\}^{-1}. \tag{10.2.12}$$

In the second formula we assume that $g(\theta)$ is positive.

A further example is

$$\begin{gathered} f(\theta) = 2^{\gamma+\delta}(1 - \cos\theta)^\gamma(1 + \cos\theta)^\delta; \qquad D(f; z) = (1 - z)^\gamma(1 + z)^\delta; \\ \gamma > -\tfrac{1}{2},\ \delta > -\tfrac{1}{2}. \end{gathered} \tag{10.2.13}$$

10.3. Further study of the representation of positive functions

The derivation of an asymptotic formula for the polynomials orthogonal on the unit circle requires some further properties of the representation defined before. In particular, we shall deal with the behavior of the function $D(z)$ on the unit circle $|z| = 1$. In this connection certain restrictive conditions on the function $f(\theta)$ are necessary.

(1) First let us consider again the case of a non-negative trigonometric polynomial $g(\theta)$ not identically zero, and let $g(\theta) = |h(z)|^2$, $z = e^{i\theta}$, be the normalized representation defined in Theorem 1.2.2. Then, by (10.2.10), we have for $0 \leqq r < 1$,

$$D(g; z) = h(z) = h(re^{i\theta}) = \exp\left\{\frac{1}{4\pi}\int_{-\pi}^{+\pi} \log g(t)\, \frac{1 + re^{i(\theta-t)}}{1 - re^{i(\theta-t)}}\, dt\right\}. \tag{10.3.1}$$

Whence

$$\begin{aligned} &\operatorname{sgn} h(re^{i\theta}) \\ &= |h(re^{i\theta})|^{-1} h(re^{i\theta}) = \exp\left\{\frac{i}{4\pi}\int_{-\pi}^{+\pi} \log g(t)\, \frac{2r\sin(\theta - t)}{1 - 2r\cos(\theta - t) + r^2}\, dt\right\} \\ &= \exp\left\{\frac{i}{4\pi}\int_{-\pi}^{+\pi} [\log g(t) - \log g(\theta)]\, \frac{2r\sin(\theta - t)}{1 - 2r\cos(\theta - t) + r^2}\, dt\right\}. \end{aligned} \tag{10.3.2}$$

Therefore, for all values of θ with the exception of the zeros of $g(\theta)$,

$$\operatorname{sgn} h(e^{i\theta}) = |h(e^{i\theta})|^{-1} h(e^{i\theta}) = \{g(\theta)\}^{-\frac{1}{2}} h(e^{i\theta})$$

$$(10.3.3) \qquad = \exp\left\{\frac{i}{4\pi}\int_{-\pi}^{+\pi} \log g(t) \cot\frac{\theta - t}{2}\, dt\right\}$$

$$= \exp\left\{\frac{i}{4\pi}\int_{-\pi}^{+\pi} [\log g(t) - \log g(\theta)] \cot\frac{\theta - t}{2}\, dt\right\}.$$

The first integral is taken in the sense of Cauchy's principal value; the second integral is absolutely convergent.

(2) This consideration can easily be extended to a positive continuous function $f(\theta)$ which satisfies certain conditions sufficient for the existence of the integrals corresponding to those in (10.3.3). Since $f(\theta)$ is continuous, defining $D(z)$ as in §10.2 we now have, uniformly for all values of θ,

$$(10.3.4) \qquad \lim_{r\to 1-0} |D(re^{i\theta})|^2 = f(\theta).$$

If the integral

$$(10.3.5) \qquad \int_{-\pi}^{+\pi} |\log f(t) - \log f(\theta)| \left|\cot\frac{\theta - t}{2}\right| dt$$

exists, then

$$(10.3.6) \qquad \int_{-\pi}^{+\pi} \log f(t) \cot\frac{\theta - t}{2}\, dt$$

exists in the sense of Cauchy's principal value. We then show the existence of the boundary values of

$$\operatorname{sgn} D(re^{i\theta}) = \exp\left\{\frac{i}{4\pi}\int_{-\pi}^{+\pi} \log f(t) \frac{2r\sin(\theta - t)}{1 - 2r\cos(\theta - t) + r^2}\, dt\right\}$$

$$(10.3.7) \qquad = \exp\left\{\frac{i}{4\pi}\int_{-\pi}^{+\pi} [\log f(t) - \log f(\theta)] \frac{2r\sin(\theta - t)}{1 - 2r\cos(\theta - t) + r^2}\, dt\right\}.$$

Indeed, if ϵ is a fixed positive number, we have

$$\lim_{r\to 1-0} \int_{|\theta - t| > \epsilon} [\log f(t) - \log f(\theta)] \frac{2r\sin(\theta - t)}{1 - 2r\cos(\theta - t) + r^2}\, dt$$

$$= \int_{|\theta - t| > \epsilon} [\log f(t) - \log f(\theta)] \cot\frac{\theta - t}{2}\, dt.$$

On the other hand, $1 - 2r\cos(\theta - t) + r^2 \geqq 2r\{1 - \cos(\theta - t)\}$; so

$$\left|\int_{|\theta - t| \leqq \epsilon} [\log f(t) - \log f(\theta)] \frac{2r\sin(\theta - t)}{1 - 2r\cos(\theta - t) + r^2}\, dt\right|$$

$$\leqq \int_{|\theta - t| \leqq \epsilon} |\log f(t) - \log f(\theta)| \frac{|\sin(\theta - t)|}{1 - \cos(\theta - t)}\, dt,$$

and the last integral is arbitrarily small with ϵ.

Therefore, under the condition mentioned,

$$\lim_{r \to 1-0} D(re^{i\theta}) = D(e^{i\theta}) \tag{10.3.8}$$

exists, and a representation analogous to (10.3.3) holds for sgn $D(e^{i\theta})$,

$$\begin{aligned} \operatorname{sgn} D(e^{i\theta}) &= \exp\left\{\frac{i}{4\pi}\int_{-\pi}^{+\pi} \log f(t) \cot\frac{\theta - t}{2}\, dt\right\} \\ &= \exp\left\{\frac{i}{4\pi}\int_{-\pi}^{+\pi} [\log f(t) - \log f(\theta)] \cot\frac{\theta - t}{2}\, dt\right\}. \end{aligned} \tag{10.3.9}$$

In this case we have $f(\theta) = |D(e^{i\theta})|^2$.

The condition (10.3.5) is satisfied if the function $f(\theta)$ fulfills the Lipschitz-Dini condition

$$|f(\theta + \delta) - f(\theta)| < L\,|\log \delta|^{-1-\lambda}, \tag{10.3.10}$$

L and λ being fixed positive constants. This is assumed in the present section. *Then* (10.3.8) *exists uniformly in* θ, *and* $D(e^{i\theta})$ *is continuous.*

The preceding exposition could be considerably simplified by using the theory of conjugate functions (see Zygmund **2**, pp. 50, 54, 55, §§3.321 and 3.45).

(3) Let m be a positive integer. Then there exists a positive trigonometric polynomial $g(\theta)$ of degree m such that

$$|f(\theta) - \{g(\theta)\}^{-1}| < P(\log m)^{-1-\lambda}, \tag{10.3.11}$$

where P is a constant, depending on the minimum and maximum of $f(\theta)$ and on L and λ. This follows by applying Theorem 1.3.2 to the function $\{f(\theta)\}^{-1}$, which satisfies the Lipschitz-Dini condition

$$|\{f(\theta + \delta)\}^{-1} - \{f(\theta)\}^{-1}| < \{\min f(\theta)\}^{-2} L\,|\log \delta|^{-1-\lambda}.$$

If $D(z)$ and $h(z)$ denote the functions defined in §10.2 which correspond, respectively, to $f(\theta)$ and $g(\theta)$, we can show that

$$|D(z) - \{h(z)\}^{-1}| < Q(\log m)^{-\lambda} \tag{10.3.12}$$

uniformly for $|z| \leqq 1$. The constant Q depends on the minimum and maximum of $f(\theta)$ as well as on L and λ.

It suffices to prove this for $|z| = 1$. An analogous inequality for the difference of $|D(z)|$ and $|h(z)|^{-1}$ is trivial (even with $(\log m)^{-1-\lambda}$). We therefore need a bound only for

$$\begin{aligned} &\operatorname{sgn} h(e^{i\theta})\{\operatorname{sgn} D(e^{i\theta}) - \operatorname{sgn}[h(e^{i\theta})]^{-1}\} \\ &\quad = \exp\left\{\frac{i}{4\pi}\int_{-\pi}^{+\pi} \{\log f(t) - \log [g(t)]^{-1}\} \cot\frac{\theta - t}{2}\, dt\right\} - 1; \end{aligned} \tag{10.3.13}$$

or, what amounts to the same thing, for

$$\int_{-\pi}^{+\pi} \{\log f(t) - \log [g(t)]^{-1}\} \cot \frac{\theta - t}{2} \, dt. \tag{10.3.14}$$

To this end, we use Theorem 1.22.1. We have

$$| g(\theta + \delta) - g(\theta) | \leqq \delta m \{\max g(\theta)\}, \tag{10.3.15}$$

so that

$$| \log g(\theta + \delta) - \log g(\theta) | \leqq \delta m \frac{\max g(\theta)}{\min g(\theta)} . \tag{10.3.16}$$

Let $E = E(\theta, m, \lambda)$ denote the set of t-values defined by the condition $| \theta - t | \leqq m^{-1} (\log m)^{-\lambda}$, and let E' be the complementary set with respect to $[-\pi, +\pi]$. On writing (10.3.14) in the form

$$\begin{aligned} &\int_E \{\log f(t) - \log f(\theta) - \log [g(t)]^{-1} + \log [g(\theta)]^{-1}\} \cot \frac{\theta - t}{2} \, dt \\ &\qquad + \int_{E'} \{\log f(t) - \log [g(t)]^{-1}\} \cot \frac{\theta - t}{2} \, dt, \end{aligned} \tag{10.3.17}$$

and using (10.3.10) and (10.3.16), we obtain for the first integral

$$\begin{aligned} O(1) \int_E | \log | \theta - t | |^{-1-\lambda} | \theta - t |^{-1} dt + O(m) \int_E | \theta - t | \left| \cot \frac{\theta - t}{2} \right| dt \\ = O[(\log m)^{-\lambda}] + O(m) O[m^{-1} (\log m)^{-\lambda}] = O[(\log m)^{-\lambda}]. \end{aligned}$$

On the other hand, (10.3.11) yields the bound

$$O[(\log m)^{-1-\lambda}] \int_{E'} \left| \cot \frac{\theta - t}{2} \right| dt = O[(\log m)^{-\lambda}]$$

for the second integral. This establishes the statement.

10.4. "Local" properties of the representation of positive functions

In this section we prove certain theorems on the representation and approximation of positive functions, important for the purposes of Chapters XII and XIII.

(1) We have the following theorem:

THEOREM 10.4.1. *Let $f(\theta)$ be integrable in Riemann's sense, and let it have the form*

$$f(\theta) = \phi(\theta) \, | (z - z_1)^{\sigma_1} (z - z_2)^{\sigma_2} \cdots (z - z_l)^{\sigma_l} |, \qquad z = e^{i\theta}, \tag{10.4.1}$$

where $0 < A \leqq \phi(\theta) \leqq B$, and $z_\nu = e^{i\theta_\nu}$ are distinct points on the unit circle, $\sigma_\nu > 0$, $\nu = 1, 2, \cdots, l$. Let $f(\theta)$ be differentiable at the fixed point $\theta = \alpha$, $a = e^{i\alpha} \neq z_\nu$, $\nu = 1, 2, \cdots, l$, and let the following be bounded near $\theta = \alpha$:

$$\frac{f(\theta) - f(\alpha) - f'(\alpha)(\theta - \alpha)}{(\theta - \alpha)^2} . \tag{10.4.2}$$

If $D(f; z)$ denotes the analytic function corresponding to $f(\theta)$ in the sense of §10.2, *there follows the existence of the limits*

$$\lim_{r \to 1-0} D(f; re^{i\alpha}) = D(f; e^{i\alpha}) = D(f; \alpha),$$
$$\lim_{r \to 1-0} D'(f; re^{i\alpha}) = D'(f; e^{i\alpha}) = D'(f; \alpha). \tag{10.4.3}$$

This is true under more general conditions (see Zygmund **2**, pp. 52, 53, §3.44, and pp. 62, 63, example 13). The following elementary argument is based on the formula (10.2.10).

If we integrate with respect to t only over a fixed arc not containing a, the corresponding limits obviously exist. If t is near α, we can write

$$\log f(t) = c + d(e^{-it} - e^{-i\alpha}) + O(1)(t - \alpha)^2, \tag{10.4.4}$$

where c and d are certain constants. But if $|z| < 1$, then

$$\frac{1}{2\pi} \int_{-\pi}^{+\pi} \{c + d(e^{-it} - e^{-i\alpha})\} \frac{1 + ze^{-it}}{1 - ze^{-it}} dt = c - de^{-i\alpha}. \tag{10.4.5}$$

Hence it suffices to show that if $\epsilon > 0$ is small enough, the function

$$\int_{-\epsilon}^{+\epsilon} O(1)(t - \alpha)^2 \frac{1 + ze^{-it}}{1 - ze^{-it}} dt, \tag{10.4.6}$$

as well as the derivative of this function with respect to z, is arbitrarily small if $z = re^{i\alpha}$, $r \to 1 - 0$. This is true for the derivative since

$$\int_{-\epsilon}^{+\epsilon} \frac{(t - \alpha)^2 dt}{|1 - re^{i(\alpha - t)}|^2} = \int_{-\epsilon}^{+\epsilon} \frac{(t - \alpha)^2}{1 - 2r \cos(\alpha - t) + r^2} dt$$
$$< \frac{1}{4r} \int_{-\epsilon}^{+\epsilon} \frac{(t - \alpha)^2}{\sin^2 \frac{\alpha - t}{2}} dt. \tag{10.4.7}$$

The argument is even simpler for the function itself, for in the last denominator we then have $|\sin\{(\alpha - t)/2\}|$ instead of $\sin^2\{(\alpha - t)/2\}$.

(2) Now we prove the following theorem:

THEOREM 10.4.2. *Let $f(\theta)$ be integrable in Riemann's sense, and assume $0 < A \leqq f(\theta) \leqq B$ if $-\pi \leqq \theta \leqq +\pi$. Also suppose $f(\theta)$ differentiable at the fixed point $\theta = \alpha$ and the expression* (10.4.2) *bounded near $\theta = \alpha$.*

If ϵ is an arbitrary positive number, there exist positive trigonometric polynomials $g_1(\theta)$ and $g_2(\theta)$ such that

$$f_1(\theta) \leqq f(\theta) \leqq f_2(\theta), \qquad f_1(\alpha) = f_2(\alpha), \tag{10.4.8}$$

where $f_1(\theta) = \{g_1(\theta)\}^{-1}$, $f_2(\theta) = \{g_2(\theta)\}^{-1}$, and

$$\int_{-\pi}^{+\pi} \frac{f_2(\theta) - f_1(\theta)}{\sin^2 \frac{\theta - \alpha}{2}} d\theta < \epsilon. \tag{10.4.9}$$

Let m and M be the lower and upper bound of (10.4.2) for $-\pi \leqq \theta \leqq +\pi$. The function $f_1(\theta)$ is greater than a positive number depending only on m, A, $f(\alpha)$, and $f'(\alpha)$; similarly $f_2(\theta)$ is less than a number depending only on M, $f(\alpha)$, and $f'(\alpha)$.

We observe that (10.4.8) implies

$$f_1(\alpha) = f(\alpha) = f_2(\alpha), \qquad f_1'(\alpha) = f'(\alpha) = f_2'(\alpha), \tag{10.4.10}$$

so that the integral in (10.4.9) exists; furthermore, (10.4.9) implies

$$\int_{-\pi}^{+\pi} \frac{\log f_2(\theta) - \log f_1(\theta)}{\sin^2 \dfrac{\theta - \alpha}{2}}\, d\theta < \epsilon', \tag{10.4.11}$$

where ϵ' is arbitrarily small with ϵ.

For the proof we apply Theorem 1.5.4 to the functions

$$\begin{aligned} p(\theta) &= \frac{\{f(\theta)\}^{-1} - c - d \sin (\theta - \alpha)}{\sin^2 \dfrac{\theta - \alpha}{2}}, & c &= \{f(\alpha)\}^{-1}, \\ & & d &= -f'(\alpha)\{f(\alpha)\}^{-2}, \\ q(\theta) &= \frac{f(\theta) - f(\alpha) - f'(\alpha) \sin(\theta - \alpha)}{\sin^2 \dfrac{\theta - \alpha}{2}}, \end{aligned} \tag{10.4.12}$$

which are both integrable in Riemann's sense. Therefore, given $\delta > 0$, there exist certain trigonometric polynomials $P(\theta)$ and $Q(\theta)$ such that

$$\begin{aligned} p(\theta) &\leqq P(\theta), & \int_{-\pi}^{+\pi} \{P(\theta) - p(\theta)\}\, d\theta &< \delta, \\ q(\theta) &\leqq Q(\theta), & \int_{-\pi}^{+\pi} \{Q(\theta) - q(\theta)\}\, d\theta &< \delta. \end{aligned} \tag{10.4.13}$$

Here $\max P(\theta)$ and $\max Q(\theta)$ are less than certain constants depending on m, A, $f(\alpha)$, $f'(\alpha)$, and on M and $f'(\alpha)$, respectively. Writing

$$g_1(\theta) = c + d \sin (\theta - \alpha) + P(\theta) \sin^2 \frac{\theta - \alpha}{2}, \qquad f_1(\theta) = \{g_1(\theta)\}^{-1}, \tag{10.4.14}$$

we find that $\{f(\theta)\}^{-1} \leqq g_1(\theta)$, or $f_1(\theta) \leqq f(\theta)$, $f_1(\alpha) = f(\alpha)$, and that

$$\int_{-\pi}^{+\pi} \frac{f(\theta) - f_1(\theta)}{\sin^2 \dfrac{\theta - \alpha}{2}}\, d\theta = \int_{-\pi}^{+\pi} f(\theta) f_1(\theta) \frac{g_1(\theta) - \{f(\theta)\}^{-1}}{\sin^2 \dfrac{\theta - \alpha}{2}}\, d\theta \leqq B^2 \delta. \tag{10.4.15}$$

Here $f_1(\theta)$ is greater than a positive constant depending on m, A, $f(\alpha)$, and $f'(\alpha)$.

On the other hand, considering the continuous function

$$G(\theta)$$

$$(10.4.16)\qquad = \frac{\left\{f(\alpha) + f'(\alpha)\sin(\theta-\alpha) + Q(\theta)\sin^2\dfrac{\theta-\alpha}{2}\right\}^{-1} - c - d\sin(\theta-\alpha)}{\sin^2\dfrac{\theta-\alpha}{2}}$$

$$= \frac{\{R(\theta)\}^{-1} - c - d\sin(\theta-\alpha)}{\sin^2\dfrac{\theta-\alpha}{2}} > \frac{R^{-1} - c - d\sin(\theta-\alpha)}{\sin^2\dfrac{\theta-\alpha}{2}},$$

we have $f(\theta) \leqq R(\theta) < R$, where R is a positive constant depending on M, $f(\alpha)$, and $f'(\alpha)$. Now according to Theorem 1.3.1, a trigonometric polynomial $S(\theta)$ can be determined such that

$$(10.4.17)\qquad \frac{R^{-1} - c - d\sin(\theta-\alpha)}{\sin^2\dfrac{\theta-\alpha}{2}} < S(\theta) < G(\theta) < S(\theta) + \delta.$$

If we write

$$(10.4.18)\qquad g_2(\theta) = c + d\sin(\theta-\alpha) + S(\theta)\sin^2\frac{\theta-\alpha}{2}, \qquad f_2(\theta) = \{g_2(\theta)\}^{-1},$$

we find that

$$(10.4.19)\qquad R^{-1} < g_2(\theta) < \{R(\theta)\}^{-1} < g_2(\theta) + \delta\sin^2\frac{\theta-\alpha}{2}.$$

Furthermore, $g_2(\theta) < \{f(\theta)\}^{-1}$. On account of the last inequality in (10.4.19) and (10.4.13) we have

$$\int_{-\pi}^{+\pi} \frac{f_2(\theta) - f(\theta)}{\sin^2\dfrac{\theta-\alpha}{2}}\,d\theta = \int_{-\pi}^{+\pi} \frac{\left\{g_2(\theta) + \delta\sin^2\dfrac{\theta-\alpha}{2}\right\}^{-1} - f(\theta)}{\sin^2\dfrac{\theta-\alpha}{2}}\,d\theta$$

$$+ \int_{-\pi}^{+\pi} \frac{\{g_2(\theta)\}^{-1} - \left\{g_2(\theta) + \delta\sin^2\dfrac{\theta-\alpha}{2}\right\}^{-1}}{\sin^2\dfrac{\theta-\alpha}{2}}\,d\theta$$

$$(10.4.20)\qquad < \int_{-\pi}^{+\pi} \frac{R(\theta) - f(\theta)}{\sin^2\dfrac{\theta-\alpha}{2}}\,d\theta + \int_{-\pi}^{+\pi} \frac{\{g_2(\theta)\}^{-1} - \left\{g_2(\theta) + \delta\sin^2\dfrac{\theta-\alpha}{2}\right\}^{-1}}{\sin^2\dfrac{\theta-\alpha}{2}}\,d\theta$$

$$< \delta + \int_{-\pi}^{+\pi} \frac{\{g_2(\theta)\}^{-1} - \left\{g_2(\theta) + \delta\sin^2\dfrac{\theta-\alpha}{2}\right\}^{-1}}{\sin^2\dfrac{\theta-\alpha}{2}}\,d\theta.$$

The last integral is arbitrarily small with δ (since $g_2(\theta) > R^{-1}$).

Combining (10.4.15) and (10.4.20), and taking δ sufficiently small, we obtain the statement above.

(3) THEOREM 10.4.3. *For the analytic functions $D(f_1; z)$, $D(f; z)$, $D(f_2; z)$, corresponding to the functions $f_1(\theta)$, $f(\theta)$, $f_2(\theta)$ of Theorem 10.4.2 in the sense of §10.2, the following inequalities hold:*

$$|D(f; a) - D(f_\nu; a)| < \epsilon', \qquad |D'(f; a) - D'(f_\nu; a)| < \epsilon', \tag{10.4.21}$$

where $a = e^{i\alpha}$, $\nu = 1, 2$; *here ϵ' is arbitrarily small with ϵ.*

The symbols $D(f; a)$, $D'(f; a)$ have the same meaning as in (10.4.3).

According to (10.2.10) we have for $|z| < 1$, $\nu = 1, 2$,

$$\begin{aligned} \log D(f; z) - \log D(f_\nu; z) &= \frac{1}{4\pi}\int_{-\pi}^{+\pi} \{\log f(t) - \log f_\nu(t)\} \frac{1 + ze^{-it}}{1 - ze^{-it}}\, dt, \\ z\frac{D'(f; z)}{D(f; z)} - z\frac{D'(f_\nu; z)}{D(f_\nu; z)} &= \frac{1}{2\pi}\int_{-\pi}^{+\pi} \{\log f(t) - \log f_\nu(t)\} \frac{ze^{-it}}{(1 - ze^{-it})^2}\, dt, \end{aligned} \tag{10.4.22}$$

and for $z = re^{i\alpha} = ra$, $r \to 1 - 0$,

$$\begin{aligned} \log D(f; a) - \log D(f_\nu; a) &= \frac{i}{4\pi}\int_{-\pi}^{+\pi} \{\log f(t) - \log f_\nu(t)\} \cot\frac{\alpha - t}{2}\, dt, \\ a\frac{D'(f; a)}{D(f; a)} - a\frac{D'(f_\nu; a)}{D(f_\nu; a)} &= -\frac{1}{8\pi}\int_{-\pi}^{+\pi} \frac{\log f(t) - \log f_\nu(t)}{\sin^2 \frac{t - \alpha}{2}}\, dt. \end{aligned} \tag{10.4.23}$$

Both integrals are absolutely convergent and arbitrarily small with ϵ (see (10.4.11)); $D(f; a)$ is a fixed number different from zero.

(4) THEOREM 10.4.4. *Let $f(\theta)$ be a function satisfying the conditions of Theorem 10.4.1. Given an arbitrary positive number ϵ, there exist positive trigonometric polynomials $g_1(\theta)$ and $g_2(\theta)$, such that on putting*

$$\begin{aligned} f_1(\theta) &= \{g_1(\theta)\}^{-1} |(z - z_1)(z - z_2) \cdots (z - z_l)|^\sigma, \qquad z = e^{i\theta}, \\ f_2(\theta) &= \{g_2(\theta)\}^{-1}, \end{aligned} \tag{10.4.24}$$

we have

$$0 \leqq f_1(\theta) \leqq f(\theta) \leqq f_2(\theta), \qquad f_1(\alpha) = f_2(\alpha), \tag{10.4.25}$$

and

$$\int_{-\pi}^{+\pi} \frac{\log f_2(\theta) - \log f_1(\theta)}{\sin^2 \frac{\theta - \alpha}{2}}\, d\theta < \epsilon. \tag{10.4.26}$$

Here σ is the least even number greater than $\max(\sigma_1, \sigma_2, \cdots, \sigma_l)$; $\max f_2(\theta)$ *is bounded from above, while* $\min\{g_1(\theta)\}^{-1}$ *is bounded from below, and both bounds are independent of ϵ.*

For the functions $D(f_1; z)$, $D(f; z)$, and $D(f_2; z)$, corresponding, respectively, to $f_1(\theta)$, $f(\theta)$, and $f_2(\theta)$, a statement similar to that in Theorem 10.4.3 *holds.*

REMARK. We can choose for σ any even number which is greater than $\max(\sigma_1, \sigma_2, \cdots, \sigma_l)$; in particular, we can choose an even number divisible by 4. This is important for certain later purposes (cf. §13.5 (2)).

Let the function $\hat{f}(\theta)$ be identical with $f(\theta)$ except in certain closed intervals around θ_ν, $\nu = 1, 2, \cdots, l$, in which $\hat{f}(\theta) = 1$. These intervals are chosen so small that they do not overlap, they do not contain α, and in each of them $f(\theta) \leqq 1$. Furthermore, let

$$\int_{-\pi}^{+\pi} \frac{\log \hat{f}(\theta) - \log f(\theta)}{\sin^2 \dfrac{\theta - \alpha}{2}}\, d\theta < \frac{\epsilon}{4}. \tag{10.4.27}$$

Then $f(\theta) \leqq \hat{f}(\theta)$ for each θ, and $f(\theta) = \hat{f}(\theta)$ in a certain neighborhood of α which can be chosen independent of ϵ provided ϵ is small enough. The function $\hat{f}(\theta) = \hat{f}(\epsilon; \theta)$ satisfies the conditions of Theorem 10.4.2 and depends on ϵ, although it has an upper bound independent of ϵ. The same is true for the upper bound $\hat{M}$ of the ratio corresponding to (10.4.2).

We now determine the trigonometric polynomial $g_2(\theta)$ such that, for $f_2(\theta) = \{g_2(\theta)\}^{-1}$,

$$\begin{gathered} f(\theta) \leqq \hat{f}(\theta) \leqq f_2(\theta); \qquad f(\alpha) = \hat{f}(\alpha) = f_2(\alpha); \\ \int_{-\pi}^{+\pi} \frac{\log f_2(\theta) - \log \hat{f}(\theta)}{\sin^2 \dfrac{\theta - \alpha}{2}}\, d\theta < \frac{\epsilon}{4}. \end{gathered} \tag{10.4.28}$$

We observe that $\max f_2(\theta)$ is less than a number independent of ϵ. (Here we use the independence of $\hat{M}$ of ϵ.)

On the other hand, let $\underline{f}(\theta)$ be identical with $f(\theta)$ except in certain closed non-overlapping intervals around θ_ν, $\nu = 1, 2, \cdots, l$, not containing α, and such that in these intervals

$$|(z - z_1)(z - z_2) \cdots (z - z_l)|^\sigma \leqq |(z - z_1)^{\sigma_1}(z - z_2)^{\sigma_2} \cdots (z - z_l)^{\sigma_l}|, \tag{10.4.29}$$

$z = e^{i\theta}$. In these intervals we define

$$\underline{f}(\theta) = \phi(\theta)\, |(z - z_1)(z - z_2) \cdots (z - z_l)|^\sigma, \qquad z = e^{i\theta}. \tag{10.4.30}$$

Furthermore, assume

$$\int_{-\pi}^{+\pi} \frac{\log f(\theta) - \log \underline{f}(\theta)}{\sin^2 \dfrac{\theta - \alpha}{2}}\, d\theta < \frac{\epsilon}{4}. \tag{10.4.31}$$

Then $\underline{f}(\theta) \leqq f(\theta)$, and $\underline{f}(\theta) = f(\theta)$ near $\theta = \alpha$. Moreover,

$$\underline{f}(\theta)\, |(z - z_1)(z - z_2) \cdots (z - z_l)|^{-\sigma}, \qquad z = e^{i\theta}, \tag{10.4.32}$$

satisfies the conditions of Theorem 10.4.2 and depends on ϵ, although it has a positive lower bound independent of ϵ; the same is true for the (not necessarily positive) lower bound $\underset{\vee}{m}$ of the ratio corresponding to (10.4.2).

We determine a positive trigonometric polynomial $g_1(\theta)$ such that

$$\{g_1(\theta)\}^{-1} \leqq f(\theta) \mid (z - z_1)(z - z_2) \cdots (z - z_l) \mid^{-\sigma}, \qquad z = e^{i\theta}, \tag{10.4.33}$$

with equality for $\theta = \alpha$, and

$$\int_{-\pi}^{+\pi} \frac{\log \{ f(\theta) \mid (z - z_1)(z - z_2) \cdots (z - z_l) \mid^{-\sigma} \} - \log \{g_1(\theta)\}^{-1}}{\sin^2 \dfrac{\theta - \alpha}{2}} \, d\theta < \frac{\epsilon}{4}. \tag{10.4.34}$$

Then, using the notation (10.4.24), we obtain

$$\int_{-\pi}^{+\pi} \frac{\log f(\theta) - \log f_1(\theta)}{\sin^2 \dfrac{\theta - \alpha}{2}} \, d\theta < \frac{\epsilon}{4}. \tag{10.4.35}$$

Addition of the inequalities (10.4.27), (10.4.28), (10.4.31), and (10.4.35) establishes (10.4.26). We observe that $\min \{g_1(\theta)\}^{-1}$ is greater than a positive constant independent of ϵ.

The assertion concerning $D(f_1 ; z)$, $D(f; z)$, and $D(f_2 ; z)$ follows as in Theorem 10.4.3.

(5) Theorem 10.4.5. *Let $f(\theta)$ be an even function which satisfies the conditions of Theorem* 10.4.1; *assume that $\phi(\theta) = \phi(-\theta)$, and that all non-real "zeros" z_ν of $f(\theta)$ occur in conjugate pairs with the same "multiplicity." Also, let $0 < \alpha < \pi$. Then the functions $f_1(\theta)$ and $f_2(\theta)$ of Theorem* 10.4.4 *can be chosen as even functions, and we have instead of* (10.4.26)

$$\int_0^{\pi} \frac{\log f_2(\theta) - \log f_1(\theta)}{(\cos \theta - \cos \alpha)^2} \, d\theta < \epsilon. \tag{10.4.36}$$

An analogous supplement can be made to Theorem 10.4.2. Previous proofs need only a slight modification. Instead of the first ratio in (10.4.12), consider

$$\frac{\{f(\theta)\}^{-1} - c - d(\cos \theta - \cos \alpha)}{(\cos \theta - \cos \alpha)^2}, \tag{10.4.37}$$

$$c = \{f(\alpha)\}^{-1}, \; d \sin \alpha = f'(\alpha) \{f(\alpha)\}^{-2}.$$

The other ratios occurring in the proof of Theorem 10.4.2 must be similarly modified. The supplements to Theorems 1.3.1 and 1.5.4 concerning even functions have to be observed. The functions $\hat{f}(\theta)$ and $\underset{\vee}{f}(\theta)$ occurring in the proof of Theorem 10.4.4 can be chosen as even functions. The inequality (10.4.36) is equivalent to (10.4.26), since the function

$$\frac{\sin^2 \{(\theta - \alpha)/2\}}{(\cos \theta - \cos \alpha)^2} \tag{10.4.37}$$

is bounded from 0 and ∞ if $0 \leqq \theta \leqq \pi$.

CHAPTER XI

POLYNOMIALS ORTHOGONAL ON THE UNIT CIRCLE

A weight function on a given curve having been assigned, we may extend the definition of polynomials orthogonal on a real interval to the more general complex domain. The corresponding polynomials are then orthogonal with this given weight function on the specified curve in the complex plane (Chapter XVI).

Of all the special instances, that of a circle is most interesting, and in the present chapter we shall consider the polynomials orthogonal on the unit circle with a given weight function. It will be seen that these polynomials possess properties which are in some respects simpler than those derived for polynomials orthogonal on a real interval. Moreover, there exists a relation between the case of the circle and that of a real, finite interval which enables us to apply certain results of this chapter to polynomials orthogonal on a real interval.

Concerning §§11.1–11.4 see Szegö **4**. See also Grenander-Szegö **1**, Chapter 2.

11.1. Definition; preliminaries

(1) Let $f(\theta)$ be a non-negative function of period 2π, integrable on $[-\pi, +\pi]$ in Lebesgue's sense, and assume

$$\int_{-\pi}^{+\pi} f(\theta)\, d\theta > 0. \tag{11.1.1}$$

We introduce the Fourier constants

$$c_n = \frac{1}{2\pi} \int_{-\pi}^{+\pi} f(\theta) e^{-in\theta}\, d\theta, \qquad n = 0, \pm 1, \pm 2, \cdots. \tag{11.1.2}$$

Obviously, $c_{-n} = \bar{c}_n$, so that the matrix of "Toeplitz' type"

$$T_n = (c_{\nu-\mu}), \qquad \nu, \mu = 0, 1, 2, \cdots, n, \tag{11.1.3}$$

is Hermitian. The corresponding Hermitian form

$$H_n = \sum_{\nu=0}^{n} \sum_{\mu=0}^{n} c_{\nu-\mu} u_\mu \bar{u}_\nu = \frac{1}{2\pi} \int_{-\pi}^{+\pi} f(\theta) \,|\, u_0 + u_1 z + u_2 z^2 + \cdots + u_n z^n \,|^2\, d\theta, \tag{11.1.4}$$

where $z = e^{i\theta}$, is positive definite and has the positive determinant

$$D_n = [c_{\nu-\mu}], \qquad \nu, \mu = 0, 1, 2, \cdots, n. \tag{11.1.5}$$

(2) Definition. *If we orthogonalize the system*[57]

[57] Cf. the last remark in §2.1 (4).

$$\{(f(\theta))^{\frac{1}{2}}z^n\}, \qquad z = e^{i\theta};\ n = 0, 1, 2, \cdots, \tag{11.1.6}$$

we obtain a system of polynomials

$$\phi_0(z), \phi_1(z), \phi_2(z), \cdots, \phi_n(z), \cdots \tag{11.1.7}$$

with the following properties:

(a) *$\phi_n(z)$ is a polynomial of precise degree n in which the coefficient of z^n is real and positive;*

(b) *the system $\{\phi_n(z)\}$ is orthonormal; that is,*

$$\frac{1}{2\pi}\int_{-\pi}^{+\pi} f(\theta)\phi_n(z)\overline{\phi_m(z)}\, d\theta = \delta_{nm}, \qquad z = e^{i\theta};\ n, m = 0, 1, 2, \cdots. \tag{11.1.8}$$

Moreover, the system $\{\phi_n(z)\}$ is uniquely determined by conditions (a) and (b). If $f(\theta)$ is an even function, that is, if $f(\theta) = f(-\theta)$, the coefficients of $\phi_n(z)$ are real.

(3) We show (see §2.2 (2)) that

$$\phi_n(z) = (D_{n-1}D_n)^{-\frac{1}{2}} \begin{vmatrix} c_0 & c_{-1} & c_{-2} & \cdots & c_{-n} \\ c_1 & c_0 & c_{-1} & \cdots & c_{-n+1} \\ \cdots & \cdots & \cdots & \cdots & \cdots \\ c_{n-1} & c_{n-2} & c_{n-3} & \cdots & c_{-1} \\ 1 & z & z^2 & \cdots & z^n \end{vmatrix} \tag{11.1.9}$$

$$= (D_{n-1}D_n)^{-\frac{1}{2}} \begin{vmatrix} c_0z - c_{-1} & c_{-1}z - c_{-2} & \cdots & c_{-n+1}z - c_{-n} \\ c_1z - c_0 & c_0z - c_{-1} & \cdots & c_{-n+2}z - c_{-n+1} \\ \cdots & \cdots & \cdots & \cdots \\ c_{n-1}z - c_{n-2} & c_{n-2}z - c_{n-3} & \cdots & c_0z - c_{-1} \end{vmatrix},$$

$$\phi_0(z) = D_0^{-\frac{1}{2}}, \qquad n = 1, 2, 3, \cdots.$$

The coefficient of z^n in $\phi_n(z)$ is

$$\kappa_n = (D_{n-1}D_n^{-1})^{\frac{1}{2}}. \tag{11.1.10}$$

The analogues of the representations (2.2.10) and (2.2.11) can also be readily derived.

(4) We pass now to considerations corresponding to those of (1), (2), and (4) of §3.1.

THEOREM 11.1.1. *Let $F(e^{i\theta})$ be a given measurable function for which*

$$\int_{-\pi}^{+\pi} f(\theta)\, |F(e^{i\theta})|^2\, d\theta \tag{11.1.11}$$

exists. The weighted quadratic deviation

$$\frac{1}{2\pi}\int_{-\pi}^{+\pi} f(\theta)\, |F(z) - \rho(z)|^2\, d\theta, \qquad z = e^{i\theta}, \tag{11.1.12}$$

where $\rho(z)$ *ranges over the set of all* π_n, *is a minimum if* $\rho(z)$ *is the nth partial sum of the Fourier expansion*

$$F(z) \sim F_0\phi_0(z) + F_1\phi_1(z) + F_2\phi_2(z) + \cdots + F_n\phi_n(z) + \cdots,$$

(11.1.13)

$$F_n = \frac{1}{2\pi}\int_{-\pi}^{+\pi} f(\theta)F(z)\overline{\phi_n(z)}\,d\theta, \qquad z = e^{i\theta}; n = 0, 1, 2, \cdots.$$

As a ready consequence, there follows Bessel's inequality

(11.1.14)
$$|F_0|^2 + |F_1|^2 + |F_2|^2 + \cdots + |F_n|^2 + \cdots \leqq \frac{1}{2\pi}\int_{-\pi}^{+\pi} f(\theta)\,|F(e^{i\theta})|^2\,d\theta.$$

In addition, Parseval's formula (that is, (11.1.14) with the equality sign) holds if one of the following sets of conditions is satisfied:

(i) $F(z)$ is regular and bounded for $|z| < 1$.
(ii) $f(\theta)$ is bounded and $F(z)$ is of the class H_2 (see §10.1).

Concerning a more general condition, see Smirnoff **2**, p. 363.
A consequence of Theorem 11.1.1 is the following:

THEOREM 11.1.2. *The polynomial* $\kappa_n^{-1}\phi_n(z)$ *minimizes the integral*

(11.1.15)
$$\frac{1}{2\pi}\int_{-\pi}^{+\pi} f(\theta)\,|z^n + a_1z^{n-1} + \cdots + a_n|^2\,d\theta, \qquad z = e^{i\theta},$$

if $z^n + a_1z^{n-1} + \cdots + a_n$ *ranges over the set of all* π_n *with the highest term* z^n. *The minimum is* κ_n^{-2}.

11.2. Example

An important special case in which the system $\{\phi_n(z)\}$ can be calculated explicitly, except for a finite number of terms, is

(11.2.1)
$$f(\theta) = \{g(\theta)\}^{-1},$$

where $g(\theta)$ is a positive trigonometric polynomial of degree m.

THEOREM 11.2. *Let* $f(\theta)$ *be defined by* (11.2.1), *and let* $g(\theta) = |h(z)|^2$, $z = e^{i\theta}$, *be the normalized representation of* $g(\theta)$ *defined in Theorem* 1.2.2. *Using the notation of* (1.12.4), *we have*

(11.2.2)
$$\phi_n(z) = z^{n-m}h^*(z) = z^n\bar{h}(z^{-1}), \qquad n = m, m + 1, m + 2, \cdots.$$

Evidently, condition (a) of the definition in §11.1 (2) is satisfied. In order to show the orthogonality, let $\rho(z)$ be an arbitrary π_{n-1}. Then, if $z = e^{i\theta}$, we have, according to Cauchy's theorem,

$$\frac{1}{2\pi}\int_{-\pi}^{+\pi} f(\theta)\phi_n(z)\overline{\rho(z)}\,d\theta = \frac{1}{2\pi}\int_{|z|=1} \{h(z)\bar{h}(z^{-1})\}^{-1} z^n\,\bar{h}(z^{-1})\bar{\rho}(z^{-1})\,\frac{dz}{iz}$$
$$= \frac{1}{2\pi i}\int_{|z|=1} \frac{z^{n-1}\,\bar{\rho}(z^{-1})}{h(z)}\,dz = 0.$$

In addition

$$\frac{1}{2\pi}\int_{-\pi}^{+\pi} f(\theta)\,|\,\phi_n(z)\,|^2\,d\theta = \frac{1}{2\pi}\int_{-\pi}^{+\pi} |\,h(z)\,|^{-2}\,|z^n\bar{h}\,(z^{-1})\,|^2\,d\theta = 1.$$

The simplest case is $f(\theta) = 1$. Then

$$\phi_n(z) = z^n, \qquad n = 0, 1, 2, \cdots. \tag{11.2.3}$$

Concerning other cases in which an explicit calculation of $\phi_n(z)$ is possible, we refer to Szegö **4**, pp. 187–188; and **12**, pp. 245–247. See also (11.5.3) and (11.5.4).

11.3. A maximum problem

Because of the similarity between this problem and that treated in §3.1 (3), we can omit details.

(1) THEOREM 11.3.1. *Let $\rho(z)$ be an arbitrary π_n subject to the condition*

$$\frac{1}{2\pi}\int_{-\pi}^{+\pi} f(\theta)\,|\,\rho(z)\,|^2\,d\theta = 1, \qquad z = e^{i\theta}. \tag{11.3.1}$$

For a fixed value of a, the maximum of $|\,\rho(a)\,|^2$ is attained if

$$\rho(z) = \epsilon\{s_n(a, a)\}^{-\frac{1}{2}}s_n(a, z), \qquad |\,\epsilon\,| = 1, \tag{11.3.2}$$

where

$$s_n(a, z) = \sum_{\nu=0}^{n} \overline{\phi_\nu(a)}\phi_\nu(z). \tag{11.3.3}$$

The maximum itself is $s_n(a, a)$.

These "kernel polynomials" $s_n(a, z)$ can be used for the representation of the partial sums of the expansion (11.1.13) in form of integrals (see (3.1.11)).

(2) THEOREM 11.3.2. *For $a \neq 0$ the polynomials* (11.3.2) *satisfy the following identity:*

$$s_n(a, z) = (\bar{a}z)^n s_n(\bar{z}^{-1}, \bar{a}^{-1}). \tag{11.3.4}$$

Furthermore,

$$s_n(0, z) = \sum_{\nu=0}^{n} \overline{\phi_\nu(0)}\phi_\nu(z) = \kappa_n\, z^n\,\bar{\phi}_n(z^{-1}) = \kappa_n\,\phi_n^*(z \tag{11.3.5}$$

where κ_n has the same meaning as in (11.1.10); *finally*

$$s_n(0, 0) = \sum_{\nu=0}^{n} |\,\phi_\nu(0)\,|^2 = \kappa_n^2 = D_{n-1}/D_n\,. \tag{11.3.6}$$

The last formula also holds for $n = 0$ if $D_{-1} = 1$.

Introducing $\rho^*(z) = r(z)$ or $\rho(z) = r^*(z)$, we have

$$(11.3.7)\qquad \frac{1}{2\pi}\int_{-\pi}^{+\pi} f(\theta)\,|\,\rho(z)\,|^2\,d\theta = \frac{1}{2\pi}\int_{-\pi}^{+\pi} f(\theta)\,|\,r(z)\,|^2\,d\theta = 1, \qquad z = e^{i\theta},$$

and, for $a \neq 0$,

$$(11.3.8)\qquad |\,\rho(a)\,|^2 = |\,a^n\bar{r}(a^{-1})\,|^2 = |\,a\,|^{2n}\,|\,r(\bar{a}^{-1})\,|^2.$$

This yields $s_n(a, a) = |\,a\,|^{2n}s_n(\bar{a}^{-1}, \bar{a}^{-1})$, that is, (11.3.4) for $z = a$, and also

$$(11.3.9)\qquad \{s_n(a, a)\}^{-\frac{1}{2}}s_n(a, z) = \epsilon[\{s_n(\bar{a}^{-1}, \bar{a}^{-1})\}^{-\frac{1}{2}}s_n(\bar{a}^{-1}, z)]^*$$

where ϵ is a proper constant with $|\,\epsilon\,| = 1$. (The symbol * refers to the variable z.) On combining this with the first result, we obtain (11.3.4). In the limiting case $a = 0$, the identity (11.3.5) arises, from which, for $z = 0$, (11.3.6) follows.

(3) THEOREM 11.3.3. *Let* $\log f(\theta)$ *be integrable in Lebesgue's sense.*[58] *Then the following limits exist:*

$$(11.3.10)\qquad \lim_{n\to\infty} s_n(a, a) = \sum_{\nu=0}^{\infty} |\,\phi_\nu(a)\,|^2, \qquad |\,a\,| < 1,$$

$$(11.3.11)\qquad \lim_{n\to\infty} s_n(a, z) = \sum_{\nu=0}^{\infty} \overline{\phi_\nu(a)}\phi_\nu(z), \qquad |\,a\,| < 1, \quad |\,z\,| < 1,$$

$$(11.3.12)\qquad \lim_{n\to\infty} \kappa_n = \kappa > 0, \qquad \lim_{n\to 0} z^{-n}\phi_n(z), \qquad |\,z\,| > 1,$$

$$(11.3.13)\qquad \lim_{n\to\infty} \phi_n(z) = 0, \qquad |\,z\,| < 1.$$

We first consider the special case $f(\theta) \geqq \mu > 0$, assuming $|\,a\,| < 1$. Let $\rho(z)$ have the same meaning as in (11.3.2). Then we have, because of Cauchy's inequality (see (7.1.4)),

$$(11.3.14)\qquad \begin{aligned} |\,\rho(a)\,|^2 &\leqq \frac{1}{1-|\,a\,|^2}\frac{1}{2\pi}\int_{-\pi}^{+\pi} |\,\rho(z)\,|^2\,d\theta \\ &\leqq \frac{\mu^{-1}}{1-|\,a\,|^2}\frac{1}{2\pi}\int_{-\pi}^{+\pi} f(\theta)\,|\,\rho(z)\,|^2\,d\theta = \frac{\mu^{-1}}{1-|\,a\,|^2}, \qquad z = e^{i\theta}; \end{aligned}$$

consequently, the same inequality holds for $n \to \infty$. Thus (11.3.10) and (11.3.11) are established in this case. For $a = 0$, (11.3.6) and (11.3.5) show that the limits (11.3.12) exist; (11.3.13) follows from the convergence of (11.3.10).

In order to prove the statement generally, we first observe that the maximum in the problem of Theorem 11.3.1 is attained for a polynomial $\rho(z)$ which is different from zero for $|\,z\,| < 1$; here again $|\,a\,| < 1$. In fact, if z_0 is a zero of $\rho(z)$, $|\,z_0\,| < 1$, then

[58] Cf. §10.2 (2).

$$\frac{1}{2\pi}\int_{-\pi}^{+\pi} f(\theta)\left|\rho(z)\frac{1-\bar{z}_0 z}{z-z_0}\right|^2 d\theta = 1; \qquad |\rho(a)|^2 < \left|\rho(a)\frac{1-\bar{z}_0 a}{a-z_0}\right|^2, \qquad z = e^{i\theta}.$$

Now assume $a = re^{i\phi}$, $0 \leqq r < 1$, $z = e^{i\theta}$; according to (1.11.3),

$$1 = \frac{1}{2\pi}\int_{-\pi}^{+\pi} f(\theta)\,|\rho(z)|^2\,d\theta \geqq \frac{1-r}{1+r}\frac{1}{2\pi}\int_{-\pi}^{+\pi} f(\theta)\,|\rho(z)|^2 \frac{1-r^2}{1-2r\cos(\theta-\phi)+r^2}\,d\theta$$

$$\geqq \frac{1-r}{1+r}\exp\left\{\frac{1}{2\pi}\int_{-\pi}^{+\pi}\log f(\theta)\frac{1-r^2}{1-2r\cos(\theta-\phi)+r^2}\,d\theta\right\}$$

$$\cdot\exp\left\{\frac{1}{2\pi}\int_{-\pi}^{+\pi}\log|\rho(z)|^2\frac{1-r^2}{1-2r\cos(\theta-\phi)+r^2}\,d\theta\right\}.$$

The last integral is Poisson's integral of the harmonic function $2\Re\log[\rho(z)]$, which is regular for $|z| < 1$. The last exponential factor is therefore equal to

$$\exp\{2\Re\log[\rho(a)]\} = |\rho(a)|^2;$$

this establishes the boundedness of $\max|\rho(a)|^2 = s_n(a, a)$.

The further formulas follow as before. Later (§12.3 (6)) we shall calculate the limits (11.3.10)-(11.3.12).

11.4. Algebraic properties

(1) Let a be fixed, $|a| < 1$. The above considerations show that the zeros of $s_n(a, z)$ lie in $|z| \geqq 1$. The same is true, of course, for $|a| \leqq 1$. From (11.3.5) we find that the zeros of $\phi_n(z)$ lie in $|z| \leqq 1$.

We now prove the following more informative statement:

THEOREM 11.4.1. *For $|a| < 1$ the zeros of $s_n(a, z)$ lie in $|z| > 1$, for $|a| > 1$ in $|z| < 1$, and for $|a| = 1$ on $|z| = 1$. The zeros of $\phi_n(z)$ lie in $|z| < 1$.*

Let z_0 be an arbitrary zero of $s_n(a, z)$. If we put

$$f_1(\theta) = f(\theta)\left|\frac{s_n(a, z)}{z - z_0}\right|^2, \qquad z = e^{i\theta}, \tag{11.4.1}$$

and consider all the linear functions $\rho(z)$ with

$$\frac{1}{2\pi}\int_{-\pi}^{+\pi} f_1(\theta)\,|\rho(z)|^2\,d\theta = 1, \qquad z = e^{i\theta}, \tag{11.4.2}$$

it is clear that $\max|\rho(a)|^2$ is attained for $\rho(z) = \text{const.}\,(z - z_0)$. It therefore suffices to discuss the case $n = 1$. From (11.1.9)

$$\begin{aligned} s_1(a, z) &= \overline{\phi_0(a)}\,\phi_0(z) + \overline{\phi_1(a)}\,\phi_1(z) \\ &= D_0^{-1} + D_0^{-1}D_1^{-1}(c_0\bar{a} - c_1)(c_0 z - \bar{c}_1), \end{aligned} \tag{11.4.3}$$

with the zero

$$z = \frac{\bar{c}_1\bar{a} - c_0}{c_0\bar{a} - c_1}. \tag{11.4.4}$$

This establishes the statement concerning $s_n(a, z)$ since $|c_1| < c_0$ (see (11.1.2)). The statement concerning $\phi_n(z)$ follows from (11.3.5).

(2) THEOREM 11.4.2. *We have the identity*

$$s_n(a, z) = \sum_{\nu=0}^{n} \overline{\phi_\nu(a)}\phi_\nu(z) = \frac{\overline{\phi_{n+1}^*(a)}\phi_{n+1}^*(z) - \overline{\phi_{n+1}(a)}\phi_{n+1}(z)}{1 - \bar{a}z} \tag{11.4.5}$$

and the "recurrence formulas"

$$\kappa_n z\phi_n(z) = \kappa_{n+1}\phi_{n+1}(z) - \phi_{n+1}(0)\phi_{n+1}^*(z), \tag{11.4.6}$$

$$\kappa_n\phi_{n+1}(z) = \kappa_{n+1}z\phi_n(z) + \phi_{n+1}(0)\phi_n^*(z). \tag{11.4.7}$$

The first identity corresponds, in some respects, to the Christoffel-Darboux formula (3.2.3). The proof can be given along the same line as the proof in §3.2 (3). As in the case of a real interval, we can characterize $s_n(a, z)$ by the equation

$$\frac{1}{2\pi}\int_{-\pi}^{+\pi} f(\theta)s_n(a, z)\overline{\rho(z)}\, d\theta = \overline{\rho(a)}, \qquad z = e^{i\theta}, \tag{11.4.8}$$

which holds if $\rho(z)$ is an arbitrary π_n. But

$$\begin{aligned} &\frac{1}{2\pi}\int_{-\pi}^{+\pi} f(\theta)\, \frac{\overline{\phi_{n+1}^*(a)}\phi_{n+1}^*(z) - \overline{\phi_{n+1}(a)}\phi_{n+1}(z)}{1 - \bar{a}z}\, \overline{\rho(z)}\, d\theta \\ &= \overline{\rho(a)}\, \frac{1}{2\pi}\int_{-\pi}^{+\pi} f(\theta)\, \frac{\overline{\phi_{n+1}^*(a)}\phi_{n+1}^*(z) - \overline{\phi_{n+1}(a)}\phi_{n+1}(z)}{1 - \bar{a}z}\, d\theta \\ &+ \frac{1}{2\pi}\int_{-\pi}^{+\pi} f(\theta)\{\overline{\phi_{n+1}^*(a)}\phi_{n+1}^*(z) - \overline{\phi_{n+1}(a)}\phi_{n+1}(z)\}\, \frac{\overline{\rho(z)} - \overline{\rho(a)}}{1 - \bar{a}z}\, d\theta, \\ &\qquad z = e^{i\theta}. \end{aligned} \tag{11.4.9}$$

The last integral vanishes, for if we write $\rho(z) - \rho(a) = (z - a)r(z)$, we have

$$\int_{-\pi}^{+\pi} f(\theta)\phi_{n+1}^*(z)\bar{z}\overline{r(z)}\, d\theta = \int_{-\pi}^{+\pi} f(\theta)\phi_{n+1}(z)\bar{z}\overline{r(z)}\, d\theta = 0, \qquad z = e^{i\theta}. \tag{11.4.10}$$

Therefore,

$$\frac{\overline{\phi_{n+1}^*(a)}\phi_{n+1}^*(z) - \overline{\phi_{n+1}(a)}\phi_{n+1}(z)}{1 - \bar{a}z} = cs_n(a, z), \tag{11.4.11}$$

where c is independent of z. Interchanging a and z and taking the conjugate complex values of both sides, we see that c is also independent of a. Writing $z = a = 0$, we obtain

$$\kappa_{n+1}^2 - |\phi_{n+1}(0)|^2 = c\sum_{\nu=0}^{n} |\phi_\nu(0)|^2, \tag{11.4.12}$$

so that $c = 1$, by (11.3.6).

Comparison of the coefficients of $\bar{a}^n$ in (11.4.5) leads to (11.4.6). By taking the reciprocal polynomials of both sides of (11.4.6), and eliminating $\phi^*_{n+1}(z)$, we find (11.4.7).

11.5. Relation to polynomials orthogonal on a real interval

(1) THEOREM 11.5. *Let $w(x)$ be a weight function on the interval $-1 \leqq x \leqq +1$, and let*

$$f(\theta) = w(\cos\theta)\,|\sin\theta|. \tag{11.5.1}$$

Further let $\{p_n(x)\}$ and $\{q_n(x)\}$ be the orthonormal sets of polynomials which belong respectively to $w(x)$ and $(1 - x^2)w(x)$ in $-1 \leqq x \leqq +1$, and $\{\phi_n(z)\}$ the orthonormal set associated with $f(\theta)$ on $z = e^{i\theta}$. Then, on writing $x = \frac{1}{2}(z + z^{-1})$, we have for $n \geqq 1$

$$\begin{aligned} p_n(x) &= (2\pi)^{-\frac{1}{2}}\left\{1 + \frac{\phi_{2n}(0)}{\kappa_{2n}}\right\}^{-\frac{1}{2}}\{z^{-n}\phi_{2n}(z) + z^n\phi_{2n}(z^{-1})\} \\ &= (2\pi)^{-\frac{1}{2}}\left\{1 - \frac{\phi_{2n}(0)}{\kappa_{2n}}\right\}^{-\frac{1}{2}}\{z^{-n+1}\phi_{2n-1}(z) + z^{n-1}\phi_{2n-1}(z^{-1})\}; \\ q_n(x) &= (2/\pi)^{\frac{1}{2}}\left\{1 - \frac{\phi_{2n+2}(0)}{\kappa_{2n+2}}\right\}^{-\frac{1}{2}}\frac{z^{-n-1}\phi_{2n+2}(z) - z^{n+1}\phi_{2n+2}(z^{-1})}{z - z^{-1}} \\ &= (2/\pi)^{\frac{1}{2}}\left\{1 + \frac{\phi_{2n+2}(0)}{\kappa_{2n+2}}\right\}^{-\frac{1}{2}}\frac{z^{-n}\phi_{2n+1}(z) - z^{n}\phi_{2n+1}(z^{-1})}{z - z^{-1}}. \end{aligned} \tag{11.5.2}$$

See Szegö **6**, pp. 204–206. The second equation follows from the first one, and similarly, the fourth follows from the third, by means of (11.4.7). The function $f(\theta)$ is even; so the polynomials $\phi_n(z)$ have real coefficients.

The constant factors in these equations are different from zero (see (11.3.6)).

These formulas, except the second one, hold also for $n = 0$.

(2) The right member of the first equation represents a π_n in x; the property of orthogonality can be expressed in the form

$$\int_{-\pi}^{+\pi} p_n(\cos\theta)\cos\nu\theta\cdot w(\cos\theta)\,|\sin\theta|\,d\theta = 0;$$

or, for $z = e^{i\theta}$,

$$\int_{-\pi}^{+\pi}\{z^{-n}\phi_{2n}(z) + z^n\phi_{2n}(z^{-1})\}\{z^\nu + z^{-\nu}\}f(\theta)d\theta = 0, \qquad \nu = 0, 1, 2, \cdots, n-1.$$

This is the case because

$$\int_{-\pi}^{+\pi}\phi_{2n}(z)(\bar{z}^{n-\nu} + \bar{z}^{n+\nu})f(\theta)d\theta = 0,$$

and $\bar{\phi}_{2n}(z) = \phi_{2n}(\bar{z})$. Moreover,

$$\int_0^{\pi} |z^{-n}\phi_{2n}(z) + z^n\phi_{2n}(z^{-1})|^2 f(\theta)\, d\theta$$

$$= \pi + \pi + \Re\left\{\int_{-\pi}^{+\pi} z^{-n}\phi_{2n}(z)\cdot\overline{z^n\phi_{2n}(z^{-1})}f(\theta)\, d\theta\right\}$$

$$= 2\pi + \Re\left\{\int_{-\pi}^{+\pi} \phi_{2n}(z)\overline{z^{2n}\phi_{2n}(z^{-1})}f(\theta)\, d\theta\right\} = 2\pi + 2\pi\frac{\phi_{2n}(0)}{\kappa_{2n}}.$$

For the third formula the proof is similar, except that it is convenient now to express the orthogonality in the form

$$\int_{-\pi}^{+\pi} q_n(\cos\theta)\frac{\sin(\nu+1)\theta}{\sin\theta}\sin^2\theta\cdot w(\cos\theta)\,|\sin\theta|\, d\theta = 0,$$

$$\nu = 0, 1, 2, \cdots, n-1.$$

A new proof for the formulas (2.6.2) and (2.6.3) can be derived from the first and third formulas in (11.5.2) by use of (11.2.2).

If in the last two equations of (11.5.2) we replace n by $n - 1$, we can express $z^{-n}\phi_{2n}(z)$ and $z^{-n+1}\phi_{2n-1}(z)$ as linear combinations of $p_n(x)$ and $(1-x^2)^{1/2}q_{n-1}(x)$, where $x = \frac{1}{2}(z + z^{-1})$. (See Szegö **18**, pp. 9–11.) These relations enable us to calculate the polynomials $\phi_n(z)$ associated with

$$(11.5.3)\quad f(\theta) = |(1-z)^\gamma(1+z)^\delta|^2 = 2^{\gamma+\delta}(1-\cos\theta)^\gamma(1+\cos\theta)^\delta,\quad z = e^{i\theta},$$

in terms of Jacobi polynomials. We find

$$z^{-n}\phi_{2n}(z) = AP_n^{(\gamma-\frac{1}{2},\delta-\frac{1}{2})}\{\tfrac{1}{2}(z+z^{-1})\} + B(z-z^{-1})P_{n-1}^{(\gamma+\frac{1}{2},\delta+\frac{1}{2})}\{\tfrac{1}{2}(z+z^{-1})\},$$

$$(11.5.4)\quad z^{-n+1}\phi_{2n-1}(z) = CP_n^{(\gamma-\frac{1}{2},\delta-\frac{1}{2})}\{\tfrac{1}{2}(z+z^{-1})\}$$

$$+ D(z-z^{-1})P_{n-1}^{(\gamma+\frac{1}{2},\delta+\frac{1}{2})}\{\tfrac{1}{2}(z-z^{-1})\},$$

where A, B, C, D are proper real constants.

CHAPTER XII

ASYMPTOTIC PROPERTIES OF GENERAL ORTHOGONAL POLYNOMIALS

The following sections deal with the asymptotic properties of polynomials orthogonal on the unit circle, or on a real, finite interval, when the degree n of these polynomials becomes infinite. In both cases the weight function will be restricted merely by certain conditions of continuity and boundedness.

Two important problems appear in connection with polynomials orthogonal on the unit circle. These are (a) the asymptotic behavior exterior to the unit circle, (b) the behavior on the unit circle itself. For a weight function identically unity, the system in question is $\{z^n\}$. This latter instance is typical to a certain extent.

The corresponding problems for polynomials orthogonal on a finite segment are (a′) the asymptotic behavior in the complex plane cut along the given segment, (b′) the corresponding question on the segment itself. (See Chapter VIII.) We give the following illustration as characteristic:

$$T_n(x) = \tfrac{1}{2}(z^n + z^{-n}), \qquad x = \tfrac{1}{2}(z + z^{-1}).$$

Problems (a) and (a′) are simpler, and relative to them comparatively general results will be obtained. It is only recently that problems (b) and (b′), which are much more difficult than (a) and (a′), have been treated by S. Bernstein and G. Szegö. We remark that the weight function conditions in case (b) are more restrictive than those in case (a). A similar comment may be made concerning (a′) and (b′).

The results of Chapter X are applied in discussing the above problems. Our investigation is first concerned with questions (a) and (a′). As regards (b′) we may state that S. Bernstein's main result is obtained in a new way which is shorter than his original argument. Then there follows Szegö's older method as applied to (b′).

12.1. Results

(1) Let G denote the class of functions $f(\theta) \geqq 0$, defined and measurable in $[-\pi, +\pi]$, for which the integrals

$$(12.1.1) \qquad \int_{-\pi}^{+\pi} f(\theta)\, d\theta, \qquad \int_{-\pi}^{+\pi} |\log f(\theta)|\, d\theta$$

exist with the first integral supposed positive. With such a function $f(\theta)$ we associated in §10.2 a uniquely determined analytic function $D(f; z) = D(z)$, regular and nonzero for $|z| < 1$ with $D(0) > 0$. The conditions for the class G imply the existence of the "geometric mean" $\mathfrak{G}(f) = \{D(0)\}^2$ of $f(\theta)$.

THEOREM 12.1.1 (Asymptotic formula for the polynomials orthogonal on the unit circle, considered for z *exterior* to the unit circle). *Let $f(\theta)$, belonging to the class G, be a weight function on the unit circle $z = e^{i\theta}$. If $\{\phi_n(z)\}$ denotes the corresponding orthonormal set of polynomials, then in the exterior of the unit circle*

$$\phi_n(z) \cong z^n\{\bar{D}(z^{-1})\}^{-1}, \qquad |z| > 1. \tag{12.1.2}$$

This holds uniformly for $|z| \geqq R > 1$.

THEOREM 12.1.2 (Asymptotic formula for the polynomials orthogonal on the segment $[-1, +1]$, considered for x *exterior* to this segment). *Let $w(x)$ be a weight function on the interval $-1 \leqq x \leqq +1$ such that $w(\cos\theta)\,|\sin\theta| = f(\theta)$ belongs to the class G. If $D(f; z) = D(z)$ denotes the analytic function corresponding to $f(\theta)$ in the sense mentioned, the orthonormal polynomials $\{p_n(x)\}$, associated with $w(x)$, possess the asymptotic formula*

$$p_n(x) \cong (2\pi)^{-\frac{1}{2}} z^n\{D(z^{-1})\}^{-1}. \tag{12.1.3}$$

Here x is in the complex plane cut along the real segment $[-1, +1]$, and $x = \frac{1}{2}(z + z^{-1})$, where $|z| > 1$. Formula (12.1.3) *holds uniformly for $|z| \geqq R > 1$.*

(2) In order to obtain the deeper asymptotic formulas valid on $|z| = 1$ and on $-1 \leqq x \leqq +1$, respectively, we must impose certain further restrictions on $f(\theta)$ and $w(x)$.

THEOREM 12.1.3 (Asymptotic formula for the polynomials orthogonal on the unit circle, considered for z *on* the unit circle). *Let $f(\theta)$ be a positive weight function on the unit circle, which satisfies the Lipschitz-Dini condition*

$$|f(\theta + \delta) - f(\theta)| < L\,|\log\delta|^{-1-\lambda}, \tag{12.1.4}$$

where L and λ are fixed positive numbers. Then we have, for $|z| = 1$,

$$\phi_n(z) = z^n\{\bar{D}(z^{-1})\}^{-1} + \epsilon_n(z) = z^n\{\overline{D(z)}\}^{-1} + \epsilon_n(z), \tag{12.1.5}$$

where $\lim_{n\to\infty} \epsilon_n(z) = 0$, uniformly for $|z| = 1$. More precisely,

$$|\epsilon_n(z)| < C(\log n)^{-\lambda}; \tag{12.1.6}$$

the positive constant C depends on L, λ, and the minimum and maximum of $f(\theta)$.

THEOREM 12.1.4 (Asymptotic formula for the polynomials orthogonal on the segment $[-1, +1]$, considered for x *on* this segment). *Let $w(x)$ be a weight function on the interval $-1 \leqq x \leqq +1$, $x = \cos\theta$, such that the function $w(\cos\theta)\,|\sin\theta| = f(\theta)$ satisfies the conditions of Theorem* 12.1.3. *Putting*

$$\operatorname{sgn} D(e^{i\theta}) = |D(e^{i\theta})|^{-1} D(e^{i\theta}) = e^{i\gamma(\theta)}, \tag{12.1.7}$$

we have uniformly on the segment $-1 \leqq x \leqq +1$ or $0 \leqq \theta \leqq \pi$, $x = \cos\theta$,

$$(1 - x^2)^{\frac{1}{4}}(w(x))^{\frac{1}{2}} p_n(x) = (2/\pi)^{\frac{1}{2}} \cos\{n\theta + \gamma(\theta)\} + O\{(\log n)^{-\lambda}\}. \tag{12.1.8}$$

The constant factor in the O-term depends only on L, λ, and the minimum and maximum of $f(\theta)$.

(3) Finally, we prove two theorems similar to Theorems 12.1.3 and 12.1.4 under conditions possessing a kind of "local" character.

THEOREM 12.1.5 (Asymptotic formula for the polynomials orthogonal on the unit circle for $|z| = 1$, and with weight function subject to a "local" condition). *Let $f(\theta)$ satisfy the conditions of Theorem* 10.4.1. *Then* (12.1.5) *holds for $z = a = e^{i\alpha}$, in the less informative form*

$$\phi_n(a) = a^n\{\overline{D(a)}\}^{-1} + \epsilon_n, \qquad \epsilon_n \to 0. \tag{12.1.9}$$

From this we obtain the following theorem of "local" character, which corresponds to Theorem 12.1.4:

THEOREM 12.1.6 (Asymptotic formula for the polynomials orthogonal on the segment $[-1, +1]$ for x on this segment, and with weight function restricted by a "local" condition). *Let $w(x)$ be integrable in Riemann's sense, and let it have the form*

$$w(x) = t(x)\,|x - x_1|^{\tau_1}\,|x - x_2|^{\tau_2} \cdots |x - x_l|^{\tau_l}, \tag{12.1.10}$$

where $0 < A \leqq t(x) \leqq B$, *and* $-1 \leqq x_1 < x_2 < \cdots < x_l \leqq 1$, $\tau_\nu > 0$, $\nu = 1, 2, \cdots, l$.[59] *Further let $w(x)$ be differentiable at the fixed point $x = \xi$, where $-1 < \xi < +1$, and $\xi \neq x_\nu$, $\nu = 1, 2, \cdots, l$, and let*

$$\frac{w(x) - w(\xi) - (x - \xi)w'(\xi)}{(x - \xi)^2} \tag{12.1.11}$$

be bounded if x is near to ξ.

Then (12.1.8) *holds in the less informative form*

$$\begin{gathered}(1 - \xi^2)^{\frac{1}{4}}(w(\xi))^{\frac{1}{2}}p_n(\xi) = (2/\pi)^{\frac{1}{2}} \cos\{n\alpha + \gamma(\alpha)\} + \epsilon_n, \\ \xi = \cos\alpha,\ 0 < \alpha < \pi,\ \lim_{n\to\infty} \epsilon_n = 0,\end{gathered} \tag{12.1.12}$$

where $\gamma(\alpha)$ has the same meaning as in (12.1.7).

12.2. Remarks

(1) Theorems 12.1.2, 12.1.4, and 12.1.6 follow readily from 12.1.1, 12.1.3, and 12.1.5, respectively. We observe that $D(z) = \bar{D}(z)$ in (12.1.3); that is, $D(z)$ is in this case a "real" function.

In Theorems 12.1.3-12.1.6 the function $D(z)$ has boundary values at the point considered (in Theorems 12.1.3 and 12.1.4 even continuous boundary values on the whole unit circle $|z| = 1$). This follows from the considerations of Chapter X.

The important function $\gamma(\theta)$, defined by (12.1.7), is completely determined

[59] For $x_1 = -1$ it suffices to assume $\tau_1 \geqq -1/2$; similarly for $x_l = +1$.

save for an integral multiple of 2π. If we choose (see (10.3.9))

$$\gamma(\theta) = \frac{1}{4\pi}\int_{-\pi}^{+\pi} [\log f(t) - \log f(\theta)] \cot \frac{\theta - t}{2}\, dt, \tag{12.2.1}$$

then $\gamma(\theta)$ is continuous. In the cases occurring in Theorems 12.1.4 and 12.1.6, $\gamma(\theta)$ can easily be expressed in terms of the weight function $w(x)$. Let

$$f(\theta) = w(\cos\theta) \mid \sin\theta \mid = W(\cos\theta). \tag{12.2.2}$$

From (10.3.9) we obtain (see S. Bernstein **2**, p. 132),

$$\begin{aligned}
\gamma(\theta) &= \frac{1}{4\pi}\int_{-\pi}^{+\pi} \{\log W(\cos t) - \log W(\cos\theta)\} \cot\frac{\theta - t}{2}\, dt \\
&= \frac{1}{4\pi}\int_{0}^{\pi} \{\log W(\cos t) - \log W(\cos\theta)\} \left\{\cot\frac{\theta - t}{2} + \cot\frac{\theta + t}{2}\right\} dt \\
&= \frac{1}{2\pi}\int_{0}^{\pi} \{\log W(\cos t) - \log W(\cos\theta)\} \frac{\sin\theta}{\cos t - \cos\theta}\, dt \\
&= \frac{1}{2\pi}\int_{-1}^{+1} \frac{\log W(\xi) - \log W(x)}{\xi - x} \left(\frac{1 - x^2}{1 - \xi^2}\right)^{\frac{1}{2}} d\xi, \qquad x = \cos\theta.
\end{aligned} \tag{12.2.3}$$

In this case

$$\gamma(-\theta) = -\gamma(\theta). \tag{12.2.4}$$

For the functions $\{(f(\theta))^{\frac{1}{2}}\phi_n(z)\}$, $z = e^{i\theta}$, $n = 0, 1, 2, \cdots$, which form an orthonormal system in the usual sense (see the Definition in §11.1 (2)), we obtain from (12.1.5) the simple asymptotic expression $z^n e^{i\gamma(\theta)} = e^{i\{n\theta + \gamma(\theta)\}}$.

(2) Theorem 12.1.1 is a direct consequence of Theorem 12.1.3 provided condition (12.1.4) is satisfied. In fact, the function $z^{-n}\phi_n(z) - \{\bar{D}(z^{-1})\}^{-1}$ is regular for $|z| > 1$ and continuous for $|z| \geqq 1$. In this special case, (12.1.2) follows in the more informative form

$$z^{-n}\phi_n(z) = \{\bar{D}(z^{-1})\}^{-1} + O\{(\log n)^{-\lambda}\}, \tag{12.2.5}$$

uniformly for $|z| \geqq 1$.

(3) Concerning Theorems 12.1.1 and 12.1.2, see Szegö **6**. Under more restrictive conditions than those in Theorem 12.1.2, Faber (**4**) proved

$$\lim_{n\to\infty} |p_n(x)|^{1/n} = |z|, \qquad |z| > 1. \tag{12.2.6}$$

This less informative statement suffices for various applications, for instance, for the purposes of §12.7 (2) and (3). Theorem 12.1.3 is new, while 12.1.4 is due to S. Bernstein (**2**). The proofs of Theorems 12.1.3 and 12.1.4, given in §12.4 and 12.5 (2), respectively, are based essentially on an idea of S. Bernstein, used in his original proof, and on another idea similar to that used in connection with the method of Liouville-Stekloff (§8.61). As mentioned, this arrangement seems to be simpler than S. Bernstein's original line of argument. Theorem 12.1.5, and Theorem 12.1.6 which is its consequence, are due to Szegö (8). The conditions required in the present treatment are slightly more general than those in the paper just cited.

S. Bernstein assumes (2, p. 132, (18))

$$|W(x+\delta) - W(x)| < L\,|\log \delta|^{-1-\lambda}; \quad -1 \leqq x \leqq +1, \; -1 \leqq x+\delta \leqq +1,$$

instead of (12.1.4). These conditions are, however, equivalent since

$$\frac{\log\{|\cos\theta_1 - \cos\theta_2|^{-1}\}}{\log\{|\theta_1 - \theta_2|^{-1}\}}$$

is bounded from 0 and ∞ if θ_1 and θ_2 are arbitrary in $[0, \pi]$, $|\theta_1 - \theta_2| \leqq \frac{1}{2}$.[60]

12.3. Proof of Theorem 12.1.1; applications

(1) Before we proceed to this proof, let us consider the special case $f(\theta) = \{g(\theta)\}^{-1}$, where $g(\theta)$ is a positive trigonometric polynomial of degree m. Let $h(z)$ have the same meaning as in §10.2 (1). According to the second formula in (10.2.12) we have $D(f; z) = \{h(z)\}^{-1}$. On the other hand, by (11.2.2)

$$\phi_n(z) = z^n \bar{h}(z^{-1}) = z^n\{\bar{D}(z^{-1})\}^{-1}, \qquad n \geqq m. \tag{12.3.1}$$

The limit relation (12.1.2) *can therefore be replaced by one of equality in this case provided* $n \geqq m$.

(2) Now let $f(\theta)$ again be an arbitrary function satisfying the conditions of Theorem 12.1.1. Let $\rho(z) = z^n + a_1 z^{n-1} + \cdots + a_n$ be an arbitrary π_n with the highest term z^n. According to Theorem 11.1.2, the minimum $\mu_n(f)$ of

$$\frac{1}{2\pi}\int_{-\pi}^{+\pi} f(\theta)\,|\rho(z)|^2\,d\theta, \qquad z = e^{i\theta}, \tag{12.3.2}$$

is κ_n^{-2}, attained for $\rho(z) = \kappa_n^{-1}\phi_n(z)$.

LEMMA. *Let* $f(\theta)$ *satisfy the conditions of Theorem* 12.1.1, *and let* $\mu_n(f)$ *have the previous meaning. Then*

$$\lim_{n\to\infty} \mu_n(f) = \mathfrak{G}(f), \tag{12.3.3}$$

where $\mathfrak{G}(f)$ *is the "geometric mean" of* $f(\theta)$.[61]

If $\rho(z)$ is any one of the polynomials considered, then $z\rho(z)$ is a π_{n+1} with the highest term z^{n+1}, and $|z\rho(z)|^2 = |\rho(z)|^2$ if $z = e^{i\theta}$. Thus $\mu_{n+1}(f) \leqq \mu_n(f)$. (This follows also from (11.3.6).) Consequently, $\lim_{n\to\infty}\mu_n(f) = \mu(f)$ exists, and $\mu(f) \geqq 0$. We must show that $\mu(f) = \mathfrak{G}(f)$.

(3) Using the inequality for the arithmetic and geometric means, we have for $z = e^{i\theta}$,

[60] If $|\theta_1 - \theta_2| = \delta$, $\delta \leqq \pi/2$, is given, the maximum and minimum of $|\cos\theta_1 - \cos\theta_2|$ are $2\sin(\delta/2)$ and $2\sin^2(\delta/2)$, respectively.

[61] Cf. §10.2.

$$\frac{1}{2\pi}\int_{-\pi}^{+\pi} f(\theta)\,|\,\rho(z)\,|^2\,d\theta \geqq \mathfrak{G}(f)\exp\left\{\frac{1}{2\pi}\int_{-\pi}^{+\pi}\log|\,\rho(z)\,|^2\,d\theta\right\}$$

(12.3.4)

$$= \mathfrak{G}(f)\exp\left\{\frac{1}{2\pi}\int_{-\pi}^{+\pi}\log|\,\rho^*(z)\,|^2\,d\theta\right\} \geqq \mathfrak{G}(f)\,|\,\rho^*(0)\,|^2 = \mathfrak{G}(f),$$

according to Jensen's theorem (see, for example, Titchmarsh **1**, p. 125). Therefore, $\mu_n(f) \geqq \mathfrak{G}(f)$, and $\mu(f) \geqq \mathfrak{G}(f)$.

On the other hand, let $T(\theta)$ be a non-negative trigonometric polynomial of degree k, not vanishing identically, and let $T(\theta) = |\,P(e^{i\theta})\,|^2$ be the corresponding normalized representation in the sense of Theorem 1.2.2. Then $\mathfrak{G}(T) = \{P(0)\}^2$. The highest coefficient of $\{\overline{P}(0)\}^{-1}P^*(z)$ is 1, so that

$$\mathfrak{G}(f) \leqq \mu(f) \leqq \mu_k(f) \leqq \frac{1}{2\pi}\int_{-\pi}^{+\pi} f(\theta)\,|\,\{\overline{P}(0)\}^{-1}P^*(z)\,|^2\,d\theta$$

(12.3.5)

$$= \{\mathfrak{G}(T)\}^{-1}\frac{1}{2\pi}\int_{-\pi}^{+\pi} f(\theta)T(\theta)\,d\theta, \qquad z = e^{i\theta}.$$

By Weierstrass' theorem, the inequality

$$\mathfrak{G}(f) \leqq \mu(f) \leqq \{\mathfrak{G}(T)\}^{-1}\frac{1}{2\pi}\int_{-\pi}^{+\pi} f(\theta)T(\theta)\,d\theta \tag{12.3.6}$$

holds for an arbitrary positive and continuous function $T(\theta)$ which has the period 2π. For the special case in which $f(\theta)$ is positive and continuous, the lemma follows from this by writing $T(\theta) = \{f(\theta)\}^{-1}$.

(4) Now in the general case assume first $f(\theta) \geqq \mu > 0$, and let ϵ be an arbitrary positive number. By Theorem 1.5.3, we can find a trigonometric polynomial $\phi(\theta)$ such that $\phi(\theta) \geqq \mu$ and

$$\frac{1}{2\pi}\int_{-\pi}^{+\pi} |\,f(\theta) - \phi(\theta)\,|\,d\theta < \epsilon. \tag{12.3.7}$$

We then have

$$\log\phi(\theta) - \log f(\theta) \leqq \mu^{-1}\,|\,\phi(\theta) - f(\theta)\,|; \tag{12.3.8}$$

hence $\mathfrak{G}(\phi) \leqq \mathfrak{G}(f)e^{\mu^{-1}\epsilon}$. In (12.3.6) we write $T(\theta) = \{\phi(\theta)\}^{-1}$. Then

$$\left|\frac{1}{2\pi}\int_{-\pi}^{+\pi} f(\theta)T(\theta)\,d\theta - 1\right| \leqq \mu^{-1}\frac{1}{2\pi}\int_{-\pi}^{+\pi}|\,f(\theta) - \phi(\theta)\,|\,d\theta < \mu^{-1}\epsilon, \tag{12.3.9}$$

so that

$$\mathfrak{G}(f) \leqq \mu(f) \leqq \mathfrak{G}(f)e^{\mu^{-1}\epsilon}(1 + \mu^{-1}\epsilon); \tag{12.3.10}$$

whence, since ϵ is arbitrary, $\mu(f) = \mathfrak{G}(f)$.

In the general case we use the obvious inequality $\mu(f) \leqq \mu(f + \epsilon) = \mathfrak{G}(f + \epsilon)$, $\epsilon > 0$, from which $\mu(f) = \mathfrak{G}(f)$ follows again. This establishes the proof of the lemma.

(5) For the proof of Theorem 12.1.1 let us now consider the function

$$D(z)\phi_n^*(z) - 1 = [D(0)\kappa_n - 1] + d_{n1}z + d_{n2}z^2 + \cdots, \tag{12.3.11}$$

which is regular for $|z| < 1$. If $r < 1$, we have

$$\begin{aligned} \frac{1}{2\pi}\int_{-\pi}^{+\pi} |D(re^{i\theta})\phi_n^*(re^{i\theta}) - 1|^2 d\theta &= \frac{1}{2\pi}\int_{-\pi}^{+\pi} |D(re^{i\theta})|^2 |\phi_n^*(re^{i\theta})|^2 d\theta \\ &\quad +1 - 2\Re\left\{\frac{1}{2\pi}\int_{-\pi}^{+\pi} D(re^{i\theta})\phi_n^*(re^{i\theta})\, d\theta\right\}. \end{aligned} \tag{12.3.12}$$

The third term is evidently $-2\Re[D(0)\phi_n^*(0)] = -2D(0)\kappa_n$. As $r \to 1 - 0$, we obtain, by use of (10.2.9),

$$\begin{aligned} \lim_{r\to 1-0} \frac{1}{2\pi}\int_{-\pi}^{+\pi} |D(re^{i\theta})\phi_n^*(re^{i\theta}) - 1|^2 d\theta &= \frac{1}{2\pi}\int_{-\pi}^{+\pi} f(\theta)\, |\phi_n^*(e^{i\theta})|^2 d\theta + 1 - 2D(0)\kappa_n \\ &= \frac{1}{2\pi}\int_{-\pi}^{+\pi} f(\theta)\, |\phi_n(e^{i\theta})|^2 d\theta + 1 - 2D(0)\kappa_n = 2 - 2D(0)\kappa_n \\ &= 2 - 2\{\mathfrak{G}(f)\}^{\frac{1}{2}}\{\mu_n(f)\}^{-\frac{1}{2}}; \end{aligned}$$

or in another form,

$$\begin{aligned} |D(0)\kappa_n - 1|^2 + |d_{n1}|^2 + |d_{n2}|^2 + |d_{n3}|^2 + \cdots \\ = 2 - 2\{\mathfrak{G}(f)\}^{\frac{1}{2}}\{\mu_n(f)\}^{-\frac{1}{2}}. \end{aligned} \tag{12.3.13}$$

In consequence of Cauchy's inequality this yields, for $|z| < 1$,

$$\begin{aligned} |d_{n1}z + d_{n2}z^2 + d_{n3}z^3 + \cdots|^2 &\leqq (|d_{n1}|^2 + |d_{n2}|^2 + |d_{n3}|^2 + \cdots)\frac{|z|^2}{1 - |z|^2} \\ &\leqq \frac{|z|^2}{1 - |z|^2}[2 - 2\{\mathfrak{G}(f)\}^{\frac{1}{2}}\{\mu_n(f)\}^{-\frac{1}{2}}], \end{aligned} \tag{12.3.14}$$

and as $n \to \infty$, the last expression tends to 0 uniformly in z for $|z| \leqq r < 1$. The same is true for $|D(0)\kappa_n - 1|^2$. Theorem 12.1.1 now follows at once from (12.3.11).

(6) As an application we calculate the limits occurring in Theorem 11.3.3. The second limit in (11.3.12) is given by (12.1.2); the special case $z = \infty$ yields

$$\lim_{n\to\infty} \kappa_n = \lim_{n\to\infty}\lim_{z\to\infty} \{z^{-n}\phi_n(z)\} = \{D(0)\}^{-1} = \{\mathfrak{G}(f)\}^{-\frac{1}{2}}. \tag{12.3.15}$$

The same formula (12.1.2) furnishes, for $|z| < 1$,

$$\lim_{n\to\infty} \phi_{n+1}^*(z) = \lim_{n\to\infty} z^{n+1}\bar{\phi}_{n+1}(z^{-1}) = \{D(z)\}^{-1}, \tag{12.3.16}$$

and a similar formula holds for $\phi_{n+1}^*(a)$, $|a| < 1$. Since $\phi_{n+1}(a) \to 0$ and $\phi_{n+1}(z) \to 0$ (see (11.3.13)), we have from (11.4.5)

$$\lim_{n\to\infty} s_n(a, z) = \sum_{\nu=0}^{\infty} \overline{\phi_\nu(a)}\phi_\nu(z) = \frac{1}{1-\bar{a}z}\frac{1}{\overline{D(a)}}\frac{1}{D(z)}. \tag{12.3.17}$$

For instance,

$$\sum_{\nu=0}^{\infty} |\phi_\nu(z)|^2 = \frac{1}{1-|z|^2}\frac{1}{|D(z)|^2}, \qquad |z| < 1, \tag{12.3.18}$$

and in particular (see (11.3.6))

$$\sum_{\nu=0}^{\infty} |\phi_\nu(0)|^2 = \lim_{n\to\infty} \kappa_n^2 = \lim_{n\to\infty} \frac{D_{n-1}}{D_n} = \frac{1}{|D(0)|^2} = \frac{1}{\mathfrak{G}(f)}. \tag{12.3.19}$$

From the last equation (12.3.15) follows again.

12.4. Proof of Theorem 12.1.3

(1) Let $g(\theta)$ be a positive trigonometric polynomial of degree m, determined as in §10.3 (3), and let $D(g; z) = h(z)$, as there, so that $D(g^{-1}; z) = \{h(z)\}^{-1}$. Let $\{\psi_n(z)\}$ be the orthonormal set of polynomials associated with the weight function $\{g(\theta)\}^{-1}$ on the unit circle. We have, by Theorem 11.2,

$$\psi_n(z) = z^n\bar{h}(z^{-1}), \qquad n \geqq m. \tag{12.4.1}$$

In close relationship with an idea of S. Bernstein (**2**, p. 158, (75)), we shall now express the polynomial $\phi_m(z)$, associated with $f(\theta)$, in terms of the polynomials $\psi_\nu(z)$, corresponding to $\{g(\theta)\}^{-1}$:

$$\begin{aligned}\phi_m(z) &= \sum_{\nu=0}^{m} \alpha_\nu \psi_\nu(z) = \alpha_m\psi_m(z) + \frac{1}{2\pi}\int_{-\pi}^{+\pi} \{g(t)\}^{-1}\phi_m(\zeta)\left\{\sum_{\nu=0}^{m-1}\overline{\psi_\nu(\zeta)}\psi_\nu(z)\right\}dt \\ &= \alpha_m\psi_m(z) + \frac{1}{2\pi}\int_{-\pi}^{+\pi} [\{g(t)\}^{-1} - f(t)]\phi_m(\zeta)\left\{\sum_{\nu=0}^{m-1}\overline{\psi_\nu(\zeta)}\psi_\nu(z)\right\}dt \\ &\quad + \frac{1}{2\pi}\int_{-\pi}^{+\pi} f(t)\phi_m(\zeta)\left\{\sum_{\nu=0}^{m-1}\overline{\psi_\nu(\zeta)}\psi_\nu(t)\right\}dt, \qquad \zeta = e^{it}.\end{aligned} \tag{12.4.2}$$

The last term vanishes because of the orthogonality of $\phi_m(z)$.

Let κ_m and $\kappa_m' = h(0)$ be the highest coefficients of $\phi_m(z)$ and $\psi_m(z)$, respectively; then $\alpha_m = \kappa_m\kappa_m'^{-1} = \kappa_m\{h(0)\}^{-1}$. Furthermore, by §12.3 (2), we have

$$\begin{aligned}\{D(0)\}^2 = \mathfrak{G}(f) \leqq \mu_m(f) = \kappa_m^{-2} &\leqq \frac{1}{2\pi}\int_{-\pi}^{+\pi} f(\theta)\,|\{h(0)\}^{-1}h^*(z)|^2\,d\theta \\ &= \{h(0)\}^{-2}\frac{1}{2\pi}\int_{-\pi}^{+\pi} |D(z)h(z)|^2\,d\theta, \qquad z = e^{i\theta};\end{aligned}$$

and because of (10.3.12) this equals

$$\{h(0)\}^{-2}\{1 + O[(\log m)^{-\lambda}]\} = \{D(0)\}^2 + O[(\log m)^{-\lambda}],$$

so that

$$\kappa_m = \{D(0)\}^{-1} + O[(\log m)^{-\lambda}], \qquad \alpha_m = 1 + O[(\log m)^{-\lambda}]. \tag{12.4.3}$$

(2) We shall next try to find a bound for

$$\max_{|z|=1} |\phi_m(z)| = M = M(m). \tag{12.4.4}$$

Using (10.3.11), we find from (12.4.2) that

$$M \leqq O(1) + O[(\log m)^{-1-\lambda}] \cdot M \cdot \max_{|z|=1} \int_{-\pi}^{+\pi} \left| \sum_{\nu=0}^{m-1} \overline{\psi_\nu(\zeta)}\psi_\nu(z) \right| dt, \quad \zeta = e^{it}. \tag{12.4.5}$$

Because of (11.4.5), the sum under the integral sign can be represented in the form

$$\sum_{\nu=0}^{m-1} \overline{\psi_\nu(\zeta)}\psi_\nu(z) = \frac{\overline{h(\zeta)}h(z) - \overline{h^*(\zeta)}h^*(z)}{1 - \bar{\zeta}z}. \tag{12.4.6}$$

We now show that

$$\int_{-\pi}^{+\pi} \left| \frac{\overline{h(\zeta)}h(z) - \overline{h^*(\zeta)}h^*(z)}{1 - \bar{\zeta}z} \right| dt = O(\log m), \qquad \zeta = e^{it}, \tag{12.4.7}$$

uniformly in z for $|z| = 1$. Indeed, the numerator is a π_m in z, which vanishes for $z = \zeta$. Therefore, the theorem of S. Bernstein (see M. Riesz **1**, especially p. 357) furnishes $O(m)$ for the integrand. Thus the contribution of the arc $|\zeta - z| \leqq m^{-1}$ is $O(1)$, while the complementary arc $|\zeta - z| > m^{-1}$ supplies

$$O(1) \int_{|\zeta - z| > m^{-1}} \frac{dt}{|1 - \bar{\zeta}z|} = O(\log m).$$

Returning to (12.4.5), we obtain

$$M \leqq O(1) + O[(\log m)^{-\lambda}]M,$$

so that $M = O(1)$. Hence, in view of (12.4.3), (12.4.1), (10.3.11), and (12.4.7), equation (12.4.2) yields

$$\phi_m(z) = \{1 + O[(\log m)^{-\lambda}]\}z^m\bar{h}(z^{-1}) + O[(\log m)^{-1-\lambda}]O(1)O(\log m). \tag{12.4.8}$$

The assertion of Theorem 12.1.3 now follows because of (10.3.12). The constants of all O-terms of this section depend only on L, λ, and the minimum and maximum of $f(\theta)$.

12.5. Asymptotic formulas for the polynomials on a finite segment; proof of Theorems 12.1.2 and 12.1.4

Theorems 12.1.2 and 12.1.4 follow from Theorems 12.1.1 and 12.1.3, respectively, almost immediately by using (11.5.2). It suffices to use the first formula.

(1) If x and z have the same meaning as in Theorem 12.1.2, we have

$$\begin{aligned} &\lim_{n\to\infty} z^{-2n}\phi_{2n}(z) = \{D(z^{-1})\}^{-1}; \qquad \lim_{n\to\infty} \phi_{2n}(z^{-1}) = 0; \\ &\lim_{n\to\infty} \phi_{2n}(0) = 0; \qquad \lim_{n\to\infty} \kappa_{2n} = \kappa > 0. \end{aligned} \tag{12.5.1}$$

Here we took into account (12.1.2), (11.3.13), and (11.3.12). If (11.5.2) is taken into consideration, this establishes Theorem 12.1.2.

(2) For the proof of Theorem 12.1.4, let n be an arbitrary integer, $m = n - 1$, and let $g(\theta)$, as well as $h(z)$, be determined as in §10.3 (3). Then by (12.1.5) and (12.1.6),

$$\phi_n(0) = \frac{1}{2\pi}\int_{-\pi}^{+\pi} \phi_n(z)\, d\theta = \frac{1}{2\pi}\int_{-\pi}^{+\pi} z^n[\{D(\bar{z})\}^{-1} - h(\bar{z})]\, d\theta$$

(12.5.2)

$$+ \frac{1}{2\pi}\int_{-\pi}^{+\pi} z^n h(\bar{z})\, d\theta + O[(\log n)^{-\lambda}], \quad z = e^{i\theta}.$$

The first integral on the right is $O[(\log n)^{-\lambda}]$, while the second vanishes, since $h(z)$ is a π_{n-1}. Therefore,

$$\phi_n(0) = O[(\log n)^{-\lambda}]. \tag{12.5.3}$$

The sequence $\{\kappa_{2n}\}$ is bounded from 0 ($\kappa_n^2 \geqq |\phi_0(0)|^2 \geqq \{\max f(\theta)\}^{-1}$, according to (11.3.6)), so that

$$\begin{aligned} p_n(x) &= (2\pi)^{-\frac{1}{2}}\{1 + O[(\log n)^{-\lambda}]\}2\Re\{z^n[\overline{D(z)}]^{-1} + O[(\log n)^{-\lambda}]\} \\ &= (2/\pi)^{\frac{1}{2}}\, |D(e^{i\theta})|^{-1} \cos\{n\theta + \gamma(\theta)\} + O[(\log n)^{-\lambda}], \end{aligned} \tag{12.5.4}$$

$$x = \cos\theta, \; z = e^{i\theta},$$

which is identical with (12.1.8). The assertion concerning the constant of the O-term is immediate.

The same result is obtained from the second formula in (11.5.2).

12.6. The asymptotic problem under "local" conditions; proof of Theorems 12.1.5 and 12.1.6

In this section essential use is made of the approximations given in Theorems 10.4.4 and 10.4.5.

(1) First the following problem will be considered:

PROBLEM. *Let λ, μ, and a be arbitrary complex numbers, and let $f(\theta)$ be an arbitrary weight function on the unit circle. We intend to determine the maximum of*

$$|\lambda\rho(0) + \mu\rho(a)|^2, \tag{12.6.1}$$

when $\rho(z)$ ranges over the set of all π_n which satisfy the condition

$$\frac{1}{2\pi}\int_{-\pi}^{+\pi} f(\theta)\, |\rho(z)|^2\, d\theta = 1, \qquad z = e^{i\theta}. \tag{12.6.2}$$

Concerning the special case $\lambda = 0$, $\mu = 1$, see §11.3.

We write

$$\rho(z) = u_0\phi_0(z) + u_1\phi_1(z) + \cdots + u_n\phi_n(z).$$

where $\{\phi_n(z)\}$ is the orthonormal set associated with $f(\theta)$. Then $\sum_{\nu=0}^{n} |u_\nu|^2 = 1$, and with the notation of (11.3.3), according to Cauchy's inequality,

$$(12.6.3)\quad \begin{aligned} |\lambda\rho(0) + \mu\rho(a)|^2 &= \left|\sum_{\nu=0}^{n} u_\nu\{\lambda\phi_\nu(0) + \mu\phi_\nu(a)\}\right|^2 \leqq \sum_{\nu=0}^{n} |\lambda\phi_\nu(0) + \mu\phi_\nu(a)|^2 \\ &= |\lambda|^2 s_n(0, 0) + 2\Re\{\bar{\lambda}\mu s_n(0, a)\} + |\mu|^2 s_n(a, a). \end{aligned}$$

The expression in the right-hand member is the desired maximum.

(2) For clarity we now write $s_n(f; a, z)$ instead of $s_n(a, z)$ and use an analogous notation for the orthonormal polynomials $\phi_n(z) = \phi_n(f; z)$, associated with $f(\theta)$, and their highest coefficients $\kappa_n = \kappa_n(f)$. Then the preceding solution of the maximum problem shows immediately that for the functions $f_1(\theta)$, $f(\theta)$, $f_2(\theta)$ of Theorem 10.4.4,

$$(12.6.4)\quad \begin{aligned} &|\lambda|^2 s_n(f_1; 0, 0) + 2\Re\{\bar{\lambda}\mu s_n(f_1; 0, a)\} + |\mu|^2 s_n(f_1; a, a) \\ &\geqq |\lambda|^2 s_n(f; 0, 0) + 2\Re\{\bar{\lambda}\mu s_n(f; 0, a)\} + |\mu|^2 s_n(f; a, a) \\ &\geqq |\lambda|^2 s_n(f_2; 0, 0) + 2\Re\{\bar{\lambda}\mu s_n(f_2; 0, a)\} + |\mu|^2 s_n(f_2; a, a). \end{aligned}$$

In particular, we have for arbitrary a

$$(12.6.5)\quad s_n(f_1; a, a) \geqq s_n(f; a, a) \geqq s_n(f_2; a, a).$$

Furthermore, we find

$$(12.6.6)\quad \begin{aligned} &|s_n(f; 0, a) - s_n(f_2; 0, a)|^2 \\ &\leqq \{s_n(f; 0, 0) - s_n(f_2; 0, 0)\}\{s_n(f; a, a) - s_n(f_2; a, a)\} \\ &\leqq \{s_n(f_1; 0, 0) - s_n(f_2; 0, 0)\}\{s_n(f_1; a, a) - s_n(f_2; a, a)\}. \end{aligned}$$

Then, by virtue of (11.3.5) and (11.3.6),

$$\begin{aligned} &|\kappa_n(f)\phi_n^*(f; a) - \kappa_n(f_2)\phi_n^*(f_2; a)|^2 \\ &\leqq \{[\kappa_n(f_1)]^2 - [\kappa_n(f_2)]^2\}\{s_n(f_1; a, a) - s_n(f_2; a, a)\}. \end{aligned}$$

Here ϕ_n^* denotes the polynomial reciprocal to ϕ_n. If $|a| = 1$, the same inequality holds for ϕ_n, that is

$$(12.6.7)\quad \begin{aligned} &|\kappa_n(f)\phi_n(f; a) - \kappa_n(f_2)\phi_n(f_2; a)|^2 \\ &\leqq \{[\kappa_n(f_1)]^2 - [\kappa_n(f_2)]^2\}\{s_n(f_1; a, a) - s_n(f_2; a, a)\}. \end{aligned}$$

Now, if n is sufficiently large,

$$(12.6.8)\quad \phi_n(f_2; a) = a^n\{\overline{D(f_2; a)}\}^{-1}$$

(see Theorem 11.2). Next, according to (12.3.15),

$$(12.6.9)\quad \lim_{n\to\infty} \kappa_n(f) = \{\mathfrak{G}(f)\}^{-\frac{1}{2}} = \exp\left\{-\frac{1}{4\pi}\int_{-\pi}^{+\pi} \log f(\theta)\, d\theta\right\};$$

and similar relations hold for f_1, f_2. By Theorem 10.4.4,

$$\{\mathfrak{G}(f_1)\}^{-1} - \{\mathfrak{G}(f_2)\}^{-1}$$
$$= \exp\left\{-\frac{1}{2\pi}\int_{-\pi}^{+\pi} \log f_2(\theta)\,d\theta\right\}\left\{\exp\left(\frac{1}{2\pi}\int_{-\pi}^{+\pi} [\log f_2(\theta) - \log f_1(\theta)]\,d\theta\right) - 1\right\}$$
$$< \{\mathfrak{G}(f)\}^{-1}(e^{\epsilon/2\pi} - 1).$$

Thus

$$\limsup_{n\to\infty}\left|\phi_n(f;a) - \left\{\frac{\mathfrak{G}(f)}{\mathfrak{G}(f_2)}\right\}^{\frac{1}{2}} a^n\{\overline{D(f_2;a)}\}^{-1}\right|^2$$
$$\leqq (e^{\epsilon/2\pi} - 1)\limsup_{n\to\infty}\{s_n(f_1;a,a) - s_n(f_2;a,a)\},$$

or (see Theorem 10.4.4)

$$\text{(12.6.10)}\qquad \limsup_{n\to\infty}|\phi_n(f;a) - a^n\{\overline{D(f;a)}\}^{-1}|^2$$
$$< \epsilon'' + (e^{\epsilon/2\pi} - 1)\limsup_{n\to\infty}\{s_n(f_1;a,a) - s_n(f_2;a,a)\},$$

where ϵ'' is arbitrarily small with ϵ. This reduces the proof of the statement to the discussion of the difference

$$s_n(f_1\,;a,a) - s_n(f_2\,;a,a).$$

(3) Let us use the abbreviation $D(f_2^{-1};z) = h(z)$. We find from (11.4.5) and Theorem 11.2 that

$$\text{(12.6.11)}\qquad s_n(f_2;a,z) = \frac{\overline{h(a)}h(z) - (\bar{a}z)^{n+1}h(a)\bar{h}(z^{-1})}{1 - \bar{a}z}$$

provided n exceeds the degree of $\{f_2(\theta)\}^{-1} = g_2(\theta)$; consequently using l'Hospital's rule, $a = e^{i\alpha}$, we obtain

$$\text{(12.6.11')}\qquad s_n(f_2;a,a) = (n+1)\,|h(a)|^2 - 2\Re\{a\overline{h(a)}h'(a)\}$$
$$= \{f_2(\alpha)\}^{-1}\left\{n + 1 + 2\Re\left[a\,\frac{D'(f_2;a)}{D(f_2;a)}\right]\right\}.$$

(4) The discussion of $s_n(f_1\,;a,a)$ is slightly more complicated. From (10.4.24) it follows that

$$\text{(12.6.12)}\qquad D(f_1\,;z) = D(g_1^{-1};z)\{(1 - \bar{z}_1z)(1 - \bar{z}_2z)\cdots(1 - \bar{z}_lz)\}^{\sigma/2}.$$

Now let $\rho(z)$ range over the set of the π_n satisfying the condition

$$\text{(12.6.13)}\qquad \frac{1}{2\pi}\int_{-\pi}^{+\pi} f_1(\theta)\,|\rho(z)|^2\,d\theta = 1, \qquad z = e^{i\theta},$$

which can be written in the form

$$\frac{1}{2\pi}\int_{-\pi}^{+\pi} \{g_1(\theta)\}^{-1} |\{(1-\bar{z}_1 z)(1-\bar{z}_2 z)\cdots(1-\bar{z}_l z)\}^{\sigma/2}\rho(z)|^2 d\theta = 1, \quad z = e^{i\theta}. \tag{12.6.14}$$

Therefore, by Theorem 11.3.1, if $l' = \sigma l/2$,

$$\begin{aligned} s_{n+l'}(g_1^{-1}; a, a) &\geqq \max |(1-\bar{z}_1 a)(1-\bar{z}_2 a)\cdots(1-\bar{z}_l a)|^\sigma |\rho(a)|^2 \\ &= |(1-\bar{z}_1 a)(1-\bar{z}_2 a)\cdots(1-\bar{z}_l a)|^\sigma s_n(f_1; a, a). \end{aligned} \tag{12.6.15}$$

Making use of the previous result, we find that

$$\begin{aligned} s_n(f_1; a, a) \leqq |(1-\bar{z}_1 a)(1-\bar{z}_2 a)\cdots(1-\bar{z}_l a)|^{-\sigma} \\ \cdot g_1(\alpha)\left\{n + l' + 1 + 2\Re\left[a\frac{D'(g_1^{-1}; a)}{D(g_1^{-1}; a)}\right]\right\}, \end{aligned} \tag{12.6.16}$$

provided $n + l'$ exceeds the degree of $g_1(\theta)$. Now from (12.6.12) we have

$$\begin{aligned} a\frac{D'(g_1^{-1}; a)}{D(g_1^{-1}; a)} &= a\frac{D'(f_1; a)}{D(f_1; a)} - \frac{\sigma}{2}\sum_{\nu=1}^{l}\frac{-a\bar{z}_\nu}{1-\bar{z}_\nu a}, \\ 2\Re\left[a\frac{D'(g_1^{-1}; a)}{D(g_1^{-1}; a)}\right] &= 2\Re\left[a\frac{D'(f_1; a)}{D(f_1; a)}\right] - \frac{\sigma l}{2}. \end{aligned} \tag{12.6.17}$$

Consequently, the important inequality

$$s_n(f_1; a, a) \leqq \{f_1(\alpha)\}^{-1}\left\{n + 1 + 2\Re\left[a\frac{D'(f_1; a)}{D(f_1; a)}\right]\right\} \tag{12.6.18}$$

holds provided n is sufficiently large. Since $f_1(\alpha) = f_2(\alpha) = f(\alpha)$, we have

$$\begin{aligned} &\limsup_{n\to\infty} \{s_n(f_1; a, a) - s_n(f_2; a, a)\} \\ &\leqq \{f(\alpha)\}^{-1} 2\Re\left\{a\frac{D'(f_1; a)}{D(f_1; a)} - a\frac{D'(f_2; a)}{D(f_2; a)}\right\} \\ &\leqq 2\{f(\alpha)\}^{-1}\left|\frac{D'(f_1; a)}{D(f_1; a)} - \frac{D'(f_2; a)}{D(f_2; a)}\right|. \end{aligned} \tag{12.6.19}$$

But this expression is arbitrarily small with ϵ, which establishes the proof of Theorem 12.1.5.

(5) Under the assumptions of Theorem 12.1.6, the function

$$f(\theta) = w(\cos\theta)\,|\sin\theta| \tag{12.6.20}$$

satisfies the conditions of Theorem 12.1.5 or of Theorem 10.4.1. Indeed, we have, if $x_\nu = \cos\theta_\nu$, $0 \leqq \theta_\nu \leqq \pi$, $e^{i\theta_\nu} = \zeta_\nu$,

$$f(\theta) = 2^{-\tau_1-\tau_2-\cdots-\tau_l-1} t(\cos\theta)\left|(z^2-1)\prod_{\nu=1}^{l}(z-\zeta_\nu)^{\tau_\nu}(z-\bar{\zeta}_\nu)^{\tau_\nu}\right|, \quad z = e^{i\theta}; \tag{12.6.21}$$

whence

$$\phi(\theta) = 2^{-r_1-r_2-\cdots-r_l-1}t(\cos\theta). \tag{12.6.22}$$

We also observe that $f(\theta)$ is differentiable at $\theta = \alpha$ and that the ratio (10.4.2) is bounded near $\theta = \alpha$.

As in §12.5, we use Theorem 11.5, particularly the first formula in (11.5.2), and find that $\kappa_{2n} = \kappa_{2n}(f)$ tends to a positive limit and $\lim_{n\to\infty}\phi_{2n}(0) = 0$. Thus

$$p_n(\xi) = p_n(\cos\alpha) = (2/\pi)^{\frac{1}{2}}\{1 + \epsilon_n\}\Re[a^n\{\overline{D(f;a)}\}^{-1}], \quad \lim_{n\to\infty}\epsilon_n = 0.$$

By use of the function $\gamma(\alpha)$ discussed in §12.2 (1) we obtain

$$\Re[a^n\{\overline{D(f;a)}\}^{-1}] = |D(f;a)|^{-1}\Re\{a^n e^{i\gamma(\alpha)}\} = \{f(\alpha)\}^{-\frac{1}{2}}\cos\{n\alpha + \gamma(\alpha)\}.$$

This establishes Theorem 12.1.6.

The validity of the asymptotic formula (12.1.9) has been extended lately by G. Freud (**3**).

12.7. Applications

(1) By means of Theorem 12.1.2 we can readily derive certain asymptotic formulas for the highest coefficients of the orthonormal polynomials $\{p_n(x)\}$.

THEOREM 12.7.1. *Let $w(x)$ be a weight function on the interval $-1 \leqq x \leqq +1$, satisfying the conditions of Theorem 12.1.2, and let*

$$p_n(x) = k_{n0}x^n + k_{n1}x^{n-1} + k_{n2}x^{n-2} + \cdots, \qquad n = 0, 1, 2, \cdots, \tag{12.7.1}$$

be the associated orthonormal system. Then as $n \to \infty$,

$$k_{n0} \cong \pi^{-\frac{1}{2}}2^n \exp\left\{-\frac{1}{2\pi}\int_{-1}^{+1}\log w(x)\frac{dx}{(1-x^2)^{\frac{1}{2}}}\right\}, \tag{12.7.2}$$

and

$$k_{n1} \cong -\pi^{-\frac{1}{2}}2^{n-1}\int_{-1}^{+1}x\log w(x)\frac{dx}{(1-x^2)^{\frac{1}{2}}} \cdot\exp\left\{-\frac{1}{2\pi}\int_{-1}^{+1}\log w(x)\frac{dx}{(1-x^2)^{\frac{1}{2}}}\right\}. \tag{12.7.3}$$

Concerning these formulas, see Shohat **2**, p. 577. If the right-hand member of (12.7.3) vanishes, we read (12.7.3) as follows: $\lim_{n\to\infty}2^{-n}k_{n1} = 0$. There is no difficulty in deriving corresponding formulas for the later coefficients $k_{n\nu}$, ν fixed. (See Problems 54, 55, 56.) From (12.7.2) we readily derive an asymptotic formula for D_n/D_{n-1}, where D_n is the determinant of Hankel's type defined by (2.2.7) (see (2.2.15)). The first asymptotic investigation of these determinants is due to Szegö (**1**, p. 517). The coefficient k_{n0} was denoted by k_n in Chapter II (see (2.2.15)).

The proof follows immediately from (12.1.3). If

$$D(z) = d_0 + d_1z + d_2z^2 + \cdots,$$

we have

$$\lim_{n\to\infty} 2^{-n}\left(\frac{1+(1-x^{-2})^{\frac{1}{2}}}{2}\right)^{-n}(k_{n0}+k_{n1}x^{-1}+k_{n2}x^{-2}+\cdots)$$

$$= (2\pi)^{-\frac{1}{2}}(d_0+d_1z^{-1}+\cdots)^{-1} = (2\pi)^{-\frac{1}{2}}\left(d_0^{-1}-\frac{d_1}{2d_0^2}x^{-1}+\cdots\right).$$

Now we use Titchmarsh **1**, p. 95 and (10.2.10). We note that on account of the mean-value theorem of Gauss,

$$\exp\left\{-\frac{1}{4\pi}\int_{-\pi}^{+\pi}\log|\sin\theta|\,d\theta\right\} = \exp\left\{-\frac{1}{4\pi}\int_{-\pi}^{+\pi}\log\left|\frac{1-z^2}{2}\right|d\theta\right\}$$

$$= \exp\left(-\frac{1}{2}\log\frac{1}{2}\right) = 2^{\frac{1}{2}}, \qquad z = e^{i\theta}.$$

Combining Theorem 12.7.1 with the formulas (3.2.2), we obtain for the coefficients A_n, C_n of the recurrence formula (3.2.1),

$$\lim_{n\to\infty} A_n = 2, \qquad \lim_{n\to\infty} C_n = 1, \tag{12.7.4}$$

provided the conditions of Theorem 12.7.1 are satisfied. If we compare the terms with x^{n-1} in (3.2.1), we obtain

$$B_n = \frac{k_{n0}}{k_{n-1,0}}\left(\frac{k_{n1}}{k_{n0}}-\frac{k_{n-1,1}}{k_{n-1,0}}\right). \tag{12.7.5}$$

Thus, under the same conditions,

$$\lim_{n\to\infty} B_n = 0. \tag{12.7.6}$$

The formulas (12.7.4) and (12.7.6) are noteworthy from the point of view of a classical theorem of Poincaré on recurrence formulas (see Blumenthal **1**, p. 16).

(2) For the further applications we need only the less informative form (12.2.6) of our asymptotic formula.

Theorem 12.7.2 (Distribution of the zeros). *Let the weight function $w(x)$ satisfy the conditions of Theorem* 12.1.2, *and let*

$$x_{1n},\, x_{2n},\, \cdots,\, x_{nn} \tag{12.7.7}$$

denote the zeros of the orthogonal polynomial $p_n(x)$. Let $x_{\nu n} = \cos\theta_{\nu n}$, $0 < \theta_{\nu n} < \pi$. If $F(\theta)$ is an arbitrary Riemann-integrable function, we have

$$\lim_{n\to\infty}\frac{F(\theta_{1n})+F(\theta_{2n})+\cdots+F(\theta_{nn})}{n} = \pi^{-1}\int_0^{\pi}F(\theta)\,d\theta. \tag{12.7.8}$$

Using the terminology of Weyl, we say that the values $\{\theta_{\nu n}\}$ are equally distributed in the interval $[0, \pi]$ (see Pólya-Szegö **1**, vol. 1, p. 70). This result is proved under slightly more restrictive conditions by Szegö (**1**, p. 531).

It is rather remarkable that the asymptotic nature of the distribution is independent of the weight function. As a consequence of (12.7.8) we obtain the following result: *Let* $[a, b]$ *be a subinterval of* $[-1, +1]$, *and let* $a = \cos \alpha$, $b = \cos \beta$, $\pi \geqq \alpha > \beta \geqq 0$. *If* $N = N(n, a, b)$ *denotes the number of the zeros of* $p_n(x)$ *in the interval* $[a, b]$, *we have*

$$\lim_{n\to\infty} \frac{N(n, a, b)}{n} = \frac{\alpha - \beta}{\pi}. \tag{12.7.9}$$

Thus the "density" of the zeros $x_{\nu n}$ towards the end-points of the interval $[-1, +1]$ becomes large.

For the proof of Theorem 12.7.2 we write (12.2.6) in the following form (see (12.7.2)):

$$\begin{aligned} \lim_{n\to\infty} n^{-1}\{\log|p_n(x)| - \log|k_{n0}x^n|\} &= \lim_{n\to\infty} n^{-1}\sum_{\nu=1}^{n}\log\left|1 - \frac{x_{\nu n}}{x}\right| \\ &= \log\left|\frac{z}{2x}\right| = \pi^{-1}\int_0^\pi \log\left|1 - \frac{\cos\theta}{x}\right| d\theta. \end{aligned} \tag{12.7.10}$$

Here $|z| > 1$. The last formula follows from Gauss's mean-value theorem, since

$$\left|1 - \frac{\zeta + \zeta^{-1}}{z + z^{-1}}\right| = \left|\frac{(\zeta - z)(\zeta - \bar{z})}{2x\bar{z}}\right|, \qquad \zeta = e^{i\theta}, |z| > 1.$$

Thus,

$$\lim_{n\to\infty} n^{-1}\sum_{\nu=1}^{n}\Re\left\{\log\left(1 - \frac{\cos\theta_{\nu n}}{x}\right)\right\} = \pi^{-1}\int_0^\pi \Re\left\{\log\left(1 - \frac{\cos\theta}{x}\right)\right\} d\theta, \tag{12.7.11}$$

so that (see Titchmarsh **1**, p. 95)

$$\lim_{n\to\infty} n^{-1}\sum_{\nu=1}^{n}(\cos\theta_{\nu n})^k = \pi^{-1}\int_0^\pi (\cos\theta)^k\, d\theta, \qquad k = 0, 1, 2, \cdots. \tag{12.7.12}$$

This establishes the statement (see Pólya-Szegö **1**, loc. cit.).

(3) Further applications are the following theorems:

THEOREM 12.7.3 (Expansion of an analytic function in terms of orthogonal polynomials). *Let the weight function* $w(x)$ *satisfy the conditions of Theorem 12.1.2, and let* $\{p_n(x)\}$ *be the associated orthonormal system of polynomials. Let* $f(x)$ *be an analytic function regular on the segment* $[-1, +1]$, *and let*

$$\begin{aligned} f(x) &\sim f_0p_0(x) + f_1p_1(x) + \cdots + f_np_n(x) + \cdots, \\ f_n &= \int_{-1}^{+1} f(x)p_n(x)w(x)\, dx, \qquad n = 0, 1, 2, \cdots, \end{aligned} \tag{12.7.13}$$

be its Fourier expansion. Let R *be the sum of the semi-axes of the largest ellipse with foci at* ± 1 *in the interior of which* $f(x)$ *is regular. Then the Fourier expansion* (12.7.13) *is convergent (with the sum* $f(x)$*) in the interior and divergent in* *the exterior of this ellipse. The convergence is uniform on every closed set lying in the interior of the ellipse. Moreover*

$$\liminf_{n\to\infty} |f_n|^{-1/n} = R. \tag{12.7.14}$$

THEOREM 12.7.4. *Let the weight function $w(x)$ satisfy the conditions of Theorem 12.1.2, and let*

$$f_0 p_0(x) + f_1 p_1(x) + f_2 p_2(x) + \cdots + f_n p_n(x) + \cdots \tag{12.7.15}$$

be an infinite series proceeding in terms of the orthonormal polynomials $p_n(x)$ associated with $w(x)$. Let

$$\liminf_{n\to\infty} |f_n|^{-1/n} = R. \tag{12.7.16}$$

Then the series (12.7.15) *is convergent in the interior of the ellipse with foci at ± 1, whose semi-axes have the sum R; it is divergent in the exterior of this ellipse. The series represents an analytic function which is regular in the interior of the ellipse and has at least one singular point on the ellipse itself. The expansion associated with this function is identical with* (12.7.15).

Thus the "ellipse of regularity" coincides with the "ellipse of convergence"; (12.7.14) and (12.7.16) are the analogues of the Cauchy-Hadamard formula (see (9.1.4)). Concerning these theorems, see Szegö **1**, p. 538; **6**, p. 193. Theorem 12.7.4 must be modified in an obvious way if $R \leqq 1$, or if $R = \infty$.

As a consequence of (12.2.6), we see that the domain of convergence of series of type (12.7.15) is always an ellipse $|z| = \text{const}$. To prove Theorem 12.7.3, we use Theorem 1.3.5. We can find a π_{n-1}, say $\rho(x)$, such that

$$|f(x) - \rho(x)| < M(R^{-1} + \epsilon)^n. \tag{12.7.17}$$

Here $\epsilon > 0$ is arbitrarily small, and $M = M(\epsilon)$ is independent of n. Thus,

$$f_n = \int_{-1}^{+1} f(x)p_n(x)w(x)\,dx = \int_{-1}^{+1} \{f(x) - \rho(x)\}p_n(x)w(x)\,dx,$$

so that

$$|f_n|^2 \leqq M^2(R^{-1} + \epsilon)^{2n} \int_{-1}^{+1} \{p_n(x)\}^2 w(x)\,dx \int_{-1}^{+1} w(x)\,dx;$$

whence $\limsup_{n\to\infty} |f_n|^{1/n} \leqq R^{-1}$. This furnishes the statement of Theorem 12.7.3 concerning the convergence. Now the last relation must be an equality, for otherwise, the expansion would be uniformly convergent in the interior of an ellipse larger than that of the ellipse of regularity, and $f(x)$ would be regular therein. This argument also furnishes the divergence in the exterior of the ellipse of regularity.

Theorem 12.7.4 can also be established without difficulty. It is quite remarkable that the domain of convergence of expansions of type (12.7.13) and (12.7.15) is independent of the weight function $w(x)$ (compare Theorem 9.1.1).

CHAPTER XIII

EXPANSION PROBLEMS ASSOCIATED WITH GENERAL ORTHOGONAL POLYNOMIALS

We shall now prove four theorems of the equiconvergent type in the sense of Chapter IX (see the introduction). Two of them, Theorems 13.1.2 and 13.1.4, deal with expansions of a preassigned function on a finite segment in terms of polynomials orthogonal on this segment. The other two Theorems, 13.1.1 and 13.1.3, are concerned with the expansion of the boundary values of an analytic function, regular in the interior of the unit circle, $|z| < 1$, in terms of polynomials orthogonal on the unit circle. In all cases the expansions in question are compared with trigonometric and power series expansions, and the weight functions are subject to conditions similar to those in the asymptotic theorems of Chapter XII. The function developed is very general and merely satisfies certain integrability conditions.

The basic idea of the method used in the proof of Theorems 13.1.1 and 13.1.2 is due to Szegö (see **9** where only the case of a segment is considered). Our present treatment of these theorems is slightly different from and more general than in Szegö **9**. The two other theorems are new. It is noteworthy that no direct use is made of the asymptotic results of the previous chapter. The methods, however, are very closely related.

After having finished the manuscript, I came into possession of three important papers of Korous (**3**, **4**, **5**).

In **3**, Korous deals with the expansion problem of Theorem 13.1.2. His conditions are of "local" character but less restrictive than those of Theorem 13.1.2. Also, his method is entirely different from that used by Szegö in **9** or from that of the present treatment.

In **4** and **5**, Korous proves two other equiconvergence theorems generalizing the Laguerre series of Theorem 9.1.5.

13.1. Results and remarks

(1) Theorem 13.1.1 (Equiconvergence theorem on the unit circle $|z| = 1$ if the weight function is subject to "local" conditions). *Let $f(\theta)$ be a weight function on the unit circle which satisfies the conditions of Theorem* 12.1.5 (= 10.4.1). *Let $F(z)$ be an analytic function regular in $|z| < 1$ and of the class H_2* (§10.1).

If $s_n(z)$ denotes the nth partial sum of the expansion of the boundary values $F(z)$, $|z| = 1$, in terms of the orthonormal polynomials $\{\phi_n(z)\}$ associated with $f(\theta)$, and if $\mathrm{s}_n(z)$ is the nth partial sum of the ordinary power series expansion of $F(z)$, we have

(13.1.1) $$\lim_{n\to\infty} \{\mathfrak{s}_n(a) - \mathbf{s}_n(a)\} = 0.$$

Here $a = e^{i\alpha}$ has the same meaning as in Theorem 12.1.5.

THEOREM 13.1.2 (Equiconvergence theorem on a finite real segment if the weight function is subject to "local" conditions). *Let $w(x)$ be a weight function on the segment $[-1, +1]$ subject to the conditions of Theorem* 12.1.6. *Let $\Phi(x)$ be an arbitrary real-valued function, measurable in Lebesgue's sense, for which the integrals*

(13.1.2) $$\int_{-1}^{+1} \{\Phi(x)\}^2 w(x)\, dx, \qquad \int_{-1}^{+1} |\Phi(x)|\, (1 - x^2)^{-\frac{1}{4}}\, dx$$

exist.

If $s_n(x)$ and $\mathbf{s}_n(x)$ denote the nth partial sums of the expansions of $\Phi(x)$ in terms of the orthonormal polynomials $\{p_n(x)\}$ associated with $w(x)$ and of the Tchebichef polynomials $\{\cos n\theta\}$, $\cos\theta = x$, respectively, we have

(13.1.3) $$\lim_{n\to\infty} \{s_n(\xi) - \mathbf{s}_n(\xi)\} = 0.$$

Here $-1 < \xi < +1$, and ξ has the same meaning as in Theorem 12.1.6.

It is remarkable that in both cases a wide class of expansions display the same convergence behavior. The expansion into a series of $\cos n\theta$ occurring in Theorem 13.1.2, is, of course, the ordinary cosine expansion of $\Phi(\cos\theta)$. A comparison of $s_n(\xi)$ with other special expansions can also be readily made. In making such comparisons, existence of certain other integrals must be required. By the use of theorems thus obtained, the customary convergence and summability theorems of the classical Fourier expansions can be easily extended to the general expansions in question. Theorem 13.1.2 can be easily extended to an arbitrary finite segment instead of the segment $[-1, +1]$.

These theorems hold under the conditions of "local" character of Theorems 12.1.5 and 12.1.6.

(2) The following theorems correspond to the conditions of "S. Bernstein's type" occurring in Theorems 12.1.3 and 12.1.4.

THEOREM 13.1.3 (Equiconvergence theorem of S. Bernstein's type on the unit circle). *Let $f(\theta)$ be a weight function on the unit circle $z = e^{i\theta}$ which satisfies the conditions of Theorem* 12.1.3 *with $\lambda > 1$. Let $F(z)$ be an analytic function, regular and bounded for $|z| < 1$. Employing the same notation as in Theorem* 13.1.1, *we have*

(13.1.4) $$\lim_{n\to\infty} \{s_n(z) - \mathbf{s}_n(z)\} = 0, \qquad |z| \leqq 1,$$

uniformly in the whole closed unit circle $|z| \leqq 1$.

The function $F(z)$ has integrable boundary values $F(e^{i\theta})$ (for almost all θ).

THEOREM 13.1.4 (Equiconvergence theorem of S. Bernstein's type on a finite real segment). *Let $w(x)$ be a weight function on the interval $-1 \leqq x \leqq +1$, $x = \cos\theta$, which satisfies the conditions of Theorem 12.1.4 with $\lambda > 1$. Let $\Phi(x)$ be an arbitrary bounded function, which is measurable in Lebesgue's sense. Employing the same notation as in Theorem 13.1.2, we have*

$$\lim_{n\to\infty} \{s_n(x) - \mathbf{s}_n(x)\} = 0, \qquad -1 < x < +1, \tag{13.1.5}$$

uniformly in the interval $[-1+\epsilon, 1-\epsilon]$, $0 < \epsilon < \frac{1}{2}$.

In the proofs of Theorems 13.1.3 and 13.1.4 essential use is made of Theorems 12.1.3 and 12.1.4, respectively.

13.2. A maximum problem on the unit circle

(1) Let $f(\theta)$ and $F(z)$ have the same meaning as in Theorem 13.1.1. We write $G(\theta) = (2\pi)^{-1}f(\theta)F(e^{i\theta})$, so that

$$\int_{-\pi}^{+\pi} |G(\theta)|\, d\theta \tag{13.2.1}$$

is convergent. In what follows we may vary $f(\theta)$, but $G(\theta)$ is supposed to be a fixed function for which (13.2.1) exists.

PROBLEM. *Let λ and μ be arbitrary complex numbers, and let $|a| = 1$. We intend to determine the maximum of*

$$\left|\lambda\rho(a) + \mu \int_{-\pi}^{+\pi} \overline{G(\theta)}\rho(z)\, d\theta\right|^2, \qquad z = e^{i\theta}, \tag{13.2.2}$$

for $\rho(z)$ ranging over the set of all π_n which satisfy the condition

$$\frac{1}{2\pi}\int_{-\pi}^{+\pi} f(\theta)\,|\rho(z)|^2\, d\theta = 1, \qquad z = e^{i\theta}. \tag{13.2.3}$$

Concerning the special case $\lambda = 1$, $\mu = 0$, see §11.3. Compare also with §12.6 (1). We write again

$$\rho(z) = u_0\phi_0(z) + u_1\phi_1(z) + \cdots + u_n\phi_n(z), \tag{13.2.4}$$

where $\{\phi_n(z)\}$ has the usual meaning. Then $\sum_{\nu=0}^{n} |u_\nu|^2 = 1$, and

$$\begin{aligned} \left|\lambda\rho(a) + \mu\int_{-\pi}^{+\pi}\overline{G(\theta)}\rho(z)\,d\theta\right|^2 &= \left|\sum_{\nu=0}^{n} u_\nu\left\{\lambda\phi_\nu(a) + \mu\int_{-\pi}^{+\pi}\overline{G(\theta)}\phi_\nu(z)\,d\theta\right\}\right|^2 \\ &\leqq \sum_{\nu=0}^{n}\left|\lambda\phi_\nu(a) + \mu\int_{-\pi}^{+\pi}\overline{G(\theta)}\phi_\nu(z)\,d\theta\right|^2 \\ &= |\lambda|^2 s_n(a,a) + 2\Re\{\lambda\bar{\mu}s_n(a)\} + |\mu|^2 H_n, \qquad z = e^{i\theta}. \end{aligned} \tag{13.2.5}$$

Here, as in Theorem 13.1.1, $s_n(a)$ is the nth partial sum of the expansion of

$2\pi G(\theta)\{f(\theta)\}^{-1}$ in terms of the polynomials $\{\phi_n(z)\}$, and

$$H_n \equiv \sum_{\nu=0}^{n} \left| \int_{-\pi}^{+\pi} \overline{G(\theta)}\phi_\nu(z)\,d\theta \right|^2, \qquad z = e^{i\theta}. \tag{13.2.6}$$

The right-hand member of (13.2.5) is the maximum in question.

(2) For clearness, as in §12.6 (2), we now write $s_n(f;\,a,\,z)$, $s_n(f;\,a)$, $\phi_n(f;\,a)$, $H_n(f)$ for the expressions involved in the previous considerations, which are associated with the weight function $f(\theta)$. We note that $s_n(f;\,a)$ is the nth partial sum of the expansion of $2\pi G(\theta)\{f(\theta)\}^{-1}$ in terms of the polynomials $\phi_n(f;\,z)$ associated with $f(\theta)$. We add that in all subsequent considerations, even if $f(\theta)$ is replaced by some other weight function, we shall keep the function $G(\theta)$ used before.

Applying Theorem 10.4.4 again, we obtain

$$\begin{aligned} &|\lambda|^2 s_n(f_1\,;\,a,\,a) + 2\Re\{\lambda\bar{\mu}s_n(f_1\,;\,a)\} + |\mu|^2 H_n(f_1) \\ &\qquad \geqq |\lambda|^2 s_n(f;\,a,\,a) + 2\Re\{\lambda\bar{\mu}s_n(f;\,a)\} + |\mu|^2 H_n(f) \\ &\qquad \geqq |\lambda|^2 s_n(f_2\,;\,a,\,a) + 2\Re\{\lambda\bar{\mu}s_n(f_2\,;\,a)\} + |\mu|^2 H_n(f_2). \end{aligned} \tag{13.2.7}$$

In particular, we have

$$s_n(f_1\,;\,a,\,a) \geqq s_n(f;\,a,\,a) \geqq s_n(f_2\,;\,a,\,a), \tag{13.2.8}$$

$$H_n(f_1) \geqq H_n(f) \geqq H_n(f_2), \tag{13.2.9}$$

$$\begin{aligned} |s_n(f;\,a) - s_n(f_2\,;\,a)|^2 &\leqq \{H_n(f) - H_n(f_2)\}\{s_n(f;\,a,\,a) - s_n(f_2\,;\,a,\,a)\} \\ &\leqq H_n(f)\{s_n(f_1\,;\,a,\,a) - s_n(f_2\,;\,a,\,a)\}. \end{aligned} \tag{13.2.10}$$

13.3. Proof of Theorem 13.1.1

(1) Bessel's inequality enables us to obtain

$$H_n(f) \leqq 2\pi \int_{-\pi}^{+\pi} |\overline{G(\theta)}\{f(\theta)\}^{-1}|^2 f(\theta)\,d\theta = \frac{1}{2\pi}\int_{-\pi}^{+\pi} |F(e^{i\theta})|^2 f(\theta)\,d\theta. \tag{13.3.1}$$

Whence (12.6.19) shows that

$$\begin{aligned} &\limsup_{n\to\infty} |s_n(f;\,a) - s_n(f_2;\,a)|^2 \\ &\leqq \frac{1}{2\pi}\int_{-\pi}^{+\pi} |F(e^{i\theta})|^2 f(\theta)\,d\theta \cdot \limsup_{n\to\infty}\{s_n(f_1;\,a,\,a) - s_n(f_2;\,a,\,a)\} \\ &\leqq \frac{1}{2\pi}\int_{-\pi}^{+\pi} |F(e^{i\theta})|^2 f(\theta)\,d\theta \cdot 2\{f(\alpha)\}^{-1}\left|\frac{D'(f_1;\,a)}{D(f_1;\,a)} - \frac{D'(f_2;\,a)}{D(f_2;\,a)}\right|. \end{aligned} \tag{13.3.2}$$

The right-hand member is arbitrarily small with the positive number ϵ occurring in Theorem 10.4.4.

(2) Next we discuss the behavior of the partial sum $s_n(f_2\,;\,a)$ as $n \to \infty$. As in §12.6 (3), we use the abbreviation $D(f_2^{-1};\,z) = h(z)$. We find that if n

is sufficiently large (see (12.6.11)),

$$
\begin{aligned}
(13.3.3)\qquad s_n(f_2; a) &= \int_{-\pi}^{+\pi} G(\theta) s_n(f_2; z, a)\, d\theta = \int_{-\pi}^{+\pi} G(\theta) \frac{h(a)\overline{h(z)} - (a\bar{z})^{n+1}\overline{h(a)}h(z)}{1 - a\bar{z}}\, d\theta \\
&= \frac{1}{2\pi}\int_{-\pi}^{+\pi} f(\theta) F(z) \frac{h(a)\overline{h(z)} - (a\bar{z})^{n+1}\overline{h(a)}h(z)}{1 - a\bar{z}}\, d\theta, \qquad z = e^{i\theta}.
\end{aligned}
$$

Comparing this latter expression with

$$
(13.3.4)\qquad \mathbf{s}_n(a) = \frac{1}{2\pi}\int_{-\pi}^{+\pi} F(z) \frac{1 - (a\bar{z})^{n+1}}{1 - a\bar{z}}\, d\theta, \qquad z = e^{i\theta},
$$

we obtain

$$
\begin{aligned}
(13.3.5)\qquad s_n(f_2; a) - \mathbf{s}_n(a) &= \frac{1}{2\pi}\int_{-\pi}^{+\pi} F(z) \frac{f(\theta)h(a)\overline{h(z)} - 1}{1 - a\bar{z}}\, d\theta \\
&- \frac{1}{2\pi}\int_{-\pi}^{+\pi} F(z) \frac{f(\theta)\overline{h(a)}h(z) - 1}{1 - a\bar{z}} (a\bar{z})^{n+1}\, d\theta, \quad z = e^{i\theta}.
\end{aligned}
$$

The second integral approaches zero as $n \to \infty$, since

$$
(13.3.6)\qquad \frac{f(\theta)\overline{h(a)}h(z) - 1}{1 - a\bar{z}}, \qquad z = e^{i\theta},
$$

has a limit at $\theta = \alpha$. The first integral can be written as follows:

$$
\begin{aligned}
(13.3.7)\qquad &\int_{-\pi}^{+\pi} F(z) \frac{f_2(\theta)h(a)\overline{h(z)} - 1}{1 - a\bar{z}}\, d\theta + \int_{-\pi}^{+\pi} F(z) \frac{f(\theta) - f_2(\theta)}{1 - a\bar{z}} h(a)\overline{h(z)}\, d\theta \\
&= \int_{-\pi}^{+\pi} F(z) \frac{f(\theta) - f_2(\theta)}{1 - a\bar{z}} h(a)\overline{h(z)}\, d\theta, \qquad z = e^{i\theta}.
\end{aligned}
$$

since (see the remark at the end of §10.1)

$$
(13.3.8)\qquad \int_{|z|=1} F(z) \frac{1 - h(a)/h(z)}{a - z}\, dz = 0.
$$

By virtue of Schwarz's inequality,

$$
\begin{aligned}
&4\pi^2 \limsup_{n\to\infty} | s_n(f_2; a) - \mathbf{s}_n(a) |^2 \\
&\leqq \int_{-\pi}^{+\pi} | F(z) |^2\, d\theta \int_{-\pi}^{+\pi} \frac{\{f_2(\theta) - f(\theta)\}^2}{4 \sin^2 \frac{\theta - \alpha}{2}} \{f(\alpha)\}^{-1}\{f_2(\theta)\}^{-1}\, d\theta \\
&\leqq \{4f(\alpha)\}^{-1} \int_{-\pi}^{+\pi} | F(z) |^2\, d\theta \int_{-\pi}^{+\pi} \frac{f_2(\theta) - f(\theta)}{\sin^2 \frac{\theta - \alpha}{2}}\, d\theta \\
&\leqq \{4\, f(\alpha)\}^{-1} \max f_2(\theta) \int_{-\pi}^{+\pi} | F(z) |^2\, d\theta \int_{-\pi}^{+\pi} \frac{\log f_2(\theta) - \log f(\theta)}{\sin^2 \frac{\theta - \alpha}{2}}\, d\theta, \quad z = e^{i\theta}.
\end{aligned}
$$

Combining this result with (13.3.2), we find $\limsup_{n\to\infty} |s_n(f; a) - \mathbf{s}_n(a)|$ to be arbitrarily small with the positive number ϵ occurring in Theorem 10.4.4. This establishes the statement.

13.4. A special case of Theorem 13.1.2

(1) In this section we consider the special case

$$w(x) = (1 - x^2)^{-\frac{1}{2}}\{\rho(x)\}^{-1}, \tag{13.4.1}$$

where $\rho(x)$ is a π_l which is positive in the interval $[-1, +1]$. The corresponding orthogonal polynomials (for $2n > l$) have been calculated in §2.6. We shall verify the validity of Theorem 13.1.2 in this special case. To this end it suffices to assume the existence of the integral

$$\int_{-1}^{+1} |\Phi(x)| (1 - x^2)^{-\frac{1}{2}} dx, \tag{13.4.2}$$

a condition which is more general than that required in Theorem 13.1.2. As in that theorem we have $\xi = \cos\alpha$, $-1 < \xi < +1$, $0 < \alpha < \pi$.

By virtue of the Christoffel-Darboux formula (3.2.3)

$$s_n(\xi) = \frac{k_n}{k_{n+1}} \int_{-1}^{+1} \Phi(x) \frac{p_{n+1}(x)p_n(\xi) - p_n(x)p_{n+1}(\xi)}{x - \xi} w(x)\, dx;$$

or

$$\frac{k_{n+1}}{k_n} s_n(\cos\alpha) = \int_0^{\pi} \Phi(\cos\theta) \frac{p_{n+1}(\cos\theta)p_n(\cos\alpha) - p_n(\cos\theta)p_{n+1}(\cos\alpha)}{\cos\theta - \cos\alpha} \cdot \{\rho(\cos\theta)\}^{-1} d\theta. \tag{13.4.3}$$

Assume now $2n > l$. Use formulas (2.6.2) and write $(2/\pi)^{\frac{1}{2}} e^{in\theta} \overline{h(e^{i\theta})} = u_n(\theta) + iv_n(\theta)$. Then

$$\begin{aligned} p_n(\cos\theta) &= u_n(\theta), \\ p_{n+1}(\cos\theta) &= \Re\{e^{i\theta}[u_n(\theta) + iv_n(\theta)]\} = u_n(\theta)\cos\theta - v_n(\theta)\sin\theta, \end{aligned} \tag{13.4.4}$$

so that

$$\begin{aligned} &\frac{p_{n+1}(\cos\theta)p_n(\cos\alpha) - p_n(\cos\theta)p_{n+1}(\cos\alpha)}{\cos\theta - \cos\alpha} \\ &= \frac{\{u_n(\theta)\cos\theta - v_n(\theta)\sin\theta\}u_n(\alpha) - u_n(\theta)\{u_n(\alpha)\cos\alpha - v_n(\alpha)\sin\alpha\}}{\cos\theta - \cos\alpha} \\ &= u_n(\theta)u_n(\alpha) - v_n(\theta)u_n(\alpha)\frac{\sin\theta - \sin\alpha}{\cos\theta - \cos\alpha} \\ &\qquad + \frac{u_n(\theta)v_n(\alpha) - v_n(\theta)u_n(\alpha)}{\cos\theta - \cos\alpha}\sin\alpha. \end{aligned} \tag{13.4.5}$$

In view of Riemann's lemma, this leads to

$$\frac{k_{n+1}}{k_n} s_n(\cos\alpha)$$

(13.4.6)
$$= \sin\alpha \int_0^\pi \Phi(\cos\theta) \frac{u_n(\theta)v_n(\alpha) - v_n(\theta)u_n(\alpha)}{\cos\theta - \cos\alpha} \{\rho(\cos\theta)\}^{-1} d\theta + o(1),$$

$$n \to \infty.$$

Now

$$\{\rho(\cos\theta)\}^{-1}\{u_n(\theta)v_n(\alpha) - v_n(\theta)u_n(\alpha)\}$$

$$= -\{\rho(\cos\theta)\}^{-1}\mathfrak{J}\{u_n(\theta) + iv_n(\theta)\}\{u_n(\alpha) - iv_n(\alpha)\}$$

(13.4.7)
$$= -\frac{2}{\pi}\{\rho(\cos\theta)\}^{-1}\mathfrak{J}\{e^{in(\theta-\alpha)}\overline{h(e^{i\theta})}h(e^{i\alpha})\}$$

$$= -\frac{2}{\pi}\mathfrak{J}\left\{e^{in(\theta-\alpha)}\frac{h(e^{i\alpha})}{h(e^{i\theta})}\right\}.$$

Repeated application of Riemann's lemma shows that $h(e^{i\alpha})\{h(e^{i\theta})\}^{-1}$ can be replaced by 1, and

$$\sin\alpha\,(\cos\theta - \cos\alpha)^{-1} = \sin\alpha\left\{2\sin\frac{\alpha+\theta}{2}\sin\frac{\alpha-\theta}{2}\right\}^{-1}$$

by $\{2\sin(\alpha - \theta)/2\}^{-1}$. Thus the right-hand member of (13.4.6) assumes the form

$$\frac{1}{\pi}\int_0^\pi \Phi(\cos\theta)\frac{\sin n(\theta-\alpha)}{\sin\dfrac{\theta-\alpha}{2}}\,d\theta + o(1)$$

$$= \frac{1}{\pi}\int_{-\pi}^{+\pi} \Phi(\cos\theta)\frac{\sin(n+\frac{1}{2})(\theta-\alpha)}{\sin\dfrac{\theta-\alpha}{2}}\,d\theta + o(1).$$

Finally $k_n/k_{n+1} = \frac{1}{2}$ when n is sufficiently large (see (2.6.5)). This establishes the statement.

(2) In the same special case we intend to calculate $K_n(\xi, \xi)$, where K_n is the kernel polynomial defined by (3.1.9). We have, by (3.2.4),

(13.4.8)
$$K_n(\xi, \xi) = \frac{k_n}{k_{n+1}}\{p'_{n+1}(\xi)p_n(\xi) - p'_n(\xi)p_{n+1}(\xi)\}.$$

Now, by (13.4.4), for a sufficiently large n,

$$\sin\alpha\, p'_n(\cos\alpha) = -u'_n(\alpha),$$

(13.4.9)
$$\sin\alpha\, p'_{n+1}(\cos\alpha)$$

$$= -u'_n(\alpha)\cos\alpha + v'_n(\alpha)\sin\alpha + u_n(\alpha)\sin\alpha + v_n(\alpha)\cos\alpha,$$

and $k_n/k_{n+1} = \frac{1}{2}$. Thus,

$$
\begin{aligned}
2 \sin \alpha \, K_n(\cos \alpha, \cos \alpha) &= \{-u_n'(\alpha) \cos \alpha + v_n'(\alpha) \sin \alpha \\
&\qquad + u_n(\alpha) \sin \alpha + v_n(\alpha) \cos \alpha\} u_n(\alpha) \\
&\qquad + u_n'(\alpha)\{u_n(\alpha) \cos \alpha - v_n(\alpha) \sin \alpha\} \\
&= \sin \alpha \, \mathfrak{I}\{[u_n(\alpha) - iv_n(\alpha)][u_u'(\alpha) + iv_n'(\alpha)]\} \\
&\qquad + u_n(\alpha)\mathfrak{I}\{[u_n(\alpha) + iv_n(\alpha)]e^{i\alpha}\};
\end{aligned}
$$

hence

$$
\begin{aligned}
\frac{2 \sin \alpha}{|u_n(\alpha) + iv_n(\alpha)|^2} K_n(\cos \alpha, \cos \alpha) &= \sin \alpha \, \mathfrak{I}\left\{\frac{u_n'(\alpha) + iv_n'(\alpha)}{u_n(\alpha) + iv_n(\alpha)}\right\} \\
&\qquad + \tfrac{1}{2} \sin \alpha + \tfrac{1}{2}\mathfrak{I}\left\{\frac{[u_n(\alpha) + iv_n(\alpha)]^2}{|u_n(\alpha) + iv_n(\alpha)|^2} e^{i\alpha}\right\}.
\end{aligned}
$$

Observing that

$$
\begin{aligned}
\mathfrak{I}\left\{\frac{u_n'(\alpha) + iv_n'(\alpha)}{u_n(\alpha) + iv_n(\alpha)}\right\} &= -\mathfrak{I}\left\{\frac{u_n'(\alpha) - iv_n'(\alpha)}{u_n(\alpha) - iv_n(\alpha)}\right\} \\
&= n - \mathfrak{I}\left[ie^{i\alpha}\frac{h'(e^{i\alpha})}{h(e^{i\alpha})}\right] = n - \mathfrak{R}\left[a\frac{h'(a)}{h(a)}\right],
\end{aligned}
$$

we get the important formula

$$
\begin{aligned}
\pi\{\rho(\xi)\}^{-1}K_n(\xi, \xi) &= n + \tfrac{1}{2} - \mathfrak{R}\left[a\frac{h'(a)}{h(a)}\right] \\
&\quad + (2 \sin \alpha)^{-1}\mathfrak{I}\left[a^{2n+1}\frac{\overline{h(a)}}{h(a)}\right], \qquad \xi = \cos \alpha,\ a = e^{i\alpha},\ 0 < \alpha < \pi.
\end{aligned}
\tag{13.4.10}
$$

This holds for sufficiently large values of n.

13.5. Preliminaries for the proof of Theorem 13.1.2

(1) Let $w(x)$ and $\Phi(x)$ have the same meaning as in Theorem 13.1.2. Write $G(x) = w(x)\Phi(x)$, so that the integral

$$
\int_{-1}^{+1} |G(x)|\, dx \tag{13.5.1}
$$

is convergent. In subsequent considerations we may vary $w(x)$, but $G(x)$ is supposed to be a fixed function for which the integral (13.5.1) exists.

(2) In what follows we apply Theorem 10.4.5 with $f(\theta) = w(\cos \theta)\,|\sin \theta|$. This function satisfies the conditions of the theorem mentioned (see §12.6 (5)). We define the functions $w_1(x)$ and $w_2(x)$ by

$$
f_\nu(\theta) = w_\nu(\cos \theta)\,|\sin \theta|, \qquad \nu = 1, 2. \tag{13.5.2}
$$

Obviously, $\xi = \cos \alpha$, $0 < \alpha < \pi$,

$$
0 \leqq w_1(x) \leqq w(x) \leqq w_2(x), \qquad w_1(\xi) = w(\xi) = w_2(\xi). \tag{13.5.3}
$$

Moreover, $w_2(x)$ is of the form (13.4.1), while $w_1(x) = k(x)u(x) = \{r(x)\}^2u(x)$. Here $u(x)$ is of the form (13.4.1), and $r(x)$ is a polynomial of degree $\sigma l/4$. (See the remark to Theorem 10.4.4.)

(3) PROBLEM. *Let λ and μ be arbitrary complex numbers, and assume that $-1 < \xi < +1$. We intend to determine the maximum of*

$$\left|\lambda\rho(\xi) + \mu\int_{-1}^{+1} G(x)\rho(x)\,dx\right|^2 \tag{13.5.4}$$

when $\rho(x)$ ranges over the set of all π_n which satisfy the condition

$$\int_{-1}^{+1} |\rho(x)|^2 w(x)\,dx = 1. \tag{13.5.5}$$

This is the problem corresponding to that of §13.2. We substitute again

$$\rho(x) = u_0p_0(x) + u_1p_1(x) + \cdots + u_np_n(x), \tag{13.5.6}$$

where $\{p_n(x)\}$ is the orthonormal set of polynomials associated with $w(x)$. Then $\sum_{\nu=0}^{n} |u_\nu|^2 = 1$, and according to Cauchy's inequality

$$\begin{aligned} \left|\lambda\rho(\xi) + \mu\int_{-1}^{+1} G(x)\rho(x)\,dx\right|^2 &= \left|\sum_{\nu=0}^{n} u_\nu\left\{\lambda p_\nu(\xi) + \mu\int_{-1}^{+1} G(x)p_\nu(x)\,dx\right\}\right|^2 \\ &\leqq \sum_{\nu=0}^{n}\left|\lambda p_\nu(\xi) + \mu\int_{-1}^{+1} G(x)p_\nu(x)\,dx\right|^2 \\ &= |\lambda|^2K_n(\xi) + 2\Re[\bar\lambda\mu s_n(\xi)] + |\mu|^2H_n. \end{aligned} \tag{13.5.7}$$

Here $K_n(\xi) = K_n(w;\xi)$ has been written for the "kernel" $K_n(\xi,\xi)$, and

$$s_n(\xi) \equiv s_n(w;\xi) = \sum_{\nu=0}^{n} p_\nu(\xi)\int_{-1}^{+1} G(x)p_\nu(x)\,dx = \int_{-1}^{+1} G(x)K_n(\xi,x)\,dx, \tag{13.5.8}$$

$$H_n \equiv H_n(w;\xi) = \sum_{\nu=0}^{n}\left\{\int_{-1}^{+1} G(x)p_\nu(x)\,dx\right\}^2. \tag{13.5.9}$$

We notice that $s_n(\xi)$ is the nth partial sum of the expansion of $\Phi(x) = \{w(x)\}^{-1}G(x)$ in terms of the polynomials $p_n(x)$ associated with $w(x)$.

The right-hand member of (13.5.7) is the maximum in question. Hence

$$\begin{aligned} &|\lambda|^2K_n(w_1;\xi) + 2\Re[\bar\lambda\mu s_n(w_1;\xi)] + |\mu|^2H_n(w_1) \\ &\geqq |\lambda|^2K_n(w;\xi) + 2\Re[\bar\lambda\mu s_n(w;\xi)] + |\mu|^2H_n(w) \\ &\geqq |\lambda|^2K_n(w_2;\xi) + 2\Re[\bar\lambda\mu s_n(w_2;\xi)] + |\mu|^2H_n(w_2). \end{aligned} \tag{13.5.10}$$

In particular

$$K_n(w_1\,;\,\xi) \geqq K_n(w;\,\xi) \geqq K_n(w_2\,;\,\xi), \tag{13.5.11}$$

$$H_n(w_1) \geqq H_n(w) \geqq H_n(w_2). \tag{13.5.12}$$

Furthermore,

$$\begin{aligned} |\, s_n(w;\,\xi) - s_n(w_2\,;\,\xi)\,|^2 &\leqq \{H_n(w) - H_n(w_2)\}\{K_n(w;\,\xi) - K_n(w_2\,;\,\xi)\} \\ &\leqq H_n(w)\{K_n(w_1\,;\,\xi) - K_n(w_2\,;\,\xi)\} \\ &\leqq \int_{-1}^{+1} \{\Phi(x)\}^2 w(x)\,dx\{K_n(w_1\,;\,\xi) - K_n(w_2\,;\,\xi)\}, \end{aligned} \tag{13.5.13}$$

if we observe Bessel's inequality. We next show that the last difference is arbitrarily small with ϵ, uniformly in n.

13.6. Proof of Theorem 13.1.2

The main tool of this proof is the special equiconvergence theorem of §13.4 (1) and the representation (13.4.10) of the kernel. The latter yields immediately the representation

$$\begin{aligned} \pi f(\alpha) K_n(w_2\,;\,\xi) &= n + \tfrac{1}{2} + \Re\left[a\,\frac{D'(f_2\,;\,a)}{D(f_2\,;\,a)}\right] \\ &\quad + (2\sin\alpha)^{-1}\Im\left[a^{2n+1}\,\frac{\overline{D(f_2\,;\,a)}}{D(f_2\,;\,a)}\right], \end{aligned} \tag{13.6.1}$$

since $\{g_2(\alpha)\}^{-1} = f_2(\alpha) = f(\alpha)$. We derive, on the other hand, an inequality for $K_n(w_1\,;\,\xi)$ analogous to that in (12.6.18). Using the notation introduced in §13.5 (2) and writing $\frac{1}{2}\sigma l = l'$, we obtain as in §12.6 (4),

$$K_{n+l'/2}(u;\,\xi) \geqq \{r(\xi)\}^2 K_n(w_1\,;\,\xi) = k(\xi) K_n(w_1\,;\,\xi). \tag{13.6.2}$$

Also, $u(x) = u(\cos\theta) = \{g_1(\theta)\}^{-1}\,|\sin\theta\,|^{-1}$, so that from (13.4.10)

$$\begin{aligned} \pi\{g_1(\alpha)\}^{-1} K_{n+l'/2}(u;\,\xi) &= n + l'/2 + \tfrac{1}{2} - \Re\left[a\,\frac{D'(g_1\,;\,a)}{D(g_1\,;\,a)}\right] \\ &\quad + (2\sin\alpha)^{-1}\Im\left[a^{2n+l'+1}\,\frac{\overline{D(g_1\,;\,a)}}{D(g_1\,;\,a)}\right] \\ &= n + l'/2 + \tfrac{1}{2} + \Re\left[a\,\frac{D'(g_1^{-1};\,a)}{D(g_1^{-1};\,a)}\right] \\ &\quad + (2\sin\alpha)^{-1}\Im\left[a^{2n+l'+1}\,\frac{\overline{D(g_1^{-1};\,a)}}{D(g_1^{-1};\,a)}\right], \end{aligned} \tag{13.6.3}$$

where, as before, $\xi = \cos\alpha$, $a = e^{i\alpha}$. Now we can use (12.6.17). Moreover from (12.6.12)

$$\frac{\overline{D(f_1\,;\,a)}}{D(f_1;\,a)} = \frac{\overline{D(g_1^{-1};\,a)}}{D(g_1^{-1};\,a)} \prod_{\nu=1}^{l} \left(\frac{1 - \bar{z}_\nu a}{1 - z_\nu \bar{a}}\right)^{\sigma/2} = a^{l'}\,\frac{\overline{D(g_1^{-1};\,a)}}{D(g_1^{-1};\,a)}\,; \tag{13.6.4}$$

also $\{g_1(\alpha)\}^{-1}k(\xi) = f_1(\alpha) = f(\alpha)$, so that from (13.6.2)

$$\pi f(\alpha)K_n(w_1;\xi) \leqq n + \tfrac{1}{2} + \Re\left[a\,\frac{D'(f_1;a)}{D(f_1;a)}\right] + (2\sin\alpha)^{-1}\Im\left[a^{2n+1}\,\frac{D(f_1;a)}{\overline{D(f_1;a)}}\right]. \tag{13.6.5}$$

On comparing this with (13.6.1), we have

$$\begin{aligned} K_n(w_1;\xi) - K_n(w_2;\xi) &\leqq \{\pi f(\alpha)\}^{-1}\left\{\Re\left[a\,\frac{D'(f_1;a)}{D(f_1;a)} - a\,\frac{D'(f_2;a)}{D(f_2;a)}\right]\right. \\ &\quad \left. + (2\sin\alpha)^{-1}\Im\left[a^{2n+1}\,\frac{D(f_1;a)}{\overline{D(f_1;a)}} - a^{2n+1}\,\frac{D(f_2;a)}{\overline{D(f_2;a)}}\right]\right\} \\ &\leqq \{\pi f(\alpha)\}^{-1}\left|\frac{D'(f_1;a)}{D(f_1;a)} - \frac{D'(f_2;a)}{D(f_2;a)}\right| \\ &\quad + \{2\pi f(\alpha)\sin\alpha\}^{-1}\left|\frac{D(f_1;a)}{\overline{D(f_1;a)}} - \frac{D(f_2;a)}{\overline{D(f_2;a)}}\right|. \end{aligned} \tag{13.6.6}$$

The left-hand side is non-negative, while the right-hand side is arbitrarily small with ϵ.

Returning to (13.5.13), we notice that $s_n(w_2;\xi)$ is the nth partial sum of the expansion of $\{w_2(x)\}^{-1}G(x) = \{w_2(x)\}^{-1}w(x)\Phi(x)$ in terms of the polynomials associated with $w_2(x)$. Since $\{w_2(x)\}^{-1}w(x) \leqq 1$, and since the second integral in (13.1.2) exists, the result of §13.4 (1) can be applied. Thus $s_n(w_2;\xi) = s_n(w_2;\cos\alpha)$ can be replaced by the nth partial sum of the corresponding Fourier expansion with an error $o(1)$, $n \to \infty$. Also $\{w_2(x)\}^{-1}w(x) = 1$ for $x = \xi$, so that the partial sum can be replaced by that of $\Phi(x)$, which is $\mathbf{s}_n(\xi)$. This completes the proof of Theorem 13.1.2.

REMARK. It is easy to show that the difference

$$H_n(w) - H_n(w_2), \tag{13.6.7}$$

occurring in (13.5.13), is also arbitrarily small with ϵ. If this fact were used, it would suffice to obtain an upper bound for the first difference in (13.6.6). This remark furnishes a slight variation of the preceding argument.

13.7. Proof of Theorem 13.1.3

(1) It is sufficient to discuss the statement for $|z| = 1$. The partial sums in question are

$$\frac{1}{2\pi}\int_{-\pi}^{+\pi} F(\zeta)s_n(\zeta, z)f(x)\,dx \quad\text{and}\quad \frac{1}{2\pi}\int_{-\pi}^{+\pi} F(\zeta)\,\frac{1-(\bar\zeta z)^{n+1}}{1-\bar\zeta z}\,dx,\quad \zeta = e^{ix}, \tag{13.7.1}$$

where $s_n(\zeta, z)$ has the usual meaning. We introduce the difference of the "kernels"

$$\Delta_n(\zeta, z) = s_n(\zeta, z)f(x) - \frac{1 - (\bar{\zeta}z)^{n+1}}{1 - \bar{\zeta}z}, \tag{13.7.2}$$

and show that

$$\lim_{n\to\infty} \int_{-\pi}^{+\pi} F(\zeta)\Delta_n(\zeta, z)\, dx = 0, \qquad \zeta = e^{ix}, \tag{13.7.3}$$

uniformly on the unit circle $|z| = 1$.

(2) First, as a consequence of the Lipschitz-Dini condition (10.3.10) assumed for $f(\theta)$, we prove that a similar condition (with the exponent $-\lambda$ instead of $-1 - \lambda$) holds for $D(f; z) = D(z)$; that is,

$$|D(e^{i(\theta+\delta)}) - D(e^{i\theta})| < L' |\log \delta|^{-\lambda}, \tag{13.7.4}$$

where L' is a positive constant. Let m be an arbitrary integer. By use of (10.3.12) we obtain

$$|D(e^{i(\theta+\delta)}) - D(e^{i\theta})| < 2Q(\log m)^{-\lambda} + |\{h(e^{i(\theta+\delta)})\}^{-1} - \{h(e^{i\theta})\}^{-1}|.$$

The last term, according to Theorem 1.22.2 is equal to $\delta O(m)$; whence the bound $O[(\log m)^{-\lambda}] + \delta O(m)$ follows. Putting $m \sim \delta^{-1} |\log \delta|^{-\lambda}$, we have $(\log m)^{-\lambda} \sim |\log \delta|^{-\lambda}$, and $\delta m \sim |\log \delta|^{-\lambda}$, which establishes the statement. This furnishes, for the function $\gamma(\theta) = \mathfrak{J}\{\log D(e^{i\theta})\}$ defined by (12.1.7), also the relation

$$|\gamma(\theta + \delta) - \gamma(\theta)| < L'' |\log \delta|^{-\lambda}, \tag{13.7.5}$$

where L'' is a positive constant. These results hold for an arbitrary $\lambda > 0$.

By the same argument the following more general inequality follows:

$$|D(z_1) - D(z_2)| < L' |\log |z_1 - z_2||^{-\lambda}, \tag{13.7.6}$$

where z_1 and z_2 are arbitrary in the unit circle $|z| \leqq 1$.

(3) Let $|z| = |\zeta| = 1, z \neq \zeta$. From (11.4.5) and Theorem 12.1.3 it is seen that

$$\begin{aligned} \Delta_n(\zeta, z) &= \frac{\{\overline{D(\zeta)}D(z)\}^{-1} - (\bar{\zeta}z)^{n+1}\{D(\zeta)\overline{D(z)}\}^{-1}}{1 - \bar{\zeta}z} |D(\zeta)|^2 \\ &\qquad - \frac{1 - (\bar{\zeta}z)^{n+1}}{1 - \bar{\zeta}z} + |1 - \bar{\zeta}z|^{-1} O[(\log n)^{-\lambda}] \\ &= \frac{D(\zeta)\{D(z)\}^{-1} - 1}{1 - \bar{\zeta}z} - (\bar{\zeta}z)^{n+1} \frac{\overline{D(\zeta)}\{\overline{D(z)}\}^{-1} - 1}{1 - \bar{\zeta}z} \\ &\qquad + |1 - \bar{\zeta}z|^{-1} O[(\log n)^{-\lambda}]. \end{aligned} \tag{13.7.7}$$

Now, according to Cauchy's theorem,

$$\frac{1}{2\pi i}\int_{|\zeta|=1} F(\zeta)\,\frac{D(\zeta)\{D(z)\}^{-1}-1}{\zeta - z}\,d\zeta = 0, \tag{13.7.8}$$

since the integrand is a function of ζ which is of the class H_1 (see the remark at the end of §10.1). Here we used the condition $\lambda > 1$. And by Riemann's lemma,

$$\lim_{n\to\infty}\int_{|\zeta|=1} (\bar{\zeta}z)^{n+1} F(\zeta)\,\frac{\overline{D(\zeta)}\{\overline{D(z)}\}^{-1}-1}{1-\bar{\zeta}z}\,dx = 0, \qquad \zeta = e^{ix}. \tag{13.7.9}$$

(Concerning the uniformity cf. below.) Therefore we obtain, if ϵ is an arbitrary positive number, $E = E(z, n, \epsilon)$ is the set on $|\zeta| = 1$ with $|\zeta - z| \geqq \epsilon n^{-1}$, and E' the complementary set, the relations

$$\int_E F(\zeta)\Delta_n(\zeta, z)\,dx = -\int_{E'} F(\zeta)\left\{\frac{D(\zeta)\{D(z)\}^{-1}-1}{1-\bar{\zeta}z} - (\bar{\zeta}z)^{n+1}\frac{\overline{D(\zeta)}\{\overline{D(z)}\}^{-1}-1}{1-\bar{\zeta}z}\right\}dx$$

$$+\, o(1) + O\{(\log n)^{-\lambda}\}\int_E |1-\bar{\zeta}z|^{-1}\,dx$$

$$= O(1)\int_{E'} |1-\bar{\zeta}z|^{-1}\,|\log|z-\zeta||^{-\lambda}\,dx$$

$$+\, o(1) + O\{(\log n)^{-\lambda}\}\int_E |1-\bar{\zeta}z|^{-1}\,dx$$

$$= O\{(\log n)^{1-\lambda}\} + o(1) + O\{(\log n)^{-\lambda}\}O(\log n)$$

$$= o(1), \qquad \zeta = e^{ix}, \quad n\to\infty.$$

On the other hand, the numerator of (11.4.5) is bounded in the present case so that Theorem 1.22.2 yields (see §12.4 (2)) $s_n(\zeta, z) = O(n)$, and hence also, $\Delta_n(\zeta, z) = O(n)$. This holds uniformly for $|z| = |\zeta| = 1$. Therefore, we have

$$\int_{E'} F(\zeta)\Delta_n(\zeta, z)\,dx = O(n)\epsilon n^{-1} = \epsilon O(1), \qquad \zeta = e^{ix},$$

where $O(1)$ is independent of ϵ and n. This establishes the statement.

(4) The uniform validity of (13.7.9) needs some explanation. According to (13.7.4) (see the remark at the end of (2)),

$$\frac{\bar{D}(\zeta)\{\overline{D(z)}\}^{-1}-1}{1-\zeta z} = K(\zeta)$$

is a function of ζ which is of the class H_1. If $\rho_n(\zeta)$ denotes the nth Cesàro mean of the partial sums of the power series expansion of $F(\zeta)$, we have

$$\int_{|\zeta|=1} \bar{\zeta}^{n+1} F(\zeta)K(\bar{\zeta})\,dx = \int_{|\zeta|=1} \bar{\zeta}^{n+1}\{F(\zeta) - \rho_n(\zeta)\}K(\bar{\zeta})\,dx = \int_{E_1} + \int_{E_1'},$$

where E_1 is the set on $|\zeta| = 1$ with $|\zeta - z| \geqq \epsilon$, and E_1' the complementary set. The integral over E_1 as $n \to \infty$ is

$$O(1) \int_{E_1} |F(\zeta) - \rho_n(\zeta)|\, dx = O(1) \int_{-\pi}^{+\pi} |F(\zeta) - \rho_n(\zeta)|\, dx$$
$$= o(1).$$

The integral over E_1' (since $F(\zeta) - \rho_n(\zeta)$ is uniformly bounded) is equal to

$$O(1) \int_{E_1'} |K(\bar{\zeta})|\, dx,$$

that is, it is arbitrarily small with ϵ.

13.8. Proof of Theorem 13.1.4

(1) We show that

$$\lim_{n\to\infty} \int_{-1}^{+1} \Phi(\xi)\Delta_n'(\xi, x)(1 - \xi^2)^{-\frac{1}{2}}\, d\xi = 0, \tag{13.8.1}$$

uniformly in $-1 + \epsilon \leqq x \leqq 1 - \epsilon$; here $\Delta_n'(\xi, x)$ is the difference of the "kernels"; that is, according to Christoffel's formula

$$\begin{aligned} \Delta_n'(\xi, x) &= (1 - \xi^2)^{\frac{1}{2}} w(\xi) \frac{k_n}{k_{n+1}} \frac{p_{n+1}(x)p_n(\xi) - p_n(x)p_{n+1}(\xi)}{x - \xi} \\ &\quad - \frac{1}{2\pi}\left\{\frac{\sin (2n+1)\dfrac{\theta+\phi}{2}}{\sin\dfrac{\theta+\phi}{2}} + \frac{\sin (2n+1)\dfrac{\theta-\phi}{2}}{\sin\dfrac{\theta-\phi}{2}}\right\}, \end{aligned} \tag{13.8.2}$$

$$\xi = \cos\phi, \quad x = \cos\theta.$$

The symbol k_n is used for the highest coefficient of $p_n(x)$.

(2) We first notice that from (11.5.2) for $n \geqq 1$

$$\begin{aligned} k_n &= (2\pi)^{-\frac{1}{2}}\left\{1 + \frac{\phi_{2n}(0)}{\kappa_{2n}}\right\}^{-\frac{1}{2}} 2^n\{\kappa_{2n} + \phi_{2n}(0)\} \\ &= (2\pi)^{-\frac{1}{2}}\left\{1 + \frac{\phi_{2n}(0)}{\kappa_{2n}}\right\}^{\frac{1}{2}} 2^n \kappa_{2n} \end{aligned} \tag{13.8.3}$$

follows so that on account of formulas (12.4.3) and (12.5.3) (see also Theorem 12.7.1)

$$\begin{aligned} k_n &= (2\pi)^{-\frac{1}{2}}\{1 + O[(\log n)^{-\lambda}]\}2^n\{[D(0)]^{-1} + O[(\log n)^{-\lambda}]\} \\ &= (2\pi)^{-\frac{1}{2}}\{D(0)\}^{-1}2^n\{1 + O[(\log n)^{-\lambda}]\}, \qquad \frac{k_n}{k_{n+1}} = \tfrac{1}{2} + O[(\log n)^{-\lambda}]. \end{aligned} \tag{13.8.4}$$

From (12.1.8) we conclude, using the notation (12.2.2), that

$$\Delta_n'(\xi, x) = \Delta_n'(\cos\phi, \cos\theta)$$

$$= \pi^{-1}(x-\xi)^{-1}\{W(x)\}^{-\frac{1}{2}}\{W(\xi)\}^{\frac{1}{2}}\{\cos[(n+1)\theta+\gamma(\theta)]\cos[n\phi+\gamma(\phi)] - \cos[n\theta+\gamma(\theta)]\cos[(n+1)\phi+\gamma(\phi)]\}$$

$$(13.8.5) \qquad - \frac{1}{2\pi}\left\{\frac{\sin(2n+1)\frac{\phi+\theta}{2}}{\sin\frac{\phi+\theta}{2}} + \frac{\sin(2n+1)\frac{\phi-\theta}{2}}{\sin\frac{\phi-\theta}{2}}\right\} + |x-\xi|^{-1}O[(\log n)^{-\lambda}]$$

$$= A_n(\phi, \theta) - B_n(\phi, \theta) + |x-\xi|^{-1}O[(\log n)^{-\lambda}].$$

The first term can be written in the form

$$A_n(\phi, \theta) = (2\pi)^{-1}\left\{2\sin\frac{\phi+\theta}{2}\sin\frac{\phi-\theta}{2}\right\}^{-1}\{W(\cos\theta)\}^{-\frac{1}{2}}\{W(\cos\phi)\}^{\frac{1}{2}}$$

$$\cdot\{\cos[n\phi+(n+1)\theta+\gamma(\phi)+\gamma(\theta)] + \cos[n\phi-(n+1)\theta+\gamma(\phi)-\gamma(\theta)] - \cos[(n+1)\phi+n\theta+\gamma(\phi)+\gamma(\theta)] - \cos[(n+1)\phi-n\theta+\gamma(\phi)-\gamma(\theta)]\}$$

$$= \left\{2\pi\sin\frac{\phi+\theta}{2}\sin\frac{\phi-\theta}{2}\right\}^{-1}\{W(\cos\theta)\}^{-\frac{1}{2}}\{W(\cos\phi)\}^{\frac{1}{2}}$$

$$(13.8.6) \qquad \cdot\left\{\sin\left[(2n+1)\frac{\phi+\theta}{2}+\gamma(\phi)+\gamma(\theta)\right]\sin\frac{\phi-\theta}{2} + \sin\left[(2n+1)\frac{\phi-\theta}{2}+\gamma(\phi)-\gamma(\theta)\right]\sin\frac{\phi+\theta}{2}\right\}$$

$$= \frac{1}{2\pi}\{W(\cos\theta)\}^{-\frac{1}{2}}\{W(\cos\phi)\}^{\frac{1}{2}}\left\{\frac{\sin\left[(2n+1)\frac{\phi+\theta}{2}+\gamma(\phi)+\gamma(\theta)\right]}{\sin\frac{\phi+\theta}{2}} + \frac{\sin\left[(2n+1)\frac{\phi-\theta}{2}+\gamma(\phi)-\gamma(\theta)\right]}{\sin\frac{\phi-\theta}{2}}\right\}.$$

(3) The integrals

$$(13.8.7) \qquad \int_{-\pi}^{+\pi}\left|\frac{\{W(\cos\theta)\}^{-\frac{1}{2}}\{W(\cos\phi)\}^{\frac{1}{2}}\cos[\gamma(\phi)\pm\gamma(\theta)]-1}{\sin\frac{\phi\pm\theta}{2}}\right|d\phi, \qquad \int_{-\pi}^{+\pi}\left|\frac{\sin[\gamma(\phi)\pm\gamma(\theta)]}{\sin\frac{\phi\pm\theta}{2}}\right|d\phi$$

exist (see (12.2.4), (12.1.4), and (13.7.5)), so that by Riemann's lemma,

$$\lim_{n\to\infty}\int_{-\pi}^{+\pi}\Phi(\cos\phi)\{A_n(\phi,\theta)-B_n(\phi,\theta)\}\,d\phi=0, \tag{13.8.8}$$

uniformly in θ. Let η be an arbitrary positive number, $E = E(\theta, n, \eta)$ the set $|\cos\phi - \cos\theta| \geqq \eta n^{-1}$, and E' the complementary set. We have

$$\begin{aligned}\int_{-\pi}^{+\pi}&\Phi(\cos\phi)\Delta_n'(\cos\phi,\cos\theta)\,d\phi\\ &=\int_E\Phi(\cos\phi)\{A_n(\phi,\theta)-B_n(\phi,\theta)\}d\phi+O[(\log n)^{-\lambda}]\int_E|\cos\phi-\cos\theta|^{-1}d\phi\\ &\quad+\int_{E'}\Phi(\cos\phi)\Delta_n'(\cos\phi,\cos\theta)\,d\phi\\ &=-\int_{E'}\Phi(\cos\phi)\{A_n(\phi,\theta)-B_n(\phi,\theta)\}\,d\phi+o(1)\\ &\quad+O[(\log n)^{-\lambda}]\int_E|\cos\phi-\cos\theta|^{-1}d\phi+\int_{E'}\Phi(\cos\phi)\Delta_n'(\cos\phi,\cos\theta)d\phi.\end{aligned} \tag{13.8.9}$$

Now

$$\begin{aligned}|A_n(\phi,\theta)|&\leqq\pi^{-1}\left\{\frac{\max W(x)}{\min W(x)}\right\}^{\frac12}\left\{2n+1+\left|\frac{\sin\{\gamma(\phi)+\gamma(\theta)\}}{2\sin\dfrac{\phi+\theta}{2}}\right|\right.\\ &\qquad\left.+\left|\frac{\sin\{\gamma(\phi)-\gamma(\theta)\}}{2\sin\dfrac{\phi-\theta}{2}}\right|\right\}\\ &=O(n)+O(1)\,|\phi-\theta|^{-1}\,|\log|\phi-\theta||^{-\lambda},\\ |B_n(\phi,\theta)|&\leqq\pi^{-1}(2n+1),\end{aligned} \tag{13.8.10}$$

so that the first term in the right-hand member of (13.8.9) is

$$O(n)\int_{E'}d\phi+O(1)\int_{E'}|\phi-\theta|^{-1}\,|\log|\phi-\theta||^{-\lambda}d\phi=\eta O(1)+(\log n)^{-\lambda+1}O(1),$$

$$n\to\infty,$$

where the bound of $O(1)$ in the term $\eta O(1)$ is independent of η. The third term is $O[(\log n)^{-\lambda}]O(\log n)$. Finally, in the last term we apply Theorem 1.22.3. Since the polynomial $p_{n+1}(x)p_n(\xi) - p_n(x)p_{n+1}(\xi)$ is uniformly bounded, $-1 \leqq x \leqq +1$, $-1 \leqq \xi \leqq +1$, we have $\Delta_n'(\xi, x) = O(n)$, uniformly in x and ξ, confined to an interval of the form $[-1 + \epsilon, 1 - \epsilon]$. Thus, for the term in question, the bound $\eta n^{-1}O(n) = \eta O(1)$ can be obtained. Here $O(1)$ is again independent of η. This completes the proof of Theorem 13.1.4.

CHAPTER XIV

INTERPOLATION

In this chapter we shall consider certain problems of interpolation related to the theory of orthogonal polynomials. In particular, we are interested in interpolations whose abscissas are the zeros of the orthogonal polynomials $p_n(x)$ associated with a given distribution of the type $d\alpha(x)$ or $w(x)dx$. We deal with the ordinary Lagrange polynomials and with the "step polynomials" (Treppenpolynome) introduced by Fejér. This topic is naturally very closely related to that of the next chapter on mechanical quadrature.

Concerning the subject matter of this and of the next chapter, see the recent monograph of Feldheim (**4**).

14.1. Definitions; problems

(1) Let $[a, b]$ be a finite or infinite interval, and let

$$S_n: x_{1n} < x_{2n} < \cdots < x_{nn}, \qquad x_{1n} \geqq a, x_{nn} \leqq b \tag{14.1.1}$$

denote a set of n distinct points of this interval. Let $l(x)$ be a π_n, not identically zero, vanishing at $x = x_{\nu n}$, $\nu = 1, 2, \cdots, n$; it is determined save for a constant nonzero factor. When there is no ambiguity, we shall write x_ν instead of $x_{\nu n}$. The polynomials

$$l_\nu(x) = \frac{l(x)}{l'(x_\nu)(x - x_\nu)}, \qquad \nu = 1, 2, \cdots, n, \tag{14.1.2}$$

are called the *fundamental polynomials* of the Lagrange interpolation corresponding to the set S_n. They have the property

$$l_\nu(x_\mu) = \delta_{\nu\mu}, \qquad \nu, \mu = 1, 2, \cdots, n. \tag{14.1.3}$$

Let $f_1, f_2, \cdots, f_n$ be arbitrary values. Then the expression

$$L_n(x) = \sum_{\nu=1}^{n} f_\nu l_\nu(x) \tag{14.1.4}$$

represents the uniquely determined π_{n-1} which assumes the value f_ν at $x = x_\nu$. This is the nth *Lagrange polynomial* corresponding to the abscissas S_n. It is readily seen that

$$l_1(x) + l_2(x) + \cdots + l_n(x) = 1. \tag{14.1.5}$$

(2) Now let S_n, $n = 1, 2, 3, \cdots$, be a sequence of sets of abscissas. If $f(x)$ is a given function defined in $[a, b]$, we can consider the sequence of the corresponding Lagrange polynomials $L_n(x)$, $n = 1, 2, 3, \cdots$, defined by (14.1.4) with $f_\nu = f(x_{\nu n})$. Various convergence and divergence properties of this sequence have been studied under proper conditions of continuity concerning $f(x)$. In

what follows we shall be interested exclusively in the case in which the abscissas S_n are the zeros of the orthogonal polynomials associated with a preassigned distribution. Different types of "convergence" can then be considered; for instance:

(a) ordinary convergence: $\lim_{n\to\infty} L_n(x) = f(x)$;

(b) convergence in mean: $\lim_{n\to\infty} \int_a^b | L_n(x) - f(x) |^2 dx = 0$;

(c) generalized convergence in mean, arising from (b) by replacing the exponent 2 by p, where p is positive;

(d) quadrature convergence: $\lim_{n\to\infty} \int_a^b \{L_n(x) - f(x)\} dx = 0$.

The last type is particularly important from the point of view of Chapter XV. Of course, a certain fixed weight function can be introduced into the integral conditions.

(3) Let a and b be finite. If only the continuity of $f(x)$ is assumed, the behavior of the Lagrange polynomials is rather irregular. Faber (**2**; see also Fejér **11**, pp. 450-453, and Marcinkiewicz **1**) proved that for a given arbitrary sequence $\{S_n\}$ there exists a continuous function $f(x)$ such that the sequence $\{L_n(x)\}$ is not uniformly convergent. S. Bernstein (**4**) has even proved the existence of a continuous function for which $L_n(x)$ is unbounded at a preassigned point x_0. According to Helly's theorem (§1.6), this is equivalent to the unboundedness of the sequence of "Lebesgue constants"

$$\sum_{\nu=1}^{n} | l_\nu(x_0) |, \qquad \text{as } n \to \infty. \tag{14.1.6}$$

In the case of the special sequence

$$x_{\nu n} = \cos(2\nu - 1)\frac{\pi}{2n}, \qquad \nu = 1, 2, \cdots, n,$$

$a = -1$, $b = +1$, that is, for the zeros of $T_n(x)$, much more is known. Grünwald (**1**) and Marcinkiewicz (**2**) proved the existence of a continuous function $f(x)$ for which the sequence of Lagrange polynomials corresponding to these $x_{\nu n}$ is everywhere divergent, even everywhere unbounded.

(4) In order to obtain convergent sequences of interpolation polynomials, it is necessary to introduce additional restrictions concerning either: (a) the function $f(x)$, particularly, restrictions concerning its modulus of continuity (see Theorem 1.3.2), or (b) the interpolation polynomial, such as conditions concerning its derivative, and the like.

We introduce the polynomials

$$\begin{aligned} h_\nu(x) &= \left\{1 - \frac{l''(x_\nu)}{l'(x_\nu)}(x - x_\nu)\right\}\{l_\nu(x)\}^2 \\ &= v_\nu(x)\{l_\nu(x)\}^2, \\ \mathfrak{h}_\nu(x) &= (x - x_\nu)\{l_\nu(x)\}^2, \end{aligned} \tag{14.1.7}$$

called the *fundamental polynomials* of the first and second kind of the "Hermite interpolation" corresponding to the set S_n. These π_{2n-1} are completely determined by the conditions

$$h_\nu(x_\mu) = \delta_{\nu\mu}, \; h'_\nu(x_\mu) = 0; \qquad \mathfrak{h}_\nu(x_\mu) = 0, \qquad \mathfrak{h}'_\nu(x_\mu) = \delta_{\nu\mu}, \qquad \nu, \mu = 1, 2, \cdots, n, \tag{14.1.8}$$

and for any given set of values f_ν, f'_ν,

$$W_n(x) = \sum_{\nu=1}^{n} f_\nu h_\nu(x) + \sum_{\nu=1}^{n} f'_\nu \mathfrak{h}_\nu(x) \tag{14.1.9}$$

represents the uniquely determined π_{2n-1} for which

$$W_n(x_\nu) = f_\nu, \qquad W'_n(x_\nu) = f'_\nu, \qquad \nu = 1, 2, \cdots, n. \tag{14.1.10}$$

(5) Again, let S_n, $n = 1, 2, 3, \cdots$, be a sequence of sets of abscissas. If $f(x)$ is a given function having a derivative in $[a, b]$, we can take $f_\nu = f(x_{\nu n})$, $f'_\nu = f'(x_{\nu n})$ and consider the corresponding "Hermite polynomials"[62] $W_n(x)$, $n = 1, 2, 3, \cdots$, defined by (14.1.9). If only the continuity of $f(x)$ is known, we may choose f'_ν arbitrarily; for instance, we may take $f'_\nu = 0$. If $f'_\nu = 0$, we call $W_n(x)$ the *step polynomials* corresponding to $\{S_n\}$. In the more general case $|f'_\nu| < A$, where A is a constant independent of ν and n, they are called *generalized step polynomials*.

The simple and generalized step polynomials $W_n(x)$ display a more regular behavior as $n \to \infty$ than do the ordinary Lagrange polynomials $L_n(x)$. They coincide with the given function $f(x)$ at the same points as do the corresponding Lagrange polynomials, but they satisfy certain additional restrictions concerning their first derivatives. Their degree is $2n - 1$ instead of $n - 1$. We shall show that for certain sequences $\{S_n\}$ the step polynomials (even the generalized step polynomials) are uniformly convergent if $f(x)$ is an arbitrary continuous function.

The step polynomials and their generalizations have been introduced and investigated by Fejér (**10**, **11**, **13**, **16**). The trigonometric analogue for $f'_\nu = 0$ had been previously considered in the simplest case of equidistant abscissas by Jackson (**4**, p. 145, Theorem VI).

For the fundamental polynomials (14.1.7) the following important relations hold:

$$\begin{gathered} h_1(x) + h_2(x) + \cdots + h_n(x) = 1, \\ \sum_{\nu=1}^{n} x_\nu h_\nu(x) + \sum_{\nu=1}^{n} \mathfrak{h}_\nu(x) = x. \end{gathered} \tag{14.1.11}$$

An important consequence of the first identity may be pointed out for the step polynomials, that is, when $f'_\nu = 0$. Let the set $\{S_n\}$ be such that

$$h_\nu(x) \geqq 0, \qquad a \leqq x \leqq b, \; \nu = 1, 2, \cdots, n. \tag{14.1.12}$$

[62] We write "Hermite polynomials" in quotation marks in order to avoid confusion with the Hermite polynomials of Chapter V.

Then (14.1.9) implies

$$\min f_\nu \leqq W_n(x) \leqq \max f_\nu, \qquad a \leqq x \leqq b,\ n = 1, 2, 3, \cdots. \tag{14.1.13}$$

We observe further that (14.1.12) is equivalent to the fact that the linear functions

$$v_\nu(x) = 1 - \frac{l''(x_\nu)}{l'(x_\nu)}(x - x_\nu) \tag{14.1.14}$$

do not vanish in the open interval $a < x < b$, or, what amounts to the same thing, that the "conjugate points"

$$x_\nu + \frac{l'(x_\nu)}{l''(x_\nu)}, \qquad \nu = 1, 2, \cdots, n, \tag{14.1.15}$$

lie outside of this interval. (See Fejér **13**, **16**.)

14.2. Fundamental polynomials of the Lagrange interpolation

(1) THEOREM 14.2.1.[63] *Let $d\alpha(x)$ be an arbitrary distribution on the interval $[a, b]$, $\{p_n(x)\}$ the associated orthonormal polynomials, and $l_1(x), l_2(x), \cdots, l_n(x)$ the fundamental polynomials* (14.1.2) *of the Lagrange interpolation corresponding to the set of zeros of $p_n(x)$. Then we have*

$$\int_a^b l_\nu(x) l_\mu(x)\, d\alpha(x) = \lambda_\nu \delta_{\nu\mu}, \qquad \nu, \mu = 1, 2, \cdots, n, \tag{14.2.1}$$

where λ_ν are the Christoffel numbers defined by (3.4.1).

This follows immediately from (3.4.1), since $l_\nu(x)l_\mu(x)$ vanishes at the zeros of $p_n(x)$ if $\nu \neq \mu$. For $\nu = \mu$ the zero x_ν is the only exception, and we have $\{l_\nu(x_\nu)\}^2 = 1$.

Also, by (14.1.8),

$$\begin{aligned} &\int_a^b h_\nu(x)\, d\alpha(x) = \lambda_\nu, && \int_a^b h_\nu'(x)\, d\alpha(x) = \int_a^b x h_\nu'(x)\, d\alpha(x) = 0;\\ &\int_a^b \mathfrak{h}_\nu(x)\, d\alpha(x) = 0, && \int_a^b \mathfrak{h}_\nu'(x)\, d\alpha(x) = \lambda_\nu, \qquad \int_a^b x\mathfrak{h}_\nu'(x)\, dx = \lambda_\nu x_\nu;\end{aligned} \tag{14.2.2}$$

$$\nu = 1, 2, \cdots, n.$$

(2) We obtain as an important consequence the following result:

THEOREM 14.2.2. *Let $K_n(x_0, x)$ have the same meaning as in* (3.1.9). *Then the following identity holds:*

$$K_{n-1}(x_0, x) = \sum_{\nu=0}^{n-1} p_\nu(x_0) p_\nu(x) = \sum_{\nu=1}^{n} \lambda_\nu^{-1} l_\nu(x_0) l_\nu(x). \tag{14.2.3}$$

[63] Erdös-Turán **1**.

Indeed, let $\rho(x)$ be an arbitrary π_{n-1}. The polynomial $K_{n-1}(x_0, x)$ is uniquely determined by the condition (3.1.12). But by (14.1.4) and (14.2.1),

$$\int_a^b \left\{\sum_{\nu=1}^n \lambda_\nu^{-1} \overline{l_\nu(x_0)} l_\nu(x)\right\} \left\{\sum_{\mu=1}^n \rho(x_\mu) l_\mu(x)\right\} d\alpha(x)$$

$$= \sum_{\nu=1}^n \rho(x_\nu) \overline{l_\nu(x_0)} = \sum_{\nu=1}^n \rho(x_\nu) l_\nu(\bar{x}_0) = \rho(\bar{x}_0),$$

and this establishes the statement.

On setting $x_0 = x = x_\nu$ in (14.2.3), we obtain a new proof of (3.4.8).

(3) In view of (14.2.1), (14.1.4) shows that

$$\int_a^b |L_n(x)|^2 d\alpha(x) = \sum_{\nu=1}^n \lambda_\nu |f_\nu|^2. \tag{14.2.4}$$

Furthermore, let $f(x)$ be an arbitrary complex-valued function for which the integrals $\int_a^b f(x) x^\nu d\alpha(x)$, $\nu = 0, 1, \cdots, n-1$, exist. Then (14.2.3) yields the identity

$$\sum_{\nu=0}^{n-1} \left| \int_a^b f(x) p_\nu(x) d\alpha(x) \right|^2 = \sum_{\nu=1}^n \lambda_\nu^{-1} \left| \int_a^b f(x) l_\nu(x) d\alpha(x) \right|^2. \tag{14.2.5}$$

Thus, according to Bessel's inequality (see (3.1.5)),

$$\sum_{\nu=1}^n \lambda_\nu^{-1} \left| \int_a^b f(x) l_\nu(x) d\alpha(x) \right|^2 \leqq \int_a^b |f(x)|^2 d\alpha(x), \tag{14.2.6}$$

provided the last integral exists.

14.3. Convergence in mean of Lagrange polynomials

(1) THEOREM 14.3.1.[64] *Let $d\alpha(x)$ be an arbitrary distribution on the finite interval $[a, b]$, and let $\{p_n(x)\}$ be the corresponding set of orthonormal polynomials. For the complex-valued function $f(x)$ let the Riemann-Stieltjes integrals*

$$\int_a^b f(x) x^n d\alpha(x), \qquad \int_a^b |f(x)|^2 d\alpha(x), \qquad n = 0, 1, 2, \cdots, \tag{14.3.1}$$

exist. Then if $L_n(x)$ denotes the Lagrange polynomial of degree $n - 1$ which coincides with $f(x)$ at the zeros of $p_n(x)$, we have

$$\lim_{n \to \infty} \int_a^b |f(x) - L_n(x)|^2 d\alpha(x) = 0. \tag{14.3.2}$$

The convergence in mean in the usual sense follows from (14.3.2) for any $f(x)$ integrable in Riemann's sense in the case where $d\alpha(x) = w(x)dx$, $w(x) \geqq \mu > 0$.

For the proof we use (14.2.4), (14.2.6), and the later Theorem 15.2.3 on quadrature convergence. Now

[64] Cf. Erdös-Turán 1 and Shohat 8.

$$
\begin{aligned}
\int_a^b |f(x) - L_n(x)|^2 d\alpha(x) &= \int_a^b |f(x)|^2 d\alpha(x) + \sum_{\nu=1}^{n} \lambda_\nu |f(x_\nu)|^2 \\
&\qquad - 2\Re \sum_{\nu=1}^{n} \overline{f(x_\nu)} \int_a^b f(x) l_\nu(x)\, d\alpha(x) \\
&\leqq \int_a^b |f(x)|^2 d\alpha(x) + \sum_{\nu=1}^{n} \lambda_\nu |f(x_\nu)|^2 \\
&\qquad + 2\left\{\sum_{\nu=1}^{n} \lambda_\nu |f(x_\nu)|^2\right\}^{\frac{1}{2}} \left\{\sum_{\nu=1}^{n} \lambda_\nu^{-1} \left| \int_a^b f(x) l_\nu(x)\, d\alpha(x)\right|^2\right\}^{\frac{1}{2}} \\
&\leqq \int_a^b |f(x)|^2 d\alpha(x) + \sum_{\nu=1}^{n} \lambda_\nu |f(x_\nu)|^2 \\
&\qquad + 2\left\{\sum_{\nu=1}^{n} \lambda_\nu |f(x_\nu)|^2\right\}^{\frac{1}{2}} \left\{\int_a^b |f(x)|^2 d\alpha(x)\right\}^{\frac{1}{2}}.
\end{aligned} \tag{14.3.3}
$$

Hence, using Theorem 15.2.3, we obtain

$$
\limsup_{n\to\infty} \int_a^b |f(x) - L_n(x)|^2 d\alpha(x) \leqq 4 \int_a^b |f(x)|^2 d\alpha(x). \tag{14.3.4}
$$

Next let ϵ be an arbitrary positive number, and $\rho(x)$ a polynomial for which (Theorem 1.5.2)

$$
\int_a^b |f(x) - \rho(x)|^2 d\alpha(x) < \epsilon. \tag{14.3.5}
$$

If $L_n(f;x)$ denotes the Lagrange polynomial of degree $n - 1$ corresponding to $f(x)$, we have

$$
f(x) - L_n(f;x) = f(x) - \rho(x) - L_n(f - \rho; x), \tag{14.3.6}
$$

provided n exceeds the degree of $\rho(x)$. This establishes the statement.

(2) In case $d\alpha(x) = (1 - x^2)^{-\frac{1}{2}}dx$, $a = -1$, $b = +1$ (that is, for the Tchebichef abscissas of the first kind) Erdös and Feldheim showed that we may even assert the validity of the following theorem:

THEOREM 14.3.2.[65] *Let p be an arbitrary positive number, and let $f(x)$ be a continuous function. Then for the Lagrange polynomials corresponding to the Tchebichef abscissas of the first kind,*

$$
\lim_{n\to\infty} \int_{-1}^{+1} |f(x) - L_n(x)|^p (1 - x^2)^{-\frac{1}{2}} dx = 0. \tag{14.3.7}
$$

It suffices to show this for even integral values of p. In the proof which is based on induction with respect to the even integer p essential use is made of the property of the Tchebichef abscissas formulated in Problem 57 (see below).

[65] Erdös-Feldheim **1**; cf. also Feldheim **2** and **3**, pp. 33-36.

Feldheim points out (**2**, p. 330) that (14.3.2) is not true in general if $d\alpha(x) = (1 - x^2)^{\frac{1}{2}}dx$, $a = -1$, $b = +1$ (that is, for the Tchebichef abscissas of the second kind), and if the exponent 2 is replaced by $p = 4$. For the same abscissas it is also *not* true that

$$\lim_{n\to\infty} \int_{-1}^{+1} | f(x) - L_n(x) |^2 \, dx = 0. \tag{14.3.8}$$

In both cases the superior limit of the integrals in question may be $+\infty$ if $f(x)$ is a properly chosen continuous function.

(3) From Theorem 14.3.1 the mean convergence in the sense of (14.3.8) follows immediately for the Jacobi abscissas, that is, for the zeros of the Jacobi polynomial $P_n^{(\alpha,\beta)}(x)$, provided $\max(\alpha, \beta) \leqq 0$ (cf. Erdös-Turán **1**); here $f(x)$ is an arbitrary continuous (or even Riemann-integrable) function.

Recently, Á. Holló (**1**) proved this mean convergence for the Jacobi abscissas with $\max(\alpha, \beta) < \frac{1}{2}$ provided $f(x)$ is continuous. The bound $\frac{1}{2}$ is the precise one, on account of the last result of Feldheim mentioned in (2). Holló also investigated the validity of

$$\lim_{n\to\infty} \int_{-1}^{+1} | f(x) - L_n(f) | \, dx = 0, \tag{14.3.9}$$

where $f(x)$ is continuous, and showed that (14.3.9) holds for $\max(\alpha, \beta) < \frac{3}{2}$. The latter bound is again the precise one, at least in the sense that for $\max(\alpha, \beta) > \frac{3}{2}$ and for a proper continuous $f(x)$, the statement (14.3.9) does not hold. (This follows from the second part of Theorem 15.4.)

14.4. Lagrange polynomials for Jacobi abscissas

For the following discussion Theorem 8.9.1 will be found invaluable, while the bounds of $P_n^{(\alpha,\beta)}(\cos\theta)$ obtained in §7.32 will also be used.

(1) Assume $\alpha > -1$, $\beta > -1$, and let $x_1 > x_2 > \cdots > x_n$ denote the zeros of the Jacobi polynomial $P_n^{(\alpha,\beta)}(x)$ in decreasing order. Here $x_\nu = \cos\theta_\nu$, $0 < \theta_\nu < \pi$. Then we may assert the following theorem:

THEOREM 14.4. *Let $f(x)$ be continuous in $[-1, +1]$ with the modulus of continuity $\omega(\delta)$.*[66] *Then the Lagrange polynomials coinciding with $f(x)$ at the zeros of $P_n^{(\alpha,\beta)}(x)$ converge uniformly to $f(x)$ in every interval $[-1 + \epsilon, 1 - \epsilon]$, where $0 < \epsilon < \frac{1}{2}$, provided that $\omega(\delta) = o(|\log \delta|^{-1})$. The same holds in the interval $[-1 + \epsilon, 1]$ if either $\alpha \leqq -\frac{1}{2}$ and $\omega(\delta) = o(|\log\delta|^{-1})$ or $\mu - \frac{1}{2} \leqq \alpha < \mu + \frac{1}{2}$*[67] *and $f(x)$ has a continuous derivative of order μ with modulus of continuity $\omega_\mu(\delta)$ satisfying the condition $\omega_\mu(\delta) = o(\delta^{\alpha-\mu+\frac{1}{2}})$, $\mu = 0, 1, 2, \cdots$.*

In case $\alpha < -\frac{1}{2}$ the Lagrange polynomials are convergent at the point $x = +1$ if $f(x)$ is an arbitrary continuous function.

There exist functions continuous in $[-1, +1]$ whose Lagrange polynomials are

[66] Cf. Theorem 1.3.2.

[67] For $\mu = 0$ the equality sign, that is, the case $\alpha = -1/2$, is excluded.

divergent (unbounded) at a preassigned point x_0, $-1 < x_0 < +1$; *the same is true at the point* $x_0 = +1$ *provided that* $\alpha \geqq -\frac{1}{2}$.

Similar statements hold in the interval $[-1, 1 - \epsilon]$ and at the point $x = -1$ if we replace α by β. The convergence is uniform in the whole interval $[-1, +1]$ provided $\max(\alpha, \beta) \leqq -\frac{1}{2}$ and $\omega(\delta) = o(|\log \delta|^{-1})$, or $\mu - \frac{1}{2} \leqq \max(\alpha, \beta) < \mu + \frac{1}{2}$ and $f(x)$ has a continuous derivative of order μ with modulus of continuity $\omega_\mu(\delta) = o(\delta^{\max(\alpha,\beta)-\mu+\frac{1}{2}})$, $\mu = 0, 1, 2, \cdots$. Again the equality sign in case $\mu = 0$, that is, the case $\max(\alpha, \beta) = -\frac{1}{2}$, is excluded.

Compare Fejér **13**, pp. 22, 24, 27; Shohat **5**, p. 146. The present results are more precise than those obtained in these papers.

(2) We shall start with a discussion of the "Lebesgue constants"

$$\sum_{\nu=1}^{n} |l_\nu(x_0)|, \tag{14.4.1}$$

where $-1 < x_0 < +1$. Let δ be a fixed positive number, $\delta < 1 - |x_0|$. Then $P_n^{(\alpha,\beta)}(x_0) = O(n^{-\frac{1}{2}})$, so that

$$\begin{aligned} \sum_{|x_\nu - x_0| > \delta} |l_\nu(x_0)| &= O(n^{-\frac{1}{2}}) \sum_{|x_\nu - x_0| > \delta} |P_n^{(\alpha,\beta)\prime}(x_\nu)|^{-1} \\ &= O(n^{-\frac{1}{2}}) \sum_{\nu=1}^{n} |P_n^{(\alpha,\beta)\prime}(x_\nu)|^{-1}. \end{aligned} \tag{14.4.2}$$

According to (8.9.2)

$$\sum_{\nu=1}^{n} |P_n^{(\alpha,\beta)\prime}(x_\nu)|^{-1} = O(1) \sum_{\nu=1}^{n} \nu^{\alpha+\frac{3}{2}} n^{-\alpha-2} + O(1) \sum_{\nu=1}^{n} \nu^{\beta+\frac{3}{2}} n^{-\beta-2} = O(n^{\frac{1}{2}}). \tag{14.4.3}$$

Consequently,

$$\sum_{|x_\nu - x_0| > \delta} |l_\nu(x_0)| = O(1). \tag{14.4.4}$$

On the other hand, assume $|x_\nu - x_0| \leqq \delta$. For a fixed ν we have (see (8.8.2))

$$l_\nu(x_0) = O(n^{-\frac{1}{2}}) \frac{P_n^{(\alpha,\beta)}(x_\nu) - P_n^{(\alpha,\beta)}(x_0)}{x_\nu - x_0} = O(1), \tag{14.4.5}$$

so that if $x_0 = \cos\theta_0$, $0 < \theta_0 < \pi$,

$$\begin{aligned} \sum_{|x_\nu - x_0| \leqq \delta} |l_\nu(x_0)| &= \sum_{n^{-1} < |\theta_\nu - \theta_0| \leqq \delta'} |l_\nu(x_0)| + O(1) \\ &= O(n^{-\frac{1}{2}}) O(n^{-\frac{1}{2}}) \sum_{n^{-1} < |\theta_\nu - \theta_0| \leqq \delta'} |\theta_\nu - \theta_0|^{-1} + O(1), \end{aligned} \tag{14.4.6}$$

where δ' is a fixed positive number. According to (8.9.1) the last expression is $O(\log n)$. The same bound holds for (14.4.1), and does so uniformly if the condition $-1 + \epsilon \leqq x_0 \leqq 1 - \epsilon$ is satisfied.

We can also show that

$$\sum_{|x_\nu - x_0| \leqq \delta} |l_\nu(x_0)| \sim \log n, \tag{14.4.7}$$

if n tends to $+\infty$ *over a proper sequence of integers*. It suffices to choose n so that $|\cos(N\theta + \gamma)| \geqq \cos \epsilon$, where N and γ have the same meaning as in Darboux's formula (Chapter VIII), and $\epsilon = \frac{1}{2} \min (\theta_0, \pi - \theta_0)$. This is obviously possible since from

$$(m + \tfrac{1}{2})\pi - \epsilon < N\theta_0 + \gamma < (m + \tfrac{1}{2})\pi + \epsilon,$$

m integral, the inequalities

$$(m + \tfrac{1}{2})\pi + \epsilon < (N + 1)\theta_0 + \gamma < (m + \tfrac{3}{2})\pi - \epsilon$$

follow. Now the formula of Darboux coupled with the previous argument and (8.9.1) gives us the desired proof.

(3) As a consequence of the last result, we may conclude by means of Theorem 1.6 (Helly's theorem) the existence of a continuous function $f(x)$ whose Lagrange polynomials are unbounded at the interior point $x = x_0$. On the other hand, let $f(x)$ have the modulus of continuity $\omega(\delta) = o(|\log \delta|^{-1})$, and let $L_n(f; x)$ be the Lagrange polynomial of degree $n - 1$ corresponding to $f(x)$. Approximate $f(x)$ by a $\pi_{n-1} = \rho(x)$ such that

$$f(x) - \rho(x) = o[(\log n)^{-1}], \qquad -1 \leqq x \leqq +1 \tag{14.4.8}$$

(Theorem 1.3.2). Then (see (14.3.6))

$$\begin{aligned} |L_n(f; x) - f(x)| &= |L_n(f - \rho; x) - \{f(x) - \rho(x)\}| \\ &= o\{(\log n)^{-1}\} O(\log n) = o(1). \end{aligned} \tag{14.4.9}$$

(4) Assume $1 - \delta \leqq x_0 \leqq 1$. On putting $\mu = n + 1 - \nu$, $\mu = O(n)$, we have, by Theorem 7.32.2 and (8.9.2),

$$\begin{aligned} \sum_{|x_\nu - x_0| > \delta} |l_\nu(x_0)| &= O(1) |P_n^{(\alpha,\beta)}(x_0)| \sum_{|x_\nu - x_0| > \delta} |P_n^{(\alpha,\beta)\prime}(x_\nu)|^{-1} \\ &= O(n^a) \sum \mu^{\beta + \frac{3}{2}} n^{-\beta-2} = O(n^{a+\frac{1}{2}}), \end{aligned} \tag{14.4.10}$$

$$a = \max (\alpha, -\tfrac{1}{2}).$$

We now pass to the determination of an upper bound for

$$\begin{aligned} \sum_{|x_\nu - x_0| \leqq \delta} |l_\nu(x_0)| &= \sum \left| \frac{P_n^{(\alpha,\beta)}(\cos \theta_0)}{\cos \theta_0 - \cos \theta_\nu} \right| |P_n^{(\alpha,\beta)\prime}(\cos \theta_\nu)|^{-1} \\ &= O(1) \, |P_n^{(\alpha,\beta)}(\cos \theta_0)| \sum \frac{\nu^{\alpha+\frac{3}{2}} n^{-\alpha-2}}{|\theta_0^2 - \theta_\nu^2|}, \end{aligned} \tag{14.4.11}$$

(since $(\theta_0^2 - \theta^2)/(\cos \theta_0 - \cos \theta)$ is bounded). In what follows we use both bounds (14.4.11). We have by (4.21.7),

$$(14.4.12)\quad \left|\frac{P_n^{(\alpha,\beta)}(\cos\theta_0)-P_n^{(\alpha,\beta)}(\cos\theta_\nu)}{\cos\theta_0-\cos\theta_\nu}\right| |P_n^{(\alpha,\beta)\prime}(\cos\theta_\nu)|^{-1} = O(1)\,|P_{n-1}^{(\alpha+1,\beta+1)}(\cos\tau)|\,\nu^{\alpha+\frac{3}{2}}n^{-\alpha-1},$$

where τ is between θ_0 and θ_ν.

Let $\theta_0 = n^{-1}\xi$, and consider first the case $\xi = O(1)$. Let $\nu = O(1)$; then (14.4.12) becomes $O(1)n^{\alpha+1}\nu^{\alpha+\frac{3}{2}}n^{-\alpha-1} = O(1)$. On the other hand, if ν is larger than a sufficiently large fixed positive number, the second expression in the right-hand member of (14.4.11) becomes (see the second bound in (7.32.5))

$$(14.4.13)\quad O(1)n^\alpha \sum \nu^{\alpha+\frac{3}{2}}n^{-\alpha-2}\theta_\nu^{-2} = O(1)\sum \nu^{\alpha-\frac{1}{2}} = O(n^{\alpha+\frac{1}{2}}),\ O(\log n),\ O(1),$$

according as $\alpha > -\frac{1}{2}$, $\alpha = -\frac{1}{2}$, or $\alpha < -\frac{1}{2}$.

Now let ξ be "large" and $\xi - \nu\pi = O(1)$, so that the number of these values of ν is bounded. Then we obtain for (14.4.12) (see the first bound in (7.32.5))

$$(14.4.14)\quad O(1)\tau^{-\alpha-\frac{3}{2}}n^{-\frac{1}{2}}\nu^{\alpha+\frac{3}{2}}n^{-\alpha-1} = O(1)(\nu/n)^{-\alpha-\frac{3}{2}}n^{-\frac{1}{2}}\nu^{\alpha+\frac{3}{2}}n^{-\alpha-1} = O(1).$$

Finally, assume that both ξ and $\xi - \nu\pi$ are "large". Then the second expression in the right-hand member of (14.4.11) becomes (see the first bound in (7.32.5))

$$(14.4.15)\quad O(\xi^{-\alpha-\frac{1}{2}})\sum\frac{\nu^{\alpha+\frac{3}{2}}}{|\xi^2-\nu^2\pi^2|} = \sum\nolimits_1 + \sum\nolimits_2 + \sum\nolimits_3,$$

where the summation is extended over $\nu\pi \leqq \xi/2$, $\xi/2 < \nu\pi \leqq 3\xi/2$, $\nu\pi > 3\xi/2$, respectively. Here we have to take into account the fact that the ratio

$$\left|\frac{\xi^2-\{\nu\pi+O(1)\}^2}{\xi^2-\nu^2\pi^2}\right|$$

has a positive lower bound. In the second sum $|\xi - \nu\pi|$ is larger than a fixed positive constant. We find now

$$\sum\nolimits_1 = O(\xi^{-\alpha-\frac{1}{2}})O(\xi^{-2})\sum_{\nu\pi\leqq\xi/2}\nu^{\alpha+\frac{3}{2}} = O(1),$$

$$(14.4.16)\quad \sum\nolimits_2 = O(\xi^{-\alpha-\frac{1}{2}})O(\xi^{\alpha+\frac{1}{2}})\sum_{\xi/2<\nu\pi\leqq3\xi/2}|\xi-\nu\pi|^{-1} = O(\log\xi) = O(\log n),$$

$$\sum\nolimits_3 = O(\xi^{-\alpha-\frac{1}{2}})\sum_{\nu\pi>3\xi/2}\nu^{\alpha-\frac{1}{2}} = O(n^{\alpha+\frac{1}{2}}),\ O(\log n),\ O(1),$$

according as $\alpha > -\frac{1}{2}$, $\alpha = -\frac{1}{2}$, or $\alpha < -\frac{1}{2}$.

Recapitulating, we obtain for the Lebesgue constant (14.4.1), the bounds $O(n^{\alpha+\frac{1}{2}})$ and $O(\log n)$, uniformly in $-1+\epsilon \leqq x_0 \leqq 1$, according as $\alpha > -\frac{1}{2}$ or $\alpha \leqq -\frac{1}{2}$. A similar result is found for the interval $-1 \leqq x_0 \leqq 1-\epsilon$ by replacing α by β. In the whole interval $-1 \leqq x_0 \leqq +1$ we obtain the bounds $O(n^{\gamma+\frac{1}{2}})$ and $O(\log n)$ for $\gamma > -\frac{1}{2}$ and $\gamma \leqq -\frac{1}{2}$, respectively, where we have $\gamma = \max(\alpha, \beta)$.

Now an argument similar to that in (3) must be used. Assume first that

$-1 + \epsilon \leqq x_0 \leqq 1$, and let $f(x)$ satisfy the conditions of Theorem 14.4. According to Theorem 1.3.3 a $\pi_{n-1} = \rho(x)$ exists such that

$$(14.4.17) \qquad f(x) - \rho(x) = \begin{cases} n^{-\mu} o(n^{-\alpha+\mu-\frac{1}{2}}) = o(n^{-\alpha-\frac{1}{2}}), \\ o[(\log n)^{-1}], & -1 \leqq x \leqq +1, \end{cases}$$

according as $\alpha > -\frac{1}{2}$ or $\alpha \leqq -\frac{1}{2}$. This establishes the statement concerning the convergence in the interval $[-1 + \epsilon, 1]$. The proof is obviously the same for $[-1, 1 - \epsilon]$ and for $[-1, +1]$.

(5) There remain to be discussed the "Lebesgue constants" for $x_0 = +1$, that is, the expressions

$$(14.4.18) \qquad \sum_{\nu=1}^{n} | l_\nu(1) | \sim n^\alpha \sum_{\nu=1}^{n} (1 - x_\nu)^{-1} | P_n^{(\alpha,\beta)\prime}(x_\nu) |^{-1}.$$

The positive zeros x_ν furnish a contribution $\sim n^\alpha \sum (\nu/n)^{-2} \nu^{\alpha+\frac{3}{2}} n^{-\alpha-2} = \sum \nu^{\alpha-\frac{1}{2}}$, which is $O(n^{\alpha+\frac{1}{2}})$, $O(\log n)$, $O(1)$, according as $\alpha > -\frac{1}{2}$, $\alpha = -\frac{1}{2}$, $\alpha < -\frac{1}{2}$. The contribution of the negative zeros is

$$n^\alpha \sum | P_n^{(\alpha,\beta)\prime}(x_\nu) |^{-1} \sim n^\alpha \sum \nu^{\beta+\frac{3}{2}} n^{-\beta-2} \sim n^{\alpha+\frac{1}{2}}.$$

And now, one more step, namely, the application of Helly's theorem, yields the desired proof of Theorem 14.4.

14.5. Preliminary discussion of the step polynomials in the classical cases

We calculate the linear functions $v_\nu(x)$ occurring in (14.1.7) for the Jacobi, Laguerre, and Hermite abscissas, respectively. We assume $\alpha > -1$, $\beta > -1$ in the first and $\alpha > -1$ in the second case.

(1) From (4.2.1) we obtain in the Jacobi case, since $l(x_\nu) = 0$,

$$(14.5.1) \qquad \frac{l''(x_\nu)}{l'(x_\nu)} = \frac{\alpha - \beta + (\alpha + \beta + 2)x_\nu}{1 - x_\nu^2}.$$

Whence

$$(14.5.2) \quad v_\nu(x) = \frac{1 - x[\alpha - \beta + (\alpha + \beta + 2)x_\nu] + (\alpha - \beta)x_\nu + (\alpha + \beta + 1)x_\nu^2}{1 - x_\nu^2}.$$

In particular,

$$(14.5.3) \qquad \begin{aligned} v_\nu(-1) &= \frac{(1 + \alpha)(1 + x_\nu) - \beta(1 - x_\nu)}{1 - x_\nu}, \\ v_\nu(+1) &= \frac{(1 + \beta)(1 - x_\nu) - \alpha(1 + x_\nu)}{1 + x_\nu}. \end{aligned}$$

The zeros x_ν are everywhere dense in $[-1, +1]$ (Theorem 6.1.1) if n is large. Thus $v_\nu(-1)$ is non-negative for each ν and n if and only if $\beta \leqq 0$. Similarly, $v_\nu(+1) \geqq 0$ if and only if $\alpha \leqq 0$. Since $v_\nu(x)$ is linear, we obtain the following result:

THEOREM 14.5. *The fundamental polynomials $h_\nu(x)$ of the first kind associated with the Jacobi abscissas, are non-negative in $-1 \leqq x \leqq +1$ for all values of ν and n, if and only if*

$$-1 < \alpha \leqq 0, \qquad -1 < \beta \leqq 0. \tag{14.5.4}$$

(2) In the Laguerre case (cf. (5.1.2), first equation)

$$v_\nu(x) = \frac{x_\nu(x_\nu - \alpha) + x(\alpha + 1 - x_\nu)}{x_\nu}; \qquad v_\nu(0) = x_\nu - \alpha. \tag{14.5.5}$$

Here $v_\nu(x)$ changes its sign for all values of α if ν and n are properly chosen.

In the Hermite case (cf. (5.5.2), first equation) we obtain

$$v_\nu(x) = 1 - 2x_\nu x + 2x_\nu^2. \tag{14.5.6}$$

14.6. Step polynomials and "Hermite polynomials" for Jacobi abscissas

We again assume $\alpha > -1, \beta > -1$ and use the same notation as in §14.4.

THEOREM 14.6. *Let $f(x)$ be continuous in $[-1, +1]$. The generalized step polynomials* (14.1.9) *$[f_\nu = f(x_\nu), |f'_\nu| < A]$ converge uniformly to $f(x)$ over every interval $[-1 + \epsilon, 1 - \epsilon]$. The same holds over the interval $[-1 + \epsilon, 1]$ provided $\alpha < 0$. The step polynomials are in general divergent at $x = +1$ if $f(x)$ is merely continuous and $\alpha \geqq 0$.*

The "Hermite polynomials" (14.1.9) *$[f_\nu = f(x_\nu), f'_\nu = f'(x_\nu)]$ converge uniformly to $f(x)$ in $[-1 + \epsilon, 1]$ if $\alpha < \frac{1}{2}$ and $f(x)$ has a continuous second derivative, or if $\mu/2 \leqq \alpha < (\mu + 1)/2$ and $f(x)$ has a continuous $(\mu + 1)$st derivative with modulus of continuity $\omega_{\mu+1}(\delta)$ satisfying the condition $\omega_{\mu+1}(\delta) = o(\delta^{2\alpha-\mu})$, $\mu = 1, 2, 3, \cdots$.*

Similar statements hold in $[-1, 1 - \epsilon]$ and at $x = -1$ if we replace α by β, and in the whole interval $[-1, +1]$ if we replace α by $\max(\alpha, \beta)$. (Cf. Shohat **5**, pp. 138–139; Szegö **14**.)

(1) We start with the discussion of the convergence for $-1 < x_0 < +1$. Here again Theorem 8.9.1 is virtually indispensable.

If $x = x_\nu$, the numerator of $v_\nu(x)$ in (14.5.2) is

$$\begin{aligned} &1 - x_\nu[\alpha - \beta + (\alpha + \beta + 2)x_\nu] + (\alpha - \beta)x_\nu + (\alpha + \beta + 1)x_\nu^2 \\ &\qquad = 1 - x_\nu^2 > 0. \end{aligned} \tag{14.6.1}$$

Therefore, $v_\nu(x_0)$ is positive if $|x_\nu - x_0|$ is sufficiently small, that is, $|x_\nu - x_0| \leqq \delta$. The same is of course true for $h_\nu(x_0)$. Furthermore, $v_\nu(x_0)$ has for these ν a positive lower bound which is independent of δ. We therefore obtain, on account of the first identity in (14.1.11),

$$\sum_{|x_\nu - x_0| \leqq \delta} |h_\nu(x_0)| = \sum_{|x_\nu - x_0| \leqq \delta} h_\nu(x_0) = 1 - \sum_{|x_\nu - x_0| > \delta} h_\nu(x_0), \tag{14.6.2}$$

and from (14.1.7) we see that

$$\sum_{|x_\nu - x_0| \leqq \delta} |\mathfrak{h}_\nu(x_0)| \leqq \delta \sum_{|x_\nu - x_0| \leqq \delta} \frac{|h_\nu(x_0)|}{v_\nu(x_0)} = O(1)\delta \sum_{|x_\nu - x_0| \leqq \delta} h_\nu(x_0). \tag{14.6.3}$$

Here the bound $O(1)$ is independent of δ.

We shall now find a bound for the corresponding sums if $|x_\nu - x_0| > \delta$. From (14.5.2) and (8.9.2) we have

$$\begin{aligned}
\sum_{|x_\nu - x_0| > \delta} |h_\nu(x_0)| &= O(1) \sum_{|x_\nu - x_0| > \delta} (1 - x_\nu^2)^{-1}\{l_\nu(x_0)\}^2 \\
&= O(1)\{P_n^{(\alpha,\beta)}(x_0)\}^2 \sum_{|x_\nu - x_0| > \delta} (1 - x_\nu^2)^{-1}\{P_n^{(\alpha,\beta)\prime}(x_\nu)\}^{-2} \\
&= O(n^{-1}) \sum_{\nu=1}^{n} (\nu/n)^{-2}\nu^{2\alpha+3}n^{-2\alpha-4} + O(n^{-1}) \sum_{\nu=1}^{n} (\nu/n)^{-2}\nu^{2\beta+3}n^{-2\beta-4} \\
&= O(n^{-1}).
\end{aligned} \tag{14.6.4}$$

Moreover, we notice

$$\begin{aligned}
\sum_{|x_\nu - x_0| > \delta} |\mathfrak{h}_\nu(x_0)| &= O(1) \sum_{|x_\nu - x_0| > \delta} \{l_\nu(x_0)\}^2 \\
&= O(1)\{P_n^{(\alpha,\beta)}(x_0)\}^2 \sum_{|x_\nu - x_0| > \delta} \{P_n^{(\alpha,\beta)\prime}(x_\nu)\}^{-2} \\
&= O(n^{-1}) \sum_{\nu=1}^{n} \nu^{2\alpha+3}n^{-2\alpha-4} + O(n^{-1}) \sum_{\nu=1}^{n} \nu^{2\beta+3}n^{-2\beta-4} \\
&= O(n^{-1}).
\end{aligned} \tag{14.6.5}$$

This yields the convergence of the generalized step polynomials of a continuous function if $-1 < x_0 < +1$. Indeed, according to (14.1.9) and (14.1.11),

$$\begin{aligned}
|W_n(x_0) - f(x_0)| &\leqq \sum_{\nu=1}^{n} |f(x_\nu) - f(x_0)|\,|h_\nu(x_0)| + A \sum_{\nu=1}^{n} |\mathfrak{h}_\nu(x_0)| \\
&\leqq \max_{|x_\nu - x_0| \leqq \delta} |f(x_\nu) - f(x_0)| \sum_{|x_\nu - x_0| \leqq \delta} h_\nu(x_0) + O(1) \sum_{|x_\nu - x_0| \leqq \delta} |\mathfrak{h}_\nu(x_0)| \\
&\quad + 2\max|f(x)| \sum_{|x_\nu - x_0| > \delta} |h_\nu(x_0)| + O(1) \sum_{|x_\nu - x_0| > \delta} |\mathfrak{h}_\nu(x_0)| \\
&= \max_{|x_\nu - x_0| \leqq \delta} |f(x_\nu) - f(x_0)|\,O(1) + \delta O(1) + O(n^{-1}) + O(n^{-1}).
\end{aligned} \tag{14.6.6}$$

The factors $O(1)$ in the last expression are independent of δ.

(2) Now assume $\alpha < 0$ and $1 - \delta \leqq x_0 \leqq 1$. The second formula in (14.5.3) shows that $v_\nu(x_0)$ and $h_\nu(x_0)$ are again positive and $v_\nu(x_0)$ is bounded from zero if $|x_\nu - x_0| \leqq \delta$, provided δ is sufficiently small. (We have $v_\nu(x_0) \geqq v_\nu(+1)$ if $\alpha - \beta + (\alpha + \beta + 2)x_\nu > 0$.) Then the analogues of (14.6.2) and (14.6.3) follow immediately. In (14.6.4) and (14.6.5) a slight modification is necessary due to the fact that in this case $P_n^{(\alpha,\beta)}(x_0) = O(n^a)$, $a = \max(\alpha, -\frac{1}{2})$. (The formulas corresponding to (14.6.4), (14.6.5) hold for arbitrary $\alpha > -1$; this remark is used in (3).) Since $a < 0$, the conclusions of (1) remain valid; this establishes the convergence of the generalized step polynomials if $\alpha < 0$ and $f(x)$

is continuous. (Of course, the same is true for the "Hermite polynomials" if $f'(x)$ is bounded.)

(3) We postpone the discussion of the step polynomials at $x = +1$ if $f(x)$ is continuous and $\alpha \geqq 0$, and we pass on to a discussion of the "Hermite polynomials" for $1 - \delta \leqq x_0 \leqq 1$ with α arbitrary but greater than -1. First we observe that the numerator of (14.5.2) vanishes for $x = x_\nu = +1$; thus we have for $|x_\nu - x_0| \leqq \delta$,

$$|v_\nu(x_0)| < (1 - x_\nu^2)^{-1}\epsilon(\delta), \tag{14.6.7}$$

where $\epsilon(\delta) \to 0$ if $\delta \to 0$. Now,

$$\begin{aligned} \sum_{|x_\nu - x_0| \leqq \delta} |h_\nu(x_0)| &= \epsilon(\delta) \sum_{|x_\nu - x_0| \leqq \delta} (1 - x_\nu^2)^{-1}\{l_\nu(x_0)\}^2 \\ &= \epsilon(\delta) \sum_{|x_\nu - x_0| \leqq \delta} (1 - x_\nu^2)^{-1}\left\{\frac{P_n^{(\alpha,\beta)}(x_0)}{x_0 - x_\nu}\right\}^2 \{P_n^{(\alpha,\beta)\prime}(x_\nu)\}^{-2}. \end{aligned} \tag{14.6.8}$$

By use of the notation and argument of §14.4 (4), we obtain for the sum above:

(a) $O(n^2)$ if $\xi = O(1)$, $\nu = O(1)$;

(b) $O(1) \sum \nu^{2\alpha-1}(\nu/n)^{-2} = O(n^{2\alpha})$, $O(n^2 \log n)$, $O(n^2)$, according as $\alpha > 1$, $\alpha = 1$, $\alpha < 1$, if $\xi = O(1)$, and ν is "large";

(c) $O(1)(\nu/n)^{-2} = O(n^2)$ if ξ is "large" and $\xi - \nu\pi = O(1)$;

(d) $O(1)\xi^{-2\alpha-1} \sum \nu^{2\alpha+3}(\xi^2 - \nu^2\pi^2)^{-2} (\nu/n)^{-2} = \sum_1' + \sum_2' + \sum_3'$ if both ξ and $\xi - \nu\pi$ are large. The summations in the last three sums are extended over the same values of ν as in (14.4.16). We have

$$\begin{aligned} \textstyle\sum_1' &= O(\xi^{-2\alpha-5}) \sum_{\nu\pi \leqq \xi/2} \nu^{2\alpha+3}(\nu/n)^{-2} = O(\xi^{-3} n^2) = O(n^2), \\ \textstyle\sum_2' &= O(1) \sum_{\xi/2 < \nu\pi \leqq 3\xi/2} (\xi - \nu\pi)^{-2}(\nu/n)^{-2} = O(\xi^{-2} n^2) = O(n^2), \\ \textstyle\sum_3' &= O(\xi^{-2\alpha-1}) \sum_{\nu\pi > 3\xi/2} \nu^{2\alpha-1}(\nu/n)^{-2} = O(n^{2\alpha}), \qquad O(n^2 \log n), \qquad O(n^2), \end{aligned} \tag{14.6.9}$$

according as $\alpha > 1$, $\alpha = 1$, or $\alpha < 1$. To these cases the bounds

$$\sum_{|x_\nu - x_0| \leqq \delta} |h_\nu(x_0)| = \epsilon(\delta)O(n^{2\alpha}),\ \epsilon(\delta)O(n^2 \log n),\ \epsilon(\delta)O(n^2) \tag{14.6.10}$$

correspond.

In order to obtain the analogous bounds for $\mathfrak{h}_\nu(x_0)$, we cancel in (14.6.8) the factor $(1 - x_\nu^2)^{-1} \sim (\nu/n)^{-2}$. Thus, we readily obtain

$$\sum_{|x_\nu - x_0| \leqq \delta} |\mathfrak{h}_\nu(x_0)| = \delta O(n^{2\alpha}),\quad \delta O(\log n),\quad \delta O(1), \tag{14.6.11}$$

according as $\alpha > 0$, $\alpha = 0$, or $\alpha < 0$.

The corresponding bounds for $|x_\nu - x_0| > \delta$ are $O(n^{2\alpha})$ (cf. the remark made in (2)). Thus, (14.1.9) and the first formula in (14.1.11) give the result

(14.6.12)

$$
\begin{aligned}
| W_n(x_0) - f(x_0) | \leqq & \max_{|x_\nu - x_0| \leqq \delta} | f(x_\nu) - f(x_0) | \epsilon(\delta) \begin{Bmatrix} O(n^{2\alpha}) \\ O(n^2 \log n) \\ O(n^2) \end{Bmatrix} \\
& + \max_{|x_\nu - x_0| \leqq \delta} | f'(x_\nu) | \delta \begin{Bmatrix} O(n^{2\alpha}) \\ O(\log n) \\ O(1) \end{Bmatrix} + \max | f(x) | O(n^{2a}) \\
& + \max | f'(x) | O(n^{2a}); \; a = \max(\alpha, -\tfrac{1}{2}).
\end{aligned}
$$

In the first term we have the alternative $\alpha > 1$, $\alpha = 1$, or $\alpha < 1$, in the second term $\alpha > 0$, $\alpha = 0$, or $\alpha < 0$. The O-expressions are independent of $f(x)$ and δ.

We now apply the usual argument. Let $W_n(f; x)$ be the "Hermite polynomial" corresponding to $f(x)$, and $\rho(x)$ an arbitrary π_{2n-1}. Then

(14.6.13)
$$W_n(f; x_0) - f(x_0) = W_n(f - \rho; x_0) - \{f(x_0) - \rho(x_0)\}.$$

Under the condition mentioned in Theorem 14.6 we can determine $\rho(x)$ (cf. Theorem 1.3.3) such that

(14.6.14)

$$
\begin{aligned}
& f(x) - \rho(x) = o(n^{-2}), \qquad f'(x) - \rho'(x) = o(n^{-1}) \quad \text{if } \alpha < \tfrac{1}{2}, \\
& f(x) - \rho(x) = o(n^{-2\alpha-1}), \qquad f'(x) - \rho'(x) = o(n^{-2\alpha}), \\
& \qquad \text{if } \tfrac{1}{2}\mu \leq \alpha < \tfrac{1}{2}(\mu + 1), \; \mu = 1, 2, 3, \cdots.
\end{aligned}
$$

This establishes the statement concerning the "Hermite polynomials."

(4) Finally, we discuss (cf. 14.5.3))

(14.6.15)
$$\sum_{\nu=1}^{n} | h_\nu(1) | = \sum_{\nu=1}^{n} \left| \frac{(1 + \beta)(1 - x_\nu) - \alpha(1 + x_\nu)}{1 + x_\nu} \right| \{l_\nu(1)\}^2.$$

The part of this sum defined by $x_\nu \geqq 1 - \delta$ is, for $\alpha \neq 0$,

(14.6.16)
$$\sim n^{2\alpha} \sum_{x_\nu \geqq 1-\delta} (1 - x_\nu)^{-2} \{P_n^{(\alpha,\beta)\prime}(x_\nu)\}^{-2} \sim n^{2\alpha} \sum_{x_\nu \geqq 1-\delta} (\nu/n)^{-4} \nu^{2\alpha+3} n^{-2\alpha-4}$$

which is of the order $n^{2\alpha}$ or 1, according as $\alpha > 0$ or $\alpha < 0$. This shows that the step polynomials (and also the generalized step polynomials) of a continuous function are in general divergent at $x = +1$ if $\alpha > 0$. (The convergence for $\alpha < 0$ has been proved in (2).) The possibility of divergence in case $\alpha = 0$ follows by choosing $f(x) = 1 - x$. The corresponding step polynomial is in fact

(14.6.17)

$$
\begin{aligned}
\sum_{\nu=1}^{n} h_\nu(1) f(x_\nu) &= (1 + \beta) \sum_{\nu=1}^{n} \frac{(1 - x_\nu)^2}{1 + x_\nu} \{l_\nu(1)\}^2 \\
&= (1 + \beta) \sum_{\nu=1}^{n} (1 + x_\nu)^{-1} \{P_n^{(0,\beta)\prime}(x_\nu)\}^{-2} \\
&\sim \sum (\nu/n)^{-2} \nu^{2\beta+3} n^{-2\beta-4} \sim 1,
\end{aligned}
$$

so that it cannot tend to $f(1) = 0$. Still simpler is the proof of the divergence of the step polynomials of $f(x) = (1 + \beta)(1 - x) - \alpha(1 + x)$ if $\alpha > 0$, since $f(1) = -2\alpha < 0$.

14.7. Step polynomials for Laguerre abscissas

Assume $\alpha > -1$, and let $x_1 < x_2 < \cdots < x_n$ denote the zeros of the Laguerre polynomial $L_n^{(\alpha)}(x)$. We prove the following theorem:

THEOREM 14.7.[68] *Let $f(x)$ be continuous for $x \geqq 0$ and $f(x) = O(x^m)$ if $x \to +\infty$; here m is an arbitrary but fixed positive number. The generalized step polynomials* (14.1.9) *($f_\nu = f(x_\nu)$, $|f_\nu'| < A$) converge uniformly to $f(x)$ over every positive interval $\epsilon \leqq x \leqq \omega$. The same holds in the interval $0 \leqq x \leqq \omega$ provided $\alpha < 0$. The step polynomials are in general divergent at $x = 0$ if $f(x)$ is continuous and $\alpha \geqq 0$.*

For the proof we shall need considerations similar to those in the Jacobi case (§14.6). In particular, we shall use Theorem 8.9.2. Some modifications in the argument are necessary due to the fact that the zeros are unbounded. The mechanical quadrature appears as an important new tool (cf. (15.3.5)).

(1) Assume $0 < \epsilon \leqq x_0 \leqq \omega$. If $v_\nu(x)$ has the same meaning as in (14.5.5), the values $v_\nu(x_0)$ and $h_\nu(x_0)$ are positive, and $v_\nu(x_0)$ is bounded from zero and infinity provided $|x_\nu - x_0|$ is sufficiently small. Thus for small δ,

$$\sum_{|x_\nu - x_0| \leqq \delta} |h_\nu(x_0)| = \sum_{|x_\nu - x_0| \leqq \delta} h_\nu(x_0) = 1 - \sum_{|x_\nu - x_0| > \delta} h_\nu(x_0), \tag{14.7.1}$$

$$\sum_{|x_\nu - x_0| \leqq \delta} |\mathfrak{h}_\nu(x_0)| = \delta O(1) \sum_{|x_\nu - x_0| \leqq \delta} h_\nu(x_0). \tag{14.7.2}$$

Here $O(1)$ is independent of δ.

If x_ν is small, $v_\nu(x_0) = O(x_\nu^{-1})$; if x_ν is large, $v_\nu(x_0) = O(x_\nu)$. Therefore (cf. (7.6.8))

$$\begin{aligned} \sum_{|x_\nu - x_0| > \delta} |h_\nu(x_0)| &= O(1) \sum_{x_\nu < x_0 - \delta} x_\nu^{-1} \left\{ \frac{L_n^{(\alpha)}(x_0)}{L_n^{(\alpha)\prime}(x_\nu)(x_0 - x_\nu)} \right\}^2 \\ &\quad + O(1) \sum_{x_\nu > x_0 + \delta} x_\nu \left\{ \frac{L_n^{(\alpha)}(x_0)}{L_n^{(\alpha)\prime}(x_\nu)(x_0 - x_\nu)} \right\}^2 \\ &= O(n^{\alpha - \frac{1}{2}}) \sum_{x_\nu < x_0 - \delta} x_\nu^{-1} \{L_n^{(\alpha)\prime}(x_\nu)\}^{-2} + O(n^{\alpha - \frac{1}{2}}) \sum_{x_\nu > x_0 + \delta} x_\nu^{-1} \{L_n^{(\alpha)\prime}(x_\nu)\}^{-2} \\ &= O(n^{\alpha - \frac{1}{2}}) \sum_{\nu=1}^{n} x_\nu^{-1} \{L_n^{(\alpha)\prime}(x_\nu)\}^{-2}. \end{aligned} \tag{14.7.3}$$

But combining (15.3.5) with (3.4.5), we find

$$\sum_{\nu=1}^{n} x_\nu^{-1} \{L_n^{(\alpha)\prime}(x_\nu)\}^{-2} = \frac{\Gamma(n+1)\Gamma(\alpha+1)}{\Gamma(n+\alpha+1)}, \tag{14.7.4}$$

which yields the bound $O(n^{\alpha - \frac{1}{2}})O(n^{-\alpha}) = O(n^{-\frac{1}{2}})$ for (14.7.3).

[68] Cf. Shohat **5**, p. 139; Szegö **14**, p. 597.

More generally, we find from (15.3.5) and (3.4.1), if m is a positive integer, $m \leqq 2n - 1$, that

$$\sum_{\nu=1}^{n} x_\nu^{m-1}\{L_n^{(\alpha)\prime}(x_\nu)\}^{-2} = \frac{\Gamma(n+1)\Gamma(m+\alpha+1)}{\Gamma(n+\alpha+1)}. \tag{14.7.5}$$

Hence the same argument as before leads to

$$\sum_{|x_\nu - x_0|>\delta} x_\nu^m \,|\, h_\nu(x_0) \,| = O(n^{-\frac{1}{2}}), \tag{14.7.6}$$

m fixed. We have also

$$\sum_{|x_\nu - x_0|>\delta} |\, \mathfrak{h}_\nu(x_0) \,| = O(n^{-\frac{1}{2}}). \tag{14.7.7}$$

Equations (14.7.1), (14.7.2), (14.7.6), and (14.7.7) establish the uniform convergence of the step polynomials in the interval $\epsilon \leqq x_0 \leqq \omega$ (cf. (14.6.6)).

(2) Now assume $\alpha < 0$ and $0 \leqq x_0 \leqq \delta$. If δ is sufficiently small, and $|\, x_\nu - x_0 \,| \leqq \delta$, both $v_\nu(x_0)$ and $h_\nu(x_0)$ are positive, and $v_\nu(x_0)$ is bounded from zero. Hence (14.7.1) and (14.7.2) are again valid. In (14.7.3) only the bound of $\{L_n^{(\alpha)}(x_0)\}^2$ must be changed. According to (7.6.11) this will be $O(n^{2a})$ where $a = \max(\frac{1}{2}\alpha - \frac{1}{4}, \alpha)$. Therefore,

$$\sum_{|x_\nu - x_0|>\delta} |\, h_\nu(x_0) \,| = O(n^{2a-\alpha}), \tag{14.7.8}$$

and the same bound holds for the sums in (14.7.6) and (14.7.7). Since the exponent $2a - \alpha = \max(-\frac{1}{2}, \alpha) < 0$, these sums tend to zero. From this the uniform convergence in $0 \leqq x \leqq \omega$ follows.

(3) The case $x_0 = 0$, $\alpha \geqq 0$, can be readily disposed of by choosing $f(x) = x - \alpha$. We have

$$\begin{aligned} \sum_{\nu=1}^{n} f(x_\nu)h_\nu(0) &= \sum_{\nu=1}^{n} (x_\nu - \alpha)^2\{l_\nu(0)\}^2 \\ &\cong n^{2\alpha}\{\Gamma(\alpha+1)\}^{-2} \sum_{\nu=1}^{n} \left(\frac{x_\nu - \alpha}{x_\nu}\right)^2 \{L_n^{(\alpha)\prime}(x_\nu)\}^{-2}. \end{aligned} \tag{14.7.9}$$

Since this expression is positive, it cannot tend to $f(0) = -\alpha$ if α is positive. If $\alpha = 0$, the last expression in (14.7.9) is 1 (cf. (14.7.5), $m = 1$), and $f(0) = 0$.

14.8. Lagrange polynomials for certain general classes of abscissas

(1) Let $x_{1n} > x_{2n} > \cdots > x_{nn}$ denote the zeros of the nth orthogonal polynomial $p_n(x)$ associated with the weight function $w(x)$ in the interval $-1 \leqq x \leqq +1$. We consider two classes A, B of weight functions characterized by the following conditions:

A. There exists a positive number μ such that

$$w(x) \geqq \mu, \qquad -1 \leqq x \leqq +1. \tag{14.8.1}$$

B. There exists a positive number μ such that

(14.8.2) $$w(x) \geqq \mu(1-x^2)^{-\frac{1}{2}}, \qquad -1 < x < +1.$$

By using this notation we prove the following statement:

THEOREM 14.8. *Let $f(x)$ be defined in $-1 \leqq x \leqq +1$. Let $\{L_n(f; x)\}$ denote the sequence of the Lagrange polynomials coinciding with $f(x)$ at the zeros $x_{\nu n}$ of the orthogonal polynomials $p_n(x)$ associated with the weight function $w(x)$, $-1 \leqq x \leqq +1$. Then $\lim_{n\to\infty} L_n(f; x) = f(x)$, uniformly in the interval $[-1, +1]$, provided that $w(x)$ belongs to A and $f(x)$ has a continuous derivative in $[-1, +1]$. The same conclusion holds if $w(x)$ belongs to B and $\omega(\delta) = o(\delta^{\frac{1}{2}})$. Moreover $\lim_{n\to\infty} L_n(f; x) = f(x)$, uniformly in the interval $[-1+\epsilon, 1-\epsilon]$, where $0 < \epsilon < 1$, provided that $w(x)$ belongs to A and $\omega(\delta) = o(\delta^{\frac{1}{2}})$.*

Compare Shohat **7**, Grünwald-Turán **1**. Here, as before, $\omega(\delta)$ is the modulus of continuity of $f(x)$ in $[-1, +1]$.

(2) We show that for the fundamental polynomials of Lagrange interpolation

(14.8.3) $$\sum_{\nu=1}^{n} |l_\nu(x)| = \begin{cases} O(n), & -1 \leqq x \leqq +1, \\ O(n^{\frac{1}{2}}), & -1 \leqq x \leqq +1, \\ O(n^{\frac{1}{2}}), & -1+\epsilon \leqq x \leqq 1-\epsilon, \end{cases}$$

where $w(x)$ belongs to A in the first and third case, and to B in the second case. From (14.8.3) the statement follows by reference to Theorems 1.3.2 and 1.3.3.

Let x be fixed, $\epsilon_\nu = \operatorname{sgn} l_\nu(x)$. We write in the case A

(14.8.4) $$\rho(t) = \sum_{\nu=1}^{n} \epsilon_\nu l_\nu(t) = \sum_{\nu=0}^{n-1} c_\nu P_\nu(t),$$

where $P_\nu(t)$ is the νth Legendre polynomial. Then

(14.8.5) $$\begin{aligned} \rho(x) = \sum_{\nu=1}^{n} |l_\nu(x)| = \sum_{\nu=0}^{n-1} c_\nu P_\nu(x) &\leqq \left\{\sum_{\nu=0}^{n-1} \frac{c_\nu^2}{\nu+\frac{1}{2}}\right\}^{\frac{1}{2}} \left\{\sum_{\nu=0}^{n-1} (\nu+\tfrac{1}{2})[P_\nu(x)]^2\right\}^{\frac{1}{2}} \\ &= \left\{\int_{-1}^{+1} [\rho(t)]^2\, dt\right\}^{\frac{1}{2}} \left\{\sum_{\nu=0}^{n-1} (\nu+\tfrac{1}{2})[P_\nu(x)]^2\right\}^{\frac{1}{2}} \\ &\leqq \mu^{-\frac{1}{2}} \left\{\int_{-1}^{+1} w(t)[\rho(t)]^2\, dt\right\}^{\frac{1}{2}} \left\{\sum_{\nu=0}^{n-1} (\nu+\tfrac{1}{2})[P_\nu(x)]^2\right\}^{\frac{1}{2}}. \end{aligned}$$

Now, according to (14.2.4),

(14.8.6) $$\int_{-1}^{+1} w(t)[\rho(t)]^2\, dt = \sum_{\nu=1}^{n} \lambda_\nu[\rho(x_\nu)]^2 = \sum_{\nu=1}^{n} \lambda_\nu \epsilon_\nu^2 = \sum_{\nu=1}^{n} \lambda_\nu = \int_{-1}^{+1} w(t)\, dt.$$

Statement (14.8.3) is readily derived by using (7.21.1) and (7.3.8).

The only essential modification of the proof in the case B is that we write

(14.8.7) $$\rho(t) = \sum_{\nu=1}^{n} \epsilon_\nu l_\nu(t) = \sum_{\nu=0}^{n-1} d_\nu T_\nu(t),$$

where $T_\nu(t)$ is Tchebichef's polynomial of the first kind.

14.9. Further results on interpolation

We indicate briefly a few more recent results on interpolation.

(1) We employ the notation of §14.8. Let the weight function $w(x)$ be continuous with a positive minimum in $[-1, +1]$ and let $\epsilon > 0$. Then

$$\max |l_\nu(x)| \to 1 \quad \text{as} \quad n \to \infty$$

where the maximum is taken in the interval $[-1 + \epsilon, 1 - \epsilon]$ and the limit relation holds for any sequence $\nu = \nu(n)$ for which the corresponding zero $x_\nu = x_{\nu n}$ is in $[-1 + \epsilon, 1 - \epsilon]$. (Erdös-Lengyel **1**.)

(2) The result of Grünwald (**1**) and Marcinkiewicz (**2**) mentioned in §14.1 (3) has been deepened by Erdös-Grünwald and Erdös. We consider the zeros of $T_n(x)$ as the set of abscissas.

Erdös-Grünwald (**1**) have shown the existence of a continuous function $f(x) = f(\cos\theta)$ the Fourier series of which is uniformly convergent and at the same time the sequence of the corresponding Lagrange polynomials $L_n(x)$ is everywhere divergent, even everywhere unbounded.

Erdös (**2**) has shown that if $x_0 = \cos(p\pi/q)$, p and q odd, there exist continuous functions $f(x)$ for which $L_n(x_0) \to \infty$. Erdös-Turán (**1**) have shown previously that this cannot hold for any other points x_0; see also Erdös **2**, p. 313.

(3) Important results have been obtained by Erdös for the "normal" sets of abscissas introduced and investigated by Fejér. They are characterized by the property that the "conjugate points" (14.1.15) lie outside of the interval $[a, b]$ of interpolation.

Fejér showed that for a normal set $x_\nu - x_{\nu-1}$ tends to zero as $n \to \infty$. Erdös-Turán (**2**) have sharpened this result and Erdös (**1**) proved that

$$x_\nu - x_{\nu-1} = \frac{\pi}{n}(1 - x_\nu^2)^{-\frac{1}{2}} + O(n^{-\frac{3}{2}})$$

provided that $x_\nu = x_{\nu n}$ is restricted to a fixed interval $[-1 + \epsilon, 1 - \epsilon]$.

Erdös (**1**) solves the following remarkable extremum problem. We consider all normal sets $\{x_{\nu n}\}$, ν and n fixed. (We follow the notation (14.1.1).) What is the minimum and maximum of $x_{\nu n}$? They are given by z_ν and $-z_{n-\nu}$ where $z_\nu = z_{\nu n}$ are the zeros of $P_n(z) + P_{n-1}(z)$.

(4) Grünwald (**2**) and Webster (**1**) have studied a certain type of "summability" for Lagrange polynomials similar to a procedure of W. Rogosinski for the partial sums of Fourier series. The abscissas used are the zeros of the Tchebichef polynomials of the first and second kind, respectively.

Grünwald (**3**) gives a survey of divergence properties of the Lagrange polynomials. This paper contains also a discussion of the convergence of the Lagrange and Hermite interpolation polynomials under the condition that the abscissas form a normal set.

(5) Balázs-Turán (**1**, **2**) and Surányi-Turán (**1**) have investigated various properties of certain polynomials of interpolation connected with ultraspherical abscissas. The novel feature of this investigation is the study of inter-

polation polynomials for which $f(x)$ and $f''(x)$ are prescribed.

(6) Erdös (3) has investigated the "Lebesgue function"

$$\sum_{\nu=1}^{n} |l_\nu(x)|,$$

see (14.1.6). Denoting the maximum of this function in the interval $-1 \leqq x \leqq 1$, by M_n, he proved the following inequality

$$M_n > (2/\pi)\log n - c.$$

This is a rather deep result. It is known that for the zeros of Tchebichef polynomials

$$M_n < (2/\pi)\log n + c.$$

Here c is a positive absolute constant.

(7) The more difficult analogous problem concerning the fundamental polynomials $\mathfrak{h}_\nu(x)$ of Hermite interpolation has been solved by Erdös-Turán (4), who prove that

$$\max_{-1 \leqq x \leqq +1} \sum_{\nu=1}^{n} |\mathfrak{h}_\nu(x)| \geqq \frac{1}{n}\left(\frac{2}{\pi}\log n - c_2 \log\log n\right).$$

This is "asymptotically best possible" as the case of Tchebichef abscissae shows.

The paper Erdös-Turán 4 also contains a proof that

$$\max_{-1 \leqq x \leqq +1} \sum_{\nu=1}^{n} |l_\nu(x)| > \frac{2}{\pi}\log n - c_3 \log\log n,$$

which is weaker than the result cited in (6) above, but preceded it in time.

(8) If $L_n(x)$ is the Lagrange interpolation polynomial of a continuous function $f(x)$ with interpolation at the zeros of $P_n^{(\alpha,\beta)}(x)$, $\alpha, \beta \geqq -\frac{1}{2}$, then

$$\lim_{n\to\infty} \int_{-1}^{1} |L_n(x) - f(x)|^p (1-x)^\alpha (1+x)^\beta dx = 0 \tag{14.9.1}$$

when $p < \min(4(\alpha+1)/(2\alpha+1), 4(\beta+1)/(2\beta+1)) = A(\alpha,\beta)$ and for $p > A(\alpha,\beta)$ there is a continuous function $f(x)$ for which (14.9.1) fails. See Askey 6.

(9) By simple methods of interpolation Egerváry-Turán 1 have proved the following beautiful identity

$$1 - [P_n(x)]^2 = \sum_{\nu=1}^{n} \frac{1-x^2}{1-x_\nu^2}\left\{\frac{P_n(x)}{P_n'(x_\nu)(x-x_\nu)}\right\}^2,$$

where the x_ν are the zeros of the Legendre polynomial $P_n(x)$. From this they derive a new proof of the inequality (7.21.1). By similar reasoning one can obtain a new proof of the inequality (7.21.3). See Egerváry-Turán 2.

CHAPTER XV

MECHANICAL QUADRATURE

The reader will recall that the Gauss-Jacobi mechanical quadrature was studied in §3.4. In the present chapter we turn to other mechanical quadrature problems which are also connected with the theory of orthogonal polynomials.

15.1. Definitions

(1) Let $[a, b]$ be a finite or infinite interval, and let

$$S_n : x_{1n} < x_{2n} < \cdots < x_{nn}, \qquad a \leqq x_{1n}, x_{nn} \leqq b, \tag{15.1.1}$$

denote a set of n distinct points in $[a, b]$. Furthermore, let

$$\Lambda_n : \lambda_{1n}, \lambda_{2n}, \cdots, \lambda_{nn} \tag{15.1.2}$$

be a set of real numbers. If $f(x)$ is an arbitrary function defined in $[a, b]$, we write

$$Q_n(f) = \sum_{\nu=1}^{n} \lambda_{\nu n} f(x_{\nu n}). \tag{15.1.3}$$

We call the numbers $x_{\nu n}$ the abscissas and the numbers $\lambda_{\nu n}$ the Cotes numbers of the "mechanical quadrature" $Q_n(f)$. Having been given the sequences of corresponding sets $\{S_n\}$ and $\{\Lambda_n\}$, $n = 1, 2, 3, \cdots$, we are interested in the convergence properties of the sequence

$$Q_1(f), Q_2(f), \cdots, Q_n(f), \cdots \tag{15.1.4}$$

associated with a given function $f(x)$.

As in Chapter XIV we write x_ν and λ_ν instead of $x_{\nu n}$ and $\lambda_{\nu n}$ when there is no ambiguity.

(2) An important special case is the following. Let the set $\{S_n\}$ be an arbitrary set of distinct numbers in $[a, b]$, and let $u(x)$ be a given non-decreasing function. We shall define the Cotes numbers λ_ν by requiring that

$$Q_n(f) = \int_a^b f(x)\, du(x) \tag{15.1.5}$$

shall hold if $f(x)$ is an arbitrary π_{n-1}. Obviously, under such a condition,

$$\lambda_\nu = \int_a^b l_\nu(x)\, du(x), \qquad \nu = 1, 2, \cdots, n, \tag{15.1.6}$$

where $l_\nu(x)$ denote the fundamental polynomials (14.1.2) of the Lagrange interpolation corresponding to the abscissas $\{S_n\}$. In this case $Q_n(f)$ is called a quadrature of the interpolatory type.

The Gauss-Jacobi quadrature is seen to be a special case of the interpolatory type when the abscissas $\{S_n\}$ are the zeros of the orthogonal polynomials associated with the distribution $du(x)$. Another remarkable case of the interpolatory type is $u(x) = x$ with $\{S_n\}$ arbitrary. The integral (15.1.5) is then the ordinary integral of $f(x)$ over the interval $[a, b]$, which is assumed to be finite in this case.

We shall use the previous notation throughout the whole chapter.

15.2. A general convergence theorem on mechanical quadrature; theorem of Stekloff-Fejér

(1) THEOREM 15.2.1.[69] *Let $[a, b]$ be a finite interval, and let the system $\{S_n\}$ (in $[a, b]$) and $\{\Lambda_n\}$ be arbitrary; let $Q_n(f)$ be defined by* (15.1.3). *Denote by $u(x)$ a non-decreasing function, and assume the "quadrature convergence"*

$$\lim_{n\to\infty} Q_n(f) = \int_a^b f(x)\,du(x) \tag{15.2.1}$$

for an arbitrary polynomial $f(x)$. Then a necessary and sufficient condition for the validity of (15.2.1) *for an arbitrary continuous function $f(x)$ is the boundedness of the sequence of the "Lebesgue constants"*

$$|\lambda_{n1}| + |\lambda_{n2}| + \cdots + |\lambda_{nn}|, \qquad n \to \infty. \tag{15.2.2}$$

This theorem is an immediate consequence of Theorem 1.6 (Helly's theorem). It is to be noted that condition (15.2.1) holds for an arbitrary polynomial $f(x)$ if the quadrature is of the interpolatory type (cf. §15.1 (2)).

(2) As an application we shall prove the following important theorem of Stekloff and Fejér:

THEOREM 15.2.2.[70] *Let $[a, b]$ be a finite interval, and let the Cotes numbers $\{\Lambda_n\}$ be non-negative. If the quadrature convergence* (15.2.1) *for an arbitrary polynomial $f(x)$ is assumed, the same convergence can be stated for an arbitrary function for which the Riemann-Stieltjes integral in the right-hand member of* (15.2.1) *exists.*

The expression (15.2.2) remains bounded in the present case, since

$$\sum_{\nu=1}^{n} |\lambda_{\nu n}| = \sum_{\nu=1}^{n} \lambda_{\nu n} = Q_n(1) \tag{15.2.3}$$

has a limit as $n \to \infty$. Therefore, (15.2.1) holds for a continuous $f(x)$. The

[69] Cf. Pólya **4**, p. 267, Theorem I.

[70] Cf. Stekloff **2**, pp. 176–179; Fejér **15**, p. 291; Pólya **4**, p. 282, d); Shohat **7**, pp. 474–476.

extension to Riemann-integrable functions is made by means of Theorem 1.5.4.

Theorem 15.2.2 holds without change if the Cotes numbers $\{\Lambda_n\}$ are non-negative only for sufficiently large values of n.

(3) A remarkable special case of Theorem 15.2.1 is obtained by considering an arbitrary quadrature of the interpolatory type (defined by a monotonic non-decreasing function $u(x)$), and by choosing for the set $\{S_n\}$ the zeros of the orthogonal polynomials $\{p_n(x)\}$ associated with a preassigned distribution $d\alpha(x)$. Here $\alpha(x)$ is in general different from $u(x)$. Then (15.2.1) holds for an arbitrary polynomial $f(x)$.

The Gauss-Jacobi quadrature is derived as a special case by taking $\alpha(x) = u(x)$. The Cotes numbers are then identical with the Christoffel numbers of §3.4 and are all positive. Applying Theorem 15.2.2, we obtain the following:

THEOREM 15.2.3. *Let $d\alpha(x)$ be an arbitrary distribution on the finite interval $[a, b]$, and let $Q_n(f)$ be the corresponding Gauss-Jacobi mechanical quadrature (that is, $x_{\nu n}$ are the zeros of the orthogonal polynomials $p_n(x)$ associated with $d\alpha(x)$ and $\lambda_{\nu n}$ the corresponding Christoffel numbers). Then the "quadrature convergence"*

$$\lim_{n\to\infty} Q_n(f) = \int_a^b f(x)\, d\alpha(x) \tag{15.2.4}$$

holds for an arbitrary function $f(x)$ for which the Riemann-Stieltjes integral in the right-hand member exists.

(4) Another remarkable special case arises upon choosing $u(x) = x$ with $\alpha(x)$ arbitrary. Such a choice permits us to assert the following theorem:

THEOREM 15.2.4.[71] *Let x_ν be the zeros of the orthogonal polynomial $p_n(x)$ associated with the arbitrary distribution $d\alpha(x)$ on the finite or infinite interval $[a, b]$. If we define the mechanical quadrature $Q_n(f)$ of the interpolatory type by the requirement in §15.1 (2) with $u(x) = x$, the corresponding Cotes numbers can be represented as follows:*

$$\lambda_\nu = -\frac{k_{n+1}}{k_n}\frac{K_n(x_\nu)}{p_n'(x_\nu)p_{n+1}(x_\nu)}, \qquad \nu = 1, 2, \cdots, n, \tag{15.2.5}$$

with

$$K_n(x) = \int_a^b K_n(\bar{x}, t)\, dt = \int_a^b \left\{\sum_{\nu=0}^n p_\nu(x)p_\nu(t)\right\} dt = \sum_{\nu=0}^n p_\nu(x) \int_a^b p_\nu(t)\, dt. \tag{15.2.6}$$

Here we use the symbols k_n and $K_n(x_0, t)$ in the sense explained in (2.2.15) *and* (3.1.9); *x_ν and λ_ν stand for $x_{\nu n}$ and $\lambda_{\nu n}$, respectively.*

For the proof we write (15.1.6) as follows:

$$\lambda_\nu = \{p_n'(x_\nu)p_{n+1}(x_\nu)\}^{-1} \int_a^b \frac{p_n(t)p_{n+1}(x_\nu) - p_{n+1}(t)p_n(x_\nu)}{t - x_\nu}\, dt. \tag{15.2.7}$$

[71] Szegö **17**, p. 94.

Comparison of this formula with equation (3.2.3) establishes the statement.

According to (3.3.6) we have

$$\operatorname{sgn} \lambda_\nu = \operatorname{sgn} K_n(x_\nu). \tag{15.2.8}$$

In particular, if $K_n(x) \geqq 0$ in $[a, b]$, the Cotes numbers are non-negative. Then the Stekloff-Fejér theorem can be applied.

The polynomial $K_n(x)$ may be characterized by the following condition:

$$\int_a^b K_n(x)\rho(x)\, d\alpha(x) = \int_a^b \rho(x)\, dx, \tag{15.2.9}$$

where $\rho(x)$ is an arbitrary π_n. In case $d\alpha(x) = w(x)\, dx$, we see that $K_n(x)$ is the nth partial sum of the expansion of $\{w(x)\}^{-1}$ in terms of the polynomials $p_n(x)$.

(5) In the next sections we consider the special cases previously mentioned, namely:

(a) $$u(x) = \alpha(x) \text{ (cf. (3))},$$

(b) $$u(x) = x \text{ (cf. (4))};$$

here the abscissas $\{S_n\}$ are the zeros of the orthonormal polynomials associated with the distribution $d\alpha(x)$ with the additional specialization that these polynomials are the classical ones.

15.3. Cotes-Christoffel numbers in the case $u(x) = \alpha(x)$ (Gauss-Jacobi quadrature) for the classical abscissas

(1) We use the representation (3.4.7) of the Christoffel numbers λ_ν and arrange the zeros x_ν in decreasing order in the Jacobi and Hermite case and in increasing order in the Laguerre case. Concerning the following results see Winston **1**. The representation (15.3.5) has already been used in §14.7.

By use of (4.3.4) and (4.21.6), the second formula (4.5.7) furnishes for the *Jacobi abscissas*

$$\lambda_\nu = 2^{\alpha+\beta+1} \frac{\Gamma(n+\alpha+1)\Gamma(n+\beta+1)}{\Gamma(n+1)\Gamma(n+\alpha+\beta+1)} (1 - x_\nu^2)^{-1}\{P_n^{(\alpha,\beta)\prime}(x_\nu)\}^{-2}, \tag{15.3.1}$$
$$\nu = 1, 2, \cdots, n; \alpha > -1, \beta > -1,$$

and, in particular, for the *ultraspherical abscissas* (cf. (4.7.1) and (1.7.3))

$$\lambda_\nu = 2^{2-2\lambda}\pi\{\Gamma(\lambda)\}^{-2} \frac{\Gamma(n+2\lambda)}{\Gamma(n+1)} (1 - x_\nu^2)^{-1}\{P_n^{(\lambda)\prime}(x_\nu)\}^{-2}, \tag{15.3.2}$$
$$\nu = 1, 2, \cdots, n; \lambda > -\tfrac{1}{2}, \lambda \neq 0.$$

For $\lambda = 0$, that is, in the "Mehler case" $w(x) = (1 - x^2)^{-\frac{1}{2}}$, we find

$$\lambda_\nu = \frac{\pi}{n}, \qquad \nu = 1, 2, \cdots, n, \tag{15.3.3}$$

that is, $\lambda_1 = \lambda_2 = \cdots = \lambda_n$. For $\lambda = 1$, that is, in case $w(x) = (1 - x^2)^{\frac{1}{2}}$, we obtain

$$\lambda_\nu = \frac{\pi}{n+1}(1 - x_\nu^2) = \frac{\pi}{n+1}\sin^2\frac{\nu\pi}{n+1}, \quad \nu = 1, 2, \cdots, n. \tag{15.3.4}$$

For the *Laguerre abscissas* we have from (5.1.10) and (5.1.14)

$$\lambda_\nu = \frac{\Gamma(n+\alpha+1)}{\Gamma(n+1)} x_\nu^{-1}\{L_n^{(\alpha)\prime}(x_\nu)\}^{-2}, \quad \nu = 1, 2, \cdots, n; \alpha > -1, \tag{15.3.5}$$

whereas for the *Hermite abscissas*, from the second identity of (5.5.10),

$$\lambda_\nu = \pi^{\frac{1}{2}} 2^{n+1} n! \{H_n'(x_\nu)\}^{-2}, \quad \nu = 1, 2, \cdots, n. \tag{15.3.6}$$

(2) In the Jacobi case the argument of §7.32 (2) shows that, for $\alpha > -\frac{1}{2}$ and $\beta > -\frac{1}{2}$, that part of the sequence $\lambda_1, \lambda_2, \cdots, \lambda_n$ corresponding to the zeros $x_\nu > x_0$ (cf. (7.32.1)) is increasing, and that part corresponding to $x_\nu < x_0$ is decreasing. The opposite is true if $\alpha < -\frac{1}{2}$, $\beta < -\frac{1}{2}$. If $\alpha > -\frac{1}{2}$, $\beta < -\frac{1}{2}$, or $\alpha < -\frac{1}{2}$, $\beta > -\frac{1}{2}$, the whole sequence in question is increasing or decreasing according as the first or second pair of relations holds.

In the ultraspherical case this argument furnishes

$$\lambda_1 < \lambda_2 < \cdots < \lambda_{[(n+1)/2]} \quad \text{if } \lambda > 0, \tag{15.3.7}$$

and

$$\lambda_1 > \lambda_2 > \cdots > \lambda_{[(n+1)/2]} \quad \text{if } \lambda < 0. \tag{15.3.8}$$

The symmetry relation (Problem 11) yields the analogous statement for the other λ_ν. The values (15.3.3) correspond to the case $\lambda = 0$. In the Legendre case we have (15.3.7).

In the Laguerre and Hermite cases we use the argument of §7.6 (1). In the Laguerre case the sequence $\lambda_1, \lambda_2, \cdots, \lambda_n$ is increasing for $x_\nu < \alpha + \frac{1}{2}$, and decreasing for $x_\nu > \alpha + \frac{1}{2}$. In the Hermite case we have ($x_1 > x_2 > \cdots$)

$$\lambda_1 < \lambda_2 < \cdots < \lambda_{[(n+1)/2]}. \tag{15.3.9}$$

(3) In the Jacobi case, if $x_\nu = \cos\theta_\nu$, and θ_ν belongs to a fixed interval in the interior of $[0, \pi]$, Darboux's asymptotic formula (cf. (8.21.10) and (8.8.1)) yields

$$\lambda_\nu = \lambda_{\nu n} \cong \frac{2^{\alpha+\beta+1}\pi}{n}\left(\sin\frac{\theta_\nu}{2}\right)^{2\alpha+1}\left(\cos\frac{\theta_\nu}{2}\right)^{2\beta+1}. \tag{15.3.10}$$

Here, for $\alpha = \beta = -\frac{1}{2}$ the symbol $\cong$ can be replaced by $=$, according to (15.3.3). The same is true for $\alpha = \beta = +\frac{1}{2}$ (see (15.3.4)) if we replace n by $n + 1$. On the other hand, if ν is fixed and $n \to \infty$, we obtain, by use of equation (8.1.1),

$$\lambda_\nu = \lambda_{\nu n} \cong 2^{\alpha+\beta+1}(j_\nu/2)^{2\alpha}\{J_\alpha'(j_\nu)\}^{-2} n^{-2\alpha-2}, \tag{15.3.11}$$

where j_ν is the νth positive zero of $J_\alpha(z)$. (Here the symbol $\cong$ can be replaced by $=$ if $\alpha = \beta = -\frac{1}{2}$.) Therefore, on account of the monotonic property proved in (2), we have uniformly in ν,

$$\lambda_\nu = O(n^{-1}) \quad \text{or} \quad O(n^{-2\alpha-2}) \qquad \text{if } 0 < \theta_\nu \leqq \pi - \epsilon, \tag{15.3.12}$$

according as $\alpha \geqq -\frac{1}{2}$, or $\alpha \leqq -\frac{1}{2}$. Here ϵ is a fixed positive number.

By use of the argument and notation in §7.32 (3), we readily see that for increasing ν, $\nu \geqq \nu_0$,

$$\phi(\theta_\nu)\left(\sin\frac{\theta_\nu}{2}\right)^{-2\alpha-1}\left(\cos\frac{\theta_\nu}{2}\right)^{-2\beta-1}\lambda_\nu \tag{15.3.13}$$

is increasing or decreasing according as $\alpha^2 \geqq \frac{1}{4}$ or $\alpha^2 \leqq \frac{1}{4}$, and $0 < \theta_\nu \leqq \delta$, where δ is a sufficiently small positive number. (If n is sufficiently large, $\phi(\theta_\nu) > 0$, $\nu \geqq \nu_0$.) Thus (15.3.13) has uniformly the "order" n if $0 < \theta_\nu \leqq \delta$. In view of (15.3.10) and (15.3.11) for $0 < \theta_\nu \leqq \pi - \epsilon$ this yields (see (8.9.1))

$$\lambda_\nu \sim \theta_\nu^{2\alpha+1} n^{-1} \sim \nu^{2\alpha+1} n^{-2\alpha-2}, \qquad 0 < \theta_\nu \leqq \pi - \epsilon; \tag{15.3.14}$$

that is to say, the ratio of these expressions is uniformly bounded from zero and infinity if ν and n are arbitrary, $0 < \theta_\nu \leqq \pi - \epsilon$. The expression in the right-hand member of (15.3.14) attains its greatest value for $\nu \sim n$ or $\nu \sim 1$ according as $\alpha \geqq -\frac{1}{2}$ or $\alpha \leqq -\frac{1}{2}$. This again furnishes (15.3.12).

Similar results can be obtained if θ_ν is confined to the interval $\epsilon \leqq \theta_\nu < \pi$.

(4) In the Laguerre case, if $\epsilon \leqq x_\nu \leqq \omega$, we obtain (cf. (8.22.1) and (8.8.4))

$$\lambda_\nu = \lambda_{\nu n} \cong \pi e^{-x_\nu} x_\nu^{\alpha+\frac{1}{2}} n^{-\frac{1}{2}}, \qquad n \to \infty, \tag{15.3.15}$$

where ϵ and ω are fixed positive numbers. On the other hand, if ν is fixed and $n \to \infty$, we find from (8.1.8)

$$\lambda_\nu = \lambda_{\nu n} \cong (j_\nu/2)^{2\alpha}\{J'_\alpha(j_\nu)\}^{-2} n^{-\alpha-1}, \tag{15.3.16}$$

where j_ν has the previous meaning.

Applying an argument similar to that in the Jacobi case to the fourth equation in (5.1.2), and taking into consideration the argument of §7.6 (2), we find that the sequences

$$\begin{gathered} \left(4n + 2\alpha + 2 - x_\nu + \frac{1 - 4\alpha^2}{4x_\nu}\right)^{-1} e^{-x_\nu} x_\nu^{\alpha+\frac{1}{2}}\{L_n^{(\alpha)\prime}(x_\nu)\}^2, \\ \left(4n + 2\alpha + 2 - x_\nu + \frac{1 - 4\alpha^2}{4x_\nu}\right) e^{x_\nu} x_\nu^{-\alpha-\frac{1}{2}} \lambda_\nu \end{gathered} \tag{15.3.17}$$

are monotonic as ν increases provided $\alpha^2 \leqq \frac{1}{4}$ and $0 < x_\nu \leqq \omega$. In case $\alpha^2 > \frac{1}{4}$ the same sequences consist of two monotonic parts ($\nu \geqq \nu_0$). On account of (15.3.15) and (15.3.16) we obtain for the second sequence (15.3.17) the uniform "order" $n^{\frac{1}{2}}$, so that

(15.3.18) $$\lambda_\nu \sim x_\nu^{\alpha+\frac{1}{2}} n^{-\frac{1}{2}}, \qquad 0 < x_\nu \leqq \omega.$$

Because of (8.9.10) we may write

(15.3.19) $$\lambda_\nu \sim \nu^{2\alpha+1} n^{-\alpha-1}, \qquad 0 < x_\nu \leqq \omega.$$

The condition $x_\nu = O(1)$ is equivalent to $\nu = O(n^{\frac{1}{2}})$.

15.4. Quadrature of interpolatory type with $u(x) = x$ for Jacobi abscissas

(1) In this section we prove the following theorem:

THEOREM 15.4. *Assume* $\alpha > -1$, $\beta > -1$, *and let the mechanical quadrature* $Q_n(f)$ *be defined by the following conditions:*

(15.4.1)
$$\text{(a)} \quad P_n^{(\alpha,\beta)}(x_\nu) = 0, \qquad \nu = 1, 2, \cdots, n,$$
$$\text{(b)} \quad Q_n(x^k) = \sum_{\nu=1}^{n} \lambda_\nu x_\nu^k = \int_{-1}^{+1} x^k \, dx, \qquad k = 0, 1, 2, \cdots, n-1.$$

Assume $\max(\alpha, \beta) \leqq 3/2$. *Then for an arbitrary function* $f(x)$ *which is continuous in* $[-1, +1]$,

(15.4.2) $$\lim_{n\to\infty} Q_n(f) = \int_{-1}^{+1} f(x) \, dx$$

holds. If the numbers α, β *are such that* $\max(\alpha, \beta) > 3/2$, *there exist continuous functions* $f(x)$ *for which* (15.4.2) *is not true.*

Notice the difference between this statement and that of Theorem 15.2.3. Apart from the special case $\max(\alpha, \beta) = 3/2$, $\alpha \neq \beta$, this theorem has been proved by Szegö (**17**, pp. 102–108). The present proof is simpler; it is based on the bounds (7.32.5) of $P_n^{(\alpha,\beta)}(x)$, and on Theorem 8.9.1 concerning the zeros of $P_n^{(\alpha,\beta)}(x)$. The notation of §14.4 is used.

(2) First assume $\alpha \leqq 3/2$, $\beta \leqq 3/2$. We must show the boundedness of the sum (15.2.2) if $\lambda_\nu = \lambda_{\nu n}$ is defined by (15.4.1). We discuss λ_ν only in case $0 \leqq x_\nu < 1$, or $0 < \theta_\nu \leqq \pi/2$. For the remaining values, (4.1.3) can be used. According to (8.9.2)

(15.4.3) $$\lambda_\nu \sim \nu^{\alpha+\frac{3}{2}} n^{-\alpha-2} \int_0^\pi \frac{P_n^{(\alpha,\beta)}(\cos\theta)}{\cos\theta - \cos\theta_\nu} \sin\theta \, d\theta.$$

We decompose the last integral into five parts I, II, III, IV, V, corresponding to the intervals

(15.4.4)
$$0 \leqq \theta \leqq \theta_1/2, \qquad \theta_1/2 \leqq \theta \leqq \theta_\nu/2, \qquad \theta_\nu/2 \leqq \theta \leqq 3\theta_\nu/2,$$
$$3\theta_\nu/2 \leqq \theta \leqq 3\pi/4, \qquad 3\pi/4 \leqq \theta \leqq \pi,$$

and we note that $(\theta^2 - \theta_\nu^2)/(\cos\theta - \cos\theta_\nu)$ is bounded in all these integrals.

Since $\theta_1 = O(n^{-1})$,

$$(15.4.5) \qquad \mathrm{I} = O(n^\alpha) \int_0^{\theta_1/2} \theta_\nu^{-2}\, \theta \, d\theta = O(n^{\alpha-2})(\nu/n)^{-2} = O(\nu^{-2} n^\alpha).$$

Further, applying (8.21.18), we obtain

$$(15.4.6) \qquad \mathrm{II} = \int_{\theta_1/2}^{\theta_\nu/2} \frac{n^{-\frac{1}{2}} k(\theta)}{\cos\theta - \cos\theta_\nu} \cos(N\theta + \gamma) \sin\theta \, d\theta + O(1) \int_{\theta_1/2}^{\theta_\nu/2} \frac{\theta^{-\alpha-\frac{1}{2}} n^{-\frac{1}{2}}}{\theta_\nu^2 - \theta^2} \, d\theta.$$

The first integrand can be written as follows:

$$(15.4.7) \qquad n^{-\frac{1}{2}} \frac{\theta^{-\alpha+\frac{1}{2}}}{\theta_\nu^2 - \theta^2} \phi(\theta_\nu, \theta) \cos(N\theta + \gamma),$$

where $\phi(\theta_\nu, \theta)$ and its partial derivative with respect to θ remain bounded in the interval considered (uniformly in ν). Therefore, integration by parts gives

$$(15.4.8) \qquad O(n^{-\frac{3}{2}}) \left[\frac{\theta^{-\alpha+\frac{1}{2}}}{\theta_\nu^2 - \theta^2} \right]_{\theta_1/2}^{\theta_\nu/2} + O(n^{-\frac{3}{2}}) \int_{\theta_1/2}^{\theta_\nu/2} \left| \frac{\partial}{\partial\theta} \left\{ \frac{\theta^{-\alpha+\frac{1}{2}}}{\theta_\nu^2 - \theta^2} \phi(\theta_\nu, \theta) \right\} \right| d\theta.$$

Here the symbol $[f(\theta)]_a^b$ means $f(b) + f(a)$. A simple discussion gives for the first term of (15.4.8) (cf. (8.9.1))

$$(15.4.9) \qquad O(n^{-\frac{3}{2}}\theta_\nu^{-\alpha-\frac{3}{2}}) + O(n^{-\frac{3}{2}}\theta_1^{-\alpha+\frac{1}{2}}\theta_\nu^{-2}) = O(\nu^{-\alpha-\frac{3}{2}} n^\alpha) + O(\nu^{-2} n^\alpha),$$

and for the second term

$$(15.4.10) \qquad O(n^{-\frac{3}{2}}) \int_{\theta_1/2}^{\theta_\nu/2} \left\{ \frac{\theta^{-\alpha+\frac{1}{2}}}{\theta_\nu^2 - \theta^2} + \frac{\theta^{-\alpha-\frac{1}{2}}}{\theta_\nu^2 - \theta^2} + \frac{\theta^{-\alpha+\frac{3}{2}}}{(\theta_\nu^2 - \theta^2)^2} \right\} d\theta.$$

We readily see that (15.4.10) can be combined with the second integral of II. On putting $\theta = \theta_\nu x$ in this integral, we obtain

$$(15.4.11) \qquad n^{-\frac{1}{2}} \int_{\theta_1/2}^{\theta_\nu/2} \frac{\theta^{-\alpha-\frac{1}{2}}}{\theta_\nu^2 - \theta^2} d\theta = \theta_\nu^{-\alpha-\frac{3}{2}} n^{-\frac{1}{2}} \int^{\frac{1}{2}} \frac{x^{-\alpha-\frac{1}{2}}}{1 - x^2} dx.$$

The lower limit of integration is $\theta_1/(2\theta_\nu) \sim 1/\nu$, so that the last integral is $O(\nu^{\alpha-\frac{1}{2}})$, $O(\log \nu)$, or $O(1)$ according as $\alpha > \frac{1}{2}$, $\alpha = \frac{1}{2}$, or $\alpha < \frac{1}{2}$. Consequently,

$$(15.4.12) \qquad \mathrm{II} = O(\nu^{-2} n^\alpha), \qquad O(\nu^{-2} n^{\frac{1}{2}} \log \nu), \quad \text{or} \quad O(\nu^{-\alpha-\frac{3}{2}} n^\alpha),$$

according as $\alpha > \frac{1}{2}$, $\alpha = \frac{1}{2}$, or $\alpha < \frac{1}{2}$.

Now by (8.8.2),

$$(15.4.13) \qquad \mathrm{III} = \int_{\theta_\nu/2}^{3\theta_\nu/2} n^{-\frac{1}{2}} k(\theta_\nu) \frac{\cos(N\theta + \gamma) - \cos(N\theta_\nu + \gamma)}{\cos\theta - \cos\theta_\nu} \sin\theta \, d\theta + O(1)\theta_\nu^{-\alpha-\frac{1}{2}} n^{-\frac{1}{2}} \theta_\nu^2.$$

Here $n^{-\frac{1}{2}}k(\theta_\nu) = O(n^{-\frac{1}{2}}\theta_\nu^{-\alpha-\frac{1}{2}}) = O(\nu^{-\alpha-\frac{1}{2}} n^\alpha)$, and the same bound holds for the second term. Furthermore,

$$\int_{\theta_\nu/2}^{3\theta_\nu/2} \frac{\sin N \dfrac{\theta - \theta_\nu}{2}}{\sin \dfrac{\theta - \theta_\nu}{2}} \sin\left(N\frac{\theta - \theta_\nu}{2} + N\theta_\nu + \gamma\right) \frac{\sin\theta}{\sin\dfrac{\theta + \theta_\nu}{2}}\, d\theta$$

(15.4.14)
$$= \int_{\theta_\nu/2}^{3\theta_\nu/2} \frac{\sin N \dfrac{\theta - \theta_\nu}{2}}{\sin \dfrac{\theta - \theta_\nu}{2}} \sin\left(N\frac{\theta - \theta_\nu}{2} + N\theta_\nu + \gamma\right) d\theta + O(1)$$

$$= \tfrac{1}{2} \sin (N\theta_\nu + \gamma) \int_{\theta_\nu/2}^{3\theta_\nu/2} \frac{\sin N(\theta - \theta_\nu)}{\sin \dfrac{\theta - \theta_\nu}{2}}\, d\theta + O(1),$$

since $|\sin\theta - \sin(\theta + \theta_\nu)/2| < |(\theta - \theta_\nu)/2|$, and $\theta_\nu\{\sin(\theta + \theta_\nu)/2\}^{-1}$ is bounded. The term with $\cos(N\theta_\nu + \gamma)$ vanishes, since the integrand is an odd function of $\theta - \theta_\nu$. The denominator in the last integral can be replaced by $(\theta - \theta_\nu)/2$. Hence (15.4.14) is bounded, and

(15.4.15)
$$\text{III} = O(\nu^{-\alpha-\frac{1}{2}} n^{\alpha}).$$

Now, as in II,

(15.4.16)
$$\text{IV} = \int_{3\theta_\nu/2}^{3\pi/4} n^{-\frac{1}{2}} \frac{\theta^{-\alpha+\frac{1}{2}}}{\theta_\nu^2 - \theta^2} \phi(\theta_\nu, \theta) \cos(N\theta + \gamma)\, d\theta + O(1) \int_{3\theta_\nu/2}^{3\pi/4} \frac{\theta^{-\alpha-\frac{1}{2}} n^{-\frac{3}{2}}}{\theta^2 - \theta_\nu^2}\, d\theta,$$

and the first integral is

$$O(n^{-\frac{3}{2}}) \left[\frac{\theta^{-\alpha+\frac{1}{2}}}{\theta^2 - \theta_\nu^2}\right]_{3\theta_\nu/2}^{3\pi/4} + O(n^{-\frac{3}{2}}) \int_{3\theta_\nu/2}^{3\pi/4} \left| \frac{\partial}{\partial\theta} \left\{ \frac{\theta^{-\alpha+\frac{1}{2}}}{\theta^2 - \theta_\nu^2} \phi(\theta_\nu, \theta) \right\} \right| d\theta$$

(15.4.17)
$$= O(n^{-\frac{3}{2}}) + O(n^{-\frac{3}{2}})\theta_\nu^{-\alpha-\frac{3}{2}} + O(n^{-\frac{3}{2}}) \int_{3\theta_\nu/2}^{3\pi/4} \left\{ \frac{\theta^{-\alpha+\frac{1}{2}}}{\theta^2 - \theta_\nu^2} + \frac{\theta^{-\alpha-\frac{1}{2}}}{\theta^2 - \theta_\nu^2} + \frac{\theta^{-\alpha+\frac{3}{2}}}{(\theta^2 - \theta_\nu^2)^2} \right\} d\theta$$

$$= O(\nu^{-\alpha-\frac{1}{2}} n^{\alpha}) + O(n^{-\frac{3}{2}}) \int_{3\theta_\nu/2}^{3\pi/4} \frac{\theta^{-\alpha-\frac{1}{2}}}{\theta^2 - \theta_\nu^2}\, d\theta.$$

On putting $\theta = \theta_\nu x$, we have

(15.4.18)
$$n^{-\frac{3}{2}} \int_{3\theta_\nu/2}^{3\pi/4} \frac{\theta^{-\alpha-\frac{1}{2}}}{\theta^2 - \theta_\nu^2}\, d\theta = n^{-\frac{3}{2}} \theta_\nu^{-\alpha-\frac{3}{2}} \int_{\frac{3}{2}}^{3\pi/(4\theta_\nu)} \frac{x^{-\alpha-\frac{1}{2}}}{x^2 - 1}\, dx = O(\nu^{-\alpha-\frac{3}{2}} n^{\alpha}),$$

so that

(15.4.19)
$$\text{IV} = O(\nu^{-\alpha-\frac{1}{2}} n^{\alpha}).$$

Finally, the second mean-value theorem gives (compare (4.21.7) and also §7.32 (2))

$$
\begin{aligned}
\mathrm{V} &= -(\cos\theta_\nu - \cos[3\pi/4])^{-1}\int_{3\pi/4}^{\theta'} P_n^{(\alpha,\beta)}(\cos\theta)\sin\theta\,d\theta \\
&= O(n^{-1})\,|\,P_{n+1}^{(\alpha-1,\beta-1)}(\cos\theta') - P_{n+1}^{(\alpha-1,\beta-1)}(\cos 3\pi/4)\,| \\
&= O[n^{\max(\beta-2,-\frac{1}{2})}] = O(n^{-\frac{1}{2}}), \qquad 3\pi/4 \leqq \theta' \leqq \pi.
\end{aligned} \tag{15.4.20}
$$

Here we used the assumption $\beta \leqq 3/2$. Hence

$$
\begin{aligned}
\lambda_\nu &= O(1)\nu^{\alpha+\frac{3}{2}}n^{-\alpha-2}(\nu^{-2}n^\alpha + \nu^{-\alpha-\frac{1}{2}}n^\alpha + n^{-\frac{1}{2}}) \\
&= O(1)\nu^{\alpha+\frac{3}{2}}n^{-\alpha-2}(\nu^{-\alpha-\frac{1}{2}}n^\alpha + n^{-\frac{1}{2}}) = O(1)(\nu n^{-2} + \nu^{\alpha+\frac{3}{2}}n^{-\alpha-\frac{5}{2}}).
\end{aligned} \tag{15.4.21}
$$

In all cases we obtain

$$
\sum_{0 \leqq x_\nu < 1} |\lambda_\nu| = O(1), \tag{15.4.22}
$$

and this establishes the boundedness of (15.2.2) and the first part of the statement.

(3) Now assume $\alpha \geqq \beta$, $\alpha > 3/2$. We assume that θ_ν lies in a fixed interval $[a, b]$, $0 < a < b < \pi$; then $\nu \sim n$, and the number of zeros θ_ν satisfying this condition is also $\sim n$. Let ϵ and ω be fixed positive numbers, $\epsilon < \min(a, \pi - b)$. Decompose the integral in question into the parts I, II, III, IV, V, corresponding to the intervals

$$
\begin{aligned}
&0 \leqq \theta \leqq \omega/n, \qquad \omega/n \leqq \theta \leqq \theta_\nu - \epsilon, \qquad \theta_\nu - \epsilon \leqq \theta \leqq \theta_\nu + \epsilon, \\
&\theta_\nu + \epsilon \leqq \theta \leqq \pi - \omega/n, \qquad \pi - \omega/n \leqq \theta \leqq \pi.
\end{aligned} \tag{15.4.23}
$$

Then,

$$
\begin{aligned}
\mathrm{I} = (1 - \cos\theta_\nu)^{-1}\int_0^{\omega/n} P_n^{(\alpha,\beta)}(\cos\theta)\sin\theta\,d\theta \\
+ O(1)\int_0^{\omega/n} |\,P_n^{(\alpha,\beta)}(\cos\theta)\,|\,\theta\theta^2\,d\theta.
\end{aligned} \tag{15.4.24}
$$

The first integral can be calculated according to (4.21.7), and we obtain (cf. (7.32.5))

$$
\begin{aligned}
\mathrm{I} &= (1 - \cos\theta_\nu)^{-1}\frac{2}{n+\alpha+\beta}\left\{P_{n+1}^{(\alpha-1,\beta-1)}(1) - P_{n+1}^{(\alpha-1,\beta-1)}\left(\cos\frac{\omega}{n}\right)\right\} \\
&\qquad + O(n^{\alpha-4}) \\
&= \frac{2}{\Gamma(\alpha)}\frac{n^{\alpha-2}}{1 - \cos\theta_\nu} + \omega^{-\alpha+\frac{1}{2}}O(n^{\alpha-2}) + O(n^{\alpha-3}).
\end{aligned} \tag{15.4.25}
$$

The bound of the term $O(n^{\alpha-2})$ is independent of ω. In the same way we find

$$
\begin{aligned}
\mathrm{V} &= (-1)^{n-1}\int_0^{\omega/n}\frac{P_n^{(\beta,\alpha)}(\cos\theta)}{\cos\theta + \cos\theta_\nu}\sin\theta\,d\theta \\
&= (-1)^{n-1}\frac{2}{\Gamma(\beta)}\frac{n^{\beta-2}}{1 + \cos\theta_\nu} + \omega^{-\beta+\frac{1}{2}}O(n^{\beta-2}) + O(n^{\beta-3}),
\end{aligned} \tag{15.4.26}
$$

where $O(n^{\beta-2})$ has a bound independent of ω. (For $\beta = 0$ the first term vanishes.)

Furthermore,

$$\text{(15.4.27)}\quad \begin{aligned} \mathrm{II} + \mathrm{IV} &= O(n^{-\frac{1}{2}}) \int_{\omega/n}^{\theta_\nu - \epsilon} \theta^{-\alpha+\frac{1}{2}}\, d\theta + O(n^{-\frac{1}{2}}) \int_{\theta_\nu+\epsilon}^{\pi-\omega/n} (\pi - \theta)^{-\beta+\frac{1}{2}}\, d\theta \\ &= \omega^{-\alpha+\frac{3}{2}} O(n^{\alpha-2}) + \omega^{-\beta+\frac{3}{2}} O(n^{\beta-2}) + O(n^{-\frac{1}{2}}). \end{aligned}$$

Here the same remark applies to the O-terms as above.

Finally, according to (8.8.2),

$$\text{(15.4.28)}\quad \begin{aligned} \mathrm{III} &= n^{-\frac{1}{2}} k(\theta_\nu) \int_{\theta_\nu-\epsilon}^{\theta_\nu+\epsilon} \frac{\cos(N\theta + \gamma) - \cos(N\theta_\nu + \gamma)}{\cos\theta - \cos\theta_\nu} \sin\theta\, d\theta + O(n^{-\frac{1}{2}}) \\ &= O(n^{-\frac{1}{2}}) \int_{\theta_\nu-\epsilon}^{\theta_\nu+\epsilon} \left| \frac{\sin N \dfrac{\theta - \theta_\nu}{2}}{\sin \dfrac{\theta - \theta_\nu}{2}} \right| d\theta + O(n^{-\frac{1}{2}}). \end{aligned}$$

The last integral is $O(\log n)$ (cf. Zygmund **2**, p. 172); whence

$$\text{(15.4.29)}\quad \mathrm{III} = O(n^{-\frac{1}{2}} \log n).$$

Now assume $\alpha > \beta$, $\alpha > 3/2$; then the previous results give

$$\text{(15.4.30)}\quad \mathrm{I} + \mathrm{II} + \mathrm{III} + \mathrm{IV} + \mathrm{V} = n^{\alpha-2}\left\{\frac{2}{\Gamma(\alpha)}\frac{1}{1 - \cos\theta_\nu} + \omega^{-\alpha+\frac{3}{2}} O(1)\right\} + o(n^{\alpha-2}),$$

where $O(1)$ is independent of ω. Hence, by (15.4.3)

$$\text{(15.4.31)}\quad \sum_{a \leqq \theta_\nu \leqq b} |\lambda_\nu| \sim n^{\alpha-\frac{3}{2}}.$$

In the more complicated case $\alpha = \beta > 3/2$ we find

$$\text{(15.4.32)}\quad \begin{aligned} &\mathrm{I} + \mathrm{II} + \mathrm{III} + \mathrm{IV} + \mathrm{V} \\ &= n^{\alpha-2}\left\{\frac{2}{\Gamma(\alpha)}\frac{1}{1 - \cos\theta_\nu} + \frac{2}{\Gamma(\alpha)}\frac{(-1)^{n-1}}{1 + \cos\theta_\nu} + \omega^{-\alpha+\frac{3}{2}} O(1)\right\} + o(n^{\alpha-2}), \end{aligned}$$

where again $O(1)$ is independent of ω. This furnishes the same result (15.4.31).

15.5. An alternative method in the ultraspherical case

In the ultraspherical case $\alpha = \beta$ the convergence theorem (15.4.2) follows from the following theorem:

THEOREM 15.5. *Let $-1 < \alpha = \beta \leqq \frac{3}{2}$. Then the Cotes numbers λ_ν, defined by (15.4.1), are positive provided that n is chosen sufficiently large. In the cases $-1 < \alpha = \beta \leqq 0$ and $\frac{1}{2} \leqq \alpha = \beta \leqq 1$ the statement is true for all values of n.*

In view of the Stekloff-Fejér theorem (Theorem 15.2.2; cf. the last remark in §15.2 (2)) this leads indeed to another proof of the convergence theorem of §15.4, even for Riemann-integrable functions. It is very probable that $\lambda_\nu > 0$ for all values of n provided $-1 < \alpha = \beta \leqq \frac{3}{2}$.

Concerning this section, refer to Szegö **17**, pp. 95–99, 109–110, where the positiveness of λ_ν is proved for $-\frac{1}{2} \leqq \alpha = \beta \leqq 0$ and for certain other special cases.

(1) We prove $\lambda_\nu > 0$ by showing that the function $K_n(x) = K_n^{(\lambda)}(x)$ of §15.2 is positive for $-1 < x < +1$; here $d\alpha(x) = (1 - x^2)^{\lambda - \frac{1}{2}}\,dx$ with $\lambda = \alpha + \frac{1}{2}$ (cf. (15.2.8)). We can assume that n is even, since the last integral in (15.2.6) vanishes if n is odd, and this assumption will be made throughout this section.

According to (4.7.15), (4.7.14), and (4.7.3), we have (cf. (1.7.3))

$$
\begin{aligned}
K_n^{(\lambda)}(x) &= 2^{3-2\lambda}\frac{\Gamma(2\lambda)}{\Gamma(\lambda + \frac{1}{2})\Gamma(\lambda - \frac{1}{2})}\sum \frac{\lambda + m}{(2\lambda + m - 1)(m+1)}P_m^{(\lambda)}(x)\\
&= \sum d_m^{(\lambda)} P_m^{(\lambda)}(x), \qquad m = 0, 2, 4, \cdots, n.
\end{aligned}
\tag{15.5.1}
$$

Let $\lambda < 2$. From a remark made in connection with (15.2.9), we have by Theorem 9.1.2, for $-1 < x < +1$,

$$
\lim_{n\to\infty} K_n^{(\lambda)}(x) = (1 - x^2)^{\frac{1}{2}-\lambda}, \tag{15.5.2}
$$

uniformly in $-1 + \epsilon \leqq x \leqq 1 - \epsilon$.

(2) First assume $0 < \lambda < \frac{1}{2}$. Then

$$
d_0^{(\lambda)} > 0, \qquad d_m^{(\lambda)} < 0, \qquad m = 2, 4, 6, \cdots. \tag{15.5.3}
$$

In addition, $P_m^{(\lambda)}(\cos\theta)$, as a cosine polynomial, has non-negative coefficients (cf. (4.9.19)). Therefore $K_n^{(\lambda)}(\cos\theta)$ attains its minimum for $\theta = 0$, and this minimum *decreases* as n increases. Now

$$
\begin{aligned}
\sum_{m=0,2,4,\cdots} d_m^{(\lambda)} P_m^{(\lambda)}(1) &= 2^{2-2\lambda}\frac{\Gamma(2\lambda)}{\Gamma(\lambda + \frac{1}{2})\Gamma(\lambda - \frac{1}{2})}\\
&\qquad \sum_{m=0,2,4,\cdots}\left(\frac{1}{2\lambda + m - 1} + \frac{1}{m+1}\right)\binom{2\lambda + m - 1}{m}\\
&= 2^{2-2\lambda}\frac{\Gamma(2\lambda - 1)}{\Gamma(\lambda + \frac{1}{2})\Gamma(\lambda - \frac{1}{2})}\\
&\qquad \sum_{m=0,2,4,\cdots}\left\{\binom{2\lambda + m - 2}{m} + \binom{2\lambda + m - 1}{m+1}\right\}\\
&= 2^{2-2\lambda}\frac{\Gamma(2\lambda - 1)}{\Gamma(\lambda + \frac{1}{2})\Gamma(\lambda - \frac{1}{2})}\sum_{m=0}^{\infty}\binom{2\lambda + m - 2}{m} = 0,
\end{aligned}
\tag{15.5.4}
$$

and this establishes the statement for $0 < \lambda < \frac{1}{2}$. If $\lambda = \frac{1}{2}$, we have $d_m^{(\lambda)} = 0$,

$m = 2, 4, 6, \cdots$, and $K_n^{(\lambda)}(x) = 1$. In the limiting case $\lambda \to 0$, we find (cf. (4.7.8))

$$\lim_{\lambda \to 0} K_n^{(\lambda)}(\cos\theta) = \frac{2}{\pi} - \frac{2}{\pi}\sum\left(\frac{1}{m-1} - \frac{1}{m+1}\right)\cos m\theta, \tag{15.5.5}$$

$$m = 2, 4, \cdots, n.$$

Therefore, this case can be discussed as before. (See Szegö **17**, p. 96.)

(3) Next, assume that $-\frac{1}{2} < \lambda < 0$. In this case $d_m^{(\lambda)} > 0$. According to (4.9.19), the coefficients of the trigonometric polynomial $P_m^{(\lambda)}(\cos\theta)$ are non-negative except for the highest coefficient, that is, that of $\cos m\theta$, which is negative, $m > 0$. Therefore, the coefficient of an arbitrary term $\cos m\theta$, $m > 0$ even, in $K_n^{(\lambda)}(\cos\theta)$ appears as a sum of the form $-u_0 + u_1 + u_2 + \cdots + u_l$ with non-negative $u_0, u_1, \cdots, u_l$ where $l = (n - m)/2$. If m is fixed, $n \to \infty$, $l \to \infty$, this expression tends to the coefficient of $\cos m\theta$ in $(1 - x^2)^{\frac{1}{2}-\lambda} = |\sin\theta|^{1-2\lambda}$. But this is negative (cf., for example, Pólya-Szegö **2**, pp. 31–32), so that the same is true for the partial sums $-u_0 + u_1 + u_2 + \cdots + u_l$. Consequently, $K_n^{(\lambda)}(\cos\theta)$ is again of the same type as in the previous case, and it attains its minimum for $\theta = 0$. The terms $d_m^{(\lambda)}P_m^{(\lambda)}(1)$ are negative, except for $m = 0$. This establishes the statement.

(4) In case $1 < \lambda \leqq \frac{3}{2}$, we start with the following identity:

$$(1 - x^2)K_n^{(\lambda)}(x) = K_n^{(\lambda-1)}(x) + \sigma_\lambda P_{n+2}^{(\lambda-1)}(x), \qquad n \text{ even}, \tag{15.5.6}$$

where σ_λ is a proper constant. We readily show (we can write $x = 1$, or compare the highest terms, cf. (4.7.9)) that $\sigma_\lambda < 0$. For later purposes we give the value of σ_λ:

$$\sigma_\lambda = \frac{2^{1-2\lambda}}{1-\lambda}\frac{\Gamma(2\lambda)}{\Gamma(\lambda+\frac{1}{2})\Gamma(\lambda-\frac{1}{2})}\frac{n+2}{2\lambda+n-1}. \tag{15.5.7}$$

To prove (15.5.6), let $\rho(x)$ be an arbitrary π_n. Then, by (15.2.9),

$$\int_{-1}^{+1}\{(1 - x^2)^{\lambda-\frac{3}{2}}[K_n^{(\lambda-1)}(x) + \sigma_\lambda P_{n+2}^{(\lambda-1)}(x)] - 1\}\rho(x)\,dx = 0, \tag{15.5.8}$$

where σ_λ must be determined so that the right-hand member of (15.5.6) vanishes for $x = 1$.

According to the result in (2), the coefficient of $\cos m\theta$, $m > 0$, in the right-hand member of (15.5.6) is again non-positive, and the minimum is attained for $\theta = 0$, or $x = 1$; whence $(1 - x^2)K_n^{(\lambda)}(x) > 0$, $-1 < x < +1$.

This argument needs only a slight modification for $\lambda = 1$. (See Szegö **17**, pp. 96–97.)

(5) We now prove $K_n^{(\lambda)}(x) > 0$, $-1 < x < +1$, for sufficiently large values of n, provided $\frac{1}{2} < \lambda < 1$, or $\frac{3}{2} < \lambda \leqq 2$.

We may assume n even, $0 < x < 1$. First let $\lambda < 1$. By (7.33.6), we have for $\theta \geqq cn^{-1}$

$$P_n^{(\lambda)}(\cos\theta) = \theta^{-\lambda}O(n^{\lambda-1}). \tag{15.5.9}$$

Now from (15.5.2), n even, we obtain

$$K_n^{(\lambda)}(\cos\theta) \geqq (\sin\theta)^{1-2\lambda} - \sum_{m=n+2,n+4,\cdots} |d_m^{(\lambda)}|\,|P_m^{(\lambda)}(\cos\theta)|. \tag{15.5.10}$$

The last sum is (cf. (15.5.1)) equal to

$$O(1)\sum m^{-1}\theta^{-\lambda}m^{\lambda-1} = \theta^{-\lambda}O(n^{\lambda-1}), \tag{15.5.11}$$

since $n^{-1} > m^{-1}$ and $\lambda < 1$; this furnishes

$$K_n^{(\lambda)}(\cos\theta) > \tfrac{1}{2}(\sin\theta)^{1-2\lambda} > 0,$$

provided n is sufficiently large and $\theta > cn^{-1}$, where c is a proper positive constant.

This argument also holds for $\lambda = 1$ with the following modification. We have (cf. (4.7.2))

$$K_n^{(1)}(\cos\theta) = \frac{4}{\pi}\sum\frac{1}{m+1}\,\frac{\sin(m+1)\theta}{\sin\theta} = \frac{1}{\sin\theta} - \frac{4}{\pi}\sum\frac{1}{m+1}\,\frac{\sin(m+1)\theta}{\sin\theta}.$$

Here m is even; in the first sum $m \leqq n$, in the second sum $m > n$. By (1.11.6) we find

$$K_n^{(1)}(\cos\theta) = (\sin\theta)^{-1} + \theta^{-2}O(n^{-1}) > 0, \tag{15.5.12}$$

provided $\theta > cn^{-1}$, $c > 0$.

In the case $\frac{3}{2} < \lambda < 2$ we use (15.5.6) and (15.5.7); taking the previous results into account, we find

$$K_n^{(\lambda-1)}(\cos\theta) > \tfrac{1}{2}(\sin\theta)^{1-2(\lambda-1)}, \qquad \sigma_\lambda P_{n+2}^{(\lambda-1)}(\cos\theta) = \theta^{1-\lambda}O(n^{\lambda-2}), \qquad \theta > cn^{-1}, \tag{15.5.13}$$

which again shows that $K_n^{(\lambda)}(\cos\theta)$ is positive provided $\theta > c'n^{-1}$, where c' is a proper positive constant. For $\lambda = 2$ we use (15.5.12). Since $\sigma_2 = -2/\pi + O(n^{-1})$, we obtain for $\theta > c'n^{-1}$,

$$\sin^2\theta\,K_n^{(2)}(\cos\theta) > (1-2/\pi)(\sin\theta)^{-1} + \theta^{-2}O(n^{-1}) + O(n^{-1})(\sin\theta)^{-1} > 0. \tag{15.5.14}$$

(6) Finally, we assume $\theta > 0$, $\theta = O(n^{-1})$, n even. According to (4.7.1), we have, $\lambda = \alpha + \frac{1}{2}$,

$$K_n^{(\lambda)}\left(\cos\frac{x}{n}\right) = \frac{2^{2-2\alpha}}{\Gamma(\alpha)}\sum\frac{\lambda+m}{(2\lambda+m-1)(m+1)}\,\frac{\Gamma(m+2\lambda)}{\Gamma(m+\lambda+\frac{1}{2})}\,P_m^{(\alpha,\alpha)}\left(\cos\frac{x}{n}\right),$$

where $m = 0, 2, 4, \cdots, n$. By (8.1.1) we obtain for $\alpha > 0$:

$$K_n^{(\lambda)}\left(\cos\frac{x}{n}\right) = \frac{2^{1-\alpha}}{\Gamma(\alpha)}\,n^{2\alpha}x^{-2\alpha}\,\frac{2x}{n}\sum\left\{\left(\frac{mx}{n}\right)^{\alpha-1}J_\alpha\left(\frac{mx}{n}\right)\right\} + \sum m^{2\alpha-1}o(1) + O(1), \qquad m = 2, 4, \cdots, n. \tag{15.5.15}$$

If $n \to \infty$, the first term of the right-hand member is

$$\cong 2^{1-\alpha}\{\Gamma(\alpha)\}^{-1}n^{2\alpha}x^{-2\alpha}f_\alpha(x),$$

uniformly in x, $0 < x \leqq x_0$, where

$$x^{-2\alpha}f_\alpha(x) = x^{-2\alpha}\int_0^x t^{\alpha-1}J_\alpha(t)\,dt = \int_0^1 t^{2\alpha-1}\{(tx)^{-\alpha}J_\alpha(tx)\}\,dt. \tag{15.5.16}$$

The second term is equal to $o(n^{2\alpha})$, $\alpha > 0$.

Obviously, $\lim_{x\to+0} x^{-2\alpha}f_\alpha(x)$ exists and is positive. We shall show that the integral function $f_\alpha(x) > 0$ for $x > 0$ provided $0 < \alpha \leqq \frac{3}{2}$.

The representation (Watson 3, p. 170, (3))

$$t^{\alpha-1}J_\alpha(t) = \frac{2^{\alpha+1}\pi^{-\frac{1}{2}}}{\Gamma(\frac{1}{2}-\alpha)}\int_1^\infty \frac{t^{-1}\sin tu}{(u^2-1)^{\alpha+\frac{1}{2}}}\,du, \qquad t > 0, \tag{15.5.17}$$

holds for $-\frac{1}{2} < \alpha < +\frac{1}{2}$. Assuming $0 < \alpha < \frac{1}{2}$, we can integrate by parts. We thus obtain the following representation, valid for $0 < \alpha < \frac{3}{2}$:

$$t^{\alpha-1}J_\alpha(t) = \frac{2^{\alpha}\pi^{-\frac{1}{2}}}{\Gamma(\frac{3}{2}-\alpha)}\int_1^\infty \frac{(tu)^{-1}\sin tu - \cos tu}{u(u^2-1)^{\alpha-\frac{1}{2}}}\,du, \qquad t > 0. \tag{15.5.18}$$

This reduces the statement to the discussion of the special case

$$(\pi/2)^{\frac{1}{2}}f_{\frac{3}{2}}(x) = \int_0^x (t^{-1}\sin t - \cos t)\,dt = \int_0^x t^{-1}\sin t\,dt - \sin x. \tag{15.5.19}$$

This function increases for $0 < x < t_1$ and decreases for $t_1 < x < t_2$, where t_1 and t_2 are the least positive roots of $\sin t - t\cos t = 0$; $t_1 < 2\pi < t_2$. Thus we need only prove $f_{\frac{3}{2}}(x) > 0$ if $x \geqq 2\pi$. We have, however, for $x \geqq 2\pi$

$$\int_0^x t^{-1}\sin t\,dt = \pi/2 - \int_x^\infty t^{-1}\sin t\,dt. \tag{15.5.20}$$

According to the second mean-value theorem the modulus of the last integral is less than $2x^{-1} \leqq 2(2\pi)^{-1}$. The left-hand member of (15.5.20) is therefore greater than $\pi/2 - 2(2\pi)^{-1} > 1$.

(7) The positivity of all the Cotes numbers λ_ν for $0 \leqq \alpha = \beta \leqq 3/2$ has been proven in Askey-Fitch 1, and for $\alpha, \beta \geqq 0$, $\alpha + \beta \leqq 1$ in Askey 5. Two inequalities of Vietoris 1 for trigonometric polynomials give the positivity of another sum which is related to a more general quadrature problem (interpolate at the zeros of $P_n^{(\alpha,\beta)}(x)$ but integrate with respect to $(1-x)^\gamma(1+x)^\delta dx$). For this and a summary of other positive sums related to quadrature problems see Askey-Steinig 1.

(8) A new proof of the divergence half of Theorem 15.4 was given by Locher 1.

CHAPTER XVI

POLYNOMIALS ORTHOGONAL ON AN ARBITRARY CURVE

In Chapter XI there were introduced certain systems of polynomials which play a rôle with respect to the unit circle similar to that which the orthogonal polynomials previously discussed do for a real interval. We shall now give a short survey of a further generalization which uses a rectifiable Jordan curve in place of either the unit circle, or a real interval.

16.1. Preliminaries; definitions

(1) Let T be a simply connected region in the complex x-plane, containing $x = \infty$ as an interior point; let the boundary C of T be a continuum consisting of a finite number of rectifiable Jordan arcs. When integrating along C, the parts of C are described in an arbitrarily fixed order; along the parts which have the character of cuts, we must integrate twice. The integrals considered will have the form

$$\int_C f(x) \, | dx \, | . \tag{16.1.1}$$

Here $f(x)$ is a Lebesgue-integrable function defined on C, and $| dx |$ is the arc element on C. We denote the total length of the boundary C by L, counting the cuts twice.

A particularly important case is that of a rectifiable Jordan curve. Another remarkable instance is that of a Jordan arc. The case of a finite interval is of this type.

(2) Let

$$x = \phi(z) = cz + c_0 + c_1 z^{-1} + c_2 z^{-2} + \cdots , \qquad c > 0, \tag{16.1.2}$$

be the analytic function, regular and univalent for $| z | > 1$, which maps $| z | > 1$ conformally onto the region T, preserving the point at infinity and the direction therein. According to a theorem of Osgood and Carathéodory (Carathéodory **1**, p. 86), the function $\phi(z)$ is continuous in $| z | \geqq 1$ and furnishes a one-to-one and continuous correspondence between the unit circle $| z | = 1$ and the boundary C of T (described in the way indicated above). The function $\phi(z)$ is uniquely determined, and the number c is called the *transfinite diameter* (*Robin's constant, capacity*) of C (cf. §16.2 (5)).

Let C be a Jordan curve, and let x_0 be a preassigned point in the interior of C. In this case we may consider also the function

$$x = \psi(z) = x_0 + d_1 z + d_2 z^2 + \cdots , \qquad d_1 > 0, \tag{16.1.3}$$

which is regular and univalent for $| z | < 1$, and which furnishes the conformal mapping of $|z| < 1$ onto the interior U of C; it carries the origin $z = 0$ into $x = x_0$ and preserves the direction at the origin; $\phi(z)$ is also uniquely determined and continuous in $|z| \leqq 1$.

For the interval $-1 \leqq x \leqq +1$ we have $\phi(z) = \frac{1}{2}(z + z^{-1})$ (cf. §1.9), while $\psi(z)$ has no sense in this case. For the circle $|x| = 1$, $x_0 = 0$, we have $\phi(z) = \psi(z) = z$.

We shall denote the inverse functions of (16.1.2) and (16.1.3) by $z = \Phi(x)$ and $z = \Psi(x)$, respectively.

A particularly simple case occurs when the boundary C of T (or of U) consists of a finite number of *analytic arcs*. Then $\phi(z)$ is analytic on $|z| = 1$, apart from a finite number of points corresponding to the corners of C. The length of any subarc of C corresponding to $z = e^{i\theta}$, $\theta_1 \leqq \theta \leqq \theta_2$, can be represented in the form

(16.1.4)
$$\int_{\theta_1}^{\theta_2} |\phi'(e^{i\theta})| \, d\theta$$

(similarly for the mapping (16.1.3)).

(3) Let C be a Jordan curve and $w(x)$ a positive and continuous weight function defined on C. Then the considerations of §10.2 can be applied to the functions $w\{\phi(e^{-i\theta})\}$ and $w\{\psi(e^{i\theta})\}$ which are both positive and continuous on the unit circle $z = e^{i\theta}$, $-\pi \leqq \theta \leqq +\pi$. Substituting into the corresponding analytic functions $D_e(z)$ and $D_i(z)$ the functions $z = \{\Phi(x)\}^{-1}$ and $z = \Psi(x)$, we obtain certain analytic functions $\Delta_e(x)$ and $\Delta_i(x)$ which have the following properties:

(a) $\Delta_e(x)$ is regular in T including $x = \infty$, $\Delta_i(x)$ is regular in U;

(b) $\Delta_e(x) \neq 0$, $\Delta_i(x) \neq 0$;

(c) $\Delta_e(\infty)$ and $\Delta_i(x_0)$ are real and positive.

Furthermore, we have

(16.1.5)
$$\lim_{x_e \to x} |\Delta_e(x_e)|^2 = \lim_{x_i \to x} |\Delta_i(x_i)|^2 = w(x),$$

where $x_e \to x$ indicates an exterior approach to the point x of C and $x_i \to x$ stands for an interior approach to x of C. The convergence is in both cases uniform with respect to x.

These considerations can be generalized by replacing the continuity of $w(x)$ by more general conditions, and also the curve C by more general point sets. If C is an arc (for instance if C is a finite segment), $\Delta_i(x)$ is meaningless.

(4) We define the scalar product of two functions $f(x)$ and $g(x)$, x on C, by the integral

(16.1.6)
$$(f, g) = \frac{1}{L} \int_C f(x)\overline{g(x)} w(x) \, |dx| .$$

We can then orthogonalize the system

(16.1.7)
$$1, x, x^2, \cdots, x^n, \cdots,$$

and this procedure will lead to a set of polynomials uniquely determined by the following requirements (cf. §§2.2, 11.1):

(a) $p_n(x)$ is a polynomial of precise degree n in which the coefficient of x^n is real and positive;

(b) the system $\{p_n(x)\}$ is orthonormal, that is,

$$\frac{1}{L}\int_C p_n(x)\overline{p_m(x)}w(x)\,|dx| = \delta_{nm}, \qquad n, m = 0, 1, 2, \cdots. \tag{16.1.8}$$

If C is a finite real segment, or the unit circle, we obtain the polynomials previously considered. In the next section we take up certain fundamental properties of the polynomials $p_n(x)$ which can be classified as follows: formal properties (minimum-maximum properties, zeros), asymptotic behavior of $p_n(x)$ if $n \to \infty$ and x is in the interior of C, asymptotic behavior of $p_n(x)$ if $n \to \infty$ and x is in the exterior of C, asymptotic behavior of $p_n(x)$ if $n \to \infty$ and x is on C.

In what follows we give only short indications of the proofs, particularly when no essential change is necessary in the arguments used for the previous special cases.

Concerning the definition and principal properties of orthogonal polynomials, see Szegö **5** and Walsh **1**, Chapter VI. Most of these properties have analogues for the polynomials orthogonal on the *interior* U of C; they are associated with the following definition of scalar product:

$$(f, g) = \frac{1}{A}\iint_U f(x)\overline{g(x)}\,d\sigma, \tag{16.1.9}$$

where A is the area and $d\sigma$ the element of area of U. These polynomials have been investigated by Carleman (**1**, pp. 20–30). Here a weight function can also be introduced.

16.2. Formal properties

(1) Let $D_n > 0$ be the determinant of the positive definite quadratic form (cf. (2.2.8) and (11.1.4))

$$\begin{aligned}\frac{1}{L}\int_C |u_0 + u_1x + u_2x^2 + \cdots + u_nx^n|^2\, w(x)\,dx \\ = \sum_{\nu,\mu=0,1,2,\cdots,n} k_{\nu\mu}u_\nu\bar{u}_\mu,\end{aligned} \tag{16.2.1}$$

$$k_{\nu\mu} = \frac{1}{L}\int_C x^\nu\bar{x}^\mu w(x)\,|dx| \tag{16.2.2}$$

(cf. (2.2.1) and (11.1.2)). The orthogonal polynomials $p_n(x)$ can be represented as follows (cf. (2.2.6), (2.2.10), and (11.1.9)):

$$p_0(x) = D_0^{-\frac{1}{2}},$$

$$(16.2.3)\quad p_n(x) = (D_{n-1}D_n)^{-\frac{1}{2}} \begin{vmatrix} k_{00} & k_{10} & \cdots & k_{n0} \\ k_{01} & k_{11} & \cdots & k_{n1} \\ \cdots & \cdots & \cdots & \cdots \\ k_{0,n-1} & k_{1,n-1} & \cdots & k_{n,n-1} \\ 1 & x & \cdots & x^n \end{vmatrix}$$

$$= \frac{(D_{n-1}D_n)^{-\frac{1}{2}}}{L^n n!} \int_C \cdots \int_C (x - x_0)(x - x_1) \cdots (x - x_{n-1})$$

$$\cdot \prod_{\substack{\nu,\mu=0,1,2,\cdots,n-1 \\ \nu<\mu}} |x_\nu - x_\mu|^2 w(x_0)w(x_1) \cdots w(x_{n-1}) \, |dx_0| \, |dx_1| \cdots |dx_{n-1}|.$$

We have (cf. (2.2.7), (2.2.11) and (11.1.5))

$$D_n = [k_{\nu\mu}]_{\nu,\mu=0,1,2,\cdots,n}$$

$$(16.2.4)\quad = \frac{1}{L^{n+1}(n+1)!} \int_C \cdots \int_C \prod_{\substack{\nu,\mu=0,1,2,\cdots,n \\ \nu<\mu}} |x_\nu - x_\mu|^2$$

$$\cdot\, w(x_0)w(x_1) \cdots w(x_n) \, |dx_0| \, |dx_1| \cdots |dx_n|.$$

(2)[72] Let $f(x)$ be a continuous function defined on C. Then the partial sums $s_n(x)$ of the Fourier expansion

$$(16.2.5)\quad \begin{aligned} &f(x) \sim f_0 p_0(x) + f_1 p_1(x) + f_2 p_2(x) + \cdots + f_n p_n(x) + \cdots, \\ &f_n = \frac{1}{L} \int_C f(x)\overline{p_n(x)} w(x) \, |dx|, \qquad n = 0, 1, 2, \cdots, \end{aligned}$$

minimize the integral

$$(16.2.6)\quad \frac{1}{L} \int_C |f(x) - \rho(x)|^2 w(x) \, |dx|$$

if $\rho(x)$ ranges over the class of all π_n. The minimum is

$$(16.2.7)\quad \frac{1}{L} \int_C |f(x)|^2 w(x) \, |dx| - |f_0|^2 - |f_1|^2 - \cdots - |f_n|^2.$$

This also yields Bessel's inequality

$$(16.2.8)\quad |f_0|^2 + |f_1|^2 + |f_2|^2 + \cdots \leqq \frac{1}{L} \int_C |f(x)|^2 w(x) \, |dx|.$$

Let C be a rectifiable Jordan curve, and let $f(x)$ be an analytic function regular in the interior of C and continuous on C. Then the equality sign in (16.2.8)

[72] Concerning the considerations of (2), (3), (4), cf. §3.1, §11.1, §11.3.

is to be taken; that is, Parseval's formula holds. This follows from Theorem 1.3.4.

Smirnoff (**1**) investigated the validity of this formula for the more general class of functions $f(x)$ for which

$$f\{\psi(z)\}D_i(z)\{\psi'(z)\}^{\frac{1}{2}} \tag{16.2.9}$$

is of class H_2 in $|z| < 1$ (cf. §10.1); here we have used the symbols previously introduced. Parseval's formula holds for this class if and only if the map function $\psi(z)$ satisfies the hypothesis

$$\log|\psi'(re^{i\theta})| = \frac{1}{2\pi}\int_{-\pi}^{+\pi} \log|\psi'(e^{it})| \frac{1-r^2}{1-2r\cos(\theta-t)+r^2}\,dt, \qquad 0 \leqq r < 1. \tag{16.2.10}$$

(See Smirnoff **1**, pp. 164–168.) According to Keldysch and Lavrentieff (**1**), there exist rectifiable curves for which the hypothesis (16.2.10) is not satisfied.

(3) Let $\rho(x)$ be an arbitrary π_n in which the coefficient of x^n equals unity. Then

$$\min \frac{1}{L}\int_C |\rho(x)|^2 w(x)\,|dx| \tag{16.2.11}$$

is attained if and only if $\rho(x) = (D_n/D_{n-1})^{\frac{1}{2}}p_n(x)$, and this minimum is D_n/D_{n-1}.

(4) Let x_0 be an arbitrary but fixed point in the complex plane, and let $\rho(x)$ be an arbitrary π_n with $\rho(x_0) = 1$. Then the minimum of (16.2.11) is attained for $\rho(x) = \{K_n(x_0, x_0)\}^{-1}K_n(x_0, x)$, where

$$K_n(x_0, x) = \overline{p_0(x_0)}p_0(x) + \overline{p_1(x_0)}p_1(x) + \cdots + \overline{p_n(x_0)}p_n(x). \tag{16.2.12}$$

The minimum is $\{K_n(x_0, x_0)\}^{-1}$. (The same result can be expressed in terms of a maximum property; see §3.1 (3) and §11.3.) The "kernel polynomials" $K_n(x_0, x)$ can be characterized by the condition

$$\frac{1}{L}\int_C K_n(x_0, x)\overline{\rho(x)}w(x)\,|dx| = \overline{\rho(x_0)}, \tag{16.2.13}$$

where $\rho(x)$ is an arbitrary π_n (cf. (3.1.12)).

Using $K_n(x_0, x)$, we can express the nth partial sum $s_n(x)$ of the development (16.2.5) in the following form (cf. (3.1.11)):

$$s_n(x) = \frac{1}{L}\int_C f(\xi)K_n(\xi, x)w(\xi)\,|d\xi|. \tag{16.2.14}$$

(5) The extremum problem of (3) can be generalized as in the case of an interval (§3.11). For instance, the problem of the "Tchebichef deviation" corresponding to (16.2.11) consists of the determination of the minimum of $\max|\rho(x)|$, x on C, when $\rho(x)$ is an arbitrary π_n in which the coefficient of x^n is 1. The polynomials which solve this problem have been investigated by Faber (**3**).

Let μ_n be the minimum in question. Then (cf. Faber, loc. cit.)

$$\lim \mu_n^{1/n} = c, \tag{16.2.15}$$

where c is the transfinite diameter defined in §16.1 (2). We shall see that a similar formula holds for the minimum of the problem in (3) (cf. (16.4.3)).

Fekete (**1**) extended the definition of the transfinite diameter to an arbitrary closed set in the complex plane by showing that for the corresponding minimum values μ_n (which have meaning if we replace the curve C by an arbitrary closed set), the limit (16.2.15) still exists.

The general transfinite diameter is a remarkable set-function which can be calculated in various cases (cf. for instance Szegö **5**, p. 254).

Concerning the extension of §3.11 (5) to an arbitrary curve, see Julia **1**.

(6) *The zeros of $p_n(x)$ lie in the least convex region containing the curve C.* See Szegö **5**, pp. 236–241; Fejér **7**. The proof can be based on an argument similar to that of §3.3 (2) (cf. also §16.4 (1), (a)).

Concerning the location of the zeros of $K_n(x_0, x)$, see Szegö **5**, pp. 241–244. See also Theorem 11.4.1 and Problems 5 and 49.

16.3. Asymptotic behavior of $K_n(x_0, x)$ in the interior of the curve C

(1) In this section we assume that C is an analytic Jordan curve and the weight function $w(x)$ defined on C is positive and continuous there. Let x_0 be an arbitrary point interior to C; denote by $z = \Psi(x) = \Psi(x_0, x)$ the inverse function of the map function $\psi(z)$ defined by (16.1.3). In the present case $\psi(z)$ is regular and univalent in a certain circle $|z| \leqq \mathrm{P}$ where $\mathrm{P} > 1$. Let $\Delta_i(x)$ have the same meaning as in §16.1 (3). We shall prove the following theorem:

THEOREM 16.3. *Let $\{p_n(x)\}$ denote the set of orthogonal polynomials defined by the conditions* (16.1.8). *Then the series*

$$K(x_0, x) = \overline{p_0(x_0)}p_0(x) + \overline{p_1(x_0)}p_1(x) + \cdots + \overline{p_n(x_0)}p_n(x) + \cdots \tag{16.3.1}$$

is convergent provided x_0 and x are arbitrary points in the interior of C. The convergence is uniform both in x_0 and x if x_0 and x are limited to a closed set in the interior of C. Furthermore, we have

$$K(x_0, x) = \frac{L}{2\pi} \{\Delta_i(x_0)\Delta_i(x)\}^{-1}\{\Psi'(x_0)\Psi'(x)\}^{\frac{1}{2}}. \tag{16.3.2}$$

In case $w(x) = 1$, see Szegö **5**, pp. 244–251. The assumptions of this theorem concerning $w(x)$ and C can be generalized (cf. Smirnoff **2**, pp. 353–356). For a Jordan arc (especially for a finite segment) these considerations have no sense. For the special case where C is the unit circle, see (12.3.17).

As a consequence of the convergence of (16.3.1) we notice that

$$\lim_{n\to\infty} p_n(x) = 0, \tag{16.3.3}$$

if x is in the interior of C (uniformly on every closed set in the interior of C).

(2) Let $\rho(x)$ be an arbitrary π_n, and let x_0 be in the interior of C. Then according to Cauchy's theorem,

$$\{\rho(x_0)\}^2 = \frac{1}{2\pi i}\int \frac{\{\rho(x)\}^2}{x - x_0}\,dx. \tag{16.3.4}$$

Consequently,

$$|\rho(x_0)|^2 \leqq \frac{1}{2\pi}\int \frac{|\rho(x)|^2}{|x - x_0|}\,|dx| \leqq \frac{\mu^{-1}L}{2\pi\delta}\frac{1}{L}\int_C |\rho(x)|^2 w(x)\,|dx|, \tag{16.3.5}$$

where $w(x) \geqq \mu$ and δ denotes the least distance of x_0 from the points x on the curve C. Taking into account §16.2 (4), we have

$$K_n(x_0, x_0) \leqq \frac{\mu^{-1}L}{2\pi\delta}. \tag{16.3.6}$$

Hence the series $K(x_0, x_0)$, and therefore (16.3.1), are convergent, as is readily shown by Cauchy's inequality.

(3) Now we show that

$$K(x_0, x_0) = \lim_{n\to\infty} K_n(x_0, x_0) \geqq \frac{L}{2\pi}\{\Delta_i(x_0)\}^{-2}\Psi'(x_0). \tag{16.3.7}$$

To this end we consider the function

$$\left(\frac{L}{2\pi}\right)^{\frac{1}{2}}\{\Delta_i(x)\}^{-1}\{\Psi'(x)\}^{\frac{1}{2}} = F(z), \tag{16.3.8}$$

which is regular in the interior of C, that is, for $|z| < 1$; $x = \psi(z)$, $z = \Psi(x)$. (The last factor is regular even for $|z| \leqq 1$.) Let r be fixed, $0 < r < 1$, and ϵ an arbitrary positive number. According to Theorem 1.3.4 we can find a polynomial $Q(x)$ such that

$$|F(rz) - Q(x)| < \epsilon, \qquad x \text{ on } C. \tag{16.3.9}$$

Writing $\rho(x) = \{Q(x_0)\}^{-1}Q(x)$, we obtain from §16.2 (4), for a sufficiently large n,

$$\{K(x_0, x_0)\}^{-1} \leqq \{K_n(x_0, x_0)\}^{-1} \leqq \frac{|Q(x_0)|^{-2}}{L}\int_C |Q(x)|^2 w(x)\,|dx|; \tag{16.3.10}$$

whence, on allowing ϵ to approach zero, we obtain

$$\{K(x_0, x_0)\}^{-1} \leqq \frac{|F(0)|^{-2}}{L}\int_C |F(rz)|^2 w(x)\,|dx|. \tag{16.3.11}$$

Now if $r \to 1 - 0$, we have

$$\{K(x_0, x_0)\}^{-1} \leqq \frac{|F(0)|^{-2}}{2\pi}\int_C |\Psi'(x)|\,|dx| = |F(0)|^{-2}, \tag{16.3.12}$$

which is equivalent to (16.3.7).

(4) Finally, we consider

$$
\begin{aligned}
J_n &= \lim_{r\to 1-0} \frac{1}{L} \int_{|z|=r} \left| K_n(x_0, x)\Delta_i(x)\{\Psi'(x)\}^{-\frac{1}{2}} - \frac{L}{2\pi}\{\Delta_i(x_0)\}^{-1}\{\Psi'(x_0)\}^{\frac{1}{2}} \right|^2 |dz| \\
&= \frac{1}{L}\int_C |K_n(x_0, x)|^2 w(x)\,|dx| \\
&\quad - \lim_{r\to 1-0} \pi^{-1} r\{\Delta_i(x_0)\}^{-1}\{\Psi'(x_0)\}^{\frac{1}{2}} \int_{|z|=r} K_n(x_0, x)\Delta_i(x)\{\Psi'(x)\}^{-\frac{1}{2}}\frac{dz}{iz} \\
&\quad + \frac{L}{2\pi}\{\Delta_i(x_0)\}^{-2}\Psi'(x_0).
\end{aligned} \tag{16.3.13}
$$

The first term equals $K_n(x_0, x_0)$; for the second term we obtain

$$
-\pi^{-1}\{\Delta_i(x_0)\}^{-1}\{\Psi'(x_0)\}^{\frac{1}{2}}\cdot 2\pi K_n(x_0, x_0)\Delta_i(x_0)\{\Psi'(x_0)\}^{-\frac{1}{2}} = -2K_n(x_0, x_0); \tag{16.3.14}
$$

so

$$
J_n = -K_n(x_0, x_0) + \frac{L}{2\pi}\{\Delta_i(x_0)\}^{-2}\Psi'(x_0). \tag{16.3.15}
$$

Therefore, $\lim_{n\to\infty} J_n = 0$ by (16.3.7). From this it follows, if $|z| < 1$, or if x is in the interior of C, that (cf. (7.1.4))

$$
\lim_{n\to\infty} K_n(x_0, x)\Delta_i(x)\{\Psi'(x)\}^{-\frac{1}{2}} = \frac{L}{2\pi}\{\Delta_i(x_0)\}^{-1}\{\Psi'(x_0)\}^{\frac{1}{2}}, \tag{16.3.16}
$$

which is the same as (16.3.2).

16.4. Asymptotic behavior of $p_n(x)$ in the exterior of the curve C

(1) We shall introduce the assumptions concerning the curve C and the weight function $w(x)$ used in §16.3. We denote by $z = \Phi(x)$ the inverse function of the map function $x = \phi(z)$, defined by (16.1.2); in the present case $\phi(z)$ is regular and univalent in the closed exterior of a certain circle $|z| = r$ where $r < 1$. Let $\Delta_e(x)$ have the same meaning as in §16.1 (3). We may then make the following assertion:

THEOREM 16.4. *Let $\{p_n(x)\}$ denote the set of orthogonal polynomials defined by the conditions* (16.1.8). *If x is in the exterior of the curve C, we have*

$$
p_n(x) \cong \left(\frac{L}{2\pi}\right)^{\frac{1}{2}}\{\Delta_e(x)\}^{-1}\{\Phi'(x)\}^{\frac{1}{2}}\{\Phi(x)\}^n. \tag{16.4.1}
$$

The ratio of these expressions as $n \to \infty$ tends to unity uniformly in every finite or infinite closed region exterior to the curve C.

In case $w(x) = 1$, see Szegö **5**, pp. 260–263.

We mention the following noteworthy consequences of (16.4.1):

(a) The zeros of $p_n(x)$ approach the closed interior of C uniformly as $n \to \infty$ (apply Theorem 1.91.3 (Hurwitz's theorem)).

(b) If x is in the exterior of C,

$$\text{(16.4.2)} \qquad \lim_{n\to\infty} \frac{p_{n+1}(x)}{p_n(x)} = \lim_{n\to\infty} \{p_n(x)\}^{1/n} = \Phi(x).$$

Cf. (12.2.6).

(c) The convergence domain of the expansion (16.2.5) of an analytic function, regular in the closed interior of C, is the interior of a level curve C_R of the conformal mapping $x = \phi(z)$ (cf. §1.3 (2)). The determination of R from the coefficients f_n is analogous to that in the case of a power series (cf. Theorems 1.3.5 and 12.7.3).

(d) Let k_n be the coefficient of x^n in $p_n(x)$. Then

$$\text{(16.4.3)} \qquad k_n \cong \left(\frac{L}{2\pi}\right)^{\frac{1}{2}} \{\Delta_e(\infty)\}^{-1} c^{-n-\frac{1}{2}},$$

where c is the transfinite diameter of C (cf. §16.1 (2); §16.2 (5)). This result can easily be expressed in terms of the determinants D_n introduced in (16.2.4).

(2) The proof of (16.4.1) is based on an argument similar to that in §16.3; here the minimum property of §16.2 (3) is used. We first show that

$$\text{(16.4.4)} \qquad \limsup_{n\to\infty} c^{-2n-1} k_n^{-2} \leqq \frac{2\pi}{L} \{\Delta_e(\infty)\}^2.$$

To this end we introduce the polynomials of Faber's type $f_n(x)$ (Faber **1**), defined as the polynomial part in the Laurent development of the function

$$\text{(16.4.5)} \qquad \left(\frac{L}{2\pi}\right)^{\frac{1}{2}} \{\Delta_e(x)\}^{-1}\{\Phi'(x)\}^{\frac{1}{2}}\{\Phi(x)\}^n = g_n(x)$$

around $x = \infty$. The function $g_n(x)$ is regular for $|z| \geqq r$, and $f_n(x)$ is clearly a π_n. Let C_r be the curve corresponding to $|z| = r$. On applying Cauchy's theorem to the ring-shaped region bounded by C_r and a large circle, we obtain, if x is on C,

$$\text{(16.4.6)} \quad g_n(x) - f_n(x) = \frac{1}{2\pi i} \int_{C_r} \frac{g_n(\xi) - f_n(\xi)}{\xi - x} d\xi = \frac{1}{2\pi i} \int_{C_r} \frac{g_n(\xi)}{\xi - x} d\xi,$$

where the integration is extended in the negative sense. Hence

$$\text{(16.4.7)} \qquad |g_n(x) - f_n(x)| < Mr^n \qquad \text{for } x \text{ on } C,$$

where M depends only on C and r. The functions $g_n(x)$ are uniformly bounded if x is on C.

Now let $\gamma_1, \gamma_2, \cdots, \gamma_m$ be arbitrary constants, $n > m$. The polynomial

$$\text{(16.4.8)} \quad \rho(x) = \left(\frac{L}{2\pi}\right)^{-\frac{1}{2}} c^{n+\frac{1}{2}}\{f_n(x) + \gamma_1 f_{n-1}(x) + \gamma_2 f_{n-2}(x) + \cdots + \gamma_m f_{n-m}(x)\}$$

is of degree n, and its highest coefficient is unity. Then, according to §16.2 (3) and (16.4.7),

$$\begin{aligned} k_n^{-2} &\leqq \left(\frac{L}{2\pi}\right)^{-1} c^{2n+1} \frac{1}{L} \int_C | f_n(x) + \gamma_1 f_{n-1}(x) + \cdots \\ &\qquad\qquad + \gamma_m f_{n-m}(x) |^2 w(x) \, | dx | \\ &= \left(\frac{L}{2\pi}\right)^{-1} c^{2n+1} \frac{1}{L} \int_C | g_n(x) + \gamma_1 g_{n-1}(x) + \cdots \\ &\qquad\qquad + \gamma_m g_{n-m}(x) |^2 w(x) \, | dx | + c^{2n} O(r^n), \end{aligned} \tag{16.4.9}$$

so that

$$\begin{aligned} &\limsup_{n\to\infty} c^{-2n-1} k_n^{-2} \\ &\qquad \leqq \frac{1}{L} \int_{|z|=1} | 1 + \gamma_1 z^{-1} + \gamma_2 z^{-2} + \cdots + \gamma_m z^{-m} |^2 w(x) \, | dz |; \end{aligned} \tag{16.4.10}$$

or

$$\limsup_{n\to\infty} c^{-2n-1} k_n^{-2} \leqq \frac{1}{L} \int_{|z|=1} | \gamma(z) |^2 w(x) \, | dz |, \tag{16.4.11}$$

where $\gamma(z)$ is an arbitrary analytic function, regular for $|z| \geqq 1$, with $\gamma(\infty) = 1$. On putting

$$\gamma(z) = \left\{ \frac{\Delta_e\{\phi(Rz)\}}{\Delta_e(\infty)} \right\}^{-1}, \qquad R > 1, \tag{16.4.12}$$

and allowing R to approach $1 + 0$, we obtain the inequality (16.4.4).

(3) Now we consider

$$\begin{aligned} J_n' &= \lim_{R\to 1+0} \frac{1}{L} \int_{|z|=R} \left| p_n(x)\Delta_e(x) \{\Phi'(x)\}^{-\frac{1}{2}} \{\Phi(x)\}^{-n} - \left(\frac{L}{2\pi}\right)^{\frac{1}{2}} \right|^2 | dz | \\ &= \frac{1}{L} \int_C | p_n(x) |^2 w(x) \, | dx | \\ &\quad - \lim_{R\to 1+0} \frac{2R}{L} \left(\frac{L}{2\pi}\right)^{\frac{1}{2}} \int_{|z|=R} p_n(x)\Delta_e(x) \{\Phi'(x)\}^{-\frac{1}{2}} \{\Phi(x)\}^{-n} \frac{dz}{iz} + 1. \end{aligned} \tag{16.4.13}$$

The second term is

$$-\frac{2}{L}\left(\frac{L}{2\pi}\right)^{\frac{1}{2}} 2\pi \lim_{x\to\infty} p_n(x)\Delta_e(x)\{\Phi'(x)\}^{-\frac{1}{2}}\{\Phi(x)\}^{-n} = -2\left(\frac{2\pi}{L}\right)^{\frac{1}{2}} \Delta_e(\infty) c^{n+\frac{1}{2}} k_n.$$

Therefore,

$$J_n' = 2 - 2\left(\frac{2\pi}{L}\right)^{\frac{1}{2}} \Delta_e(\infty) c^{n+\frac{1}{2}} k_n. \tag{16.4.14}$$

In view of (16.4.4), this implies $\lim_{n\to\infty} J_n' = 0$, which establishes the statement (16.4.1).

16.5. Asymptotic behavior of $p_n(x)$ on the curve C

Finally we have the following result:

THEOREM 16.5. *The asymptotic formula* (16.4.1) *holds uniformly outside and on the curve C provided the function $\Delta_e(x)$ is regular in the closed exterior of C; more precisely we have*

$$p_n(x) = \left(\frac{L}{2\pi}\right)^{\frac{1}{2}} \{\Delta_e(x)\}^{-1} \{\Phi'(x)\}^{\frac{1}{2}} \{\Phi(x)\}^n + O(h^n), \tag{16.5.1}$$

where $0 < h < 1$; this constant h depends on C and $w(x)$.

The same formula holds in a sufficiently small neighborhood of C in the interior of C.

(1) For the proof we use the polynomials $F_n(x)$ of Faber's type associated with

$$\left(\frac{L}{2\pi}\right)^{\frac{1}{2}} \{\Delta_e(x)\}^{-1} \{\Phi'(x)\}^{\frac{1}{2}} \{\Phi(x)\}^n = G_n(x) \tag{16.5.2}$$

(which is the "principal part" of the right-hand side of (16.5.1)) in the same way as the polynomials $f_n(x)$ defined in §16.4 (2) are associated with (16.4.5). If $0 < r < 1$ and r is sufficiently near to 1, the function $G_n(x)$ is regular for $|z| \geqq r$. We have again

$$| G_n(x) - F_n(x) | < Mr^n \qquad \text{for } x \text{ on } C, \tag{16.5.3}$$

where M depends only on C, r and $w(x)$. The functions $G_n(x)$ are uniformly bounded on C.

We write (cf. §16.4 (2))

$$\rho(x) = \left(\frac{L}{2\pi}\right)^{-\frac{1}{2}} \Delta_e(\infty) c^{n+\frac{1}{2}} F_n(x). \tag{16.5.4}$$

This is a π_n with the highest coefficient unity. Hence

$$\begin{aligned} k_n^{-2} &\leqq \left(\frac{L}{2\pi}\right)^{-1} \{\Delta_e(\infty)\}^2 c^{2n+1} \frac{1}{L} \int_C | F_n(x) |^2 w(x) \, | dx | \\ &= \left(\frac{L}{2\pi}\right)^{-1} \{\Delta_e(\infty)\}^2 c^{2n+1} \frac{1}{L} \int_C | G_n(x) |^2 w(x) \, | dx | + c^{2n} O(r^n) \\ &= \frac{2\pi}{L} \{\Delta_e(\infty)\}^2 c^{2n+1} + c^{2n} O(r^n). \end{aligned} \tag{16.5.5}$$

(The simpler nature of this argument is due to the fact that $\Delta_e(x)$ is regular on C.) On the other hand,

$$\begin{aligned} 1 &= \frac{1}{L} \int_C | p_n(x) |^2 w(x) \, | dx | = \frac{1}{2\pi} \int_{|z|=1} \left| \frac{p_n(x)}{G_n(x)} \right|^2 | dz | \\ &\geqq \lim_{z \to \infty} \left| \frac{p_n(x)}{G_n(x)} \right|^2 = \frac{2\pi}{L} \{\Delta_e(\infty)\}^2 c^{2n+1} k_n^2, \end{aligned} \tag{16.5.6}$$

so that

$$k_n^{-2} \geqq \frac{2\pi}{L} \{\Delta_e(\infty)\}^2 c^{2n+1}; \tag{16.5.7}$$

consequently

$$k_n^{-2} = \frac{2\pi}{L} \{\Delta_e(\infty)\}^2 c^{2n+1} + c^{2n} O(r^n). \tag{16.5.8}$$

(2) Now let

$$p_n(x) = \lambda_0 F_0(x) + \lambda_1 F_1(x) + \lambda_2 F_2(x) + \cdots + \lambda_n F_n(x), \tag{16.5.9}$$

where $\lambda_0, \lambda_1, \lambda_2, \cdots, \lambda_n$ are proper constants, $\lambda_\nu = \lambda(\nu, n)$; we have

$$\lambda_n = \left(\frac{2\pi}{L}\right)^{\frac{1}{2}} \Delta_e(\infty) c^{n+\frac{1}{2}} k_n, \tag{16.5.10}$$

so that λ_n is real and, by (16.5.8),

$$\lambda_n = 1 + O(r^n). \tag{16.5.11}$$

From the definition of $F_n(x)$ we conclude that

$$\begin{aligned} &\left(\frac{2\pi}{L}\right)^{\frac{1}{2}} \Delta_e(x) \{\Phi'(x)\}^{-\frac{1}{2}} \{p_n(x) - \lambda_n F_n(x)\} \\ &= \left(\frac{2\pi}{L}\right)^{\frac{1}{2}} \Delta_e(x) \{\Phi'(x)\}^{-\frac{1}{2}} \{\lambda_0 F_0(x) + \lambda_1 F_1(x) + \cdots + \lambda_{n-1} F_{n-1}(x)\} \\ &= \lambda_0 + \lambda_1 \Phi(x) + \lambda_2 \{\Phi(x)\}^2 + \cdots + \lambda_{n-1} \{\Phi(x)\}^{n-1} + \gamma_1 x^{-1} \\ &\qquad + \gamma_2 x^{-2} + \cdots \\ &= \lambda_0 + \lambda_1 z + \lambda_2 z^2 + \cdots + \lambda_{n-1} z^{n-1} + \gamma_1' z^{-1} + \gamma_2' z^{-2} + \cdots, \end{aligned} \tag{16.5.12}$$

where γ_ν, γ_ν' are certain constants. Therefore,

$$\begin{aligned} &|\lambda_0|^2 + |\lambda_1|^2 + \cdots + |\lambda_{n-1}|^2 \\ &\leqq \frac{2\pi}{L} \frac{1}{2\pi} \int_{|z|=1} |\Delta_e(x) \{\Phi'(x)\}^{-\frac{1}{2}} \{p_n(x) - \lambda_n F_n(x)\}|^2 |dz| \\ &= \frac{1}{L} \int_C |p_n(x) - \lambda_n F_n(x)|^2 w(x) |dx| \\ &= \frac{1}{L} \int_C |p_n(x)|^2 w(x) |dx| \\ &\qquad - \frac{2\lambda_n}{L} \int_C p_n(x) \overline{F_n(x)} w(x) |dx| + \frac{\lambda_n^2}{L} \int_C |F_n(x)|^2 w(x) |dx|. \end{aligned} \tag{16.5.13}$$

In the second term $F_n(x)$ can be replaced by $\lambda_n^{-1} p_n(x)$; in the third term we use (16.5.11) and (16.5.3) and obtain

$$|\lambda_0|^2 + |\lambda_1|^2 + \cdots + |\lambda_{n-1}|^2 \leqq -1 + \frac{1}{L}\int_C |G_n(x)|^2 w(x) \, |dx| + O(r^n), \tag{16.5.14}$$

or

$$|\lambda_0|^2 + |\lambda_1|^2 + \cdots + |\lambda_{n-1}|^2 = O(r^n). \tag{16.5.15}$$

Now $F_n(x) = O(1)$, uniformly if x is on C. Whence, according to Schwarz's inequality,

$$\lambda_0 F_0(x) + \lambda_1 F_1(x) + \cdots + \lambda_{n-1} F_{n-1}(x) = O(n^{\frac{1}{2}} r^{n/2}); \tag{16.5.16}$$

and from (16.5.9), (16.5.11), and (16.5.3),

$$\begin{aligned} p_n(x) &= \lambda_n F_n(x) + O(n^{\frac{1}{2}} r^{n/2}) = F_n(x) + O(n^{\frac{1}{2}} r^{n/2}) \\ &= G_n(x) + O(n^{\frac{1}{2}} r^{n/2}). \end{aligned} \tag{16.5.17}$$

This establishes the statement. The extension of (16.5.1) to the interior of C also follows immediately, since (16.5.3) holds if x is in the interior of C and sufficiently near to C.

PROBLEMS AND EXERCISES

1. We denote by $i_1 < i_2 < \cdots$ the positive zeros of Airy's function $A(x)$ (§1.81); then $i_\nu \sim \nu^{2/3}$ if $\nu \to \infty$. (Use (1.81.4), (1.81.1), and (1.71.7).)

2. Let $A(x)$ denote Airy's function (§1.81). We have, for real values of x,

$$\mathfrak{I}\left\{e^{2\pi i/3} \int_0^\infty \exp\left(-\rho^3 - \rho e^{2\pi i/3} x\right) d\rho\right\} = A(x).$$

(Develop both sides in a power series in x; see (1.81.4), (1.81.1), and (1.7.3).)

3. Let $p_n(x)$ denote the Poisson-Charlier polynomial (2.81.2); then

$$(-1)^n a^{-n/2}(n!)^{1/2} p_n(x) = \mathbf{p}_n(x)$$

satisfies, for $x = 0, 1, 2, \cdots$, the relation

$$\mathbf{p}_n(x) = \mathbf{p}_x(n).$$

4. By use of the notation (2.2.1), (2.2.7), the "kernel" polynomials $K_n(x_0, x)$ (cf. (3.1.9)) can be represented as follows:

$$K_n(x_0, x) = -D_n^{-1} \begin{vmatrix} c_0 & c_1 & c_2 & \cdots & c_n & 1 \\ c_1 & c_2 & c_3 & \cdots & c_{n+1} & \bar{x}_0 \\ \cdots & \cdots & \cdots & \cdots & \cdots & \cdots \\ c_n & c_{n+1} & c_{n+2} & \cdots & c_{2n} & \bar{x}_0^n \\ 1 & x & x^2 & \cdots & x^n & 0 \end{vmatrix}$$

(Use (3.1.12) with $\rho(x) = x^\nu$, $\nu = 0, 1, \cdots, n$.)

5. *Location of the zeros of the "kernel" polynomials* $K_n(x_0, x)$ (cf. (3.1.9)). Let a and b be finite, and let x_0 be an arbitrary non-real number. Every zero ξ of $K_n(x_0, x)$ lies in the area bounded by the interval $[a, b]$ and by the circular arc through a and b whose continuation passes through x_0. (Cf. Szegö **5**, p. 244. By (3.1.12)

$$\int_a^b \left|(x - x_0) \frac{K_n(x_0, x)}{x - \xi}\right|^2 \frac{x - \xi}{x - x_0}\, d\alpha(x) = 0.$$

In the conformal mapping $(x - \xi)/(x - x_0) = x'$, the image of $a \leqq x \leqq b$ is a circular arc; the segment bounded by this arc and its chord contains $x' = 0$.)

6. Derive from (3.2.1) the representation, $n = 1, 2, 3, \cdots$,

$$p_n(x) = p_0(x) \begin{vmatrix} A_1 x + B_1 & C_2^{1/2} & 0 & \cdots & 0 \\ C_2^{1/2} & A_2 x + B_2 & C_3^{1/2} & \cdots & 0 \\ 0 & C_3^{1/2} & A_3 x + B_3 & \cdots & 0 \\ \cdots & \cdots & \cdots & \cdots & \cdots \\ 0 & 0 & 0 & C_n^{1/2} & A_n x + B_n \end{vmatrix}$$

7. Prove the reality of the zeros of $p_n(x)$ using (2.2.9), or using Problem 6.

8. Let $x_1, x_2, \cdots, x_n$ be distinct values in $[a, b]$, and let $f(x)$ have a derivative of order $2n$ in $[a, b]$. If $H(x)$ is the π_{2n-1} satisfying the conditions

$$H(x_\nu) = f(x_\nu), \qquad H'(x_\nu) = f'(x_\nu), \qquad \nu = 1, 2, \cdots, n,$$

there exists a value $\xi = \xi(x)$ in $[a, b]$ such that

$$f(x) - H(x) = \frac{f^{(2n)}(\xi)}{(2n)!}(x - x_1)^2(x - x_2)^2 \cdots (x - x_n)^2.$$

(See A. Markoff 5, pp. 6–8. Let x be a fixed value, $x \neq x_\nu$; and to the function of z given by

$$f(z) - H(z) - \frac{f(x) - H(x)}{(x - x_1)^2(x - x_2)^2 \cdots (x - x_n)^2}(z - x_1)^2(z - x_2)^2 \cdots (z - x_n)^2,$$

apply Rolle's theorem.)

9. Let $f(x)$ have a continuous derivative of order $2n$ in $[a, b]$. By using the notation of Theorem 3.4.1 obtain

$$\int_a^b f(x)\, d\alpha(x) = \lambda_1 f(x_1) + \lambda_2 f(x_2) + \cdots + \lambda_n f(x_n) + \frac{f^{(2n)}(\xi)}{(2n)!} k_n^{-2}.$$

Here ξ is a proper value in $[a, b]$, and k_n is the highest coefficient of the orthonormal polynomial $p_n(x)$ associated with the distribution $d\alpha(x)$ (cf. (2.2.15)). (See A. Markoff 5, p. 81; use the preceding problem.)

10. Let $x_1, x_2, \cdots, x_n$ be the zeros of the orthogonal polynomial $p_n(x)$ associated with a given distribution $d\alpha(x)$ on the interval $[a, b]$, and let $\lambda_1, \lambda_2, \cdots, \lambda_n$ denote the Christoffel numbers (3.4.3). Define the scalar product of two functions $f(x)$, $g(x)$ by

$$(f, g) = \sum_{\nu=1}^{n} \lambda_\nu f(x_\nu) g(x_\nu).$$

The functions $1, x, x^2, \cdots, x^{n-1}$ are linearly independent; by orthogonalization we obtain the polynomials $p_0(x), p_1(x), \cdots, p_{n-1}(x)$ which are the same as those associated with $d\alpha(x)$.

11. If $d\alpha(x) = w(x)dx$, $w(-x) = w(x)$ and $a + b = 0$, we have for the Christoffel numbers (3.4.3)

$$\lambda_\nu = \lambda_{n+1-\nu}, \qquad \nu = 1, 2, \cdots, n.$$

12. For the Jacobi polynomial $P_n^{(-\frac{1}{2}, \frac{1}{2})}(x)$ the Christoffel numbers are

$$\lambda_\nu = \frac{2\pi}{2n + 1}(1 + x_\nu), \qquad \nu = 1, 2, \cdots, n.$$

(Cf. (4.1.8) and (15.3.1).)

13. In the special case $P_n^{(-\frac{1}{2}, -\frac{1}{2})}(x)$, the numbers $y_\nu = \cos \phi_\nu$, $0 < \phi_\nu < \pi$, of the separation theorem of §3.41 are

$$\phi_\nu = (n - \nu)\pi/n, \qquad \nu = 1, 2, \cdots, n - 1.$$

Verify the separation theorem for $P_n^{(-\frac{1}{2},-\frac{1}{2})}(x)$, $P_n^{(\frac{1}{2},\frac{1}{2})}(x)$, $P_n^{(-\frac{1}{2},\frac{1}{2})}(x)$.

14. Let $x_1, x_2, \cdots, x_n$ be the zeros of $P_n^{(\alpha,\beta)}(x)$. Then

$$x_1 + x_2 + \cdots + x_n = (\beta - \alpha)\frac{n}{2n + \alpha + \beta}.$$

(Cf. (4.21.2).)

15. *New proof of* (4.7.31). Combining the second part of (4.1.5) (polynomials in $1 - 2x^2$) with the first formula in (4.22.1), we find

$$P_n^{(\lambda)}(x) = 2^n \binom{n + \lambda - 1}{n} x^n F(-n/2, (1 - n)/2, -n - \lambda + 1; x^{-2}).$$

16. The generating functions (4.7.16) and (4.7.23) of the ultraspherical polynomials $P_n^{(\lambda)}(x)$ are identical if and only if $\lambda = \frac{1}{2}$, that is, in the Legendre case.

17. The functional equation

$$(1 - x)f'(x) = \lambda f(-x),$$

where λ is a parameter, has a polynomial solution $f(x) \not\equiv 0$ if and only if $\lambda = (-1)^n(n + 1)$ and $f(x) = \text{const.}\,\{P_n(x) + P_{n+1}(x)\}$, $n = -1, 0, 1, \cdots$; $P_{-1}(x) = 0$. (Write

$$f(x) = \sum_{\nu=-1}^{N} c_\nu\{P_\nu(x) + P_{\nu+1}(x)\},$$

where $c_{-1}, c_0, \cdots, c_N$ are constants, and use

$$(1 - x)\{P_n'(x) + P_{n+1}'(x)\} = (n + 1)\{P_n(x) - P_{n+1}(x)\},$$

which follows from (4.7.27).)

18. Show that

$$\mathbf{Q}_n'(0) = (-1)^{n/2}\frac{2\cdot 4 \cdots n}{1\cdot 3 \cdots (n - 1)}, \qquad n \text{ even},$$

$$\mathbf{Q}_n(0) = (-1)^{(n+1)/2}\frac{2\cdot 4 \cdots (n - 1)}{3\cdot 5 \cdots n}, \qquad n \text{ odd}.$$

Here $\mathbf{Q}_0'(0) = 1$, $\mathbf{Q}_1(0) = -1$. (Use the recurrence formula (4.62.13), (4.62.14), (4.62.3).)

19. The "Laplace transform"

$$f(s) = \int_0^\infty e^{-st}F(t)\,dt,$$

of Laguerre's function $F(t) = t^\alpha L_n^{(\alpha)}(t)$ is

$$f(s) = \frac{\Gamma(n + \alpha + 1)}{\Gamma(n + 1)} s^{-n-\alpha-1}(s - 1)^n.$$

(Cf. Sonin **1**, p. 42; use the method of the "generating function".)

20. Writing

$$x^{\alpha}L_n^{(\alpha)}(x)(\Gamma(n+\alpha+1))^{-1} = f(n, \alpha; x),$$

we have, for $\alpha > \beta > -1$,

$$f(n, \alpha; x) = \frac{1}{\Gamma(\alpha - \beta)} \int_0^x (x-u)^{\alpha-\beta-1} f(n, \beta; u)\, du.$$

(Kogbetliantz **22**, p. 156. Take the generating function of both sides and use (5.1.16); the resulting formula for Bessel functions can be verified by means of (1.71.1) and (1.7.5).)

21. Writing $e^{-x/2}x^{\alpha/2}L_n^{(\alpha)}(x) = f_n(x)$, we have

$$f_n(x) = \frac{(-1)^n}{2} \int_0^\infty J_\alpha\{(xy)^{\frac{1}{2}}\} f_n(y)\, dy.$$

(Hardy **1**, p. 139; use (5.1.9) and (1.71.1).)

22. Assume $x \geqq 0$, $y \geqq 0$, $\max(x, y) > 0$. Then

$$e^{-x-y} \sum_{n=0}^{\infty} \frac{L_n(x)L_n(y)}{n+1} = \int_{\max(x,y)}^{\infty} t^{-1}e^{-t}\, dt.$$

(E. R. Neumann **1**, Watson **6**. Apply Theorem 9.1.5 and use the formula

$$(n+1)\int_0^x L_n(t)\, dt = x\{L_n(x) - L_n'(x)\}$$

(cf. (5.1.2)).)

23. From (5.1.15) derive Mehler's formula

$$\sum_{n=0}^{\infty} \frac{H_n(x)H_n(y)}{n!}(w/2)^n = (1-w^2)^{-\frac{1}{2}} \exp\left\{\frac{2xyw - (x^2+y^2)w^2}{1-w^2}\right\}.$$

(See Watson **5**, Erdélyi **2**.)

24. From (5.1.9) and (5.6.1) derive the following generating function for Hermite polynomials:

$$\sum_{n=0}^{\infty} \frac{H_n(x)}{m!} w^n = (1+4w^2)^{-\frac{3}{2}}(1+2xw+4w^2)\exp\left(\frac{4x^2w^2}{1+4w^2}\right),$$

where $m = [n/2]$. (Cf. Doetsch **1**, p. 590, (7).)

25. Assume $k > -\frac{1}{2}$ and let $H_n^{(k)}(x)$ denote the orthogonal polynomials corresponding to the weight function $e^{-x^2}|x|^{2k}$ in $[-\infty, +\infty]$. Then the following differential equations are satisfied:

$$xy'' + 2(k - x^2)y' + (2nx - \epsilon x^{-1})y = 0, \qquad \epsilon = \begin{cases} 0, & n \text{ even}, \\ 2k, & n \text{ odd}; \end{cases} \quad y = H_n^{(k)}(x);$$

$$z'' + \left\{2n + 2k + 1 - x^2 + \frac{(-1)^n k - k^2}{x^2}\right\} z = 0; \qquad z = e^{-x^2/2}x^k H_n^{(k)}(x).$$

(Generalizations of (5.5.2).)

26. For fixed β, n, and x we have

$$\lim_{\alpha\to\infty} \alpha^{-n} P_n^{(\alpha,\beta)}(x) = \frac{1}{n!}\left(\frac{x+1}{2}\right)^n.$$

(Use (4.21.2).)

27. Let $\{x_\nu\}$ be the zeros of $P_n^{(\alpha,\beta)}(x)$ in decreasing order, $\alpha > -1, \beta > -1$. Then for $\nu = 1, 2, \cdots, n$

$$\lim_{\alpha\to+\infty} x_\nu = -1, \qquad \lim_{\beta\to+\infty} x_\nu = +1.$$

In the first case β and n, and in the second case α and n, are fixed. (Use Problem 26 and (4.1.3).)

28. Let $\{x_\nu\}$ be the zeros of $P_n^{(\lambda)}(x)$, $\lambda > 0$. We have for fixed n

$$\lim_{\lambda\to+\infty} x_\nu = 0.$$

(From (4.7.6) $\lim_{\lambda\to\infty} (2\lambda)^{-n} P_n^{(\lambda)}(x) = x^n/n!$; see also (5.6.3).)

29. Assume $\alpha > -1, \beta > -1$; if $\{x_\nu\}$ and $\{x_\nu'\}$ denote the zeros of $P_n^{(\alpha,\beta)}(x)$ and $L_n^{(\alpha)}(x)$ in decreasing and increasing order, respectively, we have (α, n, and ν fixed)

$$\lim_{\beta\to+\infty} \beta(1 - x_\nu) = 2x_\nu'.$$

(Use (5.3.4).)

30. For fixed n and x we have

$$\lim_{\alpha\to\infty} \alpha^{-n} L_n^{(\alpha)}(\alpha x) = \frac{(1-x)^n}{n!}.$$

(Use (5.1.6).)

31. Let $j_0 = 0 < j_1 < j_2 < \cdots$ be the positive zeros of $J_\alpha(x)$. Then $\{j_\nu\}$ is a convex sequence if $-\frac{1}{2} < \alpha < +\frac{1}{2}$, that is, $j_{\nu+1} - j_\nu$ is increasing. Furthermore $\nu^{-1}j_\nu$ is increasing. (Apply Theorem 1.82.2 to (1.8.9).)

32. With the same notation and assumptions as in Problem 31, we have

$$j_\nu > (\nu + \alpha/2 - 1/4)\pi, \qquad \nu = 1, 2, 3, \cdots.$$

(Put $n = 2\nu - 1$ in (6.3.13).)

33. We have for the factor $C_{\nu n}$, defined by (6.31.13),

$$(j_1/4)^2 < C_{\nu n} < 4.$$

Here j_1 is the least positive root of $J_0(x)$. These bounds are the best possible. (Use the increasing property of $\nu^{-1}j_\nu$; see Problem 31.)

34. Assume $\alpha > -1$; denote by $x_1 < x_2 < \cdots < x_n$ the zeros of $L_n^{(\alpha)}(x)$. If $\alpha^2 \leqq \frac{1}{4}$, the sequence $x_\nu^{\frac{1}{2}} - x_{\nu-1}^{\frac{1}{2}}$, $\nu = 2, 3, \cdots, n$, is increasing; if $\alpha^2 > \frac{1}{4}$, this is true for $x_{\nu-1} > (\alpha^2 - \frac{1}{4})^{\frac{1}{2}}$. (Apply Theorem 1.82.2 to the fourth equation in (5.1.2).)

35. With the notation of the previous problem,

$$\min (x_\nu^{\frac{1}{2}} - x_{\nu-1}^{\frac{1}{2}}) \sim n^{-\frac{1}{2}}, \qquad \max (x_\nu^{\frac{1}{2}} - x_{\nu-1}^{\frac{1}{2}}) = x_n^{\frac{1}{2}} - x_{n-1}^{\frac{1}{2}} \cong 2^{-\frac{2}{3}}3^{-\frac{1}{3}}(i_2 - i_1)n^{-\frac{1}{6}},$$
$$n \to \infty,$$

where i_1 and i_2 are the least positive zeros of Airy's function $A(x)$. (Use (8.1.8), (8.22.1), and (8.9.15).)

36. Let $x_1 > x_2 > \cdots > x_{[(n+1)/2]}$ be the non-negative zeros of the Hermite polynomial $H_n(x)$. Write

$$x_\nu = x_{\nu n} = h_n - (6h_n)^{-\frac{1}{3}} t_{\nu n}, \qquad h_n = (2n + 1)^{\frac{1}{2}}.$$

Then for fixed ν and increasing n, the numbers $t_{\nu n}$ are decreasing. Further, show that for $1 \leqq \nu \leqq (n + 1)/2$,

$$x_{\nu n} = h_n - \nu^{\frac{2}{3}} n^{-\frac{1}{6}} \rho_{\nu n},$$

where $P < \rho_{\nu n} < Q$; here P and Q are two absolute positive constants. (On writing $\xi = (6h_n)^{-\frac{1}{3}}t$ in (6.32.10), we have

$$\frac{d^2z}{dt^2} + \{t/3 - (6h_n)^{-\frac{4}{3}}t^2\}z = 0.$$

The monotonic character of $t_{\nu n}$ follows by means of Theorem 1.82.1. Furthermore, see (6.32.3),

$$i_\nu < t_{\nu n} \leqq t_{\nu, 2\nu-1} = 6^{\frac{1}{3}} h_{2\nu-1}^{\frac{4}{3}}.$$

Now apply Problem 1.)

37. Consider n unit "masses", $n \geqq 2$, at the variable points $x_1, x_2, \cdots, x_n$ in the interval $[-1, +1]$. For what position of these points does the expression

$$\prod_{\substack{\nu,\mu=1,2,\cdots,n \\ \nu<\mu}} |x_\nu - x_\mu|$$

become a maximum? (Stieltjes **4**, p. 441; the maximum position is the same as in Theorem 6.7.1, obtained by replacing n by $n - 2$ and writing $p = q = 1$. We have $(1 - x^2)P_{n-2}^{(1,1)}(x) = \text{const.}\ \{P_n(x) - P_{n-2}(x)\} = \text{const.}\ (1 - x^2)P'_{n-1}(x)$; see (4.7.27).)

38. Consider n unit "masses", $n \geqq 2$, at the variable points $x_1, x_2, \cdots, x_n$ in the interval $[0, +\infty]$, such that

$$n^{-1}(x_1 + x_2 + \cdots + x_n) \leqq K$$

where K is a fixed positive number. For what position of these points does the expression

$$\prod_{\substack{\nu,\mu=1,2,\cdots,n \\ \nu<\mu}} |x_\nu - x_\mu|$$

become a maximum? (Cf. Problem 37 and Theorem 6.7.2. We have $xL_{n-1}^{(1)}(x) = \text{const.}\ \{L_n(x) - L_{n-1}(x)\} = \text{const.}\ xL'_n(x)$; see (5.1.14).)

39. Assume $\alpha > -1$, $\beta > -1$. With the notation of (7.32.2) we have, for $n \to \infty$,

$$\max_{-1 \leqq x \leqq +1} | P_n^{(\alpha,\beta)}(x) | \cong \begin{cases} \{\Gamma(q+1)\}^{-1} n^q & \text{if } q \geqq -\frac{1}{2}, \\ \pi^{-\frac{1}{2}} n^{-\frac{1}{2}} | \alpha + \frac{1}{2} |^{-\alpha/2-\frac{1}{4}} | \beta + \frac{1}{2} |^{-\beta/2-\frac{1}{4}} | \alpha + \beta + 1 |^{(\alpha+\beta+1)/2} & \\ & \text{if } -1 < q \leqq -\frac{1}{2}. \end{cases}$$

(Cf. (7.32.2).)

40. From (7.33.10) derive

$$P_n(\cos\theta) - P_{n+1}(\cos\theta) = \theta^{\frac{1}{2}} O(n^{-\frac{3}{2}}), \qquad 0 < \theta \leqq \pi/2.$$

(Use

$$(1 - x)\{P_n'(x) + P_{n+1}'(x)\} = (n + 1)\{P_n(x) - P_{n+1}(x)\}$$

(see Problem 17) and (7.33.9).)

41. We have for $0 < \theta < \pi$

$$(\sin\theta)^{\frac{1}{2}} | \mathbf{Q}_n (\cos\theta) | < \{\pi/(2n)\}^{\frac{1}{2}}.$$

The constant $(\pi/2)^{\frac{1}{2}}$ cannot be replaced by a smaller one. (See Hobson **1**, p. 309, where the bound $(\pi/n)^{\frac{1}{2}}$ is obtained; compare Theorem 7.3.3, and Problem 18.)

42. Let $f(x)$ be an arbitrary π_n, non-negative for all real values of x, and let

$$\int_{-\infty}^{+\infty} e^{-x^2} f(x)\, dx = 1.$$

Then

$$\max f(0) = \pi^{-\frac{1}{2}} \frac{3 \cdot 5 \cdots (2m+1)}{2 \cdot 4 \cdots 2m} \cong \pi^{-1} n^{\frac{1}{2}}, \qquad m = [n/4],\ n \to \infty.$$

(Use Theorem 1.21.2, (7.71.2), (5.5.9), and (5.5.5).)

43. *A mean-value theorem for polynomials.* Let $f(x)$ be a π_{2n}. Then

$$f(b) - f(a) = (b - a) f'(\xi),$$

where ξ is a proper point in the interval

$$\tfrac{1}{2}(a + b) - \tfrac{1}{2}(b - a)x_1 \leqq \xi \leqq \tfrac{1}{2}(a + b) + \tfrac{1}{2}(b - a)x_1 .$$

Here x_1 denotes the greatest zero of the Legendre polynomial $P_n(x)$. (Tchakaloff **1**; use (3.4.1).)

44. Derive the formula of Lipschitz

$$\int_0^\infty e^{-at} J_0(bt)\, dt = (a^2 + b^2)^{-\frac{1}{2}}, \qquad a > 0,\ b > 0,$$

from the generating function of the Legendre polynomials, that is, from (4.7.23) for $\lambda = \frac{1}{2}$. (Write $w = e^{-a/N}$, $x = \cos(b/N)$, a and b fixed, $N \to \infty$, separate

$\sum_{m=0}^{\infty} P_m(x)w^m$ into $m \leqq \omega N$ and $m > \omega N$ (ω a fixed positive number), and use (8.1.1) and (7.3.8).)

45. Derive (5.4.2) from (5.1.9) by writing $x = a/N$, $w = e^{-b/N}$, a and b fixed, $b > 0$, $N \to +\infty$. (See Problem 44; use (8.1.8) and (7.6.8).)

46. *New proof of the asymptotic formula* (8.22.4) *of Hilb's type for Laguerre polynomials.* From the generating function (5.1.9) we obtain

$$e^{-x/2}L_n^{(\alpha)}(x) = \frac{1}{2\pi i}\int_{+\infty}^{(1-)} \exp\left\{-\frac{x}{2}\frac{1+w}{1-w}\right\}(1-w)^{-\alpha-1}w^{-n-1}\,dw$$

$$= \frac{1}{2\pi i}\int_{-\infty}^{(0+)} \exp\left\{-\frac{x}{2}\frac{1+e^{-z}}{1-e^{-z}}\right\}(1-e^{-z})^{-\alpha-1}e^{nz}\,dz$$

$$= \frac{1}{2\pi i}\int_{-\infty}^{(0+)} e^{nz-z^{-1}x}z^{-\alpha-1}\Phi(x, z)\,dz,$$

where $\Phi(x, z)$ is regular in $|z| < 2\pi$. If we develop $\Phi(x, z)$ in a power series in z, the resulting integrals can be reduced to Bessel functions (Watson **3**, p. 176, (1)). (This argument furnishes not only (8.22.4) for an arbitrary real α, but also an asymptotic expansion of Hilb's type whose terms are Bessel functions; see Wright **1**. This is the analogue of Szegö's argument used in **15** for Legendre polynomials.)

47. Let a be real, but different from zero. The infinite series

$$\sum_{n=1}^{\infty} n^{-\lambda}e^{ian^{1/2}}$$

is convergent if $\lambda > \frac{1}{2}$, and divergent if $\lambda \leqq \frac{1}{2}$. (For $\lambda > 0$ the convergence of the series is equivalent to that of the integral $\int_1^\infty x^{-\lambda}e^{iax^{1/2}}\,dx$.)

48. The polynomial $s_n(a, z)$ (see (11.3.3)) admits the following representation, in which the notation of §11.1 is used:-

$$s_n(a, z) = -D_n^{-1}\begin{vmatrix} c_0 & c_{-1} & \cdots & c_{-n+1} & c_{-n} & 1 \\ c_1 & c_0 & \cdots & c_{-n+2} & c_{-n+1} & \bar{a} \\ \cdots & \cdots & \cdots & \cdots & \cdots & \cdots \\ c_n & c_{n-1} & \cdots & c_1 & c_0 & \bar{a}^n \\ 1 & z & \cdots & z^{n-1} & z^n & 0 \end{vmatrix}.$$

(Cf. Problem 4.)

49. *Second proof of Theorem* 11.4.1 *on the zeros of* $s_n(a, z)$. Use the argument of Problem 5. (If z_0 is the zero in question, and $(z - z_0)/(z - a) = z'$, the image of $|z| = 1$ contains $z' = 0$.)

50. Theorem 11.3.3 on the convergence of $\sum_{n=0}^{\infty} |\phi_n(z)|^2$, $|z| < 1$, is not true if the weight function $f(\theta)$ is such that $\log f(\theta)$ is not integrable in Lebesgue's sense. (Let $f(\theta) = 0$ for $-\epsilon < \theta < +\epsilon$, and $f(\theta) = 1$ for $\epsilon \leqq \theta \leqq 2\pi - \epsilon$, and apply (16.4.2). In this case $|\phi_n(0)|^{1/n} \to R$, $R > 1$.)

51. Let $f(\theta)$ be a weight function on the unit circle, $-\pi \leqq \theta \leqq +\pi$, for which $\int_{-\pi}^{+\pi} \log f(\theta)\, d\theta$ exists. Assume $p > 0$, and let $\mu_n = \mu_n(f; p)$ be the minimum of

$$\frac{1}{2\pi} \int_{-\pi}^{+\pi} f(\theta) \mid \rho(z) \mid^p d\theta, \qquad z = e^{i\theta},$$

where $\rho(z) = z^n + \cdots$ is an arbitrary π_n with the highest term z^n. Then

$$\lim_{n\to\infty} \mu_n(f; p) = \exp\left\{\frac{1}{2\pi} \int_{-\pi}^{+\pi} \log f(\theta)\, d\theta\right\} = \mathfrak{G}(f).$$

(This is a generalization of (12.3.3); see the argument used there.)

52. Let k_{n0} be the highest coefficient of the orthonormal polynomial $p_n(x)$ associated with the distribution $d\alpha(x)$ on the finite interval $[a, b]$. Then

$$k_{n0} > 2^{2n-1}(b - a)^{-n}\left\{\int_a^b d\alpha(x)\right\}^{-\frac{1}{2}}.$$

(Cf. Theorem 12.7.1; Shohat **2**, p. 575, (24). Use the extremum property of Theorem 3.1.2, choosing $\rho(x) = 2^{1-2n}(b - a)^n T_n\{2(x - a)/(b - a) - 1\}$.)

53. Use the notation of Problem 52. Let $[a', b']$ be a subinterval of $[a, b]$ such that $\alpha(x)$ is constant in $[a', b']$. Then two positive constants A, B exist, $B > 4/(b - a)$, such that $k_{n0} > AB^n$. (Cf. Shohat **2**, p. 577. Use the extremum property of Theorem 3.1.2, and choose for $\rho(x)$ the "Tchebichef polynomial" corresponding to two disjoint segments in the sense of §16.2 (5).)

54. By use of the notation of §12.7 (1) we have, under the assumption of Theorem 12.7.1,

$$k_{n\nu} \cong \begin{cases} (2\pi)^{-\frac{1}{2}} \dfrac{(-1)^{\nu/2}}{(\nu/2)!}\, 2^{n-\nu} n^{\nu/2} d_0^{-1}, & \nu \text{ even} \\[2ex] (2\pi)^{-\frac{1}{2}} \dfrac{(-1)^{(\nu+1)/2}}{[(\nu - 1)/2]!}\, 2^{n-\nu} n^{(\nu-1)/2} d_0^{-2} d_1, & \nu \text{ odd}. \end{cases}$$

Here ν is fixed, $n \to \infty$. If $d_1 = 0$, the second formula is to be read as follows: $\lim_{n\to\infty} 2^{-n} n^{(1-\nu)/2} k_{n\nu} = 0$. (See the special cases in Shohat **2**, p. 577. Use Theorem 12.1.2 and observe that on putting

$$2^{-n}\{1 + (1 - x^{-2})^{\frac{1}{2}}\}^n = 1 + \gamma_{n1} x^{-2} + \gamma_{n2} x^{-4} + \cdots,$$

we have $\gamma_{n\nu} \cong (2^{2\nu}\nu!)^{-1}(-n)^\nu$, ν fixed, $n \to \infty$.)

55. In case of the Jacobi polynomials

$$P_n^{(\alpha,\beta)}(x) = l_{n0}^{(\alpha,\beta)} x^n + l_{n1}^{(\alpha,\beta)} x^{n-1} + \cdots$$

we have, ν fixed, $n \to \infty$,

$$l_{n\nu} \cong \pi^{-\frac{1}{2}} \frac{(-1)^{\nu/2}}{(\nu/2)!}\, 2^{n-\nu+\alpha+\beta} n^{(\nu-1)/2}, \qquad (\alpha - \beta)\pi^{-\frac{1}{2}} \frac{(-1)^{(\nu-1)/2}}{[(\nu - 1)/2]!}\, 2^{n-\nu+\alpha+\beta} n^{\nu/2-1},$$

according as ν is even or odd. In the second case we assume $\alpha \neq \beta$. (See

Geronimus **1**, p. 380; use the result of Problem 54; in this case we have $d_0 = 2^{-(\alpha+\beta+1)/2}$, $d_1 = (\beta - \alpha)\, 2^{-(\alpha+\beta+1)/2}$.)

56. With the notation and assumption of Problem 54 we have

$$\frac{k_{n2}}{k_{n0}} = -n/4 + c + \epsilon_n, \qquad \lim_{n\to\infty} \epsilon_n = 0,$$

where c is a constant. (Cf. Shohat **2**, p. 577. Use the same argument as in Problem 54.)

57. Let $l_\nu(x)$, $\nu = 1, 2, \cdots, n$, be the fundamental polynomials of the Lagrange interpolation with the zeros of $T_n(x)$ as abscissas. Then if k is even,

$$\int_{-1}^{+1} l_{\nu_1}(x) l_{\nu_2}(x) \cdots l_{\nu_k}(x)(1 - x^2)^{-\frac{1}{2}}\, dx = 0.$$

Here $\nu_1, \nu_2, \cdots, \nu_k$ are distinct integers between 1 and n. (E. Feldheim **1**; $\{T_n(x)\}^{k-1}$, $x = \cos\theta$, is a cosine polynomial not containing terms with $\cos \nu\theta$, $\nu < n$.)

58. In the ultraspherical case $\alpha = \beta$, $-1 < \alpha = \beta < 0$, we have for $-1 \leqq x \leqq +1$ (notation as in Problem 57)

$$\{l_1(x)\}^2 + \{l_2(x)\}^2 + \cdots + \{l_n(x)\}^2 \leqq |\alpha|^{-1}.$$

(For $\alpha = -\frac{1}{2}$ see Fejér **13**, p. 5. Use Problem 59 and the first identity in (14.1.11).)

59. In the ultraspherical case $\alpha = \beta > -1$ we have (cf. (14.5.2))

$$v_\nu(x) = \frac{1 - 2(\alpha + 1)xx_\nu + (2\alpha + 1)x_\nu^2}{1 - x_\nu^2} \geqq -\alpha, \quad -1 \leqq x \leqq +1.$$

60. In the Legendre case we have

$$v_\nu(x) \geqq \tan^2 \frac{3\pi}{4(2n+1)}, \qquad -1 \leqq x \leqq +1.$$

(Cf. Fejér **13**, p. 23. Use (6.6.5).)

61. Prove the following identities:

$$P'_{2n}(x) = (2n+1)xP^{(1,\frac{1}{2})}_{n-1}(2x^2-1),$$

$$xP'_n(x) = nP_n(x) + (2n-3)P_{n-2}(x) + (2n-7)P_{n-4}(x) + \cdots,$$

$$(1-x^2)P''_n(x) = -n(n-1)P_n(x) + 2(2n-3)P_{n-2}(x) + 2(2n-7)P_{n-4}(x) + \cdots.$$

62. Prove:

$$P^{(1)}_n(x) = U_n(x);$$

$$P^{(-\frac{1}{2})}_n(x) = \begin{cases} -x & \text{if } n = 1, \\ \int_x^1 P_{n-1}(t)\,dt & \text{if } n \geq 2. \end{cases}$$

63. Prove the formula:

$$\int_{-1}^{+1} P_n(x)x^{n+2\nu}\,dx = \frac{1}{2^n}\frac{(n+2\nu)!}{(2\nu)!}\frac{\Gamma(\nu+\frac{1}{2})}{\Gamma(n+\nu+\frac{3}{2})}, \quad \nu = 0, 1, 2, \cdots.$$

64. Prove the identity

$$\int_{-1}^{+1} P_n(x)e^{-i\lambda x}\,dx = i^{-n}(2\pi/\lambda)^{\frac{1}{2}}J_{n+\frac{1}{2}}(\lambda).$$

(Use Problem 63 and (1.71.1).)

65. Prove that

$$\int_{-\infty}^{+\infty} \lambda^{-\frac{1}{2}}J_{n+\frac{1}{2}}(\lambda)e^{i\lambda x}\,d\lambda = \begin{cases} (2\pi)^{\frac{1}{2}}i^nP_n(x) & \text{if } -1 < x < 1, \\ 0 & \text{if } x > 1 \text{ or } x < -1. \end{cases}$$

(Use Problem 64 applying Fourier's inversion formula.)

66. Prove the identity

$$\sum_{\nu=0}^{n}\binom{n}{\nu}\frac{P^{(\lambda)}_\nu(x)}{P^{(\lambda)}_\nu(1)}y^{n-\nu} = (1+2xy+y^2)^{n/2}\frac{P^{(\lambda)}_n\{(1+2xy+y^2)^{-\frac{1}{2}}(x+y)\}}{P^{(\lambda)}_n(1)},$$

in particular the identity

$$\sum_{\nu=0}^{n}\binom{n}{\nu}P_\nu(x)y^{n-\nu} = (1+2xy+y^2)^{n/2}P_n\{(1+2xy+y^2)^{-\frac{1}{2}}(x+y)\}.$$

(Use (4.7.23).)

67. Prove the identity

$$\sum_{\nu=0}^{n}\binom{n}{\nu}\frac{L^{(\alpha)}_\nu(x)}{L^{(\alpha)}_\nu(0)}y^{n-\nu} = (y+1)^n\frac{L^{(\alpha)}_n\{(y+1)^{-1}x\}}{L^{(\alpha)}_n(0)}.$$

(Use (5.1.9).)

68. Prove the identity

$$\sum_{\nu=0}^{n} \binom{n}{\nu} H_\nu(x)(2y)^{n-\nu} = H_n(x+y).$$

(Use (5.5.7).)

69. Let x be a parameter, $-1 < x < +1$. The zeros of the polynomial in z:

$$P_0(x) + \binom{n}{1} P_1(x)z + \binom{n}{2} P_2(x)z^2 + \cdots + P_n(x)z^n$$

are all real.

(Use Problem 66.)

70. Prove Turán's inequality

$$\{P_n(x)\}^2 - P_{n-1}(x)P_{n+1}(x) \geqq 0, \qquad -1 \leqq x \leqq +1.$$

(Turán 1, see also Szegö 22, Karlin-Szegö 1 and Csordas-Williamson 1.) If a_1 and a_2 denote the first and second elementary-symmetric function of n real numbers, we have:

$$(a_1/n)^2 \geqq a_2 \Big/ \binom{n}{2}.$$

(Use Problem 69.)

71. Derive from Problem 66 the generating series (4.10.6) and, in particular, (4.10.7). (Put $y = n/z$, $n \to \infty$; use (8.1.1).)

72. Derive from Problem 67 the generating series (5.1.16). (Put $y = n/z$, $n \to \infty$; use (8.1.8).)

73. Prove the following formula of the "Rodrigues type":

$$e^{-x}x^\alpha L_n^{(\alpha)}(x) = \frac{(-1)^n}{n!} t^{n+1} \left(\frac{d}{dt}\right)^n (e^{-1/t}t^{-\alpha-1}), \qquad xt = 1.$$

(Use Taylor's formula and (5.1.9).—In the special case $\alpha = 0$ this is due to G. Pólya, 1941.)

74. Let $\{u_n\}$ and $\{v_n\}$ be two sequences; $n = 0, 1, 2, \cdots$. One of the relations

$$u_n = \sum_{\nu=0}^{n} \binom{n+\alpha}{n-\nu} (-1)^\nu v_\nu; \quad v_n = \sum_{\nu=0}^{n} \binom{n+\alpha}{n-\nu} (-1)^\nu u_\nu$$

implies the other.

75. Using Problem 74 and (5.1.6) prove

$$\frac{x^n}{n!} = \sum_{\nu=0}^{n} \binom{n+\alpha}{n-\nu} (-1)^\nu L_\nu^{(\alpha)}(x).$$

76. Let u_n and v_n be two sequences; $n = 0, 1, 2, \cdots$. One of the relations

$$u_n = \sum_{\nu=0}^{[n/2]} \frac{(-1)^\nu}{\nu!} v_{n-2\nu}; \quad v_n = \sum_{\nu=0}^{[n/2]} \frac{1}{\nu!} u_{n-2\nu}$$

implies the other.

77. Prove the identity:

$$\frac{(2x)^n}{n!} = \sum_{\nu=0}^{[n/2]} \frac{1}{\nu!} \frac{H_{n-2\nu}(x)}{(n-2\nu)!}.$$

(Use Problem 76 and (5.5.4).)

78. Prove the following identities:

$$y'' + 2xy' + 2(n+1)y = 0, \qquad y = e^{-x^2}H_n(x);$$

$$\int_0^x e^{-t^2/2}\,dt = 6^{-\frac{1}{2}} \sum_{\nu=0}^{\infty} (-\tfrac{1}{12})^\nu \frac{H_{2\nu+1}(x)}{(2\nu+1)\nu!}.$$

79. Prove that

$$\lim_{n\to\infty} \left(\frac{y}{n}\right)^n H_n\left(\frac{n}{2y}\right) = e^{-y^2}.$$

(Use (5.5.4). Cf. P. Turán, Matematikai Lapok, vol. 5 (1954), pp. 134–137.)

80. Prove that

$$\lim_{\alpha\to\infty} \alpha^{-n/2} L_n^{(\alpha)}(\alpha^{\frac{1}{2}}x + \alpha) = (-1)^n 2^{-n/2}(n!)^{-1}H_n(2^{-\frac{1}{2}}x).$$

81. Let $\{p_m(x)\}$ be the orthonormal polynomials associated with the distribution $d\alpha(x)$ in $0 \leqq x < +\infty$. Denoting by $\xi_1, \xi_2, \cdots, \xi_k$ any zeros of $p_m(x)$ we have

$$f(t) = \int_0^\infty e^{-xt}\{(\xi_1 - x)\cdots(\xi_k - x)\}^{-1}\{p_m(x)\}^2\,d\alpha(x) > 0, \qquad t > 0.$$

(See Karlin-McGregor **1**, pp. 507–509. Since $f(0) = 0$, we have

$$e^{-\xi_k t}\frac{d}{dt} e^{\xi_k t} f(t) = \int_0^\infty e^{-xt}\{(\xi_1 - x)\cdots(\xi_{k-1} - x)\}^{-1}\{p_m(x)\}^2\,d\alpha(x) = \varphi(t),$$

$$f(t) = e^{-\xi_k t}\int_0^t e^{\xi_k \tau}\varphi(\tau)\,d\tau. \qquad \text{Induction.)}$$

82. Notation and assumption as in Problem 81, $p_m(0) > 0$ for all m. Prove that

$$\int_0^\infty e^{-xt} p_m(x) p_n(x)\,d\alpha(x) > 0, \qquad t > 0.$$

(See Karlin-McGregor **1**, loc. cit. Let $m > n$. We represent $p_n(x)$ by Lagrange's interpolation formula corresponding to the abscissas $\xi_{\mu_1}, \cdots, \xi_{\mu_{n+1}}$ chosen as in 3.3 (6); use Problem 81.)

83. Let $\{p_n(x)\}$ be the Poisson-Charlier polynomials (2.81, $\operatorname{sgn} p_n(0) = (-1)^n$). We have, $\lambda > 0$, $m \geqq n$, $j(x)$ as in (2.81.1),

$$\sum_{x=0}^{\infty} e^{-\lambda x} j(x) p_n(x) p_m(x)$$
$$= \exp\left[a(e^{-\lambda} - 1)\right] a^{\frac{1}{2}(n+m)} (m!/n!)^{\frac{1}{2}} \sum_{\nu=0}^{n} \binom{n}{\nu} (e^{-\lambda} - 1)^{n+m-2\nu} \frac{(ae^{\lambda})^{-\nu}}{(m-\nu)!}.$$

84. Let $\lambda > -\frac{1}{2}$, $\lambda \neq 0$. Using the notation (4.9.21) we have

$$\int_{-1}^{+1} (1 - x^2)^{\lambda - \frac{1}{2}} P_l^{(\lambda)}(x) P_m^{(\lambda)}(x) P_n^{(\lambda)}(x)\, dx$$
$$= \frac{\alpha_{s-l}\alpha_{s-m}\alpha_{s-n}}{\alpha_s} \int_{-1}^{+1} (1 - x^2)^{\lambda - \frac{1}{2}} \{P_s^{(\lambda)}(x)\}^2\, dx.$$

provided that $l + m + n = 2s$ is even and a triangle with sides l, m, n exists. The integral on the left is zero in every other case. (Cf. (4.7.15); cf. Hsü **1**.)

85. Let $t_n(x)$ have the same meaning as in (2.8.1); we have then

$$t_n(x) = (-1)^n t_n(N - 1 - x).$$

Here n and x run over the range $0, 1, \cdots, N - 1$. Prove also that

$$t_n(N-1) = n! \binom{N-1}{n}.$$

86. Let $p_n(x)$ have the same meaning as in (2.81.2); we have then

$$p_n(0) = (-1)^n (a^n/n!)^{1/2} = (-1)^n e^{a/2} (j(n))^{1/2}.$$

Prove also that, writing $p_n(x)/p_n(0) = c_n(x, a)$, we have

$$c_n(x; a) = c_x(n; a).$$

87. Writing $H_n(x) = H_n$ we have

$$\int_{-\infty}^{\infty} e^{-x^2} H_\alpha H_\beta H_\gamma\, dx = \pi^{1/2} \frac{2^s \alpha! \beta! \gamma!}{(s-\alpha)!(s-\beta)!(s-\gamma)!}$$

provided that $\alpha + \beta + \gamma = 2s$ is an even integer and $s \geqq \alpha$, $s \geqq \beta$, $s \geqq \gamma$. In all other cases the integral is zero.

88. Prove that

$$\lim_{n \to \infty} n! \left(-\frac{y}{n^2}\right)^n L_n^{(\alpha)}\left(\frac{n^2}{y}\right) = e^{-y}.$$

89. Prove the following identities:

$$L_n^{(\alpha)}(x) = y, \quad (e^{-x} x^{\alpha+1} y')' + n e^{-x} x^\alpha y = 0;$$

$$e^{-x/2} L_n(x) = h, \quad x h'' + h' + \left(n + \frac{2-x}{4}\right) h = 0.$$

90. Prove the identity

$$L_n^{(\alpha+\beta+1)}(x+y)=\sum_{k=0}^{n} L_{n-k}^{(\alpha)}(x)\,L_k^{(\beta)}(y).$$

91. Obtain the estimates

$$\int_0^1 (1-x)^{\mu}\,|P_n^{(\alpha,\beta)}(x)|^{p}dx \sim \begin{cases} n^{\alpha p-2\mu-2}, & 2\mu<\alpha p-2+p/2,\\ n^{-p/2}\log n, & 2\mu=\alpha p-2+p/2,\\ n^{-p/2}, & 2\mu>\alpha p-2+p/2,\end{cases}$$

where $\alpha,\beta,\mu>-1$ and $p>0$ are fixed. (See (7.34.1).)

92. Prove that

$$-\frac{1}{3}\leqq\frac{\sin(n+1)\theta}{(n+1)\sin\theta}\leqq\frac{\sqrt{6}}{9},\quad \frac{\pi}{n}\leqq\theta\leqq\pi-\frac{\pi}{n},$$

$$-\frac{1}{3}\leqq\frac{\sin(n+1)\theta}{(n+1)\sin\theta}\leqq\frac{1}{5},\quad \frac{\pi}{n}\leqq\theta\leqq\frac{\pi}{2},$$

and show that all the bounds are obtained for some θ and n. (Use (7.8.1) for $P_n^{(1/2,1/2)}(x)/P_n^{(1/2,1/2)}(1)$.)

93. Prove that

$$\frac{d^{2n}}{dx^{2n}}(1-x)^{-\alpha}(1+x)^{-\beta}>0,\qquad -1<x<1,\ \alpha,\beta\geqq 0,$$

and

$$\frac{d^{2n}}{dx^{2n}}x^{-\alpha}e^{x}>0,\qquad x>0,\ \alpha\geqq 0.$$

(Use Theorems 6.72 and 6.73. These results were conjectured by I. Joó.)

94. Prove that

$$(-1)^{k+m+n}\int_0^\infty L_k^{(\alpha)}(x)\,L_m^{(\alpha)}(x)\,L_n^{(\alpha)}(x)\,x^{\alpha}e^{-x}dx\geq 0,\ \ \alpha>-1,\ k,m,n=0,1,\cdots.$$

(Use (5.1.9).)

95. Prove that

$$p_n(x)p_m(x)=\sum_{k=|n-m|}^{n+m} a(k,m,n)\,p_k(x)$$

with $a(k,m,n)\geqq 0$ if $p_n(x)$ satisfies

$$p_1(x)p_n(x)=p_{n+1}(x)+\alpha_n p_n(x)+\beta_n p_{n-1}(x),\qquad n=1,2,\cdots,$$

where $\alpha_n\geqq 0$, $\beta_n>0$, $\alpha_{n+1}\geqq\alpha_n$, $\beta_{n+1}\geqq\beta_n$, $n=1,2,\cdots,p_0(x)=1$, $p_1(x)=x+a$. (Askey 4. Use induction. This contains Problem 94.)

96. Prove that

$$P_k(x)+P_m(x)\leqq 1+P_n(x), \qquad 0\leqq x\leqq 1,$$

when $n(n+1)=k(k+1)+m(m+1)$. Also show that

$$J_0(x)+J_0(y)\leqq 1+J_0(z)$$

when $x^2+y^2=z^2$. Grünbaum **1**, **2**, Askey **7**.

97. If $\sum_{n=0}^{\infty} a_n^2$ is finite, $\alpha>-1$, and if

$$\sum_{n=0}^{\infty}\frac{a_n n!\,L_n^{(\alpha)}(x)L_n^{(\alpha)}(y)}{\Gamma(n+\alpha+1)}\geqq 0, \qquad 0\leqq x,\, y<\infty,$$

prove that

$$a_n=\int_0^1 x^n d\mu(x),\, d\mu(x)\geqq 0. \quad \text{(Sarmonov 1.)}$$

98. If $a_{k,m,n}$ are defined by

$$\frac{1}{(1-r)(1-s)+(1-r)(1-t)+(1-s)(1-t)}=\sum_{k,m,n=0}^{\infty} a_{k,m,n}r^k s^m t^n$$

show that

$$a_{k,m,n}=\int_0^{\infty} L_k(x)L_m(x)L_n(x)e^{-3x}dx.$$

Then use Problem 84 for $\lambda=\frac{1}{2}$ and (4.5.4) to show that $a_{k,m,n}\geqq 0$. Generalize to $\int_0^{\infty} L_k^{(\alpha)}(x)L_m^{(\alpha)}(x)L_n^{(\alpha)}(x)x^{\alpha}e^{-3x}dx$, $\alpha\geqq-\frac{1}{2}$. (Szegö **26**, Askey-Gasper **3**.)

99. Prove that

$$\int_{-\infty}^{\infty}\int_{-\infty}^{\infty}\prod_{i=1}^{k}(H_{n_i}(x)\pm H_{n_i}(y))e^{-x^2-y^2}dxdy\geqq 0, \qquad k=1,2,\cdots,$$

for all choices of plus and minus signs and any choice of integers n_i. (Use (5.5.11) and Problem 87. See Ginibre **1**.)

100. Show that

$$2^n n!\sum_{k=0}^{n}(-1)^k L_k(2x^2+2y^2)=\sum_{k=0}^{n}\binom{n}{k}\,[H_k(x)]^2[H_{n-k}(y)]^2$$

and in particular that

$$2^n\sum_{k=0}^{n}(-1)^k L_k(2x^2)=\sum_{j=0}^{[n/2]}\frac{(2j)!}{j!\,j!\,(n-2j)!}\,[H_{n-2j}(x)]^2.$$

Extensions of this sum to $P_n^{(\alpha,0)}(x)$ are given in Askey-Gasper **4**.

101. Show that

$$[L_n^{(\alpha)}(x)]^2=\frac{\Gamma(n+\alpha+1)}{2^{2n}n!}\sum_{k=0}^{n}\frac{(2k)!\,(2n-2k)!}{k!\,[(n-k)!]^2\Gamma(k+\alpha+1)}L_{2k}^{(2\alpha)}(2x).$$

This formula of Howell is given in Bailey **2**.

APPENDIX

ON A SINGULAR CASE OF ORTHOGONAL POLYNOMIALS

In recent publications F. Pollaczek (**1–4**) has introduced certain remarkable generalizations of the Legendre and other classical polynomials which should be treated in this Appendix in a brief way. The polynomials of F. Pollaczek show in many respects a singular behavior. For a short treatment of this topic we refer to Bateman Manuscript Project **1**, vol. 2, pp. 218–221. Cf. also Szegö **24**.

1. Definitions and formal properties

Let a and b be real parameters, $a > |b|$. We write

$$h(\theta) = \frac{a \cos \theta + b}{2 \sin \theta} \tag{1.1}$$

and define the polynomials $P_n(x; a, b) = k_n x^n + \cdots$ by the generating function

$$\begin{aligned} f(x, w) = f(\cos \theta, w) &= \sum_{n=0}^{\infty} P_n(x; a, b) w^n \\ &= (1 - we^{i\theta})^{-\frac{1}{2}+ih(\theta)}(1 - we^{-i\theta})^{-\frac{1}{2}-ih(\theta)}, \end{aligned} \tag{1.2}$$

or, in another form:

$$f(x, w) = (1 - 2xw + w^2)^{-\frac{1}{2}} \exp\left\{(ax + b) \sum_{m=1}^{\infty} \frac{w^m}{m} U_{m-1}(x)\right\} \tag{1.3}$$

where U_{m-1} has the same meaning as in (1.12.3). The polynomials $P_n(x; a, b)$ reduce to the Legendre polynomials in the limiting case $a = b = 0$.

The following identities are easy to establish:

$$P_n(x; a, b) = (-1)^n P_n(-x; a, -b), \tag{1.4}$$

$$P_n(1; a, b) = L_n(-a - b), \qquad P_n(-1; a, b) = (-1)^n L_n(-a + b) \tag{1.5}$$

where $L_n(x) = L_n^{(0)}(x)$ is Laguerre's polynomial (Chapter V). The highest coefficient k_n of $P_n(x; a, b)$ can be obtained by replacing w by w/x in (1.3) and taking $x \to \infty$. We find

$$k_n = 2^n \binom{n + \frac{1}{2}(a-1)}{n} \cong 2^n \frac{n^{\frac{1}{2}(a-1)}}{\Gamma(\frac{1}{2}(a+1))} \quad \text{as} \quad n \to \infty. \tag{1.6}$$

The following recurrence formula holds (cf. Bateman Manuscript Project, loc. cit.):

$$\begin{aligned} nP_n(x; a, b) = {} & [(2n - 1 + a)x + b] P_{n-1}(x; a, b) \\ & - (n-1)P_{n-2}(x; a, b), \quad n = 2, 3, 4, \cdots. \end{aligned} \tag{1.7}$$

Here $P_0 = 1$, $P_1 = (2a + 1)x + 2b$.

We have the important relation of orthogonality (cf. Szegö **24**):

$$(1.8)\quad \int_{-1}^{+1} P_n(x; a, b)P_m(x; a, b)w(x; a, b)\, dx = (n + \tfrac{1}{2}(a + 1))^{-1}\delta_{nm}, \qquad n, m = 0, 1, 2, \cdots,$$

where the weight function is defined by

$$(1.9)\qquad w(\cos\theta; a, b) = e^{(2\theta-\pi)h(\theta)}[\cosh(\pi h(\theta))]^{-1}.$$

We note that

$$(1.10)\qquad w(\cos\theta; a, b) = 2\exp\{(a + b)(1 - \pi/\theta)\} \quad \text{as} \quad \theta \to +0.$$

The behavior of w is similar as $\theta \to \pi - 0$.

The following representation in terms of the hypergeometric function holds (Bateman Manuscript Project, loc. cit.):

$$(1.11)\qquad P_n(\cos\theta; a, b) = e^{in\theta}F(-n, \tfrac{1}{2} + ih(\theta); 1; 1 - e^{-2i\theta}).$$

2. Generalization

Let λ be real, $\lambda > -\frac{1}{2}$. We define the polynomials $P_n^{(\lambda)}(x; a, b)$ by the generating function

$$(2.1)\qquad \sum_{n=0}^{\infty} P_n^{(\lambda)}(x; a, b)w^n = (1 - 2xw + w^2)^{-\lambda}\exp\left\{(ax + b)\sum_{m=1}^{\infty}\frac{w^m}{m}U_{m-1}(x)\right\}.$$

For $a = b = 0$ we obtain the ultraspherical polynomials $P_n^{(\lambda)}(x)$. The case dealt with in **1**. corresponds to $\lambda = \frac{1}{2}$. The polynomials $P_n^{(\lambda)}(x; a, b)$ are orthogonal in $-1 \leqq x \leqq 1$, $x = \cos\theta$, with the weight function

$$(2.2)\qquad w^{(\lambda)}(x; a, b) = \pi^{-1}2^{2\lambda-1}e^{(2\theta-\pi)h(\theta)}(\sin\theta)^{2\lambda-1}|\Gamma(\lambda + ih(\theta))|^2.$$

Concerning a recurrence formula and a representation in terms of the hypergeometric function, see Bateman Manuscript Project, loc. cit.

3. Integral representations

The following generalizations of the Laplace integral (4.8.11) and of the Mehler integrals (4.8.6) and (4.8.7) hold (Novikoff **1**):

$$(3.1)\qquad P_n(\cos\theta; a, b) = \pi^{-1}e^{-2\theta h(\theta)}\cosh(\pi h(\theta))\int_0^{\pi}\exp\left\{2ih(\theta)\log\operatorname{ctg}\frac{t}{2}\right\} \cdot (\cos\theta + i\cos t\sin\theta)^{-n-1}\, dt$$

$$(3.2)\qquad = e^{-\theta h(\theta)}\cosh(\pi h(\theta)) \cdot \frac{2}{\pi}\int_0^{\theta}\cos\left\{(n + \tfrac{1}{2})t - h(\theta)\log\frac{\sin\frac{\theta + t}{2}}{\sin\frac{\theta - t}{2}}\right\} \cdot (2\cos t - 2\cos\theta)^{-\frac{1}{2}}\, dt$$

$$= e^{(\pi-\theta)h(\theta)} \cosh (\pi h(\theta))$$

(3.3)
$$\cdot \frac{2}{\pi} \int_\theta^\pi \sin \left\{ (n + \tfrac{1}{2})t - h(\theta) \log \frac{\sin \frac{t+\theta}{2}}{\sin \frac{t-\theta}{2}} \right\} \cdot (2 \cos \theta - \cos t)^{-\frac{1}{2}}\, dt.$$

In these formulas $0 < \theta < \pi$.

4. Infinite interval

Pollaczek (**3**) defines another remarkable class of polynomials $P_n^{(\lambda)}(x; \alpha)$ by the following generating function:

$$\sum_{n=0}^{\infty} P_n^{(\lambda)}(x; \alpha) w^n = (1 - we^{i\alpha})^{-\lambda+ix}(1 - we^{-i\alpha})^{-\lambda-ix}. \tag{4.1}$$

Here $0 < \alpha < \pi$ and $\lambda > 0$. These polynomials are orthogonal in the interval $-\infty < x < \infty$ with the weight function

$$\pi^{-1}(2 \sin \alpha)^{2\lambda-1} e^{-(\pi-2\alpha)x} |\Gamma(\lambda + ix)|^2. \tag{4.2}$$

Laguerre polynomials appear as a limiting case. Indeed, replacing x by x/α and assuming $\alpha \to 0$ we obtain

$$\lim_{\alpha \to 0} (1 - we^{i\alpha})^{-\lambda}(1 - we^{-i\alpha})^{-\lambda} \exp \left\{ \frac{ix}{\alpha} \log \frac{1 - we^{i\alpha}}{1 - we^{-i\alpha}} \right\} = (1 - w)^{-2\lambda} \exp \left(2x \frac{w}{1 - w} \right)$$

so that, cf. (5.1.9),

$$\lim_{\alpha \to 0} P_n^{(\lambda)}(x/\alpha; \alpha) = L_n^{(\beta)}(-2x), \quad \beta = 2\lambda - 1. \tag{4.3}$$

It is also clear that the polynomials in **2.** arise from $P_n^{(\lambda)}(x; \alpha)$ as follows:

$$P_n^{(\lambda)}(h(\theta); \theta) = P_n^{(\lambda)}(\cos \theta; a, b). \tag{4.4}$$

5. Asymptotic properties

(a) By means of the generating function (1.1) it is not difficult to obtain an asymptotic expression for $P_n(x; a, b)$ when $n \to \infty$. We may use Darboux's method (§8.4).

First let x be outside of the closed interval $[-1, +1]$. Writing $x = \frac{1}{2}(z + z^{-1})$, $z = e^{i\theta}$, $\Im\theta > 0$, we find

(5.1)
$$P_n(x; a, b) = \{\Gamma(\tfrac{1}{2} + ih(\theta))\}^{-1}(1 - e^{2i\theta})^{-\frac{1}{2}+ih(\theta)} \cdot e^{-in\theta} n^{-\frac{1}{2}+ih(\theta)} \left(1 + O\left(\frac{1}{n}\right)\right).$$

It is not difficult to extend this to an asymptotic expansion. Now let $-1 < x < +1$; forming the real part of the right-hand expression in (5.1), the asymptotic formula for $\frac{1}{2}P_n(x; a, b)$ arises. Finally, in view of (1.5) we find from (8.22.3)

$$P_n(1; a, b) \cong \tfrac{1}{2}\pi^{-\frac{1}{2}}e^{-\frac{1}{2}(a+b)}(a+b)^{-\frac{1}{4}}n^{-\frac{3}{4}} \cdot \exp\{2(a+b)^{\frac{1}{2}}n^{\frac{1}{2}}\}. \tag{5.2}$$

(b) Novikoff (**1**) investigated the asymptotic behavior of $P_n(\cos(tn^{-\frac{1}{2}}); a, b)$ where $t > 0$ is fixed. His principal results are as follows:

$$\begin{aligned} P_n(\cos(tn^{-\frac{1}{2}}); a, b) &= \tfrac{1}{2}\pi^{-\frac{1}{2}}(a+b-t^2)^{-\frac{1}{4}}\exp(-\tfrac{1}{2}(a+b)) \\ &\quad \cdot n^{-\frac{1}{4}}\exp\left\{n^{\frac{1}{2}}\left(\frac{\pi}{2}\frac{a+b}{t} + \lambda(t)\right)\right\}\left(1 + O\left(\frac{1}{n}\right)\right), \\ \lambda(t) &= a_1 t - \frac{a+b}{t}\operatorname{arc\,tg} a_1, \\ a_1 &= t^{-1}(a+b-t^2)^{\frac{1}{2}}, \qquad 0 < t < (a+b)^{\frac{1}{2}}; \end{aligned} \tag{5.3}$$

$$\begin{aligned} P_n(\cos(tn^{-\frac{1}{2}}); a, b) &= \pi^{-\frac{1}{2}}(t^2-a-b)^{-\frac{1}{4}}\exp(-\tfrac{1}{2}(a+b)) \\ &\quad \cdot n^{-\frac{1}{4}}\exp\left\{n^{\frac{1}{2}}\left(\frac{\pi}{2}\frac{a+b}{t}\right)\right\} \cdot \left\{\cos\left(\frac{\pi}{4} - n^{\frac{1}{2}}\mu(t)\right) + O\left(\frac{1}{n}\right)\right\}, \\ \mu(t) &= \alpha_1 t - \frac{a+b}{2t}\log\frac{1+\alpha_1}{1-\alpha_1}, \\ \alpha_1 &= t^{-1}(t^2-a-b)^{\frac{1}{2}}, \qquad t > (a+b)^{\frac{1}{2}}. \end{aligned} \tag{5.4}$$

From (5.3) and (5.4) interesting conclusions can be drawn about the "extreme" zeros of $P_n(x; a, b)$. Let us denote the zeros of these polynomials, as in Chapter VI, by $\cos\theta_\nu$ where $0 < \theta_1 < \theta_2 < \cdots < \theta_n < \pi$; $\theta_\nu = \theta_{\nu n}$. Then, for any fixed value of ν,

$$\lim_{n\to\infty} n^{\frac{1}{2}}\theta_{\nu n} = (a+b)^{\frac{1}{2}}. \tag{5.5}$$

6. Associated orthogonal polynomials

(a) We consider the system $\{\phi_n(z)\}$ of polynomials which are orthogonal on the unit circle $|z| = 1$ relative to the weight function

$$f(\theta) = w(\cos\theta)|\sin\theta|; \tag{6.1}$$

here $w(x)$ has the same meaning as in (1.9). These polynomials show also in many respects a singular behavior.

The relation to the Pollaczek polynomials can be established by means of the formulas (11.5.2):

$$\left.\begin{aligned} z^{-n}\phi_{2n}(z) \\ z^{-n+1}\phi_{2n-1}(z) \end{aligned}\right\} = \left(\frac{\pi}{2}\right)^{\frac{1}{2}}\left\{1 \pm \frac{\phi_{2n}(0)}{\kappa_{2n}}\right\}^{\frac{1}{2}} p_n(x) + \tfrac{1}{2}(z - z^{-1})\left(\frac{\pi}{2}\right)^{\frac{1}{2}}\left\{1 \mp \frac{\phi_{2n}(0)}{\kappa_{2n}}\right\}^{\frac{1}{2}} q_{n-1}(x). \tag{6.2}$$

We have $p_n(x) = \{n + \frac{1}{2}(a+1)\}^{\frac{1}{2}} P_n(x; a, b)$ [cf. (1.8)]; the system $\{q_n(x)\}$ is orthonormal with the weight function $(1 - x^2)w(x)$. Moreover, κ_{2n} denotes the (positive) highest coefficient of $\phi_{2n}(z)$. Between x and z the relation $x = \frac{1}{2}(z + z^{-1})$ holds.

(b) Let $|z| < 1$. We investigate the asymptotic behavior of the polynomials $\phi_n(z)$ as $n \to \infty$. Let $z \neq 0$, $z = e^{i\theta}$, $\Im\theta > 0$. We rewrite (5.1) in the form

$$(6.3)\qquad \begin{aligned} p_n(x) &= A(x) n^{ih(\theta)} z^{-n}\left(1 + O\left(\frac{1}{n}\right)\right),\\ A(x) &= \{\Gamma(\tfrac{1}{2} + ih(\theta))\}^{-1}(1 - z^2)^{-\frac{1}{2}+ih(\theta)}. \end{aligned}$$

Now the polynomials $\{q_n(x)\}$ can be represented in terms of $\{p_n(x)\}$ by using Christoffel's formula 2.5, taking (1.5) into account. Thus

$$(1 - x^2)q_{n-1}(x) = \text{const.}\begin{vmatrix} P_{n-1}(x; a, b) & P_n(x; a, b) & P_{n+1}(x; a, b)\\ L_{n-1} & L_n & L_{n+1}\\ L'_{n-1} & -L'_n & L'_{n+1} \end{vmatrix},$$

$$L_n = L_n(-a - b), \quad L'_n = L_n(-a + b).$$

Denoting the latter determinant by $\Delta_n(x)$ we find by an easy calculation:

$$(1 - x^2)q_{n-1}(x) = (L_{n-1}L'_n + L_nL'_{n-1})^{-\frac{1}{2}}(L_nL'_{n+1} + L_{n+1}L'_n)^{-\frac{1}{2}} \cdot\left(\frac{k_{n-1}}{k_{n+1}}\right)^{\frac{1}{2}}\{n - 1 + \tfrac{1}{2}(a+1)\}^{\frac{1}{2}}\Delta_n(x).$$

Using (1.5), (8.22.3), and (1.6) we obtain

$$(6.4)\qquad \begin{aligned} (1 - x^2)q_{n-1}(x) &= \tfrac{1}{2}A(x)n^{ih(\theta)}z^{-n-1}\\ &\cdot\left\{\left(1 + \frac{\epsilon + \epsilon'}{2}\right)z^2 + (\epsilon - \epsilon')z - \left(1 - \frac{\epsilon + \epsilon'}{2}\right)\right\}\left(1 + O\left(\frac{1}{n}\right)\right),\\ \epsilon &= (a + b)^{\frac{1}{2}}n^{-\frac{1}{2}}, \qquad \epsilon' = (a - b)^{\frac{1}{2}}n^{-\frac{1}{2}}. \end{aligned}$$

(c) From (6.3) and (6.4) we conclude, since $1 - x^2 = -(1 - z^2)^2(2z)^{-2}$,

$$(6.5)\qquad \begin{aligned} \frac{\frac{1}{2}(z - z^{-1})q_{n-1}(x)}{p_n(x)} = \frac{1}{1 - z^2}\left\{\left(1 + \frac{\epsilon + \epsilon'}{2}\right)z^2\right.\\ \left. + (\epsilon - \epsilon')z - \left(1 - \frac{\epsilon + \epsilon'}{2}\right)\right\}\left(1 + O\left(\frac{1}{n}\right)\right). \end{aligned}$$

Denoting by $k'_n = \{n + \frac{1}{2}(a+1)\}^{\frac{1}{2}}k_n$ the coefficient of x^n in $p_n(x)$ and by l_{n-1} the coefficient of x^{n-1} in $q_{n-1}(x)$, we find from (6.5) for $z \to 0$:

$$\frac{l_{n-1}}{k'_n} = \left(1 - \frac{\epsilon + \epsilon'}{2}\right)\left(1 + O\left(\frac{1}{n}\right)\right).$$

Now the first formula in (6.2) yields:

$$\left.\begin{matrix}\kappa_{2n}\\ \phi_{2n}(0)\end{matrix}\right\} = \left(\frac{\pi}{2}\right)^{\frac{1}{2}}\left\{1+\frac{\phi_{2n}(0)}{\kappa_{2n}}\right\}^{\frac{1}{2}}\frac{k_n'}{2^n} \pm \left(\frac{\pi}{2}\right)^{\frac{1}{2}}\left\{1-\frac{\phi_{2n}(0)}{\kappa_{2n}}\right\}^{\frac{1}{2}}\frac{l_{n-1}}{2^n} .$$

so that

$$\begin{aligned}\kappa_{2n} &= \pi^{\frac{1}{2}}\{(k_n')^2+l_{n-1}^2\}^{\frac{1}{2}}\cdot 2^{-n} = \pi^{\frac{1}{2}}n^{\frac{1}{2}}k_n\cdot 2^{-n+\frac{1}{2}}\left(1-\frac{\epsilon+\epsilon'}{4}\right)\left(1+O\left(\frac{1}{n}\right)\right)\\ &= \frac{(2\pi)^{\frac{1}{2}}}{\Gamma(\frac{1}{2}(a+1))}n^{a/2}\left(1-\frac{\epsilon+\epsilon'}{4}\right)\left(1+O\left(\frac{1}{n}\right)\right),\end{aligned} \tag{6.6}$$

$$\begin{aligned}\phi_{2n}(0) &= \pi^{\frac{1}{2}}\frac{(k_n')^2-l_{n-1}^2}{\{(k_n')^2+l_{n-1}^2\}^{\frac{1}{2}}}\cdot 2^{-n} = \pi^{\frac{1}{2}}n^{\frac{1}{2}}k_n\cdot 2^{-n-\frac{1}{2}}\left(\epsilon+\epsilon'+O\left(\frac{1}{n}\right)\right)\\ &= \frac{(2\pi)^{\frac{1}{2}}}{\Gamma(\frac{1}{2}(a+1))}n^{a/2}\left(\frac{\epsilon+\epsilon'}{2}+O\left(\frac{1}{n}\right)\right).\end{aligned} \tag{6.7}$$

In particular,

$$\frac{\phi_{2n}(0)}{\kappa_{2n}} = \frac{\epsilon+\epsilon'}{2}+O\left(\frac{1}{n}\right). \tag{6.8}$$

(d) We obtain from (6.2), $|z| < 1$, in view of (6.3), (6.5), (6.8),

$$\begin{aligned}\phi_{2n}(z) &= \left(\frac{\pi}{2}\right)^{\frac{1}{2}}A(x)n^{ih(\theta)}\left\{1+\frac{\epsilon+\epsilon'}{4}+\left(1-\frac{\epsilon+\epsilon'}{4}\right)\right.\\ &\qquad\cdot\left.\frac{(1+\frac{1}{2}(\epsilon+\epsilon'))z^2+(\epsilon-\epsilon')z-(1-\frac{1}{2}(\epsilon+\epsilon'))}{1-z^2}\right\}\left(1+O\left(\frac{1}{n}\right)\right)\\ &= \left(\frac{\pi}{2}\right)^{\frac{1}{2}}\frac{(1-z^2)^{\alpha(z)}}{\Gamma(\beta(z))}n^{\gamma(z)}\{(a+b)^{\frac{1}{2}}(1+z)+(a-b)^{\frac{1}{2}}(1-z)\}\left(1+O\left(\frac{1}{n}\right)\right),\end{aligned} \tag{6.9}$$

$$\begin{aligned}\phi_{2n-1}(z) &= \left(\frac{\pi}{2}\right)^{\frac{1}{2}}A(x)n^{ih(\theta)}z^{-1}\left\{1-\frac{\epsilon+\epsilon'}{4}+\left(1+\frac{\epsilon+\epsilon'}{4}\right)\right.\\ &\qquad\cdot\left.\frac{(1+\frac{1}{2}(\epsilon+\epsilon'))z^2+(\epsilon-\epsilon')z-(1-\frac{1}{2}(\epsilon+\epsilon'))}{1-z^2}\right\}\left(1+O\left(\frac{1}{n}\right)\right)\\ &= \left(\frac{\pi}{2}\right)\frac{(1-z^2)^{\alpha(z)}}{\Gamma(\beta(z))}n^{\gamma(z)}\{(a+b)^{\frac{1}{2}}(1+z)-(a-b)^{\frac{1}{2}}(1-z)\}\left(1+O\left(\frac{1}{n}\right)\right)\end{aligned} \tag{6.10}$$

where we set for abbreviation:

$$\begin{aligned}\alpha(z) &= \frac{\frac{1}{2}(a-3)+bz+\frac{1}{2}(a+3)z^2}{1-z^2},\\ \beta(z) &= \frac{\frac{1}{2}(a+1)+bz+\frac{1}{2}(a-1)z^2}{1-z^2},\\ \gamma(z) &= \frac{\frac{1}{2}(a-1)+bz+\frac{1}{2}(a+1)z^2}{1-z^2}.\end{aligned} \tag{6.11}$$

The argument of the Γ-function can be written as follows:

$$\beta(z) = \frac{a}{2}\frac{1+z^2}{1-z^2} + b\frac{z}{1-z^2} + \frac{1}{2} \tag{6.12}$$

and the real part of this expression is $>\frac{1}{2}$ for $|z| < 1$. Hence the function $1/\Gamma$ is never zero in $|z| < 1$. The only point of accumulation of the zeros of the polynomials $\{\phi_m(z)\}$ in $|z| < 1$ is the point

$$-\frac{(a+b)^{\frac{1}{2}} - (a-b)^{\frac{1}{2}}}{(a+b)^{\frac{1}{2}} + (a-b)^{\frac{1}{2}}} = \frac{(a^2-b^2)^{\frac{1}{2}} - a}{b}$$

appearing only for $m = 2n - 1$.

The exponent $\gamma(z)$ of n has a real part $> -\frac{1}{2}$. Hence the series $\sum|\phi_m(z)|^2$ is divergent for all $|z| < 1$. In particular,

$$\left.\begin{matrix}\phi_{2n}(0)\\ \phi_{2n-1}(0)\end{matrix}\right\} \doteq \left(\frac{\pi}{2}\right)^{\frac{1}{2}}\{\Gamma(\tfrac{1}{2}(a+1))\}^{-1}n^{\frac{1}{2}(a-1)}((a+b)^{\frac{1}{2}} \pm (a-b)^{\frac{1}{2}})\left(1 + O\left(\frac{1}{n}\right)\right). \tag{6.13}$$

Taking (11.3.6) into account, this yields again the main term of (6.6).

(e) Recapitulating, we may point out certain properties of the Pollaczek polynomials which indicate a rather singular behavior compared with the classical polynomials.

The weight functions $w(x)$ or $f(\theta)$ in Theorems 12.1.1 and 12.1.2, respectively, are such that $\log w(\cos\theta)$ and $\log f(\theta)$ are integrable. The weight function $w(x)$ of the Pollaczek polynomials vanishes at the endpoints $x = \pm 1$ so strongly that $\log w(\cos\theta)$ is not integrable [cf. (1.10)].

The normalized Jacobi polynomials are at $x = \pm 1$ of the order $n^{\alpha+\frac{1}{2}}$ and $n^{\beta+\frac{1}{2}}$, respectively. The orthonormal Pollaczek polynomials at $x = \pm 1$ are of order $n^{\frac{1}{4}} \exp\{2(a+b)^{\frac{1}{2}}n^{\frac{1}{2}}\}$ [cf. (5.2)].

The Toeplitz minima $\mu_n(f)$ (12.3) associated with the weight function $f(\theta)$ tend to a positive limit under the assumption of Theorem 12.1.1. In case of the polynomials discussed in **6.** this limit is zero; the weight function defines a "deterministic" process. We have in this case: $\mu_n(f) = \kappa_n^{-2} \sim n^{-a}$. Let $f(\theta) = 0$ in a certain interval, say $-\epsilon < \theta < +\epsilon$, $0 < \epsilon < \pi$ and $f(\theta) = 1$ otherwise. We have then again $\mu_n(f) \to 0$, and, more precisely, $\mu_n(f) \sim r^n$, $r < 1$; cf. (16.4.3) and Problem 50.

The orthonormal polynomials defined in Theorem 12.1.2 are asymptotically of the order $(x + (x^2-1)^{\frac{1}{2}})^n$ if x is not on the cut $[-1, +1]$). The Pollaczek polynomials are under the same condition of the order $n^K(x + (x^2-1)^{\frac{1}{2}})^n$ where K is a function of x. A similar discrepancy arises if x is on the segment $[-1, +1]$.

For the "largest" zeros $\cos\theta_\nu$, $0 < \theta_\nu < \pi$, $\theta_\nu = \theta_\nu(n)$, ν fixed, $n \to \infty$, of the Jacobi polynomials we have $\theta_\nu(n) \cong n^{-1}j_\nu$ where j_ν is the corresponding zero of an appropriate Bessel function; see (6.3.15). The similar zeros of the Pollaczek polynomials satisfy the relation $\theta_\nu(n) \cong n^{-\frac{1}{2}}(a + b)^{\frac{1}{2}}$, see (5.5); the order of magnitude is different and the constant does not depend on ν.

LIST OF REFERENCES[73]

ABEL, N. H.

*1. *Oeuvres Complètes.* Vol. 2, 1881.

ACHIESER, N. I.

1. *Über eine Eigenschaft der "elliptischen" Polynome.* Communications de la Société Mathématique de Kharkoff, (4), vol. 9 (1934), pp. 3-8.
2. *Verallgemeinerung einer Korkine-Zolotareffschen Minimum-Aufgabe.* Ibid., (4), vol. 13 (1936), pp. 3-14.

ADAMOFF, A.

1. *On the asymptotic expansion of the polynomials* $U_n(x) = e^{ax^2/2}d^n[e^{-ax^2/2}]/dx^n$ *for large values of n* (in Russian). Annals of the Polytechnic Institute of St. Petersburg, vol. 5 (1906), pp. 127-143.
2. *Expansions of an Arbitrary Function of a Single Real Variable in Series of Functions of a Preassigned Kind* (in Russian). Thesis. St. Petersburg, 1907, 191 pp.

BANACH, S.

*1. *Théorie des Opérations Linéaires.* Warszawa-Lwów, 1932.

BERNSTEIN, S.

*1. *Leçons sur les Propriétés Extrémales et la Meilleure Approximation des Fonctions Analytiques d'une Variable Réelle.* Paris, 1926.

2. *Sur les polynomes orthogonaux relatifs à un segment fini.* Journal de Mathématiques, (9), vol. 9 (1930), pp. 127-177; vol. 10 (1931), pp. 219-286.
3. *Sur une classe de polynomes orthogonaux.* Communications de la Société Mathématique de Kharkoff, (4), vol. 4 (1930), pp. 79-93. *Complément.* Ibid., vol. 5 (1932), pp. 59-60.
4. *Sur la limitation des valeurs d'un polynôme* $P_n(x)$ *de degré n sur tout un segment par ses valeurs en* $(n + 1)$ *points du segment.* Bulletin de l'Académie des Sciences de l'URSS, 1931, pp. 1025-1050.

BLUMENTHAL, O.

1. *Ueber die Entwicklung einer willkürlichen Funktion nach den Nennern des Kettenbruches für* $\int_{-\infty}^{0} [\phi(\xi)/(z - \xi)]d\xi$. Inaugural-Dissertation. Göttingen, 1898.

BOCHNER, S.

1. *Über Sturm-Liouvillesche Polynomsysteme.* Mathematische Zeitschrift, vol. 29 (1929), pp. 730-736.

BOTTEMA, O.

1. *Die Nullstellen der Hermiteschen Polynome.* Koninklijke Akademie van Wetenschappen te Amsterdam, Proceedings, vol. 33 (1930), pp. 495-503.
2. *Die Nullstellen gewisser durch Rekursionsformeln definierten Polynome.* Ibid., vol. 34 (1931), pp. 681-691.

BRAUER, A.

1. *Über die Nullstellen der Hermiteschen Polynome.* Mathematische Annalen, vol. 107 (1932), pp. 87-89.

BRUNS, H.

1. *Zur Theorie der Kugelfunctionen.* Journal für die reine und angewandte Mathematik, vol. 90 (1881), pp. 322-328.

BUELL, C. E.

1. *The zeros of Jacobi and related polynomials.* Duke Mathematical Journal, vol. 2 (1936), pp. 304-316.

[73] The asterisks indicate items not dealing with orthogonal polynomials.

Carathéodory, C.
*1. *Conformal Representation.* Cambridge, 1932.
Carleman, T.
1. *Über die Approximation analytischer Funktionen durch lineare Aggregate von vorgegebenen Potenzen.* Arkiv för Matematik, Astronomi och Fysik, vol. 17 (1923), no. 9, 30 pp.
Christoffel, E. B.
1. *Über die Gaussische Quadratur und eine Verallgemeinerung derselben.* Journal für die reine und angewandte Mathematik, vol. 55 (1858), pp. 61-82.
Cramér, H.
1. *On some classes of series used in mathematical statistics.* Comptes Rendus du Sixième Congrès des Mathématiciens Scandinaves, Stockholm, 1926, pp. 399-425.
Darboux, G.
1. *Mémoire sur l'approximation des fonctions de très grands nombres.* Journal de Mathématiques, (3), vol. 4 (1878), pp. 5-56, 377-416.
Dirichlet, G. L.
1. *Sur les séries dont le terme général dépend de deux angles, et qui servent à exprimer des fonctions arbitraires entre des limites données.* Journal für die reine und angewandte Mathematik, vol. 17 (1837), pp. 35-56.
Doetsch, G.
1. *Integraleigenschaften der Hermiteschen Polynome.* Mathematische Zeitschrift, vol. 32 (1930), pp. 587-599.
2. *Die in der Statistik seltener Ereignisse auftretenden Charlierschen Polynome und eine damit zusammenhängende Differentialdifferenzengleichung.* Mathematische Annalen, vol. 109 (1933), pp. 257-266.
Du Bois-Reymond, P.
*1. *Untersuchungen über die Convergenz und Divergenz der Fourierschen Darstellungsformeln.* Abhandlungen der Akademie München, vol. 12 (1876), pp. 1-103.
Erdélyi, A.
1. *Über eine Integraldarstellung der $M_{k,m}$-Funktionen und ihre asymptotische Darstellung für grosse Werte von $\Re k$.* Mathematische Annalen, vol. 113 (1936), pp. 357-362.
2. *Über eine erzeugende Funktion von Produkten Hermitescher Polynome.* Mathematische Zeitschrift, vol. 44 (1938), pp. 201-211.
Erdös, P., and Feldheim, E.
1. *Sur le mode de convergence pour l'interpolation de Lagrange.* Comptes Rendus de l'Académie des Sciences, Paris, vol. 203 (1936), pp. 913-915.
Erdös, P., and Turán, P.
1. *On interpolation.* I. *Quadrature- and mean-convergence in the Lagrange interpolation.* Annals of Mathematics, (2), vol. 38 (1937), pp. 142-155.
2. *On interpolation.* II. *On the distribution of the fundamental points of Lagrange and Hermite interpolation.* Ibid., vol. 39 (1938), pp. 703-724.
Euler, L.
*1. *Institutiones Calculi Integralis.* Vol. 2, sect. I, chap. X, problem 130. *Opera Omnia.* Ser. 1, vol. 12, p. 224.
Faber, G.
1. *Über polynomische Entwicklungen.* Mathematische Annalen, vol. 57 (1903), pp. 389-408.
*2. *Über die interpolatorische Darstellung stetiger Funktionen.* Jahresbericht der Deutschen Mathematiker-Vereinigung, vol. 23 (1914), pp. 192-210.
3. *Tschebyscheffsche Polynome.* Journal für die reine und angewandte Mathematik, vol. 150 (1919), pp. 79-106.

4. *Über nach Polynomen fortschreitende Reihen.* Sitzungsberichte der Bayrischen Akademie der Wissenschaften, 1922, pp. 157–178.

FATOU, P.

*1. *Séries trigonométriques et séries de Taylor.* Acta Mathematica, vol. 30 (1906), pp. 335–400.

FAVARD, J.

1. *Sur les polynomes de Tchebicheff.* Comptes Rendus de l'Académie des Sciences, Paris, vol. 200 (1935), pp. 2052–2053.

FEJÉR, L.

*1. *Sur les fonctions bornées et intégrables.* Comptes Rendus de l'Académie des Sciences, Paris, vol. 131 (1900), pp. 984–987.

*2. *Untersuchungen über Fouriersche Reihen.* Mathematische Annalen, vol. 58 (1904), pp. 51–69.

3. *Asymptotikus értékek meghatározásáról.* Mathematikai és Természettudományi Értesitö, vol. 27 (1909), pp. 1–33.

4. *Über die Laplacesche Reihe.* Mathematische Annalen, vol. 67 (1909), pp. 76–109.

*5. *Über trigonometrische Polynome.* Journal für die reine und angewandte Mathematik, vol. 146 (1915), pp. 53–82.

6. *Über Interpolation.* Nachrichten der Gesellschaft der Wissenschaften zu Göttingen, 1916, pp. 66–91.

7. *Über die Lage der Nullstellen von Polynomen, die aus Minimumforderungen gewisser Art entspringen.* Mathematische Annalen, vol. 85 (1922), pp. 41–48.

8. *Über die Summabilität der Laplaceschen Reihe durch arithmetische Mittel.* Mathematische Zeitschrift, vol. 24 (1925), pp. 267–284.

9. *Abschätzungen für die Legendreschen und verwandte Polynome.* Mathematische Zeitschrift, vol. 24 (1925), pp. 285–298.

10. *Über Weierstrass'sche Approximation, besonders durch Hermitesche Interpolation.* Mathematische Annalen, vol. 102 (1930), pp. 707–725.

11. *Die Abschätzung eines Polynoms in einem Intervalle, wenn Schranken für seine Werte und ersten Ableitungswerte in einzelnen Punkten des Intervalles gegeben sind, und ihre Anwendung auf die Konvergenzfrage Hermitescher Interpolationsreihen.* Mathematische Zeitschrift, vol. 32 (1930), pp. 426–457.

12. *Ultrasphärikus polynomok összegéröl.* Matematikai és Fizikai Lapok, vol. 38 (1931), pp. 161–164.

13. *Lagrangesche Interpolation und die zugehörigen konjugierten Punkte.* Mathematische Annalen, vol. 106 (1932), pp. 1–55.

14. *Bestimmung derjenigen Abszissen eines Intervalles, für welche die Quadratsumme der Grundfunktionen der Lagrangeschen Interpolation im Intervalle ein möglichst kleines Maximum besitzt.* Annali della Scuola Normale Superiore di Pisa, (2), vol. 1 (1932), pp. 3–16.

15. *Mechanische Quadraturen mit positiven Cotesschen Zahlen.* Mathematische Zeitschrift, vol. 37 (1933), pp. 287–309.

16. *On the characterization of some remarkable systems of points of interpolation by means of conjugate points.* American Mathematical Monthly, vol. 41 (1934), pp. 1–14.

17. *Potenzreihen mit mehrfach monotoner Koeffizientenfolge und ihre Legendre Polynome.* Proceedings of the Cambridge Philosophical Society, vol. 31 (1935), pp. 307–316.

18. *A hatványsorról és a vele kapcsolatos Legendre-féle többtagúakról.* Mathematikai és Természettudományi Értesitö, vol. 53 (1935), pp. 1–17.

19. *Bestimmung von Grenzen für die Nullstellen des Legendreschen Polynoms aus der Stieltjesschen Integraldarstellung desselben.* Monatshefte für Mathematik und Physik, vol. 43 (1936), pp. 193–209.

20. *Trigonometrische Reihen und Potenzreihen mit mehrfach monotoner Koeffizientenfolge.* Transactions of the American Mathematical Society, vol. 39 (1936), pp. 18–59.

Fejér, L., and **Szegö, G.**

*1. *Über die monotone Konvergenz von Potenzreihen mit mehrfach monotoner Koeffizientenfolge.* Prace Matematyczno-Fizyczne, vol. 44 (1935), pp. 15–25.

Fekete, M.

*1. *Über die Verteilung der Wurzeln bei gewissen algebraischen Gleichungen mit ganzzahligen Koeffizienten.* Mathematische Zeitschrift, vol. 17 (1923), pp. 228–249.

Feldheim, E.

1. *Sur l'orthogonalité des fonctions fondamentales de l'interpolation de Lagrange.* Comptes Rendus de l'Académie des Sciences, Paris, vol. 203 (1936), pp. 650–652.
2. *Sur le mode de convergence dans l'interpolation de Lagrange.* Comptes Rendus de l'Académie des Sciences de l'URSS, vol. 14 (1937), pp. 327–331.
3. *Sur l'orthogonalité des fonctions fondamentales, et sur la forte convergence en moyenne des polynomes d'interpolation de Lagrange dans le cas des abscisses de Tchebychef.* Bulletin de la Société Mathématique de France, vol. 65 (1937), pp. 1–40.
4. *Théorie de la convergence des procédés d'interpolation et de quadrature mécanique.* Mémorial des Sciences Mathématiques, vol. 95, Paris, 1939.

Friedrichs, K.

1. *On certain inequalities and characteristic value problems for analytic functions and for functions of two variables.* Transactions of the American Mathematical Society, vol. 41 (1937), pp. 321–364.

Fujiwara, M.

1. *On the zeros of Jacobi polynomials.* Japanese Journal of Mathematics, vol. 2 (1925), pp. 1–2.

Galbrun, H.

1. *Sur un développement d'une fonction à variable réelle en séries de polynomes.* Bulletin de la Société Mathématique de France, vol. 41 (1913), pp. 24–47.

Gauss, C. F.

*1. *Summatio quarumdam serierum singularium.* *Werke.* Vol. 2, pp. 9–45.
2. *Methodus nova integralium valores per approximationem inveniendi.* *Werke.* Vol. 3, pp. 163–196.

Gegenbauer, L.

1. *Über einige bestimmte Integrale.* Sitzungsberichte der mathematisch-naturwissenschaftlichen Klasse der Akademie der Wissenschaften in Wien, Abteilung IIa, vol. 70 (1874), pp. 433–443.
2. *Über die Functionen $C_n^\nu(x)$.* Ibid., vol. 75 (1877), pp. 891–905.
3. *Über die Functionen $C_n^\nu(x)$.* Ibid., vol. 97 (1888), pp. 259–270.
4. *Zur Theorie der hypergeometrischen Reihe.* Ibid., vol. 100 (1891), pp. 225–244.
5. *Das Additionstheorem der Functionen $C_n^\nu(x)$.* Ibid., vol. 102 (1893), pp. 942–950.
6. *Zur Theorie der Functionen $C_n^\nu(x)$.* Denkschriften der Akademie der Wissenschaften in Wien, Mathematisch-naturwissenschaftliche Klasse, vol. 48 (1884), pp. 293–316.
7. *Einige Sätze über die Functionen $C_n^\nu(x)$.* Ibid., vol. 57 (1890), pp. 425–480.

G(u)eronimus, J.

1. *Sur le polynôme multiplement monotone qui s'écarte le moins de zéro, dont un coefficient est donné.* Bulletin de l'Académie des Sciences de l'URSS, 1929, pp. 377–389.
2. *Sur l'écart minimal quadratique de zéro d'un polynome.* Rendiconti del Circolo Matematico di Palermo, vol. 54 (1930), pp. 298–313.
3. *On some problems of Tchebycheff.* American Journal of Mathematics, vol. 53 (1931), pp. 597–604.
4. *On a problem of M. J. Shohat.* Ibid., vol. 54 (1932), pp. 85–91.
5. *On some extremal properties of polynomials.* Annals of Mathematics, (2), vol. 37 (1936), pp. 483–517.

GOTTLIEB, M. J.

1. *Concerning some polynomials orthogonal on a finite or enumerable set of points.* American Journal of Mathematics, vol. 60 (1938), pp. 453-458.

GRONWALL, T. H.

1. *Über die Laplacesche Reihe.* Mathematische Annalen, vol. 74 (1913), pp. 213-270.
2. *Über die Summierbarkeit der Reihen von Laplace und Legendre.* Ibid., vol. 75 (1914), pp. 321-375.

GRÜNWALD, G.

1. *Über Divergenzerscheinungen der Lagrangeschen Interpolationspolynome stetiger Funktionen.* Annals of Mathematics, (2), vol. 37 (1936), pp. 908-918.

GRÜNWALD, G., and TURÁN, P.

1. *Über Interpolation.* Annali della Scuola Normale Superiore de Pisa, (2), vol. 7 (1938), pp. 137-146.

HAAR, A.

*1. *Zur Theorie der orthogonalen Funktionensysteme.* I. Mathematische Annalen, vol. 69 (1910), pp. 331-371.

2. *Reihenentwicklungen nach Legendreschen Polynomen.* Ibid., vol. 78 (1917), pp. 121-136.

HAHN, W.

1. *Die Nullstellen der Laguerreschen und Hermiteschen Polynome.* Inauguraldissertation, Berlin. Schriften des Mathematischen Seminars und des Instituts für angewandte Mathematik der Universität Berlin, vol. 1 (1933), pp. 213-244.
2. *Bericht über die Nullstellen der Laguerreschen und der Hermiteschen Polynome.* Jahresbericht der Deutschen Mathematiker-Vereinigung, vol. 44 (1934), pp. 215-236. *Nachtrag.* Ibid., vol. 45 (1935), p. 211.
3. *Über die Jacobischen Polynome und zwei verwandte Polynomklassen.* Mathematische Zeitschrift, vol. 39 (1935), pp. 634-638.
4. *Über höhere Ableitungen von Orthogonalpolynomen.* Ibid., vol. 43 (1937), p. 101.

HAMBURGER, H.

1. *Beiträge zur Konvergenztheorie der Stieltjesschen Kettenbrüche.* Mathematische Zeitschrift, vol. 4 (1919), pp. 186-222.
2. *Über eine Erweiterung des Stieltjesschen Momentenproblems.* Mathematische Annalen, vol. 81 (1920), pp. 235-319; vol. 82 (1920), pp. 120-164, 168-187.

HARDY, G. H.

1. *Summation of a series of polynomials of Laguerre.* Journal of the London Mathematical Society, vol. 7 (1932), pp. 138-139, 192.

HARDY, G. H., LITTLEWOOD, J. E., and PÓLYA, G.

*1. *Inequalities.* Cambridge, 1934.

HAUSDORFF, F.

*1. *Momentprobleme für ein endliches Intervall.* Mathematische Zeitschrift, vol. 16 (1923), pp. 220-248.

HEINE, E.

1. *Mittheilung über Kettenbrüche.* Journal für die reine und angewandte Mathematik, vol. 67 (1867), pp. 315-326.
2. *Die Fourier-Besselsche Function.* Ibid., vol. 69 (1869), pp. 128-141.
3. *Handbuch der Kugelfunctionen.* Vols. I, II. 2d edition. Berlin, 1878, 1881.

HELLY, E.

*1. *Über lineare Funktionaloperationen.* Sitzungsberichte der mathematisch-naturwissenschaftlichen Klasse der Akademie in Wien, Abteilung IIa, vol. 121 (1912), pp. 265-297

HERMITE, C.

1. *Sur les polynômes de Legendre.* Rendiconti del Circolo Matematico di Palermo, vol. 4 (1890), pp. 146-152. *Oeuvres.* Vol. 4, pp. 314-320.

2. *Sur les polynômes de Legendre.* Journal für die reine und angewandte Mathematik, vol. 107 (1891), pp. 80-83. *Oeuvres.* Vol. 4, pp. 321-326.
3. *Sur les racines de la fonction sphérique de seconde espèce.* Annales de la Faculté des Sciences de Toulouse, vol. 4 (1890), 10 pp. *Oeuvres.* Vol. 4, pp. 327-336.

HERMITE, C., and STIELTJES, T. J.
1. *Correspondance d'Hermite et de Stieltjes.* Vols. I, II. Paris, 1905.

HILB, E.
1. *Über die Laplacesche Reihe.* Mathematische Zeitschrift, vol. 5 (1919), pp. 17-25; vol. 8 (1920), pp. 79-90.

HILBERT, D.
1. *Über die Discriminante der im Endlichen abbrechenden hypergeometrischen Reihe.* Journal für die reine und angewandte Mathematik, vol. 103 (1888), pp. 337-345.

HILBERT, D., and COURANT, R.
1. *Methoden der mathematischen Physik.* Vol. 1. 2d edition. Berlin, 1931.

HILDEBRANDT, T. H.
*1. *On integrals related to and extensions of the Lebesgue integrals.* Bulletin of the American Mathematical Society, vol. 24 (1918), pp. 113-144, 177-202.

HILLE, E.
1. *A class of reciprocal functions.* Annals of Mathematics, (2), vol. 27 (1926), pp. 427-464.
2. *On Laguerre's series.* I, II, III. Proceedings of the National Academy of Sciences, vol. 12 (1926), pp. 261-265, 265-269, 348-352.
3. *Bemerkung zu einer Arbeit des Herrn Müntz.* Mathematische Zeitschrift, vol. 32 (1930), pp. 422-425.
4. *Über die Nullstellen der Hermiteschen Polynome.* Jahresbericht der Deutschen Mathematiker-Vereinigung, vol. 44 (1933), pp. 162-165.

HOBSON, E. W.
1. *The Theory of Spherical and Ellipsoidal Harmonics.* Cambridge, 1931.

HOLLÓ, Á.
1. *A mechanikus quadraturáról.* Thesis. Budapest, 1939, 23 pp.

JACKSON, D.
1. *On functions of closest approximation.* Transactions of the American Mathematical Society, vol. 22 (1921), pp. 117-128.
2. *Note on a class of polynomials of approximation.* Ibid., vol. 22 (1921), pp. 320-326.
3. *A generalized problem in weighted approximation.* Ibid., vol. 26 (1924), pp. 133-154.
4. *The Theory of Approximation.* American Mathematical Society Colloquium Publications, vol. 11, 1930.
5. *Series of orthogonal polynomials.* Annals of Mathematics, (2), vol. 34 (1933), pp. 527-545.
6. *Certain problems of closest approximation.* Bulletin of the American Mathematical Society, vol. 39 (1933), pp. 889-906.
7. *The summation of series of orthogonal polynomials.* Ibid., vol. 40 (1934), pp. 743-752.
8. *Formal properties of orthogonal polynomials in two variables.* Duke Mathematical Journal, vol. 2 (1936), pp. 423-434.

JACOB, M. M.
1. *Sullo sviluppo di una funzione di ripartizione in serie di polinomi di Hermite.* Giornale dell'Istituto Italiano degli Attuari, vol. 2 (1931), pp. 100-106, 356-368.
2. *Sur le phénomène de Gibbs dans les développements de séries de polynomes d'Hermite.* Comptes Rendus de l'Académie des Sciences, Paris, vol. 204 (1937), pp. 1540-1543.

JACOBI, C. G. J.
1. *Ueber Gauss' neue Methode, die Werthe der Integrale näherungsweise zu finden.* Journal für die reine und angewandte Mathematik, vol. 1 (1826), pp. 301-308. *Gesammelte Werke.* Vol. 6, pp. 3-11.

2. *Über die Entwickelung des Ausdrucks* $(aa - 2aa' [\cos \omega \cos \phi + \sin \omega \sin \phi \cdot \cos (\theta - \theta')] + a'a')^{-1/2}$ Ibid., vol. 26, 1843, pp. 81–87. *Gesammelte Werke*. Vol. 6, pp. 148–155.
3. *Untersuchungen über die Differentialgleichung der hypergeometrischen Reihe*. Ibid., vol. 56 (1859), pp. 149–165. *Gesammelte Werke*. Vol. 6, pp. 184–202.

JORDAN, CAMILLE.
1. *Cours d'Analyse de l'École Polytechnique*. Vol. 3, 2d edition. Paris, 1896.

JORDAN, CHARLES.
1. *Sur une série de polynomes dont chaque somme partielle représente la meilleure approximation d'un degré donné suivant la méthode des moindres carrés*. Proceedings of the London Mathematical Society, (2), vol. 20 (1920), pp. 297–325.

JULIA, G.
1. *Sur les polynomes de Tchebichef*. Comptes Rendus de l'Académie des Sciences, Paris, vol. 182 (1926), pp. 1201–1202.

KACZMARZ, ST., and STEINHAUS, H.
*1. *Theorie der Orthogonalreihen*. Warszawa-Lwów, 1935.

KELDYSCH, M., and LAVRENTIEFF, M.
1. *Sur la représentation conforme des domaines limités par des courbes rectifiables*. Annales Scientifiques de l'École Normale Supérieure, (3), vol. 54 (1937), pp. 1–38.

KLEIN, F.
1. *Ueber die Nullstellen der hypergeometrischen Reihe*. Mathematische Annalen, vol. 37 (1890), pp. 573–590. *Gesammelte Abhandlungen*. Vol. 2, pp. 550–567.

KOGBETLIANTZ, E.
1. *Sur les séries de fonctions ultrasphériques*. Comptes Rendus de l'Académie des Sciences, Paris, vol. 163 (1916), pp. 601–603.
2. *Sur la sommation des séries ultrasphériques*. Ibid., vol. 164 (1917), pp. 510–513, 626–628, 778–780; vol. 169 (1919), pp. 54–57.
3. *Sur les développements de Jacobi*. Ibid., vol. 168 (1919), pp. 992–994.
4. *Sur les séries ultrasphériques*. Ibid., vol. 169 (1919), pp. 322–324.
5. *Nouvelle observations sur les séries ultrasphériques*. Ibid., vol. 169 (1919), pp. 423–426.
6. *Sur l'unicité des développements ultrasphériques*. Ibid., vol. 169 (1919), pp. 769–770, 950–953.
7. *Sur les développements de Jacobi*. Ibid., vol. 172 (1921), pp. 1333–1334; vol. 192 (1931), pp. 915–918.
8. *Sur la sommabilité (C, δ) de développements suivant les polynomes d'Hermite*. Ibid., vol. 192 (1931), pp. 662–663.
9. *Nouvelles observations sur le système orthogonal de polynomes d'Hermite*. Ibid., vol. 192 (1931), pp. 1696–1698.
10. *Sur les séries d'Hermite et de Laguerre*. Ibid., vol. 193 (1931), pp. 386–389.
11. *Sur la convergence des séries d'Hermite*. Ibid., vol. 194 (1932), pp. 161–163.
12. *Sur les développements de Laguerre*. Ibid., vol. 194 (1932), pp. 1422–1424.
13. *Sur la série de Laguerre*. Ibid., vol. 196 (1933), pp. 523–525.
14. *Expression approchée du polynome de Laguerre* $L_n^{(\alpha)}(x)$. Ibid., vol. 196 (1933), pp. 1079–1080.
15. *Sur la détermination du saut* $D(x_0)$ de $f(x)$. Ibid., vol. 196 (1933), pp. 464–466.
16. *Über die $(C\ \delta)$-Summierbarkeit der Laplaceschen Reihe für* $1/2 < \delta < 1$. Mathematische Zeitschrift, vol. 14 (1922), pp. 99–109.
17. *Analogie entre les séries trigonométriques et les séries sphériques au point de vue de leur sommabilité par les moyennes arithmétiques*. Thèse. Paris, 1923, 65 pp.
18. *Sur la sommabilité de la série ultrasphérique à l'intérieur de l'intervalle* $(-1, +1)$ *par la méthode des moyennes arithmétiques*. Bulletin de la Société Mathématique de France, vol. 51 (1923), pp. 244–295.

19. *Recherches sur la sommabilité des séries ultrasphériques par la méthode des moyennes arithmétiques.* Journal de Mathématiques, (9), vol. 3 (1924), pp. 107-187.
20. *Recherches sur l'unicité des séries ultrasphériques.* Ibid., (9), vol. 5 (1926), pp. 125-196.
21. *Sommation des séries et intégrales divergentes par les moyennes arithmétiques et typiques.* Mémorial des Sciences Mathématiques. Vol. 51. Paris, 1931.
22. *Recherches sur la sommabilité des séries d'Hermite.* Annales Scientifiques de l'École Normale Supérieure, (3), vol. 49 (1932), pp. 137-221.
23. *Sur les moyennes arithmétiques des séries-noyaux des développements en séries d'Hermite et de Laguerre et sur celles de ces séries-noyaux dérivées terme à terme.* Journal of Mathematics and Physics, Massachusetts Institute of Technology, vol. 14 (1935), pp. 37-99.
24. *Contribution à l'étude du saut d'une fonction donnée par son développement en séries d'Hermite ou de Laguerre.* Transactions of the American Mathematical Society, vol. 38 (1935), pp. 10-47.

Korous, J.
1. *O rozvoji funkcí jedné reálné proměnné v řadu Hermiteových polynomů.* Rozpravy České Akademie, (2), vol. 37 (1928), no. 11, 34 pp.
2. *O řadách Laguerrových polynomů* (with French abstract). Ibid., no. 40, 23 pp.
3. *O rozvoji funkcí jedné reálné proměnné v řadu jistých ortogonálních polynomů* (with English abstract). Ibid., vol. 48 (1938), 12 pp.
4. *Über Reihenentwicklungen nach verallgemeinerten Laguerreschen Polynomen mit drei Parametern.* Věstník Královské České Společnosti Naúk, Třída Matemat.-Přírodověd, 1937, 26 pp.
5. *Über Entwicklungen der Funktionen einer reellen Veränderlichen in Reihen einer gewissen Klasse orthogonaler Polynome im unendlichen Intervalle.* Ibid., 1937, 19 pp.

Koschmieder, L.
1. *Über besondere Jacobische Polynome.* Mathematische Zeitschrift, vol. 8 (1920), pp. 123-137.

Kowalewski, G.
*1. *Einführung in die Determinantentheorie.* Leipzig, 1909.

Kowallik, U.
1. *Entwicklung einer willkürlichen Funktion nach Hermiteschen Orthogonalfunktionen.* Mathematische Zeitschrift, vol. 31 (1930), pp. 498-518.

Krall, H. L.
1. *On derivatives of orthogonal polynomials.* Bulletin of the American Mathematical Society, vol. 42 (1936), pp. 423-428.
2. *On higher derivatives of orthogonal polynomials.* Ibid., vol. 42 (1936), pp. 867-870.

Krawtchouk, M.
1. *Sur une généralisation des polynomes d'Hermite.* Comptes Rendus de l'Académie des Sciences, Paris, vol. 189 (1929), pp. 620-622.
2. *Sur la distribution des racines des polynomes orthogonaux.* Ibid., vol. 196 (1933), pp. 739-741.

Kronecker, L.
*1. *Zur Theorie der Elimination einer Variabeln aus zwei algebraischen Gleichungen.* Monatsberichte der Preussischen Akademie der Wissenschaften zu Berlin, 1881, pp. 535-600.

Lagrange, J. L.
1. *Oeuvres.* Vol. 1, pp. 534-539. 1867.

Laguerre, E. N.
1. *Sur l'intégrale* $\int_x^\infty x^{-1} e^{-x}\, dx$. Bulletin de la Société Mathématique de France, vol. 7 (1879), pp. 72-81. *Oeuvres.* Vol. 1, pp. 428-437.

2. *Sur l'approximation des fonctions circulaires au moyen des fonctions algébriques.* Comptes Rendus de l'Académie des Sciences, Paris, vol. 90 (1880), pp. 304–307. *Oeuvres.* Vol. 1, pp. 104–107.
3. *Sur les équations algébriques dont le premier membre satisfait à une équation linéaire du second ordre.* Ibid., vol. 90 (1880), pp. 809–812. *Oeuvres.* Vol. 1, pp. 126–132.

LANGER, R. E.
*1. *The asymptotic solutions of ordinary linear differential equations of the second order, with special reference to the Stokes phenomenon.* Bulletin of the American Mathematical Society, vol. 40 (1934), pp. 545–582.
*2. *The asymptotic solutions of certain linear ordinary differential equations of the second order.* Transactions of the American Mathematical Society, vol. 36 (1934), pp. 90–106.
*3. *On the asymptotic solutions of ordinary differential equations, with reference to the Stokes' phenomenon about a singular point.* Ibid., vol. 37 (1935), pp. 397–416.

LAWTON, W.
1. *On the zeros of certain polynomials related to Jacobi and Laguerre polynomials.* Bulletin of the American Mathematical Society, vol. 38, (1932), pp. 442–448.

LEBESGUE, H.
*1. *Sur la divergence et la convergence non-uniforme des séries de Fourier.* Comptes Rendus, de l'Académie des Sciences, Paris, vol. 141 (1905), pp. 875–877.
*2. *Leçons sur les Séries Trigonométriques.* Paris, 1906.

LEGENDRE, A. M.
1. *Exercices de Calcul Intégral sur Divers Ordres de Transcendantes.* Vol. 2. Paris, 1817.

LE ROY, É.
1. *Sur les séries divergentes et les fonctions définies par un développement de Taylor.* Annales de la Faculté des Sciences de Toulouse, (2), vol. 2 (1900), pp. 317–430.

LUKÁCS, F.
1. *Verschärfung des ersten Mittelwertsatzes der Integralrechnung für rationale Polynome.* Mathematische Zeitschrift, vol. 2 (1918), pp. 295–305.
2. *Über die Laplacesche Reihe.* Ibid., vol. 14 (1922), pp. 250–262.

MARCINKIEWICZ, J.
*1. *Quelques remarques sur l'interpolation.* Acta Litterarum ac Scientiarum Regiae Universitatis Hungaricae Francisco-Josephinae, vol. 8 (1937), pp. 127–130.
2. *Sur la divergence des polynomes d'interpolation.* Ibid., vol. 8 (1937), pp. 131–135.

MARKOFF, A.
1. *On some applications of algebraic continued fractions* (in Russian). Thesis. St. Petersburg, 1884, 131 pp.
2. *Démonstration de certaines inégalités de M. Tchébycheff.* Mathematische Annalen, vol. 24 (1884), pp. 172–180.
3. *Extrait d'une lettre.* Annales Scientifiques de l'École Normale Supérieure, (3), vol. 2 (1885), p. 183.
4. *Sur les racines de certaines équations (second note).* Mathematische Annalen, vol. 27 (1886), pp. 177–182.
5. *Differenzenrechnung.* Leipzig, 1896.

MEHLER, F. G.
1. *Bemerkungen zur Theorie der mechanischen Quadraturen.* Journal für die reine und angewandte Mathematik, vol. 63 (1864), pp. 152–157.
2. *Ueber die Entwicklung einer Function von beliebig vielen Variablen nach Laplaceschen Functionen höherer Ordnung.* Ibid., vol. 66 (1866), pp. 161–176.

3. *Ueber die Vertheilung der statischen Elektricität in einem von zwei Kugelkalotten begrenzten Körper.* Ibid., vol. 68 (1868), pp. 134–150.
4. *Ueber die Darstellung einer willkürlichen Function zweier Variablen durch Cylinderfunctionen.* Mathematische Annalen, vol. 5 (1872), pp. 135–140.
5. *Notiz über die Dirichlet'schen Integralausdrücke für die Kugelfunction $P^n(\cos\theta)$ und über eine analoge Integralform für die Cylinderfunction $J(x)$.* Ibid., vol. 5 (1872), pp. 141–144.

MEIXNER, J.
1. *Orthogonale Polynomsysteme mit einer besonderen Gestalt der erzeugenden Funktion.* Journal of the London Mathematical Society, vol. 9 (1934), pp. 6–13.
2. *Erzeugende Funktionen der Charlierschen Polynome.* Mathematische Zeitschrift, vol. 44 (1938), pp. 531–535.

MOECKLIN, E.
1. *Asymptotische Entwicklungen der Laguerreschen Polynome.* Commentarii Mathematici Helvetici, vol. 7 (1934), pp. 24–46.

MÜNTZ, C.
1. *Über die Potenzsummation einer Entwicklung nach Hermiteschen Polynomen.* Mathematische Zeitschrift, vol. 31 (1929), pp. 350–355.

MYLLER-LEBEDEFF, V.
1. *Die Theorie der Integralgleichungen in Anwendung auf einige Reihenentwicklungen.* Mathematische Annalen, vol. 64 (1907), pp. 388–416.

NEUMANN, E. R.
1. *Die Entwicklung willkürlicher Funktionen nach den Hermiteschen und Laguerreschen Orthogonalfunktionen auf Grund der Theorie der Integralgleichungen.* Inaugural-Dissertation. Breslau, 1912.
2. *Beiträge zur Kenntnis der Laguerreschen Polynome.* Jahresbericht der Deutschen Mathematiker-Vereinigung, vol. 30 (1921), pp. 15–35.

OBRECHKOFF, N.
1. *Sur la sommation de la série ultrasphérique par la méthode des moyennes arithmétiques.* Rendiconti del Circolo Matematico di Palermo, vol. 59 (1936), pp. 266–287.
2. *Formules asymptotiques pour les polynomes de Jacobi et sur les séries suivant les mêmes polynomes.* Annuaire de l'Université de Sofia, Faculté Physico-Mathématique, vol. 1 (1936), pp. 39–133.

PERRON, O.
1. *Über das infinitäre Verhalten der Koeffizienten einer gewissen Potenzreihe.* Archiv der Mathematik und Physik, (3), vol. 22 (1914), pp. 329–340.
2. *Über das Verhalten einer ausgearteten hypergeometrischen Reihe bei unbegrenztem Wachstum eines Parameters.* Journal für die reine und angewandte Mathematik, vol. 151 (1921), pp. 63–78.
3. *Die Lehre von den Kettenbrüchen.* 2d edition. Leipzig, 1929.
*4. *Algebra.* Vols. I, II, 2d edition. Berlin, 1932, 1933.

PLANCHEREL, M., and ROTACH, W.
1. *Sur les valeurs asymptotiques des polynomes d'Hermite $H_n(x) = (-1)^n e^{x^2/2} \cdot d^n(e^{-x^2/2})/dx^n$.* Commentarii Mathematici Helvetici, vol. 1 (1929), pp. 227–254.

PÓLYA, G.
1. *Sur un théorème de Stieltjes.* Comptes Rendus de l'Académie des Sciences, Paris, vol. 155 (1912), pp. 767–769.
2. *Sur un algorithme toujours convergent pour obtenir les polynomes de meilleure approximation de Tchebychef pour une fonction continue quelconque.* Ibid., vol. 157 (1913), pp. 840–843.

*3. *Über die Nullstellen gewisser ganzer Funktionen.* Mathematische Zeitschrift, vol. 2 (1918), pp. 352-383.

4. *Über die Konvergenz von Quadraturverfahren.* Ibid., vol. 37 (1933), pp. 264-286.

PÓLYA, G., and SZEGÖ, G.

1. *Aufgaben und Lehrsätze aus der Analysis.* Vols. I, II. Berlin, 1925.

2. *Über den transfiniten Durchmesser (Kapazitätskonstante) von ebenen und räumlichen Punktmengen.* Journal für die reine und angewandte Mathematik, vol. 165 (1931), pp. 4-49.

POPOVICIU, T.

1. *Sur la distribution des zéros de certains polynomes minimisants.* Bulletin de l'Académie Roumaine, vol. 16 (1934), pp. 214-217.

2. *Sur certains problèmes de maximum de Stieltjes.* Bulletin Mathématique de la Société Roumaine des Sciences, vol. 38 (1936), pp. 73-96.

RAU, H.

1. *Über die Lebesgueschen Konstanten der Reihenentwicklungen nach Jacobischen Polynomen.* Journal für die reine und angewandte Mathematik, vol. 161 (1929), pp. 237-254.

2. *Über eine asymptotische Darstellung der Jacobischen Polynome durch Besselsche Funktionen.* Mathematische Zeitschrift, vol. 40 (1936), pp. 683-692.

RIESZ, F.

*1. *Sur certains systèmes singuliers d'équations intégrales.* Annales Scientifiques de l'École Normale Supérieure, (3), vol. 28 (1911), pp. 33-62.

*2. *Über die Randwerte einer analytischen Funktion.* Mathematische Zeitschrift, vol. 18 (1923), pp. 87-95.

RIESZ, M.

1. *Eine trigonometrische Interpolationsformel und einige Ungleichungen für Polynome.* Jahresbericht der Deutschen Mathematiker-Vereinigung, vol. 23 (1915), pp. 354-368.

2. *Sur le problème des moments.* Arkiv för Matematik, Astronomi och Fysik, vol. 16 (1921), no. 12, 23 pp.; vol. 16 (1922), no. 19, 21 pp.; vol. 17 (1923), no. 16, 52 pp.

ROTACH, W.

1. *Reihenentwicklungen einer willkürlichen Funktion nach Hermiteschen und Laguerreschen Polynomen.* Inauguraldissertation, Eidgenössische Technische Hochschule Zürich, 1925.

SCHMIDT, E.

1. *Über die Charlier-Jordansche Entwicklung einer willkürlichen Funktion nach der Poissonschen Funktion und ihren Ableitungen.* Zeitschrift für angewandte Mathematik und Mechanik, vol. 13 (1933), pp. 139-142.

SCHUR, I.

1. *Über die Verteilung der Wurzeln bei gewissen algebraischen Gleichungen mit ganzzahligen Koeffizienten.* Mathematische Zeitschrift, vol. 1 (1918), pp. 377-402.

2. *Affektlose Gleichungen in der Theorie der Laguerreschen und Hermiteschen Polynome.* Journal für die reine und angewandte Mathematik, vol. 165 (1931), pp. 52-58.

SCHWID, N.

1. *The asymptotic forms of the Hermite and Weber functions.* Transactions of the American Mathematical Society, vol. 37 (1935), pp. 339-362.

SEN, D. N., and RANGACHARIAR, V.

1. *Generalized Jacobi polynomials.* Bulletin of the American Mathematical Society, vol. 42 (1936), pp. 901-908.

SHERMAN, J.

1. *On the numerators of the convergents of the Stieltjes continued fractions.* Transactions of the American Mathematical Society, vol. 35 (1933), pp. 64-87.

SHIBATA, K.

1. *On the distribution of the roots of a polynomial satisfying a certain differential equation of the second order.* Japanese Journal of Mathematics, vol. 1 (1924), pp. 147-153.

SHOHAT, J.

1. *On the polynomial of the best approximation to a given continuous function.* Bulletin of the American Mathematical Society, vol. 31 (1925), pp. 509-514.
2. *On a general formula in the theory of Tchebycheff polynomials and its applications.* Transactions of the American Mathematical Society, vol. 29 (1927), pp. 569-583.
3. *On a certain formula of mechanical quadratures with non-equidistant ordinates.* Ibid., vol. 31 (1929), pp. 448-463.
4. *On the polynomial and trigonometric approximation of measurable bounded functions on a finite interval.* Mathematische Annalen, vol. 102 (1929), pp. 157-175.
5. *On interpolation.* Annals of Mathematics, (2), vol. 34 (1933), pp. 130-146.
6. *Théorie Générale des Polynomes Orthogonaux de Tchebichef.* Mémorial des Sciences Mathématiques. Vol. 66. Paris, 1934.
7. *On mechanical quadratures, in particular, with positive coefficients.* Transactions of the American Mathematical Society, vol. 42 (1937), pp. 461-496.
8. *On the convergence properties of Lagrange interpolation based on the zeros of orthogonal Tchebycheff polynomials.* Annals of Mathematics, (2), vol. 38 (1937), pp. 758-769.

SMIRNOFF, V. J.

1. *Sur la théorie des polynomes orthogonaux à une variable complexe.* Journal de la Société Physico-Mathématique de Léningrad, vol. 2 (1928), pp. 155-179.
2. *Sur les formules de Cauchy et de Green et quelques problèmes qui s'y rattachent.* Bulletin de l'Académie des Sciences de l'URSS, 1932, pp. 337-372.

SMITH, E. R.

1. *Zeros of the Hermitian polynomials.* American Mathematical Monthly, vol. 43 (1936), pp. 354-358.

SONIN(E), N. J.

1. *Recherches sur les fonctions cylindriques et le développement des fonctions continues en séries.* Mathematische Annalen, vol. 16 (1880), pp. 1-80.
2. *On the precision of the determination of limiting values of definite integrals* (in Russian). Zapiski Akademii Nauk, vol. 69 (1892), pp. 1-30.

SPENCER, V. E.

1. *Asymptotic expressions for the zeros of generalized Laguerre polynomials and Weber functions.* Duke Mathematical Journal, vol. 3 (1937), pp. 667-675.

STEKLOFF, W.

1. *Sur les expressions asymptotiques de certaines fonctions, définies par les équations différentielles linéaires du second order, et leurs applications au problème du développement d'une fonction arbitraire en séries procédant suivant les-dites fonctions.* Communications de la Société Mathématique de Kharkow, (2), vol. 10 (1907), pp. 97-200. *Remarque complémentaire*, p. 201.
2. *On approximate evaluation of definite integrals by means of formulas of mechanical quadratures. I. Convergence of formulas of mechanical quadratures* (in Russian). Bulletin de l'Académie Impériale des Sciences, Petrograd, (6), vol. 10 (1916), pp. 169-186.

STIELTJES, T. J.

1. *Quelques recherches sur la théorie des quadratures dites méchaniques.* Annales Scientifiques de l'École Normale Supérieure, (3), vol. 1 (1884), pp. 409-426. *Oeuvres Complètes.* Vol. 1, pp. 377-394.
2. *Note à l'occasion de la réclamation de M. Markoff.* Ibid., (3), vol. 2 (1885), pp. 183-184. *Oeuvres Complètes.* Vol. 1, pp. 430-431.

3. *Sur certains polynômes qui vérifient une équation différentielle linéaire du second ordre et sur la théorie des fonctions de Lamé.* Acta Mathematica, vol. 6 (1885), pp. 321-326. *Oeuvres Complètes.* Vol. 1, pp. 434-439.

4. *Sur quelques théorèmes d'algèbre.* Comptes Rendus de l'Académie des Sciences, Paris, vol. 100 (1885), pp. 439-440. *Oeuvres Complètes.* Vol. 1, pp. 440-441.

5. *Sur les polynômes de Jacobi.* Comptes Rendus de l'Académie des Sciences, Paris, vol. 100 (1885), pp. 620-622. *Oeuvres Complètes.* Vol. 1, pp. 442-444.

6. *Sur les racines de l'équation* $X_n = 0$. Acta Mathematica, vol. 9 (1886), pp. 385-400. *Oeuvres Complètes.* Vol. 2, pp. 73-88.

7. *Sur la valeur asymptotique des polynômes de Legendre.* Comptes Rendus de l'Académie des Sciences, Paris, vol. 110 (1890), pp. 1026-1027. *Oeuvres Complètes.* Vol. 2, pp. 234-235.

8. *Sur les polynômes de Legendre.* Annales de la Faculté des Sciences de Toulouse, vol. 4 (1890), 17 pp. *Oeuvres Complètes.* Vol. 2, pp. 236-252.

9. *Sur les racines de la fonction sphérique de seconde espèce.* Annales de la Faculté des Sciences de Toulouse, vol. 4 (1890), 10 pp. *Oeuvres Complètes.* Vol. 2, pp. 253-262.

10. *Recherches sur les fractions continues.* Comptes Rendus de l'Académie des Sciences, Paris, vol. 118 (1894), pp. 1401-1403. *Oeuvres Complètes.* Vol. 2, pp. 398-401.

11. *Recherches sur les fractions continues.* Annales de la Faculté des Sciences de Toulouse, vol. 8 (1894), 122 pp.; vol. 9, 1895, 47 pp. *Oeuvres Complètes.* Vol. 2, pp. 402-566.

12. *Sur certaines inégalités dues à M. P. Tchebychef.* Article rédigé d'après un manuscript inédit. *Oeuvres Complètes.* Vol. 2, pp. 586-593.

STONE, M. H.

1. *Developments in Hermite polynomials.* Annals of Mathematics, (2), vol. 29 (1928), pp. 1-13.

*2. *Linear Transformations in Hilbert Space and their Applications to Analysis.* American Mathematical Society Colloquium Publications, vol. 15, 1932.

SZÁSZ, O.

*1. *Korlátos hatványsorokról.* Mathematikai és Természettudományi Értesítö, vol. 43 (1926), pp. 504-520.

SZEGÖ, G.

1. *A Hankel-féle formákról.* Mathematikai és Természettudományi Értesítö, vol. 36 (1918), pp. 497-538.

2. *Ein Beitrag zur Theorie der Polynome von Laguerre und Jacobi.* Mathematische Zeitschrift, vol. 1 (1918), pp. 341-356.

3. *Über Orthogonalsysteme von Polynomen.* Ibid., vol. 4 (1919), pp. 139-151.

4. *Beiträge zur Theorie der Toeplitzschen Formen.* II. Ibid., vol. 9 (1921), pp. 167-190.

5. *Über orthogonale Polynome, die zu einer gegebenen Kurve der komplexen Ebene gehören.* Ibid., vol. 9 (1921), pp. 218-270.

6. *Über die Entwickelung einer analytischen Funktion nach den Polynomen eines Orthogonalsystems.* Mathematische Annalen, vol. 82, 1921, pp. 188-212.

7. *Über die Randwerte analytischer Funktionen.* Ibid., vol. 84 (1921), pp. 232-244.

8. *Über den asymptotischen Ausdruck von Polynomen, die durch eine Orthogonalitätseigenschaft definiert sind.* Ibid., vol. 86 (1922), pp. 114-139.

9. *Über die Entwicklung einer willkürlichen Funktion nach den Polynomen eines Orthogonalsystems.* Mathematische Zeitschrift, vol. 12 (1921), pp. 61-94.

10. *Beiträge zur Theorie der Laguerreschen Polynome.* I. *Entwicklungssätze.* Ibid., vol. 25 (1926), pp. 87-115.

11. *Bemerkungen zu einer Arbeit von Herrn Fejér über die Legendreschen Polynome.* Ibid., vol. 25 (1926), pp. 172–187.
12. *Ein Beitrag zur Theorie der Thetafunktionen.* Sitzungsberichte der Preussischen Akademie der Wissenschaften, physikalish-mathematische Klasse, 1926, pp. 242–252.
13. *Koeffizientenabschätzungen bei ebenen und räumlichen harmonischen Entwicklungen.* Mathematische Annalen, vol. 96 (1927), pp. 601–632.
14. *Über gewisse Interpolationspolynome, die zu den Jacobischen und Laguerreschen Abszissen gehören.* Mathematische Zeitschrift, vol. 35 (1932), pp. 579–602.
15. *Über einige asymptotische Entwicklungen der Legendreschen Funktionen.* Proceedings of the London Mathematical Society, (2), vol. 36 (1932), pp. 427–450.
16. *Über eine von Herrn S. Bernstein herrührende Abschätzung der Legendreschen Polynome.* Mathematische Annalen, vol. 108 (1933), pp. 360–369.
17. *Asymptotische Entwicklungen der Jacobischen Polynome.* Schriften der Königsberger Gelehrten Gesellschaft, naturwissenschaftliche Klasse, vol. 10 (1933), pp. 35–112.
18. *Bemerkungen zu einem Satz von E. Schmidt über algebraische Gleichungen.* Sitzungsberichte der Preussischen Akademie der Wissenschaften, physikalish-mathematische Klasse, 1934, pp. 3–15.
19. *Über gewisse orthogonale Polynome, die zu einer oszillierenden Belegungsfunktion gehören.* Mathematische Annalen, vol. 110 (1934), pp. 501–513.
20. *Inequalities for the zeros of Legendre polynomials and related functions.* Transactions of the American Mathematical Society, vol. 39 (1936), pp. 1–17.
21. *An integral equation for the square of a Laguerre polynomial.* Journal of the London Mathematical Society, vol. 12 (1937), pp. 162–163.

TAMARKIN, J. D.
1. *On the Theory of Polynomials of Approximation.* Lecture delivered at Brown University, 1935–1936.

TCHAKALOFF, L.
1. *Sur la structure des ensembles linéaires définis par une certaine propriété minimale.* Acta Mathematica, vol. 63 (1934), pp. 77–97.

TCHEBICHEF, P. L.
1. *Sur les fractions continues.* Journal de Mathématiques, (2), vol. 3 (1858), pp. 289–323. *Oeuvres.* Vol. 1, pp. 201–230.
2. *Sur l'interpolation par la méthode des moindres carrés.* Mémoires de l'Académie Impériale des Sciences de St. Pétersbourg, (7), vol. 1 (1859), pp. 1–24. *Oeuvres.* Vol. 1, pp. 471–498.
3. *Sur le développement des fonctions à une seule variable.* Bulletin de l'Académie Impériale des Sciences de St. Pétersbourg, vol. 1 (1859), pp. 193–200. *Oeuvres.* Vol. 1, pp. 499–508.
4. *Sur l'interpolation.* Zapiski Akademii Nauk, vol. 4, Supplement no. 5, 1864. *Oeuvres.* Vol. 1, pp. 539–560.
5. *Sur les fonctions analogues à celles de Legendre.* Zapiski Akademii Nauk, vol. 16 (1870), pp. 131–140. *Oeuvres.* Vol. 2, pp. 59–68.
6. *Sur les valeurs limites des intégrales.* Journal de Mathématiques, (2), vol. 19 (1874), pp. 157–160. *Oeuvres.* Vol. 2, pp. 181–185.
7. *Sur le rapport de deux intégrales étendues aux mêmes valeurs de la variable.* Zapiski Akademii Nauk, vol. 44, Supplement no. 2, 1883. *Oeuvres.* Vol. 2, pp. 375–402.
8. *Sur la représentation des valeurs limites des intégrales par des résidus intégraux.* Ibid., vol. 51, Supplement no. 4, 1885. *Oeuvres.* Vol. 2, pp. 419–440. Acta Mathematica, vol. 9 (1886) pp. 35–56.

TITCHMARSH, E. C.

*1. *The Theory of Functions*. Oxford, 1932.

TRICOMI, F.

1. *Trasformazione di Laplace e polinomi di Laguerre. I. Inversione della trasformazione*. Rendiconti della R. Accademia dei Lincei, (6), vol. 21 (1935), pp. 232–239.

USPENSKY, J. V.

1. *On the development of arbitrary functions in series of Hermite's and Laguerre's polynomials*. Annals of Mathematics, (2), vol. 28 (1927), pp. 593–619.

VAN VEEN, S. C.

1. *Asymptotische Entwicklung und Nullstellenabschätzung der Hermiteschen Funktionen*. Proceedings, Koninklijke Akademie van Wetenschappen te Amsterdam, vol. 34 (1931), pp. 257–267.

2. *Asymptotische Entwicklung und Nullstellenabschätzung der Hermiteschen Funktionen*. Mathematische Annalen, vol. 105 (1931), pp. 408–436.

3. *Zusatz zum vorangehenden Berichte* (of W. Hahn). Jahresbericht der Deutschen Mathematiker-Vereinigung, vol. 44 (1934), pp. 236–238.

VITALI, G., and SANSONE, G.

1. *Moderna Teoria delle Funzioni di Variabile Reale. II. Sviluppi in Serie di Funzioni Ortogonali*. Bologna, 1935.

WALSH, J. L.

1. *Interpolation and Approximation by Rational Functions in the Complex Domain*. American Mathematical Society Colloquium Publications, vol. 20, 1935.

WANGERIN, A.

1. *Theorie der Kugelfunktionen und der verwandten Funktionen, insbesondere der Laméschen und Besselschen (Theorie spezieller, durch lineare Differentialgleichungen definierter Funktionen)*. Encyklopädie der Mathematischen Wissenschaften, vol. II.1.2, pp. 695–759.

WATSON, G. N.

1. *The harmonic functions associated with the parabolic cylinder*. Proceedings of the London Mathematical Society, (2), vol. 8 (1910), pp. 393–421; (2), vol. 17 (1918), pp. 116–148.

2. *Approximate formulae for Legendre functions*. Messenger of Mathematics, vol. 47 (1918), pp. 151–160.

3. *A Treatise on the Theory of Bessel Functions*. Cambridge, 1922.

4. *Notes on generating functions of polynomials*: (1) *Laguerre polynomials*. Journal of the London Mathematical Society, vol. 8 (1933), pp. 189–192.

5. *Notes on generating functions of polynomials*: (2) *Hermite polynomials*. Ibid., vol. 8 (1933), pp. 194–199.

6. *Über eine Reihe aus verallgemeinerten Laguerre'schen Polynomen*. Sitzungsberichte der mathematisch-naturwissenschaftlichen Klasse der Akademie Wien, IIa, vol. 147 (1938), pp. 151–159.

WEYL, H.

1. *Singuläre Integralgleichungen mit besonderer Berücksichtigung des Fourierschen Integraltheorems*. Inauguraldissertation, Göttingen, 1908. Mathematische Annalen, vol. 66 (1909), pp. 273–324.

WHITTAKER, E. T., AND WATSON, G. N.

1. *A Course of Modern Analysis*. 4th edition. Cambridge, 1935.

WIGERT, S.

1. *Contributions à la théorie des polynomes d'Abel-Laguerre*. Arkiv för Matematik, Astronomi och Fysik, vol. 15 (1921), no. 25, 22 pp.

2. *Sur les polynomes orthogonaux et l'approximation des fonctions continues*. Ibid., vol. 17 (1923), no. 18, 15 pp.

WIMAN, A.

1. *Über eine asymptotische Eigenschaft der Ableitungen der ganzen Funktionen von den Geschlechtern 1 und 2 mit einer endlichen Anzahl von Nullstellen.* Mathematische Annalen, vol. 104 (1931), pp. 169–181.

WINSTON, C.

1. *On mechanical quadratures formulae involving the classical orthogonal polynomials.* Annals of Mathematics, (2), vol. 35 (1934), pp. 658–677.

WRIGHT, E. M.

1. *The coefficients of a certain power series.* Journal of the London Mathematical Society, vol. 7 (1932), pp. 256–262.

YOUNG, W. H.

1. *On the connexion between Legendre series and Fourier Series.* Proceedings of the London Mathematical Society, (2), vol. 18 (1919), pp. 141–162.

ZERNIKE, F.

1. *Eine asymptotische Entwicklung für die grösste Nullstelle der Hermiteschen Polynome.* Koninklijke Akademie van Wetenschappen te Amsterdam, Proceedings, vol. 34 (1931), pp. 673–680.

ZYGMUND, A.

1. *Sur la théorie riemannienne de certains systèmes orthogonaux,* II. Prace Matematyczno-Fizyczne, vol. 39 (1932), pp. 73–117.
2. *Trigonometrical Series.* Warszawa-Lwów, 1935.

FURTHER REFERENCES

ACHIESER, N. I. (See List of References above.)

3. *Theory of Approximation.* New York, 1956.
4. *The Classical Moment Problem.* Oliver and Boyd, Edinburgh and London, 1965.

ASKEY, R.

1. *Orthogonal expansions with positive coefficients.* Proceedings of the American Mathematical Society, vol. 16 (1965), pp. 1191-1194.
2. *Jacobi polynomial expansions with positive coefficients and imbeddings of projective spaces.* Bulletin of the American Mathematical Society, vol. 74 (1968), pp. 301-304.
3. *A transplantation theorem for Jacobi series.* Illinois Journal of Mathematics, vol. 13 (1969), pp. 583-590.
4. *Linearization of the product of orthogonal polynomials,* in *Problems in analysis.* Edited by R. Gunning, Princeton University Press, Princeton, N. J., 1970, pp. 223-228.
5. *Positivity of the Cotes numbers for some Jacobi abscissas.* Numerische Mathematik, vol. 19 (1972), pp. 46-48.
6. *Mean convergence of orthogonal series and Lagrange interpolation.* Acta Mathematica Academiae Scientiarum Hungaricae, vol. 23 (1972), pp. 71-85.
7. *Grünbaum's inequality for Bessel functions.* Journal of Mathematical Analysis and Applications, vol. 41 (1973), pp. 122-124.
8. *Summability of Jacobi series.* Transactions of the American Mathematical Society, vol. 179 (1973), pp. 71-84.
9. *Certain rational functions whose power series have positive coefficients.* II. SIAM Journal on Mathematical Analysis, vol. 5 (1974), pp. 53-57.
10. *Jacobi polynomials.* I. *New proofs of Koornwinder's Laplace type integral representation and Bateman's bilinear sum.* SIAM Journal on Mathematical Analysis, vol. 5 (1974), pp. 119-124.

ASKEY, R., AND FITCH, J.

1. *Positivity of the Cotes numbers for some ultraspherical abscissas.* SIAM Journal on Numerical Analysis, vol. 5 (1968), pp. 199-201.
2. *Integral representations for Jacobi polynomials and some applications.* Journal of Mathematical Analysis and Applications, vol. 26 (1969), pp. 411-437.

ASKEY, R., AND GASPER, G.

1. *Linearization of the product of Jacobi polynomials.* III. Canadian Journal of Mathematics, vol. 23 (1971), pp. 332-338.
2. *Jacobi polynomial expansions of Jacobi polynomials with non-negative coefficients.* Proceedings of the Cambridge Philosophical Society, vol. 70 (1971), pp. 243-255.
3. *Certain rational functions whose power series have positive coefficients.* American Mathematical Monthly, vol. 79 (1972), pp. 327-341.
4. *Positive Jacobi polynomial sums.* II. American Journal of Mathematics.

ASKEY, R., AND POLLARD, H.

1. *Some absolutely monotonic and completely monotonic functions.* SIAM Journal on Mathematical Analysis, vol. 5 (1974), pp. 58-63.

ASKEY, R., AND STEINIG, J.

1. *Some positive trigonometric sums.* Transactions of the American Mathematical Society, vol. 187 (1974), pp. 295-307.
2. *A monotonic trigonometric sum.* American Journal of Mathematics.

ASKEY, R., AND WAINGER, S.

1. *Mean convergence of expansions in Laguerre and Hermite series.* American Journal of Mathematics, vol. 87 (1965), pp. 695-708.
2. *A transplantation theorem between ultraspherical series.* Illinois Journal of Mathematics, vol. 10 (1966), pp. 322-344.
3. *A transplantation theorem for ultraspherical coefficients.* Pacific Journal of Mathematics, vol. 16 (1966), pp. 393-405.
4. *A convolution structure for Jacobi series.* American Journal of Mathematics, vol. 91 (1969), pp. 463-485.

BAILEY, W. N.

1. *The generating function of Jacobi polynomials.* Journal of the London Mathematical Society, vol. 13 (1938), pp. 8-12.
2. *On the product of two Laguerre polynomials.* Quarterly Journal of Mathematics, vol. 10 (1939), pp. 60-66.

BALÁZS, J., AND TURÁN, P.

1. *Notes on Interpolation.* II. (*Explicit formulae*). Acta Mathematica Academiae Scientiarum Hungaricae, vol. 8 (1957), pp. 201-215.
2. *Notes on Interpolation.* III. (*Convergence*). Ibid., vol. 9 (1958), pp. 195-214.

BATEMAN, H.

1. *A generalization of the Legendre polynomial.* Proceedings of the London Mathematical Society, ser. 2, vol. 3 (1905), pp. 111-123.
2. *The solution of linear differential equations by means of definite integrals.* Transactions of the Cambridge Philosophical Society, vol. 21 (1909), pp. 171-196.
3. *Partial Differential Equations.* Cambridge University Press, Cambridge, 1932.

BATEMAN MANUSCRIPT PROJECT (Director: A. ERDÉLYI).

1. *Higher Transcendental Functions.* Vols. 1, 2, 3. New York-Toronto-London, 1953, 1955.

BOCHNER, S. (See List of References above.)

2. *Positive zonal functions on spheres.* Proceedings of the National Academy of Sciences, vol. 40 (1954), pp. 1141-1147.

BONAMI, A., AND CLERC, J.-L.

1. *Sommes de Cesàro et multiplicateurs des développements en harmoniques sphériques.* Transactions of the American Mathematical Society, vol. 183 (1973), pp. 223-263.

BUTLEWSKI, Z.

1. *Sur les intégrales d'une équation différentielle du second ordre.* Mathematica (Cluj), vol. 12 (1936), pp. 36-48.

CARLESON, L.

1. *On convergence and growth of partial sums of Fourier series.* Acta Mathematica, vol. 116 (1966), pp. 135-157.

COIFMAN, R., AND WEISS, G.

1. *Analyse harmonique non-commutative sur certaines espaces homogènes.* Lecture Notes in Mathematics, vol. 242, Springer-Verlag, Berlin and New York, 1971.

CSORDAS, G., AND WILLIAMSON, J.
1. *On polynomials satisfying a Turán type inequality*. Proceedings of the American Mathematical Society, vol. 43 (1974), pp. 367-372.
DAVIS, J., AND HIRSCHMAN, I. I., JR.
1. *Toeplitz forms and ultraspherical polynomials*. Pacific Journal of Mathematics, vol. 18 (1966), pp. 73-95.
DAVIS, J., AND RABINOWITZ, P.
1. (a) *Some geometrical theorems for abscissas and weights of Gauss type*. Journal of Mathematical Analysis and Applications, vol. 2 (1961), pp. 428-437. (b) *Erratum*. Ibid., vol. 3 (1961), p. 619.
DE BRUIJN, N. G.
1. *Uncertainty principles in Fourier analysis*, Inequalities, Edited by O. Shisha, Academic Press, New York, 1967, pp. 57-71.
DOETSCH, G. (See List of References above.)
3. *Handbuch der Laplace-Transformationen*. Vol. 2: *Anwendungen der Laplace-Transformationen*, 1. *Abteilung*. Basel-Stuttgart, 1955.
EAGLESON, G.
1. *A characterization theorem for positive definite sequences on the Krawtchouk polynomials*. Australian Journal of Statistics, vol. 11 (1969), pp. 29-38.
EGERVÁRY, E., AND TURÁN, P.
1. *Notes on interpolation*. V. (*On the stability of interpolation*). Acta Mathematica Academiae Scientiarum Hungaricae, vol. 9 (1958), pp. 259-267.
2. *Notes on interpolation*. VI. (*On the stability of the interpolation on an infinite interval*). Acta Mathematica Academiae Scientiarum Hungaricae, vol. 10 (1959), pp. 55-62.
ERDÉLYI, A. (See List of References above.)
3. *Asymptotic forms for Laguerre polynomials*. Journal of the Indian Mathematical Society, Golden Jubilee Volume, vol. 24 (1960), pp. 235-250.
ERDÉLYI, A., AND SWANSON, C. A.
1. *Asymptotic forms of Whittaker's confluent hypergeometric functions*. Memoirs of the American Mathematical Society, No. 25, 1957.
ERDÖS, P.
1. *On the distribution of normal point groups*. Proceedings of the National Academy of Sciences, vol. 26 (1940), pp. 294-297.
2. *On divergence properties of the Lagrange interpolation parabolas*. Annals of Mathematics, vol. 42 (1941), pp. 309-315.
3. *Problems and results on the theory of interpolation*. II. Acta Mathematica Academiae Scientiarum Hungaricae, vol. 12 (1961), pp. 235-244.
ERDÖS, P., AND GRÜNWALD, G.
1. *Über einen Faber'schen Satz*. Annals of Mathematics, vol. 39 (1938), pp. 257-261.
ERDÖS P., AND LENGYEL, B. A.
1. *On fundamental functions of Lagrangean interpolation*. Bulletin of the American Mathematical Society, vol. 44 (1938), pp. 828-834.
ERDÖS, P., AND TURÁN, P. (See List of References above.)
3. *On Interpolation*. III. *Interpolatory theory of polynomials*. Annals of Mathematics, vol. 41 (1940), pp. 510-553.
4. *An extremal problem in the theory of interpolation*. Acta Mathematica Academiae Scientiarum Hungaricae, vol. 12 (1961), pp. 221-234.
FELDHEIM E. (See List of References above.)
5. *On the positivity of certain sums of ultraspherical polynomials*. (Translated and edited by G. Szegö.) Journal d'Analyse Mathématique, vol. 11 (1963), pp. 275-284.
FREUD, G.
1. *Az Hermite-Fejér-féle interpolációs eljárás konvergenciájáról*. Magyar Tudományos Akadémia III. osztályának közleményei, vol. 5 (1955), pp. 29-47.
2. *Ortogonális polinómokról*. Dissertation. Budapest, 1956.
3. *Über die Asymptotik orthogonaler Polynome*. Académie Serbe des Sciences, vol. 11 (1957), pp. 19-32.
4. *Orthogonale Polynome*. Basel, 1969. English translation, New York, 1971.

GASPER, G.

1. *Linearization of the product of Jacobi polynomials*. I. Canadian Journal of Mathematics, vol. 22 (1970), pp. 171-175.
2. *Linearization of the product of Jacobi polynomials*. II. Canadian Journal of Mathematics, vol. 22 (1970), pp. 582-593.
3. *Positivity and the convolution structure for Jacobi series*. Annals of Mathematics, (2), vol. 93 (1971), pp. 112-118.
4. *Banach algebras for Jacobi series and positivity of a kernel*. Annals of Mathematics, (2), vol. 95 (1972), pp. 261-280.
5. *An inequality of Turán type for Jacobi polynomials*. Proceedings of the American Mathematical Society, vol. 32 (1972), pp. 435-439.
6. *Nonnegativity of a discrete Poisson kernel for the Hahn polynomials*. Journal of Mathematical Analysis and Applications, vol. 42 (1973), pp. 438-451.
7. *Projection formulas for orthogonal polynomials of a discrete variable*. Journal of Mathematical Analysis and Applications, vol. 45 (1974), pp. 176-198.

GATTESCHI, L.

1. *Approssimazione asintotica degli zeri dei polinomi ultrasferici*. Rendiconti di Matematica e delle sue Applicazioni, (5), vol. 8 (1949), pp. 399-411.
2. *Limitazione degli errori nelle formule asintotiche per le funzioni speciali*. Rendiconti del Seminario Matematico dell'Università e del Politecnico di Torino, vol. 16 (1956-1957), pp. 83-94.

GINIBRE, R.

1. *General formulation of Griffiths' inequalities*. Communications on Mathematical Physics, vol. 16 (1970), pp. 310-328.

GRENANDER, U., AND SZEGÖ, G.

1. *Toeplitz Forms and their Applications*. Berkeley-Los Anageles, 1958.

GRÜNBAUM, F. A.

1. *A property of Legendre polynomials*. Proceedings of the National Academy of Sciences, vol. 67 (1970), pp. 959-960.
2. *A new kind of inequality for Bessel functions*. Journal of Mathematical Analysis and Applications, vol. 41 (1973), pp. 115-121.

GRÜNWALD, G. (See List of References above.)

2. *On a convergence theorem for the Lagrange interpolation polynomials*. Bulletin of the American Mathematical Society, vol. 47 (1941), pp. 271-275.
3. *On the theory of Interpolation*. Acta Mathematica, vol. 75 (1943), pp. 219-245.

HAHN, W. (See List of References above.)

5. *Über Orthogonalpolynome, die q-Differenzengleichungen genügen*. Mathematische Nachrichten, vol. 2 (1949), pp. 4-34.

HARTMAN, P., AND WINTNER, A.

1. *On non-conservative linear oscillators of low frequency*. American Journal of Mathematics, vol. 70 (1948), pp. 529-539.

HIRSCHMAN, I. I., JR.

1. *Variation diminishing transformations and orthogonal polynomials*. Journal d'Analyse Mathématique, vol. 9 (1961), pp. 177-193.
2. *The strong Szegö limit theorem for Toeplitz determinants*. American Journal of Mathematics, vol. 88 (1966), pp. 73-95.

HORTON, R.

1. *Expansions using orthogonal polynomials*. Ph.D. Thesis, University of Wisconsin, Madison, 1973, 60 pp.

HSÜ, H.

1. *Certain integrals and infinite series involving ultraspherical polynomials and Bessel functions*. Duke Mathematical Journal, vol. 4 (1938), pp. 374-383.

HUA, L. K.

1. *Harmonic Analysis of Functions of Several Complex Variables in the Classical Domains*, Translations of Mathematical Monographs, vol. 6 (1963), American Mathematical Society, Providence, R. I.

HUNT, R.

1. *On the convergence of Fourier series*, Orthogonal Expansions and their Continuous Analogues, Edited by D. Haimo, Southern Illinois University Press, Carbondale, Illinois, 1968, pp. 235-255.

KARLIN, S., AND MCGREGOR, J. L.

1. *The differential equations of birth-and-death processes, and the Stieltjes moment problem*. Transactions of the American Mathematical Society, vol. 85 (1957), pp. 489-546.
2. *The Hahn polynomials, formulas and an application*. Scripta Mathematica, vol. 26 (1961), pp. 33-46.
3. *Classical diffusion processes and total positivity*. Journal of Mathematical Analysis and Applications, vol. 1 (1960), pp. 163-183.

KARLIN, S., AND STUDDEN, W. J.

1. *Tchebycheff Systems with Applications in Analysis and Statistics*. Interscience Publishers, New York, 1966.

KARLIN, S., AND SZEGŐ, G.

1. *On certain determinants whose elements are orthogonal polynomials*. Journal d'Analyse Mathématique, vol. 8 (1960), pp. 1-157.

KOORNWINDER, T.

1. *The addition formula for Jacobi polynomials*. I. *Summary of results*. Indagationes Mathematicae, vol. 34 (1972), pp. 188-191.
2. *The addition formula for Jacobi polynomials and spherical harmonics*. SIAM Journal on Applied Mathematics, vol. 25 (1973), pp. 236-246.
3. *Jacobi polynomials*. II. *An analytic proof of the product formula*. SIAM Journal on Mathematical Analysis, vol. 5 (1974), pp. 125-137.
4. *Jacobi polynomials*. III. *An analytic proof of the addition formula*. SIAM Journal on Mathematical Analysis, vol. 6 (1975), pp. 533-543.

LOCHER, F.

1. *Norm bounds of quadrature processes*. SIAM Journal on Numerical Analysis, vol. 10 (1973), pp. 553-558.

LORCH, L.

1. *The Lebesgue constants for Jacobi series*. I. Proceedings of the American Mathematical Society, vol. 10 (1959), pp. 756-761.
2. *The Lebesgue constants for Jacobi series*. II. American Journal of Mathematics, vol. 81 (1959), pp. 875-888.
3. *Comparison of two formulations of Sonin's theorem and of their respective applications to Bessel functions*. Studia Scientiarum Mathematicarum Hungarica, vol. 1 (1966), pp. 141-145.

LORCH, L., MULDOON, M. E., AND SZEGO, P.

1. *Higher monotonicity properties of certain Sturm-Liouville functions*. III. Canadian Journal of Mathematics, vol. 22 (1970), pp. 1238-1265.
2. *Higher monotonicity properties of certain Sturm-Liouville functions*. IV. Canadian Journal of Mathematics, vol. 24 (1972), pp. 349-368.

LORCH, L., AND SZEGO, P.

1. *Higher monotonicity properties of certain Sturm-Liouville functions*. Acta Mathematica, vol. 109 (1963), pp. 55-73.
2. *Higher monotonicity properties of certain Sturm-Liouville functions*. II. Bulletin de l'Académie Polonaise des Sciences. Série des Sciences Mathématiques, Astronomiques et Physiques, vol. 11 (1963), pp. 455-457.

MAGNUS, W., AND OBERHETTINGER, F.

1. *Formeln und Lehrsätze für die speziellen Funktionen der mathematischen Physik*. Berlin, 1948.

MAKAI, E.

1. *Über die Nullstellen von Funktionen, die Lösungen Sturm-Liouville'scher Differentialgleichungen sind*. Commentarii Mathematici Helvetici, vol. 16 (1944), pp. 153-199.
2. *On a monotonic property of certain Sturm-Liouville functions*. Acta Mathematica Academiae Scientiarum Hungaricae, vol. 3 (1952), pp. 165-172.

3. *On systems of polynomials orthogonal in two intervals*. Publicationes Mathematicae, vol. 2 (1952), pp. 222-228.

MAKAI, E., AND TURÁN, P.

1. *Hermite expansions and distribution of zeros of polynomials*. Magyar Tud. Akadémia Mat. Kutató Int. Közl. vol. 8 (1963), pp. 157-163.

MUCKENHOUPT, B.

1. *Asymptotic forms for Laguerre polynomials*. Proceedings of the American Mathematical Society, vol. 24 (1970), pp. 288-292.
2. *Mean convergence of Hermite and Laguerre series*. I. Transactions of the American Mathematical Society, vol. 147 (1970), pp. 419-432.
3. *Mean convergence of Hermite and Laguerre series*. II. Transactions of the American Mathematical Society, vol. 147 (1970), pp. 433-460.
4. *Equiconvergence and almost everywhere convergence of Hermite and Laguerre series*. SIAM Journal on Mathematical Analysis, vol. 1 (1970), pp. 295-321.
5. *Mean convergence of Jacobi series*. Proceedings of the American Mathematical Society, vol. 23 (1969), pp. 306-310.

MUCKENHOUPT, B., AND STEIN, E.

1. *Classical expansions and their relation to conjugate harmonic functions*. Transactions of the American Mathematical Society, vol. 118 (1965), pp. 17-92.

NEWMAN, J., AND RUDIN, W.

1. *Mean convergence of orthogonal series*. Proceedings of the American Mathematical Society, vol. 3 (1952), pp. 219-222.

NOVIKOFF, A.

1. *On a special system of orthogonal polynomials*. Dissertation, Stanford University, 1954.

OLVER, F. W. J.

1. *A paradox in asymptotics*. SIAM Journal on Mathematical Analysis, vol. 1 (1970), pp. 533-534.

PEETRE, J.

1. *The Weyl transform and Laguerre polynomials*. Le Matematiche, vol. 27 (1972), pp. 301-323.

POLLACZEK, F.

1. *Sur une généralisation des polynomes de Legendre*. Comptes Rendus de l'Académie des Sciences, Paris, vol. 228 (1949), pp. 1363-1365.
2. *Systèmes de polynomes biorthogonaux qui généralisent les polynomes ultrasphériques*. Ibid., vol. 228 (1949), pp. 1998-2000.
3. *Sur une famille de polynomes orthogonaux qui contient les polynomes d'Hermite et de Laguerre comme cas limites*. Ibid., vol. 230 (1950), pp. 1563-1565.
4. *Sur une généralisation des polynomes de Jacobi*. Mémorial des Sciences Mathématiques, vol. 131 (1956).

POLLARD, H.

1. *The mean convergence of orthogonal series*. I. Transactions of the American Mathematical Society, vol. 62 (1947), pp. 387-403.
2. *The mean convergence of orthogonal series*. II. Transactions of the American Mathematical Society, vol. 63 (1948), pp. 355-367.
3. *The mean convergence of orthogonal series*. III. Duke Mathematical Journal, vol. 16 (1949), pp. 189-191.

ROOSENRAAD, C. T.

1. *Inequalities with orthogonal polynomials*. Ph.D. Thesis, University of Wisconsin, Madison, 1969, 52 pp.

ŠAPIRO, R. L.

1. *The special functions connected with the representations of the group $SU(n)$ of class one with respect to $SU(n-1)$ $(n \geqq 3)$*. Izvestija Vysših Učebnyh Zavedeniĭ, Matematika, vol. 71 (1968), pp. 97-107, (Russian).

SARMANOV, I. O.

1. *A generalized symmetric gamma correlation*. (Russian; translation in) Soviet Mathematics, Doklady, vol. 9 (1968), pp. 547-550.

SCHMEISSER, G.
1. *Optimale Schranken zu einem Satz über Nullstellen Hermitescher Trinome*. Journal für die reine und angewandte Mathematik, vol. 246 (1971), pp. 147-160.
SCHOENBERG, I., AND SZEGÖ, G.
1. *An extremum problem for polynomials*. Compositio Mathematica, vol. 14 (1960), pp. 260-268.
SEIDEL, W., AND SZÁSZ, O.
1. *On positive harmonic functions and ultraspherical polynomials*. Journal of the London Mathematical Society, vol. 26 (1951), pp. 36-41.
SHOHAT, J. A., HILLE, E., AND WALSH, J. L.
1. *A Bibliography on Orthogonal Polynomials*. Washington, 1940.
SURÁNYI, J., AND TURÁN, P.
1. *Notes on interpolation*. I. (*On some interpolatorical properties of the ultraspherical polynomials*). Acta Mathematica Academiae Scientiarum Hungaricae, vol. 6 (1955), pp. 67-79.
SZÁSZ, O. (See List of References above.)
2. *On the relative extrema of ultraspherical polynomials*. Bollettino della Unione Matematica Italiana, series 7, vol. 5 (1950), pp. 125-127.
3. *On the relative extrema of the Hermite orthogonal functions*. Journal of the Indian Mathematical Society, vol. 25 (1951), pp. 129-134.
SZEGÖ, G. (See List of References above.)
22. *On an inequality of P. Turán concerning Legendre polynomials*. Bulletin of the American Mathematical Society, vol. 54 (1948), pp. 401-405.
23. *On the relative extrema of Legendre polynomials*. Bollettino della Unione Matematica Italiana, (3), vol. 5 (1950), pp. 120-121.
24. *On certain special sets of orthogonal polynomials*. Proceedings of the American Mathematical Society, vol. 1 (1950), pp. 731-737.
25. *Ultrasphaerikus polinomok összegéröl*. Matematikai és Fizikai Lapok, vol. 45 (1938), pp. 36-38.
26. *Über gewisse Potenzreihen mit lauter positiven Koeffizienten*. Mathematische Zeitschrift, vol. 37 (1933), pp. 674-688.
27. *On some Hermitian forms associated with two given curves of the complex plane*. Transactions of the American Mathematical Society, vol. 40 (1936), pp. 450-461.
SZEGÖ, G., AND TURÁN, P.
1. *On the monotone convergence of certain Riemann sums*. Publicationes Mathematicae (Debrecen), vol. 8 (1961), pp. 326-335.
THORNE, R. C.
1. *The asymptotic expansion of Legendre functions of large degree and order*. Technical Report, Office of Naval Research, California Institute of Technology, 1956.
TRICOMI, F. G. (See List of References above.)
2. *Sul comportamento asintotico dell'n-esimo polinomio di Laguerre nell'intorno dell'ascissa 4n*. Commentarii Mathematici Helvetici, vol. 22 (1949), pp. 150-167.
3. *Sul comportamento asintotico dei polinomi di Laguerre*. Annali di Matematica, (4), vol. 28 (1949), pp. 263-289.
4. *Sugli zeri dei polinomi sferici ed ultrasferici*. Ibid., (4), vol. 31 (1950), pp. 93-97.
5. *Vorlesungen über Orthogonalreihen*. Berlin-Göttingen-Heidelberg, 1955.
TURÁN, P.
1. *On the zeros of the polynomials of Legendre*. Časopis pro Pěstováni Matematiky a Fysiky, vol. 75 (1950), pp. 113-122.
2. *On Descartes-Herriot's rule*. Bulletin of the American Mathematical Society, vol. 55 (1949), pp. 797-800.
3. *Remark on a theorem of Erhard Schmidt*. Mathematica, vol. 2 (1960), pp. 373-378.
VIETORIS, L.
1. *Über das Vorzeichen gewisser trigonometrischer Summen*. Sitzungsberichte der mathematisch-naturwissenschaftlichen Klasse der Akademie der Wissenschaften in Wien, vol. 167 (1958), pp. 125-135.

VITALI, G., AND SANSONE, G.

See List of References above. Third edition. Bologna, 1952.

WEBSTER, M. S.

1. *A convergence theorem for certain Lagrange interpolation polynomials*. Bulletin of the American Mathematical Society, vol. 49 (1943), pp. 114-119.

WIDOM, H., AND WILF, H.

1. *Small eigenvalues of large Hankel matrices*. Proceedings of the American Mathematical Society, vol. 17 (1966), pp. 338-344.

ZYGMUND, A.

See List of References above. Second edition. New York, 1952.

INDEX

The numbers refer to pages

GHIJ–AMS–898765